W9-AQP-569

# Shakespeare's Lives

# Shakespeare's Lives

## New Edition

S. SCHOENBAUM

CLARENDON PRESS · OXFORD
1991

Oxford University Press, Walton Street, Oxford OX2 6DP
Oxford New York Toronto
Delhi Bombay Calcutta Madras Karachi
Petaling Jaya Singapore Hong Kong Tokyo
Nairobi Dar es Salaam Cape Town
Melbourne Auckland
and associated companies in
Berlin Ibadan

Oxford is a trade mark of Oxford University Press

Published in the United States
by Oxford University Press, New York

© S. Schoenbaum 1991

British Library Cataloguing in Publication Data
(Data available)
ISBN 0–19–818618–5

Library of Congress Cataloging in Publication Data
Schoenbaum, S. (Samuel), 1927–
Shakespeare's lives/S. Schoenbaum.—New ed.
p.    cm.
Includes bibliographical references and index.
1. Shakespeare, William, 1564–1616—Biography.   2. Shakespeare,
William, 1564–1616, in fiction, drama, poetry, etc.   3. Dramatists,
English—Early modern, 1500–1700—Biography—History and criticism.
4. Biography (as a literary form)   I. Title.
PR2894.S3   1991
822.3'3—dc20
ISBN 0–19–818618–5

Typeset by Latimer Trend & Company Ltd, Plymouth
Printed in Great Britain by
Bookcraft Ltd
Midsomer Norton, Avon

*To*

*Kenneth Muir*

# Preface

THE idea of writing this book first entered my mind at Stratford-upon-Avon on 1 September 1964. I had gone there to attend an international conference honouring the quatercentenary of Shakespeare's birth, and that day, after the learned papers on the bear in *The Winter's Tale* and on *Hamlet* without words had been read and discussed by the assembled experts, I wandered down to the Avon, speckled white with swans, and entered (for the first time) the splendid Collegiate Church of the Holy Trinity. It was late afternoon; the tourists had departed. Although outside the late summer sun still shone brilliantly, I could barely make out the monument and bust in the shadows of the north wall of the chancel.

As I stood there I thought of the pilgrims, many thousands strong, who had looked up as I now did and pondered the inconceivable mystery of creation. Keats had been there; Boswell also, and the Irelands, father and son, and Washington Irving. The scholars had come too: Malone, who dimmed his name by having the bust painted white; Halliwell-Phillipps, who had loved Stratford but in the end turned his back on the town. The mad folk had been drawn, moth-like, to the Shakespearian flame. Poor Delia Bacon had tiptoed through the portal after dusk, a candle in one hand and a lantern in the other; ignoring the gravestone malediction, she came to dig for wonderful secrets beneath the slab, but, frightened by ghosts, she fled before dawn. Yeats, it seems, resisted the lure of monument and Birthplace when he visited Stratford in 1901, but few have shown a similar resolution. I thought of these vistors and others, and it occurred to me that some interest might attach to a little book narrating the quest for knowledge of Shakespeare the man; a book describing the different, sometimes opposing, ideas of him which people over the centuries had entertained. It would be called Shakespeare's Lives. Despite all that had been written on this great subject, there had not been such a book, only a few popular studies (useful in their own way, and sometimes entertaining) of such related subjects as the Shakespeare cult and Shakespeare's reputation.

Widening inexorably in scope, the project came to fill my days, and sometimes to haunt my nights. It occurred to me that I must try to find out how the various documents—the marriage-licence bond, the Belott-Mountjoy deposition, the will—came to light. The formal Lives, from Rowe to Rowse, would of course occupy much of my space, but what of the accretions of biographical notes in eighteenth-century editions, the bits and pieces of information in newspapers, magazines, and miscellanies, the prefaces to innumerable popular collections, the encyclopaedia articles from which ordinary readers formed their impressions of the National Poet? These, surely, could not be ignored. From the nineteenth century onwards, critics hunted for biographical revelations in the works, especially the Sonnets: I would have to confront this daunting mass of material. The representations of Shakespeare are in their own way biographical documents; belief in the various icons, all but two doubtful or spurious, would furnish curious evidence of human credulity deserving a place in my pages. The invention of biographical data by means of forged papers would also comprise part—a fascinating part—of the story. And then one must reckon with the amateurs, the eccentrics, the cranks with theories. Of these the worst would be the heretics, alert to conspiracies, who saw a sinister plot to take away the plays from their true progenitor, Bacon or Marlowe or some Earl or other, and bestow them instead on the Stratford boor. I did not relish this aspect of my assignment, although I knew that Mark Twain and Freud rubbed shoulders with less celebrated schismatics.

I embarked on my work without any preconceived theme or thesis, for these I distrust as strait-jackets, stifling insight and reducing complex intellectual phenomena to simplistic formulae. But I quickly recognized the truth of the observation that biography tends towards oblique self-portraiture. How much must this be so with respect to Shakespeare, where the sublimity of the subject ensures empathy and the impersonality of the life-record teases speculation! I remember once mentioning this pattern to the late John Crow in the familiar columned portico of the British Museum, and he reminded me that Desmond McCarthy had said somewhere that trying to work out Shakespeare's personality was like looking at a very dark glazed picture in the National Portrait Gallery: at first you see nothing, then you begin to recognize features, and then you realize that they are your own. Not everyone, I discovered however, had come to a similar realization. In any event, the biographers' recurring self-identification with their subject supplied me, if not with a thesis, at least with a leitmotif.

The phenomenon made it necessary for me to acquaint myself with the lives and personalities of the biographers. Here, for the first time, I found the printed sources insufficient. Although I was not surprised to be unable to locate memoirs of Robert Bell Wheler or Nathan Drake or other forgotten worthies, I discovered with astonishment the neglect of such eminent Victorians as Charles Knight, the great popularizer; Halliwell-Phillipps, most considerable of nineteenth-century scholars on the biographical side and a fascinating character; and J. Payne Collier, the famous forger whose intriguing career spanned most of a century. Others had been written about without scholarly rigour. James Prior was no doubt an excellent naval surgeon, and his *Life of Malone* (1860) qualifies as an agreeable exercise in piety, but it is not an authoritative study of one of the ornaments of the Johnson circle and one of the greatest of all Shakespearian scholars. Yet a more up-to-date biography had not superseded Prior's, despite quantities of untapped manuscript material at the Bodleian Library, the British Library and elsewhere. The most recent book on William-Henry Ireland, whose forgeries constitute a unique chapter in literary history, misses out on important printed evidence as well as manuscripts in at least half a dozen public and private collections. These establish definitively for the first time such basic facts as the date of the forger's birth and the identity of his mother; they reveal certain aspects of his career in a new light and affect our understanding of the father's role in the deception.

Having determined to consult primary materials wherever possible, I made full use of the excellent resources of the Newberry Library in Chicago near which I then lived, but soon found it necessary to embark upon a long odyssey. For a period of two years my university arranged my lecture schedule in such a way as to enable me to commute twice or three times a month for a few days at the Houghton Library in Cambridge or the New York Public Library or the Folger Shakespeare Library in Washington. In California I visited the Francis Bacon Foundation in Pomona and enjoyed three stays, two of them extended, within the agreeable confines of the Huntington Library in San Marino. The Beinecke Library of Yale University furnished me with several valuable items, as did the Morgan Library in New York. Dr Louis Marder of Evanston, Illinois, gave me free access to his distinguished collection of Shakespeariana; Mrs Donald F. Hyde (now the Viscountess Eccles) more than once received me with gracious hospitality at Four Oaks Farm in New Jersey, where I was delighted to find an important Ireland manuscript, as well as to examine the bellows portrait of Shakespeare. In Great Britain I have worked at the British Library, University College, London, the University of London Library, the Royal Society of Literature, the Victoria and Albert Museum, the Bodleian Library, the Records Office and Shakespeare Centre at Stratford-upon-Avon, the Birmingham Shakespeare Library, the Shakespeare Institute of the University of Birmingham, the Edinburgh University Library, and the National Library of Scotland.

These researches—made possible by leaves from Northwestern University, a Fellowship from the John Simon Guggenheim Memorial Foundation, and a grant-in-aid from the Huntington Library—have proved rewarding beyond expectation. Take, for example, the Collier affair. I found correspondence at the Huntington. Dr Arthur Freeman, now of Bernard Quaritch Ltd., Antiquarian Booksellers, and of London, showed me an entertaining letter in his possession, and Dr Marder placed at my disposal an unpublished manuscript by C. M. Ingleby, Collier's chief antagonist. The Folger supplied me with the malefactor's unpublished autobiography and the twelve surviving volumes of his manuscript diary, the last of which contains his repentance. There too I came upon more Ingleby papers. Sir Frederic Madden's voluminous manuscript Journal, at the Bodleian, furnished new information about the circumstances leading to Collier's fall. The same pattern has repeated itself again and again. Quite by chance I discovered, tucked away in a copy of an obscure nineteenth-century monograph, a letter in which Thomas Campbell recants the most famous of all biographical interpretations of literature: his inspired identification of Prospero abjuring his magic with Shakespeare bidding the stage farewell. At the Huntington I examined the papers of the Wallaces, who came to London from Nebraska and made the most spectacular Shakespearian biographical recoveries of the twentieth century. A postmark on an envelope addressed to Charles Wallace helped me to understand the acrimonious feud which developed between him and the distiguished French scholar Albert Feuillerat. Perhaps there is a moral here about the utility of preserving the apparently trivial.

While pursuing the biographers I have tried not to lose sight of their subject. My book may perhaps be most appropriately viewed as a novel species of Shakespearian biography, with the protagonist gradually emerging from the mists of ignorance and misconception, to be seen through a succession of different eyes and from constantly shifting vantage points. The sources for such a book as this are of course overwhelmingly numerous. I expect that I have found my way through perhaps a million pages of printed and manuscript material. Much of this has not left a trace on the pages which follow. Why fatigue readers with every third-rate Victorian biography, every handbook designed for the edification of undergraduates, every scrap from the Baconian rubbish heap? Nor have I seen fit to record such unilluminating controversies as that over the supposititious Shakespearian annotations of a copy of Halle's *Chronicle*, as described by Alan Keen and Roger Lubbock in *The Annotator* (1954), although surmises about the poet's early life entered their speculations. I have not, however, knowingly passed over any substantive biographical contribution, and I have found room for much that is representative or engagingly idiosyncratic.

Some of the documents on which I touch—by Keats, Carlyle, De Quincey, and others—have a literary interest in their own right, but the majority do not challenge esteem as monuments of unageing intellect. The interest of some writings, such as those of Delia Bacon, is mainly clinical. On the assumption that few readers will experience an irresistible compulsion to examine for themselves my less well-known sources, I have quoted more generously than is my wont in order to give the flavour of the material.

Writings in other than the mother tongue presented a problem. For a time I toyed with the notion of inviting authorities in other countries to furnish brief accounts of Shakespearian biography in Russia, Scandinavia, and elsewhere, for a series of appendices; but such supplementary matter would have swollen a book already not short, and so I decided against this course. Instead

I have dealt briefly myself with some of the major German and French figures, especially those (e.g., August Wilhelm Schlegel) who have notably influenced the English biographical tradition. Brandes, being of exceptional importance, is discussed at some length. I regret not having space to treat, except by implication, the theory and practice of literary biography in different periods, although that subject interests me greatly. One must impose some limits upon what one attempts.

A passionate interest in the lives and achievements of people has sustained me in seeing through to completion a project which may, I believe, be fairly described as not easy. Much of the really significant work on Shakespeare in this century has been impersonal and technical—I am thinking particularly of the achievements in analytical bibliography which have revolutionized our understanding of Shakespeare's text. These contributions are beyond praise. Still, when literary pursuits have become so specialized and objective (indeed, at times, mathematical), when the electronic computers have been placed at the service of scholarship, I believe it more than ever necessary from time to time to demonstrate the continuity of an older kind of literary study: one which did not disdain to offer its fruits to those outside the academy's privileged walls. I like to think of this book belonging to that tradition and as affirming, in its own modest way, our common humanity.

For this revised edition, almost twenty years after the fact, I have both abridged and enlarged an original text of more than 800 pages. Much of the initial survey of the biographical facts—the matter of the first section, Materials for a Life—is present in curtailed form: the information, in greater detail, is conveniently available in my *William Shakespeare: A Compact Documentary Life*, a revised edition of which with a new postcript has recently (New York and Oxford, 1987) been issued. Less space is now given to the section I called Deviations, and concerned with the genesis and varied manifestations of anti-Stratfordianisn; the mad and affecting Delia Bacon remains, as does Dr Freud, but some other curious and wayward folk no longer recline on my couch. I have also—somewhat more reluctantly—made do without Shakespeare as a character in plays and novels. I have, however, added a Recent Lives chapter appraising the work of the past two decades, new information—and hypotheses—showing no sign of abatement after the passage of more than four centuries; scholarship, as I am fond of saying, is process. In my discussion, in this section, of Anthony Burgess's *Shakespeare* (1970) I have seen fit to include some material on *Nothing Like the Sun* (1964), which had appeared in my first edition.

Some of the relevant items I have already taken up in reviews or in the postcript to the revised edition of my *Compact Documentary Life*, and I have resisted the temptation to rephrase for no other reason that to say things a little differently. Here and there in the text itself, I have corrected errors and made stylistic refinements. So I am luckier than another terrified reviser, poor Dencombe in Henry James's 'The Middle Years', who yearns in vain for a second chance.

S.S.

*Washington, DC*
*December 1988*

*Note*: For the convenience of ordinary readers I have modernized the spelling and punctuation, and expanded contractions, of material prior to 1700, although I have not tampered with proper names, as these forms sometimes have a special interest. Footnote references to old-spelling texts are furnished for the benefit of those who may wish to consult versions without editorial intervention. After 1700 the conventions of spelling and punctuation present few problems, and so I have reproduced these passages—the great majority of my quotations—unaltered. For citations from Shakespeare I have throughout used the *The Complete Works*, ed. Stanley Wells and Gary Taylor (1986).

# Acknowledgements

FOR permission to quote from unpublished manuscript sources I am obliged to the following: the Hyde Collection, Somerville, New Jersey (Ireland's *Confessions*); Dr Arthur Freeman, London (Letters of Collier and Halliwell-Phillipps); Dr Louis Marder, Evanston, Illinois (Letter of Thomas Campbell, Ingleby's *Supplement*); Mr Stuart B. Schimmel, New York City (Ireland marginalia); Dr H. Stopes-Roe, Birmingham, England (Letters of Charlotte Carmichael Stopes); Henry W. and Albert A. Berg Collection of the New York Public Library, Astor, Lenox and Tilden Foundations (Mark Twain notation); the Folger Shakespeare Library, Washington, DC (various manuscripts); Harvard College Library (*Ireland's Shaksperian Fabrications*, letter of Halliwell-Phillipps); Henry E. Huntington Library and Art Gallery (Wallace papers, letter of J. Payne Collier, Ireland marginalia); the Trustees of the National Library of Scotland (Letters of E. K. Chambers, Halliwell-Phillipps, Ritson, and Dover Wilson); the Pierpont Morgan Library (Letters of Malone and Ritson); the Shakespeare Centre, Stratford-upon-Avon (Collier notation); the Shakespeare Institute, Birmingham, England (Sisson lecture); the Director of the University of London Library (Cowell's *Reflections*); the Editorial Committee of the Yale Editions of the Private Papers of James Boswell (letter of Edmund Burke, letter of Stratford town clerks to Malone).

In addition to the above, I have made full use of the rich manuscript holdings of the British Library, the Bodleian Library, the Shakespeare Birthplace Record Office, the Birmingham Shakespeare Library, and the University of Edinburgh Library. The staffs of all these institutions have been uniformly helpful. For particular courtesies I must mention Sandra Rickards (Folger), Mary Isabel Fry and Anne Hyder (Huntington), Ian Willison (British Library), Robert Bearman (Birthplace Record Office), W. R. N. Fredrick and Angela Letch (Birmingham Shakespeare Library), and Alan Bell (National Library of Scotland). At the Royal Society of Literature I was graciously received by the Secretary, J. M. Patterson.

Although I have consulted numerous manuscripts, most of my sources have been published works. I owe an enormous debt to that body of scholarly writings which it is my task to deal with critically in these pages. Specific obligations are indicated in my notes. I wish that there I might have been able to cite the sources consulted for every statement, but to do so would have been to lengthen intolerably an already long book. Perhaps here I may mention a few outstandingly helpful works. Chief among these has been E. K. Chambers's *William Shakespeare; A Study of Facts and Problems*. For more specialized purposes, Mark Eccles's *Shakespeare in Warwickshire*, Hyder

Edward Rollins's New Variorum Edition of the *Sonnets*, and Halliwell-Phillipps's scrap-books at the Folger have proved invaluable. In addition to the authorities cited in my section on the schismatics, I have found very useful, as a finding list, Joseph S. Galland's *Digesta Anti-Shakespeareana*, available in typescript at Northwestern University. The Shakespeare portraits, genuine and supposititious, present technical problems calling for expert knowledge; here I am indebted principally to the writings of M. H. Spielmann, although his conclusions have sometimes had to be modified in the light of more recent information. (I tried unsuccessfully to trace Spielmann's long, unfinished monograph on the Shakespeare iconography, which he mentioned early in the century in a letter to Sidney Lee. Only many years after *Shakespeare's Lives* had first appeared was I able to track down the Spielmann typescript to the Houghton Library of Harvard University.) Where biographies and memoirs of my subjects exist—as they do for Davenant, Sir Thomas Phillipps, Delia Bacon, Samuel Butler, Strachey, and a number of others—I have consulted them with much profit.

I have used a very small amount of my material previously in articles. For permission to reuse it here I wish to thank The Times Newspapers Ltd., the Brown University Press, and Macmillan & Co. of Canada.

To the John Simon Guggenheim Memorial Foundation I am grateful for a Fellowship which enabled me to complete not only the research for this work but also the composition of it. My appointment as a Fellow of the Research Center for Arts and Humanities (directed by Professor John Fuegi) of the University of Maryland for the autumn 1988 term, facilitated the expeditious completion of this revised edition.

Thanks for a variety of favours are due to the following: the Viscountess Eccles, R. N. Alexander, Sydney Anglo, Giles E. Dawson, Mark Eccles, G. Blakemore Evans, R. A. Foakes, Levi Fox, Terence Hawkes, Stephen and Eva May Heathcote, Samuel Hynes, Robert Kirsch, Arthur Miller, P. I. C. Payne, John W. Velz, Stanley Wells, and George H. Wiedeman.

Judi Bertman made an excellent beginning on the typing of my manuscript. Her work was continued and brought to completion by Margaret Buth, who coped with a succession of drafts, and throughout applied herself to the task with a devotion beyond the call of duty. During the last stages David Daniell of University College, London, provided a much needed additional pair of sharp eyes to check quotations.

The Oxford University Press in New York and Oxford, and the Clarendon Press in Oxford, have given all that an author can hope for in the way of encouragement and support. I must particularly mention Leona Capeless and John Bell.

It is no small favour for a busy scholar to take time away from his own commitments to read a colleague's typescript—especially if that typescript comes to over a thousand pages. Although adequately forewarned, Professors

Geoffrey Bullough and Kenneth Muir unhesitatingly came to my assistance; their suggestions have made my book better than it would otherwise have been, but they are in no way responsible for its faults.

Since setting forth the above acknowledgements twenty years ago, during my Northwestern days, a number of friends, colleagues, and well-wishers have departed the scene. Here I must mention Whitney Blake, John R. B. Brett-Smith, Geoffrey Bullough, D. M. Davin, Richard Ellmann, Alfred Harbage, Charlton Hinman, James G. McManaway, Dorothy Mason, J. C. Maxwell, James M. Osborn, and Gordon N. Ray. While they lived they made the biographer's task lighter by generously lending their good offices and often formidable expertise. I mourn their passing with a special poignance.

Happily, not all is loss. Diane M. Clark was a loyal secretary coping with a difficult task. While pursuing postgraduate studies at the University of Maryland, Vincent Pollet found time to learn some of the hazards of the game in the course of performing checks, at my direction, on this text of the revised version of *Shakespeare's Lives*. Whether a computer is capable of having a nervous breakdown I am not sure (I think of Hal in Stanley Kubrick's *2001: A Space Odyssey*), but I have discovered that some of our technological marvels—namely the Kerzweil Optical Scanner—can make mistakes, so I am especially grateful to Julie Keenan for riding herd on the gadgetry and otherwise contributing much to the rectification of accidentals. Michelle Green, also a postgraduate student at my university, performed helpful labours on the final typescript and Trudi Bellardo supplied the index. At Oxford University Press Kim Scott Walwyn gave me further welcome encouragement and I am also much indebted to the desk-editing department, and particularly to Robert Mark Ritter, for so adroitly readying this abridged and updated version of *Shakespeare's Lives* for publication; no easy task. My thanks also go to Betsy Welsh and Jean Miller at the Folger Shakespeare Library for their help.

My most personal debt is to my wife Marilyn. She endured this project, as she has the others, with sympathy and good humour.

S.S.

*Washington, DC*

# Contents

PART VI: DEVIATIONS

PART VII: THE TWENTIETH CENTURY

# List of Illustrations

# PART I

*Materials for a Life*

# 1

## *Prologue*

HE died in rainy April. That was on the 23rd, according to tradition the day
he entered upon his fifty-third year. About the circumstances of his passing
there would afterwards be conjecture. Two days later (so the registers of Holy
Trinity Church tell us) he was buried—'full seventeen foot deep, deep enough
to secure him',* according to the unlikely testimony of William Hall, rector of
Acton in Middlesex and prebendary of St Paul's, who journeyed to Stratford
in 1694, there to visit 'the ashes of the Great Shakespear'.[1] His remains lie
beneath the floor of the chancel next to the north wall. Only those who had
made their mark in the world were interred in this way; ordinary folk lie
beneath moss-green headstones in the churchyard on either side of the long
avenue, lined with lime trees, that leads to the entrance. In the churchyard
had been laid to rest his mother and father; despite the fact that John
Shakespeare had held high civic office. In time the poet's younger daughter
Judith would join them there.

On the stone slab covering the grave, which bears no name, these words
were carved:

> Good friend for Jesus' sake forbear,
> To dig the dust enclosed here!
> Blessed be the man that spares these stones,
> And cursed be he that moves my bones.

By the mid-eighteenth century the gravestone had sunk beneath the level of
the floor and was so decayed that the town fathers in their wisdom in time
replaced it with another bearing the identical legend.

Possibly Shakespeare himself composed the inscription; at any rate, the
suggestion that he did would be made some seventy-five years after his death.
Who is the 'good friend' so solemnly forewarned? Not, surely, the casual
passer-by, but rather the sexton, who sometimes had to dig up an old grave
in the parish church in order to make room for the newly deceased. The bones
thus uncovered would be thrown upon others in the charnel-house, which
stood adjoining the north wall, just a few feet away from Shakespeare's
grave. (Perhaps the place was in his mind's eye when he made Juliet protest
to Friar Laurence: 'chain me with roaring bears, | Or hide me nightly in a
charnel-house, | O'er-cover'd quite with dead men's rattling bones, | With
reeky shanks and yellow chapless skulls.') At Stratford this edifice has long

---

* A grave that deep is unlikely with the river so close by.

since been pulled down, but the door still stands. Hall on the same occasion saw the charnel-house, and thus described it to his friend Edward Thwaites, the noted Anglo-Saxon scholar: 'There is in this church a place which they call the bone-house, a repository for all bones they dig up, which are so many that they would load a great number of wagons.'[2] The malediction (about which in future years there would be much discussion) effectively accomplished its purpose. No sexton has desecrated Shakespeare's grave, nor have his bones been disturbed by modern enthusiasts—although some have sought to do so—intent in this curious fashion to illuminate with the light of science the concealing darkness of the poet's mystery.

Of a celebrated artist of this century, Douglas Goldring has remarked:

... the distillation of Modigliani's quality as a man ... is found in the masterpieces he has left behind him. They are more really *himself* than anything that has been recorded about the personality who produced them. Perhaps I have now got to the root of the matter. An artist should be judged not by his extravagances, intoxications, quarrels, vehement and silly letters, inability to be a bourgeois husband, vagaries, lapses from being a 'perfect gentleman', and so forth, but solely by the extent of his achievement.[3]

No doubt this is so; yet the biographer of Modigliani will not regret that he can draw upon his subject's vehement and silly letters, or the recollections of extravagances and quarrels by those who knew him in Montmartre and Montparnasse. For Shakespeare, of all artists the one about whom we most wish to know, such testimony is totally lacking. Little wonder that in time he might be regarded as everything and nothing; behind his face and words, 'only a bit of coldness, a dream dreamt by no one'. Instinctively, it would be said, he became proficient at simulating *someone*: on stage, as an actor he played many parts, in his imagination he became Caesar and Macbeth and other heroes. Before or after his death, 'he stood face to face with God, he said to Him, "I who, in vain have been so many men, want to be one man— myself." The voice of the Lord answered him out of the whirlwind, "I too have no self; I dreamed the world as you dreamed your work, my Shakespeare, and among the shapes of my dream are you, who, like me, are many men and no one."' So Jorge Luis Borges writes of Shakespeare.[4]

But he *was* someone, mortal not divine, who left notices of himself; not indeed intimate records, but evidences of a life spent in customary pursuits among men. Retrieved by wearisome research from the obscurity of provincial muniment rooms and parish churches, the records alienate interest by their stark formality—these wills and conveyances, subsidy rolls, fines and legal actions, and registers of christenings, marriages, and burials. Yet their information about Shakespeare's ancestry and immediate family supplies a background, however shadowy, to the dramatist's own career. And if the notices of Shakespeare himself contain nothing to satisfy the curiosity of a Keats as to the position in which the poet sat when he began 'To be or not to be', if (more importantly) they fail to lay bare the wreathed trellis of a working brain—they nevertheless possess a pattern and significance of their own.

# 2

# *The Shakespeares of Warwickshire*

ENOUGH Shakespeares appear in old records to keep any drowsy genealogist awake. Before 1400 we encounter them in Cheshire, Cumberland, Essex, Gloucester, Kent, Nottingham, Staffordshire, Warwickshire, and Yorkshire, as well as in Ireland. The ancient Shakespeares waxed and multiplied above all in the county of Warwickshire, north of the winding Avon where lay thick woods and scattered farms. After the fifteenth century they crossed into Cambridgeshire, Derbyshire, Leicestershire, and London. The spelling of the name assumed fantastic variations: Shakespey, Schacosper, Scakespeire, Saxper, Chacsper, Schaftspere, Shakstaf, and over seventy others. The disyllable carries martial associations, and it is not surprising that early commentators proposed an heroic etymology. Richard Verstegan, writing in 1605 'Of the Surnames of Our Ancient Families', remarks that 'Breakspear, Shakspear, and the like, have been surnames imposed upon the first bearers of them, for valour and feats of arms'.[5] As if to belie a derivation so pleasing to patriotic sentiment, actual records begin inauspiciously with William Saksper of Clopton in Kiftesgate Hundred, Gloucestershire, who in 1248 was hanged for robbery. Of all the early Shakespeares only Adam, son and heir of Adam of Oldediche, and his kin are known to have held (and therefore probably to have acquired) land by military service; his name appears in 1389 in the records of Baddesley Clinton.

Other early documents tell us that Shakespeares rented houses and tilled the land. They paid taxes, collected customs, witnessed wills, sued and were sued; they served as jurymen, archers, and bailiffs, and earned their livelihoods as coopers, turners, shoemakers, and weavers. At least one despaired: a John Shakespeare of Balsall hanged himself in his house on 23 July 1579. Others embraced the consolation of religion: an Isabella Shakspere was prioress of Wroxall Abbey in the early sixteenth century; at the same convent, in 1525, the sub-prioress Jane Shakspere told her beads. Most interest attaches to the Shakespeares of Balsall and Wroxall and two other villages situated in the old forest of Arden not many miles from

Stratford: Baddesley Clinton and Rowington. All the Warwickshire Shake-speares traced before 1500 resided in these hamlets. From one of them, Richard Shakespeare, who was almost certainly the poet's grandfather, made his way to Snitterfield, a small community about four miles northeast of Stratford, some time before 15 April 1529. On that date he was fined twopence for not travelling six miles to attend the manor court at Warwick.

A husbandman, he tended land on several manors in Snitterfield. The notices of him relate mainly to the small affairs of a life spent close to the soil. On one occasion every tenant was required to 'make his hedges and ditches betwixt the end of the lane of Richard Shakespere and the hedge called Dawkins hedge'. In 1535 he was fined for overburdening the common pasture with his cattle; and again, in 1560, for not ringing his swine and for allowing his stock to run loose in the meadows. The house he rented abutted on the High Street, and behind it had land that stretched down to the brook which flowed, then as now, through Snitterfield. This property belonged to Robert Arden, a gentleman of worship from the nearby hamlet of Wilmcote; we shall be hearing of him again. In the winter of 1560–61 Richard Shakespeare died, leaving an estate formally valued (probably undervalued) at £38.17s.* He had not been badly off.

Two sons survived him. The younger, Henry, never moved from Snitter-field, but farmed there and in neighbouring Ingon in the parish of Hampton Lucy. He got into scrapes. In 1574 he was fined for drawing blood in a fight; he was fined again in 1583 for not wearing a cap to church on Sundays and holidays. In October 1596, shortly before his death, the authorities penalized him for not labouring with his team to mend the Queen's highways, and 'for having a ditch between Reed Hill and Burman in decay for want of repairing'. He incurred debts. For one obligation, in 1586, his brother John stood surety, and when Henry failed to pay, the creditor brought an action against John Shakespeare, who at the time could ill afford to make good the £10. In Stratford in 1591 Henry was imprisoned in a plea of trespass in the case, and in 1596 was arrested for debt. Again jailed, he was helpless to prevent his surety, William Rounde, from marching to his house and taking away two oxen that Henry had bought but (characteristically) not paid for. Yet, after he died in December 1596, it was testified that there was money in his coffers, a mare in his stable, and plentiful corn and hay in his barn. This was William Shakespeare's Uncle Harry. John, the more stable elder brother, was granted

---

* To suggest even a rough modern equivalent for such figures would be misleading; ratios (pre-devaluation) of fifteen or twenty to one, sometimes used in the past, are now justly out of favour. Prices fluctuated widely during Elizabethan times, and the buying power of money by modern standards is difficult to ascertain. Income, however, may give some clue to the difference between then and now. In the period from 1561 to 1629 a building craftsman in southern England earned from 10d. to 12d. a day, while his labourer made from 6d. to 8d. (E. H. Phelps Brown and Sheila V. Hopkins, 'Seven Centuries of Building Wages', *Economica*, NS 22 [1955], 195–205). In *The State of England, Anno Dom. 1600*, Thomas Wilson regards as well-to-do a yeoman with from £300 to £500 yearly from lands and leases.

administration of his deceased father's estate on 10 February 1561. This John was the poet's father. Richard Shakespeare may have had a third son, Thomas, but we do not have enough evidence to be sure.

The bond naming John administrator of the estate describes him as a husbandman of Snitterfield. He may well have been brought up to the plough on his father's land, but by 1561 he no longer farmed in Snitterfield. Instead he followed the thriving glover's trade in Stratford: on market and fair days glovers had the chief standing in the town, at the picturesque timber and plaster structure called the High Cross. About John Shakespeare's occupation confusion would in aftertimes arise: he would be described as a butcher and also (probably because of his subsidiary business transactions) as a dealer in wool; in 1572 he twice appeared in the Exchequer court to answer charges of illegally purchasing wool. But he is termed a glover in a suit in 1556, and again in 1586, when he stood bail in Coventry for Michael Pryce, a tinker indicted for felony. In another action, in 1573, he is cited as a 'whittawer'— one who cured and softened the white skins used for gloves. Still other records show him speculating in barley, timber, and wool. Early biographers would have difficulty in tracing his career because there lived in Stratford another John Shakespeare, a corvisor or shoemaker, who crops up in the town annals from 1584 until 1596.

The first reference to the father of the National Poet occurs in April 1552, when he was fined a shilling for making a dungheap (*sterquinarium*) before his house in Henley Street—the same house, it may well be, that future ages would venerate as the Birthplace.[6] Evidently John Shakespeare prospered in his adopted community. In 1556 he bought two more dwellings: one, with garden, on Henley Street, adjoining the house he already occupied, and the other, with garden and croft, in Greenhill Street. At some time carpenters connected the Henley Street structures by inside doorways, and they thus became a single, imposing, close-timbered building.

From the records we may deduce that it was in 1557 that John Shakespeare married Mary Arden, youngest daughter of Richard Shakespeare's Snitterfield landlord. In his will, drawn up in November 1556 when Mary was still single, Robert Arden left her ten marks in money and the freehold estate of Asbies: a cottage and nearly sixty acres of land, with 'the crop upon the ground sown and tilled as it is'. Of the wedding no notice survives, but usually the ceremony took place in the bride's parish. Mary Arden hailed from the parish of Aston Cantlow, where the registers do not commence until 1560.

The newly married couple had a daughter Joan, baptized on 15 September 1558. (She must have died young, for her parents gave the same first name to another child christened in 1569.) Margaret Shakespeare was born in 1562 and buried the following year. The parish registers of Holy Trinity record the baptism of the Shakespeares' first son, William, on 26 April 1564: 'Gulielmus filius Johannes Shakspere'. Tradition would assign the birth itself to 23 April, the day of St George, England's patron saint. One biographer has, indeed, reminded us that such a birthday for the National Poet would be 'especially appropriate'; he no doubt is right, but other considerations usually determine these events.* Nor can the historian, alas, attach any great weight to the legend, of late origin, that the first nightingale sings each year in Stratford groves on 23 April.

The infant survived the plague that gripped Stratford in the summer of 1564 and carried off (according to estimate) nearly 250 souls in half a year. Soon he had the company of brothers and sisters. The Shakespeares produced four other offspring whose christenings are entered in the church register: Gilbert (13 October 1566), Anne (28 September 1571), Richard (11 March 1574), and Edmund (3 May 1580). Anne was buried in 1579, and Gilbert, about whom little is known, in 1612. Joan grew up to marry William Hart, a hatter of Stratford; she outlived her celebrated brother by thirty years. Edmund became an actor in London and died young, in 1607. He lies in the chancel of St Saviour's Church (now Southwark Cathderal), not far from the Globe playhouse.

As his family multiplied, so John Shakespeare waxed in importance. Stratford recognized his good parts. He arbitrated suits and served as a juror; minor offices fell to him. Meanwhile he sued fellow townsmen and was sued by them; he contributed to the poor relief and was fined for not keeping his gutters clean. By degrees he achieved local eminence. At some time before 1561 John Shakespeare became one of the chief burgesses, or members of the town council, and was thereby entitled to have his children educated without charge at the King's New School of Stratford-upon-Avon, which had been endowed more than two centuries previously.

Apparently John could not himself write, for he signed documents with a cross or made his mark: a pair of compasses emblematic of the glover's trade. That he executed written instruments in this way does not, however, definitively prove his illiteracy, for others who were able to sign their names followed the same custom; but no signature of John survives, and the natural inference is that, having been reared in a country village without a school, he never learned to write. Despite this liability he had charge of borough property and looked after the finances when he served as one of the two chamberlains of Stratford from 1561 to 1563. In 1565 John was elected one of the fourteen aldermen, a position that privileged him to wear in public a black cloth gown trimmed with fur. Soon afterwards his name appears in records with the honorific prefix 'Master', reserved for men of high social

---

* Later it would frequently be assumed that he was born on the 23rd on the unwarranted assumption that baptism customarily took place three days after birth. The *Prayer Book* of 1559 merely prescribed baptism not later than the next Sunday or other holy day following birth. In 1564, 23 April fell on a Sunday; if Shakespeare was born then, he should have been baptized by the 25th, St Mark's Day.

standing or learning. Thus we are enabled to distinguish him from the corvisor of Stratford who bore his name.

In 1568 John Shakespeare received the ultimate honour his town could confer upon him: he became High Bailiff—mayor, we would say. The glories of the station have been described by the most devoted modern student of Stratford's past, Edgar I. Fripp. The Bailiff and his deputy the High Alderman, he writes, 'were escorted from their houses to the Gild Hall by the serjeants bearing their maces before them. They were waited on by these buff-uniformed officers once a week to receive instructions, and accompanied by them through the market on Thursdays, through the fair on fair-days, about the parish-bounds at Rogation, and to and from church on Sundays. At church they sat with their wives in the front pew on the north side of the nave. At sermons in the Gild chapel they had their seats of honour'.[7] The Bailiff presided at council meetings in the guild-hall and, as a justice of the peace, at the monthly sessions of the Court of Record. During John's tenure itinerant actors—the Queen's Men and the Earl of Worcester's Servants—played in the guild-hall and received their rewards of 9s. and 12d. respectively; presumably the Bailiff sat in the front row. (In 1573 the Earl of Leicester's Men, led by James Burbage, played in Stratford—did the nine-year-old William see them?)

After John Shakespeare's term expired in 1571, he was elected High Alderman and deputy to the new Bailiff, Adrian Quiney; he continued as justice of the peace. Sometime around 1576 he initiated a petition for a grant of arms but failed to follow through. The reason is not far to seek: John Shakespeare had fallen on hard times.

Troubles closed in on him. He stopped attending council meetings. His brother aldermen in January 1578 eased his financial burden by reducing his tax for equipping soldiers from the town, but still he did not pay it; in November they exempted him from contributing towards the poor relief. Instead of buying more land and houses, John sold or mortgaged part of what he owned. Thus in 1578 he conveyed, for £40 in ready cash, his wife's Asbies inheritance to his brother-in-law, Edmund Lambert of Barton-on-the-Heath, to whom he already owed money. When the £40 fell due two years later, John did not pay it, and acrimonious litigation eventually followed; he never did recover the estate. Through the deaths of her sisters Joyce and Alice, Mary Shakespeare and her husband had come into a small share in two houses and a hundred acres in Snitterfield; this they sold in 1579 for a mere £4. John also had personal enemies to contend with: in 1582 he petitioned for sureties of the peace against four townsmen 'for fear of death and mutilation of his limbs'. The records do not reveal the source of the bad feeling. In 1586, after ten years of absence from the council, he was finally replaced as alderman; 'Mr. Shaxpere', the Corporation books sadly note, 'doth not come to the halls when they be warned, nor hath not done of long time'. He also avoided church. The two documents of 1592 citing him, along with eight others, for this infringement of Her Majesty's laws have been interpreted as offering evidence of John Shakespeare's recusancy, but notes appended to them make clear that he kept to his house 'for fear of process for debt'. In those times sheriff's officers could make arrests on Sundays, and the church was a likely place to find someone.

Against this background of declining family fortunes, William Shakespeare passed his formative years. But the father's difficulties can be exaggerated— and would be by biographers. When the glover of Stratford died in 1601 he still owned land and dwellings, including the large double-house in Henley Street. He was buried on 8 September. No will has come to light. The register of Holy Trinity records the burial of his widow seven years later, on 9 September 1608. In this family the women outlasted the men.

Patient investigation has yielded some information about the masters of the Stratford grammar school in Shakespeare's time, but records for pupils at the school during the sixteenth century have long since vanished. Scholarship has instead had to content itself with reconstructing the curriculum, about which a good deal may be inferred from the analogy of other Elizabethan grammar schools.

After William Shakespeare's christening the next event of which we have record is his marriage. The Episcopal Register of the diocese of Worcester for 27 November 1582 contains an entry for a marriage licence 'inter Willel-mum Shaxpere et Annam Whateley de Temple Grafton'. Presumably the clerk, who made mistakes with other entries, erred in recording the bride's name. Perhaps he copied it carelessly from the applicant's allegation stating upon oath the names, addresses, and occupations of the parties (as well as other information). The name Whateley slightly resembles Hathaway, or Hathwey, and a William Whateley, vicar of Crowle, was before the Consis-tory Court the same day as plaintiff in a tithe dispute. Had Parson Whateley interrupted the scribe while he was entering the day's licences? The reference to Temple Grafton, about five miles west of Stratford, presents another puzzle, for the bride came from another hamlet. Perhaps the wedding took place at Temple Grafton, where the register does not begin until 1612. The priest there, old John Frith, was regarded as 'unsound of religion'—he was a Catholic—but tolerated because of his advanced age. He would make Temple Grafton a good choice if the couple were Catholic; elsewhere in Warwickshire the rites of the Old Faith were suppressed. But these are suppositions, mere guesses; we know for certain only that William Shakespeare did not marry Anne Whateley of Temple Grafton, but Anne Hathaway of Stratford parish.

The bride's name appears correctly in the bond, or obligation, issued the next day:

The condition of this obligation is such that if hereafter there shall not appear any lawful let or impediment by reason of any precontract, consanguinity, affinity or by any other lawful means whatsoever, but that William Shagspere on the one party, and Anne Hathwey of Stratford in the diocese of Worcester, maiden, may lawfully solemnize matrimony together, and in the same afterwards remain and continue like man and wife according unto the laws in that behalf provided; and, moreover, if there be not at this present time any action, suit, quarrel, or demand moved or depending before any judge ecclesiastical or temporal for and concerning any such lawful let or impediment; and, moreover, if the said William Shagspere do not proceed to solemnization of marriage with the said Anne Hathwey without the consent of her friends; and also if the said William do upon his own proper costs and expenses defend and save harmless the right reverend Father in God, Lord John Bishop of Worcester and his officers for licencing them the said William and Anne to be married together with once asking of the banns of matrimony between them and for all other causes which may ensue by reason or occasion thereof; that then the said obligation to be void and of none effect or else to stand and abide in full force and virtue.[8]

The bond required two financially responsible citizens as sureties. On this occasion Fulk Sandells and John Rychardson, described as husbandmen of Stratford, stepped forward and agreed to indemnify the Bishop's chancellor, Richard Cosin, and his registrar Robert Warmstry, £40 if any legal actions resulted from the marriage.

Many Hathaways lived in the Stratford environs, but Anne was probably the eldest daughter of Richard Hathaway, a farmer of Shottery, a village lying about a mile west of Holy Trinity, within Stratford parish. The Hathaways, who were not badly off, dwelt in a stout farmhouse of two storeys, with stone fireplaces and heavy oak beams supporting the ceiling. This house, which belonged to the Earl of Warwick, the Lord of the Manor of Old Stratford, was called Hewlands Farm; it stands to this day* and is visited by innumerable tourists who know it as Anne Hathaway's Cottage. Anne is probably the Agnes to whom Richard Hathaway bequeathed ten marks to be paid to her on her wedding day (the names Anne and Agnes were often used interchangeably). He did not live to attend the ceremony, but was buried in the churchyard of Holy Trinity a few days after he made his will on 1 September 1581. Of Anne Hathaway we know almost nothing—not even the day of her christening, for she was born before the commencement of the Stratford baptism records. But the brass marker on her grave indicates that she was her husband's senior by seven or eight years, for it says that she was 'of the age of .67. years' when she died on 6 August 1623.

The circumstances of the marriage would, centuries later, fan controversy. Heads would shake gravely over the single asking of the banns, which normally were pronounced three times on successive Sundays or holy days. It would be noted that the young husband was a minor, not yet nineteen, and the bride a mature woman: perhaps a woman of the world. The bondsmen

---

* It suffered damage from fire in November 1969, however. The conflagration, which gutted about a third of the property, was the work of an arsonist, although apparently not a Baconian.

were allies of *her* family: Richard Hathaway named Sandells as one of the overseers of his will, and described him as a trusty friend and neighbour; Rychardson witnessed the instrument. Why was the bridegroom's family not represented in the bond? Did the Shakespeares withhold consent? The union produced a child in six months. Had the future Bard, still a mere boy, fallen prey to the lures of a siren of Shottery? Was the match a shotgun affair brought about by the consequences of (as one writer quaintly puts it) 'antenuptial fornication'?

Diligent research would show, however, that William could not have obtained a licence without the express willingness of his father, and that, when the groom was a minor, the bondsmen were customarily friends or kindred of the bride chosen to safeguard her interests. It would be pointed out that banns were prohibited from Advent Sunday (2 December in 1582) until 13 January: without a special dispensation the couple would have had to wait two months to wed, and Anne's condition made such a delay awkward. This information provides no clue, of course, to Shakespeare's mood, of which the absence of testimony is total.

Where they lived during the first years of matrimony is not recorded; probably in John Shakespeare's house in Henley Street. They had three children, all born before William attained his majority. The first, Susanna, was baptized on 26 May 1583. There followed twins, Hamnet and Judith, christened on 2 February 1585. The parish register records the burial of the poet's only son on 11 August 1596. With him died Shakespeare's hopes of preserving the family name according to the common way of mankind.

Evidently the Shakespeares named their twins after Hamnet Sadler, a young baker of Stratford, and his wife Judith Staunton of Longbridge. The Sadlers had lived in the same town for many generations, and in 1598 they gave the name William to one of their fourteen children, seven of whom died young. Hamnet Sadler witnessed the dramatist's will in 1616, and was remembered in it.

About Susanna Shakespeare a fair amount of information has trickled down. In 1606 Susanna and twenty other persons in Stratford were accused of failure to receive Communion on Easter Sunday. (The discovery of the Gunpowder Plot in the previous year had led to new regulations against those, especially Catholics, who denied the authority of the Church of England.) When, after some delay, Susanna appeared in court to answer the charge, it was dismissed. The episode would seem to suggest that she had Catholic leanings, but it is a fact that the next year, at the age of twenty-four, she married John Hall, a man with Puritan sympathies. The wedding, held on 5 June 1607, is recorded in the register of Holy Trinity.

The son of a Middlesex physician, Hall received BA and MA degrees from Queens' College, Cambridge, and had been practising medicine in Stratford since around 1600. Dr Hall treated worthies not only of Warwickshire but also of the adjacent counties: his patients included the Earl and Countess of Northampton, whose seat lay forty miles distant at Ludlow Castle, and the poet Drayton, a frequent visitor to Clifford Chambers in nearby Gloucestershire. We do not know whether he ministered to his father-in-law, for his neatly kept medical notebooks begin after Shakespeare's death. John and Susanna Hall had one child, Elizabeth, christened on 21 February 1608. On one occasion when she fell ill (she was in delicate health as a girl) her father treated her successfully with a fomentation of aqua vitae and spices.

Susanna Hall could sign her name to legal documents, and was lauded as witty beyond her sex; she was her father's favourite—so we gather from his will. Her sister Judith presumably had less wit than her sister, for she never learned to sign her name. In 1611 she twice made her mark as witness to a deed for the sale of a house belonging to Elizabeth Quiney and her eldest son Adrian. Mrs Quiney's husband Richard, a draper, was the son of the High Bailiff who went to London with John Shakespeare on Corporation business in 1572. Richard Quiney himself served two terms as Bailiff of Stratford, and was (as we shall see) on friendly terms with the poet. Judith must have known the Quineys well, for on 10 February 1616 their third son, Thomas, married her. She was then thirty-one, a bit ripe for the marriage market, and he twenty-six. The ceremony took place during the prohibited season which stretched that year from 27 January until 7 April, and the couple did not trouble to purchase the costly special wedding licence. As a result they were twice summoned before the Consistory Court, but they failed to show up. Thomas was excommunicated, and perhaps Judith too, though probably not for long. Judith's first offspring, Shakespeare Quiney, was baptized in the church in November 1616.

They had meanwhile moved into a house called The Cage, at the corner of High Street and Bridge Street. In the upper half they set up a vintner's shop and also dealt in tobacco. Things did not go very well for them. He was fined paltry sums for swearing and for suffering townsmen to tipple in his house. Once he was in danger of prosecution for vending adulterated and unwholesome wine. In 1633 kinsmen took over his lease to The Cage—a lease that he had tried to sell—in trust for his wife and progeny. (All three children died young.) Around 1652 Quiney abandoned his family and went to London; it is not clear that he ever returned to Stratford. During the present century The Cage has been restored: in the very house where Thomas Quiney once dispensed his questionable wines, hungry tourists for a time consumed Wimpy hamburgers. It is today a tourist information bureau.

# 3

## The Burgher of Stratford

WHILE raising a family, Shakespeare was also building an estate. The dry records of property transactions and the like chronicle his rise to material well-being. He became a man of substance in the town of his birth.

In 1597 he bought New Place. The second largest dwelling in Stratford, it would remain Shakespeare's permanent residence until his death. This 'pretty house of brick and timber'—so Leland, King Henry VIII's antiquary, described it—stood at the corner of Chapel Street and Chapel Lane, opposite the Guild Chapel. According to modern calculations, New Place had a frontage of over sixty feet on Chapel Street, and a depth of seventy feet along Chapel Lane; the northern gable rose up twenty-eight feet. There were two barns and two gardens, the larger of which attracted notice in the next century for the vines (probably grapevines) growing there. A fine levied five years after the purchase refers also to two orchards—did Shakespeare plant them? In winter ten fireplaces heated the mansion. No wonder Stratford folk spoke of New Place as The Great House.

Built in the fifteenth century by a member of the prominent Clopton family, New Place eventually passed into the hands of William Underhill. He had the misfortune to be poisoned by his son and heir Fulk, who was duly executed for parricide in 1599. The house that Shakespeare bought had suffered decades of neglect; hence, one gathers, the low purchase price of £60 paid by the poet. Apparently he made repairs: the load of stone that 'Mr. Shaxpere' sold the Corporation in 1598 was quite likely left over from work done on the mansion. By then he had already settled into New Place with his family, for by February 1598 he was storing grain in his new barns. An inventory made by the town lists him as having ten quarters (eighty bushels) of corn and malt for brewing. Only two residents in Chapel Street ward are credited with more.

In the seventeenth century New Place must have been a splendid house. Distinguished guests—even royalty—stopped there. During the Civil War, when Queen Henrietta Maria journeyed across England to join her husband in Oxford in July 1643, she passed two nights at The Great House as the guest of Lady Bernard. It has long since disappeared with little trace. Today only a few foundation stones remain, but a pen-and-ink drawing of New Place sketched from memory in the eighteenth century has preserved for posterity the essential features of that imposing structure of three storeys and five gables.

Upon the death of John Shakespeare in 1601, the poet presumably came into possession of the double house in Henley Street. There his mother continued to live, as did his married sister, Joan Hart, and her family.

On 1 May 1602 Shakespeare paid the goodly sum of £320 in cash for 107 acres of arable land, with grazing rights for sheep and cattle in the common pasture. This large tract lay in Old Stratford, a farming area about a mile and a half north of the town. Tenants, Thomas and Lewis Hiccox, tilled the soil. Shakespeare bought this freehold estate from William Combe of Warwick and his nephew John Combe. By lending money this John became the richest man in Stratford; of all the dwellings in the town only his mansion, The College House, rivalled New Place. The old bachelor sued many debtors, but when he died in 1614 he left £20 for the poor, as well as £60 for the erection of a suitably impressive monument to his own memory in Holy Trinity, and bequests totalling over £1500. He did not forget the poet, to whom he left £5. About his personal relations with Shakespeare the records are silent, but in the next century traditions would make good the deficiency by furnishing colourful anecdotes.

Also in 1602 Shakespeare acquired a cottage and a garden of a quarter-acre on the south side of Chapel Lane, opposite the garden of New Place. The cottage, one gathers, provided accommodation for Shakespeare's gardener or some other servant. Thus the estate grew.

The dramatist made his largest investment on 29 July 1605. On that day Ralph Hubaud, a former sheriff of Warwickshire, sold Shakespeare his half interest in the lease of tithes of 'corn, grain, blade, and hay' from Old Stratford, Bishopton, and Welcombe, and (with certain exclusions) the 'tithes of wool, lamb, and other small and privy tithes' from Stratford parish. The purchaser agreed to pay rents amounting to £22 a year, and to collect the tithes himself or have them collected in his behalf. They brought him £60 a year net. Ralph Hubaud died soon after the transaction. An inventory of his estate made on 31 January 1606 shows a debt of £20 'Owing by Mr. Shakespre'.

He engaged in minor legal wrangling. In 1608 he sued John Addenbrooke in the Stratford Court of Record to recover a debt of £6. The case dragged on for almost a year. Eventually the jury awarded Shakespeare his claim plus 24s. costs and damages. In the same court a few years earlier the attorney William Tetherton had pleaded against Shakespeare's neighbour Philip Rogers, an apothecary and tobacconist who had bought twenty bushels of malt and (in addition) borrowed 2s. from the owner of New Place, but returned only 6s. of the entire debt. Such litigation was commonplace in those times; the same Rogers was sued for debt by others, including two clergymen. Shakespeare also paid his share for civic improvement. In 1611 he contributed 'towards the charge of prosecuting the bill in Parliament for the better repair of the highways and amending divers defects in the statutes already made.' His name, inserted in the margin, follows those of the chief alderman, the steward, and the other aldermen. Only a few of the seventy-one names listed are, like Shakespeare's, preceded by the honorific 'Master'.

The land and houses that Shakespeare acquired and took pains to conserve bear witness to his material success. Along with his measure of worldly prosperity he early procured the emblem of status. When the College of Heralds in 1596 granted a coat of arms to John Shakespeare, they were probably acting on an application presented by the son in his father's name; so John's decayed fortunes would lead one to expect. (The fees were considerable.)[9] Henceforth the records speak of William Shakespeare, gentleman.

By revealing a man concerned with money, real estate, and social position, these documents, however deficient in human interest, add another dimension to our understanding of the creator of *Hamlet*. Shakespeare's prosaic interests would recommend him in a later age to partisans of the Commercial Spirit, but would alienate those nurtured on the romantic idea of the Poet. The records also show that Shakespeare's provincial roots ran deep; not for him the Bohemian squalor to which the University Wits succumbed. Here Marlowe provides the opposing image: the brilliant iconoclast who, after settling himself in London, had little to do with the Canterbury of his birth. Marlowe died young, and Shakespeare became a burgher.

# 4

# *The Player and Playwright of London*

ABOUT the origins of his professional career that made possible his rise to eminence in his home town, the records are frustratingly mute. They do not tell us when, or under what circumstances, Shakespeare embarked for the first time on the high road leading from Stratford to the capital. Nor do we know the name of the acting troupe to which he first attached himself, or the capacity in which he served them at the outset. About all these matters traditions, legends, and conjectures would in time flourish like the green bay tree. Records do, however, exist for Shakespeare in London once he established himself there. They include comments about him as a man—all the more precious by reason of their scantiness—made by those who knew him personally; these are considered elsewhere in this narrative, apart from the impersonal documents that now occupy us.

The formal records tell us of his activities as actor, playwright, and shareholder in a theatrical company. The earliest official notice of his name appears in the Declared Accounts of the Treasurer of the Royal Chamber for 15 March 1595. Along with William Kempe and Richard Burbage he is listed as one of the payees for the performance of 'two several comedies or interludes' before the Queen by the Lord Chamberlain's Men during the previous Christmas season. This early, Shakespeare is linked in a position of responsibility with the company's two leading actors. (Theatrical history in the early 1590s is complicated. The Lord Chamberlain's troupe emerged in 1594 from the remnants of the old Lord Strange's Servants. Did Shakespeare begin his career with the latter organization, or with the rival Pembroke's Men? We can only guess.) The poet's name heads the printed list of the principal comedians who acted Jonson's *Every Man in His Humour* in 1598, and appears also in the cast list of the same dramatist's *Sejanus His Fall* (1603); in the First Folio, Shakespeare is given pride of place in the table of 'the Principal Actors in all these Plays'. None of these records indicates the parts he took: tradition would furnish more precise information. In 1598 Shakespeare's name graces the title-pages of editions of *Richard II*, *Richard III*, and *Love's Labour's Lost*, although these were not the first of his plays to be published. He became a theatrical entrepreneur: an inventory of 16 May 1599 describes the newly erected playhouse as in the occupation of William Shakespeare and others ('in occupacione Willielmi Shakespeare et aliorum'); another inventory, in 1601, singles out 'Richard Burbage and William Shackespeare, Gent.' for mention as tenants of the Globe.

The names of Shakespeare and his celebrated fellow actor come together again at around the same time in a rather different kind of record. In an entry in his *Diary* dated 13 March 1601 (1602 according to the modern calendar), John Manningham, then enrolled at the Middle Temple, records a story told him by his fellow-student Edward Curle:

Upon a time when Burbidge played Richard III there was a citizen grew so far in liking with him that, before she went from the play, she appointed him to come that night unto her by the name of Richard the Third. Shakespeare, overhearing their conclusion, went before, was entertained and at his game ere Burbidge came. Then, message being brought that Richard the Third was at the door, Shakespeare caused return to be made that William the Conqueror was before Richard the Third.[10]

So that the point may not be missed, Manningham helpfully adds: 'Shakespeare's name William'. Whether the story is true we cannot say, although clearly Manningham knew that Burbage played Richard and that he had an association with Shakespeare at the Globe. The anecdote would embarrass some future biographers.

Ten days after his arrival in London to begin his reign in 1603, King James I instructed the Keeper of the Privy Seal, Lord Cecil, to prepare for Shakespeare's company a formal Patent under the Great Seal of England. Two days later, on 19 May, the Patent was issued. It licenses and authorizes these our servants Lawrence Fletcher, Will[ia]m Shakespeare, Richard Burbage, Augustine Phillipps, John Heninges, Henrie Condell, Will[ia]m Sly, Rob[er]t Armyn, Richard Cowly, and the rest of their associates, freely to use and exercise the art and faculty of playing comedies, tragedies, histories, interludes, morals, pastorals, stage-plays and such others, like as they have already studied or hereafter shall use or study, as well for the recreation of our loving subjects as for our solace and pleasure when we shall think good to see them, during our pleasure . . . within their now usual house called the Globe . . . And to allow them such former courtesies as hath been given to men of their place and quality. And also what further favour you shall show to these our servants for our sake we shall take kindly at your hands.[11]

Thus did the Chamberlain's company receive royal patronage and become the King's Men. In an historic move the King's Men acquired in 1608 the lease to the enclosed Blackfriars Theatre, although they did not abandon the Globe. From various documents we can determine Shakespeare's financial interests in the two playhouses associated with his troupe. Eventually scholars would attempt, on the basis of elaborate arithmetical calculations, to determine Shakespeare's actual earnings. Their results vary, but all agree that the professional income of the actor-playwright-entrepreneur was ample.

He required residences in London as well as in Stratford—bachelor accommodation, presumably, for there are no references to his wife and children in the metropolis. From tax assessments on personal property in the *Pipe Rolls* we learn that at some time before October 1596 Shakespeare was living in the parish of St Helen's, Bishopsgate, near the Shoreditch play-houses, The Theatre and the Curtain, where the Chamberlain's Men were then performing. His goods were valued at the not inconsiderable sum of £5. In November 1596 a writ indicates that by then the playwright had migrated across the Thames to the Surrey Bankside. By 1599, the tax records show, Shakespeare was dwelling in the Liberty of the Clink in Southwark, where the newly built Globe Theatre stood.

In 1604, and perhaps for a time before and after, 'one Mr Shakespeare . . . lay in the house' of Christopher Mountjoy, a French Huguenot tiremaker (that is, manufacturer of women's ornamental headdresses). His large dwelling—with shop on the ground floor and lodgings above—was situated at the corner of Monkwell and Silver streets, near St Olave's Church, in Cripplegate ward within the north-west corner of the city walls. Not very far off stood St Paul's Cathedral. There in the Churchyard, centre of the book trade, Shakespeare could examine new volumes on the stalls, and in the Cathedral itself, in Duke Humphrey's walk, he might observe the great human scene, for St Paul's was a magnet drawing Londoners from all walks of life.

In 1613, Shakespeare was paid 44s. for devising the *impresa*—the emblem and motto—for the shield carried by the Earl of Rutland in the tilt at Court on the King's Accession day, 24 March; the great Burbage, an amateur artist, painted Shakespeare's invention. The same month we find the poet adding to his estate by purchasing a 'dwelling house or tenement with th' appurtenances' in the Blackfriars. It stood on the west side of Puddle Hill, at the bottom of which ran the narrow creek of the Thames known as Puddle Dock; the winter playhouse of the King's Men stood within easy distance. For this property Shakespeare paid Henry Walker, 'citizen and minstrel of London', £80 in cash, and entered into a mortgage of £60 as security for the balance. Walker conveyed the house in joint ownership to Shakespeare and three others: William Johnson, John Jackson, and John Heminges. Johnson was the host of the Mermaid; Heminges was of course Shakespeare's actor-friend. The identity of Jackson is less certain, for the name, then as now, was common; the deed merely describes him as a gentleman of London. These men acted as trustees in Shakespeare's interest: he alone paid the purchase money, and to him and his heirs and assigns went all the rights in the property. It is more likely that Shakespeare bought the Blackfriars Gate-house (as it is called) for investment purposes, rather than for use as a residence; by this time he had retired to Stratford. So much we may gather from the Belott-Mountjoy suit, where the witness is cited as 'William Shakespeare of Stratford upon Aven in the county of Warwick, gentleman'. Occasionally he would journey to London to attend to business or take part in the activities of his troupe—*Henry VIII* belongs to this phase—but for all practical purposes he had returned home, there to pass his remaining days.

# 5
# *End of the Line*

HE did not have a great many days left. In January of either 1615 or 1616 he summoned his solicitor, Francis Collins of Warwick, to draw up a last will and testament. An able attorney employed also by the Combes, this Collins had served Shakespeare before in his property transactions (as when the poet bought his interest in the Stratford tithes in 1605). The will must have been dictated after July 1614, when John Combe died, for in it Shakespeare names not John but the usurer's nephew and heir Thomas, to whom he bequeaths his sword. Shakespeare was still, according to the conventional formula employed on such occasions, 'in perfect health and memory, God be praised', and he signed the third (and last) sheet of the instrument in a firm hand. On 25 March 1616 he recalled Collins to make revisions necessitated by Judith's marriage in the previous month. By then his health and memory were less perfect: his pen trembled as he signed the first two sheets, which Collins (or his clerk) had apparently recopied, and he could not supply the name of his ten-year-old nephew Thomas Hart—the lawyer had to leave a blank space.

But in his final testament, which has been justly described as 'a characteristic will of a man of property in the reign of James I', Shakespeare was able to dispose of his worldly belongings, as his wisdom dictated, for the benefit of the living. To the poor of Stratford he left £10. His little godson William Walker (aged eight) received 20s. in gold. Shakespeare remembered local acquaintances whose names and histories scholars would one day rescue from oblivion for no other reason than that these men were his friends: William Reynolds, Anthony Nash, and Anthony's brother John. These three received 26s. 8d. each to buy memorial rings. Hamnet Sadler, a lifelong friend, was remembered in the same way. So too were London colleagues from the King's Men days: the dramatist's fellows Richard Burbage, John Heminges, and Henry Condell. The last two, especially, would not forget so worthy a friend. Nowhere in his testament does he mention any lord or great man, as the world measures greatness: a circumstance rather uncooperative with the snobbish urge to place Shakespeare on familiar terms with courtiers and aristocrats.

Mainly he thought of his family. There was his last surviving sister Joan Hart, whose husband would be dead in less than three weeks. To her Shakespeare willed the use during her lifetime of the Henley Street residence where she was living; also £20 in cash and his wearing apparel. Her three sons—William, Michael, and the unspecified Thomas—inherited £5 each. To his younger daughter Shakespeare bequeathed £150 in cash: £100 as her marriage portion, to be paid within twelve months, and £50 more upon surrendering to Susanna her interest in the Chapel Lane cottage. Within three years from the date of the will another £150 would fall to her or (in the event of her decease) her issue. And lastly Judith was left Shakespeare's broad silver-gilt bowl.

Above all he looked after Susanna and her offspring. Elizabeth Hall, then eight, was left Shakespeare's plate. To her mother went the bulk of the estate:

All that capital messuage or tenement with the appurtenances in Stratford aforesaid called the New Place, wherein I now dwell, and two messuages or tenements with the appurtenances situate, lying, and being in Henley Street . . . and all my barns, stables, orchards, gardens, lands, tenements, and hereditaments whatsoever, situate . . . within the towns, hamlets, villages, fields, and grounds of Stratford upon Avon, Old Stratford, Bushopton, and Welcombe, or in any of them, in the said county of Warwick; and also all that messuage or tenement with the appurtenances wherein one John Robinson dwelleth, situate . . . in the Blackfriars in London, near the Wardrobe; and all other my lands, tenements, and hereditaments whatsoever.[12]

The 'hereditaments whatsoever' would include Shakespeare's miscellaneous investments, such as his tithe interests. All this was for Susanna to have and to hold.

The full import of the will emerges from the passage that comes next, which reveals the testator's intention of keeping intact, hopefully through his elder daughter's male line, the estate that with the endeavours of art and with business acumen he had over a lifetime built up. After Susanna's death the tenements, appurtenances, and hereditaments bequeathed to her were to pass 'to the first son of her body lawfully issuing and to the heirs males of the body of the said first son lawfully issuing, and for default of such issue to the second son of her body lawfully issuing and to the heirs males of the body of the said second son lawfully issuing'. This progresses all the way down to 'the heirs males of the bodies of the said fourth, fifth, sixth, and seventh sons lawfully issuing . . .; and for default of such issue the said premises to be and remain to my said niece Hall and the heirs males of her body lawfully issuing; and for default of issue, to my daughter Judith and the heirs males of her body lawfully issuing.' Appropriately the Halls were named as executors of the will.

About his wife Shakespeare was silent, except for an interlineated addition in the third sheet, following the instructions for the entailing of the estate: 'Item, I give unto my wife my second-best bed with the furniture'—that is, the bed furnishings (hangings, linen, etc.). Does the meagreness of the bequest betray indifference or derision? Or had this object sentimental associations that did not attach to the best bed, which was perhaps reserved for overnight guests at New Place? Long afterwards someone would point out that, according to English law, a widow was entitled to one third of her husband's goods and real property, and that there was no need for this provision to be rehearsed in the will.

Under what circumstances the poet died we do not know, although half a century later an explanation would be offered. On Thursday, 25 April—one month to the day after he had drawn up his will—Shakespeare was buried. We do not know which of his fellows attended the service on that early spring day, or in what terms the preacher eulogized the departed—was it mentioned that this good citizen of Stratford wrote plays? Later, much later, tales would circulate about what happened on that solemn occasion. Shakespeare's widow lived to see the monument erected in her husband's honour by 1623. That August she died and was buried alongside him in the chancel of Holy Trinity Church.

The elaborate programme for holding together the estate failed. No son lawfully issued from the body of Susanna Hall, nor (for that matter) did she bear another daughter. She died in 1649, at the age of sixty-six, and was laid to rest in the chancel of Holy Trinity, next to the grave of her husband, whom she had outlived by fourteen years. Their daughter Elizabeth came into possession of New Place and the rest of the entailed estate. On 22 April 1626 (it would later be asked, was the day chosen because it was her grandfather's birthday?) she had married Thomas Nash, eldest son of the Anthony remembered by Shakespeare in his will; he died in 1647, and was buried in Holy Trinity to the right of the poet. Two years later, on 5 June 1649, the widow married John Bernard at Billesley, a hamlet about four miles from Stratford. Elizabeth had no children by either match. After her interment at Abington on 17 February 1670 the houses on Henley Street passed into the hands of her cousins the Harts; the rest of the Shakespeare estate was sold following her husband's decease four years later. Judith Quiney's children had long since died: Shakespeare at six months, Richard at eleven years, Thomas at nineteen—the last two within a month of one another in 1639. The Stratford register records the burial of Judith on 9 February 1662, but the actual site of her grave is not known. Thus Shakespeare's direct line was extinct by 1670.

Even today the biographical records are widely scattered. They are preserved in the Shakespeare Birthplace Trust Records Office, the College of Heralds, the London Guildhall Library, the Public Record Office, the British Museum, the Worcestershire Record Office, Belvoir Castle, the Folger Shakespeare Library, and elsewhere. If these documents offer no clues to the anxieties, crises, and triumphs of Shakespeare the man—if they in no way illuminate the dark corners of genius—nevertheless, on a more prosaic level, they offer much for the biographer to ponder concerning the poet's origins and family, his ties to the town in which he achieved such prestige, and his remarkably thoroughgoing professional commitment as actor, playwright, and entrepreneur. Much patient sleuthing has gone into unearthing the facts we possess concerning Shakespeare; literally millions of documents have been turned over in the quest.

In the seventeenth century the Shakespeare documents were unknown, and the absence of concrete facts gave free rein to the growth and dissemination of the apocryphal stories that it is the fate of great men to inspire. These legends, the accomplishment of the myth-makers, will be part of our story. So too will be speculation on the gaps in the record, once it became available: the matter of the mysterious Lost Years, for instance, between the time Shakespeare's twins were christened in Stratford and the dramatist's first appearance in London. Several of the documents themselves would arouse fierce contention, especially the will, with its reference—seemingly an afterthought—to a second-best bed. All these developments will complicate and enrich our narrative.

# 6
## Greene and Chettle

To the laconic notations entered in their records by municipal or church authorities and the formal contracts drawn up by attorneys, we may add the more humanly rewarding comments made by those who knew Shakespeare—or knew of him—while he lived, or who read and were impressed by his writings. Almost everyone spoke well of him.

A striking exception is our first word in print regarding Shakespeare: the notorious attack by Robert Greene in 1592 in a pamphlet whose title sufficiently indicates its contents: *Greene's Groats-worth of Wit, Bought with a Million of Repentance. Describing the Folly of Youth, the Falsehood of Makeshift Flatterers, the Misery of the Negligent, and Mischiefs of Deceiving Courtesans.* This edifying work was, the title-page further informs us, *Written before his death and published at his dying request.*

Greene had lived riotously, sustaining himself by the prolific production of pamphlets, amorous or cony-catching, and by the plays, among them the romantic and spectacular *Friar Bacon and Friar Bungay*, with which he graced the London stage. Now desperately ill and desperately poor, he lay infested with lice in his squalid lodgings. His old companions—Nashe and the rest—did not visit him in his extremity. Thus isolated, he recalled the follies of his youth and reproached himself for abandoning his wife. With envious bitterness he brooded about the base-born actors who prospered by mouthing the lines that he, a Cambridge BA, had provided them. In such a condition he wrote his *Groatsworth of Wit*.

'The swan sings melodiously before death,' Greene remarks in his preface, 'that in all his lifetime useth but a jarring sound. Greene though able enough to write, yet deeplier searched with sickness than ever heretofore, sends you his swan-like song.'[13] The narrative recounts the adventures of Roberto, an impoverished graduate who is hired to write plays by a gorgeously attired actor who has grown rich on the labours of scholars. Roberto thus becomes 'an arch play-making poet'; he takes up with dissolute companions, he learns the sharp practices of the underworld, he drinks and wenches. Abruptly the fictional persona—in any event a thin disguise—is dropped, and Greene directly offers his readers the fruits of his repentance in the form of a series of moral imperatives. This is followed by a letter to three of his 'fellow scholars about this city': Marlowe and (probably) Nashe and Peele. In the course of it occurs the famous outburst against Shakespeare:

Base-minded men all three of you, if by my misery you be not warned; for unto none of you (like me) sought those burrs to cleave, those puppets (I mean) that spake from our mouths, those anticks garnished in our colours. Is it not strange that I, to whom they all have been beholding: is it not like that you, to whom they all have been beholding, shall (were ye in that case as I am now) be both at once of them forsaken? Yes, trust them not: for there is an upstart crow, beautified with our feathers, that with his *Tiger's heart wrapt in a player's hide*, supposes he is as well able to bombast out a blank verse as the best of you; and being an absolute *Johannes Factotum*, is in his own conceit the only Shake-scene in a country.[14]

That Greene has singled out Shakespeare for attack is evident not only from the punning reference to Shake-scene, but also from the parody of a line in *3 Henry VI* (I. iv. 137): 'O tiger's heart wrapt in a woman's hide!' By altering one word Greene has accused Shakespeare of cruelty.

It is not his only charge. The rest, being obscure in the best Elizabethan tradition, would in later days evoke seemingly endless debate. The passage as a whole is directed against the actors ('Puppets . . . that spake from our mouths') who live off the dramatists. Greene was in trouble with the players: he had made the mistake of selling the same play to two different companies. Shakespeare is an actor, and is attacked as such; the phrase about the 'upstart crow, beautified with our feathers' continues the idea expressed in 'anticks garnished in our colours'. Furthermore, this mere actor has the audacity to set himself up as a universal genius (*Johannes Factotum*) who can, by turning out stilted and bombastic verse, rival his superiors and deprive them of employment. Many would interpret the passage in this way.

But does it perhaps harbour a more sinister accusation? The crow may not be the bird taught to imitate its betters that derives ultimately from Aesop, Martial, and Macrobius; Greene may instead allude to the third *Epistle*, in which Horace uses the image of the crow divested of its plundered lustre (*furtivis nudata coloribus*) in connection with the idea of plagiarism. Is Greene then maliciously suggesting that Shakespeare appropriated his work? This would be the second interpretation. It would give rise to the view that Shakespeare began his literary career as a Johannes Factotum in the sense of a Jack of all trades who—in addition to acting—revised and adapted the plays of others, including Greene. About the whole issue controversy would in time become so wearying that one twentieth-century commentator would be impelled to declare, 'This passage from Greene has had such a devastating effect on Shakespearian study that we cannot but wish it had never been written or never discovered'.[15]

It caused turmoil in its own day. Nashe denounced the *Groatsworth of Wit* as 'a scald, trivial, lying pamphlet'.[16] Its author being dead, the brunt of displeasure fell on the hapless Henry Chettle, whose misfortune it was to have assumed responsibility by preparing a fair copy of the work for the printers from Greene's illegible scrawl. In a prefatory address to his *Kind-Heart's Dream* (printed in December 1592), Chettle confesses that Greene's 'letter written to divers play-makers' has been 'offensively by one or two of them taken'. The grieved parties were, the context makes clear, Marlowe and Shakespeare. In a celebrated passage Chettle comments sharply on the former and apologizes to the latter:

With neither of them that take offence was I acquainted, and with one of them I care not if I never be. The other, whom at that time I did not so much spare as since I wish I had, for that as I have moderated the heat of living writers, and might have used my own discretion (especially in such a case, the author being dead)—that I did not, I am as sorry as if the original fault had been my fault, because myself have seen his demeanour no less civil than he excellent in the quality [i.e. acting] he professes. Besides, divers of worship have reported his uprightness of dealing, which argues his honesty, and his facetious grace in writing, that approves his art.[17]

The first complainant, whose learning Chettle reverences, appears to have offered threats ('him I would wish to use me no worse than I deserve'), and Chettle reveals that he struck out from Greene's letter an aspersion which, 'had it been true, yet to publish it was intolerable'—could this have been an accusation against Marlowe of homosexuality? He has added nothing of his own, 'for I protest it was all Greene's, not mine nor Master Nashe's, as some unjustly have affirmed'.[18]

It is a glowing tribute to Shakespeare, this first glimpse we get of him as a man, and all the more effective for being set off by the evocation, however brief, of the turbulent Marlowe, to whom no apology is forthcoming. The reference to testimonials from 'divers of worship'—did Shakespeare bear them in hand when he visited Chettle or were they sent along to him afterwards?—indicates that the poet had supporters in high places. The defence of his 'uprightness of dealing' would seem to imply that he took Greene's innuendoes as impugning Shakespeare's honesty.

# 7

# *Reputation*

ALLUSIONS to Shakespeare around this time focus on the writings rather than the writer, and so they have not the immediate biographical pertinence of the Greene–Chettle affair; but they evidence the beginnings of an impact upon an audience on the part of an author in process of establishing himself. In later ages biographers concerned with this phase of Shakespeare's career would sometimes tend to impose on the past the values of their own day, and to exaggerate the significance of early citations. Not all references to Tarquin and Venus necessarily imply familiarity with *The Rape of Lucrece* and *Venus and Adonis*. Mentions of Shakespeare's works (despite enthusiastic claims by some students) are far from numerous in the last decade of the sixteenth century, and for the most part are tucked away in obscure publications put out by the lesser literati. They show that Shakespeare had begun to make his mark early, but they do not permit us to generalize safely about the scope of his appeal or to project upon times past our own idolatry. To this category of allusion belong the reference by 'W. Har.' (William Harvey?) to 'You that have writ of chaste Lucretia' in *Epicedium* (1594), Richard Barnfield's tribute to Shakespeare's 'honey-flowing vein' in *Poems in Divers Humours* (1598), and the antiquary John Weever's verses in *Epigrams in the Oldest Cut, and Newest Fashion* (1599) praising the 'honey-tongu'd' creator of rose-cheeked Adonis and proud lust-strung Tarquin.

Francis Meres's commonplace book, *Palladis Tamia, Wit's Treasury*, furnished, in 1598, the most effusive early recognition of Shakespeare. The title-page of the 1634 edition describes it, reasonably enough, as *A Treasury of Divine, Moral, and Philosophical Similes and Sentences, Generally Useful*. Much interest attaches to the section labelled 'A comparative discourse of our English poets, with the Greek, Latin, and Italian Poets', in which Meres heaps praise on Sidney and on *The Faerie Queene*, as well as on many other authors and works. But it is for his remarks on Shakespeare that he is chiefly remembered. 'The sweet, witty soul of Ovid lives in mellifluous and honey-tongued Shakespeare', Meres remarks, 'witness his *Venus and Adonis*, his *Lucrece*, his sugared sonnets among his private friends, &c.'[19] And he goes on to compare Shakespeare with the most notable Romans:

As Plautus and Seneca are accounted the best for comedy and tragedy among the Latins, so Shakespeare among the English is the most excellent in both kinds for the stage; for comedy, witness his *Gentlemen of Verona*, his *Errors*, his *Love Labour's Lost*, his *Love Labour's Won*, his *Midsummer's Night Dream*, and his *Merchant of Venice*; for tragedy, his *Richard the Second*, *Richard the Third*, *Henry the Fourth*, *King John*, *Titus Andronicus*, and his *Romeo and Juliet*.[20]

He is praised as one of the best for lyric poetry, for tragedy, and for comedy, and as one of 'the most passionate among us to bewail and bemoan the perplexities of love'.[21] If the Muses spoke English, they would choose Shakespeare's 'fine filed phrase'. Clearly Shakespeare had in Meres an ardent admirer, and it is not surprising that his tribute, regarded out of context, should mislead later students about the nature of Shakespeare's contemporary reputation. In point of fact Meres in the *Palladis Tamia* refers to Drayton more often than to Shakespeare, and he offers homage to 125 English writers, painters, and musicians. Although he has been accused of intelligence, the inclusiveness of the listings does not inspire confidence in Meres's powers of critical discrimination.

Whatever its limitations, *Palladis Tamia*, recovered in the eighteenth century, would prove of inestimable value. Students concerned with the chronology of Shakespeare's writings would find in it *termini ad quem* for no fewer than eleven plays, and definite evidence that some, at least, of the Sonnets were in circulation by 1598. The citation of *Love's Labour's Won* would tease future generations of scholars. And the very enthusiasm of Meres's praise would reveal much about the predispositions of those who sought to reconstruct Shakespeare's standing as a writer in the closing years of the old Queen's reign.

From the beginning of the new century until his death the scattered references to Shakespeare, many of them from manuscript sources, hardly add up to a very impressive testimonial to his contemporary reputation. They do show that he had some admirers, including the esteemed headmaster of Westminster, William Camden, sufficiently impressed by his writings to take the pains to set down their appreciation. They testify also to his popularity as a dramatist. But where one would look for the highest praise, from his fellow playwrights, it is (with one famous exception) withheld. We sense too some condescension towards the author who is a mere actor, and not, by the standards of the day, a learned man.

The playwrights say little. Fletcher, who succeeded Shakespeare as principal dramatist for the King's Men and during the period of transition (it seems likely) collaborated with him in the writing of at least one play, not once mentions his great colleague. In his preface to *The White Devil* (1612), John Webster mentions those other men by whose worthy labours he would wish his own to be judged; Chapman, Jonson, Beaumont, Fletcher, 'and lastly (without wrong last to be named) the right happy and copious industry of Master Shakespeare, Master Dekker, and Master Heywood'. A few years later the author of that delicious parody, *The Knight of the Burning Pestle*, addressed a verse-letter to Jonson in which he speaks of Shakespeare as representative of fecund Nature untutored in the doctrine and discipline of Art:

> here I would let slip
> (If I had any in me) scholarship,
> And from all learning keep these lines as [cl]ear
> As Shakespeare's best are, which our heirs shall hear
> Preachers apt to their auditors to show
> How far sometimes a mortal man may go
> By the dim light of Nature.[22]

Here Beaumont first suggests that popular early conception of Shakespeare as Fancy's child, warbling his native woodnotes wild.

His death evoked no great outpouring of homage. That was reserved for his rival Jonson, who was accorded, six months after he expired, an entire volume of eulogy. In the 1640 edition of Shakespeare's *Poems*, it is true, an

anonymous versifier declared that 'every eye that rains a shower for thee, |
Laments thy loss in a sad elegy'.[23] But the flow does not seem to have been so
very copious, for only a few noted his passing. William Basse urged
'Renowned' Spenser, 'learned' Chaucer, and 'rare' Beaumont to make room
in Westminster Abbey 'For Shakespeare in your threefold, fourfold tomb'[24]—
a conceit at which Jonson was, by implication, to sneer in his own lines to
Shakespeare's memory:

> I will not lodge thee by
> Chaucer, or Spenser, or bid Beaumont lie
> A little further, to make thee a room:
> Thou art a monument without a tomb.[25]

In *The Praise of Hempseed*, published in 1620, John Taylor extolled Shakespeare (whom he may have known personally) as excelling in art and
surviving immortally on paper; but he was not moved to write an elegy, as he
later did for Jonson. The rest was silence, until in 1623 there appeared the
noble volume of plays that his companions in the King's Men, Heminges and
Condell, assembled 'without ambition either of self-profit or fame: only to
keep the memory of so worthy a friend and fellow alive, as was our
Shakespeare'.[26]

In a famous passage of their preface, they recall his happy facility: 'His
mind and hand went together, and what he thought, he uttered with that
easiness that we have scarce received from him a blot in his papers'. Such
untrammelled fluency of composition seems to have left a deep impression on
Shakespeare's fellow artists. So we may gather from the celebrated rejoinder
left by Jonson in his papers, and published after his death in *Timber, or
Discoveries*. It provides a warm personal description of Shakespeare all the
more impressive for being wrung from Jonson, a laborious writer, while in a
mood of irritable protest:

I remember the players have often mentioned it as an honor to Shakespeare, that in
his writing (whatsoever he penned) he never blotted out line. My answer hath been,
'Would he had blotted a thousand!' Which they thought a malevolent speech. I had
not told posterity this but for their ignorance, who chose that circumstance to
commend their friend by wherein he most faulted; and to justify mine own candor, for
I loved the man, and do honor his memory on this side idolatry as much as any. He
was, indeed, honest, and of an open and free nature; had an excellent phantasy, brave
notions, and gentle expressions, wherein he flowed with that facility that
sometime it was necessary he should be stopped. '*Sufflaminandus erat*', as Augustus
said of Haterius. His wit was in his own power; would the rule of it had been so, too! . . .
But he redeemed his vices, with his virtues. There was ever more in him to be praised
than to be pardoned.[27]

The remarks 'on our fellow countryman Shakespeare', *De Shakespeare
nostrati*, belong to a private record. In public Jonson offered unstinting—or

almost unstinting—praise. His eulogy in the First Folio is justly esteemed one of the most splendid occasional poems of the language. In it a great dramatist and a self-praiser not easily moved to speak well of another offers homage to his colleague and supreme competitor. That the most erudite poet of the age should express a qualification about Shakespeare's learning need provoke no surprise: yet Jonson's casual observation that the dramatist had 'small Latin, and less Greek', perhaps deriving from Minturno's *Arte Poetica* ('poco del Latino e pochissimo del Greco'), would be deeply meditated by future generations of scholars. In a revealing passage Jonson, while acknowledging the gifts bestowed on his subject by Nature, goes on to emphasize the role of Art:

> Yet must I not give Nature all; thy Art,
>     My gentle Shakespeare, must enjoy a part.
> For, though the poet's matter Nature be,
>     His Art doth give the fashion; and that he
> Who casts to write a living line must sweat
>     (Such as thine are) and strike the second heat
> Upon the Muses' anvil, turn the same
>     (And himself with it) that he thinks to frame,
> Or for the laurel he may gain a scorn;
>     For a good poet's made as well as born.[28]

Thus does Jonson vindicate Shakespeare from the charge of facility, which evidently still rankles him. But what remains with the reader is not Jonson's implied self-defence of his own creative way, nor the peculiarly Jonsonian interpretation of Shakespeare's purpose as being (did he say this merely for the sake of the pun?) to shake a lance at ignorance in 'his well turned, and true-filed lines'. Rather we think of gentle Shakespeare, the sweet swan of Avon, the applause, delight, and wonder of the stage, the poet who is a monument without a tomb and who wrote not for an age but for all time.

Jonson's panegyric, to which Heminges and Condell gave pride of place, far outshines the other contributions to the preliminary matter of the Folio. Yet such an acknowledgement pays no great compliment to the eulogist, for he has scant competition. Commendatory verses were supplied by just three other admirers, and these can only be described as comparative nonentities. (The Second Folio in 1632, however, boasts Milton's sonnet.) By contrast, the 1651 octavo edition of *Comedies, Tragi-comedies, with Other Poems* by William Cartwright—the now justly forgotten author of *The Royal Slave* and *The Ordinary*—appeared with fifty-seven poems honouring the writer. Such are the vagaries of reputation.

# 8

# The Shakespeare Canon

IN addition to commendatory poems, the Heminges and Condell Folio carried a dedication to the Earl of Pembroke and to the Earl of Montgomery. The volume also contains thirty-six plays. Eighteen of them reached print for the first time in this way. They represent the achievement of more than half a lifetime of unparalleled creation; the better half, for they include *Julius Caesar*, *Macbeth*, *Antony and Cleopatra*, and *The Tempest*. To these glories we may add *Othello*, which also appeared some time after the playwright's death, in a quarto edition issued in 1622. Apparently he died neither knowing, nor caring, about the ultimate fate of works that posterity would value beyond all other accomplishments of the literary imagination. The plays published during his lifetime, even those that were not unauthorized, he failed even to proofread; so the gross errors, left uncorrected, tell us. Yet he could speak eloquently of the fame that (as King Henry of Navarre remarks in the opening words of *Love's Labour's Lost*) 'all men hunt after in their lives', and in his Sonnets he proclaims more than once the power of his mighty rhyme to outlast marble and the gilded monuments of princes. Such indifference, in the face of such knowledge, constitutes part of the enigma that his biographers would have somehow to confront and for which, in due course, they would offer explanations.

'There are two ways in which we examine the personal life of an artist,' George Seferis has said in connection with the poetry of Cavafy: 'one is by means of anecdotes, surprises, jokes, medical reports; the other is by humbly trying to see how the poet incorporates his perishable life in his work.' In later ages all the plays would come under minute scrutiny on the part of scholars seeking in this way to arrive at an understanding of Shakespeare's values, his philosophical ideas, his political convictions and religious beliefs, his likes and aversions: anything that might shed light on his buried life and on the development of his mind and art. Earlier students would grope in the dark, ignorant even of the approximate dates of Shakespeare's writings; for in the 1623 Folio the plays are printed without regard for chronological sequence, *The Tempest*, one of the last in point of actual composition, being given pride of place. At all times the task would be complicated by the nature of the dramatic medium itself, which does not permit the creator to appear in his own guise but only in the manifold masks of the personages with whom he peoples his plays. Many would carry on the pursuit with only the dimmest awareness of the methodological burdens they had assumed; some still do.

The dramatist might have provided some help, but chose not to. Towards the quartos printed while he lived he maintained a public aloofness; the plays issued from the presses unequipped with any dedication to a patron or commendatory verses by the poet's friends or prefatory word by the author for his readers—that paraphernalia which in other writers of the period sometimes provides welcome guidance. From Shakespeare we have no such enlightening document as the one in which his contemporary, John Webster, excoriated the audience—'those ignorant asses'—which had spurned *The White Devil*, answered the detractors who had sneered at his slowness of composition, and enumerated those dramatists by whose light he wishes to be judged. The only play by Shakespeare to bear a preface is the 1609 quarto of *Troilus and Cressida*, but this has been supplied by the officious publisher, whose only apparent motivation was to market his wares. Nor did Shakespeare ever furnish an Induction, such as those written by Jonson for *Every Man in his Humour* and *Bartholomew Fair*, in which the dramatist overtly states his prejudices and principles. Even Shakespeare's few prologues and epilogues are unpolemical; Jonson had used his for vigorous attack and self-defence. The man keeps his mask always firmly in place; apart from the works themselves there is only silence.

Little wonder, this being the case, that the critical imagination should devise ingenious methods for piercing behind the expressionless eyes of the poet's mask. Centuries after the Folio was published, the verbal images to be found in the plays would be assembled, sifted, and systematically examined. On the basis of such tediously garnered evidence a receptive public would be given a lively picture of the poet's personality, temperament, and mind; even his physical appearance:

The figure of Shakespeare which emerges is of a compactly well-built man, probably on the slight side, extraordinarily well co-ordinated, lithe and nimble of body, quick and accurate of eye, delighting in swift muscular movement. I suggest that he was probably fair-skinned and of a fresh colour, which in youth came and went easily, revealing his feelings and emotions.[29]

Little wonder too that Shakespeare's writings should be ransacked for covert allusions to his personal circumstances, or that individual characters—indeed, at times entire plays—should be read as personal allegory. The nineteenth century would find irresistible the temptation to equate Prospero with Shakespeare. 'The whole play', James Russell Lowell would write of *The Tempest*,

indeed, is a succession of illusions, winding up with those solemn words of the great enchanter who had summoned to his service every shape of merriment or passion, every figure in the great tragi-comedy of life, and who was now bidding farewell to the scene of his triumphs. For in Prospero shall we not recognize the Artist himself—

'That did not better for his life provide
Than public means which public manners breeds
Whence comes it that his name receives a brand',—

who has forfeited a shining place in the world's eye by devotion to his art, and who,
turned adrift on the ocean of life in the leaky carcass of a boat, has shipwrecked on
that Fortunate Island (as men always do who find their true vocation) where he is
absolute lord, making all the powers of Nature serve him, but with Ariel and Caliban
as special ministers? Of whom else could he have been thinking, when he says,—

'Graves, at my command,
Have waked their sleepers, oped, and let them forth,
By my so potent art'?[30]

Possibly, we may irreverently reply, he was thinking of Ovid's Medea, whose
incantatory speech Shakespeare draws upon throughout Prospero's valedic-
tory; the relevant passage in Golding's translation, which the dramatist
knew, goes: 'I call up dead men from their graves.'

While deprecating such fanciful interpretations, Edward Dowden would
contribute, only half-facetiously, one more fanciful than all the rest. Miranda, according to Dowden, represents Art; Ariel, 'the imaginative genius of
poetry but recently delivered in England from long slavery to Sycorax';
Caliban, the grosser appetites and passions; and Ferdinand, Shakespeare's
young colleague and possible collaborator, John Fletcher. Prospero's departure from the island of course signifies Shakespeare's farewell to the stage:
'He returns to the dukedom he had lost, in Stratford upon Avon.'[31] How
indifferent to sentimental yearnings that, having declared the revels ended,
Shakespeare should several years later return to London to try his hand at yet
another revel, *Henry VIII*, from which, alas, patient investigation has been
unable to deduce any intimate revelation to rival those yielded by *The
Tempest*!

Into such dark forests of allegorization would the featureless impersonality
of the book of his plays seduce future generations. Fortunately Shakespeare
himself looked after the non-dramatic writings he published as a young man.
His plays he sold to a company whose property they became, to be acted
before multitudes; but his long narrative poems, *Venus and Adonis* and *The
Rape of Lucrece*, were intended for a sophisticated reading public, and hence
qualified as literature. Accordingly he seems to have seen them through the
press personally—they were carefully proof-read—and furnished them with
the prefatory epistles his plays so conspicuously lack.

From the dedication of *Venus and Adonis* we learn for the first time of
Shakespeare's patron, Henry Wriothesley, third Earl of Southampton and
Baron of Titchfield. The phrasing of the dedication, formal and self-deprecating,
is such as we might expect from an inexperienced writer of country (although
not necessarily humble) origin addressing a nobleman of the realm with whom
he was not on any close personal footing:

Right Honourable, I know not how I shall offend in dedicating my unpolished lines to your Lordship, nor how the world will censure me for choosing so strong a prop to support so weak a burden; only if your Honour seem but pleased, I account myself highly praised, and vow to take advantage of all idle hours till I have honoured you with some graver labour. But if the first heir of my invention prove deformed, I shall be sorry it had so noble a godfather; and never after ear so barren a land, for fear it yield me still so bad a harvest. I leave it to your honourable survey, and your Honour to your heart's content, which I wish may always answer your own wish and the world's hopeful expectation.

Your Honour's in all duty,<br>
William Shakespeare.[32]

In 1593, when the first edition of *Venus and Adonis* appeared, Shakespeare was twenty-eight, and the lord he addressed with such deferential reserve was nineteen. The brief dedication would inspire longer commentary. On the other hand, there could scarcely be any question about the 'graver labour'. That must be *The Rape of Lucrece*, printed the following year. Once again he dedicates his work to the young nobleman:

To the Right Honourable Henry Wriothesley, Earl of Southampton and Baron of Titchfield.

The love I dedicate to your Lordship is without end; whereof this pamphlet, without beginning, is but a superfluous moiety. The warrant I have of your honourable disposition, not the worth of my untutored lines, makes it assured of acceptance. What I have done is yours; what I have to do is yours; being part in all I have, devoted yours. Were my worth greater, my duty would show greater; meantime, as it is, it is bound to your Lordship, to whom I wish long life still lengthened with all happiness.

Your Lordship's in all duty,<br>
William Shakespeare.[33]

Scholars would, of course, place the two dedications alongside one another for minute comparison and proceed diligently to read between the lines. Some would discern in the second greater assurance and a more familiar tone, even (a few would say) the expression of a deep devotion. Had the poet been welcomed into the Southampton circle? A pleasant aristocratic thought.

If the two narrative poems hold interest for the chronicler of Shakespeare's life by reason of their dedications, that interest seems slight when placed in balance with the claims made on the biographical imagination by Shakespeare's most celebrated body of non-dramatic writing, the Sonnets. They constitute a sustained outburst of passionate lyricism before which the mask at last falls—or seems to fall. Again and again the poet uses the pronoun denied him, in his own person, in the plays; we find ourselves in the presence of the naked 'I'.

Although (as Meres affirms) the Sonnets were evidently intended only for manuscript circulation, they appeared in 1609 in an unauthorized edition printed (so the title-page tells us) 'By G. Eld for T.T.'.* From an entry in the Stationers' Register on 20 May 1609 we learn that the initials stand for Thomas Thorpe. It is this Thorpe who furnished the tantalizingly enigmatic dedication that would ultimately provoke controversy which resounds to this day:

TO . THE . ONLIE . BEGETTER . OF
THESE . ENSUING SONNETS .
M$^r$. W.H. ALL . HAPPINESS .
AND . THAT . ETERNITY
PROMISED .
BY .
OUR . EVER-LIVING . POET .
WISHETH .
THE . WELL-WISHING .
ADVENTURER . IN .
SETTING .
FORTH .          T.T.[34]

Who is this Mr W.H.? Is he the 'onlie begetter' of the poems in the sense of being the sole source of their inspiration? Does he have any connection with the H.W. in Henry Willobie's *Willobie His Avisa* (1594), who, 'being suddenly infected with the contagion of a fantastical fit, at the first sight of *A*, pineth a while in secret grief; at length not able any longer to endure the burning heat of so fervent a humour, bewrayeth the secrecy of his disease unto his familiar friend W.S.'?[35] Is he perhaps William Herbert, one of the dedicatees of the First Folio? Or (a more prosaic possibility) did W.H. merely acquire the Sonnets for the illicit edition—*begetter* in this context meaning *procurer*—and in this way earn the publisher's fulsome, if ambiguous, gratitude? The patron of Shakespeare's previous poems had, in Sir William Hervey (or Harvey), a stepfather with the right initials. Unfortunately, the initials are commonplace, a circumstance which has stimulated rather than discouraged conjecture. About the mysterious Mr W.H. many books and articles would be written offering solutions to a riddle that to this day remains unsolved.

The Sonnets themselves reveal—or appear to reveal—in fascinating detail their creator's innermost feelings about himself, his friend, and his mistress, about his profession, about beauty, time, death, and the world. From them we learn the true nature of his commitment to his art. The public reticence of the playwright gives way, in true Elizabethan fashion, to the lyric poet's rapturous awareness of the power of language, despite time, to eternize what it celebrates. Most of the Sonnets are addressed to a high-born young man, 'beauteous and lovely', of around twenty. The poet's most usual epithet for him is 'love' or 'my love'. He has a woman's face and a woman's gentle

---

* Two sonnets, 138 and 144, had found their way into print earlier. Two others, 2 and 100, exist in manuscript versions not identical with Eld's edition.

disposition; but, despite his feminine nature, he is equipped with a penis that rules out, so the poet says, the possibility of their love's consummation. Other sonnets speak of the poet's humiliating infatuation with a young woman of unfashionably dark beauty—black eyes and brows—and darkly intense sexual appetite. Not content with being his mistress, she seduces the beautiful youth before the poet can free himself from enslavement to her. We hear also of another poet whose spirit is 'by spirits taught to write | Above a mortal pitch'; with 'the proud full sail of his great verse,' he sings the same youth's praises.

Who are these personages that have come to be known as the Fair Youth, Dark Lady, and Rival Poet? Can scholarship, by painstaking attention to external circumstances and the internal evidence of the poems, produce convincing historical identifications? The Sonnets, the only intimate document connected with the poet, would thus seem to bring the biographer to the threshold of the knowledge—some of it perhaps forbidden—that is the object of his tireless quest. All this will be part of our story.

To the corpus of thirty-seven plays (thirty-eight if one includes *Sir Thomas More*), the two narrative poems, and the Sonnets may be added a few miscellaneous or doubtful pieces. *The Passionate Pilgrim*, printed in 1599 as '*By W. Shakespeare*', consists of twenty poems, of which several have been identified as the work of other writers. One injured party, Thomas Heywood (two of his verse epistles had been purloined), complained angrily, and let it be known that Shakespeare was 'much offended' with the stationer, William Jaggard, who 'altogether unknown to him, presumed to make so bold with his name'.[36] Apparently Shakespeare protested effectively, for Jaggard removed his name from the title-page. In Thorpe's 1609 edition of the Sonnets, these are followed by *A Lover's Complaint*, a poem of some three hundred lines, with a clear attribution to Shakespeare; but because of the unauthorized nature of the publication, the ascription would not in general win favour. Some critics, however, would convince themselves of the genuineness of the *Lover's Complaint*, and go on to read it as autobiography.

More certainly Shakespeare's is *The Phoenix and the Turtle*, included in 1601, in a strange collection celebrating the love of Sir John Salisbury of Lleweni for his wife Ursula, natural daughter of Henry, fourth Earl of Derby; a love that has found fruition in the birth of their daughter Jane. Under the circumstances it is odd that Shakespeare should lament a union 'Leaving no posterity'. Being enigmatic, the poem would inspire fanciful interpretations. Among the more curious would be the theory that in *The Phoenix and the Turtle* the turtle-dove stands for Shakespeare, and the phoenix for Anne Whateley or Beck, the unfortunately deceased nun who was 'the one and only woman he really loved',[37] or—even quainter—the suggestion that the lyric describes 'the failure of Shakespeare's poetical aspirations' because of his unfortunate over-indulgence in wine.[38]

These speculations of the fantastics are symptomatic of a hunger from which more rational spirits would suffer too. Because Shakespeare is, of all, the artist about whom curiosity is most intense, it is not surprising that many in search of revelations would delve into his works and find what is not there. For others, wedded to a more rigorous scholarly discipline, the *œuvre* would be the point of embarkation and return for that exploration of Shakespeare's mind and art which it would be one of the principal tasks of critical biography to achieve.

What, then, can be said of the biographical evidence which has been laboriously surveyed: the portraits, the legal (or at any rate official) documents, the allusions by contemporaries, the author's creations? Reduced to essentials, they tell us that this man was extremely valuable to his theatrical company, that he had gentle manners and was affectionately regarded by his fellows, that for a time he had a patron, that to some extent (how much is debatable) he revealed himself in his sonnets, that he maintained his ties with the town of his birth, that he prospered and invested wisely in property, that his 'excellent and sweetly composed' poems (so they were described by the 1640 publisher) pleased readers, and that his plays delighted popular and courtly audiences. He did not in his own day inspire the mysterious veneration that afterwards came to surround him and his works. No playwright in that day did, and certainly no actor. If any dramatist came close to enjoying adulation, it was Jonson, whose writings exemplified the critical doctrines subscribed to by the intelligentsia, who were more inclined than ordinary folk to put down their impressions in writing.[39] This state of affairs helps to account for the limitations of the record, and also (as we shall see) for some later misinterpretations of that record.

'The writing a life is at all times, and in all circumstances the most difficult task of an historian', wrote John Dryden in his *Life of Lucian*:

and notwithstanding the numerous tribe of biographers, we can scarce find one, except Plutarch, who deserves our perusal, or can invite a second view. But if the difficulty be so great, where the materials are plentiful and the incidents extraordinary, what must it be when the person that afford the subject denies matter enough for a page?[40]

About Shakespeare we know more than about Lucian, for whom we do not have even the dates of birth and death. Yet, if it would be an exaggeration to say that the materials concerning the dramatist deny matter enough for a page, they hardly amount to a great deal; and they are fraught with perplexities for the biographer. His problems are compounded by new accretions to the record later in the century by those who did not know Shakespeare when he lived but who may yet have been in touch with an authentic tradition. Others contributing to the story were to be the artificers of myth. Where tradition, embodying probable or possible truth, leaves off and legend begins is not always easily determined. Dryden's point loses none of its force.

To the transmitters of the tradition and the makers of myths we now turn.

PART II

# Shakespeare of the Legends;
# The First Biographers

# 1
## Shakespeare's Epitaphs

> . . . done are Shakespeare's days;
> His days are done, that made the dainty plays,
> Which made the globe of heav'n and earth to ring.

THUS wrote the Trinity College, Cambridge, poet Hugh Holland in the elegiac sonnet carried on the sixth preliminary leaf to the First Folio. Disappointed in his hopes for preferment, Holland (according to Fuller) 'grumbled out the rest of his life in visible discontentment';[1] a strain of lament must have come naturally to him. But Shakespeare had indeed passed away, and with him the supreme moment of the English stage. That something so extraordinary had taken place, immediately succeeding generations did not for the most part suspect; yet his posthumous reputation grew. Shortly after the publication of the First Folio, one anonymous enthusiast penned this epitaph on the last end page of what is now Folio no. 26 of the Folger collection:

> Here Shakespeare lies whom none but death could shake;
> And here shall lie till judgement all awake;
> When the last trumpet doth unclose his eyes,
> The wittiest poet in the world shall rise.[2]

An anonymous writer in 1630 describes Stratford, in passing, as 'a town most remarkable for the birth of famous William Shakespeare'.[3]

The times favoured the proliferation of apocryphal anecdotes and episodes. Recollections at second hand now begin to appear in the commonplace books that men in those days kept for their own amusement and edification. There are scraps of fact and conjecture. Whether certain of the traditions are true reports or not, there is no way of determining; these have vexed generations of scholars. Some suggestions are oddities, as those dealing with the epitaphs that Shakespeare supposedly wrote. Others concern the composition of his plays, and in one instance at least provide a bit of verifiable information. Still others supply gossip, sometimes sensational, about the circumstances of his life and death. With the passage of time a vast accretion of legend would surround like a nimbus the blurred outlines of the Bard of men's idolatry. Fortunately there would also get under way a rational attempt to find out the actual facts of the poet's career. Towards the end of the seventeenth century the antiquaries set about assembling materials for their brief lives. These trends culminate in 1709 in the first full-dress biography, Nicholas Rowe's *Some Account of the Life, &c., of Mr. William Shakespear*, which prefaces the first volume of his edition of the *Works*.

In the seventeenth century, Shakespeare came to enjoy esteem as a writer of mortuary verses. One finds an occasional epitaph in the plays: that, for example, in *Pericles*, celebrating the maiden of Tyrus, 'the king's daughter, | On whom foul death hath made this slaughter'; or the two inscriptions left on the gravestone by the sea's edge where lies Timon, 'who alive all living men did hate. | Pass by and curse thy fill, but pass | And stay not here thy gait.' But these scarcely prepare us for the sepulchral epigrams, serious or mocking, that came to cluster around Shakespeare's name. These attributions illustrate the pitfalls confronting the biographer who must distinguish between the probable, the possible, and the preposterous.

A manuscript collection of verses compiled in mid-century by Nicholas Burgh, a Poor Knight of Windsor, links Shakespeare's name with Jonson's in this curious exercise by which the wits amused themselves:

> Mr. Ben. Johnson and Mr. Wm. Shake-speare being merry at a
> tavern, Mr. Jonson having begun this for his epitaph:
>> Here lies Ben Johnson that was once one
> he gives it to Mr. Shakespear to make up, who presently
> writes:
>> Who while he liv'd was a slow things [*sic*],
>> And now being dead is nothing.
>> > > Finis.[4]

A few years afterwards the Archdeacon of Rochester, Thomas Plume, independently describing the same occasion, furnishes a full and improved version of the epitaph:

>> Here lies Ben Johnson, who was once one,
> This he made of himself. Shakspear took the pen from him
> and made this:
>> Here lies Benjamin, with short hair upon his chin,
>> Who while he lived was a slow thing; and now he's
>> > buried is no thing.[5]

These verses appropriately twit Jonson's notorious slowness of composition ('His wits are somewhat hard bound', Dekker's Tucca sneers at him in *Satiromastix*: 'the punk his Muse has sore labor ere the whore be delivered').[6] Had not Jonson, after all, censured Shakespeare for his facility? As the century progressed a number of anecdotes would accumulate around the dramatists' relations with one another.

In another manuscript, dating from somewhere about 1630 to 1640, Shakespeare is credited with memorializing a less well-known personage:

>> When God was pleas'd (the world unwilling yet),
>> Elias James to nature paid his debt,
>> And here reposeth; as he liv'd he died,
>> The saying in him strongly verified,

> 'Such life, such death.' Then, the known truth to tell,
> He liv'd a godly life, and died as well.[7]

There would be speculation about the identity of Elias James—did he, it would be asked, belong to the Jameses of Stratford and the outlying hamlets?—until Hotson's discovery that he was a wealthy and celibate brewer whose establishment at Puddle Wharf stood some 150 yards from the Blackfriars Theatre.[8] His sister-in-law was the wife of one John Jackson, and it is intriguing that someone of that name acted as a trustee for Shakespeare in his purchase of the Blackfriars Gate-house. James died in September 1610—before the dramatist's retirement to Stratford. These facts support the possibility that he may indeed have celebrated, in dismayingly pedestrian verses, the memory of the godly brewer.

Another pair of lapidary tributes, these associated with the tomb of the Stanleys in Tong Church in Shropshire, appear twice in manuscript: first, among the 'Epitaphs Laudatory' in a miscellany 'in a handwriting of the early part of the reign of Charles I',[9] and then in a manuscript collection of monumental inscriptions appended in 1664 by Sir William Dugdale to his *Visitation of Shropshire*. 'On the northside of the chancel of Tonge Church, in the county of Salop', Dugdale writes,

stands a very stately tomb, supported with Corinthian columns. It hath two figures of men in armour thereon lying—the one below the arches and columns, and the other above them—and this epitaph upon it: 'Thomas Stanley, Knight, second son of Edward, Earl of Derby', etc. These following verses were made by William Shakespeare, the late famous tragedian:

*Written upon the East End of the Tomb*

> Ask who lies here, but do not weep;
> He is not dead, he doth but sleep.
> This stony register is for his bones,
> His fame is more perpetual than these stones;
> And his own goodness, with himself being gone,
> Shall live when earthly monument is none.

*Written upon the West End Thereof*

> Not monumental stone preserves our fame,
> Nor sky-aspiring pyramids our name;
> The memory of him for whom this stands
> Shall outlive marble and defacers' hands:
> When all to Time's consumption shall be given,
> Stanley, for whom this stands, shall stand in heaven.[10]

Whether the verses constitute one epitaph or two, or which member—or members—of the family they commemorate—remains unclear.

Could the Stanleys thus commemorated be the uncles of Ferdinando

Stanley, Lord Strange, fifth Earl of Derby, who was the patron of Lord Strange's Men—a company with which Shakespeare may have had an association in the early 1590s? Such an identification is tempting, but Sir Edward, who died in 1609, had no connection with Tong. His nephew of the same name did; but this Sir Edward died in 1632, too late to be the subject of the epitaph, if (as is very questionable) Shakespeare composed it. Such are the genealogical mazes through which scholars would have to thread their way.

Lines on King James, not yet deceased, appear unascribed beneath the frontispiece to the 1616 edition of his works—e.g. 'Of more than earth can earth make none partaker | But knowledge makes the king most like his maker'—but were attributed to the leading dramatist of the troupe of which the monarch was nominally patron in at least two seventeenth-century manuscripts and a no longer extant printed broadside.

The epitaphs for which Shakespeare came to be most renowned in the seventeenth century pay honour—if that is the word—to the memory of his old Stratford acquaintance John Combe. We have already encountered Combe: together with his uncle William, he sold Shakespeare land in Old Stratford in 1602, and in his will, made in 1613, he left the poet £5 as a token. Apparently they were friends; Shakespeare bequeathed his sword to Combe's nephew and heir Thomas. By lending money at the customary rate of ten in the hundred, John Combe became the richest man in Stratford. The legal records testify to his industrious pursuit of defaulters. He died a bachelor in 1614, releasing his 'good and just debtors' of one shilling in the pound, and leaving £20 to the poor of the town. Among the 'divers select epitaphs and hearse-attending epods' annexed to Richard Brathwait's *Remains after Death*, printed in 1618, the following unascribed doggerel verses—apparently not offered as Brathwait's own—are included:

AN EPITAPH

Upon one John Combe of Stratford-upon-Aven, a notable usurer, fastened upon a tomb that he had caused to be built in his lifetime:*

> Ten in the hundred must lie in his grave,
> But a hundred to ten whether God will him have.
> Who then must be interr'd in this tomb?
> 'Oh,' quoth the Devil, 'my John a Combe.'[11]

Similar rhymes, but on an unnamed usurer, had appeared earlier in *The More the Merrier* (1608) by 'H.P.', and in Camden's *Remains Concerning Britain* (1614), among a group of 'conceited, merry, and laughing epitaphs, the most of them composed by Master John Hoskins when he was young'; later in the century other versions substituted names—Stanhope, Spencer, or Pearse—for Combe.

* Combe, however, did not have the tomb built in his lifetime, for in his will he allocated £60 for its construction.

In 1634 the touring Lieutenant Hammond, after pausing over Shakespeare's tomb, went on to note another, 'of an old gentleman a bachelor, Mr
Combe, upon whose name the said poet did merrily fan up some witty and
facetious verses, which time would not give us leave to sack up'.*[12]
Hammond does not quote these verses, but the same Nicholas Burgh who
reports Shakespeare's epitaph on Jonson does:

On John Combe, a covetous rich man, Mr. Wm. Shakspear writ this at his request
while he was yet living, for his epitaph:

Who lies in this tomb?
'Hough', quoth the Devil, 'Tis my son John A Combe.'

Finis.

but being dead, and making the poor his heirs, he after writes this for his epitaph:

Howe'er he lived, judge not;
John Combe shall never be forgot
While poor hath memory; for he did gather
To make the poor his issue, he their father,
As record of his tilth and seed
Did crown him in his latter deed.†

Finis; W. Shak.[13]

Essentially the same epitaph as that in Brathwait's collection is quoted in
1673 by Robert Dobyns, who claims to have seen it himself on Combe's
monument. He also informs us that 'Since my being at Stratford the heirs of
Mr. Combe have caused these verses to be razed, so that now they are not
legible.'[14] (But can we trust the report? Brathwait, it will be recalled, had said
that the epitaph was fastened to the tomb, not carved on it.)

Elaboration was perhaps inevitable; Aubrey in 1681 tells how the epitaph
came to be written: 'One time as he was at the tavern at Stratford super
Avon, one Combes, an old rich usurer, was to be buried. He makes there this
extemporary epitaph . . .'[15] But this is pale beside the rich development of the
episode in Rowe's 'Account':

it is a Story almost still remember'd in that Country, that he had a particular Intimacy
with Mr. *Combe*, an old Gentleman noted thereabouts for his Wealth and Usury: It
happen'd, that in a pleasant Conversation amongst their common Friends, Mr. *Combe*
told *Shakespear* in a laughing manner, that he fancy'd, he intended to write his
Epitaph, if he happen'd to out-live him; and since he could not know what might be
said of him when he was dead, he desir'd it might be done immediately: Upon which
Shakespear gave him these four Verses.

*Ten in the Hundred lies here ingrav'd,*
*'Tis a Hundred to Ten, his Soul is not sav'd:*

* A comb is a measure for grain; hence the punning references to fanning and sacking up.
† Burgh makes Sir John Spencer the subject of the lines 'Ten in the hundred lies under this stone; |
It's a hundred to ten to the Devil he's gone,' which, however, he leaves unattributed.

> *If any Man ask, Who lies in this Tomb?*
> *Oh! ho! quoth the Devil, 'tis my John-a-Combe.*

But the Sharpness of the Satyr is said to have stung the Man so severely, that he never forgave it.[16]

In the late eighteenth century the story was further enhanced by John Jordan, the wheelwright of Tiddington, near Stratford, who in his spare hours educated himself in local antiquities and matters Shakespearian, and served as a guide to tourists who were attracted in increasing numbers to Shakespeare's birthplace. Jordan gave a local habitation to the meeting of Shakespeare and Combe: it took place at 'the sign of the Bear, in the Bridge street'. Shakespeare's epitaph upon Combe was, according to this version, greeted with a burst of loud laughter, 'perhaps from the justness of the idea, and the hatred all men have to the character of a miser and usurer'.[17]

This flowering has a sequel in the contribution made in 1740 by Francis Peck, a Leicestershire parson and antiquary, who reports an epitaph on another Combe:

Every body knows *Shakespeare's* epitaph for *John a Combe*. And I am told he afterwards wrote another for *Tom a Combe*, alias *Thin-Beard*, brother of the said *John*; & that it was never yet printed. It is as follows.

> 'Thin in *beard*, & thick in purse;
> 'Never man beloved worse:
> 'He went to th' grave with many a curse:
> 'The Devil & He had both one nurse.'

This is very sour.[18]

So sour that it has not been digested—except by Jordan, who felicitously joined together the epitaphs for the Combe brothers in a single anecdote. When at the Bear 'the violence of the mirth' had subsided, Shakespeare (in response to requests) composed extempore the 'thin in beard' epigram, thus affording 'no small diversion amongst the convivial meeting'. But, Jordan sadly concludes, 'it is said the severity of this satire made so deep an impression upon the two brothers that they never forgave the author of the epitaphs'.[19] We shall be hearing again about this Jordan.

# 2

# *Relics and Associations*

As if to compensate for their shadowy and doubtful origins, traditions would, over the centuries, attach themselves to the particular: to actual places and tangible objects. Tradition marked the pew in the Guild Chapel where sat Stratford's most illustrious son. In 1877 the master of the free school, like others before him, proudly displayed Shakespeare's desk to visitors: 'William was a studious lad', he pointed out, '& selected that corner of the room so that he might not be disturbed by the other boys.' (As early as 1835 Mrs Phoebe Dighton, by appointment fruit and flower painter to Her Majesty the Queen, included in her *Relics of Shakespeare* a lithograph of his desk as it was 'Still to be seen in the Grammar School where he received his Education'.) The poet who left behind no personal papers nevertheless favoured posterity with innumerable baubles, gewgaws, and toys to mock apes: his pencil case, walking-stick, gloves, brooch, table, spoon, and wooden salt cellars, not to mention a jug in the form of a large coffee pot, with a neat head of the owner, in his fortieth year, engraved on the silver lid. The proprietor of the Shakespeare Hotel took pleasure in showing his visitors Shakespeare's clock, before parting with it at a public auction in 1880; not to be outdone, the nearby Falcon Inn boasted the shovel-board at which he delighted to play. Stratford's most eminent antiquary, R. B. Wheler (about whom more hereafter), cherished the gold ring with the initials W. S., tied together by a true-love knot, which was found in a field near Holy Trinity on 16 March 1810, and purchased by him the same day. How many of Shakespeare's chairs could pilgrims view, sit on, and hack to bits! New Place and the house on Henley Street had their Bardic seats; in nearby Shottery, Anne Hathaway's cottage boasted Shakespeare's Courting Chair; Bidford too, as we shall see, possessed its ancient stick of furniture with legendary associations.

In the early nineteenth century, Mary Hornby, custodian of the Birthplace, imposed upon the credulous with her bogus treasures. With her 'frosty red face, lighted up by a cold blue anxious eye, and garnished with artificial locks of flaxen hair, curling from under an exceedingly dirty cap' (so Washington Irving described her), this garrulous harridan showed off the shattered stock of the dramatist's matchlock, his tobacco box, the sword with which he played Hamlet, a 'curious piece of carving' representing David slaying Goliath, and—among other curiosities—'a Gold embroidered Box' presented to Shakespeare by the King of Spain in return for a goblet of great value. The presence of these relics must be regarded as miraculous, for when Samuel Vince visited Stratford in the summer of 1787, the only Shakespearian item remaining there was the poet's chair, 'fixed in one of the Chimnies'; so Vince tells us in the manuscript itinerary of his tour preserved in the Folger Shakespeare Library.

Traditions supplied the poet with a scattered circle of acquaintance; traditions pin-pointed where he set quill to paper and how he spent his leisure hours. Friends at Dursley gave Shakespeare shelter between the time of his departure from Stratford and his appearance in London; so believed John

Henry Blunt, amateur archaeologist and rector of Beverton in the nineteenth century. How else to account for the fact that a pathway in the wood near the town was known traditionally as 'Shakespeare's Walk'? At Clifford, near Stratford, he enjoyed the hospitality of Sir Henry Rainsford and his lady, 'beautiful, and of a gallant structure of body'. The poet often passed the time at Clopton House, also near Stratford, and on numerous occasions he visited the Combe family at the Welcombe Lodge, situated near the Warwick Road. At Elizabeth's Court he formed an intimacy with the powerful family of the Prices, on whom he frequently called. A mere player, he nevertheless stood on terms of easy familiarity with royalty: on the hanging stage of the old Town Hall in Leicester, Shakespeare read his plays before the Queen, but mainly he read to her at Court, for she took 'especial delight' in inviting him thither. From *The Shakespeare Almanack* for 1870 (price one penny) we learn that he wrote *Twelfth Night* amid the straight-lined walks and green alleys of New Place; in the same garden, according to another source, he played at bowls. Shakespeare wrote his Sonnets in a small room over the porch of a half-timbered dwelling at Rowington called 'Shakespeare's Cottage'. He found the scenery for *A Midsummer Night's Dream* in the sylvan surroundings of Gum Slade, Sutton Park. Sometimes he retired to study in the large library of the old moated manor-house at Radbrook in Warwickshire. But he must have written his plays mainly in his London residence at 134 Aldersgate Street; called the Half-Moon Hotel in Shakespeare's day, this fine specimen of Elizabethan domestic architecture was regrettably pulled down in the nineteenth century.

Other traditions cluster around taverns. According to an anecdote reported by the great eighteenth-century editor Edward Capell, Shakespeare diverted himself at an alehouse in Wincot with a fool who belonged to a neighbouring mill. Another report holds that Shakespeare often met his friends at the Greyhound in Stratford. From still a different source we learn that the Three Pigeons in Brentford was the inn favoured by Shakespeare and Jonson. The poet also frequented the Devil Tavern, presumably in London, and he sought relaxation in one of the wooden chambers of the Red Lion in the Edgware Road. On other occasions Shakespeare and his friends enjoyed their potations at an old house known as the sign of the Boar in Eastcheap. The legend to a nineteenth-century engraving of the Falcon Tavern on the Bankside informs us that this establishment was 'celebrated for the daily resort of SHAKSPEARE, and his Dramatic Companions'. Yet another tradition maintains that the conviviality took place at a little tavern called the Globe, near Blackfriars— perhaps conflating public house with public theatre. On the road from Stratford to London, Shakespeare sometimes stopped for the night at The Olde Shipe Inne in the straggling village of Grendon Underwood, immortalized in the distich, 'Grendone Underwoode— | The dirtiest towne that ever stoode'; in the nineteenth century, visitors would be escorted up the old oak staircase, with its quaint balustrades, to the gabled third storey, where in a room with a curious little oval window the great man had slept.

Thus, in the absence of facts, did traditions cater to the requirements of local pride and the curiosity of multitudes. As a whole they make a contribution to the Shakespeare-Mythos—in E. K. Chambers's term (see p. 519)—rather than to biography proper. Yet historians have not been able to dismiss them all out of hand, for in some may reside a kernel of truth.

# 3

# *Traditions of Writing and Acting*

OF greater intrinsic significance are the traditions that tell us something about his professional life. From a letter of about 1625 we first learn of the controversy stirred by production of the *Henry IV* plays. The correspondent is Dr Richard James, a devoted admirer of Jonson, and the scholar and antiquary who served as Sir Robert Cotton's librarian. Writing to Sir Harry Bourchier, he relates that a 'young gentle-lady of your acquaintance' asked him how Sir John Falstaff 'could be dead in the time of Harrie the Fifth and again live in the time of Harrie the Sixth to be banished for cowardice'.[20] (In *1 Henry VI* a messenger reports that Sir John Falstaffe has 'play'd the coward', taking to his heels from a battle without 'having struck one stroke'.) James describes how he defended the valour and learning of the historical Falstaff— or Fastolfe—and how he explained the discrepancy about which the puzzled young gentlewoman inquired:

. . . in Shakespeare's first show of Harrie the Fifth, the person with which he undertook to play a buffoon was not Falstaffe but Sir Jhon Oldcastle, and that offense being worthily taken by personages descended from his title (as per-adventure by many others also who ought to have him in honorable memory), the poet was put to make an ignorant shift of abusing Sir Jhon Falstophe, a man not inferior of virtue. . . .[21]

That the dramatist originally called his greatest comic creation Oldcastle is corroborated by clues from the plays themselves: a botched pun ('my old lad of the castle'), a metrically deficient line, a surviving speech prefix 'Old'. But the clearest testimony of all is the author's humble Epilogue disclaimer: '. . . Falstaff shall die of a sweat, unless already a be killed with your hard opinions. For Oldcastle died a martyr, and this is not the man.' In 1599 a quartet of dramatists from the rival Admiral's Men pooled their inconsiderable talents to produce *The First Part of the True and Honourable History of the*

*Life of Sir John Old-castle, the Good Lord Cobham.* 'It is no pampered glutton we present', the Prologue smugly maintains,

> Nor aged councellor to youthful sin,
> But one whose virtue shone above the rest,

and concludes with a patent slam at Shakespeare's version of history:

> . . . let fair Truth be grac'd,
> Since forg'd invention former time defac'd.[22]

By a nice irony *Sir John Oldcastle* in time came to be attributed to Shakespeare, and was duly included in the Third Folio of his works in 1664.

The historical Oldcastle was a distinguished and tragic figure, a Lollard hanged and burned for his heretical faith and afterwards enrolled by John Foxe in his *Book of Martyrs*. Protest of the slur on a revered name was most likely entered by William Brooke, seventh Lord Cobham, or his successor—after his death in 1597—Henry, eighth Lord Cobham; they were lineally descended, on the mother's side, from Oldcastle. The Brookes were powerful figures at Court. William, a Privy Councillor and Lord Chamberlain, could have ordered the change directly or acted through the Master of the Revels. His son Henry, the brother-in-law of Sir Robert Cecil and intimate of Ralegh, was an adversary of Essex and of Shakespeare's patron Southampton. No matter that he was mocked by the Essex faction as 'my lord Fool'; he had successfully withstood their opposition to his appointment, in 1596, as Lord Warden of the Cinque Ports. Identification of the Cobham who remonstrated depends on the dating of *1 Henry IV*, which may have been produced either before or after the death of William. Some scholars would in after years argue that the dramatist's choice of a name was 'an unlucky accident' dictated by his source; others, that it was deliberately provocative.[23] In *1 Henry IV* in the Oxford edition of *The Complete Works* (1986), the editors restore Sir John's original surname for the first time in printed texts, although there is some reason to believe—as Wells and Taylor point out—that the name 'Oldcastle' was sometimes used, as the dramatist intended, even after the play had been produced.

The Oldcastle–Falstaff imbroglio is an interesting episode, and would have still greater interest were a royal personage involved. Such exalted intervention is obligingly conceived by Rowe in an instance of that embroidering to which traditions are subject: '. . . this Part of *Falstaff* is said to have been written originally under the Name of *Oldcastle*; some of that Family being then remaining, the Queen was pleas'd to command him to alter it; upon which he made use of *Falstaff*'.[24] Such dramatic heightening we shall be encountering more than once in this narrative.

The *Henry IV* plays did not furnish the only occasion when Shakespeare's company had to worry about the sensibilities of the Brookes. In *The Merry*

*Wives of Windsor* the jealous Ford, told that his wife has stirred Falstaff's lust, visits the Knight disguised (in the 1602 Quarto) as Master Brooke. The dramatist perhaps hit on the name through a process of verbal association with Ford, and it yields the inevitable pun: 'Such Brookes are welcome to me, that o'erflows such liquor' (II. ii. 146–7). In the Folio the word-play is sacrificed, like the old lad of the castle, to a name change; *Brooke* becomes *Broome* throughout. When the alteration was made we can only guess; perhaps during rehearsal, when someone realized that Brooke was the name of the same Lord Cobham who had lately been angered by the stage exploitation of Oldcastle, and that it would not do to repeat the offence.[25]

The same play provided matter for another seventeenth-century tradition involving the Queen. In the prefatory epistle to *The Comical Gallant: or The Amours of Sir John Falstaffe* (1702), his unsuccessful adaptation of *The Merry Wives of Windsor*, John Dennis wrote that the original play 'pleas'd one of the greatest Queens that ever was in the World. . . . This Comedy was written at her Command, and by her direction, and she was so eager to see it Acted, that she commanded it to be finished in fourteen days. . . .'[26] In his prologue Dennis again refers to the fourteen days of composition time, but two years later he improved upon the anecdote by contracting the fortnight to ten days.[27] In 1709 the story found a place in Rowe's 'Account', where an explanation is given for the Queen's directive: 'She was so well pleas'd with that admirable Character of *Falstaff*, in the two Parts of *Henry* the Fourth, that she commanded him to continue it for one Play more, and to shew him in Love.'[28] The following year Charles Gildon, the hack who earned unenviable fame by being pilloried by Pope in *The Dunciad*, enhanced the legend a bit more. 'The *Fairys* in the fifth Act', he wrote, 'makes a Handsome Complement to the Qneen [*sic*], in her Palace of *Windsor*, who had oblig'd him to write a Play of *Sir John Falstaff* in Love, and which I am very well assured he perform'd in a Fortnight; a prodigious Thing, when all is so well contriv'd, and carry'd on without the least Confusion.'[29]

Although the two-week composition period sounds like legendary exaggeration, the known facts and likelihoods lend plausibility to the tradition that *The Merry Wives of Windsor* was written at the Queen's command—a command that may have been motivated by Shakespeare's failure to deliver on his promise, in the epilogue of *2 Henry IV*, to reintroduce Falstaff in *Henry V*. (The Queen may have seen the *Henry IV* plays acted at Court during the Christmas season of 1596–7.) The carelessnesses of composition in *The Merry Wives of Windsor*, frequently noted by critics, are suggestive of haste, and the duped lover of the play bears little resemblance to the cunning and witty 'hill of flesh' depicted previously. According to the title-page of the 1602 Quarto, the play was performed before Elizabeth. The Windsor setting—and the comedy is unique among Shakespeare's plays in having a contemporary English backdrop—would render it especially appropriate for production at the Castle. Such production would help to account for the otherwise rather extraneous compliment to Windsor and its occupant by Mistress Quickly as the fairy in Act V, sc. v.

Royal acknowledgement of Shakespeare is the basis for yet another tradition that first manifests itself in an advertisement for an early eighteenth-century edition of Shakespeare. 'That most learn'd Prince, and great Patron of Learning, King *James* the First', the anonymous publicist reveals, 'was pleas'd with his own Hand to write an amicable Letter to Mr. *Shakespeare*; which Letter, tho now lost, remain'd long in the Hands of Sir *William D'avenant*, as a credible Person now living can testify.'[30] The tradition was known to the antiquary Oldys, who identified the credible person as John Sheffield, Duke of Buckingham.[31]

At the same time that accounts of royal favour spring up, we get our first Neglected Artist story. The same Oldys (of whom we shall have more to say) is reported to have noted down that Shakespeare received only £5 for *Hamlet*. The report sorts oddly with earlier rumours that the dramatist assembled a great fortune—according to one source, he lived at the rate of a thousand pounds a year; according to another he left a sister two or three hundred pounds per annum.[32]

Other traditions reveal—or purport to reveal—the playwright at work. Oldys had it from the old actor Bowman who had it from Sir William Bishop that Shakespeare in part modelled Falstaff, as a gesture of pique, upon a Stratford citizen 'who either faithlessly broke a contract or spitefully refused to part with some land, for a valuable consideration, adjoining to Shakespeare's, in or near that town'.[33] The Stratford charnel-house naturally enough inspired spectral associations. From Gildon's continuation of Langbaine we learn that Shakespeare was moved to write the Ghost scene in *Hamlet*, I. iv, 'at his House which bordered on the Charnel-House and Church-Yard'.*[34] In another passage he adds the appropriate detail that the scene was composed 'in the midst of the Night'.[35] Thus Shakespeare is given suitable inspiration; with 'oppressed and fear-surprised eyes', like his own dumbfounded sentinels, he sees the ghost stalk before him. A pity that his Stratford house was a half mile from Holy Trinity and that, in any event, he probably did most of his writing in London. (One cannot, however, place greater trust in the 'known tradition', reported in 1747 by William Guthrie in *An Essay upon English Tragedy*, 'that Shakespear shut himself up all night in Westminster-Abbey when he wrote the scene of the ghost in Hamlet'.) The same literalism that requires a nocturnal setting, preferably the charnel-house but at least a church, for the composition of supernatural scenes, is responsible for the assumption—expressed in 1818 in Fussell's *Journey Round the Coast of Kent*—that Shakespeare wrote Act IV of *King Lear* as he actually

---

* A related tradition, reported by Dodd (*The Beauties of Shakespear*), holds that the playwright wrote the scene 'in a charnel-house'.

stood in the presence of the stupendous chalk cliff visible on the left as one leaves St Martin's Priory on the road from Dover to Folkestone. This despite the fact that Edgar describes an imaginary precipice.

Now and then we hear about the dramatist's relations with the characters he created and with the actors who personified them. According to Dryden, writing in his Defence of the *Epilogue to the Second Part of the Conquest of Granada* (1672), the author of *Romeo and Juliet* said himself of his Mercutio 'that he was forced to kill him in the third act, to prevent being killed by him'. Gildon tells of being assured on good authority that Iago was acted by a famous comedian, and that his creator 'put several words, and expressions into his part (perhaps not so agreeable to his Character) to make the Audience laugh, who had not yet learnt to endure to be serious a whole Play'.[36] The denigration of Shakespeare's audience had begun.

A more curious motivation for dramatic composition is reported by Lewis Theobald. Offering *Double Falsehood, or The Distrest Lovers* as his adaptation of a play originally written by Shakespeare to a sceptical public, he writes in his preface that 'There is a Tradition (which I have from the Noble Person, who supply'd me with One of my Copies) that it was given by our Author, as a Present of Value, to a Natural Daughter of his, for whose Sake he wrote it, in the Time of his Retirement from the Stage.'[37] Who the noble individual was, Theobald does not say, but it is a most dubious tradition, reflecting as it does a curious incomprehension of the nature of a playwright's business arrangements with his company, which would not leave him with transferable property rights in playbooks. The unfortunate love child—did she exist— would have benefited little from such a bequest. Based on the story of Cardenio in Part One of Cervantes's *Don Quixote, Double Falsehood* is a tragicomedy successfully produced at Drury Lane in 1727, and more than once thereafter revived. On 9 September 1653 Humphrey Moseley, a London publisher, had entered in the Stationers' Register—along with a number of other plays—one entitled '*The History of Cardenio*, by Mr Fletcher and Shakespeare'. Earlier allusions indicate that the King's Men indeed owned a play on this theme at the time that Shakespeare and Fletcher were collaborating for the company. Conceivably, then, Theobald's *Double Falsehood* derives, however remotely, from the drama registered by Moseley.

Inevitably, legend would attempt to fathom the mystery of how a rustic with small Latin and less Greek contrived to write so profoundly. It was reported by someone intimately acquainted with Shakespeare (so the story goes) that he was supplied with predigested historical source information for his plays by an assistant, 'one of those chuckle-pated Historians'. Thus furnished, the dramatist would shape the raw material 'as his beautiful Thoughts directed', leaving it to his willing associate to see to it that everything was grammatical. 'How do you think, Reading could have assisted him in such great Thoughts?' our anonymous informant asks. 'It would only have lost Time. When he found his Thoughts grow on him so fast, he could have writ for ever, had he liv'd so long.'[38] Thus condescension and admiration comfortably sustain one another.

Seemingly more plausible information derives from the traditions concerning Shakespeare's acting career. The First Folio states that he was one of the principal actors in his own works but is silent on the parts he took. Rowe, compiling material for his 'Account', made inquiries and was able to learn that 'the top of his Performance was the Ghost in his own *Hamlet*'.[39] From the *Annual Register* for 1767 we discover, alas, that in this part Shakespeare failed. In his notes Oldys recorded that one of Shakespeare's younger brothers used to come up to London to see the dramatist perform in the plays he had written. This brother (Oldys says) survived into the Restoration, and was made much of by the actors of the latter age; 'greedily inquisitive', they tried to draw out from him information about his celebrated relation's acting career. But he was old, and could remember little except:

faint, general, and almost lost ideas he had of having once seen him act a part in one of his own comedies, wherein being to personate a decrepit old man, he wore a long beard, and appeared so weak and drooping and unable to walk, that he was forced to be supported and carried by another person to a table, at which he was seated among some company, who were eating, and one of them sung a song.[40]

It is a pleasant story, spoiled only by the fact that none of Shakespeare's three brothers lived on into the Restoration.

This recollection of Shakespeare in the part of Adam is improved upon by Edward Capell in a note to *As You Like It*; for in Capell the poet's kin is not necessarily a brother:

A traditional story was current some years ago about Stratford,—that a very old man of that place,—of weak intellects, but yet related to Shakespeare,—being ask'd by some of his neighbours, what he remember'd about him; answer'd,—that he saw him once brought on the stage upon another man's back; which answer was apply'd by the hearers, to his having seen him perform in this scene the part of Adam.[41]

Capell mentions another tradition, that Shakespeare was not an especially gifted actor,* and he goes on to adduce support for this allegation by a descent into that literalism which sometimes afflicts him as a critic. Did not the poet, in Sonnet 37, complain of being 'made lame by fortune's dearest spite'? And in Sonnet 89 he writes: 'Speak of my lameness, and I straight will halt.' Metaphor dwindles into statement; Shakespeare the actor, in Capell's view, becomes peculiarly suited to decrepit parts by reason of 'an accidental lameness, which . . . befell him in some part of life; without saying how, or when, of what sort, or in what degree; but his expressions seem to indicate—

---

* Confirmation of such a tradition comes from *Historia Histrionica: An Historical Account of the English Stage . . . in a Dialogue, of Plays and Players* (London, 1699), in which the anonymous author (James Wright?) remarks in passing that '*Shakespear* . . . was a much better poet, than Player' (p. 4).

latterly'.[42] We shall be hearing more about this supposed lameness, which will become a curious part of our narrative.

# 4

# *Shakespeare and the Vintner; Shakespeare versus Jonson*

MORE entertaining, if not necessarily more reliable, traditions furnish instances of the poet's extemporaneous wit. One story tells how Shakespeare purchased a quantity of wine but failed to pay for it, despite repeated applications on the vintner's part. Amused with the poet's company and conversation, the wine merchant at length offered to forgive the debt if his customer satisfactorily answered four questions: (1) What pleases God best? (2) What pleases the Devil best? (3) What pleases the world best? (4) What pleases me best? To which Shakespeare at once replied:

> God is best pleas'd when men discard their sin,
> Satan's best pleas'd when they persist therein.
> The world's best pleas'd when you do sell good wine,
> And you're best pleas'd when I do pay for mine.

What choice had the merchant but to write off the debt?[43]

A greater spur to wit than any vintner was the dramatist's presumed arch-rival Jonson. That anecdotes about their confrontations should come into being in the period following Shakespeare's death need occasion no surprise. Interest was keen in both poets, and especially in Jonson. His eulogy in the Folio attests to a personal relationship with his subject. The two men differed from one another temperamentally as well as in their conceptions of their vocation; yet their names are frequently brought into conjunction in writings of the period.[44] Moreover, Jonson had looked down his nose at Shakespeare's flowing invention and the limitations of his classical education. People remembered the statement about small Latin and less Greek, and they savoured Jonson's sally at the author who never blotted a line. From such beginnings legends are destined to grow.

We have already seen Shakespeare and Jonson locking wits over an epitaph. Another meeting is recounted by Sir Nicholas L'Estrange, who, ensconced in his purchased baronetcy, assembled anecdotes, most of them indecent. In *Merry Passages and Jests* (1629–55), he offers a merry tale belonging to the inoffensive minority: on the authority of one Mr Duncomb he tells how Shakespeare stood godfather to one of Ben's children. After the christening the host observed his guest to be downcast:

Johnson came to cheer him up, and asked him why he was so melancholy. 'No, faith, Ben,' says he, 'not I, but I have been considering a great while what should be the fittest gift for me to bestow upon my godchild, and I have resolved at last.' 'I prithee, what?' says he. 'I' faith, Ben; I'll e'en give him a dozen good latten spoons, and thou shalt translate them.'[45]

(For the modern reader the jest may call for a gloss: latten was a brass-like alloy.) The story appealed—Archdeacon Thomas Plume set it down in his notebooks, but as he has it, the infant is Shakespeare's, and Jonson delivers the devastating stroke of repartee.

Word of still another exchange has come down to us from Oldys, as reported by Steevens in his 1778 edition of Shakespeare. According to this story, the alleged motto of the Globe, *Totus mundus agit histrionem* (adapted from the *Fragmenta* of Petronius Arbiter: 'quod fere totus mundus exerceat histrionem'), furnished on one occasion the anvil from which the poets sent sparks of their wit flying. Their impromptu verses are thus reported:

> *Jonson.*
> If but *stage actors* all the world displays,
> Where shall we find *spectators* of their plays?
> *Shakespeare.*
> Little, or much, of what we see, we do;
> We're all both *actors* and *spectators* too.[46]

This encounter is not more probable than the others; for Oldys scrawled variant versions of the couplets on the title-page of his copy of Langbaine's *Account of the English Dramatic Poets*, without mention of any speakers. Later it would be suggested that the verses were Oldys's invention, and served up as Jonson's and Shakespeare's by Steevens in a deliberate hoax, of which that great scholar was not (as we shall see) incapable.

There are other reports. We hear in time of a convivial 'tavern-club' that attracted from the Court lords drawn by the 'wit and conversation' of the two poets.[47] But the evocation of the wit-combats between Jonson and Shakespeare that enrich our literary folklore we owe largely to 'the great Tom Fuller' (as Pepys termed him), who describes the contests in a celebrated passage from his account of Shakespeare.

A divine blessed with retentive memory and prodigious energy, Fuller enraptured throngs with his preaching and made money for the stationers by winning a large audience for his numerous publications, providing good counsel and encyclopaedic information. His *Church History of Britain*, published in an enormous folio of over thirteen hundred pages, testifies to the grandiosity of his projects. But Fuller's crowning achievement is the *History of the Worthies of England* (1662)—'his beloved book . . . the darling of his soul'[48]—the first attempt, assembled without the help of an amanuensis, at a dictionary of national biography. The book is notable for the author's avowed purpose to entertain as well as to inform. 'I confess the subject is but dull in itself', he declares,

to tell the time and place of men's birth and deaths, their names, with the names and number of their books, and therefore this bare skeleton of Time, Place, and Person must be fleshed with some pleasant passages. To this intent I have purposely interlaced (not as meat, but as condiment) many delightful stories. . . .[49]

To prepare the compilation Fuller travelled about the country—horseback riding was his sole exercise—garnering his materials: he consulted documents, looked at buildings and places, and interviewed local inhabitants and relatives of his worthies. Yet Fuller's sketch of Shakespeare, the first attempt at a formal biography, is remarkably deficient in both meat and condiment.

Indeed, it hardly offers a bare skeleton. Fuller has little to report beyond the indisputably correct information that Shakespeare was born in Stratford. The year of the poet's death eludes him, as the blank space reserved for it in the text pathetically testifies. Following lines laid down by Jonson, he instances Shakespeare as the poet who is born, not made: the naturally smooth diamond, unpolished by any learning. Where facts are unavailable, fancy fills the gap. Hence it is that in his concluding paragraph Fuller sees the two poets together in his mind's eye:

Many were the wit-combats betwixt him and Ben Johnson, which two I behold like a Spanish great galleon and an English man-of-war. Master Johnson (like the former) was built far higher in learning; solid, but slow, in his performances. Shake-spear, with the English man-of-war, lesser in bulk but lighter in sailing, could turn with all tides, tack about and take advantage of all winds, by the quickness of his wit and invention.[50]

*I behold*, Fuller writes; it is not tradition that shapes this passage, but the imagination of the biographer operating upon tradition.

For an actual tradition we turn to Rowe, who furnishes an attractive but unsubstantiated account of Shakespeare's early kindness to the man who would become his most formidable rival:

His Acquaintance with *Ben Johnson* began with a remarkable piece of Humanity and good Nature; Mr. *Johnson*, who was at that time altogether unknown to the World, had offer'd one of his Plays to the Players, in order to have it Acted; and the Persons into whose Hands it was put, after having turn'd it carelessly and superciliously over, were just upon returning it to him with an ill-natur'd Answer, that it would be of no service to their Company, when *Shakespear* luckily cast his Eye upon it, and found something so well in it as to engage him first to read it through, and afterwards to recommend Mr. *Johnson* and his Writings to the Publick. After this they were profess'd

Friends; tho' I don't know whether the other ever made him an equal return of Gentleness and Sincerity.[51]

There is little going for Rowe's story, however great its sentimental appeal, except the fact—about which clearly not much can be made—that Shakespeare acted in Jonson's first play for the Lord Chamberlain's company, *Every Man in his Humour*. To accept the tradition requires an act of faith, which would be forthcoming in an age of incipient bardolatry. 'For the honour of literature,' an early nineteenth-century commentator would assert with some vehemence, 'for the respect and veneration which I bear towards these great poets, I trust this tradition, so honourable to both, is founded in truth.'[52]

More influential than Rowe's account of Shakespeare introducing Jonson to the stage are the dark implications of his last few words, which conjure up a picture of a gentle and open Shakespeare extending himself on behalf of a surly and suspicious Jonson. The disparagement of Jonson is deliberate, and quickly becomes open. Rowe goes on to speak of his 'Proud and Insolent' nature, of his inability to look but 'with an evil Eye upon any one that seem'd to stand in Competition with him'.[53] The commendations of the Folio testimonial harbour undercurrents of reserve, 'insinuating his Uncorrectness, a careless manner of Writing, and want of Judgment'.[54] Rowe is here enlarging upon the enormously influential Dryden's characterization of Jonson's eulogy, in the *Discourse concerning the Original and Progress of Satire*, as 'an insolent, sparing, and invidious panegyric'.[55] Rowe then recounts a conversation that purportedly took place at some unspecified date, between Suckling, Davenant, Emdymion Porter, a Mr Hales of Eton (i.e., John Hales), and Jonson. Suckling, a warm advocate of Shakespeare, defended him against the detractions of Jonson, who made the familiar charges of Shakespeare's want of learning and ignorance of classical letters. Hales, who had listened in silence, finally erupted against Jonson, telling him '*That if Mr. Shakespear had not read the Antients, he had likewise not stollen any thing from 'em; (a Fault the other made no Conscience of) . . .*'[56] The same episode is mentioned by Dryden's Neander at the conclusion of his encomium of Shakespeare in the *Essay of Dramatic Poesie* (1668), but without the direct confrontation between Hales and Jonson. It is repeated, with embellishments, by Gildon in 1694 in his *Reflections on Mr. Rymer's Short View of Tragedy*. Perhaps Rowe, like Gildon before him, had the story from Dryden himself. As to Dryden's authority we can only conjecture; perhaps it was Davenant,[57] who was (as we shall see) a fountainhead of traditions concerning Shakespeare.

From these beginnings—Jonson's qualifications about Shakespeare in the Folio tribute, the epitaphs and epigrams, the wit-combats imagined by Fuller, Rowe's anecdotes—would emerge the myth of Jonson's antagonism towards Shakespeare, a hostility engendered by pride and malevolence. If Jonson truly honoured the memory of his beloved, asked the strolling player John Roberts in 1729, why did *he* not edit the volume of Shakespeare's plays and purge them of the absurdities foisted upon them by the printers and actors? But perhaps it is just as well he failed to—'We shou'd have seen as many Injuries from his prejudic'd Pen, as from all the Abuses, which now stand charg'd on the Players.'[58] 'It was, and is, a general opinion', Spence quoted Pope as saying, around 1728, 'that Ben Jonson and Shakespeare lived in enmity against one another,' although the actor Betterton had assured him that 'there was nothing in it.'[59]

In subsequent times Jonson's writings would be scrutinized for covert disparagement of Shakespeare. Late in the eighteenth century George Chalmers would hit upon the fifty-sixth epigram: 'Poor poet-ape, that would be thought our chief, | Whose works are e'en the frippery of wit. . . .' Surely this thief who 'takes up all, makes each man's wit his own' must be Shakespeare. The charge would be answered by Jonson's editor, Gifford, with heavy but efficient irony: 'Mr. Chalmers will *take it on his death* that the person here meant is Shakspeare! Who can doubt it? For my part, I am persuaded, that GROOM IDIOT in the next epigram is also Shakspeare; and, indeed, generally, that he is typified by the words "fool and knave," so exquisitely descriptive of him, wherever they occur in Jonson.'[60] Nevertheless, dark suspicions of Jonson's evil eye would not so easily succumb to rational scorn.

# 5
# *Shakespeare and the Davenants*

IF the myth-makers were to elaborate a drama of full-scale hostilities between Jonson and Shakespeare, that development is still not so sensational as the rumours that came to circulate around the connection between Shakespeare and William Davenant.

Born a decade before the dramatist's death, Davenant led a flamboyant life. A royalist, his military adventures included participation in the Army Plot to march on London and crush Parliament. The conspiracy was discovered, and Davenant imprisoned; later he was freed on bail, which he cheerfully forfeited. During the Civil Wars he commanded a packet boat that ran munitions through the Roundhead blockade. After the King's execution he was captured in the Isle of Wight as he prepared to set sail for the New World, and held by Parliament as a scapegoat exception to its act of amnesty. In prison he worked on his long heroic poem, *Gondibert*, in celebration of love and valour. He was freed after two years in the Tower. Davenant's amorous exploits were more dangerous: he contracted syphilis, and the resulting disfigurement after mercury treatment—he lost his nose—provided the wits with matter for scurrilous jests. Davenant fared better matrimonially, taking as the second of his three wives Lady Anne Cademan, the widow of his distinguished physician, a lady undeterred by her prospective bridegroom's medical history. So too Davenant's literary career prospered. He succeeded Jonson first as deviser, with Inigo Jones, of Court masques, and then as poet laureate after the great man's death in 1637. He wrote plays and managed a troupe, the King and Queen's Young Company (also known as Their Majesty's Servants) at the Phoenix Theatre in Drury Lane. When the playhouses were shut down during the Interregnum, he staged an historic series of operatic performances, first in the hall at the back part of Rutland House, and afterwards at the old Phoenix. With the triumphant return of the Stuarts in 1660, he was granted, along with Killigrew, a monopoly on theatrical activities. As governor of the Duke's company, he introduced actresses on the professional London stage and produced those adaptations of Shakespeare that fastidious later critics were to deplore, but which satisfied the Restoration palate. In this way Davenant maintained a continuity between the old drama and the new, and manifested his peculiar affinity for Shakespeare.

He was born in late February or early March of 1606. His father John was a merchant vintner, and broker as well (a middleman operating between merchants of all kinds), who ran a successful business and was a model of domestic and civic sobriety; in 1622 he became mayor of the City of Oxford. John Davenant's wife, Jennet, who reared their seven offspring, was reputedly a beauty, famed too for her cleverness and agreeable conversation. (Not until a century after Jennet's death do we encounter the remarkable claim that she was a daughter of John Florio, the writer, translator, and Italian tutor of the Earl of Southampton to whom Shakespeare dedicated his two long narrative poems.) Together the Davenants kept the unpretentious tavern known in those days simply as The Taverne and displaying as its sign the familiar vintner's bush. Later, in 1666, it would be rechristened The Crown and become famous in literary history by that name. The structure, a two-storey house with twin gables, fronted on the Corn Market (it is now no. 3 on the east side of Cornmarket Street), only a few yards from the High Street, and was thus close by to the great highway that ran from Warwick-shire to London. Shakespeare would pass through Oxford on his way back and forth from Stratford, and on occasion his company acted there: indeed, on 9 October 1605 the King's Men performed before the mayor and the whole corporation. The Taverne offered no lodging or stabling, but these were available at the prosperous Cross Inn bounding it on the north, with which it shared a common courtyard. Of course an occasional guest could have spent the night upstairs at The Taverne. There the best bedroom had a great fireplace; the walls were decorated with an interlacing pattern of vines and flowers, and on a painted frieze along the top appeared a pious exhortation:

> . . . In the morning early
> Serve God devoutly;
> Fear God above all thing,
> And [honour Him] and the King.*

This room assumes a special interest in view of a scandalous report that got under way in the late seventeenth century. This rumour held that Shakespeare stopped at The Taverne and there shared his bed with Mistress Davenant. The fruit of this union, according to the story, was the future laureate.

The tale was first set down, perhaps around 1680, by John Aubrey, who knew Oxford well. There he had met Anthony Wood (1632–95) of Merton College and agreed to supply that disagreeable character with material for his monumental *Athenae Oxonienses: An Exact History of All the Writers and Bishops Who Have Had Their Education in the Most Ancient and Famous University of Oxford from . . . 1500, to the End of the Year 1690*. Thus did Aubrey's *Brief Lives* originate. Easy-going and totally disorganized, he was nevertheless peculiarly suited for this undertaking; for he had an alert mind and an insatiable curiosity about people: their appearance, manners, and sayings, the quirks and scandals of their private lives. By birth a member of the landed gentry, Aubrey inherited a good fortune, but being incompetent in his dealings with money (as with women) he lost everything—even, at last, his beloved books, which he was compelled to sell. He had little choice but to utilize his talent for conviviality and become a professional guest. In this capacity he drank too much and slept too little, but he gathered his precious data; Wood said of him that he would rather break his neck rushing downstairs than miss getting a story from a departing guest. The next morning, bleary-eyed with hangover, he would commit his anecdotes helter-skelter to paper. 'I have, according to your desire', he wrote to Wood, 'put in writing these minutes of *Lives*, tumultuarily, as they occurred to my thoughts or as occasionally I had information of them. They may easily be reduced into order at your leisure by numbering them with red figures, according to time and place, &c.'[61] In his one factual addition to Aubrey's account of Davenant, Wood notes that the father, however reserved and melancholy he afterwards may have been reputed, was 'yet an admirer and lover of plays and playmakers, especially Shakespeare'. We now know that John Davenant was

---

* The bracketed words are conjecturally supplied (cf. 1 Peter 2: 17). The lettering first came to light during alterations in November 1927, and soon thereafter the chamber began being shown to visitors as The Painted Room. Information about The Taverne is conveniently summarized in the biographies of Davenant by Harbage (London, 1935) and Nethercot (Chicago, 1938).

living in London in the 1590s, in a parish which lay opposite the Bankside playhouses, so he was in a position to attend performances there.

Eventually Aubrey fell out with Wood, and afterwards that insufferable recluse of Postmasters' Hall wrote of him in his journal:

He was a shiftless person, roving and maggoty-headed, and sometimes little better than crazed. And being exceedingly credulous, would stuff his many letters to A.W. with fooleries and misinformations, which sometimes would guide him into the paths of error.[62]

Wood's charge has some truth, but it is also true that Aubrey recorded scrupulously what was told him, and that he had an eye for the vivid detail. ('If ever I had been good for anything 'twould have been a painter,' he once mused. 'I could fancy a thing so strongly, and have so clear an idea of it.'[63]) He is an invaluable source of information about the great and near-great of his own and the previous age.

He knew Davenant. He was also acquainted with Davenant's elder brother Robert, the parson who as a child (so he told Aubrey) had been given a hundred kisses by Shakespeare. The gossip about the laureate's mother, which perhaps came to Aubrey via Samuel Butler, appears as follows in his notes for the life of Davenant:

Mr. William Shakespeare was wont to go into Warwickshire once a year, and did commonly in his journey lie at this house in Oxon, where he was exceedingly respected. . . . Now Sir William would sometimes when he was pleasant over a glass of wine with his most intimate friends, e.g. Sam. Butler (author of *Hudibras*), &c., say, that it seemed to him that he writ with the very spirit that Shakespeare, and was ['seemed' above] contentended [sic] enough to be thought his son. He would tell them the story as above, in which way his mother had a very light report, whereby she was called a whore.[64]

The last sixteen words have been scored through, perhaps by Wood.[65]

Later it would be suggested that Davenant was jesting in his cups, or that he was misunderstood rather than misconceived. Actually he had claimed *poetical* kinship with Shakespeare, as others prided themselves on being called the Sons of Ben; he would not have casually defamed his parents and sacrificed his legitimacy in order to maintain descent from a writer who, after all, had not yet been deified. But there is no question how Aubrey, with his taste for salacious gossip, interpreted the incident, and similar rumours were spread by others who did not have access to Aubrey's notes. In his 1698 edition of Langbaine, Gildon reports that Shakespeare much frequented the Davenant tavern 'whether for the beautiful mistress of the house, or the good wine, I shall not determine'.[66] Thomas Hearne, a keeper of the Bodleian and an antiquary steeped in Oxford lore, kept a diary, and in 1709 he noted in it, as a local tradition, that Shakespeare was Davenant's godfather, and that probably he begot the child, who was named for him. He goes on to record an anecdote about young Davenant returning home from school and meeting a reverend divine. 'Child, whither art thou going in such haste?' the clergyman demanded. 'O sir', the boy replied, 'my godfather is come to Town, and I am going to ask his blessing.' To which the doctor of divinity responded: 'Hold child, you must not take the name of God in vain.'[67] The learned Hearne was presumably unaware that, long before being applied to Davenant and Shakespeare, the jest had formed part of the *Wit and Mirth* published by John Taylor in 1629.

The story that Sir William was more than Shakespeare's poetical offspring was common during Davenant's lifetime, and he seemed to relish being thus regarded; so Pope told Spence in May 1730.[68] At a brilliant dinner party given by the Earl of Oxford shortly after a monument to Shakespeare was belatedly erected in Westminster Abbey in 1741, that event naturally turned conversation to Shakespeare. Pope (who had played a part in the campaign for the monument) was present, and again mentioned the traditions about Shakespeare's sojourns at The Taverne. He went on to repeat, with a new flourish or two, the anecdote about Davenant and his godfather that Hearne had privately set down. Little Davenant, in Pope's version, was 'so fond also of Shakespeare, that whenever he heard of his arrival, he would fly from school to see him', and consequently was almost breathless from 'heat and hurry' when stopped for questioning, this time by 'an old townsman'. The story, Pope told one of the guests, the antiquary Oldys, was given him by the actor Betterton. It is a choice bit of scandal, and Oldys wondered aloud why Pope had not used it in the preface to his edition of Shakespeare. With the Augustan elegance of which he was master, Pope replied, 'There might be in the garden of mankind such plants as would seem to pride themselves more in a regular production of their own native fruits, than in having the repute of bearing a richer kind by grafting.'[69] Despite this fastidiousness, he did not hesitate to repeat the story in company. A year or two later we find him mentioning it to Spence; on this occasion the schoolboy's interlocutor was 'a head of one of the colleges (who was pretty well acquainted with the affairs of the family) . . .'[70]

Thus sown, rumours about Davenant's illustrious parentage would dig deep roots. A former prompter at Drury Lane, William Chetwood, put the gossip in print for the first time in his *General History of the Stage* in 1749, and moreover took the trouble to study the frontispiece engraving by William Faithorne (from a lost portrait by John Greenhill) of Davenant, crowned with the bays as poet laureate, in the 1673 folio of the *Works*. 'The Features', he solemnly concluded, 'seem to resemble the open Countenance of *Shakespear*, but the want of a Nose gives an odd Cast to the Face.'[71] The story is referred to in the anonymous *British Theatre* in 1750, and retold with Johnsonian eloquence in *The Lives of the Poets*. Late in the eighteenth century Walter Whiter, in his *Specimen of a Commentary*, would give a pleasingly bardolatrous explanation for the emergence of so indelicate a story: 'The truth is, that the *great beauty* and *sprightly wit* of the lady would easily afford a subject of scandal to the censorious, when the *melancholy gravity* of her husband was contrasted with the pleasantry, the accomplishments, and the genius of the *gentle Shakspeare*.'[72] But few read Whiter, and anyway salacious gossip is not easily put down.

The story would be endlessly recapitulated. Sir Walter Scott came upon it in the Variorum, and found a use for it in *Woodstock*:

'Why, we are said to have one of his [Shakespeare's] descendants among us—Sir William D'Avenant', said Louis Kerneguy; 'and many think him as clever a fellow.'

'What!' exclaimed Sir Henry—'Will D'Avenant, whom I knew in the North, an officer under Newcastle, when the Marquis lay before Hull?—why, he was an honest cavalier, and wrote good doggrel enough; but how came he a-kin to Will Shakspeare, I trow?'

'Why', replied the young Scot, 'by the surer side of the house, and after the old fashion, if D'Avenant speaks truth. It seems that his mother was a good-looking, laughing, buxom mistress of an inn between Stratford and London, at which Will Shakspeare often quartered as he went down to his native town; and that out of friendship and gossipred, as we say in Scotland, Will Shakspeare became godfather to Will D'Avenant; and not contented with this spiritual affinity, the younger Will is for establishing some claim to a natural one, alleging that his mother was a great admirer of wit, and there were no bounds to her complaisance for men of genius.'

'Out upon the hound!' said Colonel Everard; 'would he purchase the reputation of descending from poet, or from prince, at the expense of his mother's good fame?—his nose ought to be slit.'

'That would be difficult,' answered the disguised Prince, recollecting the peculiarity of the bard's countenance.

'Will D'Avenant the son of Will Shakspeare?' said the knight, who had not yet recovered his surprise at the enormity of the pretension . . . I never heard such unblushing assurance in my life!—Will D'Avenant the son of the brightest and best poet that ever was, is, or will be?'[73]

The pretension would be much debated, and there would in aftertimes be some curious transmutations of the legend.

We do not know whether Shakespeare actually stood godfather to Sir William—much less, begot him—or, for that matter, whether young Davenant ever caught a glimpse of the great man at The Taverne, if indeed he stopped there. But there can be little doubt that from an early age Davenant worshipped Shakespeare. He demonstrated his allegiance in a juvenile ode 'In Remembrance of Master William Shakespeare', in which the dead master is celebrated in elegiac strains:

> Beware (delighted poets!), when you sing
> To welcome Nature in the early spring,
>   Your·num'rous feet not tread

> The banks of Avon; for each flower
> (As it ne'er knew a sun or shower)
> Hangs there the pensive head. . . .[74]

Later it would be thought that the precocious author poured out this threnody when he was only ten or eleven, on the occasion of Shakespeare's death; but that is probably an exaggeration. In the preface to his adaptation of *The Tempest*, Dryden acknowledged that Shakespeare was a poet held by Davenant in 'particularly high veneration, and whom he first taught me to admire'.[75] He is said to have cherished (as already noticed) a letter which King James wrote to Shakespeare; tradition holds too that he owned the so-called Chandos portrait of Shakespeare.* If he was the swan of Avon, Davenant would be the swan of Isis.

The producer of his own adaptations of a number of his great predecessor's plays, Davenant stands at the centre of the transmission of theatrical traditions. Thus, for the part of Henry VIII, Betterton, we learn, received instruction from Davenant, 'who had it from Old Mr. *Lowen*, that had his Instructions from Mr. *Shakespear* himself'.[76] John Lowin, usually cast as comedian or villain, had acted with the King's Men since 1603; he was still with them in 1635, when he spoke the Prologue to Davenant's *Platonic Lovers* at the Blackfriars. It was also said that Betterton was taught his celebrated interpretation of Hamlet 'in every Particle' by Davenant, 'having seen Mr. *Taylor* of the *Black-Fryars* Company Act it, who . . . [was] Instructed by the Author Mr. *Shaksepeur*'.[77] But Taylor did not join the King's company until 1619; legends are sometimes indifferent to fact.

More central to our purposes is Davenant's responsibility for two traditions associated with problems crucial to Shakespearian biography. One concerns the playwright's relations with Southampton. 'There is one Instance so singular in the Magnificence of this Patron of *Shakespear's*', Rowe remarks in his 'Account', 'that if I had not been assur'd that the Story was handed down by Sir *William D'Avenant*, who was probably very well acquainted with his Affairs, I should not have ventur'd to have inserted, that my Lord *Southampton*, at one time, gave him a thousand Pounds, to enable him to go through with a Purchase which he heard he had a mind to.'[78] The other tradition has to do with Shakespeare's Lost Years. Davenant is the ultimate source of the story that the dramatist began his career in London by holding horses outside the theatre.[79] This story, it was said, Davenant told to Betterton, who passed it along to Rowe, who mentioned it to Pope, who then repeated it to Dr Thomas Newton, the editor of Milton; Newton in turn gave it to an unnamed gentleman, who communicated it to Robert Shiels, Johnson's amanuensis, who delivered it to print in 1753 in *The Lives of the Poets of Great Britain and Ireland, to the Time of Dean Swift*.[80] Thus are traditions handed down. There is,

---

* On the Chandos portrait, see below, Part IV, Chapter 4.

however, something murky about the windings of this one, for neither Rowe
nor Pope included it in their accounts of Shakespeare. The story achieved its
apotheosis when related by Dr Johnson in his edition of Shakespeare.[81]

# 6

# *Personal Legends; Youth and Education*

FOR biographers the most vitally interesting traditions have to do with the
events of his life. These obligingly fill in the voids in knowledge that tease
speculation. They shed light, or appear to do so, on the scenes of Shakespeare's boyhood and youth, on his departure from Stratford, his activities
during the unknown years, the manner of his death, and the fate of his
papers. We are even given glimpses of his appearance and carriage and
reassuring words about his moral character. A tradition among 'the Veterans
of the Theatre of the last century' held that Shakespeare was a fat man. So an
eighteenth-century newspaper reported, but one suspects that Aubrey is
better deserving of credit. 'He was a handsome well shaped man', he records:
'very good company, and of a very ready and pleasant smooth wit.'[82] Also:

the more to be admired q[uia] he was not a company keeper, lived in Shoreditch,
wouldn't be debauched, and if invited to, writ: he was in pain.[83]

A later tradition illustrates his largesse to the needy: Shakespeare, we learn,
relieved a widow with numerous family who was involved in a ruinous
lawsuit. Sometimes these reports are disproved by more reliable sources, but
usually the traditions supplement the documentary records, in some instances so colourfully that biographers, responding to myths stranger than
truth, would find them irresistible.

    The first informant to mention Shakespeare's father is Archdeacon Plume
who around 1657 cites Sir John Mennis as having once seen him in his shop,
'a merry cheeked old man' who said, 'Will was a good honest fellow, but he
durst have cracked a jest with him at any time.'[84] Aubrey, who makes the not
quite correct assertion that John Shakespeare was a butcher, records that
William had a childhood friend in Stratford, also a butcher's son, and every
bit as clever as he; but unfortunately this lad died young. Thus was posterity
denied a second Immortal Bard. As a boy Shakespeare occasionally took a
turn at his father's craft, and (so some of the Stratford neighbours told
Aubrey) 'when he killed a calf, he would do it in a high style, and make a
speech.'[85] This anecdote—patently ludicrous if taken at face value—would

receive the solemn approbation of some learned judges, who would favour it with an interpretation it was doubtfully intended to bear. (May not the anecdote represent a filtered down and inevitably distorted recollection of a Stratford mumming play performed by the boys at Christmas time—some such play as the one noted in this context by Douglas Hamer: the performance at Christmas 1521 before the young Princess Mary: 'Item, pd to a man at Windsor, for killing a calf before my Lady's Grace behind a cloth.') Although a glover would not be permitted to sell raw meat, they would ask, might he not himself likely slaughter the deer and sheep and goats whose skins he would then proceed to cure and whiten?[86] (The answer is that probably he would not.) And is not the killing of the calf maybe—as the Revd James Raine of Durham would first ask in 1837—a confused reference to the pantomimic exercise by that name in the repertories of wandering players?[87] The Court accounts of Princess Mary show that during Christmas 1521 an entertainer at Windsor was paid 'for killing of a calf before my lady's grace behind a cloth'. So much the erudition of commentators would reveal, but there would be no way of ascertaining whether such a performance lay behind Aubrey's report, which clearly has another meaning.

From Rowe we learn that Shakespeare attended the free school in Stratford and there acquired his rudimentary Latin. 'But the narrowness of his Circumstances,' Rowe continues, 'and the want of his assistance at Home, forc'd his Father to withdraw him from thence, and unhappily prevented his further Proficiency in that Language.'[88] Regrettably, no records have been preserved for the pupils who attended the King's New School of Stratford-upon-Avon in Shakespeare's day, but there is every likelihood that the future poet studied there. Support for Rowe's remark that Shakespeare was taken out of school is perhaps to be found in the evidence of his father's straitened finances during these years. But Rowe's information that William was one of ten offspring does not inspire confidence: in fact there were eight, of whom several died young. Thus John Shakespeare was not burdened by the large family which Rowe postulates, and which would rationalize his guess about the withdrawal of the eldest son from school. The line of reasoning appears to have been set in motion out of a desire to account for the poet's small Latin.

After Shakespeare gave up his formal education, according to Rowe, he followed his father's way of life. That Rowe describes John Shakespeare as a wool dealer is perhaps sufficient to arouse scepticism regarding this supposed period of apprenticeship; but the father did have subsidiary business interests which included wool-dealing, and it is not inconceivable that young William helped out in his father's shop. A less plausible tradition is mentioned by a Mr Dowdall, who visited Stratford in April 1693 and wrote to his cousin about his experience. At Holy Trinity Church he viewed Shakespeare's monument and gravestone, and the clerk who showed him round said that 'this Shakespear was formerly in this town bound apprentice to a butcher, but that he run from his master to London'.[89] Even this suggestion would not be entirely scorned by later biographers abhorring a vacuum. Not too dissimilar is Aubrey's conjecture that Shakespeare came to London when he was about eighteen, and became an actor in one of the playhouses. But Aubrey is frankly guessing and it is not a very good guess. Shakespeare was at that age in the process of getting married; and he was probably still in Stratford when he turned twenty-one: his wife then produced twins, and although the husband need not have attended their christening, his presence was presumably required for their conception in 1584.

# 7

## *Shakespeare the Deer-Poacher*

THE occasion on which Shakespeare left Stratford and struck out on the high road for London was a critical one for the future dramatist. How he came to make this move is explained in the most celebrated of Shakespearian legends. Like so much other information about the poet, it was first made public by Rowe. Shakespeare, according to him, was leading a settled life until

> an Extravagance that he was guilty of, forc'd him both out of his Country and that way of Living which he had taken up; and tho' it seem'd at first to be a Blemish upon his good Manners, and a Misfortune to him, yet it afterwards happily prov'd the occasion of exerting one of the greatest *Genius*'s that ever was known in Dramatick Poetry. He had, by a Misfortune common enough to young Fellows, fallen into ill Company; and amongst them, some that made a frequent practice of Deer-stealing, engag'd him with them more than once in robbing a Park that belong'd to Sir *Thomas Lucy of Cherlecot*, near *Stratford*. For this he was prosecuted by that Gentleman, as he thought, somewhat too severely; and in order to revenge that ill Usage, he made a Ballad upon him. And tho' this, probably the first Essay of his Poetry, be lost, yet it is said to have been so very bitter, that it redoubled the Prosecution against him to that degree, that he was oblig'd to leave his Business and Family, in *Warwickshire*, for some time, and shelter himself in *London*.[90]

Several decades previous to Rowe's Life the Revd William Fulman, a scholar of Corpus Christi College, Oxford, had assembled notes on the lives of various poets (among them Shakespeare) for whom he jotted down only an approximate date of birth. When Fulman died in 1688, his papers came into the hands of the Revd Richard Davies, vicar of Sapperton in Gloucestershire, about forty miles south-west of Stratford. 'Mr. Davies looked red and jolly', according to an entry by Wood in his diary in 1692, 'as if he had been at a fish dinner at Corpus Christi College and afterwards drinking, as he had been.'[91] Sometime before his own death in 1708, the genial divine added—in a single fragmentary sentence—a less detailed and more violent version of the poaching story: 'much given to all unluckiness in stealing venison and rabbits, particularly from Sir —— Lucy, who had him oft whipped and sometimes imprisoned and at last made him fly his native country to his great advancement; but his revenge was so great that he is his Justice Clodpate and calls him a great man and that in allusion to his name bore three louses rampant for his arms.'[92] Of special interest is the fact that the episode is here recorded independently of Rowe.

By Clodpate,* Davies means Shallow: he is referring to the opening dialogue of *The Merry Wives of Windsor*. Slender has just said that Shallow— like Sir Thomas Lucy a country justice of the peace—writes himself 'Armigere', or esquire, and he goes on to add: 'All his successors gone before him hath done't, and all his ancestors that come after him may. They may give the dozen white luces in their coat.' Also present is the Welsh parson, Sir Hugh Evans, who interjects the predictable quibble: 'The dozen white louses do become an old coat well. It agrees well passant: it is a familiar beast to man, and signifies love.' A luce, as Shallow goes on to note, is a 'fresh fish'— actually a species of pike. Now, the Lucys of Charlecote had punningly adopted the fish as their heraldic device; they bore the arms *Vair, three luces hauriant argent*. On a Lucy tomb at Warwick the three luces on the coat are repeated four times—thus providing the dozen to which Slender refers. In the play Shallow has come to Windsor from Gloucestershire to make a Star Chamber matter out of a poaching expedition on his estate. He complains of venison 'ill killed', and angrily upbraids Falstaff for beating the justice's men, killing his deer, and breaking open his lodge. Is Shakespeare in an obscurely allusive passage settling an old score? It is an intriguing possibility.

The bitter lost ballad to which Rowe refers would duly turn up. Malone tells of consulting an unreliable manuscript *History of the Stage*, perhaps written by Chetwood between 1727 and 1730, and there reading of how Joshua Barnes, professor of Greek at Cambridge University, alighted some forty years previously at a Stratford inn and there (such was his admiration for Shakespeare's genius) gave an old woman a new gown for the following stanzas:

> Sir Thomas was too covetous,
>   To covet so much deer,
> When horns enough upon his head
>   Most plainly did appear.

---

* A popular stage figure, Clodpate appears as the country justice in Thomas Shadwell's *Epsom Wells* (1673), and is described by the author as 'a publick, spirited, politick, discontented Fop, an immoderate hater of *London*, and a lover of the Country above measure, a hearty true *English* Coxcomb'.

> Had not his worship one deer left?
> What then? He had a wife
> Took pains enough to find him horns
> Should last him during life.[93]

One may wonder whether the learned professor received a good pennyworth.

To this doubtful report we may add the independent testimony of Capell. His eventual authority is a Thomas Jones, who dwelt in Tardebigge, a village in Worcestershire some eighteen miles from Stratford; here lived Mary Hart, a granddaughter of Shakespeare's nephew. When in his nineties, Jones, who died in 1703,* remembered hearing the deer-poaching story from several of the elders of Stratford. Their account agreed with Rowe's except for an added detail: 'the ballad written against sir Thomas by Shakespeare was stuck upon his park gate, which exasperated the knight to apply to a lawyer at Warwick to proceed against him.'[94] Jones wrote down the first stanza, all that he could recall. Thomas Wilkes, the grandfather of the man who brought it to Capell, transmitted it by memory to his son, who also recorded it. Capell prints it thus:

A Parliamente member a Justice of Peace, | At Home a poore Scarecrow at London an Asse. | If Lowsie is Lucy as some Volke Miscalle it | Then Lucy is Lowsie whatever befalle it | He thinkes himselfe great | Yet an Asse in his State | We allowe by his Eares but with Asses to mate | If Lucy is Lowsie as some Volke miscalle it | Sing [O] Lowsie Lucy whatever befalle it.[95]

The same stanza was transcribed by Oldys, who had it from a relative of the acquaintance of 'a very aged gentleman living in the neighbourhood of Stratford, (where he died fifty years since,),' who had heard of Shakespeare's transgression from several of the old people of the town.[96] In an ominous early example of fabrication, John Jordan announced in 1790 the discovery of the entire ballad, which on one occasion he claimed to have found in a private drawer of an old chest; on another, among some deeds given to a tailor for conversion into measure strips. Jordan himself dabbled in verse, and the Lucy stanza is not unworthy of his own Muse.

The deer-poaching legend has thus come down from several autonomous sources—Rowe, Davies, Barnes, Jones—all (it would seem) stemming ultimately from Stratford gossip.[97] We are even told that it was in Stratford that folk pronounced *lousy* like *lucy*. It is a satisfying story—an exciting sequence of theft, discovery, punishment, and escape. What a dramatic scene: the sensitive, abashed young man, red in face and hand, being confronted in the Great Hall at Charlecote by the imperious Sir Thomas, justice of the Queen's

---

* The burial register of Tardebigge lists no Thomas Jones in 1703 or thereabouts, but does record an Edward Jones that year.

peace, sheriff of Warwickshire and Worcestershire, and member of Parliament; the same Lucy who in 1584 had unsuccessfully introduced into Commons a bill to make poaching a felony. Artists would in fact visualize the scene in steel engravings which, framed in rosewood, adorned the corridors of country inns in the nineteenth century. The offence itself is attractively venial, associated with youthful daring and high spirits, and in Elizabethan times well esteemed as a gentlemanly crime.[98] In a passage in *Titus Andronicus* that may reflect personal knowledge, Shakespeare asks: 'What! hast not thou full often struck a doe and borne her cleanly by the keeper's nose?' The aftermath, moreover, stirs empathy: Shakespeare torn from the bosom of his family and the town where he had been born and bred. A remote figure is made humanly accessible. But is the story true?

The Great House stands four miles upstream from Stratford, and on the west side the Avon's waters lave the terrace steps. Since the twelfth century the Lucys had held the manor of Charlecote. In Elizabethan times Sir Thomas, a slight, spare man, had rebuilt the family mansion in one of the best uses to which he put his wife's fortune. There Queen Elizabeth had twice slept on goosefeathers in the Great Bed Chamber. Charlecote lay two miles across the fields from Snitterfield, whence Shakespeare's father had come to Stratford, and where an old uncle still resided; the boy may have wandered that way. Charlecote, however, had no park in the legal sense—an enclosed tract of land prescribed for the keeping of beasts of the chase—but instead a free warren stocked with 'beasts and fowls of warren': rabbits, hares, pheasants, and wood-partridge. Late in the eighteenth century this inconvenient fact would be realized, and the scene of the poaching shifted to Fulbrook, two miles north of Charlecote and midway between Stratford and Warwick. In 1828 the then master of Charlecote, a descendant of Sir Thomas, would tell a distinguished visitor, Sir Walter Scott, that Shakespeare hid the buck in a barn, no longer standing, which belonged to an estate some distance away—presumably at Fulbrook—where the Lucys resided when the trespass took place. Pilgrims to the Shakespeare country were still being shown Shakespeare's Barn in the last quarter of the nineteenth century, and a farmer of Fulbrook triumphantly displayed a sideboard made from the very bedstead to which the captured deer-stealer had been bound. For the transfer we are indebted to the ever-helpful Jordan. At Fulbrook the Lucys did indeed once own a deer preserve that attracted poachers, but unfortunately for legend, the park came into the family's possession in 1615.

Still, the shift of scene is unnecessary. Davies mentions rabbits as well as deer; moreover, if fallow deer would not come under the heading of beasts of warren, roe deer would. So the episode could have taken place at Charlecote after all. The game law of 1563 then in effect did not prescribe whipping for such an offence; but, if caught with a dead deer within a paling or fence, the poacher would be haled before the justice of the peace, who could order him to pay damages for as much as three times the amount and sentence him to prison for as long as three months. And Rowe speaks also of prosecution for libel. Thus, even if Shakespeare had not committed a felony, he would nevertheless be in serious trouble.

But other factors, less conducive to romance, suggest themselves. Is it reasonable to suppose that Shakespeare would have waited some fourteen years to give vent to his animosity in allusive dialogue that a London audience could hardly be expected to comprehend? And the satirical portrait seems far removed from the historical personage, who saw such days as Shallow has not seen. Sir Thomas was zealous in the pursuit of recusants, not poachers, and was otherwise an amiable man. He was neither the stupid and parsimonious bachelor of *2 Henry IV* and *The Merry Wives of Windsor* nor (if we may be guided by the sentiments he expressed in his epitaph for his wife) the cuckold of the ballad. The circumstantial touch of the reference to luces is by no means conclusive. A popular heraldic device, they appeared on the arms of Justice Gardiner, the Gascoignes, the Way and Geddes family, and the Earl of Northampton—not to mention the Worshipful Company of Stockfish-mongers. Shakespeare, moreover, had treated with great respect an ancestor of the Lucys of Charlecote, Sir William Lucy, in *1 Henry IV*. Could the legend possibly have originated, long after the dramatist's death, among Strat-fordians who read *The Merry Wives of Windsor*, recollected a local jest about luces and louses, and interpreted the passage in accordance with their own resentment of a powerful local family? But in that case how *are* we to interpret the passage in the play? It is a difficult case.

For a long time it would not seem difficult. The story would be marvellously embroidered. It would be said that a farmhouse on a spot of rising ground in the park at Charlecote called Daisy Hill was formerly the keeper's lodge; thither Shakespeare was conveyed and held prisoner at the time of the charge. Someone would suggest that Shakespeare stole the buck to celebrate his wedding day; according to other gossip, he not only took the game but also seduced the keeper's daughter. On the other hand, some theorized that Shakespeare was no poacher but a popular hero asserting the rights of free forestry, and therefore incurring the wrath of the lord of the manor. It would be said that, after having had Shakespeare whipped, the infuriated knight put him, like Kent, in the stocks. Another report would hold that Sir Thomas, intending not to punish the malefactor but only to frighten and reclaim him, told the latter's family, whereupon the exasperated poet composed the ballad and thus aroused Lucy's fury. Late in the eighteenth century we learn that Sir Thomas's anger drove Shakespeare 'to the extreme end of ruin', from which he was rescued by the gracious intercession of the Queen.[99] A still later report holds that the Earl of Leicester prevailed upon Sir Thomas to relent in his persecution of Shakespeare. In modern times most (but not all) responsible scholars would reject the entire episode as traditionary romance.

# 8
## *Shakespeare the Toper*

THE lusty yeoman who poached deer would presumably be able to hold his pint of nut-brown ale, so we need not be surprised that there should develop the quaint legend of Shakespeare the tippler. The story of the celebrated crab-apple tree, with which this legend is associated, makes a dubious late appearance, and was apparently fostered by local guides. Sir Hugh Clopton, the then owner of New Place, seemed to think it true, and repeated it often to Joseph Greene; but the master of the free school regarded the story as some wag's invention, and anyway (so he wrote to the Hon. James West in 1758) 'rather too mean to appear in print with other Memoirs of so great a Bard'. The Revd Mr Greene's pious wish notwithstanding, the tale shortly afterwards made its first published appearance. An anonymous traveller, writing in *The British Magazine* in 1762, tells of putting up at the White Lion in Stratford, and being taken by the landlord to the village of Bidford, about seven miles below Stratford, where he was shown, in a hedge, the crab-tree called Shakespeare's Canopy because the poet had slept under it one night. It seems that the poet was a mighty drinker, and, having heard that the men of Bidford were similarly renowned, he went there one day to try his skill with them. Asking a shepherd for the Bidford drinkers, he was told that they were absent, although the sippers, who might be sufficient for him, were at home. Thus rebuffed, Shakespeare had no choice but to make his lodging under the tree.[100]

The story is retold, with amplifications, by the ubiquitous Jordan. He informs us that 'Our Poet was extremely fond of drinking hearty draughts of English Ale, and glory'd in being thought a person of superior eminence in that proffession if I may be alowed the phrase.'[101] In this version Shakespeare and his companions learn that the topers have gone to Evesham fair on a similar errand, and the Stratford swillers stay on to do battle with the lesser order of sippers, at whom they duly scoff. The result is ignominious defeat for the Shakespeare faction, who become 'intollerable intoxicated'. After sleeping it off under a crab-tree by the side of the road, Shakespeare refuses his friends' entreaties to resume the contest, saying that he has drunk with

> Piping Pebworth, Dancing Marston,
> Haunted Hillborough, Hungry Grafton,
> Dadgeing Exhall, Papist Wicksford,
> Beggarly Broom, and Drunken Bidford.[102]

These appellations would stick to the eight villages named in the bacchanalian doggerel.*

A scene for the revelry would be found in a large building at Bidford once called the Falcon Inn; not only would the room be pointed out to admiring visitors, but also the actual chair in which Shakespeare sat. The Canopy itself would suffer a sad fate, for it would fall victim to the rapacity of the souvenir-hunters. They literally tore the crab-tree to bits. On 4 December 1824 the remains were dug up and carted to the Revd Henry Holyoakes of Bidford Grange. 'For several years previously', we learn from a Stratford antiquary's unpublished remembrance, 'the Branches had entirely vanished from the further depradations of pious votaries; & the stock had mouldered to touchwood, the roots were rotten, & the time worn remains totally useless.'[103]

# 9

# *Legends of Shakespeare's Maturity*

THESE are Warwickshire legends; eventually, whatever the circumstances, he left Stratford. 'Such wind as scatters young men through the world,'

> To seek their fortunes further than at home,
> Where small experience grows,

carried him from his drowsy childhood town to a world of larger experience, in much the same way as the happy gale blows Petruchio to Padua from old Verona. Aubrey, as we have seen, thought Shakespeare might have headed directly to London, where, still in his teens, he was swept up into theatrical life. Unworried by such trivialities as consistency, Aubrey in another passage makes a quite incompatible suggestion, and with greater conviction. 'Though as Ben. Johnson says of him, that he had but little Latin and less Greek,' Aubrey records, 'he understood Latin pretty well, for he had been in his younger years a schoolmaster in the country.'†[104] In a marginal note for this bit of information, he cites his authority: 'from Mr. Beeston'. Aubrey is unique in his time in thus challenging Jonson's famous dictum about Shakespeare's classical attainments. The proposal that he passed the Lost

---

* Not to be outdone as a nursery of bardic jingles, the nearby Cotswolds have foisted upon Shakespeare an even more depressing quatrain:

> Dirty Gretton, dingy Greet,
> Beggarly Winchcomb, Sudely sweet;
> Hartshorn and Wittington Bell,
> Andoversford and Merry Frog Mill.

† See Epilogue, p. 506.

Years as a rural schoolmaster is not especially glamorous; it would have nothing like the impact of the deer-poaching tradition, and is still ignored by some biographers. For some inexplicable reason, however, it would come to appeal marvellously to academic investigators. They would find support for the possibility in the early plays themselves: the intimate knowledge of pedagogues and school-texts in *Love's Labour's Lost*, the Plautine character of *The Comedy of Errors*, and the Senecan character of *Titus Andronicus*—the other side of the same academical coin. True, Shakespeare's own grammar-school education would not have qualified him to take charge of a school, despite the fact that the excellent Stratford institution was guided by a Bachelor of Arts; one master was a fellow of Corpus Christi; another, of St John's College, Oxford. But he would not require a university degree for the post of usher or *abecadarius* in a country school.

Whether or not he for a time (in Jonson's phrase) swept his living from the posteriors of little boys, he had by 1592 made enough of a mark as a dramatist to arouse the envious malice of Greene. As to how he got launched in the metropolis the documentary records are silent, but again tradition speaks. According to Dowdall, the runaway butcher-boy was 'Received into the playhouse as a serviture, and by this means had an opportunity to be what he afterwards proved'; Rowe merely says that he was taken in by the company 'at first in a very mean Rank'.[105] This is too meagre to satisfy curiosity. In a note appended to the reprint of Rowe's *Life* in his 1765 edition of Shakespeare, Dr Johnson furnishes ampler information, which he describes as having been related to Pope by Rowe (although, mysteriously, neither availed himself of it). It is worth quoting in full:

In the time of *Elizabeth*, coaches being yet uncommon, and hired coaches not at all in use, those who were too proud, too tender, or too idle to walk, went on horseback to any distant business or diversion. Many came on horseback to the play, and when *Shakespear* fled to *London* from the terrour of a criminal prosecution, his first expedient was to wait at the door of the play-house, and hold the horses of those that had no servants, that they might be ready again after the performance. In this office he became so conspicuous for his care and readiness, that in a short time every man as he alighted called for *Will. Shakespear*, and scarcely any other waiter was trusted with a horse while *Will. Shakespear* could be had. This was the first dawn of better fortune. *Shakespear* finding more horses put into his hand than he could hold, hired boys to wait under his inspection, who, when *Will. Shakespear* was summoned, were immediately to present themselves, *I am* Shakespear's *boy, Sir*. In time *Shakespear* found higher employment, but as long as the practice of riding to the play-house continued, the waiters that held the horses retained the appellation of Shakespear's *Boys*.[106]

Note how the supposed aftermath of the supposed deer-stealing is here presented as melodramatic fact. Johnson's charming story of successful early capitalist enterprise, originating as we have seen with Davenant, is of course pure nonsense. (Like other such nonsense, it would in time assume a pleasing specificity. The playhouse at whose door Shakespeare waited would be identified as the Red Bull in Clerkenwell, in the north-west suburbs; no matter that it came into existence after the composition of *Othello* and *King Lear*.)

Of Shakespeare after he became the leading dramatist for the Lord Chamberlain's Men and, subsequently, the King's Men, tradition is remarkably uncommunicative. There is the fantastic rumour that he earned £1,000 per year. Scattered reports filter down about his writing and acting; these we have already noticed. From the unreliable David Lloyd in 1665[107] we learn that Fulke Greville, Lord Brooke, who held important posts under Elizabeth, wished to be known to posterity as 'Master' to Shakespeare—and to Jonson as well; but the story totally lacks corroboration.* According to Aubrey, Shakespeare would together with Jonson gather the 'humours of men daily wherever they came'; once, on Midsummer Night, he lay at Grendon in Buckinghamshire, on the road from London to Stratford, and there found the model for the Constable in *A Midsummer Night's Dream*[108]—presumably Aubrey is thinking of Dogberry in *Much Ado About Nothing*, but that rather spoils a Midsummer Night's legend. These are slim pickings, especially when one considers that these were the years of Shakespeare's achievement and recognition, when one would expect information to be more eagerly sought after and passed along than during his period of obscurity. But legends thrive best in the dark.

Sometimes one cannot very clearly tell to which phase of Shakespeare's career a legend should be assigned, particularly if the story itself is trivial. An instance is the anecdote passed on in the earlier eighteenth century 'to a gentleman at Stratford by a person then above eighty years of age, whose father might have been contemporary with Shakspeare'.[109] It reached Sir Hugh Clopton, who in May 1742 told it to the actor Macklin and the great Garrick as they sat in sunshine at New Place under the spreading mulberry tree which (legend has it) Shakespeare himself had planted. According to this story, a blacksmith with carbuncled face thus accosted Shakespeare as he leaned over a mercer's door:

> Now, Mr. Shakspeare, tell me, if you can,
> The difference between a youth and a young man.

To which the poet unpremeditatedly replied:

> Thou son of fire, with *thy face like a maple*,
> The same difference as between a scalded and a coddled apple.†[110]

* As Chambers notes, Lloyd's *bona fides* is impugned by Wood, who reproaches him in *Athenae Oxonienses* as an 'impudent plagiary' and a 'false writer and meer scribbler' (iv. 352); see *William Shakespeare*, ii. 250.

† Halliwell-Phillipps quotes a letter written in 1788 by Malone, in which he makes John Combe the subject of the comparison (*Life of William Shakespeare*, 243).

The lines, as Malone recognized, are a variation on one of *Tarlton's Jests* (1638): 'Gentleman, this fellow, with this face of maple, | Instead of a pippin, hath thrown me an apple.'

Less indeterminate is the tradition related by the same Sir Hugh to Joseph Greene, the curate and schoolmaster of Stratford. In his times of pleasantry Shakespeare and his children would together exercise their talents by writing little epigrams on familiar subjects. Great pleasure was afforded the poet (Greene remarks with satisfaction) when he could trace in his offspring 'some petty display of that genius which God & Nature had bless'd himself with'. Perhaps Susanna was witty above her sex after all. The epigrams, we learn, decorated the panes of the New Place windows when the house returned to the Cloptons, but they have since regrettably disappeared without trace.

# 10
## *Legends of Death and Burial*

OF Shakespeare during the years of his retirement, tradition says nothing, apart from Theobald's unlikely tale of the origin of *Double Falsehood*. But about the causes of his death speculations would naturally float about. According to one tradition the poet fell into a decline not long after his retirement from the stage—never mind that, according to Rowe, he passed his last years in 'Ease, Retirement, and the conversation of his Friends'.[111] Greater authority attaches to an account of Shakespeare's final illness and death preserved in the notebooks of the Revd John Ward, the first individual in Stratford to take an antiquarian interest in the poet.

Born thirteen years after Shakespeare's death, Ward took his MA degree at Oxford. Although a student of divinity, he immersed himself in pharmacy, anatomy, and other subjects connected with medicine while he was at Oxford, and later in London. Thus he qualified himself to minister to men's bodies as well as to their souls. In this double capacity he came in 1662 to Stratford, where he was appointed vicar of the parish. His commonplace-books, numbering seventeen volumes, testify to his piety, humanity, and lively curiosity: he resolved, in one entry, 'to study Arabic and Saxon, with a further perfection in the Hebrew tongue'.[112] It is scarcely surprising, there-fore, that Ward should interest himself in the great man who had resided in the town a half-century earlier. 'Remember to peruse Shakespeare's plays,' he reminds himself, 'and be much versed in them, that I may not be ignorant in that matter.'[113] When Ward settled in Stratford, relatives, friends, and neighbours who knew the poet were still alive and could have imparted priceless information. There was Thomas Hart, for example, Shakespeare's nephew, who dwelt in the ancestral house on Henley Street. An entry in Ward's notebooks refers to 'Mrs Queeny'—presumably Shakespeare's daughter Judith, who married Thomas Quiney and died at the age of seventy-seven in 1662. It is a pity that Ward records so little, but he does make the following crucial statement: 'Shakespeare, Drayton, and Ben Jonson had a merry meeting, and it seems drank too hard, for Shakespear died of a fever there contracted.'[114]

Elsewhere Ward goes badly off-base in estimating Shakespeare's earnings, but he correctly notes that the dramatist had two daughters, one of whom married the physician Hall. Such a convivial meeting as he describes does not strain credulity. Jonson was associated with Shakespeare, and drank deeply. Drayton, it is true, was noted for his temperance, but he could have made a third: a Warwickshireman, he often visited the village of Clifford Chambers, nearby to Stratford, and he was once treated by Dr Hall. A possible occasion for celebration was the wedding of Judith Shakespeare in February 1616. Nevertheless, some scholars would later express scepticism. They would object that Shakespeare could scarcely have died of alcoholic poisoning in an age in which ale and wine rather than spirits were the preferred intoxicants, and they would question the cause-and-effect relationship between the drinking party and the fever. But Ward says nothing about alcoholic poisoning, and one may contract a fever at a party, as elsewhere. In another passage Ward remarks that there was much fever (typhoid?) at Stratford in certain seasons. Although he can be clinically precise, he is not of course offering a medical diagnosis of Shakespeare's fatal malady, but merely jotting down something he has heard. Nor does he record it as a fact; *it seems*, Ward says.

The story of Shakespeare's death is another tradition, and if it cannot be trusted as fact, it remains a possibility. Like other such anecdotes, it would breed variants. 'The tradition runs', we learn from the *Quarterly Review*, for April 1864, 'that he caught his death through leaving his bed when ill, because some of his old friends had called on him.' Another version holds that the friends included 'play-fellows', and they came to fetch him for a carouse. Of later date is the suggestion of W. L. Rushton, president of the Mersey Bowmen, that the merry meeting passed not in drinking but in shooting with the long bow—*the Vicar's Diary notwithstanding* [italics Rushton's]. According to this hypothesis, Shakespeare died of congestion of the lungs developing out of a cold he contracted while engaged in his favourite exercise.[115]

Of his funeral, legend says little. In the nineteenth century a flurry of excitement would be generated by the purported inscription on the red sandstone slab covering the grave of Edmund Heldon in the Fredericksburg area in Virginia, near a place called Potomac Creek: 'Here lies the body of

Edward [*sic*] Helder, practitioner in Physic Chyrugery. Born in Bedfordshire, England, in the year of our Lord 1542—was contemporary with, and one of the Pallbearers of William Shakespeare. After a brief illness his spirit ascended in the year of our Lord 1618, aged 76.'[116] A Dr Helder did exist, but his spirit, it turned out, ascended not in 1618 but in 1678. References to this lapidary intelligence would circulate in the newspapers for some twenty years before one Moncure D. Conway reported in *Harper's New Monthly Magazine* for January 1886 that he had inspected the cemetery and satisfied himself that no such tombstone inscription ever existed. Another story passed down tells of a lady who heard the funeral sermon preached that day in April 1616, and remembered the vicar bursting into tears and lamenting, before the large and solemn congregation, that Shakespeare had not been bred a divine.

After death and burial, he would be memorialized at Holy Trinity. In due course the unimposing bust of Shakespeare—actually a half-length figure—would inspire a high-Victorian Royal Academy artist of Pre-Raphaelite leanings, Henry Wallis (1830–1916), to evoke with meticulous realism the sculptor at work on the effigy in his studio, with adoring onlookers gazing and oblivious children at play alongside; in the background the Avon and church spire. Never mind that the actual statue was almost certainly chiselled in Southwark. Today the small oil canvas (the artist's own replica of a larger painting), entitled 'A Sculptor's Workshop, Stratford-upon-Avon, 1617', hangs in the Royal Shakespeare Theatre Picture Gallery, hard by Holy Trinity.

# 11
## *Shakespeare's Faith*

ABOUT the faith in which Shakespeare ended his life a solitary seventeenth-century report has come down. It is furnished by the same Richard Davies who first set down the sensational deer-poaching story. In the terse concluding sentence of his laconic notes on the poet, Davies writes: 'He died a papist.'[116] Few would accept the statement at face value; it marks the starting-point for a tortuous debate nourished by few facts and an abundance of inferences. Scholars would measure Davies's assertion against the evidence of the plays themselves, the religious implications of Shakespeare's will, and what might be deduced about the doctrinal beliefs and church affiliations of his father, who would be variously described as a Catholic, a Puritan, and a conforming Protestant.

On the subject of John Shakespeare's faith, the wheelwright of Stratford is questionably informative after his own inimitable fashion. In June 1784 Jordan offered to *The Gentleman's Magazine* the Spiritual Last Will and Testament of John Shakespeare, 'wrote in a fair and legible hand';[117] but that journal refused it. The significance of this remarkable document was not at once universally apprehended. 'This testament', reads an item in the popular press for February 1790, 'is no farther remarkable, than in proving that John Shakespear was a Butcher, and that he bequeathed all his acquirements in that profession to his son William.' In fact the testament, since lost, is not a will in the usual sense but, rather, a declaration of faith. In fourteen articles, it follows the model composed during a time of plague by St Charles Borromeo, Cardinal Archbishop of Milan, and presumably disseminated in England by Jesuit missionaries. The testator describes himself as 'an unworthy member of the holy Catholic religion', craves the sacrament of extreme unction, asks the Virgin Mary to be his chief executrix, and beseeches all his 'dear friends, parents, and kinsfolks' to succour him after his death by celebrating the mass (a demanding request, for the Act of 1581 imposed a prison sentence of one year and a fine of two hundred marks for saying mass, and a fine of one hundred marks for hearing it). Accepting his death however it befalls him, the testator bequeaths his soul 'to be entombed in the sweet and amorous coffin of the side of Jesus Christ', and begs that 'this present writing of protestation' be buried with him.

The document, originally consisting of six small leaves stitched together, was, according to Jordan, found by the master bricklayer Joseph Moseley on 27 April 1757 between the rafters and the tiling of the western Shakespeare house in Henley Street, which he was retiling. The house was then owned by Thomas Hart, the fifth lineal descendant of the poet's sister, Joan Hart. Moseley passed along the will to Jordan, who made a transcript of the complete document in his notebook. The original reached Malone around 1789 from Mr Payton of Shottery, an alderman of Stratford, via James Davenport, the vicar. The document that Malone saw lacked the first page.

It would stir sceptical inquiry. Was this the declaration of the poet's father? If so, must he not, because of the reference to his parents, have put his signature to it as a young man? (Later it would be pointed out that the word *parents* may, like the Latin *parens*, merely signify *relatives*.) Or is the will, in whole or in part, a fraud? Was Jordan capable of such a fabrication? Or was the document genuine, but with false signatures inserted in the blank spaces left for the testator's name; Jordan had, we know, experimentally forged the poet's signature. There would be much controversy. The publisher of *The Gentleman's Magazine* rejected the document as spurious. Malone published it in his 1790 edition of Shakespeare, where he says it had been found twenty— not thirty-three—years earlier; later he experienced second thoughts, and in 1796 declared that 'clearly . . . it could not have been the composition of any one of our poet's family'.[118] Jordan himself did not help matters by informing Malone on 19 March 1790 that the will was given him in June 1785—the year after he offered it for publication. Jordan also said that Moseley never showed it to the owner of the house; yet Hart informed Malone that he remembered the discovery of the document. In his letter to *The Gentleman's Magazine* Jordan remarked that he was given the testament six days previously, but to Malone he intimated that he waited for some time before sending it off: he asked Moseley to look for the missing page (which Jordan afterwards produced), and, during the long interval that followed, showed the document to several persons, including the Revd Joseph Greene, the antiquarian rector of Welford, who pronounced it a fabrication. The discrepancies in Jordan's testimony are hardly reassuring (he *was* a rogue), but they do not prove the testament a fraud.

How miraculous that it should re-emerge during renovation of the Birthplace house over 150 years after being drawn up! And what a pity that it should afterwards disappear; for disappear it did, without trace. Modern scholarship might have determined, from the watermark, where the paper was manufactured, and also helped to date the document. From the handwriting we could very likely discover whether the text was transcribed in England (where the secretary script was favoured) or on the Continent (where italic was preferred). It would be interesting also to know if John Shakespeare's name was entered in the same hand as the rest. And did he affix to the testament his characteristic glovers' mark? These questions must remain forever unanswered. But in the twentieth century much would be learned about the Borromeo formulary (an authentic English text, printed in 1638 and agreeing closely with the version published by Malone, would come to light in the Folger Library), and the Spiritual Last Will and Testament of John Shakespeare would not want champions.*

* See the authoritative article by J. G. McManaway, 'John Shakespeare's "Spiritual Testament"', *Shakespeare Quarterly*, 28 (1967), 197–205. McManaway concludes that 'the five leaves of Spiritual Testament sent to Malone must have been genuine, even though the identity of the testator cannot now be established' (p. 205).

John Shakespeare's Spiritual Last Will and Testament occupies a chapter in my *William Shakespeare: A Documentary Life*, 41–6, in which Carlo Borromeo's *Testament of the Soule* (1638) is reproduced in its entirety in facsimile. A single opening of the *Testament* is reproduced in my chapter on the same theme in *William Shakespeare: A Compact Documentary Life* (1977; rev. edn., 1987, 45–54).

# 12

# Legends of Books and Manuscripts; the Malediction

EVEN greater would be the curiosity stirred by the fate of the poet's own books and manuscripts. Copies of Bacon's *Essays*, Florio's *Montaigne*, a prayer book, and other interesting volumes, all bearing Shakespeare's presumed signature, would duly appear; also a map of Cambridge with the inscription, 'W. Shakspeare. The gift of my beloved friend B. Jonsone at Londone, April 17, 1593.' About the manuscripts rumours would circulate. In 1729 Roberts, the strolling player, lamented that '*Two* large *Chests* full of this GREAT MAN's *loose Papers* and *Manuscripts*, in the Hands of an ignorant *Baker* of WARWICK, (who married one of the Descendants from *Shakespear*) were carelessly scatter'd and thrown about, as Garret Lumber and Litter, to the particuler Knowledge of the late *Sir William Bishop*, till they were all consum'd in the generall Fire and Destruction of that Town.'[119] A fire did indeed cause much destruction in Warwick in 1694, but no descendant of Shakespeare lived there. Variations on this legend reached Jordan: he notes an 'improbable tale' of two chests of Shakespeare's manuscripts coming into the hands of Bishop and being destroyed in a Stratford conflagration.[120] This Sir William Bishop (1629–1700) resided in Bridgetown, a hamlet partly within the town; but, as Jordan observed, there was no great fire at Stratford after 1614. Jordan also mentions the tradition that the papers were once in the hands of a baker (now metamorphosed into a baker of Stratford), only to reject it as false: Shakespeare had no descendant in that trade.

Other reports raised or dashed hopes. In 1742 Sir Hugh Clopton told his visitor Macklin of 'an old tradition' that Lady Bernard carried with her from Stratford many papers belonging to her grandfather.[121] A different legend, of later origin, whispered that Shakespeare's 'play-writings' had been destroyed by his elder daughter Susanna. Rumours too sprang up that the manuscripts were removed to Clopton House, and there obliterated by a philistine tenant. After the poet's deification, it was wondered whether Mrs Hall had not bestowed the poet's manuscripts on Queen Henrietta Maria when she sojourned at New Place in 1643. Perhaps, then, they had been destroyed, along with the King's 'most secret letters', by Cromwell's army, or (if they escaped that fate) had been carried abroad by the Queen when she went into exile. A diligent search of the environs of her places of exile would be recommended.

The Shakespeare-Mythos pursues the poet beyond life to the grave. In the seventeenth century, visitors to Holy Trinity Church were struck by the malediction carved on Shakespeare's gravestone, which lies between that of Anne Shakespeare (alongside the north wall where the monument stands) and that of Susanna Hall. Dobyns, Davies, Dowdall, Hall, and Roberts all paused over the epitaph; Rowe quoted it in his Life. The tradition that Shakespeare himself wrote the lines first appears in Dowdall, and is repeated by others. Almost a century later, in 1777, an anonymous visitor to Stratford mentions being told by the guide that the poet composed the verses for himself in order to prevent having his bones thrown into the charnel-house which then stood adjacent to the chancel. To Dowdall we owe the last of our legends, which concerns the aftermath of the curse. In his gossipy letter to his cousin he remarks that the aged clerk, upwards of eighty, who escorted him around the church told him that fear of the curse had frustrated the earnest desire of Shakespeare's wife and daughters to be interred with him.[122]

# 13

## *First Biographical Notices*

'HARD has been the Fate of many a Great Genius,' reflected Dr James Welwood in his preface to Rowe's translation of Lucan's *Pharsalia*, 'that while they have conferr'd Immortality on others, they have wanted themselves some Friend, to Embalm their Names to Posterity.'[123] Shakespeare might have found such a friend in his fellow dramatist Thomas Heywood, who for many years had in hand a vast work containing 'The Lives of all the poets modern and foreign'. We hear about it in 1614, and again in 1635, but it has vanished, and scholars since Malone's day have lamented the loss. In the event, Shakespeare was dead nearly half a century before the first biographical notices began to appear. A civil war had intervened, an age had passed; almost all who had known him were gone. His sister Joan Hart, who might have recalled his early life, outlasted the poet by thirty years. His first daughter lived until 1649, and his second survived into the Restoration. Thomas Combe, to whom Shakespeare left his sword, died in Stratford in 1657, while his elder brother William lived on there until 1667. All these people might have been interviewed, and imparted intimate details about Shakespeare the man; but no one sought them out. Documentary records existed but had passed from view, and in their place had begun to flourish that curious garden of anecdotes, legends, and myths which would eventually put forth such exotic blossoms. Ultimately this heritage would present problems: it would be difficult, sometimes impossible, to determine where tradition leaves off and fancy or deceit begins.

Such perplexities did not, however, disturb the tranquillity of Shakespeare's earliest biographers, who bestirred themselves little over either fact or legend. The more considerable Lives produced in this period concern themselves with piety, power, and crime, as reflected in the careers of divines, statesmen, and gallows-birds; literary biography, like the profession of letters itself, hardly yet existed. For the first sketches of Shakespeare's career we must look to those accommodating compilations in which he figures merely as one of many.

The first published biography (if it can be called that) occupies half a folio page of Fuller's *Worthies*. As we have noticed, it is impressively uninformative, even the date of Shakespeare's death—plainly visible on his monument—escaping the antiquary who sedulously combed the countryside for data. Yet Fuller's vague generalities constituted a primary source for several accounts which followed. William Winstanley, in *The Lives of the Most Famous English Poets* (1687), eked out his four-page memoir with William Basse's elegy (mysteriously ascribed to 'one') and some verses from *Venus and Adonis*, as well as with a catalogue of Shakespeare's forty-eight plays, including such interesting items as *Lorrino* and *Oldraste's Life and Death*; but mainly he contented himself with plagiarizing Fuller—even to the extent of reproducing such non-information as the blank space for the year of Shakespeare's death. Thus does this memorialist, reputedly a barber before he turned biographer, justify the contemptuous observation of Anthony Wood, who remarked that Winstanley abandoned his razor but kept his scissors to clip from his predecessors.

In *An Account of the English Dramatic Poets* (1691), Gerard Langbaine, a much more considerable historian, quotes Fuller on Shakespeare's natural genius which, like Cornish diamonds, did not require the polishing of art. Most of Langbaine's entry for the dramatist, however, is given over to a catalogue of the forty-six plays he attributes to him, with (as is the compiler's method throughout) as much information as he has been able to uncover about their sources. But Langbaine does provide some important biographical facts at the close of his account: he gives a date, 23 April 1616, for the poet's decease; he mentions that Shakespeare's widow and his daughter Susanna who married John Hall lie buried with him in Holy Trinity Church; he describes the monument and quotes the epitaph inscribed upon it, as well as the malediction on the gravestone.

Others too trace their descent from Fuller. Pope Blount's Life in his *Remarks on Poetry* (1694) cites him by way of Langbaine, whom he quotes at length; he also furnishes generous extracts from the commentaries of Dryden and others. Strictly biographical information is restricted to the brief opening paragraph consisting of two sentences which tell us where Shakespeare was born and when he died. Jeremy Collier's paragraph on Shakespeare in *The Great Historical, Geographical, Genealogical and Poetical Dictionary* (1701) is a terse condensation, duly accredited, of Fuller.

To these unsatisfactory beginnings we may add Edward Phillips's equally unsatisfactory Life, a single paragraph in *Theatrum Poetarum* (1675). Phillips commends Stratford for having produced so excellent a poet, lets it be known that Shakespeare was an actor before becoming a playwright, and remarks with wonder on the phenomenon that a writer with no extraordinary learning could please with 'a certain wild and native Elegance', and command 'an unvulgar style'.[124] (All these early memorialists dwell on the limitations of Shakespeare's learning; Jonson's strictures had made their impact.) Meagre as it is, Phillips's sketch benefited Winstanley, who, as was his custom, utilizes him without acknowledgement.

In *The Lives and Characters of the English Dramatic Poets* (1698), Charles Gildon's entry for Shakespeare consists mainly of a catalogue of the works *à la* Langbaine, his only novel contribution being the story of the charnel-house inspiration for the ghost scene in *Hamlet*.

There is more substance to the random jottings set down by Aubrey around August 1681; these (as we have already seen) at least possess life and particularity. Aubrey gathered material from a variety of sources. Newly arrived in Oxford in 1642, he talked there with a Mr Joseph Howe who claimed to know the model for Dogberry. He looked at Dugdale's *Warwick-shire* for the Shakespeare epitaph, and in *Discoveries* for Jonson's outburst about the blotting of lines. He interviewed Sir William Davenant, and also the laureate's clerical brother Robert. Most important, Aubrey sought out and questioned old Beeston in his house at Hoglane in Shoreditch. But it would be a while before Aubrey's notes passed into the mainstream of Shakespearian biography. He left his sketches chaotic and unpublished at his death. Warton first printed extracts from the Shakespeare jottings, but out of the way, as a footnote in his *Life and Literary Remains of Ralph Bathurst* (1761). Malone planned an edition of Aubrey but never followed through, and not until 1898 did Clark reduce the notes to order and bring out the first edition of the *Brief Lives*.

# 14
## *Nicholas Rowe*

IGNORANCE of the basic facts of Shakespeare's life, even among the presumably learned, in the eighteenth century is illustrated by the *Magna Britannica* (1721–31), 'A New Survey of Great Britain'. In the brief section on Stratford, Shakespeare is singled out as the only lay worthy deserving mention. '*William Shakespear* an eminent Poet, but no scholar . . .,' begins the paragraph allotted to him by the antiquarian compiler Thomas Cox, and it ends: 'He dy'd and was buried in this Town in 1564.' Thus Cox, who was no poet, shows himself to have been no great shakes as a scholar either. In truth, he has little excuse, for the first important memoir of Shakespeare had already been published.

His predecessors' largely desultory endeavours put into appropriate relief the accomplishment of Nicholas Rowe, the poet laureate and playwright who attempted a Shakespearian flight in his *Jane Shore* but won greater acclaim for the sentimental indulgences of *The Fair Penitent*. The first critical editor of Shakespeare, in 1709, is also, for all practical purposes, the first biographer. As editor, Rowe has obvious enough deficiencies, the most glaring of which is his choice of the derivative Fourth Folio of 1685 for his copy text.* But he accomplished much. Rowe introduced a number of accepted emendations, modernized spelling and punctuation, completed the Folio division of the plays into acts and scenes, prefixed *dramatis personae* lists to all the plays, and supplied scene locations. In short, twenty-four years after the last Folio, he made Shakespeare accessible to a wide public in a convenient, illustrated octavo edition similar in format to other collections of authors popular at that time. For the first volume he supplied, by way of a preface, 'Some Account of the Life, &c. of Mr. William Shakespear'.

Slight by modern standards—it occupies forty pages of large type—this essay is notably ambitious and substantial by comparison with what had gone before. Fortunately, Rowe shares in the curiosity aroused by circumstances in the lives of men of achievement. 'How fond do we see some People of discovering any little Personal Story of the great Men of Antiquity,' he writes in his opening paragraph,

their Families, the common Accidents of their Lives, and even their Shape, Make and Features have been the Subject of critical Enquiries. How trifling soever this Curiosity

---

* The British Library has recently acquired a trial edition, dated 1708, of a few pages of *The Tempest*, which seem to indicate that Rowe originally intended to follow the text of the Second Folio, of which he owned a copy. This question is now under study by Mr R. J. Roberts of the Library.

may seem to be, it is certainly very Natural; and we are hardly satisfy'd with an Account of any remarkable Person, 'till we have heard him describ'd even to the very Cloaths he wears. As for what relates to Men of Letters, the knowledge of an Author may sometimes conduce to the better understanding his Book: And tho' the Works of Mr. *Shakespear* may seem to many not to want a Comment, yet I fancy some little Account of the Man himself may not be thought improper to go along with them.[125]

Thus does Rowe launch himself on the first attempt at a connected biography of Shakespeare.

In an engaging letter to Rowe, John Dennis described him as 'a Gentleman, who lov'd to lie in Bed all Day for his Ease, and to sit up all Night for his Pleasure'.[126] Yet Rowe laboured diligently on his edition of Shakespeare, and he took evident pains to gather material for his Life. In the 'Account' he mentions his attempts, which met with little success, to inquire into the dramatist's acting career. A press notice, dated 17 March 1708, announced that 'a very neat and correct edition of Mr. William Shakespear's Works . . . is now so near finished as to be published in a month; to which is designed to be prefixed an account of the life and writings of the said author, as far as can be collected'.[127] Any gentlemen having by them materials that might further the project were invited to pass them on, as a favour to the editor, to his publisher, Jacob Tonson. Apparently, information was slow in arriving, for publication of the edition was delayed for over two months. Rowe had to depend mainly on what the great Shakespearian actor Betterton, celebrated for his Hamlet, could find for him. Sometime in 1708 (according to Malone) Betterton, now in his old age, made a pious journey to Stratford 'on purpose to gather up what Remains he could of a Name for which he had so great a Value'.[128] Later it would be questioned whether such an expedition in fact took place, the scepticism being attributed to Betterton's fellow actor Bowman, who married the former's adopted daughter; but Rowe is explicit on this point, and there is no reason to doubt him. He handsomely acknowledges his obligation to Betterton for 'the most considerable part of the Passages' dealing with the life.

At Stratford the actor became the first investigator to delve into the records for information about Shakespeare. He examined the parish registers in Holy Trinity Church, but without an antiquary's skill. Apparently he confused Shakespeare's father with the John Shakespeare who was a shoemaker of Warwick, and he failed to deduce that the daughter Joan christened 15 September 1558 must have died before the baptism of a second Joan in 1569; thus he made William one of ten children. How he arrived at the conclusion that the dramatist had three daughters, of whom Judith (rather than Susanna) was the eldest, is less easily determined.

The misinformation yielded by Betterton's consultation of the records duly found its way into Rowe's pages, and affected larger issues, such as the fancied hardships endured by Shakespeare as a member of a large family suffering financial embarrassment. Errors and misjudgements of other kinds also mar Rowe's 'Account'. As we have already observed, he describes the father (understandably enough) as a wool dealer. Uninformed about *The Rape of Lucrece*, which he does not mention or reprint, Rowe cites *Venus and Adonis* as 'the only Piece of his Poetry which he ever publish'd himself'.[129] Although aware that his footing is precarious, he nevertheless insists on following Dryden and identifying Shakespeare with '*pleasant* Willy . . . *dead of late*' in Spenser's *Tears of the Muses*, published in 1591; it would remain for twentieth-century scholarship to associate Willy, more plausibly, with Richard Wills or Willey, the learned author of *De Re Poetica*, who died around 1579.[130] Rowe accepts, as do all the rest but Aubrey, received opinion regarding the inferiority of Shakespeare's learning: the dramatist understood French (clear from *Henry V*) but had only slight Latin—although the editor expresses puzzlement that an uneducated author should have written a comedy based on the *Menaechmi* of Plautus: it would remain for Johnson to point out that an English version was available to Shakespeare.* On the vital question of the chronology of the works, Rowe confesses himself at sea, but he begs leave to question Dryden's early date for *Pericles*, and notes that sometimes (as with *Henry V* and *Henry VIII*) the limits of composition can be narrowed down on the basis of internal references. So far as the Elizabethan world that gave birth to Shakespeare is concerned, Rowe sees it, from his high vantage point in a more polished and enlightened age, as little removed from barbarism: 'We are to consider him as a Man that liv'd in a State of almost universal License and Ignorance.'[131] The theatre that preceded Shakespeare produced not a single worthy play. Thus Rowe did nothing to initiate the enormous task, essential to Shakespearian biography, of reconstructing the historical, cultural, and theatrical context in which the poet lived and worked.

Elsewhere Rowe repeats the now familiar information about Shakespeare's monument and the inscription on the gravestone, and he quotes in full Jonson's remarks in *Discoveries*. But he also provides more novel facts: for the first time in any source we hear that Shakespeare as a boy attended the free school in Stratford, that while yet very young he married 'the Daughter of one *Hathaway*, said to have been a substantial Yeoman in the Neighbourhood of *Stratford*', and that he spent the last years of his life in prosperous retirement in the town of his origin. We also see Shakespeare as a person, although here Rowe gives no source for his impressions:

Besides the advantages of his Wit, he was in himself a good-natur'd Man, of great sweetness in his Manners, and a most agreeable Companion; so that it is no wonder if with so many good Qualities he made himself acquainted with the best Conversations of those times. . . . His exceeding Candor and good Nature must certainly have inclin'd

---

* In manuscript, however, for Warner's translation did not achieve print until 1595.

all the gentler Part of the World to love him, as the power of his Wit oblig'd the Men of the most delicate Knowledge and polite Learning to admire him.[132]

Is it too fanciful to speculate that perhaps this author, like so many biographers of Shakespeare after him, is gazing into his own mirror and finding there his subject's reflection? For he emphasizes those qualities in the poet for which he was himself cherished. 'He had a natural Sweetness and Affability', William Ayre said of Rowe, 'that it was impossible not to be obliged by something in the Tone of his Voice so soft and winning, that every Body us'd to be sorry when he left off speaking. He was of an open communicative Temper, and discours'd very freely; not lavish of Speech, but exceeding generous: No Man was better belov'd by all Degrees and Stations.'[133] At any rate, Rowe in this passage parts company with Dryden, who had declared, in his *Essay on the Dramatic Poetry of the Last Age*, that no playwright in that time, except Jonson, 'had been conversant in courts'.[134]

Traditions bulk large in Rowe's account, and if he sometimes uses such disclaimers as 'he is said to . . .' or 'if I had not been assur'd . . . I should not have ventur'd to have inserted', ordinarily he makes no attempt to distinguish between legendary anecdotes and ascertainable facts. Thus we are informed in some detail about the poaching at Charlecote, and of the relevant allusion in *The Merry Wives of Windsor* to the Lucy coat of arms. We learn that the Queen ordered Shakespeare to alter the name of Oldcastle to Falstaff, and, more pleasantly, that she commanded her favourite playwright to portray the fat knight in love. (Rowe is the first to mention in print the deer-stealing escapade and the Oldcastle directive.) And we are told too of a different sort of command: how usurious old John Combe invited him to compose the epitaph with which he was afterwards so displeased. Whatever the merits—and they vary—of these traditions, credit is due Rowe for setting them down; otherwise some at least would have perished, as Dr Johnson recognized in his life of Rowe.

Rowe is also the first to integrate biography with a critical evaluation of the poet. Disappointed in his quest for concrete details about the life, he turns to the book. 'This is what I could learn of any Note, either relating to himself or Family,' he sadly concludes: 'The Character of the Man is best seen in his writings.'[135] Behind Rowe the critic stand two formidable spokesmen of the Restoration: Dryden, whose comments on Shakespeare he echoes, elaborates, and sometimes refutes; and Rymer, whose strictures in the *Short View of Tragedy* he tactfully rebuts without appearing to invite controversy. The picture of the dramatist that emerges from Rowe's discussion sums up the late seventeenth-century position, with the supposed faults that disturbed the Restoration minimized. Ignorant of the ancients, living by the mere light of nature, Shakespeare cannot be judged by laws of which he had no knowledge. His mastery did not reside in structure—his fables are deficient—and he stands justly accused of allowing into his tragedies that admixture of comedy that spawned the genre of tragicomedy which was 'the common Mistake of that Age'. But ignorance is also liberation: adherence to the rules would only have inhibited Shakespeare's fire, impetuosity, and admirable extravagance. He excelled in the variety of his characters, which included the incomparable Falstaff and Shylock; in the liveliness of his images (what could surpass Patience on her monument?); and in his capacity to arouse that terror which is the proper passion of tragedy. In a word, as Dryden said of him, Shakespeare possessed genius, 'alone . . . a greater virtue . . . than all other Qualifications put together'.[136] Of the 'Account' as a critical document it is difficult to resist Johnson's considered judgement that it 'cannot be said to discover much profundity or penetration'.[137] Yet it epitomizes well the outlook of an age, and, along with the strictly biographical reporting, was to exert an influence beyond its time. Rowe earned the £36 10s. his publisher paid him for the six volumes.

# 15

## *The Tribe of Nicholas*

TONSON reprinted Rowe's edition in 1714 with the inscription on the Stratford monument (omitted from Rowe's 'Account') reproduced on the verso of the title-page to the first volume. It early received the tribute that hackwork pays to original research: plagiarism. In *The Poetical Register* (1719) Giles Jacob, without acknowledgement, tends to follow Rowe word for word although, by a process inevitable in rehashes, what was tentative or qualified hardens into positive assertion. If Rowe thought it 'highly probable' that Spenser represented Shakespeare as pleasant Willy, 'Mr. SPENSER [according to Jacob] speaks of him in his *Tears of the Muses*, not only with the Praises due to a good Poet, but even lamenting his Absence with the Tenderness of a Friend.'[138] The story of Southampton's unlikely gift of £1,000 to his young protégé assumes the status of fact, with accretions of circumstantial detail: Shakespeare 'had the Honour to meet with many great and uncommon Marks of Favour and Friendship from the Earl of Southampton . . . to whom he Dedicated two Poems, *Venus and Adonis*, and *Tarquin and Lucrece*. For the Former of which Dedications, that Noble Lord gave him a Thousand Pounds, which uncommon Bounty Mr. Shakespear gratefully acknowledg'd in the Dedication to the Latter.'[139] In such a way do traditions flourish and take on the particularities of truth.

The Life became standard for the century. Pope placed Rowe's 'Account' after his own preface to his 1725 edition, where it is followed by first publication of the College of Arms manuscript (transmitted to Pope by John Anstis, Garter King of Arms) of the heraldic grant to John Shakespeare. But Pope did not print Rowe's Life as he found it; instead he altered it in accordance with his own taste and the facts as he understood them. He brought together all the biographical material in a single section, and added a couple of footnotes. At the same time he made excisions. Gone are Rowe's observations on the testimony, of the plays with respect to Shakespeare's knowledge of Latin and French; the reference to Dryden's opinion of the early date of *Pericles*; the statement that *Venus and Adonis* is the only poem published by Shakespeare himself (an error glanced at by Sewell in his preface to volume seven of Pope's edition); the identification of Spenser's 'pleasant Willy' with the dramatist; the strictures on what Rowe regarded as Jonson's reserved, somewhat envious attitude towards Shakespeare; and other passages concerning Rymer and the Davenant-Dryden *Tempest*. Pope's tamperings are thus substantial, and it is curious that subsequent eighteenth-century editors should have reprinted this considerably tidied up Life under the misapprehension that Rowe himself had made the alterations for the 1714 reprint, which in fact provides a text identical with the original. Later comments on Rowe's 'Account' are, then, actually comments on Pope's version; Steevens, for example, in 1785 claims that he has 'printed [the preface] from Mr. Rowe's second edition, in which it had been abridged and altered by himself after its appearance in 1709'.[140]

(Before turning from Pope to subsequent eighteenth-century editors, we should not overlook his most influential contribution to Shakespearian biography; a contribution made by the poet rather than the editor. In *The First Epistle of the Second Book of Horace, Imitated* (1737), Pope writes:

> Shakespear, (whom you and ev'ry Play-house bill
> Style the divine, the matchless, what you will)
> For gain, not glory, wing'd his roving flight,
> And grew Immortal in his own despight.[141]

Here Pope, like so many others, admires Shakespeare in his own image: the translator of the *Iliad* and *Odyssey* undertook those labours for profit as well as fame, and made a fortune with *his* roving flight. A succession of Victorians who cheerfully embraced the profit motive would quote Pope's lines with satisfaction.)

While not reprinting Rowe in his 1733 edition of Shakespeare, Theobald in his own preface made ample (and acknowledged) use of the biographical passages in the 'Account', to which he refers readers for further particulars of the dramatist's life. But Theobald, who knew Rowe's preface in its original form, rejects with Pope the identification of Shakespeare with Spenser's dead

Willy, and he adds references to the licence granted by James to the King's Men in 1603 (first printed in Rymer's *Foedera* in 1727), the history of New Place,* and the tradition—which he distrusts—of the two chests of manuscripts regrettably burned while in the possession of the ignorant baker of Warwick.

With or without acknowledgement, Rowe continued to dominate Shakespearian biography in the mid-century. *The Life of Mr. William Shakespear, Whose Monument Was Lately Erected in Westminster-Abbey, at the Expence of the Publick*, published anonymously in 1743, is no more than a catchpenny attempt to capitalize on popular interest in the Bard by reprinting Pope's Rowe under an altered title. The revised version of Rowe (along with Pope's preface and the grant of arms) was included by Hanmer in his elaborately illustrated but otherwise dismal edition of 1744. Bishop Warburton, introducing his 1747 edition of the *Works*, sneers at Rowe for 'giving us meagre Account of the Author's Life, interlarded with some common-place Scraps from his Writings';[142] but he reprints it with no additions of his own.

Johnson in 1765 found room for all the prefaces of his predecessors, and in his own contribution justifies the retention of Rowe's Life: 'though not written with much elegance or spirit; it relates however what is now to be known, and therefore deserves to pass through all succeeding publications'.[143] The record, then, had not grown very much in half a century; but Johnson follows the 'Account' with, in addition to the grant of arms, two more accretions: the episode of the horse-holding at the playhouse door, first published in 1753, and the dramatist's will, as 'Extracted from the Registry of the Archbishop of *Canterbury*'. The last item is of course of the most signal importance.

It does not appear for the first time in Johnson's edition. In 1747 Joseph Greene had come upon a copy of Shakespeare's will in Stratford; precisely where we do not know—perhaps among the parish records laid away in Holy Trinity, of which he was vicar; perhaps among Borough documents kept in a room attached to the free school, of which he was headmaster. This transcript of the will, now at the Shakespeare Centre in Stratford, belongs to the first half of the seventeenth century and was probably made shortly after the death of Dr Hall in 1635.[144] In any event Greene seems to have been disappointed rather than gratified by his find. On 17 September he wrote to James West, Secretary to the Treasury, and afterwards President of the Royal Society:

The Legacies and Bequests therein, are undoubtedly as he intended; but the manner of introducing them, appears to me so dull and irregular, so absolutely void of y$^e$ least

---

* To Theobald we owe the suggestion that Shakespeare renamed his property (formerly the Great House) after 'having repair'd and modell'd it to his own Mind'; but it is called 'the Newe Place' in a survey of 1590, as J. M. C. Bellew pointed out in *Shakespeare's Home at New Place, Stratford-upon-Avon* (London, 1863), 52.

particle of that Spirit which Animated Our great Poet; that it must lessen his Character as a Writer, to imagine yᵉ least Sentence of it his production

The only Satisfaction I receive in reading it, is to know who were his Relations, and what he left them, which perhaps may just make you also amends for yᵉ trouble of perusing it.[145]

The will, as transcribed from the original reposited in Doctors' Commons, first saw print in the article on Shakespeare, signed 'P' (by Philip Nichols), in Part I of the sixth volume of *Biographia Britannica* in 1763. With interesting consequences, 'P' misread Shakespeare's bequest, 'I give unto my wife my second-best bed with the furniture'; in *Biographia Britannica* it is 'my brown best bed, with the furniture'. Malone scornfully attributed the alteration to Theobald, whose abilities he held in no very high esteem, but this seems rather hard on the scholar who died three years before the discovery of the will. (And, anyway, what justification had Malone for complaint? He fell into the same error himself when he first examined the will at Doctors' Commons. 'In the Register of his Will which is in the office . . .', Malone complacently noted in his copy of Langbaine now in the Bodleian Library, 'they have made a mistake, and call it "my *second* best bed".')*

That Johnson had not gone to the Archbishop of Canterbury's registry but had taken the easier course of following 'P' is suggested not only by his retention of the brown bed but also of other misreadings, including 'boxes' for 'bowl' and 'Bushaxton' for 'Bushopton', a hamlet north-west of Stratford. Subsequent editors cribbed Johnson and reproduced the item about the 'brown best bed'. Thus the question of the implications of the bequest as regards the poet's relations with his wife—a question which would afterwards engender impassioned debate—was for a time happily averted.

The year that saw the appearance of Johnson's *Shakespeare* also witnessed the publication of an editorial landmark of a very different order: Bishop Percy's *Reliques of Ancient English Poetry*, which supplied the hauntingly evocative old balladry—the strains of *Chevy Chase* and *Sir Patrick Spens* and the rest—to an audience wearying of the neo-classic mode. In his enlarged preface 'On the Origin of the English Stage' to Book II of the fourth edition, in 1794, Percy is apparently the first to suggest that Shakespeare as a boy of eleven may have attended the magnificent entertainments for the Queen at Kenilworth, a few miles distant from Stratford. Perhaps the Warwickshire lad laughed with Elizabeth at the Hock Tuesday interlude performed by players from Coventry, and '. . . if our young bard afterwards gained admittance into the castle to see a Play, which the same evening, after supper, was there "presented of a very good theme, but so set-forth by the actors' well-handling, that pleasure and mirth made it seem very short, though it lasted two good hours and more," we may imagine what an impression was made

---

* Shakespeare's will is now in the Public Record Office. The buildings of Doctors' Commons, which stood in Knightrider Street near St Paul's Cathedral, were pulled down in 1861.

on his infant mind'.[146] Later biographers would eagerly follow up on the Bishop of Dromore's intriguing speculation.

Shakespeare's peregrinations also interested William Guthrie, who in his *General History of Scotland* (1768) informs his readers that King James in 1599 asked Elizabeth to send him a company of English comedians. 'I have great reason to think', Guthrie declares, 'that the immortal Shakespear was of the number.'[147] Whatever his reasons, he does not share them with his readers, and his assertion that the King of Scotland requested players of the English Queen is an unwarranted elaboration of a passage in Spottiswood's *History of the Church of Scotland*. But others who followed Guthrie would ponder whether the author of *Macbeth* travelled into those distant parts north of Perth, where Birnam Hill rose and where stood Glamis Castle. Not until the next century, however, would it be proposed that Shakespeare visited Italy.

Other bits and pieces of information reached the public as edition succeeded edition. George Steevens, who in 1773 reprinted Rowe and all the other prefaces and ancillary materials, added an annotated transcription, made by James West, of extracts of baptism, marriage, and burial entries from the Stratford parish registers. Transcript and commentary leave something to be desired: Hamnet Shakespeare becomes Samuel in the records of both baptism and burial, the shoemaker of Warwick continues to be confused with the poet's father, and the date of Shakespeare's birth is categorically stated as 23 April 1564, although this is not a fact but merely a plausible inference from the christening entry and the wording of the monumental inscription, '*obiit anno . . . aetatis 53*'. In his 1778 edition Steevens furnished the complete text of the patent granted by James to the King's Men in 1603, extracted from Rymer's *Foedera*. With the will he gave facsimiles of the three Shakespeare signatures, but apparently did not make a fresh transcription: Bushaxton and the brown best bed remain. Steevens added, too, the Stationers' Register entries of the plays, and, without notable conviction, *Additional Anecdotes* he had found in Oldys's manuscript remains; these are, he says, 'for aught we know to the contrary, as well authenticated as any of the anecdotes delivered down to us by Rowe'.[148] Among them are the old townsman's witticism about Davenant's godfather, the recollection (purportedly by the dramatist's younger brother) of Shakespeare in the part of Adam, and the suggestion of the Stratford original for Falstaff. All this material—including Rowe's 'Account'—was reprinted in the 1785 edition prepared by Isaac Reed.

About Oldys some further word is appropriate. The illegitimate son of Dr William Oldys, the Chancellor of Lincoln and Advocate of the Admiralty, he early developed a taste for antiquarian studies. In these pursuits he was perhaps abetted rather than hindered by a compulsive temperament: 'He was so particular in his habits', according to John Taylor, whose father befriended Oldys in his later years, 'that he could not smoke his pipe with ease till his chair was fixed close to a particular crack in the floor.'[149] In 1736 he made his reputation with an ambitious *Life* of Ralegh, 'newly compil'd, from Materials more ample and authentick than have yet been publish'd'. His *British Librarian*, published anonymously in 1737, helped to establish interest in rare old books. The Earl of Oxford the next year appointed Oldys his literary secretary, and he thus came to know the glittering circle that included Pope. But Oxford died in 1741, and for fourteen years thereafter Oldys was reduced to servile employment as a bookseller's drudge, turning out to order pamphlets, prefaces, editions, essays, and biographical notices. Osborne engaged him, with Johnson, to compile the *Harleian Miscellany*. He made his voluminous contribution—twenty-two articles in all, including a Life of the actor Alleyn—to the *Biographia Britannica*. Meanwhile he continued to annotate his copy of Langbaine, acquired in 1727, which would benefit future students so greatly. Appointed Norroy King of Arms in the College of Arms in 1755, and thus delivered from poverty, he surrounded himself with books in his large room in the west wing of the college.

There he laboured on his most daunting task: a Life of Shakespeare undertaken for Walker, a bookseller in the Strand; a biography which was to have (according to Taylor) 'ten years of the life of Shakspeare unknown to the biographers and commentators'.[150] Oldys's work habits—if one can trust his memorialist Yeowell—were curious: 'His notes were written on slips of paper, which he afterwards classified and reposited in small bags suspended about his room.'[151] These methods did not ensure efficiency. Perhaps drink was a factor: his acquaintance Captain Grose says that he was almost continuously intoxicated during his latter years, and that at Princess Caroline's funeral 'he was in such a situation as to be scarcely able to walk, and actually reeled about with a crown on a cushion, to the great scandal of his brethren'.[152] In any event, the great Shakespeare project came to nothing. Chalmers reports, on the authority of Dr Ducarel, 'who knew him well', that Oldys 'had by him, at the time of his death, some collections towards a "Life of Shakespeare," but not digested into any order, as he told the doctor a few days before he died'.[153] Taylor, his executor, paid back the bookseller his advance of twenty guineas. Posterity would have to be satisfied with the notes rescued by Steevens.

Thus, as the century wore to a close, Rowe's 'Account' in the Pope version, insufficient and outdated, remained the standard authority. Yet Edward Capell did not reprint it in his great edition of 1767, and at the end of his introduction he entered a plea for a different kind of biography to preface such an enterprise. 'How much is it to be wish'd', he expostulates, 'that something . . . worthy to be intitl'd—a Life of SHAKESPEARE, could accompany this relation, and compleat the tale of those pieces which the publick is apt to expect before new editions?'[154] Instead readers were offered only 'an imperfect and loose account of his father, and family; his own marriage, and the issue of it; some traditional stories,—many of them triffling in themselves, supported by small authority, and seemingly ill-grounded; together with his life's final period as gather'd from his monument'; such is 'the full and whole amount of historical matter that is in any of these writings; in which the critick and essayist swallow up the biographer. . . .'[155] The circumstances of the poet's private life would always defeat curiosity, but for the student of literature the public self is more important, and moreover promises rewards for the investigator. Great questions remained to be answered: When did he begin writing for the stage, and with which play? What is the chronological sequence of his works? For which playhouses were they composed? Clues to the answers reside in the plays themselves.

'A Life thus constructed', the editor writes in his final paragraph,

interspers'd with such anecdotes of common notoriety as the writer's judgment shall tell him—are worth regard; together with some memorials of this Poet that are happily come down to us . . . such a Life would rise quickly into a volume; especially, with the addition of one proper and even necessary episode—a brief history of our Drama, from it's origin down to the Poet's death: even the stage he appear'd upon, it's [*sic*] form, dressings, actors should be enquir'd into, as every one of those circumstances had some considerable effect upon what he compos'd for it: The subject is certainly a good one, and will fall (we hope) ere it be long into the hands of some good writer; by whose abilities this great want may at length be made up to us, and the world of letters enrich'd by the happy acquisition of a masterly '*Life of* SHAKESPEARE'.[156]

The passage is prophetic of the path investigation would follow. Capell's challenge would be taken up by the greatest of eighteenth-century Shakespearians, Edmond Malone, who would furnish the monument for which his eminent predecessor called.

# PART III

*Edmond Malone*

# 1
## *The Rise of Bardolatry*

By Malone's time Shakespeare was enthroned as the undisputed monarch of English letters. It had not always been that way. Although he had never wanted for devotees—Dryden is the prime early example—the Restoration and earlier eighteenth century felt reservations, most powerfully expressed in Rymer's respected and influential *Short View of Tragedy*. Shakespeare, the neoclassicists objected, ignored the Unities, and he violated decorum by his excursions into grossness, his introduction of low characters into the stately scene of tragedy, and his mingling, in the same work, of comic buffoonery with high seriousness (the grave-diggers' scene in *Hamlet* offended on all three counts, and was appropriately castigated). The dramatist, furthermore, made do without the Chorus, which was indispensable, as the example of the Ancients plainly showed; he committed outrageous puns; he wrote in blank verse rather than rhymed. 'It seems no wonder', John Upton could write as late as 1746, 'that the masculine and nervous Shakespeare . . . should so little please our effeminate tast.'[1]

But a transformation was in progress. The labours of a great succession of editors ensured a vast popular audience for the poet. 'Were Shakespeare to revisit this globe', a writer in *The Universal Magazine* could declare near the end of the century, 'the first thing that would surprise him would be, to learn that above one hundred and fifty thousand pounds have lately been devoted toward splendid editions of his works.'[2] The Shakespeare enthusiasts (of whom Upton himself is not the least eloquent) defended him against the dogmatizing critics. Rowe's point—that the poet must not be condemned for violating injunctions of which he knew nothing—was taken up by the greatest poet of the age. 'To judge . . . of *Shakespear* by *Aristotle*'s rules', Pope wrote, 'is like trying a man by the Laws of one Country, who acted under those of another.'[3]

To this endorsement by the artist was added, forty years later, the approbation of the most esteemed critic the century produced. Johnson devoted much of his preface to answering the strictures of the neoclassic denigrators, dwelling especially on the great issues of decorum and the Unities. The complaints against Shakespeare are formed on narrow principles, 'the petty cavils of petty minds'. Above all writers he is the poet of nature; his drama, which mirrors life, transcends the rules of criticism. 'The work of a correct and regular writer', he sums up in a famous passage, is a garden accurately formed and diligently planted, varied with shades, and scented with flowers; the composition of *Shakespeare* is a forest, in which oaks extend their branches, and pines tower in the air, interspersed sometimes with weeds and brambles, and sometimes giving shelter to myrtles and to roses; filling the eye with awful pomp, and gratifying the mind with endless diversity. Other poets display cabinets; of precious rarities, minutely finished, wrought into shape, and polished unto brightness. *Shakespeare* opens a mine which contains gold and diamonds in unexhaustible plenty, though clouded by incrustations, debased by impurities, and mingled with a mass of meaner minerals.[4]

The note of reservation, while muted, still persists in the final clause.

It would be forever stilled in the next generation. The whole edifice of neo-classical authority was shaken by the translators of the *Poetics*, who pointed out that Aristotle gives no support to the unity of time, that he does not mention place, and that in any event he is not prescribing but referring to the customary practice of the playwrights of his day. An historical yardstick was applied to Shakespeare's puns, which were now seen as reflecting the curious taste of his time. Moreover (some would own) they are good puns: 'I remember but very few', Morgann remarks, 'which are undoubtedly his, that may not be justifyed; and if *so*, a greater instance cannot be given of the art which he so peculiarly possessed of converting base things into excellence.'[5] The dramatist's blank verse, it was now claimed, has all the variety and modulation of musical expression; rhyme, on the other hand, destroys the illusion of reality.

By the end of the century the great leap has been made to the other side of idolatry: the poet is celebrated with the full orchestration of panegyric. 'Shakespeare', declares the formidable Ritson, 'is *the God of the writers idolatry*',[6] and the Revd Martin Sherlock thus begins his *Fragment on Shakespeare*:

It is she [Nature] who was thy book, O Shakspeare. . . . Thou wert the eldest son, the darling child, of Nature; and, like thy mother, enchanting, astonishing, sublime, graceful, thy variety is inexhaustible. Always original, always new, thou art the only prodigy which Nature has produced. Homer was the first of men, but thou art more than man. The reader who thinks this elogium extravagant is a stranger to my subject. . . .

The merits of this poet are so extraordinary, that the man, who should speak of them with the most rigid truth, would seem to the highest degree extravagant. . . . I will therefore say, because a more certain truth was never said; *Shakspeare possessed, in the highest degree of perfection, all the most excellent talents of all the writers that I have ever known.*[7]

On such hyperbole the next century could scarcely improve, although it would try.

# Farmer and Shakespeare's Learning

THE question of how to interpret Shakespeare's disregard for the Unities and decorum is bound up with the debate over the extent of his learning. Was he in fact ignorant of the laws he violated? Some, as we have seen, followed the traditional estimate of the poet's small Latin and less Greek; to this number we may add Pope, who, despite evasiveness elsewhere, at one point firmly declares that Shakespeare worked 'without assistance or advice from the Learned, as without the advantage of education or acquaintance among them: without that knowledge of the best models, the Ancients, to inspire him with an emulation of them'.[8] But others defended his learning in the ancient—and modern—languages. Among these are Gildon, Sewell, Theobald, Warburton, and Upton. In the first formal treatise on the subject, Peter Whalley argued that 'Shakespeare was more indebted to the Ancients than is commonly imagined'; he not only knew Greek and Latin, but also 'arrived to a Taste and Elegance of Judgment, particularly in the Latter'.[9] Zachary Grey, in his Critical, Historical, and Explanatory Notes (1754), took a less moderate stand: 'As to his ignorance in the Greek and Latin tongues, though that point has been more than once discussed, and much said on both sides of the question; I cannot but think from his exact imitation of many of the antient poets and historians, (of which there were no tolerable translations in his time,) that his knowledge in that respect cannot reasonably be call'd in question.'[10] The show-piece of evidence is Hamlet, 'in many places . . . an exact translation of Saxo Grammaticus', but Grey bolsters his case with numerous parallels, some of them curious: the First Murderer's fairly nondescript line in Macbeth, 'We are men, my liege,' is traced to Terence's Heautontimorumenos ('Homo sum; humani nihil a me alienum puto').[11] The issue—the education of the poet—is as vital to biography as to criticism, and in Richard Farmer's casually brilliant Essay on the Learning of Shakespeare (1767) it produced one of the age's most remarkable monographs.

The Essay was the only work that Farmer ever published. A bachelor, his most impassioned love affair was with the Cambridge college, Emmanuel, of which he eventually became Master, and which returned his devotion. There the great Johnson spent a 'joyous evening' with him in February 1765, and they remained friends. Farmer was also on cordial terms with Steevens and Malone. Years later Henry Gunning of Christ's College recalled how Farmer genially applauded performances of the Norwich players at the local theatre in the company of the great Shakespearians of the age—George Steevens,

Isaac Reed, Edmond Malone—and afterwards adjourned to Emmanuel Parlour; the Shakespeare Gang, they were called. Of Farmer it was said that he enjoyed three things above all else—old port, old clothes, and old books—and that there were three things he could not be persuaded to do: rise in the morning, go to bed at night, and settle an account. To such externals as his personal appearance he was happily indifferent; the author of a sketch of Farmer in the *Encyclopaedia Britannica* was distressed to meet him in Canterbury, whither he had gone for an appointment with the Archbishop, 'dressed in stockings of unbleached thread, brown breeches, and a wig not worth a shilling'.[12] Twice he declined a bishopric offered him by Pitt in reward for his Tory steadfastness. His most delighted hours were spent over good wine and amiable conversation in the common room of Emmanuel, and it is a wonder, given his indolent nature, that he could tear himself away long enough to compose, at the age of thirty-one, the study on which his reputation rests.

In his *Essay* Farmer sets out to demonstrate Shakespeare's lack of learning in classical and modern languages. For this task he has the incalculable advantage over his predecessors of a wide and intimate familiarity with the literature of Shakespeare's day; he had not assembled his famous library in vain. Parallel after parallel between the dramatist and the classics he traces to a known English source or to a work already translated, and which therefore need not have been consulted in the original. (So committed did Farmer become to this mode of argument that he later fell back on it even when he could not put his finger on a precise source; thus when Steevens, in correspondence, pointed out to him a fragment of Cicero used by Shakespeare, Farmer blandly replied, 'I do not believe, it will make any Converts to the *Learning of our Poet*'; adding vaguely: 'I am pretty sure, I remember it in one of the old Plays, or *Chronicles*, which he follow'd on this Subject.'[13]) With ease Farmer demonstrates that Shakespeare read Plutarch not in the Greek but in North's version, which (for example) the dramatist versifies for Volumnia's great speech in Act V of *Coriolanus*. So too Farmer is able to show Shakespeare's reliance on other English sources: Holinshed's *Chronicles*, Sidney's *Arcadia*, Painter's *Palace of Pleasure*, Golding's *Ovid*. He refutes Shakespeare's claim to the epistles of Paris and Helen, translated from Ovid, which Gildon printed in the supplementary volume to Rowe, by citing the crucial passage in the *Apology for Actors* that reveals Heywood's authorship. Individual expressions credited to the poet belong to the common language of his time. In such a fashion did Shakespeare acquire the Greek, Latin, and Italian to which his writings bear witness. Farmer even goes so far as to suggest that Shakespeare had little French; in *Henry V* it is 'extremely probable' that the 'French ribaldry' was inserted by another hand. The conclusion is firmly stated: 'He remembered perhaps enough of his *school-boy* learning to put the *Hig, hag, hog*, into the mouth of Sir *Hugh Evans*; and might pick up in the Writers of the time, or the course of his conversation a familiar phrase or two of *French* or *Italian*: but his *Studies* were most demonstratively confined to *Nature* and *his own Language*.'[14]

Maybe Farmer's arguments did not crush all opposition, but without excessive modesty he was inclined to regard his own performance as 'an Answer to every thing that shall hereafter by written on the subject'.[15] He managed to have the last word. A spiteful notice appeared in the *Critical Review* for January 1767, but the evidence there presented to refute the *Essay* merely provided its author with new ammunition which he deployed in the enlarged second edition published later the same year. There, in his preface, he remarks with gentle irony that 'The very FEW, who have been pleased to controvert any part of his Doctrine, have favoured him with better manners, than arguments.'[16] When Colman (one of those whom Farmer had taken to task) ventured to express scepticism in the appendix to the second edition of his *Translation of Terence*, he was similarly demolished—or so it seemed—this time in a letter to Steevens which the latter printed, along with the *Essay*, in his edition of 1773. Farmer's performance was certainly dazzling, and among those who recognized its power was the author's greatest contemporary. '"Dr. Farmer," said Johnson [to him], "you have done that which never was done before; that is, you have completely finished a controversy beyond all further doubt." "I thank you," answered Dr. Farmer, "for your flattering opinion of my work, but still think there are some critics who will adhere to their old opinions; certain persons that I could name." "Ah!" said Johnson, "that may be true: for the limbs will quiver and move when the soul is gone."'[17]

The limbs did quiver. Colman in his appendix had asked, what does Johnson say about an apparent borrowing from Ovid in *The Taming of the Shrew*? Years later, in 1780, Boswell put the question to Johnson, who replied irritably, 'Sir, let Farmer answer for himself: I never engaged in this controversy. I always said, Shakespeare had Latin enough to grammaticise his English.'[18] The great debate was in fact far from settled, although it would no longer be claimed that Shakespeare was a learned man in the sense of being accomplished in Latin and Greek literature. Farmer did not entirely avoid the pitfalls into which his opposition had stumbled, and which he so relished discovering. Like them he adopted an extremist position; like them he laid down an hypothesis and afterwards sought arguments in its support. The availability of translations does not preclude consultation of the originals. Moreover, for several French and Italian sources—Belleforest's *Histoires tragiques*, Ser Giovanni's *Il Pecorone*, Cinthio's *Epitia* and *Hecatommithi*—Shakespeare did not have access to translations. Farmer's theory, of interpolation, resorted to for *Henry V*, was popular in his day but would ultimately be discredited. His most serious limitation, however, is his failure to enquire into the amount of Latin Shakespeare would have acquired as part of his grammar-school training. Farmer had not denied him *any* Latin; the essential question concerns degree. Such an investigation of the social context would not be undertaken until late in the nineteenth century (although Malone would make a beginning) and not accomplished with the full panoply of scholarship until the twentieth, when Farmer's ninety small pages would be answered by T. W. Baldwin's fifteen hundred large ones.

Farmer's aim, sometimes misunderstood, was not to denigrate but to honour his subject. '*Shakespeare*', he declared, 'wanted not the Stilts of Languages to raise him above all other men.'[19] And of his own performance he concluded, 'Upon the whole, I may consider myself as the *Pioneer* of the *Commentators*: I have removed a deal of *learned Rubbish*, and pointed out to them *Shakespeare's* track in the ever-pleasing *Paths of Nature*.'[20] He is beholden neither to Homer nor to Virgil nor to any other great antecedent. Thus even the disparagement of Shakespeare's learning constitutes an act of homage to the poet who now held unchallenged sway.

# 3
# *The Stratford Jubilee*

KINGS must be crowned, and it is fitting that Shakespeare's belated coronation should have been held in the town of his birth. The Stratford Jubilee (as its sponsors called it) took place, oddly, not on the bicentenary of the poet's birth in 1764 but five years later, and not, as one might expect, in April when he was born but in irrelevant September. Uncomprehending townsmen confused the forthcoming Jubilee with the Jew bill that had aroused political passions fifteen years earlier. For Londoners unable to attend, the actor Samuel Foote, darling of the town for his satirical topicalities, furnished 'The Devil's Definition' after the fact:

A Jubilee, as it has lately appeared, is a public invitation, urged by puffing, to go post, without horses, to an obscure borough, without Representatives, governed by a Mayor and Aldermen who are no Magistrates, to celebrate a great poet (whose own works have made him immortal,) by an ode without poetry, music without melody, dinners without victuals, and lodgings without beds; a masquerade, where half the people appeared bare-faced; a horserace, up to the knees in water, fireworks extinguished as soon as they were lighted, and a gingerbread amphitheatre, which, like a house of cards, tumbled to pieces as soon as it was finished.[21]

The post of Steward of the Jubilee went to David Garrick, the foremost promoter and interpreter of Shakespeare in the theatre of his day. The English

Roscius held the Bard in mysterious veneration. On his estate in Hampton he erected a temple to Shakespeare, and in it he placed the Roubiliac statue of the poet, plump with inspiration, for which Garrick himself had sat. In later years, at Abington Abbey in Northamptonshire, hallowed by association with the dramatist's last lineal descendant, Lady Bernard, he would plant a mulberry tree, just as his immortal predecessor had done (according to tradition) in the garden of New Place. 'The action', gravely reflects the author of the *Histrionic Topography*, 'might appear ostentatious, if intended for public discussion.'

Despite Garrick's special intimacy with the National Poet, Stratford's choice did not please all. 'If this Jubilee is meant to be a serious Meeting in Honour of the greatest Poet ever born in any Nation, or in any Age,' querulously asked Steevens, under the pseudonym of Zingis, in the *Public Advertiser* (23 August 1769), 'why were not literary Men placed at the Head of it?. . . . Are Men of Learning the most insufficient Preservers of the Reputation of a Poet? Shakespeare, 'tis true, wrote chiefly for the Stage, but does it follow from thence that he is entitled only to histrionic Honours?'* To Garrick's credit, however, it must be conceded that Shakespeare received almost as much attention as himself at the Jubilee.

For the occasion there rose on the banks of the Avon an octagonal wooden amphitheatre on the classical model but judiciously modified, along the lines of the Rotunda at Ranelagh, to please contemporary taste. Seventy feet in diameter, with a stage spacious enough for a hundred performers and a dance floor that could hold a thousand, the Rotunda was lined with crimson velvet draperies and illuminated by splendid chandeliers two months in the moulding by 'Gibby' Johnson, the Boxkeeper at Drury Lane. Here was consumed the 327-pound turtle cooked by Gill of Bath; here took place the performances and the costumed ball.

The most celebrated Shakespearian authority of the age did not attend; Johnson's absence was noted and regretted. On the mend after illness, he preferred to stay comfortably ensconced with the Thrales at Brighton while the Jubilee was in progress. Nor did the other intellectuals make the journey. But Boswell, delighted to defer treatment for venereal infection, was there, and as he set foot in that drowsy borough of 2,287 souls he experienced (he tells us) such emotions as stirred Cicero in Athens. On the first morning, Wednesday, 5 September, the thirty cannon roared, bells rang through all Stratford,† and the serenaders—fantastically garbed actors—sang, to the accompaniment of clarinets, flutes, hautboys, and guitars: 'Let beauty with

---

* The scholars' choice for Steward was Johnson. Martha Winburn England discusses this minor tempest in *Garrick's Jubilee* (Columbus, Ohio, 1964), 20–4, a recent and entertaining account of the Jubilee. In a prodigal display of the fortuitous duplication of scholarly effort, the quatercentenary produced two other able books on the first Shakespeare celebration: Johanne M. Stochholm's *Garrick's Folly* and Christian Deelman's *The Great Shakespeare Jubilee*.

† The din was made by a battalion of ringers hired for three guineas.

the sun arise, | To Shakespear tribute pay.' During the public breakfast in the Town Hall the county militia played Dibdin's 'Warwickshire', and Boswell joined in the chorus, 'The Will of all Wills was a Warwickshire Will.' Even the deer-poaching escapade made for local glory: 'The thief of all thieves was a Warwickshire thief.' At Holy Trinity Church the bust was hung with garlands, and, as a religious observance, Dr Thomas Arne conducted the full Drury Lane orchestra, with soloists and chorus, in his oratorio *Judith*. Then, bedecked with the Jubilee Ribbon in all the colours of the rainbow (symbolizing the poet's universal genius) and the Jubilee Medal bearing a portrait of Shakespeare and the motto 'We shall not look upon his like again', the company, led by musicians, proceeded to the Birthplace on Henley Street. There they sang the lyric Garrick had composed for the occasion:

> Here Nature nurs'd her darling boy,
> From whom all care and sorrow fly,
>   Whose harp the Muses strung:
> From heart to heart, let joy rebound,
> Now, now, we tread enchanted ground,
>   Here Shakespeare walk'd and sung!

That night in the Rotunda there was more singing, and a ball, after which Dominico Angelo, special-effects wizard and pyrotechnist extraordinary, lit up the night sky with symbolic rainbows of fireworks. So ended the first day, bright with promise of greater glories to follow.

But the next morning the rains came, first in a drizzle, then a torrent. The procession of Shakespearian characters, with a satyr-drawn triumphal chariot containing Melpomene, Thalia, and the Graces, was called off. Cancelled too was the crowning of the Bard. There were murmurs of complaint that the managers of the Jubilee had not provided any awning or covering of some sort in anticipation of such an accident. Now two thousand revellers crowded into the Rotunda built to accommodate half that many. Dr Arne led the musicians in Garrick's *Ode to Shakespear*. During the encore of Mrs Baddeley's solo, 'Thou soft-flowing Avon . . .', Garrick in a sudden gesture flung open the doors to reveal the swollen waters surging against the flimsy structure. The effect (according to one observer) was 'irresistible, electrical'. The crowd laughed, but laughter turned to tears as Mrs Baddeley, that beautiful insinuating creature, went on to sing in her haunting soprano:

> Thou soft-flowing *Avon*, by thy silver stream,
> Of things more than mortal, sweet *Shakespear* would dream,
> The fairies by moonlight dance round his green bed,
> For hallow'd the turf is which pillow'd his head.
>
>   .   .   .   .   .   .   .   .   .
>
> Flow on, silver *Avon*, in song ever flow,
> Be the swans on thy bosom still whiter than snow,

Ever full be thy stream, like his fame may it spread,
And the turf ever hallo'd which pillow'd his head.

Afterwards the throng was moved again, more tumultuously, by Garrick's Jubilee Oration, concluding with the line, 'We shall not look upon his like again,' uttered (reports the hostile witness Dibdin) with a 'manner and effect . . . which beggared all description.'[22] Then the actor drew on to his hands the gloves which, according to tradition, Shakespeare himself had worn on the stage. So violent was the applause that several benches collapsed, and Lord Carlisle providentially missed death from a falling door.

That night there was a masquerade, at which Corsican Boswell, resplendent in his scarlet breeches and grenadier cap embroidered with *Viva la Libertà* in gold letters, danced the minuet with water coming over his shoetops. Some departing merry-makers regrettably fell into flooded ditches; the more prudent stayed on until daybreak, when they retreated over planks stretching from the entrance to their waiting carriages. Several avowed that the deluge was the judgement of God on the idolatry of the Jubilites.

The next morning was anticlimactic. Many left early. The chief event was the Jubilee Sweepstakes run by five colts on the deluged Shottery Meadows course. John Pratt, a groom, won the cup, valued at fifty guineas, with Shakespeare's arms. He owned that he knew very little about Master Shakespeare's plays but would none the less treasure the memento. That night the last ball took place: Mrs Garrick gave delight with her graceful minuet; at 4.00 a.m. the Master of Ceremonies yielded up the insignia of the Steward's office. Thus ended the Stratford Jubilee: absurd, commercialized, and also somehow touching. Reservations about Shakespeare reflecting the aesthetic principles of the preceding age had yielded to the raptures of enthusiasm; the monarch celebrated at Stratford clearly ruled by divine right. The Jubilee festivities quickly had a predictable fall out in other venues. In the pleasure gardens of Marylebone, where in an early nineteenth-century church Robert Browning was to marry Elizabeth Barrett of nearby Wimpole Street, Samuel Arnold had before September produced an 'Ode in Honour of Shakespeare', with words by Francis Gentleman—a Dublin-born actor—followed by Signor Bigari's 'very elegant' transparent Temple of Apollo, and fireworks and illuminations: fancy dresses, but no masks allowed;[23] and that October the pageant designed for Stratford was mounted by Garrick at Drury Lane, where it enthralled throngs for ninety performances.

The Jubilee greatly stimulated the relics industry. We have noted the Bardic gloves dramatically donned by Garrick. In the Festival year his brother George purchased in neighbouring Shottery, at the cottage claimed for Anne Hathaway, another pair of Shakespeare's gloves, and also one of his inkstands. As though in answer to a tourist's prayer, there turned up the poet's chair, his shoehorn, his seal ring, and a dog 'spoted like a leper', which was descended from Shakespeare's coach dog. The distinguished, if not always accurate, antiquary James West, who had a house three miles from Stratford, prided himself on owning the wooden bench on which Shakespeare most enjoyed being seated, and the earthen half-pint mug from which the poet took draughts of ale each Saturday afternoon at his local.[24] 'Lately I passed through Stratford-on-Avon in Warwickshire, Shakespeare's birthplace,' wrote Georg Lichtenberg on 18 October 1775: 'I saw his house and sat on his chair, from which people are beginning to cut away pieces. I made them cut me out some of it for one shilling.'[25] The Hon. John Byng in 1785 bought a slice of the oak chair 'the size of a tobacco-stopper', and also the lower crossbar. Yet in 1790 this article of furniture was, remarkably, still able to support the Princess Czartoryska, who came from Poland to the Birthplace in the same spirit of adoration that carried others to the shrine of the Lady of Loretto. So enamoured did she become of the chair on which the Matchless Bard had pressed his sacred posterior that four months later she dispatched her secretary to obtain it at any price; he paid twenty guineas, and received with the treasure a certificate of authenticity on stamped paper. This romantic episode moved at least one son of Stratford to premeditated song. In a manuscript panegyric now in the Shakespeare Birthplace Record Office, Poet Jordan wrote:

> Yes! Princess *Czartoryska* has been here
> And after sent Orloski for the chair
> Where 'tis reported once, the poet sate
> Ere he had dream't of his superior state.

In the fullness of time there would be added to the relics exhibited in Stratford a Spanish card and dice box presented to Shakespeare by the Prince of Castile, remains of the matchlock (also Spanish) with which the poet shot the deer in Charlecote Park, and 'A gold Tissue Toilet or Table Cover' presented to him by 'his friend and admirer, Queen Elizabeth'.

At least one potentially lucrative attraction went unexploited. New Place, an obvious shrine for cultists, had been bought in 1753 by the Revd Francis Gastrell, retired vicar of Frodsham. His manuscript diary of a journey through Scotland, made some years later, reveals Gastrell as an unamiable man—crusty and excessively squeamish about dirt—but also pious, responsive to fine monuments, ruins, and prospects, and sufficiently humane to be shocked by the cruelty of the Scottish clergy with respect to the baptizing of bastard children. For his treatment of his Shakespearian properties, however, his name would live in infamy. Objecting to the payment of full taxes for a dwelling in which he lived only part of the time, he expressed his disapproval rather forcibly by having the structure pulled down, after which, amid curses and execrations, he departed Stratford never to return. A half-century earlier the house had been rebuilt on neoclassic lines by the previous owner, so the loss to Shakespeare enthusiasts was less than the benevolent divine might have wished, or than for a while was thought.* While at New Place, Gastrell also hired a carpenter to cut down and chop into firewood the celebrated mulberry tree, ostensibly because of the dampness it caused in his house, but actually because it attracted the devout. Thomas Sharpe, a watchmaker who knew the value of such sacred objects, bought the wood, and over a period of forty years, carved from it more curios and useful articles than one would expect a single mulberry capable of yielding. In such circumstances, however, a miracle is perhaps not surprising.†

The Jubilee also inspired a biography, short in length and equally short on facts, of 'the great poet of nature, and the glory of the British nation', published in the August number of the *London Magazine*, which was hawked in Stratford as a souvenir. To a considerable extent cribbed from the article on Shakespeare in the *Biographia Britannica*, it would in turn become a source for future memoirs. The Festival prompted another periodical publication of greater significance: in July 1769 the *Gentleman's Magazine* published the first picture of the Birthplace, a handsomely detailed engraving from a drawing supplied by Richard Greene (1716–93), a surgeon and apothecary of Lichfield, and the brother of Joseph Greene, the master of Stratford grammar school; all in the family. That Shakespeare drew his first breath in the dwelling on Henley Street must for long have been traditional knowledge, for the 'house where Shakespeare was born' is casually identified in the map of Stratford made around 1759. But the article in the *Gentleman's Magazine* first brought the information before the general reading public. 'I do not know', the magazine's correspondent allowed, 'whether the apartment where the incomparable Shakespeare first drew his breath, can, at this day, be ascertained, or not.'[26] Garrick, however, with splendid intuition was able to identify the room, and during the celebration a Thomas Becket set up shop there for the sale of Jubilee books.

The Stratford celebrations contributed to the emergence of Shakespeare as full-fledged culture hero. He received the ultimate popular accolade of becoming the protagonist of a jestbook, *Shakespeare's Jests, or the Jubilee Jester* (*c*.1769). For the most part the Funny Stories, Droll Adventures, and Smart Repartees of which the title-page boasts are cruelly uncomical. They turn on stale puns or feeble rejoinders; they impart homely wisdom. Some are

---

* Until well into the nineteenth century it was believed that Sir Hugh Clopton had merely beautified New Place in 1720; later some thought that he had demolished it. The truth lies somewhere between these extremes, but there can be no doubt that Clopton 'improved' the house by equipping it with a Georgian façade.

† Modern sceptics may take pause from Sharpe's deathbed affidavit. On 14 October 1799, in the presence of the mayor and a Justice of the Peace for the borough, he took a solemn oath upon the four Evangelists, in the presence of Almighty God, that the mulberry tree which 'growed' in Shakespeare's garden was indeed planted by him, and that all the many curious toys and useful articles worked therefrom came from no other source.

exceedingly gross. The novelty lies in the choice of a hero for a minority of the anecdotes: 'the Warwickshire Wag', as he is called, or (more often and more familiarly) 'our wag'. We see Shakespeare in the coffee-houses that were a feature of eighteenth-century rather than Elizabethan London, or with 'a set of merry companions, over a chearful bowl', or reeling home from the tavern. A wild young nobleman desires leave to toast the devil; 'with all my heart replies our wag, I have no objection to any of your lordship's *friends*'. 'The Countess of Essex being at supper (the Monday-night before Shrove-Tuesday) where Shakespear was present, it having struck twelve o'clock: well, says she, it is now Shrove-Tuesday; come, gentlemen, set up your *cocks, and I will knock them down*. True, replied our wag, there is none that disputes your Ladyship's abilities.' 'Shakespear being asked the reason why people of the first rank generally gave alms to the poor blind more than to those of philosophers, replied, there was a probability of their being *blind*, but never of their being *philosophers*.'[27] Some of the stories display ignorance of—or indifference to—the most elementary facts of the dramatist's life. 'A man that had *never been married*, earnestly persuaded the Warwickshire Wag to take a wife, proving to him, that marriage, so far from being derogatory to the character of a poet, was, on the contrary, but just and necessary. Give me, therefore, says Shakespear, *one of your daughters*.'[28] Only one anecdote has any bearing on an actual, if doubtfully valid, biographical tradition; it dimly adumbrates, on the coarsest public-house level, a supposed rivalry in which the Bard demonstrated his wit at the expense of a learned adversary. 'Shakespear seeing Ben Johnson in a necessary-house, with a book in his hand reading it very attentively, said he was sorry his memory was so *bad*, that he could not *sh-te without a Book*.'[29] Thus was Shakespeare the man made accessible to a non-literary public.

Among the more sophisticated, too, curious elaborations on the Shake-speare–Mythos began to flourish. At Exeter in the 1790s a society of gentlemen and scholars pledged to conversation in which learning was mixed with decent gaiety was edified by an account of 'the vagabond Shakespeare'. The Lucy story takes on a more dramatic coloration as the gulf widens between the status of the victim and of his oppressor. 'How little did Sir Thomas Lucy suppose', the speaker (identified only as 'T.O.') exclaims, 'when in the pomp and plentitude of magisterial power, he saw an idle youth stand trembling before him, or heard of his precipitate flight from the terrors of judicial authority, that he should be transmitted down to posterity, by the same disorderly youth, under the humiliating appellation of "Robert Shallow, esq. justice of peace and *coram!*"'[30]

In this context of burgeoning bardolatry and vigorous but unsystematized and unfulfilled inquiry, Edmond Malone made his contribution to Shake-speare scholarship.

# 4
## *Malone: The Making of a Scholar*

TODAY the portrait of Malone painted by Sir Joshua Reynolds in 1774 hangs on a great retractable screen in the basement storeroom of the National Portrait Gallery in London, but the curious visitor may view it upon request. The subject is in his early prime. His complexion, a fresh pink, suggests vigorous good health. The features reveal an evident intellectual: the broad expanse of forehead; the brown eyes, bright and full, gazing fiercely at some point beyond the picture's frame; the long, finely chiselled nose; the firmly set lips, with more than a trace of primness; the forceful thrust of the chin. It is the face of a serious man, and also of a man not to be trifled with. 'Mr. Malone was quite a gentleman in his manners', his friend John Taylor recalled, 'and rather of a mild disposition, except when he had to support the truth, and then there were such firmness and spirit in what he said as could hardly be expected from one so meek and courteous; but he never departed from politeness and respect.'[31] Malone was in his early thirties when he sat for Reynolds; already on intimate terms with Reynolds and other distinguished men of the age, he had not yet published the first work for which his own name would be celebrated.

Descended from an ancient Irish family that for generations had made the law their hereditary pursuit, he was born in Dublin on 4 October 1741. His grandfather, Richard, gained fame and wealth at the Irish Bar, and purchased the splendid family seat of Baronston on the banks of Lough Iron. His four sons all practised the law in the same place, along with their father, giving the family a virtual monopoly on the courts of the district. Edmond—the scholar's father—had a successful career: he attracted numerous clients, sat for the Irish House of Commons for Granard from 1760 to 1766, and was in the latter year made judge of the Court of Common Pleas. It is hardly surprising that his second son Edmond seemed destined to follow the ancestral profession.

As a child Malone attended the highly regarded school in Molesworth Street, Dublin, run by Dr Ford. There the boys excelled in the performing of plays; so much so that fashionable Dublin attended their representations, and the great Macklin himself purportedly directed several productions. Here Edmond, as a juvenile actor, enjoyed his first taste—from which he never recovered—of Shakespeare and the stage. One of his schoolmates was Robert Jephson, the future dramatist, who became a lifelong friend. Malone went on to Trinity College, Dublin, where he made other warm and lasting associations, wrote verses that impressed his contemporaries, and displayed those characteristics of the scholar by means of which he would later effect a revolution in English literary studies. 'His pursuits were various', writes his first memorialist,

but they were not desultory. He was anxious for general information, as far as it could be accurately obtained; but had no value for that superficial smattering which fills the world with brisk and empty talkers. When sitting down to the perusal of any work, either antient or modern, his attention was drawn to its chronology, the history and character of its author, the feelings and prejudices of the times in which he lived; and any other collateral information which might tend to illustrate his writings, or acquaint us with his probable views and cast of thinking.[32]

In his studies he was sustained by affectionate ties to an harmonious and closely knit family. In 1760 his father wrote to his 'Dear Neddy': 'Continue, therefore, my dear child, the same course of industry you are in, in order to qualify you to get your own bread, and to make your own way in the world. God Almighty always blesses the diligent and industrious.'[33]

After taking his BA at Trinity, Malone diligently and industriously continued his legal studies at the Inner Temple. He was also making his first exhilarated acquaintance with the great city whose attractions he would ultimately find irresistible. In the Grecian Coffee-house in the Strand he rubbed shoulders with the glamorous literary and theatrical society of the day. Of less than middling stature, he created a favourable impression by his urbanity of temper, kindliness, and social ease; later he would be celebrated for his hospitality. His Irish friend, Edmund Southwell, introduced him in the autumn of 1765 to Dr Johnson and thus initiated a friendship that was to last until the latter's death. Malone wrote about the meeting to his friend Chetwood, rusticating in a country parsonage in provincial Ireland, and received a breathless reply: 'How happy are you who can sweeten even confinement with the company of men and works of genius! I envy you your intimacy with the editor of Shakspeare, and the opportunities you have by your situation in London of collecting books.'[34] In London too he met another great Shakespearian editor, the difficult and spiteful Steevens, who was said to have only three friends, one of whom was himself.* Steevens befriended Malone and published his work, but later, unable to brook a rival editor, severed relations with a man whose reputation threatened to outstrip his own.

In 1767, after some travel in France, Malone returned to Dublin, was called to the bar, and for some years practised law on the Munster circuit with indifferent rewards. He contributed articles critical of the government to Irish newspapers and periodicals, and embarked upon his first substantial literary project: an edition of Goldsmith. He was also in the midst of a protracted and

---

* The other two were Dr Farmer and John Reed. See Samuel Parr, *Works*, ed. John Johnston (London, 1828), viii. 128.

tormenting love affair which he could neither extricate himself from nor resolve by matrimony; why has yet to be revealed.* (In later years he confided to his friend, the artist Farington, 'that contrary to what His appearance bespoke He had lived a life of much anxiety from being disappointed in hopes & wishes which related to domestic union (marriage) that must now no longer be looked for'.)[35] His delightful elder sister urged him to dispel his melancholy by having a fling. 'I wish you would partake of all sorts of diversions', she wrote to him; 'for though I do not expect they can afford you in your present situation any amusement at the time, yet I believe dissipation is the best remedy against low spirits; though I must confess I do not think it the pleasantest.'[36] Meanwhile Malone's father died, leaving him his proper-ties in Shinglas, Co. Westmeath, and in Cavan, which eventually yielded him about £800 a year. He was now independent, if not rich, and after toying with the idea of a political career, he abandoned Dublin and the law to migrate to England in the autumn of that year of revolution, 1776.

In London he set up as a man of letters, and in his bachelor quarters at 7 Marylebone Street, and (after 1779) 55 Queen Anne Street—later renamed Foley Place—he entertained with elegance and warmth. His excellent family credentials gave him entrée to the best literary and social circles. Horace Walpole showed him favour; he became a member of the Literary Club frequented by the great men of the day: Johnson, Garrick, Gibbon, Burke, Warton. Such company found him congenial rather than brilliant; 'Malone', Boswell said of him, 'is respectable and gentlemanlike rather than shining.'[37] His brother Richard, who had succeeded to the great house at Baronston, tried to lure him back to Dublin: the professional competition, once so formidable, had scattered, a seat in Parliament might be arranged for. 'I am convinced that were you to return here', Richard wrote to him in 1777, 'you could not fail of the most rapid success in the profession; and really wish—could you in any sort reconcile it with your other schemes of happiness—that you would once more adventure in a pursuit, which there is no doubt would now be attended with the greatest advantages to your fortune, if that is a matter you consider as any object.'[38]

It was not. He was already cramming his head with all such reading as was never read, 'the reading [in Farmer's judgement] necessary for a comment on Shakespeare'. Steevens in 1777 lent him his copy of Langbaine's *Account*, the annotations to which Malone laboriously transcribed in full, and, over the years, supplemented with his own notes; it became a great repository of knowledge about the early stage. Malone was also familiarizing himself with

---

* Prior, *Malone*, 34. See also Malone's letter to Caulfield, 8 January 1782: 'There is little chance of getting over an attachment that has continued with unabated force for thirteen years; nor at my time of life is the heart very easily captivated by a new object. . . . I am a very domestic kind of animal, and not at all adapted for solitude, and indeed have been peculiarly unfortunate.' [*The Manuscripts and Correspondence of James, First Earl of Charlemont*, Historical Manuscripts Commission, 12th rep., app., pt. 10; 13th rep., app., pt. 8 (London, 1891–4), i. 394.]

documents: 'I went with my friend Mr. Steevens', he noted in his *Langbaine* on 24 September 1776, '. . . to see Shakespeare's *original* Will.'

For his first Shakespearian project he began at the beginning and tackled the great uncharted sea of the poet's chronology. It was a daring change of occupation for the erstwhile barrister, and he did not undertake it lightly. Fearing ridicule, he wrote to his lifelong friend James Caulfeild, Earl of Charlemont, whose taste and learning he respected, and from him received the necessary reassurance. 'That some wise ones may smile at your lucubrations I doubt not. But let them smile. . . .', Charlemont advised him:

For my own part, I will never be laughed out of my amusements till they shall be proved hurtful to society, but will boldly proceed in those pursuits which though they cannot be deemed the fruits of literature, may at least be stiled its flowers. . . . but your undertaking, my dear Ned, needs not any such apology. The history of man is on all hands allowed to be the most important study of the human mind, and what is your chronological account of the writings of Shakspeare other than the history of the progress of the greatest genius that ever honoured and delighted human nature?[39]

In 1778 Steevens, still his friend, included in the second edition of his *Shakespeare* Malone's *An Attempt to Ascertain the Order in which the Plays of Shakespeare Were Written.*

The *Attempt* does not pretend to Johnson's imaginative acuteness, or to make the kind of ingenious demonstration with which Farmer disarmed critics. No aura of excitement clings to the enterprise: the barrister's tone is judicious, impartial. He presents in summary form his proposed chronology, and then proceeds through the plays, one by one, evaluating the evidence for date of composition. Malone is aware, as he acknowledges at the close of his essay, that to some his project will appear 'a tedious and barren speculation'. If imagination is operating it is in the fact of the undertaking. No one had previously set himself the task which Malone now confronted, and no one else—including Johnson, who was inattentive to minute accuracy[40]—had the equipment to see it through.

In his work he was aided by two recent and notable additions to Shakespearian knowledge contributed by Thomas Tyrwhitt (1730–86). One of the most learned men of the day—he reportedly knew 'almost every European tongue'—Tyrwhitt played a major role in the exposure of the Chatterton forgeries, and, equally at home in classical and modern philology, won lasting acclaim for his great editions of Chaucer and the *Poetics* of Aristotle. His *Observations and Conjectures upon Some Passages of Shakespeare*, printed anonymously in 1766, exemplifies the current rage for emendation of texts, but in it Tyrwhitt digresses to draw attention to Meres's *Palladis Tamia* of 1598, with its crucial lists of comedies and tragedies. Malone was supplied with a *terminus ad quem* for eleven plays, and confronted with interpreting Meres's reference to *Love's Labour's Won*. Equally important, Tyrwhitt, in a

note communicated to Steevens and published in volume six of the latter's
1773 edition of Shakespeare, brought to light the first reference to Shakespeare on the London stage: Greene's deathbed onslaught, with its parody of
a line from *3 Henry VI*. The 1773 *Shakespeare* also furnished the Stationers'
Register entries which Steevens had extracted and assembled, and which
Malone would find invaluable.

His subject leads him to speculate on a directly biographical question:
when, and under what circumstances, did Shakespeare begin his theatrical
career? From the absence of reference to his plays in Webbe's *Discourse of
English Poetry* (1586), Puttenham's *Art of English Poesy* (1589), and Harington's *Apology for Poetry* (1591), Malone deduces that the poet had not yet
attracted attention as 'the new prodigy of the dramatick world'. He conjectures that Shakespeare began writing for the stage around the middle of
1591. As for the circumstances, Malone suggests that someone from
Stratford may have arranged an introduction in London: perhaps the popular
actor Thomas Greene, whose name turns up in Stratford records, or Michael
Drayton, a Warwickshireman who was making a reputation for himself as a
literary man in the early 1590s. The Greene identification is unfortunate, and
illustrates the dangers posed to scholars by commonplace names. References
to Greene the actor do not begin to appear until 1603. We do not know his
origins, but the Stratford Greene, who may have been Shakespeare's kinsman
by blood or matrimony, belonged to the Inner Temple and was steward and
town clerk of Stratford from 1603 until 1617. Later Malone would come up
with a more plausible theory of Shakespeare's entry into theatrical life.

To fix the dates of the plays Malone relies on a variety of criteria: the title-
pages of the early quartos, the Stationers' Register entries, records of Court
performance, internal allusions, and references and imitations in contemporary writings. His most striking methodological innovation is his discovery
that Shakespeare's changing verse technique tells us something about
chronological succession. 'A mixture of rhymes with blank verse, in the same
play, and sometimes in the same scene', he notes,

is found in almost all his pieces, and is not peculiar to Shakspeare, being also found in
the works of Jonson, and almost all our ancient dramatick writers. It is not, therefore,
merely the use of rhymes, mingled with blank verse, but their *frequency*, that is here
urged, as a circumstance which seems to characterize and distinguish our poet's
earliest performances. In the whole number of pieces which were written antecedent
to the year 1600, and which, for the sake of perspicuity, have been called his *early
compositions*, more rhyming couplets are found, than in all the plays composed
subsequently to that year; which have been named his *late productions*. Whether in
process of time, Shakspeare grew weary of the bondage of rhyme, or whether he
became convinced of its impropriety in a dramatick dialogue, his neglect of rhyming
(for he never wholly disused it) seems to have been *gradual*. As, therefore, most of his
early productions are characterized by the multitude of similar terminations which

they exhibit, whenever, of two early pieces it is doubtful which preceded the other, I am disposed to believe, (other proofs being wanting) that play in which the greater number of rhymes is found, to have been first composed.[41]

Immediately, however, he recognizes, with the integrity of thought which distinguished all his scholarship, that the criterion is fallible, for the three *Henry VI* plays, which are among Shakespeare's earliest compositions, have few rhymes. Later students, aware that rhyme prevails less in early Elizabethan drama than Malone thought, would see other problems. Statistics can mislead, especially if ordinary couplets are not separated from alternate rhymes, sonnets, doggerel rhymes, and songs and sporadic jingles, which may serve special purposes. A dramatist's choice of rhyme over blank verse is always conscious, and may have a deliberately experimental end, as, for example, in *A Midsummer Night's Dream*, which is fantastic, or *Romeo and Juliet*, which is romantic; both plays abound in rhyme. Later, modulations in the blank-verse line itself would attract attention: variations of stress within the unit, percentages of redundant final syllables, proportions of run-on lines. To these more sophisticated tests—which, although they measure less premeditated characteristics, are not *necessarily* more reliable—Malone points the way.

Like so many pioneer efforts, Malone's chronology would be modified in the light of subsequent investigation. Some of his dates are clearly unacceptable. He gives *The Winter's Tale* much too early a place in the canon, 1594, thinking it is *perhaps* (the italics are Malone's) *A Winter Night's Pastime* entered on the Stationers' books on 22 May 1594. But he is uneasily aware that the play is not mentioned by Meres, and that, except for the Chorus, it is without rhymes. Malone also assigns too early a date to *Hamlet* (1596), on the basis of a misdating (not his own) of marginalia by Gabriel Harvey and misinterpreted allusions to the previous tragedy, perhaps by Kyd, on that subject. Too early, again, is the assignment of *Henry VIII* (1601), although Malone knows Sir Henry Wotton's letter referring to the play as new in 1613. Unaware of the Court performance on 1 November 1604, he attributes *Othello*, without conviction, to 1611, and, on the basis of a supposed allusion to Parliamentary undertaking, *Twelfth Night* to 1614. He is also wrong, if not quite so spectacularly, on other plays: *Richard III* (1597), *Cymbeline, King of Britain* (1604), *The Taming of the Shrew* (1606), and *Julius Caesar* (1607). Sometimes, when he can find no evidence, he throws up his hands in despair and assigns a play to a year simply because that year would otherwise be blank, and Shakespeare must have been continuously employed. Such is the case with *Coriolanus* (1609) and *Timon of Athens* (1610), for which objective evidence is still depressingly scant. Malone's date for *Coriolanus* is, however, a lucky guess. Whether or not we see, with one modern scholar, an allusion (III. i. 96–7) to Hugh Middleton's project, begun in February 1609, to bring clean water into London by channels from streams in Hertfordshire, that year is on general grounds plausible.[42]

Malone is also right on a number of other plays. The work of the first period is correctly specified: *Titus Andronicus* (which, however, he denies Shakespeare), the *Henry VI* trilogy, *Two Gentlemen*, *Midsummer Night's Dream*, *Romeo and Juliet*, and *Comedy of Errors*. From the mention of the Duke's aversion to crowds in *Measure for Measure*, Malone shrewdly infers an allusion to a similar disposition in James, and dates the play—rightly—soon after that monarch's accession. He knows that *The Tempest* was written after the shipwreck of Sir George Somers off the Bermudas in July 1609. Although Malone dates *All's Well* too early (1598), his identification of that play with *Love's Labour's Won*—here he is following Farmer—anticipates later opinion.

Whether right or wrong, Malone sets forth in every instance the data on which he has based his conclusion. In the case of *Henry VIII*, for example, the reader is furnished with all the evidence for a later date than Malone's; he is also given Malone's reasons, which he may or may not accept, for rejecting that evidence—but, and this is most important, he is in a position to form his own conclusion. Malone never claims infallibility. He expressly states that inquiry, at this stage, cannot produce certain results, but only a better notion of the chronology than is to be found in the various editions of Shakespeare. The enduringly admirable characteristics of Malone's scholarship thus appear in his first Shakespearian venture: his candour; his rejection of impressionism in favour of method, which is clearly expounded; his wide reading; his refusal to push the interpretation of evidence beyond legitimate bounds. Such qualities, in combination, are perhaps rarer than they should be.

Because he had refrained from dogmatism at the outset, he was able afterwards to reconsider and revise his conclusions without embarrassment. This he did when he came to republish the *Attempt* in his 1790 edition of Shakespeare. Some modifications are minor, involving merely a year or two. But, on stylistic grounds, he moves the *Shrew* back a dozen years to 1594, which is more in accord with today's thinking. He advances *The Winter's Tale* a decade to 1604, still too early but a step in the right direction. (He changed his mind twice more. In the emendations and additions included in the second volume of his edition, he shifted the play to 1613; among his posthumous papers, a revised version of the *Attempt* assigns *The Winter's Tale* to 1610 or 1611—exactly the finding of modern scholarship). At the same time he provides his public with the first notice of a document of prime biographical interest: Chettle's defence of Shakespeare's character in the preface to *Kind-Heart's Dream*. This Malone quotes and discusses. (Not until long afterwards would it be realized that the 'divers of worship' who, according to Chettle, vouched for Shakespeare's excellences were presumably not noble lords—noblemen they would refer to as 'divers of honour';

'worship' applied to gentlemen.[43]) Later he recognized that he had erred with *Hamlet*, and in the final version of the *Attempt*, published posthumously in 1821, moved the play to 1600, a date now generally accepted.

# 5

# *Malone: The* Supplement *and Other Writings*

IN the interval between the two editions of the *Attempt* Malone laboured prodigiously. In 1780 he brought out his *Supplement to the Edition of Shakespeare's Plays Published in 1778 by Samuel Johnson and George Steevens*, consisting of additional observations (including, as an addendum to one of Steevens's notes, Malone's first essay at an historical account of the Elizabethan stage) and editions of the poems and of the apocryphal plays of the Third Folio. Both the apocrypha and the poems had been edited previously, yet Malone's exertions on the latter not only advanced textual studies but also contributed powerfully, if indirectly, to Shakesperian biography.

The great succession of eighteenth-century editors, from Rowe to Capell, had failed to include the poems in their collected *Shakespeares*. Theobald promised an edition, and one was announced as ready for the press in the *Grub-Street Journal* (6 June 1734); but none appeared. Capell prepared an edition, but he too did not publish it. Thorpe's exceedingly rare 1609 quarto of the Sonnets, which Capell could never put his hands on, had dropped from sight until it was reprinted, with the *Lover's Complaint* from the same volume, in 1711 by Bernard Lintot. The Sonnets appear in volume two of *A Collection of Poems . . . Being all the Miscellanies of Mr. William Shakespeare, Which Were Publish'd by Himself in the Year 1609. and Now Correctly Printed from Those Editions*. To this misleading information about Shakespeare's role in the publication of the Sonnets, Lintot adds his misconception of the nature of the cycle itself, which he describes as *One Hundred and Fifty Four Sonnets, All of Them in Praise of His Mistress*. The 1609 quarto was also reprinted, without introduction or notes, by Steevens in volume four of his *Twenty of the Plays of Shakespeare* in 1766. But all the widely disseminated and influential editions prior to Malone derive from Benson's piratical 1640 collection of *Poems: Written by Wil. Shake-speare. Gent.* Lintot based his text on Benson, which also underlies two more important editions: Gildon's unauthorized supplement to Rowe in 1709, and the additional volume (seven) to Pope's

*Shakespeare,* provided in 1725 by the physician turned literary hack, George Sewell.

These editions follow Benson in omitting the dedication to Mr W.H. and in rearranging the Sonnets in a jumbled and misleading order. Out of the 146 sonnets are manufactured 72 'poems' with such inspired titles as 'The glory of beautie' or 'The benefit of Friendship' or 'Happinesse in content'. The idea of a cycle is wholly destroyed. Gildon and his successors also follow Benson in transforming the sex of the person addressed; thus 'friend' in 104 and 'boy' in 108 become 'love', and the concluding lines of 101 alter 'him' and 'he' to 'her' and 'she'.

These changes deceived Gildon, unaware of the existence of the 1609 edition, into conjecturing that 'these Poems being most to his [Shakespeare's] Mistress it is not at all unlikely, that she kept them by her till they fell into her Executors Hands or some Friend, who would not let them be any longer conceal'd'.[44]

Malone restored the dedication, the sequence of 1609, and the masculine references. In co-operation with Steevens, he furnished full annotation. His note to the dedication makes the important point that 120 of the Sonnets are addressed to a man (the dedicatee, he believes) and the remaining 28 to a lady. Malone reports, only to reject, Farmer's suggestion that Mr W.H. is the author's nephew William Hart: he could have been at most twelve years old in 1598, when Meres refers to Shakespeare's 'sugared sonnets'. Instead Malone favours, without notable conviction, Tyrwhitt's candidate, William Hughes, on the basis of a pun on *hew* in the twentieth sonnet. The rival poet he imagines to be Spenser. Thus tentatively begins the great debate over the *dramatis personae* of the Sonnets.

A particular passage in the Sonnets prompted a remarkable exchange between Malone and Steevens: an exchange with far-reaching consequences for Shakespearian biography. In glossing the opening lines of Sonnet 93, 'So shall I live supposing thou art true | Like a deceived husband,' Malone reports Oldys's observation that this poem and the preceding 'seem to have been addressed by Shakspeare to his beautiful wife on some suspicion of her infidelity'.[45] Malone points out that these sonnets are not addressed to a female, but wonders whether Oldys might not have come upon some other evidence in his researches: after all, Shakespeare neglected his wife in his will, and in his plays he portrayed the emotion of jealousy with such exquisite feeling as to arouse suspicion that he was himself perplexed with doubts (although perhaps not in the extreme). Previous biographers had not worried much about the poet's conjugal relations, nor (when they did evince curiosity) had they necessarily assumed his disaffection with Anne. In *The Modern Universal British Traveller*, which antedates Malone's *Supplement* by one year, the 'Biography of Warwickshire' confidently informs us that Shakespeare 'lived very happy' with his wife, and, after he made some money

minding horses, fetched her to London. This intelligence is no less trustworthy than the same biographer's description of Shakespeare's bride—'a
young lady, the daughter of one Hatchway'.[46]

Malone's speculation, offered as mere conjecture, elicited from Steevens a
famous rejoinder on the limitations of Shakespearian biography:

As all that is known with any degree of certainty concerning Shakespeare, is—*that he
was born at Stratford upon Avon,—married and had children there,—went to London, where
he commenced actor, and wrote poems and plays,—returned to Stratford, made his will, died,
and was buried,*—I must confess my readiness to combat every unfounded supposition
respecting the particular occurrences of his life.[47]

Yet, as the rest of the note shows, Steevens is not unwilling to do some
supposing himself.

To discredit Malone's conjecture, he offers his own interpretation of the
will. The poet had provided for his wife in times of health and prosperity, or
knew that her father had done so, and the bed he left her perhaps had
peculiarly tender associations; may, indeed, have been the bridal bed. The
plays must be rejected as sources of information about the author's emotional
life: 'No argument . . . is more fallacious than that which imputes the success
of a poet to his interest in his subject.' But Malone stuck to his guns, and in
his own rejoinder to Steevens he returned again to the will. In a passage that
would become profoundly influential—more influential than any other single
statement he would ever make—Malone reconstructed the circumstances
underlying the bequests: 'His wife had not wholly escaped his memory; he
had forgot her,—he had recollected her,—but so recollected her, as more
strongly to mark how little he esteemed her; he had already (as it is vulgarly
expressed) cut her off, not indeed with a shilling, but with an old bed.'[48] The
issue of the second-best bed was now clearly stated, and in the opposing
views of two great scholars the battle lines of the future were drawn.

To Malone recognition is therefore due as the first critical editor of the
Sonnets. The others had neglected the poems because they thought little of
them. Some say that 'they are not valuable enough to be reprinted', Gildon
remarks in his defence of the Sonnets, 'as was plain by the first Editors of his
Works [i.e. Heminges and Condell] who wou'd otherwise have join'd them
altogether'.[49] When in 1793 Steevens issued his fifteen-volume edition of
Shakespeare, he refused to make room for the Sonnets, with an explanation
that reflects the alien sensibility of another age, as well as the editor's
grudging admiration for the achievement of Malone, with whom he had
broken:

We have not reprinted the Sonnets, &c. of Shakspeare, because the strongest act of
Parliament that could be framed, would fail to compel readers into their service;
notwithstanding these miscellaneous Poems have derived every possible advantage
from the literature and judgement of their only intelligent editor, Mr. Malone, whose

implements of criticism, like the ivory rake and golden spade in Prudentius, are on this occasion disgraced by the objects of their culture.—Had Shakspeare produced no other works than these, his name would have reached us with as little celebrity as time has conferred on that of Thomas Watson, an older and much more elegant sonnetteer.[50]

But the Sonnets were now available in a form that made apparent their autobiographical relevance. Soon they would be closely scrutinized, and the identity of the dedicatee would become a subject for endless speculation.

Several other accomplishments of the 1780s deserve brief notice. At the Bodleian Library Malone recovered the epitaph, ascribed to Shakespeare, celebrating the godly life and death of Elias James. 'What think you?' he asked Charlemont on 20 July 1781. 'Genuine or not? It does not sound to me quite Shakspearian; and yet this may be mere whim, for we have no similar composition of his to compare it with, it being remarkable that he has not left a single line, monumental or commendatory, on any of his contemporaries, a circumstance which, I fear, will weigh against the authenticity of these lines.'[51] While making no pretence to a medievalist's training Malone, in *Cursory Observations on the Poems Attributed to Thomas Rowley* (1782), has little difficulty in showing—on grounds of versification, anachronisms, imitation of more recent authors, and handwriting and paper—that the poems purportedly written by Rowley, the fifteenth-century priest of Bristol, originated not in some miraculously discovered chest but in the head of Thomas Chatterton. This demonstration foreshadows Malone's later and even more decisive role in unmasking forgeries that more directly concern this narrative. In 1783 he published two appendices to his *Supplement* to the Johnson–Steevens *Shakespeare*, revising and expanding some of his earlier notes. By furnishing the third edition of the Johnson–Steevens *Shakespeare* in 1785 with annotations indiscreetly controverting his mentor's earlier notes, Malone made a permanent rift with Steevens inescapable. These efforts on a lesser scale were followed by one of his most ambitious and influential studies, *A Dissertation on the Three Parts of King Henry VI* (1787).

The *Dissertation*, dealing with the origin and nature of dramatic texts, is more pertinent to study of the Shakespeare canon than to the present undertaking.[52] Yet, because certain assumptions about Shakespeare's career underlie the *Dissertation*—assumptions which, in turn, it tends to reinforce— we may here pause over it briefly. Malone argues that Shakespeare did not write *1 Henry VI*, or at most contributed only a scene or two, and that the dramatist moulded *2* and *3 Henry VI*, with many revisions and additions, out of two earlier plays, *The First Part of the Contention of York and Lancaster* and *The True Tragedy of Richard Duke of York*, by Greene or Peele or both. Thus Malone endorses the view, which for long would hold sway, that Shakespeare started out as a playwright by refurbishing the works of established authors. Greene's complaint about the 'upstart crow' forms, Malone acknowledges, 'the chief hinge of my argument'. But Greene is ambiguous, and Malone is wrong; the *Contention* and *True Tragedy* are not independent plays but corrupt versions—memorial reconstructions—of 2 and 3 *Henry VI*. Not until the twentieth century would this complex and fascinating textual puzzle be worked out, and until then Malone's conclusions, brilliantly deduced from a false premiss, would be accepted as dogma by almost everyone.

The *Dissertation* appeared during the eight years, from 1782 to 1790, when Malone was engaged on his monumental ten-volume edition of Shakespeare, a task which (he complained to Warton on 23 November 1785) hardly left him time for eating or sleeping. Still, during the same period he managed to find time for another project which, although unrelated to Shakespeare studies, should not pass unnoted in any account of Malone. Early in 1785 he and Boswell became intimate friends. Johnson had died the previous December, and immediately Boswell had been mentioned as the most appropriate person to do the definitive life. But he decided first to create a book out of the journal of his tour to the Hebrides with Johnson. The work did not progress, however, until after a fateful dinner with Malone on 30 March 1785, when they talked until past midnight. It was with Malone's collaboration that Boswell was able to see his enterprise through. The manuscript discovered in this century in the croquet box at Malahide Castle reveals more revisions in Malone's hand than in Boswell's. When the latter published his book late that year, he dedicated it, in affection and gratitude, to the friend who had unstintingly helped him. For the next six years Boswell went over his journals covering the twenty-one years of his association with Johnson, and gathered collateral materials for the biography. Again Malone rendered powerful support. '[T]he revision of my *Life of Johnson* by so acute and knowing a critick as Mr. Malone', Boswell wrote to Temple on 28 November 1789, 'is of most essential consequence, especially as he is *Johnsonianissimus*.'[53] When the *Life* finally appeared in 1791, Boswell declared, in his prefatory advertisement, 'I cannot sufficiently acknowledge my obligations to my friend Mr. *Malone*, who was so good as to allow me to read to him almost the whole of my manuscript, and made such remarks as were greatly for the advantage of the Work.'[54] In addition to such tangible assistance, Malone gave Boswell moral encouragement to persevere through anxiety, dissipation, and despair, and to bring to a triumphant conclusion the greatest of English literary biographies.

# 6

## *Malone: The 1790* Shakespeare

MALONE published his *Shakespeare* on 29 November 1790. Expectations, we learn from Boswell, had for some time run high among Shakespearian enthusiasts, and he did not disappoint, although some grumbled (with cause, Malone acknowledged) about the smallness of the type. As a textual critic he has, to be sure, since been largely superseded by the great advances in bibliographical method; but he showed convincingly why the later Folios should be rejected, and he wisely chose as his copy-texts the earliest quartos and the First Folio. He claims to have restored 1654 original readings; the age of conjectural criticism and capricious innovation—Malone's terms—was over. What here chiefly concerns us, however, is his contribution to the elucidation of Shakespeare's life.

Malone examined the records, transcribed the documents, and weighed the evidence. In his searches he had assistance from John Kipling, keeper of the rolls in Chancery, and Richard Clarke, Registrar of the diocese of Worcester. Above all, the Revd James Davenport, vicar of Stratford, made inquiries for Malone in the town and environs. With the materials assembled, it would not have been difficult, he remarks in his preface, to have composed a new biography more ample and reliable than Rowe's meagre Life, but the information has been collected at very different times, and so is dispersed through the various sections of the prefatory volumes to the edition. Drily set forth in notes, appendices, and treatises, it marks nevertheless the greatest single step forward, to that time, in knowledge of Shakespeare's life and theatrical milieu.

Much of his material Malone presents as a running commentary on Rowe's essay. From the books of the coporation of Stratford he is able to reconstruct, for the first time, the career of John Shakespeare. We learn that in 1569 the poet's father filled the office of High Bailiff, and that in the former part of his life he was comfortably off but afterwards found himself in reduced circumstances: Malone notes the remission in 1569 of his weekly aldermanic tax of 4*d.*, and that in 1586 he was replaced, 'for Mr. Shakspere doth not come to the halls, when they be warned, nor hath not done of long time'. Malone also supplies information about the dramatist's mother and the ancient family of Arden. He then goes on to speculate that on leaving school 'Shakspeare was placed in the office of some country attorney, or the seneschal of some manor court.' (Later he came upon Aubrey's brief life, with its reference to Shakespeare as a schoolmaster in his younger years, but, unwilling to relinquish his theory, he now suggested that the poet was law-teacher rather than schoolmaster, and that he might have become 'sufficiently conversant with conveyances to have taught others the forms of such legal assurances as are usually prepared by country attorneys; and perhaps spent two or three years in this employment before he removed from Stratford to London'.[55] Malone was not a barrister for nothing.) He distrusts with good reason Chetwood's account of the song about Sir Thomas Lucy, his deer and his horns, with which Joshua Barnes, the Cambridge professor of Greek, was purportedly serenaded at a Stratford inn; but he makes it available to his readers. Next Malone reports a stage tradition (first brought to light in his 1780 *Supplement*) that Shakespeare's first employment in the theatre was that of call-boy, or prompter's attendant. Continuing to follow the course of the dramatist's professional career, he cites the passage in the rare old play of *The Return from Parnassus* in which Shakespeare gives Jonson 'a purge that made him bewray his credit'.

In the passages dealing with Shakespeare's later life, Malone evaluates the evidence regarding the poet's earnings in an admirably reasoned note. He rejects as a gross exaggeration Gildon's statement that Shakespeare left an estate valued at £300 a year; he considers £200—still a comfortable income—more likely, and in his reckoning takes into account the worth of the land holdings in Bishopton and Welcombe (mentioned in the grand-daughter's will), New Place and the other Stratford houses, and the Blackfriars property.

Of New Place he provides historical particulars and the first view of the house as it stood in Shakespeare's time, engraved 'From a Drawing in the Margin of an Ancient SURVEY, made by Order of Sir George Carew, (afterwards Baron Carew, of Clopton, and EARL of TOTNESS) and found near Stratford upon Avon, in 1786.' New Place (as we have seen) no longer existed, the Revd Francis Gastrell having in 1759 razed the dwelling substantially rebuilt a half-century earlier. Nor was the recovered survey available: Malone had to depend on the undependable Jordan for a copy. One detail in the latter's sketch troubled the editor. 'Before Shakspeare's house', he wrote to Jordan on 25 March 1790, 'there were certainly no *iron* rails, that not being the mode of his time. I suppose what you have represented are intended for wooden pales, and perhaps, on a larger scale, were more distinct.'[56] On the whole, though, Malone was satisfied. He did not know that George Vertue, the antiquary and celebrated engraver, had sketched New Place. In 1737 Vertue accompanied the Earl of Oxford on a tour of Oxford, returning to London via Stratford, where Vertue talked with a descendant of the poet's sister, Shakespeare Hart, born in 1670. Perhaps the engraver obtained from Hart a description of the vanished house he knew so well; in any event Vertue made a rapid sketch preserved in a small quarto volume now in the British Library. It depicts a handsome structure of three storeys and five gables (Jordan's picture shows only three gables). Above the drawing Vertue wrote, 'This Something by memory and yᵉ description of Shakespears House. . . .' On the same sheet he included a plan showing the gate and a building on either side of the court fronting the main house. For a century and a half after Malone's time, Vertue's two drawings eluded notice; they were finally published in 1952.[57] Such is the force of tradition, or ignorance, that they have not entirely supplanted Malone's engraving, which is still reproduced.

From the vicar of Stratford Malone obtained information about the tradition of the celebrated mulberry tree, of which he prints the first detailed account. He makes public the destruction wrought by the Revd Mr Gastrell, who, desiring to be 'damn'd to everlasting fame', cut down the tree 'to save himself the trouble of shewing it those whose admiration of our great poet led them to visit the poetick ground on which it stood'. Malone goes on to report

that Mr. Hugh Taylor . . . who is now eighty-five years old . . . says, he lived when a boy at the next house to New-Place; that his family had inhabited the house for almost three hundred years; . . . that this tree (of the fruit of which he had often eaten in his younger days, some of its branches hanging over his father's garden,) was planted by Shakspeare; and that till this was planted, there was no mulberry-tree in that neighbourhood. Mr. Taylor adds, that he was frequently, when a boy, at New-Place, and that this tradition was preserved in the Clopton family, as well as in his own.[58]

Thus did the legend, like the tree, grow. Shakespeare, Malone conjectures, planted the mulberry in the spring of 1609, when many thousands of the trees were imported from France by order of King James.

Everywhere in his commentary Malone clears a path through accumulated error and breaks new ground. With good reason he rejects the epitaph supposedly written on Thomas Combe, 'Thin in beard, and thick in purse.' The more famous epitaph on John Combe, which (Rowe says) gave that usurious gentleman such disquiet, presents a more difficult case, and Malone, with the thoroughness that is his hallmark as a scholar, made a special investigation. He consulted Combe's will, and caught out Brathwait, who had first published the epitaph with the assertion that it was fastened on the tomb built by Combe in his lifetime: in the will Combe is concerned to have a tomb erected to him *after* his death. Malone, moreover, discovered that Combe named Shakespeare as one of his beneficiaries, and from this evidence he concludes, not unreasonaly, that the two men lived on amicable terms with one another. He was equally persistent in his attempt to track down information about the malady which terminated Shakespeare's life, but his researches led nowhere: hopefully he examined the private notebook of Dr Hall, who presumably attended the poet in his last illness, only to find that the earliest recorded case is dated 1617. In other notes Malone discusses the Shakespeare portraits (with similar diligence he went to Stratford with the

Droeshout engraving and minutely compared it with the bust in the monument); he prints for the first time the epitaph on Elias James, and those on the Stanleys, as noted by Dugdale in his *Visitation of Shropshire*. He corrects Rowe's statement that Shakespeare had three daughters, of whom Judith was the eldest. He traces the Shakespeare line to its extinction with the death of Lady Elizabeth Bernard. He supplies a transcript of the will of John Hall, a summary of the will of Thomas Nash, who married Hall's daughter Elizabeth, and the text of Elizabeth's will. With regret he reports that, 'after a very careful search', he could not find the will of Susanna.

Other important documents and records follow the Rowe Life. Malone gives the Oldys anecdotes, with corrective notes. He prints a selection, larger than West's, of entries of baptisms, marriages, and burials, 'extracted with great care from the Registers of Stratford'; the period covered extends from the birth of Joan Shakespeare in 1558 to the death of George Hart in 1778. West is corrected: Samuel Shakespeare becomes Hamnet Shakespeare. John, however, is not disengaged from the troublesome son of the Warwick shoemaker; instead Malone conjectures that the poet's father married three times. The remaining documents include the first accurate text of Shakespeare's will, with 'Bushaxton' replaced by 'Bishopton' and the 'brown best bed' by the 'second best bed'. Malone also printed, for the first time, the text of the Blackfriars Gate-house mortgage, found in 1768 by Albany Wallis, an attorney and antiquarian hobbyist, among the title-deeds of the Revd Mr Fetherstonhaugh, of Oxted in Surrey, whose family bought the Blackfriars property in 1667. Acquired by Garrick, the deed came into the possession of his widow and was communicated to Malone on 13 February 1788 through the good offices of Horace Walpole.

These invaluable materials Malone supplements with his article on Shakespeare in the section on the actors in *An Historical Account of the Rise and Progress of the English Stage*, which opens part ii of the first volume of the edition. Here Malone is still troubled by the entries for John Shakespeare in the Stratford registers. He knows that Rowe errs in stating that John had ten children, and he now believes that the three children born between March 1588 and September 1591—Ursula, Humphrey, and Philip—belong to another John Shakespeare. Thus far later scholarship would bear him out, but he takes the wrong tack in supposing that William had an elder brother John born before the register commenced, and twice married. Malone is on firmer, if nevertheless controversial, ground in offering, for the first time, the Spiritual Last Will and Testament of the poet's father, found twenty years previously at the Birthplace; later, however, he would have misgivings about it. The testament is followed by a judicious appraisal of the strengths and limitations of Aubrey as an antiquary, and the first exact transcript of his entire sketch of Shakespeare.

The *Historical Account of the English Stage* in which these miscellaneous bits and pieces are inserted occupies 331 pages, and thus qualifies as the longest sustained narrative that Malone had yet produced. It is also, for its time, an amazing achievement. It shows the scholar attaining that knowledge of the Shakespearian context which the best later biographers would recognize as essential to their task. It also reveals Malone as sufficiently a man of his own time to share its limitations of taste: on his first page he dismisses as contemptible all plays produced before 1592—'the titles are scarcely known, except to antiquaries; nor is there one of them that will bear a second perusal'.[59] In a note he lists the works on which he has based his sweeping censure; these include *The Arraignment of Paris*, *The Spanish Tragedy*, *Friar Bacon and Friar Bungay*, and the entire Marlowe canon. (Elsewhere in his writings Malone expresses a similar disdain for Jonson, and for the masques approved by the 'wretched taste' of the period.) When the younger Boswell came to reprint Malone's history of the stage in the Third Variorum, as it would become known, he could not forbear taking exception to the judgement levelled by the scholar he idolized: it was because Malone had constantly before him the example of Shakespeare that he undervalued that matchless writer's predecessors. Boswell, however, was writing in 1821, over a decade after Lamb had extolled the Elizabethans in his *Specimens of English Dramatic Poets*; the Romantic age had begun, the sensibility had changed.

But in his *Account of the English Stage* Malone presents himself not as a critic but as a theatre historian, and his is the first history on that subject deserving of the name. For he has gone back to the primary sources. He has read the old tracts; he has studied the old plays, and gleaned from them what inductions, prologues and epilogues, and stage directions can tell him about theatrical practice. He has consulted documents in the Remembrancer's Office in the Exchequer, and in the Lord Chamberlain's Office. He has seen manuscript notes of Lord Stanhope, Treasurer of the Chamber to King James. He has examined contemporary maps of London, and Stow's chronicle of the city (and its continuation by Howe), as well as Camden's *Annales*. From an old chest, unopened for 130 years, he has retrieved the priceless Office-book of Sir Henry Herbert, Master of the Revels in the time of James I and Charles I. 'This discovery is so much beyond all calculation or expectation', he exulted to Warton, 'that I will not despair of finding Shakespeare's pocket book some time or other.'[60] This, alas, did not turn up, but just as Malone's edition was issuing from the press 'a large folio volume of accounts kept by Mr. Philip Henslowe, who appears to have been proprietor of the Rose Theatre near the Bankside in Southwark', was found at Dulwich College; Malone thus became the first to publish extracts from Henslowe's *Diary*, the most valuable single document relative to the early stage.

In his account, which is meticulously documented, Malone gives a brief sketch of the medieval drama, based mainly on Warton's *History of English Poetry*, and then goes on to the larger issues of Elizabethan stage history that chiefly interest him: the number, size, and whereabouts of the playhouses; their structure, apparatus, and accommodations; the identity of the companies; admission prices, and the time and duration of performances; the incomes of playwrights; and the careers of the actors named in the First Folio. He makes mistakes (he thinks Kyd had a university education), and he is vague where later scholars would be exact: he does not know when the first regular theatre rose in England, nor when the Globe was built, although he does not think it could have been much before 1596. He thinks that the Globe was hexagonal on the outside and a rotunda within, and that it was so called because of its circular form; but later, characteristically, he reconsidered, and decided that the name derived only from the theatre's sign and motto, *Totus mundus agit histrionem*. He knows that the Globe catered to a public audience, and the Blackfriars to a select clientele, and he conjectures rightly that the King's Men used the former in the summer and the latter as their winter headquarters. He demonstrates that movable scenery was not employed in the Elizabethan theatre, and he points to the existence of an upper stage; but he does not raise the question of the existence of an inner stage, which would so much vex later students. He guesses that the imperfect and mutilated texts of one or two of Shakespeare's plays have resulted from shorthand transcription during performance. Subsequent historians would modify and supplement Malone's researches, but he put the investigation of English stage history on a sound footing, and those who came after built on his work. He richly deserves the tribute he received from the most eminent statesman of the day. 'You have taken infinite pains', Burke wrote to him in 1790, '& pursued your Enquiries with great sagacity.'[61] In the same year Burke publicly expressed his esteem for Malone by dedicating to him his *Reflections on the Revolution in France*.

The grand dimensions of Malone's investigations contract when he turns to the detection of a minor fraud in his essay on Shakespeare, Ford, and Jonson which rounds off the prefatory matter to the edition. Charles Macklin, the great actor celebrated for having redeemed Shylock from travesty, had in the 1748 published a letter in the *General* (afterwards the *Publick*) *Advertiser* prompted by the imminent benefit revival, for Mrs Macklin, of Ford's *The Lover's Melancholy*. In it Macklin gives an account of Ford, whom he describes as 'an intimate and a professed admirer of Shakspeare'. In a second letter to the same paper, Macklin claims to have once had in his possession a Caroline pamphlet bearing the quaint title, *Old Ben's Light Heart Made Heavy by Young John's Melancholy Lover*. According to Macklin this pamphlet (since very regrettably lost in passage from Ireland) brought together all the slurs and abuse testifying to Jonson's '*ill-nature* and *ingratitude* to Shakespeare, who first introduced him to the *theatre and fame*'. What especially infuriated old Ben was public acceptance of young John's play in the very week in which his own *New Inn* was damned; for 'Ford was at the head of the partisans who supported *Shakspeare's fame* against *Ben Jonson's invectives.*' In revenge Jonson accused Ford of having stolen *The Lover's Melancholy* from Shakespeare's papers, with the connivance of Heminges and Condell, who helped him with the revisal. Malone republished Macklin's second letter—as Steevens had before him—as a note to Jonson's elegy on Shakespeare. For some time, however, he had entertained doubts as to the authenticity of its information: he had talked with Macklin, who held to his story; he had advertised in the newspapers, offering in vain a reward to anyone producing a copy of the pamphlet. His suspicions at length fully aroused, he brought to bear on the letter the full resources of his historical knowledge and powers of logical analysis.

First he studied the title. In 1631, the year in which the pamphlet was supposedly printed, Jonson was fifty-seven, and Ford forty-five; so that, if Ben is to be accounted *old*, John can scarcely be regarded as *young*. Equally damaging is the fact that such of the pretended extracts as are true are already familiar, and could have been lifted from such accessible sources as Rowe's Life and Pope's preface to Shakespeare; whereas those extracts that are new are demonstrably false. Sir Henry Herbert's Office-book, moreover, reveals that *The New Inn* was produced not in the same week with *The Lover's Melancholy* but two months previously. Malone concludes that Macklin wrote his letter to excite public interest in the benefit by giving the revival a spurious Shakespearian interest. It is under these circumstances that the first Shakespearian forgery (unless one puts Theobald's *Double Falsehood* in the same class) originated.

When Malone wrote, Macklin was still alive and in his nineties. With exquisite tact the editor refrains from charging the veteran actor with anything graver than a harmless *jeu d'esprit*—although he has, it is true, behaved 'contrary to the Statute of Biography, and other wholesome laws of the Parnassian Code, in this case made and provided, for the security of the rights of authors, and the greater certainty and authenticity of dramatick history'.[62] The lightness of touch, even occasional facetiousness, and the evident wish to save Macklin as much as possible from embarrassment, lend by contrast an unexpected power to Malone's concluding sentences, alight with the flames of scholarly dedication:

Let the present detection be a lesson to mankind in matters of greater moment, and teach those whom higher considerations do not deter from invading the rights or property of others by any kind of fiction, to abstain from such an attempt, from the *inefficacy* and *folly* of it; for the most plausible and best fabricated tale, if properly examined, will crumble to pieces, like 'the labour'd mole', loosened from its foundations by the continued force of the ocean; while simple and honest truth, firm and self-dependent, will ever maintain its ground against all assailants,—

'As rocks resist the billows and the sky.'[63]

The warning would go unheeded, and the prophecy would be fulfilled. In a few years Malone would take the centre of the stage as the remorseless prosecuting attorney in the most sensational of all cases of Shakespearian forgery.

# 7

# *Stratford Visitors*

THE great edition published, Malone did not desist from Shakespearian labours. In the spring of 1793 he was in Stratford, where, with the vicar's permission, he, out of misguided neoclassical zeal, brought the bust 'back to its original state, by painting it a good stone-colour'.[64] By 1746, we learn from the memoranda of Joseph Greene, the monument had 'through length of years and other accidents become much impaired and decayed', and two years later it was restored and beautified. Malone flattered himself on having performed 'a public service' (so he described his action to Charlemont), but his intervention was ill-advised—the bust in its original state had been painted rather than stone-coloured—and he is not entirely undeserving of the epigram inserted by a pilgrim some years later in a Stratford *Visitor's Book*:

> Stranger, to whom this monument is shewn,
> Invoke the Poet's curse upon Malone;
> Whose meddling zeal his barbarous taste betrays,
> And daubs his tombstone, as he mars his plays![65]

No other act of Malone's did such lasting damage to his reputation as his tampering with the Stratford bust. There is a late nineteenth-century memorandum (by Halliwell-Phillipps) in the Bodleian to the effect that, for the very few who respected Malone's memory, thousands accepted the judgement of the *Visitors' Book* rhyme, which found its way into Wheler's *Guide to Stratford-upon-Avon* (1814) and was thence widely dispersed. Victor Hugo, in exile on Guernsey, remembered the desecration. 'An imbecile, Malone', he wrote in his *William Shakespeare*, 'made commentaries on his plays, and, as a logical sequence, whitewashed his tomb.'[66] (The bust suffered desecration again in 1973, when vandals dislodged it from its niche by chipping away some of the plinth. At the time the police, according to a local press report, speculated that these miscreants 'were seeking valuable manuscripts written by the Bard'. These were presumably unforthcoming.)

While in Stratford, Malone met with that self-appointed custodian of Shakespearian lore, John Jordan, whom we have already encountered several times. 'One of those humble geniuses, to whom a little learning, if not a dangerous thing, proved almost useless'—so a fellow Stratfordian sized up Jordan, who had received no education beyond his seventh year. This 'very honest fellow' and 'civil inoffensive creature' (as another contemporary describes him) did not scruple to impose upon the credulity of his betters with forged signatures of Shakespeare imitated from facsimiles. Yet a witness unduped by Jordan's wiles could nevertheless affirm that in everyday affairs he was scrupulously honest; of such contraries did his nature consist. As a wheelwright he fared poorly: his brother usurped the business, and he was reduced to doing journeyman work—during the summer, fifteen or sixteen hours a day for nine shillings a week—in the employ of his loathed sibling. To Malone he poured out his heart in grandiosely platitudinous phrases: 'alas I am unoticed by the World, oppressed with affliction and Wreck'd with despair, the anchor of hope has tottaly forsook me, and I am dashed with the Waves of a boundless sea of trouble, Sorrow, and misery, which brings to my mind an expression [of] Shakespere'.[67] What is remarkable about this outpouring is that it should have Malone's ear. Clearly the Stratford wheelwright had aspirations beyond his station.

Fancying himself a poet, and as such feeling a special affinity for his town's most celebrated son, he published in 1777 a poem on the legends surrounding the nearby Welcombe Hills. It met with little success, but by immersing himself in Shakespeariana, the Stratford Poet (as locals referred to him) did achieve recognition and employment as a *cicerone* to pilgrims visiting Bardic shrines. He had his quirks: after quarrelling with the owners of the Birthplace, he denounced the house as 'a most flagrant and gross imposition, invented purposely with a design to extort pecuniary gratuities from the credulous and unwary'. Instead Jordan championed the Brook House, on the banks of the Avon between the Bancroft and Southern Lane, and in a letter to the *Gentleman's Magazine* of April 1808 he had the impudence to describe that dwelling as the house 'in which it is generally admitted that Shakspeare was really born'. (Somewhat inconsistently, in a Shakespeare Birthplace Trust manuscript, he describes John Shakespeare's Spiritual Testament as having been found in a Henley Street house 'in which his son was born'.) Jordan completed his *Original Collections on Shakspeare and Stratford-on-Avon* around 1780, and, a decade later, *Memoirs and Historical Accounts of the Families, of Shakespeare (alias Shakespere) and Hart*. But these were so ill-digested and poorly arranged that no publisher would take them.

Lacking a scholar's equipment, Jordan contributed to the Shakespeare–Mythos by inextricably blending traditions, oral and written, with his own naïve embellishments. In the *Original Collections* he offers the improbable picture of a 'libertine, sportive young Shakespear, the leader of a loose rabble in a country town, oppressed probably with debt, and incumbered prematurely with a family'.[68] Jordan's genealogy for the poet consists mainly of inferences from documents and records he found in the Johnson–Steevens *Shakespeare*, but he furnishes him with a great-great-grandfather John and a cousin Anthony who resided at Hampton-in-Arden in Warwickshire. Malone listened to Jordan (who apparently sent him the manuscript of the *Original Memoirs*), corresponded with him, and raised a subscription to relieve his poverty;* but he did not believe all that he was told.

That summer in Stratford, Jordan had a more credulous audience in the Irelands, father and son, who passed a week there in his hands. Samuel Ireland, the elder, was an engraver who enjoyed some success. An Honorary Member of the Royal Academy, represented in an Academy Exhibition in 1784, he had won a popular following with his *Picturesque Tour Through Holland, Brabant, and Part of France* (1790), succeeded by similar volumes on the Thames (1792) and the Medway (1793), profusely illustrated with lifeless but realistically detailed and decorative engravings of landscapes, buildings, and monuments. Now Ireland was preparing his *Picturesque Views on the Upper, or Warwickshire Avon, from Its Source at Naseby, to Its Junction with the Severn at Tewkesbury*, eventually to be published in 1795. Through the drowsy summer, accompanied by his quiet and slow-witted son William-Henry, then eighteen,† he followed the river's meandering course.

Given the setting, it was natural that Shakespeare should never be far from his thoughts. In the park at Fulbrook, on a spot called Daisy Hill, he sketched a farmhouse which (he had heard) was anciently the keeper's lodge where the poet was supposedly conveyed for deer poaching; the picture of it which he included among his picturesque views was the first such to appear. In Stratford a mood of exaltation came over Ireland; for he adored the Matchless Bard with mystical fervour, and was unceasing in his praises, to which his son listened in silence. Ireland was delighted to meet Jordan. This tall, muscular man with weather-beaten face, peasant features (beetle-brow and

---

* Malone's kindly interest in Jordan had a predictably diuretic effect on the latter's Muse; his 'Panegyric on Some Worthy Men' contains the following lines:

> And ever honour'd be the dear MALONE.
> Long may he live with ev'ry wish possess'd
> And with encreasing joys be ever blest
> Ye pow'rs Coelestial hear my ardent pray'r
> And let him never be perplex'd with care.

(Shakespeare Birthplace Record Office MS 16, fo. 16)

† Later, after notoriety had overtaken him, William-Henry would claim that he was born in 1777; he would thus have been sixteen at the time of the Stratford expedition. His falsification, which has won uncritical acceptance in works of reference (e.g., *DNB*, *Shakespeare Companion*, *Shakespeare Encyclopedia*), served the twin functions of exaggerating Ireland's precocity and extenuating his guilt. His modern biographers, Mair and Grebanier, have preferred 1775, and their judgement is vindicated with satisfying precision by the Ireland family Bible, not hitherto consulted by students. According to his father's entry, S[am] William-Henry entered the world at one o'clock in the afternoon of Wednesday, 2 August 1775.

extensive mouth), and fuzzy black hair looked every bit the rustic repository of folk knowledge.

With Jordan the pair toured the town and neighbourhood. He took them to the shop run by that Sharpe, now advanced in years, who had acquired Shakespeare's mulberry; here Ireland bought, for 'an adequate price', a goblet and other knick-knacks carved from that sacred object. At Holy Trinity Church the Shakespearian ecstasy entered the boy as well; 'it would be impossible for me to describe the thrill', he recalled years afterwards, 'which then took possession of my soul'.[69] While the father sketched the monument, and also the effigy of John Combe, William-Henry pushed open the door leading to the charnel-house, and stared with awe at the vast heap of human bones. Together, father and son mused on the poet's gravestone malediction designed to prevent a similar fate from overtaking his own remains, and on the grave-diggers' scene in *Hamlet*. At the Birthplace they met Hart the butcher, descended from Shakespeare's younger sister Joan, who informed Samuel that 'he well remembered having, with other boys, dressed themselves as Scaramouches (such was his phrase) in the wearing apparel of our Shakspeare'.[70] Ireland made a meticulous sketch of the kitchen, later the basis for an engraving in his volume. On the grounds where New Place once stood, he was 'by no means reserved in the curses he heaped on the unfeeling leveller of the Mulberry Tree'.[71]

At Shottery, a mile from Stratford, Jordan introduced the visitors (and consequently the public at large) to the sturdy oak-beamed farmhouse standing at Hewlands Farm that was thenceforth to be known as Anne Hathaway's Cottage. The house, still in the possession of descendants of the Hathaways, was indeed probably the ancestral home of Shakespeare's wife, for her grandfather was described in local records as holding a house, with a half-yardland, called 'Hewland'. There Ireland bought a bugle purse and an old oak chair described to him as Shakespeare's courting chair. Nearby stood an ancient bed which aroused his acquisitive lust; but the venerable owner, who had slept in it since her childhood, unreasonably refused to part with it at any price. (Ireland did not give up easily; after the old lady died, he asked Jordan to find out what they would now take for the bed, '& let me know as soon as you can'.)

The visitor sketched the thatched house, and in his *Picturesque Views* published the first delineation of Anne Hathaway's Cottage. He would be the second to provide his readers with an engraving of the Elizabethan New Place, made, he relates, from a drawing of 1599 by Robert Treswell and found in Clopton House in 1786, but since lost. The view is a variation of Malone's: not surprising, for it derives from the same source, Jordan, who in his inimitable fashion supplied the two men with different pedigrees.

Jordan also gave his client the story of Shakespeare's convivial encounter with the sippers of Bidford and the night under the crab-tree, along with the text of the song the poet sang the next morning. This material, which Ireland copied down, would be duly incorporated in his book, although this is not a first: the anecdote, minus the circumstantial detail and the song, appeared anonymously in the *British Magazine* of June 1762, and Jordan's version (submitted by 'M.E.') in the *Gentleman's Magazine* in 1794. No one, however, would anticipate Ireland in furnishing a grateful public with a picture of Shakespeare's crab-tree, which is certainly a fine specimen.

However satisfying these acquisitions of information, anecdotes, and relics, Ireland yearned with all the passion of his bardolatry for manuscript remains that would shed light on Shakespeare's life and work. In response to persistent inquiries he learned, from some of the town's oldest inhabitants, that at the time of the great fire in Stratford, manuscripts were removed from New Place to Clopton House, about a mile distant from Stratford. Thither the party repaired, only to be told by the occupant, a churlish gentleman–farmer named Williams, 'By G–d I wish you had arrived a little sooner! Why, it isn't a fortnight since I destroyed several baskets-full of letters and papers, in order to clear a small chamber for some young partridges which I wish to bring up alive: and as to Shakspeare, why there were many bundles with his name wrote upon them. Why it was in this very fireplace I made a roaring bonfire of them.' 'My G–d!' cried Ireland, starting from his chair and clasping his hands in response to the incredible revelation, 'Sir, you are not aware of the loss which the world has sustained. Would to heaven I had arrived sooner!' Then Williams called in his wife and asked her about the baskets of papers. 'Yes, my dear,' the old lady replied; 'I do remember it perfectly well! and, if you will call to mind my words, I told you not to burn the papers, as they might be of consequence.' After expressing regrets, Ireland asked to inspect the chamber in question, where, however, he found only partridges.[72] Was Williams enjoying a philistine practical joke at the expense of an intrusive antiquary and transparently naïve bardolator? Many years later Ireland's son recalled (in a manuscript note) the 'plain gentlemanly farmer' as being 'to all appearance incapable of inventing a premeditated falsehood of this description'; and Malone, informed of the episode by Jordan, took the matter seriously enough to complain to Williams's landlord. But the reliable Stratford antiquary Wheler, in an unpublished document, reports that Williams confessed with a laugh that he had deliberately perpetrated a hoax upon his trio of credulous visitors.[73] In any event, Ireland's hunger for original Shakespearian manuscripts would soon be satisfied by the industry of young William-Henry who, as a dutiful child, was anxious to secure his father's content.

# 8
## *William-Henry Ireland: The First Impositions*

AT the Ireland house at 8 Norfolk Street in the Strand, the trophies of Stratford—Shakespeare's courting chair and the mulberry souvenirs—joined a curious collection that included a number of rare books and tracts (among them several Shakespearian quartos and a First Folio), paintings by Rubens and Van Dyck, and pictures by Hogarth; but also the cere cloth of a mummy at Rotterdam, hair from the heads of Edward IV and Louis XVI, and various articles of clothing or objects belonging to the great: a bit of Wyclif's vestment, Oliver Cromwell's buff leather jacket, part of a cloak belonging to Charles I, a garter worn by James II at his coronation, a crimson velvet purse presented to Anne Boleyn by Henry VIII, Sir Philip Sidney's cloth jacket embroidered with silk knotting, and a pocket knife used by Addison to pare his fruit. These books and relics Ireland acquired at stalls and junk-shops around the Strand and Chancery Lane, and from the homes of the great who needed quick cash; in turn he sold them, discreetly, to the cultivated and well-to-do who visited him on Norfolk Street. But collecting was more to Ireland than a gentlemanly way of earning his livelihood; it had the force of mania. Greedy of acquisition, he revelled in the sheer delight of possession. It is in this light that one must interpret his disappointment at Clopton House.

Under Ireland's roof in London lived, in addition to his son, a daughter Jane (an older daughter had married) and his housekeeper Mrs Freeman. When young, Mrs Freeman had, it appears, been kept by the Earl of Sandwich; now she was Ireland's mistress and the mother of his children, facts which his moral sensibilities led him to conceal—so successfully that one modern biographer has been led to create an imaginary Mrs Ireland whose marriage was left unrecorded because of Mrs Freeman's influence. Not surprisingly, young William-Henry wondered about his origins when not retreating into fantasies of knights and monasteries, or building pasteboard theatres. His initiation into the mysteries of poetry occurred on an excursion up the Thames, when his father pointed out Pope's villa and, having expatiated on the great man, patted the boy on the head and sighed, 'I fear you will never shine such a star in the hemisphere of literary fame';[74] it was an episode that William-Henry was never able to push out of his mind, for in the year before his death he recalled it in his marginalia. Eager to win approval from a parent who regarded him at best with indifference, at worst as a simpleton, he began, when old enough, to frequent the book-stalls, and was delighted when his father expressed astonishment at some rarity the boy had procured through chance or diligence. After dinner in this household of impeccable surface respectability, the chairs would be arranged in a circle, and the family would read aloud one of Shakespeare's plays, with Samuel in the leading part, and the children and Mrs Freeman taking the other roles. At times like these, Ireland would discourse enthusiastically on the passages that took his fancy, and he lost no opportunity to extol that poet who was a god among men. Although silent, the boy 'paid the greatest attention to every statement made by Mr. Ireland; thus gradually imbibing a similar fondness and veneration for every thing that bore a reference to the mighty father of the English stage.'[75]

On one occasion, when the Shakespearian diet was varied by a reading from Herbert Croft's epistolary novel *Love and Madness*, conversation turned to Chatterton, the subject of a long digression in the novel. Enthralled, young William-Henry later sought out himself the passage in the book. Thus he became acquainted with 'our Bristol Shakespeare', as described by an author so sympathetic that he could say of the word *forgery*: 'For Chatterton's sake, the English language should add another word to its Dictionary; and should not suffer the same term to signify a crime for which a man suffers the most ignominious punishment, and the deception of ascribing a false antiquity of two or three centuries to compositions for which the author's name deserves to live for ever.*[76] *Love and Madness* made a great impression: 'the fate of Chatterton so strongly interested me', William-Henry wrote after everything had collapsed, 'that I used frequently to envy his fate, and desire nothing so ardently as the termination of my existence in a similar cause.'[77] At the time of his disgrace, when he had every reason to be filled with regret, he would nevertheless be drawn to Chatterton's Bristol. There he would climb to the turret of St Mary Redcliff Church and peer into the empty old chests in which Chatterton had pretended to discover his Rowley poems; there he would minutely question Chatterton's sister about the circumstances of his child-hood—any remarkable symptoms he displayed—and about his appearance and character. He would affirm his reverence for the Marvellous Boy, but he would stop short at emulating the latter's destiny of arsenic poisoning—if indeed Chatterton had, as was thought, committed suicide: another theory holds that his death was accidental, the consequence of an overdose of medication for venereal disease; still another, that he died after ingesting oysters that fatal August. It was a bad month for oysters as well as for Chatterton.

* Years later William-Henry offered his own extenuation of the crime of forgery: 'Had Mr. S. Ireland at his age committed a fabrication of this species the word *crime* would justly apply to the whole transaction', he defended himself in marginalia preserved in the Huntington Library; 'but when it is considered that a boy was the writer without any vicious intention but whose only desire was to gratify a parent I should conceive that the case was widely different.'

If the example of Chatterton influenced the son, so too did Stratford leave an everlasting imprint on the father. 'Mr. S. Ireland's predilection for the name of Shakspeare seemed also to have increased by this visit to the birth-place of our mighty dramatist,' William-Henry recollected, a decade later: 'his encomiums were unceasing; and he would frequently assert, that such was his veneration for the bard that he would willingly give half his library to become possessed even of his signature alone.'[78] The son was then employed as a conveyancer's clerk in the chambers of one Mr Bingley of New Inn, to whom the elder Ireland had articled him as preparation for a legal career, for which William-Henry was ill-suited. When the boy began work there, Bingley kept a hackney-writer in constant attendance, and also a messenger, Foster Powell; but the hackney-writer was let go, and the messenger died. Thus William-Henry had many hours on his own, and found scope for his designs. His father's Shakespearian obsession prompted the young clerk to ransack all the ancient deeds scattered about the chambers, in hope of finding one with the poet's signature. This failing, he scouted the stalls of vendors of old paper and parchment, again without success.

His first experiment in counterfeiting involved a quarto tract, a set of prayers, dedicated to Queen Elizabeth, and bound in vellum with the Queen's arms stamped in gold on the cover. Ireland diluted some ink with water, and, on a piece of old paper, forged a dedicatory epistle to Elizabeth; then thrust it between the cover and the endpaper, which had come unstuck. Before presenting this prize to his father, he visited a bookbinder, Laurie, who ran a shop in New Inn Passage, a couple of minutes' walk from Bingley's chambers. There, in the presence of two journeymen, Laurie examined the letter (which the youth, with a smile, said he had devised to test his father's expertise), and avowed that it certainly *did* look antique. But one of the journeymen, studying the manuscript, declared that he could provide a mixture with a much closer resemblance to old ink. This concoction, blended from three different liquids used by bookbinders for marbling the covers of calf bindings, Ireland would employ for all his subsequent fabrications. It was a dark brown fluid that dried to a light tone; but when the paper to which it had been applied was held before a fire, the ink gradually darkened once again. Hence the scorched appearance of the Ireland forgeries, for, fearful of discovery in Bingley's chambers, the counterfeiter sometimes brought the paper too close to the flames. William-Henry paid the journeyman for his trouble, rewrote the dedicatory letter, and presented the tract to his father, who did not question its authenticity. Nor did he wonder why the Queen should have so oddly interposed the epistle between unglued paper and cover. The path was now clear for the major deceptions.

From his father's bookshelf William-Henry took down the volume of the Malone *Shakespeare* with the facsimiles of the poet's signatures, and made tracings of them. The text of the Blackfriars mortgage deed, in the same source, furnished a model for legal phraseology. (Later, in his confessions, he would say that he used the Johnson–Steevens edition, but his motto—'the whole truth, and nothing but the truth'—is perhaps best received as unintended irony. Malone, to whom understandably he would give credit for nothing, first printed the Blackfriars document in his *Shakespeare*. Before revising his recollection Ireland acknowledged his debt in a manuscript addressed to Albany Wallis and also in an unpublished memoir that now forms part of the Hyde collection.) When alone in the conveyancer's office he cut off a piece of parchment from an old rent-roll, and placed before himself a deed from the reign of James I. Then, imitating the penmanship as faithfully as possible, he drew up an agreement, dated 14 July 1610, between 'William Shakespeare of Stratford on Avon in the County of Warwick Gent but now residynge in London and John Hemynge of London Gent of thone Pte and Michael Fraser and Elizabeth hys Wife of the othere Pte'. On this document, with his tracings before him, he entered Shakespeare's name, and, with his left hand, that of Michael Fraser. Now he was faced with his most difficult problem: attaching the seals, which in Jacobean times were fashioned from malleable wax and stamped upon narrow pieces of parchment that hung from the deed beneath the signatures. The old seals he had hoped to use crumbled when heated. At length he hit upon the expedient of heating a knife, with which he managed to cut in two an ancient seal without its disintegrating; he then scooped out a hollow on the side opposite to that bearing the impression, inserted the parchment strip, and filled in the remaining cavity with fresh, heated wax. Then, to make the colour uniform, he rubbed the seal with soot and coal ashes.

Having prepared the deed, he had now to prepare his father for the revelation, and to anticipate questions about the provenance of the document. So he fashioned a tale, and on 2 December 1794 related it to Ireland Senior. Some ten days previously, he announced, he had dined at the house of Mr Mitchell the banker, an acquaintance of Samuel's, and had there met a propertied young gentleman, very rich, who gathered from the lad's conversation that he had 'a great predilection for every thing like antiquity'. (Later, carelessly, William-Henry would say that the encounter had taken place in a coffee-house.) The gentleman disclosed that he possessed many old papers which had been in the family for a century and a half; some might interest the youthful antiquary, and if so he was welcome to them. A date was appointed for a visit to the gentleman's chambers, but suspecting that he was being made fun of, William-Henry failed to appear. Some mornings later, however, he had second thoughts as he passed the gentleman's residence, and he went up. After mild reproof, and an apology on the visitor's part, he was shown a great chest (he did not say such a chest as Chatterton had recourse to) standing at the rear of the room. Instantly he began rummaging through the bundles of papers within, and in a few hours uncovered a deed with Shakespeare's signature. This he showed to the gentleman, who remarked, with well-bred equanimity, that 'it was certainly a very curious instrument, but that having promised me every thing I should find worthy my notice, he would not be worse than his word.'[79] There were provisos, however: the deed must not be removed until he had had an opportunity to examine it, which would take ten days, for he had affairs to attend to at his country house; and, secondly, his identity and address must be kept secret. These whims of an aristocrat the boy of course swore to honour.

Samuel Ireland had now to endure a fortnight of suspense. The quaintly formal exchange of civilities that took place between father and son when the forged document was at last presented to him, on 16 December, is best described in the donor's own words:

It was about eight o'clock, being after my evening's attendance at chambers, that I presented the deed in question. Mr. S. Ireland's family were present; and, if I mistake not, another person;—the fact being precisely as follows:—I had placed the deed within my bosom; when, after informing Mr. Ireland that I had a very great curiosity to show him, I drew it forth and presented it, saying—'There, sir! what do you think of that?' Mr. Ireland, opening the parchment, regarded it for a length of time with the strictest scrutiny: he then examined the seals; and afterwards proceeded to fold up the instrument; and on presenting it to me he replied—'I certainly believe it to be a genuine deed of the time.' Returning it immediately into Mr. Ireland's hand, I then made answer—'If you think it so, I beg your acceptance of it.'[80]

Touched, the father insisted upon presenting his son with a copy of Stokes's *Vaulting Master*; it is scarcely surprising that in his excitement he should have forgotten his previous offer of half his library for a Shakespeare signature.

Two days later Samuel Ireland took the deed to the Heralds' Office, where the Heralds agreed as to its authenticity but were unable to decipher the devices on the seals. He thereupon sent for his young acquaintance Sir Frederick Eden, then under thirty. Sir Frederick was chairman of the Globe Insurance Company and a literary dilettante; more relevant to Ireland's needs, he made a hobby of old seals. He examined the document, emphatically affirmed its validity, and furthermore pointed out that the impression on the seal under Shakespeare's signature represented a quintain—a specially constructed post for tilting at with a lance. Obviously the device bore an analogical relationship to the name Shakespeare, 'and from that moment the Quintin was gravely affirmed to be the seal always used by our monarch of the drama'.[81] William-Henry was impressed, for he had never before heard of a quintain.

Antiquaries now flocked to Norfolk Street to inspect the deed. Soon they began to hint that the same source might yield other Shakespearian treasures; the hint became a suggestion, and the suggestion pressure. Ireland Senior, for the first time attentive to his son, pleaded with him to renew his search, and even taunted him for allowing the priceless opportunity to go by.

In his enthusiasm he was merely expressing, more urgently than others, the hunger of the age for knowledge of Shakespeare's life. Malone, through assiduous application, had uncovered much, but these documents and records left untouched the central core of the mystery: the character and spirit and daily life of the greatest of poets. Where were the papers that alone could provide the knowledge which all who loved literature craved? Boaden, decades later, expressed well what many in Ireland's generation felt: 'It was a subject of infinite surprise to the admirers of Shakspeare's genius, to observe from age to age, that while discoveries, very material to our knowledge of the period in which he lived, occasionally occupied the press, yet that with respect to himself little could be known; and all the effusions that friendship or business must have poured from his pen during a town life, and the reasonable produce of his retirement from a mind so essentially active ALL, as if collected together in one mass destroyed BY AUTHORITY, had vanished away, and were entirely lost to posterity.'[82] Yet Shakespeare had not lived in obscurity, nor had he died poor, but was surrounded by love and admiration. The great of the age—Elizabeth, James, Essex, Southampton, and the two Pembrokes—had shown him favour; Jonson and his circle were intimate associates. Nevertheless not a single letter signed by him had come down. Surely Shakespeare's manuscript remains had been collected by 'some affectionate hand'? Surely a day would come when 'a rich assemblage of Shakspeare papers would start forth from some ancient repository, to solve all our doubts, and add to our reverence and our enjoyment'.[83] Even the prudent Malone shared this illusory hope. The prospectus he issued in 1795 for a new edition of Shakespeare declared, in what was in effect a plea, 'He has always thought that much Information might be procured, illustrative of the history of this extraordinary Man, if Persons possessed of ancient Papers would take the trouble to examine them, or permit others to peruse them.' Perhaps the manuscripts had been inherited and passed down by Lady Bernard, who appointed Edward Bagley of London her executor; or perhaps some of the papers had come into the hands of the daughters (by a former wife) of Sir John Bernard, who died in 1674; maybe others were among the effects of a descendant of Heminges.

In this context of optimistic expectation William-Henry Ireland obliged curiosity by producing the Shakespeare Papers. By so doing he created a personal history for the poet, as conceived by a day-dreaming child of the eighteenth century, at that time when the neoclassic temper had already largely yielded to romanticism; a personal history fabricated by a youth regarded as stupid by his elders, but which many mature men of the age would be able to reconcile with the Shakespeare of their sentimental imagination.

Ireland's second document—more daring than the mortgage deed—ventured upon Shakespearian phrasing and handwriting, but kept within the conventions of a legal transaction. This item is a promissory note purportedly given by the dramatist to John Heminges, with a receipt of payment affixed to it with wax and bearing Heminge's name as signed by Ireland with his left hand:

One Moneth from the date hereof I doe promyse to paye to my good and Worthye Freynd John Hemynge the sume of five Pounds and five shillings English Monye as a recompense for hys greate trouble in settling and doinge much for me at the Globe Theatre as also for hys trouble in going downe for me to statford Witness my Hand

Wm Shakspere

September the Nynth 1589

Received of Master W^m Shakspeare the Sum of five Pounds and five Shillings good English Money thys Nynth Day of October 1589

Jn° Hemynge[84]

Such phrases as 'hys greate trouble' or 'hys trouble' were apparently congenial to the author, for they would recur. The reference to the Globe Theatre, not constructed until 1599, is regrettable. Ireland's orthography has not yet taken its Chattertonian flight, but the persistent 'y''s augur well in this respect. It is odd that Shakespeare should misspell the name of his native town; odd too, for that matter, that he should compose a promissory note, for none from this period has survived. Also curious, for a professional writer, is the total aversion to punctuation, a deficiency even more striking in the longer documents to follow. However puerile, the note and receipt fulfilled their function of demonstrating that 'Shakespeare, in addition to his other good qualities, was very punctual in all pecuniary transcations.'[85]

# 9

# *Ireland: The Major Impositions*

GRADUALLY the subject of the fabrications extended his dominion over the fabricator's consciousness. With the third Shakespeare counterfeit, produced two days after the promissory note, Ireland passed from ordinary technical duplication—the forger's prosaic craft—to inspired composition. As usual an occasion prompted the experiment. His father's friend, the Hon. John Byng who was the proud possessor of a crossbar from Shakespeare's chair, would often dwell on Southampton's bounty to the poet, and express a longing for documentary information about the sum donated by His Lordship. From this hint Ireland, with his tracings of the Shakespeare signatures before him, fabricated an exchange between patron and recipient. That afternoon he committed his thoughts to paper unpremeditatedly, just as they entered his mind; but Byng would have to remain content with his crossbar, for the counterfeiter prudently refrained from mentioning figures, in the event that they would one day be contradicted by a genuine document. If Ireland's Muse is liberated in his latest venture, so too is his orthography, although this would become still more luxuriant:

> Mye Lorde
> Doe notte esteeme me a sluggarde nor tardye for thus havynge delayed to answerre or rather toe thank you for your greate Bountye I doe assure you my graciouse ande good Lorde that thryce I have essayed toe wryte and thryce mye efforts have benne fruitlesse I knowe notte what toe saye Prose Verse alle all is naughte gratitude is alle I have toe utter and that is tooe greate ande tooe sublyme a feeling for poore mortalls toe expresse O my Lord itte is a Budde which Bllossommes Bllooms butte never dyes itte cherishes sweete Nature ande lulls the calme Breaste toe softe softe repose Butte my goode Lorde forgive thys mye departure fromme my Subjectte which was toe retturne thankes and thankes I Doe retturne O excuse mee mye Lorde more at presente I cannotte
>
> > Yours devotedlye and withe due
> > respecte
> > W^m Shakspeare

Southampton's reply, addressed 'To the Globe Theatre forre Mast^r William Shakspeare', reads:

> Deare William
> I cannotte doe lesse than thanke you forre youre kynde Letterre butte Whye dearest Freynd talke soe muche offe gratitude mye offerre was double the Somme butte you woulde accepte butte the halfe thereforre you neede notte speake soe muche onn thatte Subjectte as I have beene thye Freynd soe will I continue aughte thatte I canne doe forre thee praye commande mee ande you shalle fynde mee
>
> Julye the 4                                    Yours
>                                                Southampton

Not until he came to fold up the first letter and address it did Ireland realize that, if it was sent to Southampton, it would be unlikely to find a place among Shakespeare's papers. But this difficulty was easily overcome; he added, as a superscription, 'Copye of mye Letter toe hys grace offe Southampton'. William–Henry was less careful about Southampton's hand, which he merely tried to have contrast as much as possible with Shakespeare's; hence the wretched scrawl made with his left hand. He did not then know that manuscripts in Southampton's small, tidy penmanship existed. Meanwhile, however, the documents accomplished their naïve purpose. In showing Southampton's generosity with money, and Shakespeare's generosity of spirit in accepting only half the proffered gift, they brought to light another aspect of the character of an eighteenth-century gentleman born two hundred years before his time.

Ireland's next project was grander. The recent publication, in Malone's *Shakespeare*, of John Shakespeare's Spiritual Last Will and Testament had encouraged speculation that the son shared the father's apparent Catholicism; speculation which was reinforced by allusions to Purgatory on the part of the Ghost in *Hamlet*. As the forger was a staunch Anglican who righteously felt 'the most rooted antipathy to every thing like superstition and bigotry', he would banish forever the idea of Shakespeare's Catholicism 'by making the profession of faith appear to be written by a sincere votary of the protestant religion'.[86] The Profession, for which Ireland Senior had been prepared in the customary way by advance notice of the wonders it would hold, contains the following remarkable apostrophe:

O Manne whatte arte thou whye considereste thou thyselfe thus gratelye where are thye greate thye boasted attrybutes buryed loste forre everre inne colde Deathe. O Manne whye attemptest thou toe search the greatenesse offe the Almyghtye thou doste butte loose thye labourre more thou attempteste more arte thou loste tille thye poore weake thoughtes are elevated toe theyre summite ande thence as snowe fromme the leffee Tree droppe ande disstylle themselves tille theye are noe more O God Manne as I am frayle bye Nature fulle offe Synne yette greate God receyve me toe thye bosomme where alle is sweete contente ande happynesse alle is blysse where discontente isse neverre hearde butte where oune Bonde offe freyndshippe unytes alle Menne Forgive O Lorde alle oure Synnes ande with thye grete Goodnesse take usse alle to thye Breaste O cherishe usse like the sweete Chickenne thatte under the coverte offe herre spreadynge Wings Receyves herre lyttle Broode and hoeveynge oerre themme keepes themme harmlesse ande in safetye

The word *leffee*, after to come under heavy attack by Ireland's opponents, was merely a careless slip for *leafless*; but at first nobody noticed.

The father was impressed to know that William-Henry had memorized this sublime piece of writing, and repeated it morning and evening in his prayers. Nevertheless he decided to consult authorities. Thus he called in the Revd Dr Samuel Parr, a well-known and pugnacious pedagogue, and Joseph Warton, the editor of Pope described by the *Dictionary of National Biography* as 'a learned and sagacious critic'. For the first time the youth had to face the terrors of expert interrogation. He answered the questions put to him about the discovery of the manuscripts and the concealment of the gentleman's identity, then sat motionless while his father recited the Profession of Faith, with the two guests fiercely intent on each syllable. The reading done, Warton turned to Samuel and, raising his spectacles on to his forehead, exclaimed energetically, 'Sir, we have had very fine passages in our church service, and our litany abounds with beauties; but here, sir, here is a man who has distanced us all!'[87] The sweete Chickenne had come home to roost. Fully conscious of the irony of the situation, and of the implications of the praise, William-Henry absented himself from his father's study, and, cooling his head against the window-pane in the dining-room, felt a new thought possess him like a flame. He was a genius.

The genius grew reckless. In a delirium of exhilaration he heightened his father's cupidity by promising undeliverable wonders from the gentleman's country house: a seal of cornelian stone set in gold and bearing the intaglio of the quintain; two uncut First Folios of Shakespeare's plays; a full-length portrait of the poet in black draperies, holding a long fringed glove, painted on board from life. Now the boy would be pestered unmercifully for these concrete relics that would eradiacte any scruples about the authenticity of the Shakespeare Papers.* Thus he was driven to mollify Samuel by providing an earnest of treasures to come.

He turned out a crude pen-and-ink sketch of the poet, based on Droeshout, with an equally crude background of Shakespeare signatures, arms brandishing spears, antic faces, and shields with the Shakespeare arms. But his father, who understood draftsmanship better than old documents, dismissed the work as ridiculous and childish. Stung by this rebuff, and fearful lest doubt respecting one document extend itself to others, William-Henry the next day redeemed himself by producing the letter from Shakespeare to Richard Cowley in which the sketch had been enclosed:

Worthye Freynd
Havynge alwaye accountedde thee a Pleasaynte ande wittye Personne ande oune whose Companye I doe muche esteeme I have sente thee inclosedde a whymsycalle conceyte whiche I doe suppose thou wilt easylye discoverre butte shoudst thou notte whye thenne I shalle sette thee onne mye table offe *loggerre heades*
                                                                Youre trewe Freynde
Marche
nynthe

                                            W^m Shakspeare


The letter allayed doubt and prompted minute scrutiny of the drawing for clues to the interpretation of the 'whymsycalle conceyte', which was however so profound as to elude detection.

The forger had less difficulty in imposing on the believers an old colour drawing he had picked up in Butcher Row. It depicted, on one side, a young man in Jacobean costume; on the other, a bearded old Dutchman with hands in his pockets. Ireland retouched the face of the gallant along lines suggested by the Folio engraving, and improved the backdrop by the addition of

---

* These, Samuel Ireland wrote to Jordan on 23 March 1795, were now being prepared for the public, and he added, with characteristic indiscretion: 'We shall likewise have a whole length Portrait of him in oil as large as life.'

Shakespeare's initials, the titles of a few of his plays, and his arms (with the spear unfortunately reversed and incorrect colours). By incorporating a knife and pair of scales in the picture on the reverse, he implied that the figure was Shylock. The enthusiasts of Norfolk Street identified the young man as Shakespeare in the part of Bassanio, thus fitting another piece into the jigsaw of the great man's life, and it was conjectured that the drawing once graced the green-room of the Globe Theatre. Mr Hewlitt of the Common Pleas office, an authority on old handwriting, detected with the aid of magnifying glasses faint traces of the signature of John Hoskins, a painter of the time of James I. Nobody else could see the name, but it shows very clearly in the facsimile eventually published by Ireland Senior.

These artistic endeavours did not distract the forger from his literary productions. Some of the documents unveiled in 1795, it is true, are minor and prosaic: agreements between Shakespeare and the players Condell and Lowin, by which their services are retained at the fee of a guinea a week; a note recording payment to Shakespeare of the extraordinary sum of fifty pounds for 'oure Trouble' in playing before Leicester in his house; another acknowledging the reimbursement for 'oure greate Trouble' and expenses 'inne gettynge alle inne orderre forre the Lord Leycesterres Comynge'. One of the Leicester receipts William-Henry dated 1590—a mistake, for that noble-man was already dead, as Samuel Ireland instantly recognized when the document was presented to him. Panic-stricken, William-Henry offered to burn the paper at once, but his father refused: perhaps the receipt was dated on some later occasion, or confused with another. Finally they agreed that the date should be torn off. 'This is the only instance', William-Henry later recalled (in the Hyde manuscript), 'wherein I conceive that Mr. Ireland was absolutely blind for the business was then so palpable that had any one taxed me boldly at that moment I could not have resisted but made a full disclosure. If Mr. I——— suspected the Truth he ought to have checked the business, if he did not his silence was willing and he was the most believing of all believers.'[88] *If Mr. I——— suspected the Truth*: thus does the son raise the overwhelming question. At this remove of time we must share his bafflement, although Samuel Ireland's complicity in the removal of the date reveals him as a less ingenuous dupe than his modern biographers have surmised.*

It did not occur to William-Henry that Shakespeare could scarcely have established himself as a dramatist so early, but he was conscious that the sheer mass of documents would argue their validity. Ireland had heard that small slips of paper, such as the Leicester memoranda, were in this period kept

---

* Significantly William-Henry recast this episode for his published *Confessions*, excluding his father altogether and making himself the discoverer of the date of Leicester's death. Malone, reading the *Confessions*, knew better and scrawled this note alongside the passage: 'This whole statement is *false*. The father was privy to the tearing off the date; as this writer has acknowledged *under his hand*' (British Library copy). Had Malone seen the Hyde manuscript?

tied together in bundles, so, in his quest for verisimilitude, he tore off a loose-hanging segment from a decayed tapestry in the House of Lords, and unravelled it for string.

These papers, which discover Shakespeare as an efficient business adminis-trator, the youth now began to supplement with volumes from the drama-tist's library, with his own annotations. One such bit of marginalia, in a tract giving a circumstantial account of the execution of Guy Fawkes, testifies to the poet's humanity: 'hee hadd beene intreatedd bye hys freynde John Hemynges to attende sayde executyonne, butte thatte he lykedde notte toe beholde syghtes of thatte kynde.'[89] Another note, in the text of a sermon preached by Lancelot Andrewes, illustrates Shakespeare's critical acumen:

Of Poetrye I ha reade muche goode ande muche badde Playes I ha reade manye goode manye badde badde Sermonnes toe I ha reade o bothe Qualytyes butte nere dydde I before reade soe muche offe badde withe soe lyttelle goode synce William Shakspeare hathe beene mye name ande soe fare thee welle goode Byshoppe o Chychesterre.[90]

Meanwhile William-Henry's father importuned him for Shakespeare's *Holinshed*, which the boy said he had seen at the country house of Mr H. (as he now called him), but regrettably all his wanderings among the book-stalls did not procure him a copy with margins wide enough to accommodate Shakespeare's annotations. On his own, however, he produced a catalogue of Shakespeare's library, compiled by the owner, with the dates of most of the items considerately provided. Shakespeare presumably prepared the cata-logue—now at University College, London, Library—late in life, for it includes two books from 1613. A hole obliterates the date of the *Shepheardes Calenderre*—Ireland had brought the paper too close to the fire.

Byng, whose curiosity about Southampton had inspired William-Henry, also wondered whether the papers might not contain the letter which King James was said to have written to Shakespeare. William-Henry did not dare undertake anything so obvious, and, besides, he was unacquainted with James's hand; so instead he produced a letter from the Queen. Poor Byng: he never got quite what he wanted. Elizabeth's cosy epistle, addressed—as though for delivery by the postman—to 'Master William Shakspeare atte the Globe bye Thames', acknowledges receipt of some 'prettye Verses' by 'goode Masterre William', and reminds him to bring his best actors to Hampton to amuse her, for 'the lorde Leiscesterre wille bee withe usse'. That the poet and his monarch were on terms of such intimacy when he was hardly twenty-four is heartwarming knowledge; moreover, it disposes of traditions about Shakespeare's humble beginnings in the theatre. 'We think it clearly proves', declared Boaden, then a Believer, in the *Oracle*, 'that all this degrading nonsense of him holding horses, etc., will be found utterly fictitious, and that this great man was the Garrick of his age, caressed by everyone great and illustrious.' Like so many others he did not pause to ask how a letter should be addressed to Shakespeare at a playhouse not built until ten years after Leicester's death.

But the most affecting recovery was a love letter addressed to Anna Hatherrewaye, and accompanied by a lock of Shakespeare's hair—short, straight, and wiry, as in the Droeshout portrait—tied with thick woven silk removed from a royal patent of the time of Henry VIII or Elizabeth, which Ireland had purchased from Yardley, a vendor of old parchment in Clare Market. In this ardent outpouring the soul of the poet speaks:

Dearesste Anna

As thou haste alwaye founde mee toe mye Worde moste trewe soe thou shalt see I have stryctlye kepte mye promyse I praye you perfume thys mye poore Locke with thye balmye Kysses forre thenne indeede shalle Kynges themmeselves bowe ande paye homage toe itte I doe assure thee no rude hande hathe knottedde itte thye Willys alone hathe done the worke Neytherre the gyldedde bawble thatte envyronnes the heade of Majestye noe norre honourres moste weyghtye wulde give mee halfe the joye as didde thysse mye lyttle worke forre thee The feelinge thatte dydde neareste approache untoe itte was thatte whiche commethe nygheste untoe God meeke and Gentle Charytye forre thatte Virrtue O Anna doe I love doe I cheryshe thee inne mye hearte forre thou arte ass a talle Cedarre stretchynge forthe its branches ande succourynge the smallere Plants fromme nyppynge Winneterre orr the boysterouse Wyndes Farewelle toe Morrow by tymes I wille see thee tille thenne

Adewe sweete Love

Thyne everre
W$^m$ Shakspeare

Enclosed with the letter and the lock was a poem, the quality of which may be sufficiently indicated by a single stanza (there were five in all):

> Is there inne heavenne aught more rare
> Thanne thou sweete Nymphe of Avon fayre
> Is there onne Earthe a Manne more trewe
> Thanne Willy Shakspeare is toe you.

This discovery stirred much enthusiasm for the delicacy of the passion revealed. Samuel was delighted to find in the letter an anticipation of the passage in *3 Henry VI* about the cedar protecting low shrubs from winter's powerful wind. Small quantities of hair were taken from the sacred lock and set in gold rings to adorn the fingers of the faithful.

Meanwhile Ireland Senior was seeing his *Warwickshire Avon* through the press, and it naturally occurred to him to announce, in such a volume, the manuscript treasures that he intended shortly to lay before the public. Prudently, however, he decided to send proofs of his Preface to Mr H.—a move which initiated a bizarre correspondence between Ireland and the imaginary gentleman. For William-Henry, who had already assumed the identity of the Immortal Bard, it was no great trick to join the wealthy landed gentry; Mr H. now became the outlet for all his megalomaniac fantasies and the intercessor between himself and an aloof parent. With incredible fool-hardiness, William-Henry made no attempt to disguise his normal hand. Even more incredibly, his father never recognized his own son's penmanship.

The exchanges show the correspondents at cross purposes. Samuel pressed his practical interests: discreetly reminding Mr H. of the promised full-length portrait of Shakespeare; requesting more documents, 'not only papers but pictures drawings &c' that would shore up the evidence for authenticity; seeking, above all, to meet the mysterious benefactor who alone could allay doubts regarding the provenance of the papers. For his part, Mr H. preferred to dwell on the virtues of his youthful protégé. 'It may appear strange', he confessed with an impulsive familiarity which he excused,

that a Young Man like myself shou'd have thus form'd a friendship for one whom he has so little knowledge of but I do assure you D*r*. *Sir* without flattery he is the young Man after *my own Heart* & in whom I wou'd confide & even consult on the nicest affair.[91]

In succeeding letters Mr H. became more familiar still. He advised Ireland on how William-Henry should wear his hair ('with flowing locks'); he tried to persuade Samuel—as William-Henry in his own person could not—that his offspring, far from being a blockhead, possessed marvellous literary gifts:

He tells me he is in general look'd upon as a young man that scarce knows how to write a good Letter you yourself shall be the judge by what follows—I have now before me part of a *Play* written by *your Son* which for stile & greatness of thought is equal to any one of Shakspeares. . . . He has chosen the subject of *Wm* the Conqueror & tells me he intends writing a series of Plays to make up with Shakspeares a compleat history of the Kings of England. . . . Mr. I—upon my honour & Soul I would not scruple giving £2000 a Year to have a son with such extraordinary faculties If at 20 he can write so what will he do hereafter. The more I see of him more I am amaz'd—If your *Son* is not a second Shakspeare I am not a *Man*.

Lest the point be missed, he subjoins a postscript that would have aroused suspicion in anyone less willingly credulous than Ireland: 'Your *Son* is brother in Genius to Shakspeare & is the only Man that ever walkd with him hand in hand.'[92] With the letter Mr H. enclosed, for Ireland's perusal, a speech from William-Henry's play—no doubt in the same hand as the covering encomium. In his reply the proud possessor of Shakespeare's amatory verses protested in astonishment that 'it is the first specimen of my Sons poetical talent I have ever yet seen'. But Samuel soon returned to his obsessive theme. 'If it is your wish to remain unknown to the Public—may I', he pleaded, 'without intrusion on your friendship request to have an interview on the most private ground imaginable.'[93] If the son could be trusted with the momentous secret, why not his father? It is perhaps unnecessary to add that no interview took place.

Young Ireland lived in constant danger, but somehow—for a time—

survived. A visitor dropped the Shakespeare–Fraser deed on the father's mahogany desk, breaking the seal and revealing the two waxes used; but William-Henry instantly tied the two parts together with black silk. Thus it was displayed—luckily no one thought to inspect it—until the forger could frame an excuse to remove the deed to the chambers, where he cemented the parts firmly together. To keep up his production he required paper. At a bookshop in St Martin's Lane the owner, Mr Verey, let him, for five shillings, snip out all the flyleaves from his old folios and quartos. William-Henry had heard of the prevalence of jug watermarks in Elizabethan times, and so he selected sheets with this mark, but he cunningly interspersed them with blank leaves to avert suspicion. Then his ink gave out, and at a time when the Shakespeare Papers were the subject of public curiosity and journalistic comment, he had to confront Laurie's journeyman again and give him a shilling for a fresh supply. He took impulsive risks. The letter from Elizabeth was fabricated in the presence of the charwoman; he thrust the completed document into her hands and asked whether she would not have thought it very old. Yes, she laughed, adding *'that it was very odd I could do such unaccountable strange things'*.[94] A graver threat was posed by his friend Montague Talbot, another young conveyancer's clerk, who shrewdly sensed what William-Henry was up to, but had no proof because of the youth's watchfulness. One day, however, while Ireland sat at his desk in front of the window, occupied with his afternoon forgery, Talbot doubled over outside underneath the window, then darted unobserved into the office, and grabbed Ireland's arm as he tried to conceal the unfinished document. Now Talbot as well as the charwoman knew, and others—Verey, Laurie, the journeyman— had reason to suspect him.

But nobody gave him away, and William-Henry kept up his furious production. He actually began to supply the manuscript plays he had promised his father. First came the *Tragedye of Kynge Leare*, which he adapted from the Second-Jaggard-Quarto purportedly of '1608', but actually published in 1619, and recently acquired by Samuel. In Ireland's version the play is happily purged of the obscurities and ribaldries foisted on it by the players. *Leare* was followed by a few leaves from *Hamblette*, in which no fair thought lies between Ophelia's legs; Ireland was a highly moral forger. But his most ambitious enterprise was an entire historical tragedy, previously unknown, on the subject of Vortigern and Rowena. This he passed along to his father in batches of a few pages, and to save time he did not bother to feign an antique hand: the gentleman, Samuel was informed, refused to part with the original before a complete transcript was made. By a pleasant coincidence there hung in the study on Norfolk Street an engraving by Ireland himself of Vortigern and Rowena.

In early February 1795 Ireland threw open his doors to the public, and a procession of visitors—not just a trickle of antiquarian acquaintances—filed into the study. So great was the press that Samuel was forced to limit admission to ticket-holders (subscribers to a forthcoming edition of the Papers) and their guests. Some came away sceptical. Henry Bate Dudley, the Fighting Parson of Grub Street, sneered in his *Morning Herald* (17 February) at the 'antique MELANGE of *love letters!—professions of faith!—billets-doux!— locks of hair!—*and *family receipts!—*The only danger, as to *faith in the discovery*, seems to be from the indiscretion of *finding too much!*' The Marquis of Salisbury left Norfolk Street sagaciously shaking his head, but held his peace. Others responded with suitable veneration. Boswell examined the manuscripts admiringly, after which, finding himself thirsty, he requested a tumbler of warm brandy and water. Thus refreshed, he redoubled his praises of the Papers. Then, rising up from his chair, he said, 'Well, I shall now die contented, since I have lived to witness the present day'; and, kneeling before the Papers, he went on: 'I now kiss the invaluable relics of our bard: and thanks to God that I have lived to see them!'[95] After reverently kissing a volume containing some of the documents, he left. Three months later he was dead.

Of all those who with trembling awe held the Shakespeare Papers in their hands, none became more enthusiastic a partisan than Francis Webb, secretary at the College of Heralds. 'These papers', he wrote to a friend,

bear not only the Signature of his hand; but the Stamp of his Soul, & the traits of his Genius—his Mind is as manifest, as his hand. . . . They [the Papers] exhibit him full of Friendship, Benevolence, Pity, Gratitude, & Love. The milk of human kindness flows as readily from his Pen, as do his bold & sublime descriptions.—Here we see the Man, as well as the Poet.[96]

A copy of Webb's letter forms part of the Samuel Ireland collection at the British Library. Did he show it to his son?

To Norfolk Street also came the awesome Joseph Ritson, not yet mad. William-Henry felt a dread he had never before experienced as he stood by while the great scholar—sharp-featured, with piercing eye—studied the Papers. His questions were pointed and laconic. After completing the examination he departed without a word, and Ireland knew he knew. 'The Shakspeare papers, of which you have heard so much, & which I have carefully examined', Ritson wrote to a friend in Edinburgh, 'are, I can assure you, a parcel of forgeries, studiously & ably calculated to deceive the public. . . .'[97] But publicly he said nothing, and many more trusted than doubted. Indeed, at the instigation of the assertive Parr, a distinguished band of the faithful put their names to a Certificate of Belief. Boswell (who again fell to his knees) signed; so too did Byng; Isaac Heard, Garter King-of-Arms; the Whig leader Lauderdale; the young Duke of Somerset; Sir Herbert Croft, author of *Love and Madness*; Henry Pye, unembarrassed at being Poet Laureate; and a dozen others, including Parr.

Several of the visitors to Norfolk Street suggested that, if a descendant of Shakespeare were found, he might well lay claim to the papers. This possibility distressed William-Henry: 'It would be hard indeed', he considered, 'that my own productions should go into the hands of an utter stranger.'[98] Hence his stratagem to show that Shakespeare had been closely associated with an Ireland, to whom he had incurred a weighty obligation and to whom, therefore, he might reasonably bequeath his papers. It had recently come to light that in December 1604 Henry Walker, 'citizen and minstrel of London', had leased his Blackfriars Gate–house—a property purchased by Shakespeare a decade later—to William Ireland, 'citizen and haberdasher of London'. In the conveyance of 10 March 1613 the gate-house is described as 'now or late being in the tenure or occupation of one William Ireland or of his assignee or assigns'. The yard where the property had stood was called, in Samuel's time, Ireland Yard. But who would have guessed that this person—whose middle name, oddly enough, was Henry— had saved Shakespeare's life? Let the dramatist's Deed of Gift, written in his own hand on 25 October 1604, tell the story of how he came to leave his manuscript plays, the property of his company, to William Ireland and his heirs forever:

[O]nne or abowte the thyrde daye of the laste monethe beyng the monethe of Auguste havynge withe mye goode freynde Masterre William Henrye Irelande ande otherres taene boate neare untowe myne house afowresayde wee dydd purpose goynge upp Thames butt those thatte were soe toe connducte us beynge muche toe merrye throughe Lyquorre they didd upsette oure fowresayde bayrge alle butte myeselfe savedd themselves bye swimmyng for though the Waterre was deepe yette owre beynge close nygh toe shore made itte lyttel dyffyculte for themm knowinge the fowresayde Arte Masterre William henrye Irelande notte seeynge mee dydd aske for mee butte oune of the Companye dydd answerre thatte I was drownynge onn the whyche he pulledd off hys Jerrekynne and jumpedd inn afterre mee withe muche paynes he draggedd mee forthe I beynge then nearelye deade and soe he dydd save mye life.

On this moist tale the *Morning Herald* drily commented: 'The *swimming* reasons given in a paper of yesterday in favour of the authenticity of certain *musty manuscripts*, shew to what Dangers we may expose ourselves by *wading* too far in pursuit of an *object*.'

Accompanying the Deed of Gift was a rude pen-and-ink sketch of the Jacobean Ireland's house, underneath which was written 'Viewe o mye Masterre Irelands house bye the whyche I doe showe thatte hee hath falselye sayde inne tellynge mee I knewe notte howe toe showe itte hymme onne Paperre ande bye the whyche I ha wonne fromme [him] the Summe o 5 shyllynges,' signed 'W. Shakspeare'. The last item in this package, presented in a stiff parchment cover with leather strings, was a paper bearing the arms of Ireland and Shakespeare (reversed), linked together by a chain, and, underneath, these tributary lines:

> Givenne toe mye mouste worthye
> ande excellaunte Freynde Masterre
> William Henrye Irelande inne
> Remembraunce of hys havynge
> Savedde mye life whenne onne
> Thames
> > William Shakspeare
>
> > Inne life wee
> wille live togetherre
> > Deathe
> > shalle forre a lytelle
> > parte usse butte
> > > Shakspeares Soule restelesse
> > > inne the Grave shalle uppe
> Agayne ande meete hys freynde hys
> > Ireland
> Inne the Bleste Courte of Heavenne
>
> O Modelle of Virretue Charytyes sweeteste
> Chylde thye Shakspeare thanks thee
> Norre Verse norre Sygh norre Teare canne
> paynte mye Soule norre saye bye
> halfe howe muche I love thee
> > Thyne
> > W$^m$ Shakspeare

A postscript follows: 'Keepe thys forre mee ande shoulde the Worlde prove sowerre rememberre oune lives thatte loves the stylle.'

This adolescent outburst of maudlin emotion is affecting in its revelation of the forger's pathetic identification with his subject. Equally pathetic is the attempt on the part of an illegitimate youth to provide himself with a lineage—he told his father that Mr H. had found among the family papers documents tending to prove that William-Henry was the direct descendant of the Ireland named in the Deed of Gift. But the audacity of the whole performance exceeds even its pathos. Looking back years later, William-Henry himself professed amazement that it should have passed muster. 'I am absolutely astonished', he says in a manuscript comment on the Deed of Gift, 'that even credulity itself should have been duped by this flagrant document which was in itself sufficient to have overturned the whole mass of evidence, even had it been attended with circumstances a thousand times more convincing than the flimsy tale avouched by myself to stamp the whole with the Signet of Truth.'[99] But his elders, bobbing aloft on a tide of deception, rejoiced in the latest fantasy, which was palaeographically more accomplished than its predecessors. Sir Isaac Heard even suggested that Samuel should couple his arms with those of Shakespeare. 'To give an adequate idea of my feelings on this occasion', William-Henry wrote afterwards, 'is utterly impossible.'[100] His success fed his vanity. Years later he would still rejoice, in his marginalia, on having hoodwinked the experts—'I leave it to the world to judge', he crowed, 'how far the *Science & lucubrations* of the *Critic & Antiquary* are to be relied upon, when, I being then at the age of 17; they were the dupes to my juvenile effusions.'[101] As for the father, this latest revelation brought him to a 'paroxysm of madness', so extreme was his elation.[102]

But the last treasure yielded by the marvellous chest exceeded in interest even the Deed of Gift. A Deed of Trust, dated 23 February 1611, it fleshes out Shakespeare's will—after all, only an aridly legalistic document concerned with a second-best bed but not with the fate of the immortal dramas. The deed is a utilitarian masterpiece. It validates the existence of the chest, love letters to Anne Hathaway, and *Vortigern*; it explains the mysterious secrecy of Mr H., and his equally mysterious generosity to the youth he had casually befriended. In the deed, Shakespeare, having 'founde muche wickedness amongste those of the lawe and not liking to leave matterrs at theyre wills', names a trusty friend, Heminges, who will execute his instructions; but some matters he passes over in silence, in case in future he should make a will. Thus the true will is anticipated, and William-Henry need not rehearse its provisions, which knowledge in any event he scarcely commands, for he does not even know—or remember—that Shakespeare had more than one daughter. To his 'deare Daughterr' (unnamed) he leaves in the deed £20.7s., as well as a suit of black silk and his favourite ring, given him by Southampton. His wife receives £180, clothing, and 'mye lytelle Cedarr Trunke in wyche there bee three Ryngs oune lytell payntyng of myselfe in a silverr Case & sevenn letterrs wrottenn to her before oure Marryage'. From the larger oaken chest in the Globe playhouse Heminges is to remove the manuscript plays and distribute them among the actors Cowley, Lowin, and Burbage. Among these manuscripts is the as yet unwritten *Henry VIII*, but poets have prophetic souls, as is also shown by the poet's foreknowledge of eighteenth-century textual theory, which enables him to refer to 'mye altered Playe of Titus Andronicus' and 'Mye Gentlemenn of Verona alterrd'. In a quaint and endearing touch little Jonas Greggs is given 15 shillings for 'yᵉ troble he hathe hadd inn goynge often tymes withe letterrs toe yᵉ Globe'. Heminges too is to be rewarded, with five manuscript plays (including 'mye newe Playe neverr yette imprynted called Kynge Hʸ. vii') and money for a gold ring, in recompense for 'yᵉ troble hee will have in seeynge thys mye deede ryghtlye executedd'.

The dramatic heart of the document lies in the remaining instructions to Heminges:

And as there wille stille remayne in hys hands 287$^1$. & 14 shyllyngs I furtherr orderr hym to brynge up thatt Chylde of whom wee have spokenn butt who muste nott be named here & to doe same I desyre hym toe place owte s$^d$. Moneye in y$^e$ beste waye he cann doe tylle s$^d$. Child shall be of Age fytenn toe receyve s$^d$. Moneye & withe whatte shall comm uppon s$^d$. Moneye soe toe Instructe hym as aforesayde I allso orderr Masterr Hemynge toe selle mye three howses inn y$^e$ Borowghe & toe putt oute y$^e$ Moneye comynge from same forr s$^d$. Childe. I allsoe gve toe s$^d$. Chylde y$^e$ eyghte Playes thatt bee stylle inne s$^d$. Cheste as allso mye otherr Playe neverr yett Impryntedd called Kynge Vorrtygerne thys as allso y$^e$ other eyghte toe bee whollye forr y$^e$ benyfytte of s$^d$. Chylde as well y$^e$ pryntynge as playinge same and shoulde I chaunce write more as bye Gods helpe & grace I hope toe doe I herebye give $^{ye}$ Profytts of evry Kynde comynge fromm anye suche newe playes orr otherr Wrytyngs unntoe s$^d$. Chylde & hys heires for everre trustynge toe mye freynde John Hemynges honorr and allso onn hys promys of beynge clouse of speeche inn thys laste Matterr

How comes it, then, that the chest is in the possession of Mr H.? Is his ancestor that Heminges who betrayed his trust by failing to honour the clauses of the deed? Had this infamous Heminges also been entrusted with the Deed of Gift, and was he delinquent with that, appropriating property left to the Jacobean Ireland? Was Shakespeare's by-blow, after the default, raised as his own by the William-Henry Ireland who had saved the dramatist's life? Was, finally, William-Henry of Norfolk Street a descendent of the Matchless Bard, truly 'a second Shakspeare'? The possibilities for intriguing inference are mind-boggling, but no one need wonder any longer why Mr H. preferred to keep this skeleton in his closet, and why he was so bountiful to Samuel's son.

Thus, with the Deed of Trust, did the youth complete his romantic biography of Shakespeare, and appease the appetite of his seniors, grave citizens and literary, for those intimate revelations that eluded the industry of the greatest Shakespearians of the age. The bookbinder's elixir and antique paper concealed in the window-seat in the conveyancer's chambers had accomplished their task well. Royal recognition came to Samuel Ireland in the form of a summons to Carlton House for an audience with the Prince of Wales. ('Thanks to heaven', William said long afterwards; 'I was spared the crucifixion of that ordeal.') At 12.30 in the afternoon on 30 December 1795, the Prince's representative arrived at Norfolk Street to escort Ireland. As they were about to leave, his friend, Albany Wallis—discoverer of the Blackfriars Gate-house mortgage deed—entered breathless with brutal elation. 'I have here something to shew you', he said, pointing to his pocket, 'that will do your business for you—& knock up your Shakspeare papers.' He had just discovered a true Heminges signature, as different as could be from the childish left-handed scribble on William-Henry's receipt. There was nothing to do but to go on to Carlton House for an interview lasting over two hours, at which the Prince declared that the papers certainly looked authentic. Samuel perhaps found the occasion less gratifying than otherwise he might have.*

Before he returned home, the news had reached William-Henry. He rushed off to see Wallis, who lived nearby at the bottom of Norfolk Street, and made a mental record of Heminges's neat hand. Then he ran home, and, agitated and in a sweat, told his father that he would go at once and inform the gentleman. Actually, of course, he went to the office at New Inn, and there forged four new receipts with Heminges's signature. These he tied together in a bundle with a number of others, and hastened back to Wallis. Less than half an hour had elapsed. The gentleman, William-Henry reported, had smilingly removed the notes from the drawer of his writing-table and told him to show these to the sceptic. With a friend, Wallis compared the signatures, and found that they corresponded with his own find. How then to account for the odd hand on the first receipt? The gentleman had explained that there were two Jacobean Heminges, one connected with the Globe playhouse, and the other with the Curtain, although to a lesser extent with Shakespeare and the Globe. It was a bit confusing, even in those days, and so they were distinguished from one another as tall John Heminges of the Globe and short John Heminges of the Curtain; that first receipt had been signed by short John. Wallis was taken in, or seemed to be. Imminent disaster had been converted into vindication: no one after all could have produced forgeries at such short notice. Later William-Henry, having had a second look at the Heminges signature, prudently rewrote the receipts with greater fidelity to the hand, and destroyed his originals.

All now seemed well. At Drury Lane preparations moved forward for the première of *Vortigern*. The documents had just been published in a lavish folio volume, with colour illustrations and numerous facsimiles, bearing the imposing title, *Miscellaneous Papers and Legal Instruments under the Hand and Seal of William Shakspeare: including the Tragedy of King Lear and a Small Fragment of Hamlet, from the Original MSS. in the Possession of Samuel Ireland, of Norfolk Street*. This venture William-Henry had instinctively opposed, but assent was wrung from him by his father, whose rage (as recalled in the Hyde manuscript) exceeded all endurance.

Not all the fecund forger's inventions appear in this extraordinary collection. The lot of papers sold after Ireland's decease to a Mr Dent included Shakespeare's haggling correspondence with the printer William Holmes about the sale to him of a play, the dramatist's memorandum noting that Hys Grace o Southamptonne had ordered him to repeat the part of Richard II, and several letters to Heminges: Shakespeare begs his friend 'to speake toe Masterre Johnsonne who hathe treatedde mee mouste hawttylye', confesses

---

* Mair, followed by Grebanier, misdates the interview as the 28th. In his *Confessions* Ireland conceals the conjunction of this event with Wallis's visit, but Ireland Senior records it (British Library MS Add. Add. 30346, fo. 170).

that 'onne cannotte bee wise atte alle tymes evenne soe was itte withe myselfe atte Bitteforde', and complains of 'myne olde syckness'. Judging from the content and trembling hand, the olde syckness was a good deal worse when Shakespeare scrawled the document that is now MS Douce e. 8 in the Bodleian Library. It begins: 'As I nowe fynde mye houre of sycknesse faste comynge onn I doe orderre thatte thys atte mye dethe thys lyttell booke of Prayerre bee givenne toe mye wife ande atte herre dethe thatte she leave itte toe mye daughterre [which one?] thatte so itte maye keepe inne oure house. . . .' This inspiringly pious effusion was composed at Stratforde on Avon at 20 Minutes afterr twoe the mornynge in 1616 (odd, in view of the precision with respect to the hour, that month and day are unspecified). These sacred mementoes, and others, are absent from the volume.[103]

At four guineas a copy, the *Miscellaneous Papers and Legal Instruments* attracted 122 subscribers, among them Glasgow College and New College, Oxford, and such individual purchasers as Boswell, George Chalmers, Richard Brinsley Sheridan, the venerable Macklin, a number of knights, and a sprinkling of lords, including Malone's old friend the Earl of Charlemont.

The big guns of Shakespearian scholarship had, it is true, withheld their endorsement. Parr had urged the celebrated Richard Farmer to come down from Cambridge to view the documents and pass on them; but he declined. Steevens too refused to make the pilgrimage to Norfolk Street. A few months after his visit, Ritson wrote again to his Scottish friend:

I am since told that he [Samuel Ireland] has not only considerably augmented the collection which I saw, by the addition of playhouse-accounts & tracts from Shakspeares library, but has likewise occasionally varied the relation of his becoming possessed of it. However, as I had not the slightest doubt as to the fabrication or forgery of everything he shewed me, my curiosity was never tempted to repeat the visit. I take the whole scheme to have been executed within these 3 or 4 years,—since the publication, that is, of Malones edition of Shakspeare;—and by, or under the direction of, some person of genius and talents, which ought to have been better employed. It appeared to me, at the time, that Ireland himself was the dupe of this imposture; but whether he be still ignorant of its real nature and design, I cannot be quite so positive. . . . The most remarkable circumstance, perhaps, in this iniquitous business, and that which is, apparently, best calculated to promote its success, is— that the parchment & seals of the deeds are indisputably ancient & authentic; so that the original writing must have been entirely effaced.[104]

But publicly Ritson characteristically maintained his silence. The Irelands could bask in the knowledge that each in his own way had served the Immortal Bard and himself well. They had yet, however, to reckon with Malone.

# 10
## *Malone's* Inquiry

WHEN word of the sensational Ireland discoveries reached Malone, surrounded in his study by a hundred documents bearing directly or indirectly on Shakespeare, he was naturally interested—and at once suspicious. His judgement, always cool, warned him against pressing in among the visitors to Norfolk Street. Any inspection there would be cursory as well as public, and if he could not at once pronounce the papers fraudulent, it would be assumed that they had his sanction. Then if, after later study, he published an exposure, it would be thought that he had changed his mind out of envy or faction. Accordingly he sought to view the papers privately, on neutral ground. On 1 February 1795, using the artist Ozias Humphry as intermediary, he approached Thomas Caldecott, a Shakespeare enthusiast who knew Ireland. It was a bad choice: Caldecott fancied himself a scholar—he was preparing an incompetent edition of *Hamlet*—and harboured for Malone all the resentment of the muddled amateur for the successful professional.* So it was with relish that he informed Humphry that the papers 'w$^d$ not be removed into the house of any person whatever, unless he Mr. I. should be requested to wait on his Majesty w$^{th}$. them', and that Samuel had no intention of showing them 'to any Commentator or Shakspere monger whatever'.[105] Undeterred, Malone wrote to the Hon. John Byng: 'If you can get your friend to bring L$^d$. Southamptons letter & the answer to it, and the articles between Condelle, and Hemynge &c to your house tomorrow at three o'Clock, I will produce a facsimile of Lord Southamptons hand writing which will at once ascertain the matter but I beg my name may not be mentioned, let it be only a *gentleman*.'[106] But Byng did mention his name, and Ireland, sneering that the disguise of a gentleman would be inappropriate for Malone,

---

* Caldecott's resentment was perhaps not focused exclusively on Malone: he is presumably the 'T.C.' who on 30 November 1797 addressed a four-page letter to John Mander in Witney, Oxfordshire, detailing the Ireland affair and referring to George Chalmers (see pp. 167–8): 'He is a very silly Coxcomb, and an execrably bad Writer; but he has not spared his pains, and was possessed of extraordinary sources of Information. Pre-eminent above his other Follies is the extravagant conceit, that Shakespeare's Sonnets were addressed to a Lady and that Lady, Q. Elizabeth.' In a note T.C. adds, 'since the above was written, another cumbrous and tiresome volume has appeared, in which this Absurdity is repeated, and laboured with a most perverse industry'. The letter—a flyleaf insert to a bound volume of nine printed titles spawned by the Ireland controversy (e.g. G. M. Woodward's *Familiar Verses, from the Ghost of Willy Shakespeare to Sammy Ireland*) is at the Folger (PR 2950, B5, copy 2, cage). I give the text, along with commentary, in 'The Ireland Forgeries: An Unpublished Contemporary Account'; originally published in Tokyo in 1980 in a Festschrift honouring Professor Jiro Ozu, the article was subsequently reprinted in a volume of my selected essays, *Shakespeare and Others* (Washington, DC, 1985), 144–53.

forbade any but a public inspection. Byng was embarrassed; 'I will be more careful for the future,' he promised.[107]

Thus rebuffed, Malone had to wait until Christmas Eve, 1795, when Ireland published the *Miscellaneous Papers and Legal Instruments*, with its facsimiles. This volume was all Malone needed. 'The truth is', he would boast in his *Inquiry*, 'that a single perusal of it was sufficient; and in one hour afterwards the entire foundation of the Letter I am now writing [his work took the form of an epistle to Charlemont] was laid, and all the principal heads of objection briefly set down.'[108] On 10 January 1796 he settled down to what he thought would be a pamphlet, but the project grew and grew. The public, stirred up by the numerous newspaper accounts of the Shakespeare Papers and the forthcoming production at Drury Lane, eagerly awaited the thunderbolt to be hurled by the most authoritative scholar of the age. Publication, however, was more than once postponed: 'The expanding of the topicks, and the minute examination of authorities', he afterwards explained, 'necessarily required some time.'[109] On February 16th the *Chronicle* carried the following announcement, in which Malone makes known his position unequivocally:

### Spurious Shakespeare Manuscripts

Mr. Malone's detection of this Forgery has been unavoidably delayed by the Engravings having taken more time than was expected, and by some other unforeseen circumstances; but it is hoped it will be ready for Publication by the end of this month.

But February passed without the book. The newspapers jeered at the delays, but Malone worked on, aloof to publicity, and even his aloofness rankled. 'Mr. Malone', Boaden commented in the *Oracle* on 9 March, 'talks rather peevishly of the "*meretricious* and *undesirable* celebrity of a newspaper". We trust no writer will henceforward offend this fastidious gentleman, with any thing so irksome to his feelings as *diurnal praise*.'

Malone's time-consuming thoroughness enabled others to enter the lists before him. Boaden, converted from faith to disbelief, perhaps under the influence of his friend Steevens, published *A Letter to George Steevens, Esq.* (dated 11 January 1796) attacking the Papers. Ineffectual as a critic, and devoting too many of his pages to a comparison of the two *Lears*, he yet makes some telling points. In a note to Ireland's *helas*, he observes: 'By this curious mode of writing the interjection, one might be tempted to believe that Shakspeare had received a French Education at the College of St. Omers.'[110] The 'vicious and fantastic orthography' of the Ireland *Lear* finds its only parallel in Chatterton. Boaden points out that Leicester could not have passed the 'holydayes' with Elizabeth after 1585, when Shakespeare was a mere twenty-one; the letter, moreover, is addressed to the Globe, which (according to Boaden) did not exist until 1596. Lowin, described in one of Ireland's receipts as playing before Leicester, was twelve when that lord died. Anachronistic terms enter into the letter to Southampton; the Profession of Faith displays 'the puerile quaintness and idiomatic poverty of a Methodist rhapsody'.[111]

On 23 January Boaden followed up his broadside by publishing the following letter to Shakespeare in the *Oracle*:

Wee werre mightylye tickledde at the per fourmaunce offe thye pleasaunte Commedie offe the MERYE WYFFES.—Thye wytte couthe welle Maisterre William—the Gygantesse that didde lye underre Mounte Pelion was excellente—We give thee nottice thatte wee shalle *drinke Tea* withe thee bye Thames Tomorrowe, thou Monarche offe the *Globe*. Inne thye Hamblette wee perceive thou diddest oure biddinge twittinge mye Lorde of Leicesterre thatte he was fatte and scante of breathe—Write mee whatte thou thinkeste offe the lankye Ladde Southamptowne thye friende.

P.S. More offe oure virgin beautye.

> Thynne everre toe commaunde,
> ELIZABETH, R.

*Greenwyche,*
*Julye Ninethe, 1580.*

The annotations that follow pillory with devastating effect the fatuities of the Ireland forgeries: 'Here must be some mistake in all this, notwithstanding the *Water-mark*. The *Merry Wives* was not then written.—ELIZABETH had no *Tea to drink*—the *Globe* was not then built—*Hamlet* was not then in existence—and, although LEICESTERRE was certainly fat, SOUTHAMTON [sic] was then a child.'

Other newspapers carried similar additions to the Shakespeare Manuscripts yielded by old trunks everywhere. A reader in Brompton provided the *Herald* with a verse addressed by the poet to Anne Hathaway, '*On herre cruelle neglecte offe mee.*' An answer to Shakespeare's love letter, printed by the *Chronicle* (28 April), poignantly concludes: 'My Maydenne name once flunge away, | Noe longerre ANNA *Hath herre waye!*' The *Telegraph* for 14 January produced an epistle from Shakespeare to his cheesemonger ('Thee cheesesse you sentte mee werre tooee sweattie, and tooe rankee inn flauvorre, butte thee redde herringges werre addmirabblee.'); also another addressed in amicable mood to his great rival:

Tooo Missteerree BEENJAAMMIINNEE JOOHNNSSONN.

DEEREE SIRREE,

Wille you doee meee theee favvourree too dinnee wythee meee onn Friddaye nextte, attt twoo off theee clockee, too eattee sommee muttonne choppes andd somme poottaattoooeesse.

> I amm, deerree Sirree,
> Yourre goodde friendde,
> WILLIAMME SHAEKSPERE.

The dates to these letters, 12 January 1687 and 27 January 1658 (respectively), are an endearing touch.

Meanwhile, Ireland had prevailed on one of the believers, Francis Webb, to defend the faith, which he did under the name of 'Philalethes' in *Shakspeare's Manuscripts, in the Possession of Mr. Ireland*, a pamphlet issued almost simultaneously with Boaden's. The documents, Webb argues, 'reciprocally illustrate and confirm each other'. That they are Elizabethan is indicated by the paper, the antiquity of which experts have confirmed. Why would a contemporary of Shakespeare have perpetuated forgeries that would lie hidden and profitless for two hundred years? Webb does not entertain the possibility that they are modern fabrications on old paper. The young forger must have relished the judgement of an elder who could declare himself satisfied 'that no human wisdom, cunning, art, or deceit, if they could be united, are equal to the task of such an imposture', or who could discern in these relics 'the same strong marks of his [Shakespeare's] original genius, as those with which his acknowledged writings are deeply impressed'.[112] In February, Walley Chamberlain Oulton in his anonymous pamphlet, *Vortigern under Consideration*, ingeniously suggested as 'not at all unlikely', all the world being a stage, that Shakespeare used the term *Globe* as a metaphorical appellation for any theatre. The Love Letter is 'a precious relic' substantiated by the lock of hair which Boaden fails to mention. The Profession of Faith is inspiring, the drawings are too crude for a forger to have invested his time in, and the Shakespeare Library, a laborious undertaking, could only have been completed by the poet. Boaden was again attacked by J. Wyatt in his *Comparative Review of the Opinions of Mr. James Boaden*, which twitted him unmercifully for changing sides. In the same month, February, G. M. Woodward, the caricaturist, entered the fray in high spirits with *Familiar Verses, from the Ghost of Willy Shakspeare to Sammy Ireland*, which contains the following engaging lines:

> Samples of hair, love songs, and sonnets *meete*,
> Together met by *chaunce* in *Norfolk street*;
> Where, fruitful as the vine, the tiny elves
> Produce *young manuscripts* for SAMMY's shelves.
> Dramas in embrio leave their lurking holes,
> And little VORTIGERNS start forth in shoals.[113]

These exchanges maintained public interest in the Shakespeare Papers at a high level, but that they struck contemporaries as inconclusive is indicated by the detachment, as late as March, of the *Analytical Review*, which after quoting extracts from the documents, concluded that the claims for them would 'undergo an impartial examination at the bar of an intelligent and candid public'. The world awaited Malone.

He finished his work on 2 March, a little over two months after he commenced writing. He applied himself unremittingly, in the last days working from morning to night at the printing house. The book, which comes to over four hundred pages, bears the imposing title, *An Inquiry into the Authenticity of Certain Miscellaneous Papers and Legal Instruments, Published Dec. 24, M DCC XCV. and Attributed to Shakspeare, Queen Elizabeth, and Henry, Earl of Southampton: Illustrated by Fac-similes of the Genuine Hand-writing of That Nobleman, and of Her Majesty; A New Fac-simile of the Hand-writing of Shakspeare, Never before Exhibited; and Other Authentick Documents.*

The *Inquiry* appeared on 31 March, two days before the first performance of *Vortigern*. In those two days it sold five hundred copies. The *Inquiry* reveals that Malone had not cut himself off completely from the law. He is the prosecuting attorney proceeding by laws of evidence in a literary tribunal, and his address to the jury—Lord Charlemont and the British public—burns with righteous anger. Malone's indignation is not difficult to appreciate. His cause is truth, which as a scholar he had always held sacred. For the Papers that are subject to his onslaught he has only scorn: they are puerile, spurious rubbish. Because they are trash they debase with foreign alloy Shakespeare's reputation and true writings, and rob him of 'that good name and reputation which to all men of sensibility is dearer than life itself'.[114] They are an affront to English scholarship, and to the public which has been duped by them.

He begins with Mr H. In his preface to the *Miscellaneous Papers*, Ireland had partially explained the gentleman's generosity to a casual acquaintance by revealing that William-Henry, in rummaging through the documents, found some deeds that established Mr H.'s title to a considerable estate. In what county does this property lie? Malone asks. Has any suit been instituted as a result of the discovery? It is not idle curiosity to seek the name of the original owner of the papers, for every new fact or circumstance supplies the means to corroborate or disprove the contested point, and the withholding of evidence carries the presumption of falsehood; he cites Blackstone and quotes an eloquent passage from Gilbert's *Law of Evidence*. Malone, however, needs no more for his case than the Papers themselves, and he proceeds to examine them, in their sequence of publication, with respect to orthography, vocabulary and phraseology, the dates furnished or deducible by inference, and handwriting.

He has little difficulty in disposing of the 'laboured and vicious deformity' of the spelling, which represents the practice of no period. In his experience with several thousand manuscripts from Henry IV's reign onward, he has not once encountered (for example) *ande* or *forre*. Such forms as *ourself, yourself*, etc., were written as two words in Elizabethan manuscripts, but in the Ireland papers they appear as modern disyllables. Malone notes the absurdity of the double consonants of *bllossommes* and *bllooms*, and of *Anna Hatherrewaye* for *Anne Hathaway*, as the name appears in parish registers. In a comparative table he lists twenty-five words from Elizabeth's supposed letter to Shakespeare alongside the actual forms taken by these words in four documents in the Queen's hand. It is curious too that Her Majesty should consistently misspell the name of Leicester (*Leycesterre* in Ireland) to whom she had an attachment for thirty years. But Malone's most compelling argument—or so it would seem—involves the spelling of the name Shakespeare itself, and here he apparently catches the forger, red-handed, following one of the three signatures reproduced in the Johnson–Steevens *Shakespeare*. Steevens had traced these signatures from Shakespeare's will, in Malone's presence, in 1776, and the latter had since then adopted the spelling of the third signature, *Shakspeare*. But in 1793 an anonymous correspondent informed him that there was no *a* in the second syllable, and that this signature, like the others, should read *Shakspere*. Malone re-examined the will, and agreed. The forger had therefore trapped himself by imitating an error. It is a stunning bit of detection, but unfortunately Malone was right the first time. Although the signature is poor, the last three tremulous letters perhaps having been added to expand a contraction, there can be little doubt that the spelling is *Shakspeare*. At the time, however, Malone's point made a great impression.

Although this argument must now be rejected, the cumulative force of the evidence is none the less staggering. Malone independently makes Boaden's points about Leicester's death, the date of the Globe, and forms of address for an earl (he anticipated anticipation, and regarded it merely as confirmatory). Elsewhere, with respect to diction, he cites a number of words from the Ireland papers as either not used in Shakespeare's time, or not used in the sense employed; among these are *accede*, *whimsical*, *upset* (as in the upset of a boat), *composition* (for a written piece), and *view* (as a delineation of a house or object; the usual Elizabethan term is *prospect*). The discussion of handwriting is even more devastating. By minute analysis Malone demolishes Elizabeth's signature, for which alone the forger had used a model, and which the Papers' defenders had accepted with as much faith as if they had seen her writing it herself. He places facsimiles of the Queen's handwriting alongside the pretended letter and facilitates comparison by furnishing the genuine and spurious alphabets. So too, in facsimile, John Heminges's true autograph is placed beside the forgery and we are shown two letters in Southampton's neat hand, which contrasts overpoweringly with Ireland's scribble, the 'miserable scrawl of a paralytick man of four-score'.

Malone's common sense and vast knowledge of old documents, stage history, and Elizabethan customs come into play in a variety of ways. He is alert to the inconsistency in the Deed of Gift, which, although supposedly executed in October, refers to 'the last monethe beyng the monethe of Auguste'. He observes the absence of references to the Blackfriars in documents purportedly dating from the period when that theatre was principally used by the King's Men. He notes that a dramatist retained no title to the pieces he supplied to his company, that manuscript plays of the period (unlike the Ireland *Leare*) did not have line numbering, and that (also unlike the *Leare*) they were written on both sides of the page. In receipts, sums would be expressed in Roman rather than Arabic numerals. It was not the custom to baptize infants with two Christian names; hence the unlikelihood of a Jacobean William-Henry Ireland. Endorsements on legal documents give the regnal year in Latin—not English, as in the forgeries. Malone shrewdly notes the absence of any reference in Ireland's preface to the full-length portrait of Shakespeare and the two uncut folios, about which rumours had circulated from Norfolk Street. He raises pertinent questions: Why did not the faithless Heminges produce for his own profit at the Globe and Blackfriars the virgin plays—*Vortigern*, *Henry VII*—that had come into his hands, and why did he exclude them from the First Folio? How did all these papers come together in the same mysterious chest? Heminges might conceivably have the leases, deeds, etc., as well as the plays, but surely not the Love Letter, the lock of hair, and the exchange with Southampton; the widow, to whom the latter items would belong, would not have the former. 'I HAVE now done', Malone concludes his presentation of the evidence, on a note of triumphant disdain; 'and I trust I have vindicated Shakspeare from all this "imputed trash," and rescued him from the hands of a bungling impostor, by proving all these Manuscripts to be the true and genuine offspring of consummate ignorance and unparalleled audacity.'[115]

In the course of the *Inquiry* Malone touched upon several issues of Shakespearian biography independent of his immediate purpose. They have considerable interest. Of John Shakespeare's Spiritual Last Will and Testament, which he had presented to the world in 1790, he says that he has 'since obtained documents that clearly prove it could not have been the composition of any one of our poet's family; as will be fully shewn in his Life'.[116] These documents he never produced, and Boswell (the biographer's son), who went through Malone's papers after his death, could find no trace of them; he guessed that they concerned another John Shakespeare with whom the playwright's father had been confused. 'From a paper now before me', Malone says in another passage, 'which formerly belonged to Edward Alleyn, the player, our poet appears to have lived in Southwark, near the Bear-Garden, in 1596. Another curious document in my possession . . . affords the strongest presumptive evidence that he continued to reside in Southwark to the year 1608.'[117] But these papers, which he hoped to incorporate in a biography of Shakespeare, never saw publication, and they have since passed from view. Still another newly recovered document he prints for the first time as an appendix to the *Inquiry*: the conveyance (dated 10 March 1612/13) of the Blackfriars Gate-house property from Walker to Shakespeare. This is the counterpart to the mortgage deed, and was executed the previous day. Like the mortgage, it was found by Albany Wallis among the Fetherstonhaugh deeds. Wallis, whose role in the controversy is ambiguous—he shared the confidence of the Irelands while assisting Malone—communicated the manuscript to Malone, who had known of it since 1783, when he referred to it in a manuscript note in his copy of Langbaine. Purchased by the Corporation of London in 1843, it is now in the Guildhall Library.

The *Inquiry* has its faults, at least some of which may be attributed to haste of composition. Malone's irony is more laboured than elegant, and he can be so heavy-handed as to say, when comparing a counterfeit with a genuine document, 'The resemblance in the present case is that of a weatherbeaten alehouse sign in a country village to a portrait by Titian or Sir Joshua Reynolds.'[118] Angered by the reference to 'the gyldedde bawble thatte envyronnes the head of Majestye' in the Love Letter, he launches into an irrelevant denunciation (which, incidentally, delighted Burke) of the forger's presumed anti-royalist persuasion—although he makes the just point that the crown was for Shakespeare not a gyldedde bawble but a golden circle. Later commentators would point to minor inaccuracies of fact, and Malone corrected himself in several letters to the *Gentleman's Magazine*. Then, too, a memory that was (as the younger Boswell remarked) not especially tenacious betrays him into sweeping general statements to which exceptions would be found. *Ande* is an unusual spelling for the period, but it exists; *forre* is rarer still, but not unknown; *accede* appears in 1611 in Florio's *New World of Words*. It was not customary for Elizabethans to have two Christian names, but a few did. And so on. At one time Malone planned to publish a corrected second edition of his *Inquiry*; although he abandoned the project, his working copy, voluminously annotated, may be consulted at the British Library. His occasional lapses fail to shake the edifice of argument. This the public accepted, as the columns of the *Gentleman's Magazine* and other periodicals testify.

The wretched Samuel, stung by Malone's references to impostures and an impostor, protested his personal innocence in *Mr. Ireland's Vindication of His Conduct, Respecting the Publication of the Supposed Shakspeare MSS. Being a Preface or Introduction to a Reply to the Critical Labors of Mr. Malone* (November, 1796). He quotes letters from his son and Talbot, prints the Certificate of Belief, and laments 'the pecuniary losses and the consumption of time, which these transactions have led me into'.[119] The next year he published *An Investigation of Mr. Malone's Claim to the Character of Scholar, or Critic, Being an Examination of His Inquiry into the Authenticity of the Shakspeare Manuscripts, &c.*, based on material largely supplied by Caldecott. The pamphlet scores some points with respect to orthography and Christian names, but the effort is on the whole ineffectual (how could it be otherwise?). In his *naïveté* or desperation, Ireland is even driven to dredge up tall and short John Heminges. Anyway, as he admits himself, the controversy is dead. *Vortigern* had been literally howled off the stage after Kemble had declaimed, in his most sepulchral tone, the unfortunate line in Act V: 'And when this solemn mockery is o'er—.' William-Henry had confessed his misdeeds, with a swaggering plea for forgiveness, in *An Authentic Account of the Shakspearian Manuscripts, &c.* (1796). In the Westminster Forum, held on Brewer Street, a public debate was announced for 9 January 1797 on the question, 'Do the Shakespearean Manuscripts, the Play of Vortigern and Rowena, and the Apology of Mr. Ireland Jun. exhibit stronger Proofs of Authenticity, flagrant Imposition, or the Credulity of Persons of Genius?' (admittance sixpence; no political remarks permitted). Although the management invited the Irelands and Dr Parr, as well as other literary gentlemen, to attend, there is no reason to believe that they did, or that the debate caused any stir whatever. Meanwhile sales of Malone's *Inquiry* dwindled. Samuel had stopped distribution of his Folio. The Ireland bubble had burst.

It is the hard lot of deceivers not to be believed even when they speak truth: neither the public nor his father would accept William-Henry's revelations at face value. Samuel simply did not think his son clever enough to perpetrate such an imposition; a letter addressed to an unknown correspondent on 26 June 1797 shows that he preferred to regard William-Henry as a thief. 'I believe you are . . . informed that he has publicly declared himself y$^e$ author of all y$^e$ Mss—poetry, plays, Law deeds &c &c', Samuel wrote, 'no part of w$^{ch}$ declaration do I believe nor does y$^e$ world. Till something further and more satisfactory does appear—we must be separated—and I fear that will be for ever—my opinion is—that they have been stolen—& that he is afraid to declare y$^e$ truth for fear of Consequences—.'[120]

Students of this curious episode in literary history have been understandably moved to compassion for the Irelands in their disgrace, although for this writer sympathy for William-Henry is mitigated by his heedlessness—or is it simply unawareness?—of his father's sufferings. For casting Malone in the role of vindictive persecutor, however, as Mair has done in *The Fourth Forger*, there is no justification.* Vindictiveness was alien to the temper of a man valued for his kindness and generosity by a circle that included The Stratford Poet as well as the biographer of Johnson. 'His heart was warm', the younger Boswell recalls, 'and his benevolence active.'[121]

It is true that the forger was only a youth of twenty, that the relationship between parent and child was, especially towards the end, invested with an almost unendurable pathos, and that the ruin and calamity which befell both after exposure are terrible to contemplate. Ireland Senior received the bulk of the abuse. In one newspaper he was compared with Abraham offering up his son Isaac as a sacrifice. The public turned a deaf ear to his protestations, his clients disappeared, Gillray savagely caricatured him in a print entitled

---

* It may be granted that Malone's professional etiquette does not at all times conform with ours: in public and private he heaped scorn on Capell, denouncing him (in his marginalia) as a blockhead and an absurd coxcomb, yet he did not hesitate to pillage Capell's text without acknowledgement. The late James M. Osborn exaggerates his hero's virtues when he claims that 'Malone would rather have been caught without his wig than not to have properly acknowledged his authorities' (*John Dryden: Some Biographical Facts and Problems* [New York, 1940], 54).

'Notorious Characters, No. 1.' Ritson, it is true, anticipates later students in exonerating the father. 'You will be surprised to hear . . .', he wrote to an unidentified correspondent on 1 December 1796, 'that all the plays, deeds, letters, & papers of every description which have been produced by Ireland, owe their existence solely to his son Samuel alias William-Henry, a boy of 19, in whom no talents of any kind were ever before discovered, even by the father himself, who has, in fact, been the completest of all possible dupes to the astonishing artifices of this second Chatterton.'[122] But few others could bring themselves to forgive the elder Ireland, and characteristically Ritson did so only in private. Poor Samuel: his worst fault, apart from his greed and self-serving credulity, was his inability to conceive that his son, who above all wanted recognition from him, had any talent whatever.

But how was Malone to know that a half-crazed youth was dreaming a wild Shakespearian dream, or that in his heart of hearts the youth's father venerated the Matchless Bard? The *Inquiry* is directed at the Papers, not their inventor, upon whose identity the author refuses to speculate. About Ireland as a man, or about the character of the son, he says nothing; Malone stands aloof from personalities. Not that he did not privately hold convictions from the outset. On 29 December 1795 he wrote to Charlemont: 'The editor, a Mr. Ireland, a broken Spitalfields weaver, aided by his son, an attorney's clerk, are without doubt the inventors, though to avoid being pelted in the newspapers by such men, I shall leave that matter in uncertainty, and merely confine myself to prove the forgery, let it come from where it may.'[123] The fate that overtook the pair resulted not from harsh treatment at Malone's hands, but from the fact of discovery.

The father expired, of diabetes and despair, in July 1800, protesting to the end his faith in the genuineness of the Shakespeare Papers. Under such circumstances death is a release, and so his daughter Jane regarded it. 'The Death of my ever to be lamented Father has indeed been a severe stroke!' she wrote to Mrs Byng, 'but the reflection, he now enjoys those blessings (peace, & tranquillity, to which his mind was a stranger when living) ought to reconcile me to the separation, awful as it is!!'[124] His son was early able to draw a bitter lesson from his experience; on 28 October 1796 he informed Byng, '[I] now am left to little better than a state of Starvation I plainly see what the World is & little am I astonished at poor Chatterton's fate.'[125] The rest of life was anticlimax. William-Henry left home, pawned his belongings, sold forged documents as curios, elaborated on his confession, and ended his days living from hand to mouth as a Grub Street hack, never pardoned by the literary community which had been tricked into rhapsodizing over the effusions of a dull-witted young man. To the last, amid squalor and desperation, he retained his high spirits; he could pride himself on the fact that he was Shakespeare Ireland, and had had his moment of glory. His contemporaries he would impress as being rather like a roistering Restoration cavalier 'with something gleaming out of his eyes that in the height of his hilarity forbad you to trust him'.[126]

# 11
## *Postscript: George Chalmers*

THE Ireland affair had a strange aftermath in the two volumes published by the Scottish antiquary and chief clerk in the Earl of Liverpool's office, George Chalmers: *An Apology for the Believers in the Shakspeare-Papers, Which Were Exhibited in Norfolk-Street* (1797) and *A Supplemental Apology for the Believers in the Shakspeare-Papers: Being a Reply to Mr. Malone's Answer, Which Was Early Announced, but Never Published* (1799). A signer of the Certificate of Belief, Chalmers (according to rumour) had in hand a work arguing for the authenticity of the Shakespeare Papers, when Malone published his crushing exposure. Chalmers was stung by the latter's scorn for 'credulous partizans of folly and imposture', and he was also understandably reluctant to lose the fruits of his industry. With the economy for which his nation is celebrated, he salvaged his demonstration by converting it into a defence of his credulity and an onslaught against the scholar who had embarrassed him. In his learned, cranky disquisition, Chalmers attacks venomously and at length (the books come to over six hundred pages each) Malone's self-confidence, methodo-logy, and specific exhibits of evidence; an assault to which, as Chalmers's second title implies, his victim did not deign to reply. Steevens, to whom Chalmers addressed an admiring prefatory letter in his *Supplemental Apology*, saw no discredit to the Detector in his contempt for a notorious cheat, and thought the rebuttal folly.

So too it appeared to others. The *Supplemental Apology* stirred satirical mirth through the autumn of 1799 in the columns of the *Morning Chronicle*, where it was recommended to all sleepless men and women. The following epigram graced the September 4th number:

> GEORGE, 'tis odd you cannot rest,
> Since you rummag'd IRELAND's Chest.
> Think of your Office and your Head—
> Sure, you've enough of Scraps and *Lead!*[127]

Nevertheless, Chalmers along the way makes a contribution of his own to Elizabethan studies in his discussion of stage history, punctuation, and the Shakespeare chronology. If his views are at times sufficiently eccentric—

improving on Malone, he places *Merry Wives* before the *Henry IV* plays, and *A Midsummer Night's Dream* after *Hamlet*—he none the less performs a service in printing for the first time a number of documentary records, including disbursements for plays performed before the Queen, and further extracts from the Stationers' Register. He anticipates modern lines of investigation by expressing an interest in the grammars, dictionaries, and critical treatises that may have had a formative influence on Shakespeare. And in the section 'Of the History of the Stage' he prints for the first time Augustine Phillips's will with its bequest of a thirty-shilling gold piece 'to my fellow Willm̄ Shakespeare'.

Chalmers's most novel theories concern the Sonnets. Taking a hint from Elizabeth's acknowledgement of Shakespeare's 'prettye Verses' in the forged letter, he embarks on a close study of the cycle, and arrives at the conclusion that in these poems Shakespeare addresses the Queen, thus urging upon her the duties of marriage and procreation. That Elizabeth was then in her sixties does not bother Chalmers, who does not pretend to be a biologist. Nor is he disturbed by the fact—first pointed out by Malone in 1788—that most of the Sonnets are directed to a man: 'Elizabeth was often considered as a man', he argues, and termed a prince, rather than a princess, by Drant, Spenser, Ascham, and Bacon.[128] Chalmers cannot conceive that 'Shakspeare, a husband, a father, a moral man, addressed a hundred and twenty, nay, a hundred and twenty-six *Amourous* Sonnets to a *male* object!'[129] He even quotes the bawdiest of the sonnets, 20, to support his thesis, and solemnly notes that 'To *prick* is often used by Shakspeare for to *mark*.'[130] As regards the onlie begetter of the ensuing sonnets, Mr W. H., Chalmers makes the suggestion, to be echoed by later commentators without awareness of his precedence, that 'W. H. was the *getter* of the manuscript, imperfect as it was, from which the Sonnets were printed inaccurately.'[131] His theory about the cycle itself has understandably won less support. 'When a writer has once determined that all Shakspeare's Sonnets must relate to the same subject, and must be addressed to the same person', the *Monthly Review* critic mused, 'he will violate every rule of language in order to maintain his position.'[132] Absurd as that position is, Chalmers has initiated a significant and ominous trend in offering the first autobiographical reading of the Sonnets. With him, as the distinguished Variorum editor of the Sonnets remarks without mirth, the fun beings.

# 12
## Malone: The Posthumous Life

OF all the eighteenth-century editors of Shakespeare, Malone was uniquely qualified by temperament, interests, and experience to provide for his age an authoritative documentary life of Shakespeare. His great edition behind him, he was still in his prime, with the energy and leisure to persevere with the task. And how desperately needed that biography was! Largely through Malone's achievement the inadequacies of Rowe's essay were now recognized. There are not more than eleven biographical facts mentioned in it, according to Malone's analysis, and of these critical examination reveals eight to be mistaken and one doubtful; the remaining two, records of baptism and burial, were furnished by the Stratford parish register. Malone exaggerates, but not a great deal. Yet Rowe, for want of a successor, remained standard. Malone's expedient in 1790 of reprinting the preface with copious notes, corrective and supplementary, was cumbersome and inefficient, as it did not permit him to use all the materials he had gathered. Recognizing the need for a biography on a grander scale, he alludes in his own preface to his intention of producing one.

With the introductory volumes of his edition set up in type, and the opportunity for revisions or additions past, he manifested his continuing preoccupation by besieging Jordan with queries throughout 1790. He also mentioned fresh discoveries: he had come upon a record showing the transfer from Hercules Underhill to Shakespeare, in 1602, of 'one messuage, two barns, two gardens, and two orchards, with the appurtenants in Stratford upon Avon'.[133] Apparently this is a translation from the Latin of the fine (mistakenly thought by Malone to be a purchase) levied on New Place when Underhill came of age that year.[134] His other find is that Shakespeare bought some land in Old Stratford and Stratford-upon-Avon from William Combe and his nephew John in 1614. Again there is some confusion, for this transaction—not previously known—took place in 1602, although the property did become the subject of enclosure controversy in 1614.

After 1790 Malone desisted from his Shakespearian labours for a couple of years, and his correspondence with Jordan came to a halt. But in April 1793 he was writing to another Stratford acquaintance and scholarly informant, the Revd Dr Davenport, about his plans,

the first work I mean to set about is the 'Life of Shakspeare,' for which I have a good many materials already in print, which must be woven together and brought into a connected narrative, with the addition of some information obtained too late for my

8vo. edition. Preparatory to it, will you allow me, after a long interval, once again to resume our Shakspearian disquisition—

> 'Age cannot wither it, nor custom stale
> Its infinite variety.'[135]

He also renewed correspondence with Jordan, in November, for the same reason. In those more casual times ancient records were guarded less zealously than today, and Davenport could accommodate his friend by arranging for the loan of the books of the Corporation of Stratford. These Malone pored over avidly. Meanwhile, he spared no pains in his search for the record of Shakespeare's marriage, and he found out what he could about the great Stratford fire of 1614. In December 1794 he informed Davenport that he had completed half of the life of Shakespeare, bringing him 'to the door of the London theatre', and had transcribed a fair copy for the press. The next section would be 'a History of the prevailing manners of the English World when he first came on the town'.[136] For this task he had spent three or four months reading the vast quantity of material he had brought together, and making extracts.

But the essay did not materialize. 'In the meanwhile', he confessed,

the Life is at a dead stand still: but I hope soon again to resume it, and to finish it in a few months. It will make, I imagine, 'justum volumen'; and at least will have the advantage of novelty and curiosity, for I think I shall be able to overturn almost every received tradition about this extraordinary man.[137]

The next year, in a prospectus for a new edition of Shakespeare, he announced that the first volume would comprise 'an entirely NEW LIFE OF SHAKSPEARE (COMPILED FROM ORIGINAL AND AUTHENTICK DOCUMENTS,) Which is now nearly ready for the Press'. In 1796, in the *Inquiry*, he speaks of having 'amassed such an accumulation of materials for a more regular Life of our poet, as have exceeded my most sanguine expectations, and are now swelled to such a size as to form a considerable volume'.[138] But that volume, despite his claim, was nowhere near completion. 'A few years ago I executed about half of it', Malone wrote to Bishop Percy in June 1802, 'and have it by me fairly transcribed for the press. I am resolved *this summer* to sit down steadily to it, and not to make any excursion from London, but to work *doggedly* at it, till it is done.'[139] Again he was over-optimistic.

Yet he could not resist following up any lead, however slight, that might bring new documents, perhaps personal documents, his way. In his mind he ruminated over the possible sources. Before her death in 1670 Lady Bernard, Shakespeare's granddaughter, had appointed as her executor a London citizen, Edward Bagley, who must have come into possession of cabinets and coffers 'undoubtedly' containing some of the poet's papers. Had the Bagley estate, Malone wondered, descended to any person still living? And did the instrument executed by Shakespeare when he bought some property from

Ralph Hubaud (brother of Sir John Hubaud of Ipseley in Warwickshire) remain unnoticed among the title-deeds of the inheritors of that gentleman's estate? And what of the dramatist's worthy friend and colleague, John Heminges, who left behind a son and four married daughters—did priceless theatrical documents languish among the possessions of one of these families? A rumour reached Malone's ears that letters from Shakespeare to the Lord Treasurer awaited discovery among the Dorset Papers, and, although sceptical, he was characteristically unwilling to reject it out of hand. 'As I have long observed that—"omnis fabula fundatur in historia", or, in other words, that half the stories running round the world, are partly true and partly false', he wrote on 30 November 1802, probably to Sir Nathaniel Wraxall (the letter is unaddressed):

I fear that the Petition of *Sarah* Shakspere which you mention, has given rise to the tale concerning the Poet's Letters.—Yet I have a *bastard* kind of hope that they may really exist, in consequence of the dates assigned, which struck me as they did you, as probable and consistent with the time of the Ld Treasurer Dorset's death. . . . I am . . . strongly inclined to think, that these letters, if they indeed exist, and are dated in 1606, or about that time, will turn out letters of thanks to the Treasurer, for some bounty transmitted through his hands by King James, in return for the tragedy of *Macbeth*.[140]

Thus, in the dawn of the new century, hope had not yet been abandoned that genuine Shakespeare papers would come to light.*

Meanwhile the Life remained unwritten. The three 'great books', the folio, and the Chamberlain's accounts borrowed from the Stratford Corporation gathered dust in his study a dozen years. On 2 December 1805 Hunt and Hobbes, the town clerks, wrote to him pleading for the return of the records, and threatening to apply, if necessary, to the King's Bench to get them restored. 'M<sup>r</sup>. Hunt has himself', they wrote, 'and several others have called upon you repeatedly for these papers, and you have as repeatedly promised they should be carefully returned, and the Corporation do conceive they are by no means well treated.'[141] He yielded up the three huge volumes of proceedings of the Corporation from the earliest times until 1650, and on 28 December expressed his sense of shame to Davenport for having held on to them for so long. '[T]he truth is', he wrote, 'I was still in hopes that I should bring my work, which has been delayed by a thousand unforeseen circumstances, to a conclusion; and I was very desirous, while it should be passing

---

* Rumours are with difficulty dispelled. Reports of Shakespeare letters among the Dorset Papers persisted for many years. At the Shakespeare Birthplace Record Office I have come upon a fragment of a letter from the Duchess of Dorset, dated 19 June 1810, in which she says that she knows nothing of such papers at Knole; a Mr Chalmers, she faintly recalls, had searched there but with what success she does not know, and certainly Wraxall had found nothing. In 1825 it was reported in the press that two original letters of Shakespeare had been discovered among the Dorset Papers in 1803, and that shortly thereafter a third letter had been turned up in the same archive. This time Wraxall broke his silence and repeated in the *Courier* what he had said to Malone a generation earlier: between 1797 and 1800 he had reviewed all the family documents at Knole without coming upon anything Shakespearian.

through the press, to have before me the original documents from which my extracts were made, as I well know how very difficult it is to be minutely accurate in transcribing very old and often scarcely legible papers.'[142] He begged to keep the remaining materials until the publication of his biography. It is 'a formidable work, but if I can but live to finish it, I shall think nothing of the labour. I hope to put it to the press about the middle of summer.'[143] Summer came and went, and still he did not complete it. In May 1807 Malone informed Percy that a third of the *Life*, his 'favourite object', was still to be written, with all the materials for it ready. The next year, in June, the same third remained undone; a 'thousand interruptions and avocations' had held him up—'but I mean to stay in town all this summer, and hope to do a great deal between this time and Christmas, and to go to the press early in next year'.[144] But he deluded himself: he would not live to conclude his great work.

What happened? At this remove it is not possible to penetrate to the underlying reason for Malone's failure to finish, in the twenty years remaining to him, what should have been his magnum opus; one can only record the overt circumstances. He enjoyed research, which daily rewarded his ardour with some new discovery, but (so he confessed to Charlemont) found the business of arranging and putting his materials in proper form 'an anxious and laborious business'; yet more than once he showed himself capable of bringing great tasks to completion. Other demands were made upon his industry. The Ireland case was one. Being the sort of man who is chosen as an executor, he was called upon more than once to perform this office of kindness. For his friend Reynolds, who died in 1792, he superintended a collected edition of the artist's writings, which was published in 1797, with a biographical memoir by Malone. He was distracted too by new scholarly undertakings. Contemplating an edition of Aubrey, he spent the summer of 1797 in the Ashmolean Museum transcribing the *Brief Lives*. But mainly a vast new project absorbed his energies: an edition of Dryden's prose works, to be prefaced by a full and detailed biography. When Jordan enjoyed Malone's hospitality in London in June 1799, '. . . that truly great, good, and honorable gentleman . . . made a very satisfactory apology' to the poetical wheelwright

for not returning me an answer while I was at Stratford, by both assuring me and *showing* me that his time is wholly employed in the publication of the works of Dryden. He has postponed the 'Life and Works' of our Immortall Poet till the others are published; but he has not declined nor given it up, as he convinced me by shewing me the manuscript copy of the genealogy of the Shakspeare family.[145]

In 1800 Malone published, in four volumes, *The Critical and Miscellaneous Prose Works of John Dryden, Now First Collected: with Notes and Illustrations; an Account of the Life and Writings of the Author, Grounded on Original and*

*Authentick Documents; and a Collection of His Letters, the Greater Part of Which Has Never before Been Published.* 'On reviewing the received accounts of his Life and Writings', Malone remarks in his Advertisement, 'I found so much inaccuracy and uncertainty, that I soon resolved to take nothing upon trust, but to consider the subject as wholly new; and I have had abundant reason to be satisfied with my determination on this head; for by inquiries and researches in every quarter where information was likely to be obtained, I have procured more materials than my most sanguine expectations had promised.'[146]

The life of Dryden is the greatest triumph of his scholarly method as applied to the problems of biography.* After its publication he returned to Shakespeare, who occupied him until the very end; but the gradual failure of his vision prevented him from making normal progress. His eyes were never very good, and he had worsened them by studying faded manuscripts by candlelight. On 25 May 1812 he succumbed, at the age of seventy, to a wasting disease. On his deathbed he asked Boswell, the biographer's son, who had known and loved him from infancy, to complete the *Shakespeare*.

This request Boswell honoured faithfully, and in 1821 he published, in twenty-one volumes, the edition of *The Plays and Poems of William Shakspeare* that has come to be known as the Third Variorum. The second volume is given over to a life of Shakespeare. Although with its several appendices, it comes to seven hundred pages, the Life is in some ways deeply disappointing. The portion of the book which Malone left ready for the press amounts to less than half of the whole, and takes Shakespeare only to the point of his arrival in London and entry upon the stage: Malone does not even get to Greene's attack. His narrative stops short of consideration of 'Shakspeare in his higher character of a poet', but as Malone's forte was not aesthetic criticism, this deprivation may be borne with better than the others. Nor has he, in these early sections, occasion to say much about the personal or moral character of his subject. 'I cannot but lament', Boswell writes, 'that much has unquestionably been lost, which, had he lived to superintend this edition himself, he would have furnished.'[147]

Of the 287 pages that Malone completed, more than a third—112—are given over to a discussion of two poems by Spenser, *The Tears of the Muses* and *Colin Clout's Come Home Again*. In 1790 Malone had taken up the issue of whether 'pleasant Willy' in the former is Shakespeare, and arrived at no settled conclusion; now he rejected the identification. In *Colin Clout* he is concerned with an obscurely allusive passage dealing with contemporary writers. The following passage especially absorbs his interest:

> And there though last not least is *Aetion*,
> A gentler shepheard may no where be found:

---

* For an appreciative analysis of Malone's achievement, see Osborn, *John Dryden*, 39–71.

> Whose *Muse* full of high thoughts invention,
> Doth like himselfe Heroically sound.        (ii. 444–7)

Aetion, Malone concludes, is Shakespeare. Clearly he is moved by a desire to link him with the greatest non-dramatic writer of the era, and to show that Shakespeare was 'duly appreciated by his illustrious and amiable contemporary; who, in talents and virtues, more nearly resembled Shakspeare than did any writer of that age; and who, we find, at a very early period of our great poet's dramatick life, had a high and just sense of his transcendent merits'.[148] Malone denies any need to apologize for his long disquisition, but surely he had committed a colossal blunder of judgement. With respect to *The Tears of the Muses* he has laboriously disproved an already discountenanced supposition; Pope, after all, had excised pleasant Willy from Rowe's preface. The attempt to identify Aetion in *Colin Clout* with Shakespeare is a failure, as must be any effort to pin down with certainty so ambiguous a reference. Scholars today regard Drayton as more likely to have been in Spenser's mind, but this too is guesswork.

Yet this fragment of a biography has redeeming excellences. On the Shakespeare genealogy Malone has brought to bear all his incomparable gifts for patient, minute investigation. He has corresponded with anyone who was able to supply information, he has travelled everywhere: he has searched the parish registers not only of Stratford, but also of the neighbouring villages, he has retrieved wills from the Consistory Office in Worcester and elsewhere, he has studied documents in the College of Arms. He succeeds in dispelling the mist of confusion and obscurity that had for so long surrounded the poet's family. Malone is interested even in Shakespeare's remote maternal ancestors—for example, Sir John Arden, the elder brother of the grandfather of Robert Arden, the father of Shakespeare's mother. He traces back the Ardens through four generations, and speculates interestingly on how the Shakespeares came to have that name.

Revising his earlier notes, he is able to show, once and for all, that John did not have ten (or eleven) children, but eight, of whom five lived to maturity; and that he was married not twice or three times, but only once. He correctly identifies the source of the confusion, the shoemaker John Shakespeare, whom he shrewdly deduces—from the sum he paid to the Company of Shoemakers and Saddlers for his freedom—not to have been a native of Stratford. From a document giving an account of proceedings in the bailiff's court in Stratford, Malone discovers that Shakespeare's father was not a wool dealer but a glover, and he has some acute observations on the status of this craft in Elizabethan England. He is able to supplement, from the Stratford borough records, the information he offered in 1790 on John's straitened financial condition in later life. The tradition that John Shakespeare was a woolman possibly derives, Malone suggests, from the fact that a window in the Birthplace bore, on stained glass, the arms of the Merchant of the Staple. It is an interesting speculation, but wrong: the stained glass was introduced long after John's time, by the poet's great grandson Shakespeare Hart, a glazier, who removed it from a window of the Guild Chapel, where Dugdale in the seventeenth century had remarked its presence. The tradition more probably stems from John Shakespeare's collateral dealings in wool, upon which light has been shed in the present century by Leslie Hotson.

Malone's conjecture that William was named after William Smyth, the mercer of Stratford, or William Smith, the haberdasher, is idle, the name being too common; but he also proposes more plausibly that the poet christened his only son after his friend Hamnet Sadler. He notes that there is no direct evidence that Shakespeare was born on 23 April. Only once in dealing with the complex family relationships does he go seriously astray: when he rejects Joseph Greene's suggestion—quite correct—that Shakespeare's wife belonged to the Hathaways of Shottery. Instead Malone prefers Luddington, a village two miles from Stratford, where some Hathaways were tenants of Arden relations early in Elizabeth's reign. Having retrieved the will of Bartholomew Hathaway of Shottery, Malone acknowledges that Anne might have been this Hathaway's sister; but as an argument against the possibility, he cites the fact that she is nowhere mentioned in the document. But Bartholomew *was* Anne's brother, as Malone should have gathered from the provision naming John Hall of Stratford, Shakespeare's physician son-in-law, as an overseer. It is one of the rare instances where Malone, with all the facts before him, has failed to make an obvious connection.

He repeats, with modifications, his earlier speculation that Shakespeare was initiated in the forms of law, imagining entertainingly that the poet's curiosity led him to attend the fortnightly sessions at Stratford of the Court of Record, where the Bailiff presided. But Malone admits that professional habits suggested this notion to him. Elsewhere he is on firmer ground. On the history of Stratford and the organization of its local government he offers much curious information. He ingeniously estimates, on the basis of burial entries in the parish register, the population of the town in Shakespeare's time; he records the names of all the bailiffs from the time their charter was granted until 1615; he draws attention to the plague that carried off more than a seventh of the inhabitants in the last six months of the year of Shakespeare's birth. Malone has an able passage on the free school, and on the sort of education the poet received there: the schooltexts he used, the degree of proficiency he attained in Latin ('a competent, though perhaps not a profound knowledge'). As regards Shakespeare's abilities as an actor, Malone rejects the view, passed down by Rowe, that he was not especially gifted. Malone cites Chettle and Aubrey, and the grasp of the performing art revealed by the plays; but he grants that the poet did not take parts of the first order.

His attitude towards traditions, which confront the biographer with some of his most vexing problems, is notably judicious:

There is, certainly, a great difference between traditions; and some are much more worthy of credit than others. Where a tradition has been handed down, by a very industrious and careful inquirer, who has derived it from persons most likely to be accurately informed concerning the fact related, and subjoins his authority, such a species of tradition must always carry great weight along with it.[149]

Nevertheless, Malone relaxes his own standards in presenting as a fact the traditionary information that Shakespeare was Davenant's godfather. He performs, however, a great service in providing the first scholarly treatment of the popular deer-poaching story. In so doing, Malone brings forward the earliest reference to the episode: the jottings of Richard Davies, which he had found in the archives of Corpus Christi College in 1792. He forcefully rejects the tradition, noting that the statutes did not permit whipping (alleged by Davies) for such an offence, and that no deer park existed at Charlecote in Shakespeare's youth. The spiteful ballad on Lucy, attributed to the poet, is a modern innovation. Despite Malone's scepticism, the story (as we shall see) would persist; reason inevitably yields to romance. Malone is similarly destructive in his treatment of the tradition of the horse-holding by Shakespeare and Shakespeare's boys at the theatre door, made famous by Johnson's retelling. At the only time in his career when he might have been available for this occupation, Shakespeare was not a mere boy but a husband and father, and the son of a former Bailiff of Stratford, who, whatever his financial difficulties, presumably retained some London connections. And is it likely that William suffered such mean employment when he might have turned for aid to the solid citizens of his native town? There was Sadler, the substantial baker; Quiney, probably related to the rich cloth-worker, Bartholomew Quiney, who had settled in London; Richard Field, son of the Stratford tanner, and now an eminent London printer. Malone has a novel alternative theory, and a more plausible one. He points out that Warwick's Men and Leicester's Men wore the livery of noblemen living within a few miles of Stratford, and that these companies frequently played there; the Queen's Men too performed in the town in 1587, and perhaps also the year before. Malone conjectures that Shakespeare became acquainted with some of the principal actors who visited Stratford, and that he there first made up his mind to profess the quality. Not until the twentieth century would it be discovered that in the summer of 1587 the Queen's Men lacked a player: for William Knell, who had taken the part of Prince Henry in *The Famous Victories of Henrvy V*, was killed in a fight that June. No evidence, however, exists of any Elizabethan troupe ever having recruited while on the road.

That Malone's life of Shakespeare is available to us in any form we owe to Boswell. He coped heroically with an impossible task. The completed fragment of the biography he printed as he found it. Malone's notes for the section on Elizabethan manners and customs were left in such disorder that his editor did not dare attempt to arrange them for publication. Instead he followed up the last completed segment, on Shakespeare as an actor, with Malone's essay on the chronology of the plays in its final revised form, although he confesses that the biographer might have preferred to keep it distinct. This long section is succeeded by the memoranda Boswell was able to rescue from Malone's scattered papers, presented as far as possible in the author's own words. Irrecoverably lost is the account which Malone promised of Shakespeare's brother Edmund; Boswell could find only the entry for his burial copied from the Stratford register. He filled in the gaps in the biography as best he could with the notes that Malone had appended to Rowe's Life in 1790. The result is a creditable and fairly consecutive performance, although there are minor carelessnesses (section xiii, for example, mysteriously follows directly upon the heels of section x).

One can only feel gratitude for what Boswell managed to salvage. The notes include a discussion of Shakespeare's relations with his patron Southampton, in which Malone dismisses, as 'totally unworthy of credit', the anecdotes of a gift to the poet of £1,000 to enable him to make a purchase. With respect to Shakespeare's will, Malone makes the important point that the declaration of religious principles at the outset affords no support for the notion that the dramatist was a Catholic; others have since made the same observation, without always crediting Malone for it. Boswell reproduces the tale of Shakespeare in search of the topers of Bidford in Jordan's own words, as communicated to Malone. Of much greater interest is the first public appearance of a new documentary find: Richard Quiney's letter of 25 October 1598, addressed 'To my loving good friend and countryman Mr. Wm Shackespere', and requesting a loan of £30. This 'very pretty little *relick*, about *three inches long by two broad*' (so he describes it in a letter to Percy) Malone had found in a bundle of Quiney letters while rummaging for ten days in 1793 through some three thousand documents in the Stratford archives.[150] Lastly, we may mention a contribution made by Boswell himself with the aid of materials left by his master. To the text of Shakespeare's will in the appendix, he adds a note on the omission of the wife from the first draft, as originally devised, and the interlineated provision of the second-best bed, which have 'created a suspicion that his affections were estranged from her either through jealousy or some other cause'.[151] Boswell anticipates the prevailing modern view by supposing that Shakespeare provided for Anne in his lifetime, and that in those days such a bequest as the bed carried no reproach: he has found among Malone's Adversaria the will of Sir Thomas Lucy the younger, who in 1600 left his second son Richard his second-best horse, but no land, because this his father-in-law had already promised him.

Despite its fragmentary character and eccentricity of proportion, Malone's life of Shakespeare is a major achievement. One need only compare it with Rowe's Life a century earlier to be made aware of the quantum leap forward in Shakespeare studies—a leap for which Malone is largely responsible. He had limitations and made mistakes; but he found out more about Shakespeare and his theatrical milieu than anyone before or since, and, recognizing his fallibility, corrected himself in the light of new information which his own sleuthing had unearthed. He is the greatest of Elizabethan scholars; others in the new century would build on his foundations. In an age that craved revelations of the kind that only an Ireland could furnish, not everybody, it is true, appreciated Malone's minute and painstaking labours. Take Alexander Chalmers, for example. For his short memoir of Shakespeare (originally published in 1805) this prolific editor and popular biographer merely collected the *disjecta membra* scattered through the Johnson–Steevens edition, but in his 1823 revision he incorporated carping references to Malone's posthumous Life. The latter's rejection of the deer-stealing legend Chalmers regarded as a positive deprivation—'We have lost the old tradition, with all its feasible accompaniments', he laments, 'but have got nothing in return.' As a man Chalmers won esteem for his warmth and affection; others reacted more hostilely to Malone. But only the irresponsible could ignore his findings. He had erected a new foundation for the study of Shakespeare's life.

# PART IV

## *The Earlier Nineteenth Century*

# 1

# *Romantics*

ONE age impinges upon another. In 1821, when James Boswell the younger published the Third Variorum with Malone's life of Shakespeare, Keats died. The Romantic critics—Lamb, Hazlitt, Coleridge—had effected a transformation in the way that Shakespeare was read and understood. Influence from the Continent, and especially from Germany, had modified English responses. Goethe, Herder, the brothers Schlegel, and Schelling (among others) had made their pronouncements on the genius they discovered to be universal, and their voices were heard. Goethe saw the superlative clarity of the dramatist's vision. To Schelling, Shakespeare is no longer, as during the *Sturm und Drang* movement, Nature's divine and savage child, but the conscious artist who in his poems deliberately elaborates and objectifies personal emotion. None of these critics, in either England or Germany, wrote biographies or biographical sketches of the poet who stirred them so profoundly. The dry raw materials of scholarship—parish registers, mortgages, wills—failed to excite their interest. Instead they concentrated their attention on the *œuvre*.

Yet no age indulged a greater passion for biography—capsule lives in the *Penny Magazine* and *Chambers's Journal*, biographical dictionaries and encyclopaedias catering to every level of educational attainment, 'libraries' of book-length biographies (Eminent Women, Men Worth Remembering, Lives Worth Living), specialized compilations devoted to particular vocations or classes (eccentrics, Welshmen, boy princes, virtuous wives).[1] All these flourished in the century after the elder Boswell's Life of Johnson. The period encouraged those works—unknown to any age before or since—of pious homage to the departed that Gladstone described as 'a reticence in three volumes'. It was the heyday of biographical criticism, when, by a curious inversion of priorities, men read the letters for the sake of the lives. They felt that in Goethe's, 'as in every man's writings, the character of the writer must lie recorded': 'his opinions, character, personality . . . with whatever difficulty, are and must be decipherable in his writings'.[2] Thus spake Carlyle.

The National Poet could scarcely escape the fierce embrace of the enthusiasts. They extended the frontiers of Shakespearian biography by first scrutinizing the Sonnets, and then the plays, for revelations of their creator's mind and soul. Assuming that 'by a comprehensive view of the informing spirit, the final scope and tendency of his works . . . we can ascertain the actual direction of his mind', Hartley Coleridge determined to his satisfaction that Shakespeare was a gentleman and a kindly, sincere, and decisive Tory.[3] By 1839 an obscure American cleric, Jones Very, could argue for a view of Shakespeare's personages as the 'necessary growths or offshoots' of his own character.[4] 'In Claudio's reflections on death, the poet unconsciously lays bare the texture of his own mind', and in Hamlet—'such a son as Shakspeare would have made'—he shows the wrestling of his own soul, passionately committed to life, with the great antagonist Death.[5] Other critics and scholars would take their lead from more literary sources than did Very; the most powerful formative influence on his thought was the Spirit, without whose guidance (as Dr Channing reported) he ventured only with reluctance from his chair to the fireplace. It was left to Very's Transcendental acquaintance Emerson to make, at mid-century, the classic pronouncement. 'Shakspeare', he wrote in his *Representative Men*, 'is the only biographer of Shakspeare.'

A German led the way. August Wilhelm Schlegel, in an article for Schiller's *Horen* (1796), announced his startling perception of the chief value of the Sonnets: their apparent inspiration by an actual friendship and love.[6] In his *Lectures on Dramatic Art and Literature*, delivered in Vienna in 1808 'to a brilliant audience of nearly three hundred individuals of both sexes', Schlegel gave the idea fuller statement:

It betrays more than ordinary deficiency of critical acumen in Shakspeare's commentators, that none of them, so far as we know, have ever thought of availing themselves of his sonnets for tracing the circumstances of his life. These sonnets paint most unequivocally the actual situation and sentiments of the poet; they make us acquainted with the passions of the man; they even contain remarkable confessions of his youthful errors.[7]

It is in his Sonnets that the poet alludes to the public means which public manners breeds and to the vulgar scandal stamped upon his brow. In the plays the self-deceptions and semi-self-conscious hypocrisies of the personages lead Schlegel to discern, on the part of the artist, 'a certain cool indifference, but still the indifference of a superior mind, which has run through the whole sphere of human existence and survived feeling'.[8] The insight anticipates Keats.

The *Lectures* became available to an English-speaking audience in John Black's translation of 1815. Schlegel's name and his theme would echo in subsequent criticism. Others seem to have made a similar discovery on their own: 'There is extant a small Volume of miscellaneous poems, in which Shakespeare expresses his own feelings in his own person,' Wordsworth wrote in the 'Essay Supplementary to the Preface' to his *Poems* (1815). And in 'Scorn Not the Sonnet' of 1827 he formulated the phrase that would have a haunting effect: 'With this key, Shakespeare unlocked his heart.' 'Did he?' retorted Browning in 'House', forty years later; 'If so, the less Shakespeare he!' But by then the critical tide was not to be stemmed.

The key was, however, brushed aside by Wordsworth's associate, the most brilliant and influential of the Romantic critics. To Coleridge, 'Shakespeare's poetry is characterless; that is, it does not reflect the individual Shakespeare.' Chaucer we know well; of Shakespeare we know nothing. Yet Coleridge has his own vision of Shakespeare. Amid the wreckage of his private life, the Lake Poet formulated an ideal image of transcendent genius. The questionable tradition, reported by Oldys, of Shakespeare in the part of old Adam initiates not an historical enquiry into the playwright as an actor but, rather, an ecstatic reverie:

Great dramatists make great actors. But looking at him merely as a performer, I am certain that he was greater as *Adam*, in 'As you Like it,' than Burbage, as *Hamlet*, or *Richard the Third*. Think of the scene between him and *Orlando*; and think again, that the actor of that part had to carry the author of that play in his arms! Think of having had Shakespeare in one's arms! It is worth having died two hundred years ago to have heard Shakespeare deliver a single line. He must have been a great actor.[9]

Shakespeare was a patriot. He loved children and nature, and respected kings, physicians, and priests. He possessed an exquisite sense of beauty. If gross passages occur in his plays, the age is accountable; anyway, they contain not a single vicious sentiment. At all times he kept to the high road of life: 'With him there were no innocent adulteries; he never rendered that amiable which religion and reason taught us to detest; he never clothed vice in the garb of virtue.'[10] Although endowed with all manly powers, and indeed more than a man, Shakespeare yet had 'all the feelings, the sensibility, the purity, innocence, and delicacy of an affectionate girl of eighteen'.[11] What though of the Sonnets which, whatever emotions they express, they are not those ordinarily associated with virginal girls of eighteen? Although Coleridge allowed that men in a moral state might entertain a love for one another aloof from appetite, homo-erotic passion aroused his disgust. Writing in 1803 for the future guidance of his son Hartley, Coleridge proclaimed his inability to find anywhere in Shakespeare 'even an allusion to that very worst of all possible vices'. (Apparently he had read *Troilus and Cressida* carelessly.) The way in which he reconciled the Sonnets with his scheme is indicated by a remark, made shortly before his death, from the *Table Talk*: 'It seems to me that the sonnets could only have come from a man deeply in love, and in love with a woman, and there is one sonnet which, from its incongruity, I take to be a purposed blind.'[12] Was he thinking of Sonnet 20, with its phallic reference?

In Coleridge (as in the Germans) the Shakespeare who warbled native woodnotes wild gives way to the deliberate master:

Shakspeare, no mere child of nature; no automaton of genius; no passive vehicle of inspiration possessed by the spirit, not possessing it; first studied patiently, meditated deeply, understood minutely, till knowledge, become habitual and intuitive, wedding itself to his habitual feelings, and at length gave birth to that stupendous power, by which he stands alone, with no equal or second in his own class; to that power which seated him on one of the two glory-smitten summits of the poetic mountain, with Milton as his compeer not rival.[13]

But Coleridge will not yield everything to Art; Shakespeare was the child of Nature too:

That gift of true Imagination, that capability of reducing a multitude into unity of effect, or by strong passion to modify series of thoughts into one predominant thought or feeling—those were faculties which might be cultivated and improved, but could not be acquired. Only such a man as possessed them deserved the title of *poeta* who *nascitur non fit*—he was that child of Nature, and not the creature of his own efforts.[14]

Does Coleridge contradict himself? Very well, he contradicts himself; he contains multitudes.

The gigantic, myriad-minded *persona* that emerges from his scattered remarks—the remarks of a disciple of poetry who detested the stage representation of Shakespeare—bears little resemblance even to an extraordinary mortal occupied, as most mortals must be, with the bread and cheese of daily life: that life in the course of which a man wrote plays to please a theatrical company and audiences, acted in grease-paint on the stage with other actors, shared management responsibilities, made purchases, and retired in prosperity to a sleepy provincial town. Those future biographers nursed on Coleridgian criticism would have somehow to bridge the chasm between the profound self revealed by the writings and the everyday self chronicled in the records. But he furnished later biographers of the spirit with a remarkably suggestive lead; for Coleridge is the first to attempt to classify the writings according to phases of the artist's development. In his manuscripts he speaks of Five Epochs, and in a lecture at the Crown and Anchor Tavern on 25 February 1819 he outlined Five 'Aeras'. From this seed would in time spring Dowden's Four Periods.

Keats, who died young, naturally has less to say about Shakespeare than Coleridge, and as he was not a professional critic or lecturer, his remarks, more casual and programmatic, are scattered through his private correspondence. Yet Shakespeare exerted the most profound influence on his sensibility. Like so many others Keats made the pilgrimage to Stratford, signed the visitors' book, added his name to the countless signatures that blackened the Birthplace walls, and was struck by the simple statue in Holy Trinity. His appreciation of the visit, his companion Bailey observed, 'was of that genuine, quiet kind which was a part of his gentle nature; deeply feeling what he truly enjoyed, but saying little'.[15] A portrait of Shakespeare (by whom we do not know) hung over his desk; he lovingly scored, marked, and annotated his edition of the *Dramatic Works*; Shakespeare, he fancied, was the good genius who presided over him.

Three of Keats's remarks about Shakespeare as man and artist stand out especially. For Keats, the playwright has the imaginative faculty to understand and translate into poetry the random and petty circumstances of daily existence. 'A Man's life of any worth is a continual allegory', he wrote to George and Georgiana Keats in February 1819, '—and very few eyes can see the Mystery of his life—a life like the scriptures, figurative—which such people can no more make out than they can the hebrew Bible. Lord Byron cuts a figure—but he is not figurative—Shakspeare led a life of Allegory; his works are the comments on it—.'[16] Later the same year, in a letter to Sarah Jeffrey, he reflected on the English way of ill-treating their writers—the finest in the world—during their lifetime, and celebrating them after their death; he cites Shakespeare as an instance: 'The middle age of Shakspeare was all couded [clouded?] over; his days were not more happy than Hamlet's who is perhaps more like Shakspeare himself in his common every day Life than any other of his Characters.'[17] But his most piercing insight came earlier, in December 1817, as he walked back to Hampstead from the Christmas pantomime at Drury Lane with Charles Armitage Brown and Charles Dilke. They talked about various subjects, and suddenly Keats thought of that annihilation of self on the poet's part—his denial of the 'egotistical sublime'—which enables him to make his mind a 'thoroughfare for all thoughts' and to feel truth as beauty. A few days later he wrote to his brothers George and Tom:

several things dovetailed in my mind, & at once it struck me, what quality went to form a Man of Achievement especially in Literature & which Shakespeare pos[s]essed so enormously—I mean *Negative Capability*, that is when man is capable of being in uncertainties, Mysteries, doubts, without any irritable reaching after fact & reason.[18]

The lesser poet, such as Coleridge (Keats's example is not very felicitous), cannot remain satisfied with half-knowledge. 'This pursued through Volumes would perhaps take us no further than this, that with a great poet the sense of Beauty overcomes every other consideration, or rather obliterates all consideration.'[19]

Keats's statement of Shakespeare's negative capability is more astonishing for its phrasing than for any quintessential originality. Others in the period (we have already noted Schlegel) were commenting on the poet's detachment, his denial of the self. A few weeks after Keats's letter, on 27 January, Hazlitt in his lecture 'On Shakspeare and Milton' at the Surrey Institution observed that Shakespeare

was the least of an egotist that it was possible to be. He was nothing in himself; but he was all that others were, or that they could become. He not only had in himself the germs of every faculty and feeling, but he could follow them by anticipation, intuitively, into all their conceivable ramifications, through every change of fortune, or conflict of passion, or turn of thought. He had 'a mind reflecting ages past,' and present:—all the people that ever lived are there. . . . He had only to think of anything in order to become that thing, with all the circumstances belonging to it.[20]

That night Keats sat in the audience.

The same year Keats's brother Tom died, and the poet moved into the house of his new friend Charles Armitage Brown at Wentworth Place in Hampstead; the house now preserved as the Keats Museum. Brown, eleven years his companion's senior, was to be described long afterwards by his son as 'methodical, hospitable, kind hearted, and very cool in the presence of danger'. He looked after Keats, borrowed money to lend to him, and nursed him when he was ill. Keats addressed his last, heart-breaking letter to Brown. He is best remembered as the poet's friend and posthumous memorialist, and also as the author of a single volume of criticism: *Shakespeare's Autobiographical Poems*, published in 1838. Although Keats was by then dead fifteen years, some have assumed that the book owes much to conversation with him. Perhaps it does; we know that Brown was haunted for the rest of his days by the memory of the genius who had died young. It is also true, however, that he developed new loyalties. *Shakespeare's Autobiographical Poems* is dedicated to Walter Savage Landor, with whom he had discussed the subject in Florence as early as 1828. 'I have made a strange discovery into the character of Shakespear,' Brown proudly revealed to Leigh Hunt in a letter dated 1 June 1830. 'A good deal is written 60 pages, and so delighted and convinced was Landor, that he used to visit me nearly every morning to hear how I was going on. . . . Then his [Shakespeare's] Sonnets,—there I have made a discovery, a key to the whole, which Landor declares incontrovertible,—it is a discovery of which I am almost ashamed, as no one else discovered it before me. . . . Landor swears, out and out, that no man ever understood Shakespear like myself.'[21]

Brown's title is characteristically inexact, describing not the book as a whole but one section of a loosely related series of essays amounting to a biography of sorts. The facts of the poet's life, as Brown understood them from the most recent secondary sources, provide matter for three chapters. Shakespeare is oddly made one of six children, and the family's financial setbacks are exaggerated: Brown refers to John's 'ruined trade' and the difficulties of being the eldest son of 'a large and an extremely impoverished family'. Familiar with Malone's Life, he vigorously rejects the deer-poaching tradition, only to muddy the waters he has himself cleared by suggesting that Shakespeare in *The Merry Wives of Windsor* attacked Sir Thomas Lucy's son, supposedly a jealous guardian of game, and was prosecuted for the libel. Brown is misguidedly positive that Nashe's famous passage, usually applied to Kyd, on the shifting companions who busy themselves with the endeavours of art actually refers to Shakespeare, who was therefore a lawyer's clerk and the author, at the age of twenty-four, of a first draft of *Hamlet*. Other chapters bear such titles as 'His Learning', 'He Never Was a Flatterer', and 'Did He Visit Italy?'

Himself in love with Italy, Brown is apparently the first to speculate on whether Shakespeare visited that land. Not surprisingly, he decides in the affirmative, and he goes so far as to plot Shakespeare's itinerary: from Venice the poet went through Padua, Bologna, and Florence, to Pisa. Maybe too he journeyed a little out of his way to see Rome, and perhaps he also stopped at Verona, the scene of *Romeo and Juliet*; but here Brown concedes having no grounds for such a supposition. His 'evidence' for Shakespeare's Italian travels reveals his amateur status as a scholar. Act I, sc. i, of *The Taming of the Shrew* is set in '*A public place*', a fact that prompts Brown to make the following learned comment: 'For an open place or a square in a city, this is not a home-bred expression. It may be accidental; yet it is a literal translation of *una piazza publica*, exactly what was meant for the scene.'[22] Brown fails to realize that the place settings are the interpolations of later editors.

Brown's chief contribution, however, is the fullest autobiographical reading yet essayed of the Sonnets. 'With this key', his title-page proclaims, echoing Wordsworth, '. . . every difficulty is unlocked, and we have nothing but pure uninterrupted biography.' Brown also remembered Schlegel's remarks on the Sonnets a generation earlier, and wondered why they had not spurred inquiry. He sees the cycle as comprising six self-contained poems in the sonnet stanza: the poet seeks to persuade his friend to marry (1–26), forgives him for having been seduced by the Dark Lady (27–55), complains of the Fair Youth's coldness and warns him of life's decay (56–77), remonstrates with him for preferring another poet's praises and upbraids him for faults which may harm his character (78–101), excuses to him his own silence while denying the charge of inconstancy (102–26), and addresses his mistress on her unfaithfulness (127–52). The two final sonnets, on Cupid and a nymph of Diana, belong apart.

Except for his insistence on treating the Sonnets as six integrated poems, Brown's reading seems straightforward enough to the modern reader. The novelty resides in the very weight given to the sequence as personal revelation. Critics have continually lamented that we know almost nothing of Shakespeare's life, Brown reminds his readers; yet in the second of the six poems, in which the writer is torn between friendship and his sense of sexual betrayal, we have described—by Shakespeare himself—a most intimate experience, with many attendant circumstances that lay bare his character. Such tales we seek in vain in the biographies of other great men; here, at last, we witness the workings of Shakespeare's passions. Seeing his beloved friend yield to the lures of the dark beauty who has captivated him, the philosophical poet explodes in a fit of ungovernable rage—only to be instantly disarmed when the friend expresses sorrow for his lapse. Once the animal in him has yielded to reason, the poet is not merely forgiving, 'but kind, affectionate, seeking excuses for the wrong he had endured, and heart-struck at the recollection of his resentment'.[23] He is gentle Shakespeare after all.

Brown, who does not claim to be a scholar, establishes an attractively personal relationship with his readers. It is perhaps best illustrated by the passage in which he confronts the delicate question of Shakespeare's culpability in taking a mistress while his Penelope remained behind in Stratford. 'For myself', he owns,

I confess I have not the heart to blame him at all,—purely because he so keenly reproaches himself for his own sin and folly. Fascinated as he was, he did not, like other poets similarly guilty, directly or by implication, obtrude his own passion on the world as reasonable laws. Had such been the case, he might have merited our censure, possibly our contempt. On the contrary, he condemned and subdued his fault, and may therefore be cited as a good rather than as a bad example. Should it be contended that he seems to have quitted his mistress more on account of her unworthiness than from conscientious feelings, I have nothing to answer beyond this: I will not join in seeking after questionable motives for good actions, well knowing, by experience, that when intruded on me, they have been nothing but a nuisance to my better thoughts.[24]

Malone, brought up in a more austere school, would not have expressed himself thus.

Nor would he, or other critics of his age, have sought the writer's inner life in his art. It did not occur to them that his works were (in Carlyle's words) 'so many windows, through which we see a glimpse of the world that was in him'.[25] Carlyle, for whom history itself is the essence of innumerable biographies, typifies a new enthusiasm. The tradition of disinterested scholarly investigation of the lives of men would be maintained, and new facts about Shakespeare would increase the meagre store; but the seer of Cheyne Row inveighs powerfully against the Dryasdust plodders. 'How', Carlyle thunders, 'shall that unhappy Biographic brotherhood, instead of writing like Index-makers and Government-clerks, suddenly become enkindled with some sparks of intellect, or even of genial fire; and not only collecting dates and facts, but making use of them, look beyond the surface and economical form of a man's life, into its substance and spirit?'[26] Not unexpectedly, Carlyle fixed on Shakespeare as one of his subjects, and, true to his own visionary ideal, created a polemic—it is no biographical sketch in the usual sense of the term—memorably devoid of facts and dates.

The Shakespeare indistinctly discernible amid the rhetorical excesses of Carlyle's lecture of 12 May 1840 is the Poet as Hero. Unlike Dante, deep and fierce, this Shakespeare is 'wide, placid, far-seeing'. He has always been a spring in which men discover, Narcissus-like, their own reflection, and so we need feel no surprise that Carlyle, who came of Ecclefechan peasant stock, should seize on the myth that Shakespeare was a 'poor Warwickshire Peasant', and in turn help to propagate it. How fortunate for posterity that a vindictive squire prosecuted him for deer-stealing, and thus uprooted a lad for whom native woods and skies might otherwise have sufficed. True, this rustic happened to be 'the greatest intellect who, in our recorded world, has left record of himself in the way of Literature', but Carlyle has in mind instinctual, *unconscious* intellect (the italics are his). Shakespeare is a voice of nature, Carlyle proclaims; the critic thus takes his place in a biographical tradition reaching back to Beaumont and the seventeenth century. But the new note that we encountered in Charles Armitage Brown is also sounded in 'The Poet as Hero'. We are offered a Shakespeare who battled against misery, and overcame. 'Doubt it not', the author assures us, 'he had his own sorrows: those *Sonnets* of his will even testify expressly in what deep waters he had waded, and swum struggling for his life;—as what man like him ever had not to do?'[27] But perhaps his greatest torment, to which the imperfections of his work testify, is that he had to endure the constraints of the market-place: 'Alas, Shakspeare had to write for the Globe Playhouse: his great soul had to crush itself, as it could, into that and no other mould.'[28] Still, the career of the Stratford peasant was triumphant: he contrived to escape Sir Thomas Lucy's treadmill, received some kind glances from a noble lord, and rose to be a theatre manager 'so that he could live without beggary'.

Like Carlyle (the mirror again) the poet is prophet. If Dante was the melodious priest of the Catholicism of the Middle Ages, Shakespeare is 'the still more melodious Priest of a *true* Catholicism, the "Universal Church" of the Future and of all times'.[29] Not recognized as a god in his own lifetime, he represents for Carlyle and his fellow hero-worshippers the archetypal Superman-Poet, more precious to an England in the heyday of imperialism than any Indian empire. Thus did the bardolatry first naïvely celebrated at the Stratford Jubilee of 1769 achieve its imaginative apotheosis in the rhapsody of a Victorian romantic.

# 2

# *Wheler and Drake*

To turn from Coleridge, Keats, and Carlyle to Robert Bell Wheler and Nathan Drake is to move from the inspired poet-critic and seer to the antiquary and scholarly amateur. No one would lay claim that these men, who led such quiet lives, are the unacknowledged legislators of mankind. Yet one stirs with pleasure the dust of oblivion that has settled upon their writings, which make a contribution to Shakespearian biography more substantive as well as more prosaic than that of their famous contemporaries.

A true son of Stratford, Wheler was born there, educated at the free school, and articled to his father, who practised law in the town that then numbered twenty-four hundred souls. Following in his parent's footsteps professionally, Wheler passed his days in bachelor tranquillity in the house of his birth, Avon Croft 2, part of the mansion once owned by the powerful Cloptons in Old Town, near Holy Trinity. When he died he was buried in the yard, just left of the path leading to the porch of the venerable church which (so he tells us) he was much in the habit of contemplating and admiring. Wheler seems to have absented himself from Stratford only once in his lifetime, when he passed a month in London at the time of his formal admission to the Bar. His leisure hours he dedicated to his twin passions: Stratford and the poet who had conferred glory upon it. Whatever information he could acquire went into his voluminous notes, which admit of such curiosa as the fact that Martha Hart, buried 17 May 1802, was no descendant of Shakespeare's sister but 'a Hell Lane Venus, more familiarly called Pat Hart'.

In 1806 Wheler celebrated his majority by publishing a volume the contents of which are sufficiently indicated by the title: *History and Antiquities of Stratford-upon-Avon: Comprising a Description of the Collegiate Church, the Life of Shakspeare, and Copies of Several Documents Relating to Him and His Family, Never before Printed; with a Biographical Sketch of Other Eminent Characters, Natives of, or Who Have Resided in Stratford. To Which Is Added, a Particular Account of the Jubilee, Celebrated at Stratford, in Honour of Our Immortal Bard.* Like all of Wheler's few publications, the book was published, with endearing amateurishness, in Stratford; the engraving of Shakespeare's monument appears upside down in the Newberry Library copy. The eight plates that embellish the *History and Antiquities* were all engraved from Wheler's own studiously precise drawings. The Stratford scenes they depict include, in addition to the monument, the Birthplace and New Place.

For his brief life of Shakespeare, Wheler depends shamelessly and without acknowledgement on Rowe. Mostly he follows the early memoir word for word, although now and then he incorporates later information. There is the inevitable hardening of rumour into purported fact: 'One singular instance of the munificence of this patron of Shakspeare [Southampton] is, that he gave him, at one time, a thousand pounds, to enable him to compleat a purchase which he heard he had a mind to.'[30] The size of the sum had at least given Rowe pause, and he had specified Davenant as the source of the tradition. On the basis of this Life alone, Wheler would hardly merit a place in this chronicle. But in an appendix he publishes for the first time several new documents. The first is the unexecuted counterpart of the conveyance of the Old Stratford freehold to Shakespeare by William and John Combe in 1602. There follow transcripts (not entirely accurate) of two previously unknown writs, dated 15 March and 7 June 1609 and issued by the Stratford Court of Record, which had jurisdiction over cases involving debts up to a limit of £30.

The first order cites John Addenbrooke, a gentleman of Stratford, for a debt of £6, plus 26 shillings in costs and damages, recovered against him by 'Willielmo Shackspere generoso'. The second writ is issued against Thomas Horneby, Addenbrooke's surety, to show cause why he should not make good the sum defaulted by Addenbrooke. Little excitement attaches to the case, and the records are uninformative: who Addenbrooke and Horneby were, and what dealings Shakespeare had with them, remain undiscovered. Yet these documents do show Shakespeare's continuing Stratford involvement while his professional life lay in London, and their substance—his strenuous efforts to recover an inconsiderable debt—would prove disturbing to Victorian idolaters. Wheler also prints the texts of several indentures that shed light on the subsequent history of the Shakespeare estate, after New Place and other properties had passed into the hands of his granddaughter Elizabeth. In another section Wheler gives a compendious history of New Place, in which he refers casually (on the basis of papers in his possession) to Shakespeare's acquisition of the property from the Underhill family in 1597. The date of the purchase, as we have seen, was not known to Malone, and Wheler is apparently the first to mention it.

For his antiquarian labours Wheler early won modest fame as Stratford's leading authority on Shakespeare's life. 'There is great Room for a comprehensive attempt to elucidate the Passages forming the Groundwork of the received Biography of this wonderful Writer,' a well-wisher wrote to him in 1814. 'If *your* Leisure would allow I should think it an undertaking most congenial to your Inclination; and one for which your Investigations have peculiarly qualified you.'[31] But such an undertaking was not in fact congenial to his inclination; instead, that year, Wheler published *A Guide to Stratford-upon-Avon.* An abridgement of his first book, the *Guide* has a few new features: an account of the Birthplace, a revised biographical sketch of Shakespeare, the first attempt at a life of Dr Hall, and pen portraits of several local characters with Shakespearian associations: John Combe, Joseph Greene, and John Jordan.* In his pages on the dramatist, Wheler again produces some scraps of fresh information, this time from the diary of Thomas Greene, the Town Clerk, dealing with the proposed enclosure of Welcombe discussed in 1614 and 1615. Just how Greene was related to the poet we do not know, but in these memoranda he refers to conversation and correspondence with his 'Cousin Shakspeare'. Wheler also reveals for the first time Shakespeare's purchase in 1605, for £440, of half of a leasehold interest in a parcel of Stratford tithes. Thus the documentary record grew.

That record would be enhanced two decades later by Wheler's most striking single contribution to Shakespearian biography, although he did not

---

* Of special interest is the annotated interleaved copy of the *Guide* in the Shakespeare Birthplace Record Office.

himself actually make the discovery. In a letter to Sylvanus Urban, as the editor of the *Gentleman's Magazine* styled himself, Wheler reported that Henry Clifton of Worcester had passed along to him a document recently brought to light in the Consistorial Court of that city. (Ten years would pass before a grateful public would learn the name of the finder: Sir Thomas Phillipps, the bibliomaniac who became the unwilling father-in-law of the century's most distinguished contributor to Shakespearian biography, James Orchard Halliwell.) This record is the bond of 28 November 1582, according to which Fulke Sandells and John Rychardson of Stratford stood surety that William Shakespeare would lawfully solemnize matrimony with Anne Hathaway, a maiden of the same town. Thus the identity of the poet's bride, first published by Rowe, is unequivocally confirmed, and the proximate date of the marriage revealed. Wheler subjoined the complete text of the document, which was printed with his letter in the September 1836 number. Southey might deride the 'exquisite inanities' of the *Gentleman's Magazine*, but for long it furnished an indispensable outlet for Shakespearian lore.

Wheler's last publication (apart from a new abridgement of the *Guide*) is an *Historical and Descriptive Account of the Birth-place of Shakspeare*, illustrated with a plan and nine lithographs by C. F. Green. This slight work, which comes to only thirteen pages, furnishes the first connected history of the property, and a detailed and scrupulous description of it as it stood in the early nineteenth century. Halliwell-Phillipps, who thought it the best account of the Birthplace by anyone, reprinted the pamphlet in 1863 without the illustrations.

That year Halliwell-Phillipps also published a handlist of Wheler's collections, which his sister Anne had presented to the town that he loved so devotedly. The assemblage of a lifetime of antiquarian zeal, they include such curiosities as the massy gold ring that Haydon hysterically described to Keats on 4 March 1818: 'I shall certainly go mad!—In a field at Stratford upon Avon, in a field that belonged to Shakespeare; they have found a gold ring and seal with the initial thus—*a true WS Lover's Knot between*; if *this* is not Shakespeare who is it?—a true lovers Knott!!—I saw an impression to day, and am to have one as soon as possible—As sure as you breathe, & that he was the first of beings the Seal belonged to him—Oh Lord!—'[32] The ring, alas highly dubious, eventually became part of the Shakespeare Birthplace Trust, along with Wheler's bona fide acquisitions, including the record of the New Place purchase in 1597, the assignment of the interest in the Stratford tithes in 1605, and the Welcombe documents. In the Birthplace Record Office, moreover, may today be found thirty-four volumes of his manuscripts and scrapbooks; bits and pieces of every description—old documents, modern transcripts, drafts of letters, tokens, coins, pedigrees, engraved portraits and views, letters from Malone and other scholars, drawings and papers by Jordan. The prize of the collection is Wheler's unpublished repository of local lore, *Collectanea de Stratford*, in 536 quarto pages in Wheler's minute and marvellously legible hand.

Wheler's contemporary Nathan Drake was, like him, a professional man, in this case a physician. A Yorkshireman with a medical degree from Edinburgh, he eventually settled as a general practitioner in Hadleigh in Suffolk. The public at large, however, best knew him as a literary amateur of cultivated and discriminating taste. He was elected an honorary fellow of the Royal Society of Literature. Drake found a ready audience for his collections celebrating the hours and seasons: *Winter Night, Literary Hours, Evenings in Autumn, Noontide Leisure, Mornings in Spring.* In one of his essays he describes the rapture with which as a boy he hung over the pages of Spenser, Milton, Thomson, and Gray. But Shakespeare was his chief passion.

For thirty years Drake gave most of his leisure hours to study of the dramatist. Living in a small provincial market town away from the great libraries and primary documents—without the resources of a Malone or Wheler—he chose to produce a work of synthesis assimilating the uncoordinated knowledge of Shakespeare and the Shakespearian context made available by Malone, Wheler, and eighteenth-century scholarship. *Shakspeare and His Times* appeared in 1817 in two huge folio volumes comprising fourteen hundred pages. Drake had dedicated himself to the work with love, 'with incessant labour and unwearied research', and also with painful anxiety lest it prove an unworthy tribute to the Immortal Bard.

The novelty of Drake's contribution lies in the fact that he is the first to view Shakespeare's life as a challenge to the art of literary biography, and to confront in terms of that art the overwhelming problem posed by the meagreness of the personal records. Drake's solution is best expressed in his own words. His strategy is the 'blending with the detail of manners, &c. such a portion of criticism, biography, and literary history, as should render the whole still more attractive and complete':

In attempting this, it has been his [the author's] aim to place Shakspeare in the foreground of the picture, and to throw around him, in groups more or less distinct and full, the various objects of his design; giving them prominency and light, according to their greater or smaller connection with the principal figure.

More especially has it been his wish, to infuse throughout the whole plan, whether considered in respect to its entire scope, or to the parts of which it is composed, that degree of unity and integrity, of relative proportion and just bearing, without which neither harmony, simplicity, nor effect, can be expected, or produced.[34]

Because Drake sought a synthesis, it will be well to consider his achievement as a whole, including those portions that lie outside the biographical domain.

Despite his intention to keep Shakespeare in the foreground, the nature of Drake's material is such that the Times tend to engulf the Life. The first four chapters, comprising sixty-seven pages, take Shakespeare up to the time of his departure from Stratford. There follows a survey, occupying 332 pages, of country life during the Elizabethan period—a 'slight sketch', Drake describes it. Contextual material predominates throughout, and for these sections Drake read prodigiously in spite of his isolation and medical responsibilities. He knows all the standard secondary sources, and he has also profited from modern reprints: Holinshed, *The Gull's Hornbook*, the *Ancient British Drama*, the plays of Massinger and Jonson, *The Progresses and Public Processions of Queen Elizabeth*. Drake studs his pages with quotations from his varied sources, and also with apposite passages from Shakespeare's works to ensure that the object of his study does not entirely drop from view. These chapters concerned with social history provided early nineteenth-century readers with a rough equivalent of such modern guides as *Shakespeare's England*. Drake dwells on how people dressed and ate, the shelters they placed over their heads, their recreations and ceremonials, the daily round of their lives. He notes when forks were first introduced into England (1611), he investigates learnedly whether Falstaff's sack was dry or sweet (it was sweet, he decides), and he reports curious customs—for example, the practice, 'still kept up in some parts of the north', of presenting children at Easter with boiled eggs stained with various colours.

But Drake's great passion is for the supernatural. One of his most ambitious chapters reviews country superstitions, and when at length he turns from Shakespeare's times to the plays, his preoccupation reasserts itself: *A Midsummer Night's Dream* prompts an excursion into fairy mythology; *Hamlet* inspires a dissertation on apparitions; and *The Tempest*, one on magic and enchantment. Thus did the romantic impulse assert itself in the sober physician of Hadleigh.

Part II, on Shakespeare in London, includes a section on metropolitan life that parallels the view of rural customs in the first book, including an elaborate survey of the state of letters at the time, which carefully enumerates and grades more than 230 poets. An analogous later section provides listings and brief critiques of Shakespeare's predecessors and contemporaries on the stage.

As a critic, Drake is a not especially brilliant representative of the Romantic school; his heaviest debt is to Schlegel, to whom he pays tribute as the greatest of Shakespearian critics. A scholarly amateur, he is out of touch with the new sensibility: he refers to Byron but once, and not at all to Wordsworth or Coleridge. For Drake, as for Schlegel, Shakespeare's plays must be judged not by neoclassical standards but by the canons of the Romantic drama which Shakespeare largely created and which transcends the glories of the Greek theatre. Drake tends to moralize the plays, seeking and finding in them many 'just and useful maxims'. An impressionist using the vocabulary of enthusiasm, he is prone to indiscriminate eulogy: he sees no falling off in the

Falstaff of *The Merry Wives of Windsor*, and regards Parolles as second only to
the fat knight as a comic creation. We remind ourselves with difficulty that
*Shakspeare and His Times* appeared in the same year as Hazlitt's *Characters of
Shakespear's Plays*. In the section on the poets, Donne receives short shrift;
Drake, speaking for a 'more refined age', finds the greatest of the metaphysi-
cals deficient in harmony of versification and in simplicity of thought and
expression. Less sweeping than Malone in his dismissal of the earlier
Elizabethan playwrights, he recognizes among them a 'few feeble lights'
shedding a kind of twilight: *Tamburlaine*, several passages excepted, is 'a
tissue of unmingled rant, absurdity and fustian', but *Edward II* and *Doctor
Faustus* evoke praise. Drake's knowledge of Elizabethan literature does give
him an advantage over contemporary Shakespearians. Equipped with a
yardstick of critical relativism, he can show how the faults that the
commentators point to in Shakespeare characterize the age. In Drake's pages
the Sonnets, despised by Steevens and lukewarmly defended by Malone,
receive the attention that is their due.

   In the strictly biographical chapters Drake's chief contribution is the
ordering of previously disparate materials in a consecutive narrative. His
chief source is Isaac Reed's 1803 edition of *The Plays of William Shakspeare*,
known as the First Variorum. Reed, like other early editors, brings together in
his preliminary volumes the contributions of predecessors. Here Drake found
Rowe's *Account* with Malone's annotations, the Oldys anecdotes, the extracts
from the Stratford parish registers, the texts of documents, the prefaces of
previous editors from Pope to Malone, Farmer's essay on Shakespeare's
learning along with Colman's rejoinder, and the Stationers' Register entries
for the poet's works. But Drake also knows all the other essential books that
touch upon his subject: Ireland's *Picturesque Views*, Malone's *Inquiry*, Chalmers's *Apology* and *Supplemental Apology*, Wheler's *History and Antiquities* and
*Guide to Stratford*. His assimilation of this material results in a quintessential
early nineteenth-century life of Shakespeare reflecting the current state of
knowledge—and ignorance.

   If Drake's critical sensibility is limited, so too is his equipment as a scholar.
He presents suppositions as facts (Shakespeare is stated categorically to have
been born on 23 April), and he does not always get his verifiable facts
straight: he cites Fuller's *Worthies*, correctly, as enumerating Simon and
Edward Ardern as sheriffs of Warwickshire in the eleventh and sixteenth
years of Elizabeth's reign, but he wrongly gives those years as 1562 and
1568 rather than as 1569 and 1574. With the perplexing questions that
Malone would do so much to elucidate in his unfinished Life, Drake sheds
little new light: John Shakespeare was a probable Catholic, a wool stapler
(and very likely also a butcher) who had, it seems, eleven children and three
wives. As regards Shakespeare's education, Drake with amiable moderation
steers a middle course between Farmer, who would grant the dramatist only

*Hig, hag, hog* and a familiar phrase or two of French or Italian, and those extremists who saw him as a learned man.

Traditional anecdotes, which for Malone provided such a challenge to historical enquiry, find hospitable lodging in Drake's pages. We hear of the poet's celebrated pursuit of the Bidford inebriates, of the drunken blacksmith with a face like a maple, of the wit-combats with Jonson. So too Drake accepts the deer-poaching episode. From Ireland, who had it from Jordan, he takes the version that makes Fulbrook the setting and Daisy Hill the site of Shakespeare's confinement in the keeper's lodge. Indeed, we are told, Shakespeare's woodland pictures and descriptions of wounded deer derive from his own wanderings in the shades of Fulbrook. It is a pity that Drake is unaware that Fulbrook was disparked shortly before Shakespeare's birth and not bought by the Lucys until the year preceding his death. The bitter ballad about the knight with his cuckold's horns is accepted as authentic, although Drake knows the doubts surrounding it and (more damagingly) the testimonial to his wife's virtue that Lucy had carved on their tomb in the church at Charlecote. He dismisses, however, the legend of the horse-holding at the theatre door on grounds similar to those which Malone provided in his *Life*.

Rowe's statement that Shakespeare was received into his company 'in a very mean Rank' is interpreted as signifying that the poet began his stage career as an actor performing second-rate parts. Drake follows 'the most sagacious critics' in deducing, from Greene's attack, that Shakespeare made his start as a dramatist by correcting and improving the plays of others. From the works themselves Drake tries to reconstruct Shakespeare's library, an enterprise for which his wide reading in Elizabethan literature has prepared him. The long section on the chronology follows Malone's format and takes into account Chalmers's modifications; here Drake's principal contribution is the unfortunate argument that Shakespeare's first play, composed in 1590, was *Pericles*. He conveniently omits *Titus Andronicus* altogether, for it is too 'disgusting' to be Shakespeare's. With *Twelfth Night*, according to Drake, Shakespeare terminated his playwriting career in 1613 and withdrew from London to Stratford. The brief last section, on Shakespeare in retirement, chronicles the great Stratford fire of 1614, the death of John Combe, the marriage of Shakespeare's daughters, the debate over enclosure (the particulars of which Drake found in Wheler), and the making of the will. Inevitably, the second-best bed proves troublesome, but Drake is satisfied that the poet previously made ample provision for his wife. 'We may, at least, rest satisfied,' he concludes, 'as well from the known integrity of Shakspeare, as from the humanity of his disposition, that nothing harsh or unjust had been committed by him on this occasion.'[35]

The passage just quoted fairly illustrates the conception of Shakespeare formulated by the gentle and devout physician. Throughout, a pleasant halo of bardolatry surrounds the character of the poet. Not only traditions but also the works themselves bespeak 'the gentleness, the benevolence, and the goodness of his heart'. The epithets *worthy, gentle,* or *beloved* connected with him testify to his felicitous temper and the sweetness of his manners. Like Chaucer he possessed a mind wholly cheerful and serene. This happy state he owed to his moral virtues and religious convictions—for do not his writings 'breathe a spirit of pious gratitude and devotional rapture'? (The 'scandalous surmise' that the dramatist enjoyed the charms of his landlady when he sojourned at The Taverne in Oxford is alluded to without comment.) Shakespeare had an exquisite taste for beauty in nature and art, and he was a delighted enthusiast of music, painting, and sculpture.

His inspirational character is marred, however—or so it would seem—by the Sonnets, which Drake is the first biographer to take seriously as a source of personal revelation. The ardour with which Shakespeare addresses one of the same sex does not disturb Drake, for he knows that in the literature of the period the languages of love and friendship were 'mutually convertible'. Rather it is the presence of the Dark Lady, with whom the poet—a husband and a father!—entered into a liaison that Drake finds agitating. Who she was is not worth the inquiry, 'for, a more worthless character, or described as such in stronger terms, no poet ever drew'. Would that these twenty-two poems had never seen print. Still, the biographer is able to resolve the problem by concluding with '*the most entire conviction*' (emphasis Drake's), that the temptress was an imaginary creation introduced solely to express 'the contrarieties, the inconsistencies, and the miseries of illicit love'.[36] The poems yield a wholesome moral after all, and the paragon remains unsullied.

Drake does not, however, deny the reality of the Fair Youth—far from it. The identification he proposes would alone ensure him a significant place in this chronicle. Accepting Chalmers's inspiration that the onlie begetter, Mr W. H., merely procured the poems, Drake suggests that the young man who is the focus of the first 126 sonnets is the Earl of Southampton.[37] He is an obvious enough candidate, but, remarkably, nobody had previously declared for him, at least not publicly. Drake not only nominates Southampton, but he also does some electioneering. He cites the dedication of *The Rape of Lucrece*:

The love I dedicate to your Lordship is without end. . . . The warrant I have of your honourable disposition, not the worth of my untutored lines, makes it assured of acceptance. What I have done is yours; what I have to do is yours, being part in all I have, devoted yours. Were my worth greater my duty would show greater. . . .

—and notes how the opening lines of Sonnet 26 express, in similar phraseology, the poet's sense of his subject's exalted merit and his own unworthiness:

> Lord of my love, to whom in vassalage
> Thy merit hath my duty strongly knit,
> To thee I send this written embassage
> To witness duty, not to show my wit;

> Duty so great which wit so poor as mine
> May make seem bare in wanting words to show it. . .

The parallel, which Drake proclaims to have been 'hitherto unnoticed', was in fact first pointed out by Capell in a note that, as Boswell observes with no great charity, probably suggested to Drake his theory.[38] He reinforces the identification by citing somewhat similar lines in Sonnet 110 (ll. 8–12).

To the objection that the first seventeen sonnets urge the youth to marry, whereas Southampton from 1594 to 1599 required no encouragement to pay court to Elizabeth Vernon, the cousin of Essex, Drake replies that the Queen opposed the match, and on two occasions the Earl deferred to his monarch by abandoning the affair. The lover's frame of mind at this time, Drake conjectures, was '*that if he could not marry the object of his choice, he would die single*'.[39] (Is it ignorance or prudery that keeps Drake from mentioning that Elizabeth Vernon became pregnant by the Earl in 1598?) The identification, Drake claims, provides a key to four otherwise inexplicable circumstances: the expostulatory tone of the poet, who wishes to see the patron to whom he is ardently attached married to *someone*; the fact that only the first seventeen sonnets urge matrimony, for the Earl and Mistress Vernon were secretly wed in 1598; the withholding of publication until after the Queen's death; and Jaggard's failure to include any of the poems in *The Passionate Pilgrim* in 1599, when the angry monarch had the newly-weds imprisoned. The poet's reverence and homage for his friend would be misapplied in connection with William Hart or William Hughes or, for that matter, anyone of similar station. In Sonnet 78 he speaks of the youth as his literary patron. To whom *can* the Sonnets apply 'if not to Lord Southampton, the bosom-friend, the munificent patron of Shakspeare, the noble, the elegant, the brave, the protector of literature and the theme of many a song'?[40] Drake anticipates the objection that, if (as he believes) Shakespeare wrote the cycle between the years 1594 and 1609, how comes it that he can address the friend in Sonnet 126 as 'my lovely boy'? In 1609 the Earl was thirty-six. The answer is to be found in the 108th sonnet, in which the poet resolves to consider the 'sweet boy' as endowed with perpetual youth: all will remain 'as when first I hallowed thy fair name'. Other objections would in time be raised, and other candidates would have their champions. But Drake's argument would early win converts, and to this day many biographers continue to identify the Fair Youth with the passionate young nobleman to whom Shakespeare dedicated the first heir of his invention.

That *Shakspeare and His Times* met with an enthusiastic response should occasion no surprise. Nares, who wrote a long review for the *Gentleman's Magazine*, felt that Drake had made 'perfectly clear' that the youth celebrated in the Sonnets is 'undoubtedly' the Earl of Southampton. For Drake's achievement as a whole he has only praise:

no work has hitherto appeared, and we may venture almost to pronounce that none can in future be produced, in which so much of agreeable and well-digested information on this subject will be found, as in this masterly production of Dr. Drake. That it is the result of much study, and many hours devoted to research in every possible line from which the materials could be drawn, is evident from the most casual inspection of these Volumes, which will at once indulge and greatly extend the desire already prevalent, of being informed of every thing material which can illustrate the life, the writings, or the genius, of our inimitable Dramatist.[41]

To Drake credit is due for devising a structure that could accommodate the synthesis for which a century of Shakespearian scholarship supplied the materials, and one can only admire the systematic energy which enabled him to complete his monumental and self-appointed task.

# 3
# *Boaden and the Sonnets*

WORKS with as ample an embrace as *Shakspeare and His Times* do not appear often in any age: they arise when the accumulation of knowledge over a considerable period creates the hunger for a summing up. At the other end of the scale flourish the specialized monographs that swell the annual bibliographies. These were not unknown to the landscape of eighteenth-century scholarship; witness the exchanges in the debate over Shakespeare's learning, capped by Farmer's crushing *Essay* of 1767, and Malone's attempt to ascertain the chronology of the plays. In the next century, however, the monograph really came into its own. These studies deal with every phase of the poet's background, life, and work; even with the orthography of his name. Their multitudinous existence complicates the historian's task, but he dare not ignore them.

One such contributor, James Boaden, a cultivated amateur of scholarship, is today even less well known than Drake; unlike that worthy, he has been denied by posterity even the tribute of a paragraph in those handy reference compilations, *A Shakespeare Companion* and *The Reader's Encyclopedia of Shakespeare*. Yet he is not undeserving of recognition. We have met Boaden before: while editor, in his early thirties, of the *Oracle* he won attention for being successively partisan and opponent of the Shakespeare Papers. In the course of a long and useful life, he composed several novels, gave the stage half a dozen successful plays, produced biographies of Kemble, Mrs Siddons, and Mrs Jordan, and reared nine children. Engaging as a man, he seems to have found time to cultivate the art of conversation. Not until he entered his seventies did he publish the works that now concern us.

His *On the Sonnets of Shakespeare: Identifying the Person to Whom They Are Addressed; And Elucidating Several Points in the Poet's History* appeared in 1837. This little volume represents an expansion of a two-part article that Boaden contributed to the *Gentleman's Magazine* in 1832, and which he reprinted at the suggestion of J. Payne Collier, newly famous as a Shakespearian scholar and some years later to be infamous in the same capacity. Drake had proposed Southampton as the Fair Youth; Boaden countered with William Herbert, third Earl of Pembroke. In the first instalment of his article, in which Herbert is not mentioned, Boaden demolished (so he thought) the pretensions of the other claimants, notably Southampton. To him credit is due for first conceiving and enunciating a major theory about the Sonnets; one which has distinguished adherents to this day.

Boaden argues effectively. The initials of the dedication do not fit Southampton, christened Henry Wriothesley; Boaden dismisses as tortured the efforts to read *begetter* as *procurer*—only the eternized recipient of the poems merits being wished 'all happiness and that eternity promised by our ever-living poet'. Southampton, twenty-one in 1594, was not young enough to be celebrated as a lovely boy, and in any case he was not lovely as represented in his portrait. And, if Southampton is intended, how comes it that the poet fails to allude to the dramatic events in the courtier's life during the turbulent years when the Sonnets were written: Southampton's achievement as a captain at Cadiz and in the Azores, his imprisonment after marriage to Elizabeth Vernon, his conspiratorial association with the rebellious Essex, his trial, judgement, and sentence, and his imprisonment by Elizabeth and release by James? (The Earl's afflictions when in disgrace were taken note of, as Boaden observes, in verses addressed to him by Daniel.) Since none of these episodes is so much as hinted at in the Sonnets, they *cannot*—the italics are Boaden's—apply to Southampton.

Herbert, on the other hand, has the right initials as well as the requisite station. The editors of the First Folio dedicated their book to him and to his brother, the Earl of Montgomery, both of whom had rewarded the playwright, while he lived, with much favour; Southampton, although still alive, they do not mention. In 1594 Herbert was fifteen, and thus (as in the poems) a boy: 'in the vernal blossom of existence', as Boaden puts it. He had beauty, if Van Dyke's portrait of him in maturity is any index. Boaden quotes fully Wood's characterization of Pembroke ('He was not only a great favourer of learned and ingenious men, but was himself learned, and endowed to admiration with a poetical genie. . . .') and the sketch of him by Clarendon in his *History of the Rebellion*, which a leading Shakespearian in this century strangely describes as 'strangely neglected'.[42] Clarendon pictures the Earl as an incorruptible man, esteemed by the Court, a great lover of his country, and of religion and justice, and liberal towards worthy men who needed encouragement or support. At the same time he had infirmities of character: 'He indulged to himself the pleasures of all kinds, almost in all excesses.' The delights to which this righteous sensualist, who for long evaded matrimony, was most 'immoderately given up' were those of the bed. This portrayal does not contradict that of Shakespeare's youth, who is much addicted to pleasure and pursued by women. Because the Sonnets were personal and Pembroke was a statesman at the time of their publication, his identity, Boaden argues, was deliberately obscured by the use of his initials in the dedication rather than his full name.

Boaden is the first to identify the Rival Poet as Samuel Daniel, who was brought up at Wilton, the seat of the Pembrokes. In 1602 he addressed his *Defense of Rime* to Herbert: in Sonnet 82 Shakespeare jealously alludes to 'The dedicated words which writers use | Of their fair subject, blessing every book'. In *The Rape of Lucrece* the poet imitated Daniel's *Complaint of Rosamond*; he 'equally founded' his Sonnets on those of Daniel to Delia. The mysterious references to the rival being taught to write by spirits and conversing with an 'affable familiar ghost' were suggested by the astrologer Dr Dee, who was patronized by the Pembrokes; at Wilton, Daniel would have been 'within the very lime-twigs of the Necromancer's spell'.

Boaden's arguments for the identity of the Rival Poet, although considered, would carry much less weight than those for the Fair Youth; but here too, in the perennial fashion of the controversialist, he overstates his case. Mistaking a plausible hypothesis for a demonstration, he concludes that 'the question to whom Shakespeare's Sonnets were addressed, is now decided'.[43] Despite the vocabulary of confidence, nothing was decided. In the decades to follow, enormous quantities of ink would flow in inconclusive skirmishes between the Pembrokists and the Southamptonites, and other candidates would be proposed. Boaden, moreover, damages his own presentation by maintaining, as a certainty, the unlikely hypothesis (although even today it has its partisans) that the dedicatory epistle to the Folio signed by Heminges and Condell was actually written by Jonson; he is here following Steevens at his most perverse.[44] These qualifications do not lessen one's admiration for the essay as a whole: sensible, informed, and cogently argued, it won from the first considerable support.

An immediate convert, Charles Armitage Brown, put forward in *Shakespeare's Autobiographical Poems* arguments for Herbert similar to Boaden's but he curiously neglected to mention the work that had anticipated him. Like so many others since, Brown was troubled by the form of address of the dedication, which he recognized as inappropriate to a nobleman. This problem he settled to his own satisfaction by reasoning that 'the title, "Mr." was not improperly applied to the eldest son of an Earl, there not having been, at that period, any grander title of courtesy'.[45] Boaden's hypothesis was accepted as proved by the poet Thomas Campbell in his 1838 edition of the *Dramatic Works*. On the Continent support came early from the distinguished German Shakespearian Hermann Ulrici who, in *Shakspeare's dramatische Kunst* of 1839, avowed himself persuaded that Boaden had fully established his case. The first dissent was offered by Charles Knight in 1841. Boaden had maintained that the Sonnets were published with Pembroke's probable sanction; Knight now demanded, 'Would Lord Pembroke have suffered himself to be styled "W.H., the only begetter of these ensuing Sonnets"— plain Mr. W.H.—he, a nobleman, with all the pride of birth and rank about him—and represented in these poems as a man of licentious habits, and treacherous in his licentiousness?'[46] The issue was fairly joined, and debate would proceed uninterrupted to the present day.

# 4

# *Various Pictures*

THE contents of Boaden's other treatise (published earlier, in 1824) are summarized by its enormous title: *An Inquiry into the Authenticity of Various Pictures and Prints, Which, from the Decease of the Poet to Our Own Times, Have Been Offered to the Public as Portraits of Shakspeare: Containing a Careful Examination of the Evidence on Which They Claim to Be Received; by Which the Pretended Portraits Have Been Rejected, the Genuine Conformed and Established. Illustrated by Accurate and Finished Engravings, by the Ablest Artists, from Such Originals as Were of Indisputable Authority*. Boaden allots pride of place to the authenticated likenesses, the Droeshout engraving and the Stratford bust, then goes on to consider the Chandos portrait, the Felton, the Janssen, and others. These pictures have always intrigued Shakespeare lovers, avid for a veritable image flattering to their own preconceptions. As we have seen, a portrait of Shakespeare, by whom unknown, deeply impressed Keats. To the biographer these representations are pertinent documents, to be studied and interpreted. As the century wore on, claims would be made on behalf of newly recovered portraits—the Ely, the Stratford, the Ashbourne—and subsequent monographs would take into account the extended terrain. But Boaden's slim volume provides us with a suitable occasion to survey the state of Shakespeare portraiture up to his time.

'No picture within the last hundred years has been more frequently copied', Boaden remarks of the Chandos portrait.[47] The copyists include distinguished original artists—Sir Godfrey Kneller, Sir Joshua Reynolds, the sculptor Roubiliac—as well as an assortment of lesser portraitists and engravers well enough esteemed in their own day: Vertue, Houbraken, Duchange. Some made copies of copies. Unconcerned about strict fidelity to the original, they reversed the head, altered the expression, changed the costume. Kneller's copy hung in Dryden's study, a source of inspiration to the poet, who thanked the artist in his *Fourteenth Epistle* of 1694:

> Shakspeare, thy gift I place before my sight;
> With awe I ask his blessing as I write;
> With reverence look on his majestic face,
> Proud to be less, but of his godlike race.

In 1779 Capell presented his treasured version, by Ranelagh Barret from the Kneller copy, to his beloved Trinity College, Cambridge. Privileged to view the original at the seat of the Duke of Chandos, Malone procured His Lordship's permission to have Ozias Humphry make a faithful drawing of it, which now hangs in the Folger Shakespeare Library in Washington, DC. Eventually Malone would possess no fewer than three copies of the canvas, including Reynolds's. Rowe chose the Chandos portrait, in a free rendering by van der Gucht, as the frontispiece to the 1709 *Works*, and one variant or another of this picture—rather than of the Droeshout engraving—adorns (with the exception of Pope) the apostolic succession of eighteenth-century editions. Horace Walpole, who once planned to marry the owner, regarded the Chandos as 'the only original picture of Shakspeare'.[48] It has always been the favourite likeness of Shakespeare.*

Painted in oil on coarse English canvas, the oval portrait measures eighteen inches in width by twenty-two inches in height. It has been more than once retouched over the years. Against a background of rich gravy, the subject, a man apparently in his early forties, wears a black silk doublet; the plain white lawn collar stands open; white strings sewn over the collar show through the beard and hang loosely down. His complexion is swarthy. A 'distinctly Italian type', declares more than one authority; 'a decidedly Jewish physiognomy', is another, and repeated, verdict (it would be conjectured that the portrait shows Shakespeare made up for the part of Shylock). The hair—dark brown verging on black—falls crisply away in a profusion of curls from the massive expanse of forehead. The full, grey-brown eyes, edged with red,

* Of the van der Gucht frontispiece to Rowe's *Works*, David Piper remarks in his admirable *The Image of the Poet: British Poets and Their Portraits* (Oxford, 1982): 'Modest enough, too modest, a bit stiff, in that swirling baroque setting with trumpeting fame winged aloft, and the doting Muses as supporters. All the business may divert attention from the portrait to its setting, the allegorical framework, and it announces, indeed, a curious counterpoint which runs right through the English eighteenth-century interpretation of Shakespeare's image. In some insular critical appreciation of the work in the same period, even up to Johnson and beyond, there appears a faintly defensive strain, as critics, aware of Aristotelian canons of form, had to agree that Shakespeare did not abide by the rules as did the great French tragedians' (49–52). Piper goes on to note that van der Gucht lifted his design blatantly from the 1660 Rouen edition of Pierre Corneille, 'the most severely classical of all French tragedians'.

gaze thoughtfully out. To one expert the mouth is 'somewhat lubricious', the lips 'wanton'; another finds the mouth 'combining at once the expression of healthy life with slight melancholy and delicate irony'. The moustache is full; a small beard adorns the pointed chin and ascends in a thick fringe to the hair. In the right ear a gold ring glitters. Could this be Shakespeare of the Midlands?

The Chandos portrait can be traced back farther than most others put forward as genuine, but the earlier history consists mainly of shadowy and contradictory traditions. Oldys records that 'old Cornelius Jansen' was the artist, then adds: 'Others say, that it was done by Richard Burbage, the player.' Elsewhere he ascribes the picture to 'John Taylor, the player'. According to a communication in the *Critical Review* for 1770, this Taylor bequeathed the canvas to Davenant. If the Chandos portrait was painted from life (as most believers think), it could not very well have been by Janssen, who probably never saw England until two years after Shakespeare's death. Reynolds felt that the picture did not in the least resemble Janssen's performances. No John Taylor acted with the King's Men, although the company did have a Joseph Taylor who became a member after Burbage's death in 1619. This Taylor died intestate, and so could not have willed the portrait to Davenant. An artist by the name of John Taylor did, however, flourish in the seventeenth century: Malone reports seeing at Oxford a portrait by him dated 1655. If he is the Chandos artist, he must have had a long career. But Malone is wrong about the date: our John Taylor died in June of 1651, and was laid to rest at the parish church of St Bride in Fleet Street. He had been a leading member of the worshipful Company of Painter-Stainers, and was represented in the customary group painting—a genre of which the most celebrated exemplar is Rembrandt's painting of the Syndics of the Draper's Guild—which has survived and hangs to this day in the Company's Court Room in Little Trinity Lane in London. Cornelius Janssen (or Johnson) is reputedly the artist. Taylor cuts a handsome middle-aged figure with his pointed beard and encircling ruff. During the course of a long career he presented at least five apprentices of the Company for their freedom.

In notebook entries made in 1719, George Vertue—a credible witness—is the first to furnish documentary evidence about the early history of the Chandos portrait. According to Vertue, Betterton told Robert Keck, of the Inner Temple, that the picture 'was painted by John Taylor, a player who acted for Shakespear & this John Taylor in his will left it Sir William Davenant, & at death of Sir Will Davenant Mr. Betterton bought it, in whose possession it now is . . .'. Betterton was convinced that the artist was one John Taylor, having presumably been told so before the latter's death in 1651. A John Taylor was a leading member of the Company of Painter-Stainers, serving as Upper Warden in 1635–6, and as Master in 1643–4; he paid the substantial sum of seven pounds for declining to serve a second term as Upper Warden in 1639–40. His arms were among those embellishing the windows of Painters-Stainers' Hall before the Great Fire of 1666 destroyed old London.[49]

That Davenant once owned the canvas does not strain credulity. One legend holds that the copy made by Kneller for Dryden sometime between 1683 and 1692 was commissioned by Davenant, but as the latter had by then been dead for two decades, the report may be received with scepticism. Davenant died insolvent, the administration of his effects going to his principal creditor. The actor Betterton bought the Chandos portrait at a public sale, and while in his possession it was engraved for Rowe's edition. Betterton too died indigent and without a will. The picture then purportedly came into the possession of Mrs Elizabeth Barry, the actress, who afterwards sold it for forty guineas to Robert Keck; Steevens, aware of Mrs Barry's propensities, conjectured without chivalry that 'somewhat more animated than canvas, might have been included, though not specified, in a bargain with an actress of acknowledged gallantry'.[50] But the legend that Mrs Barry owned the Chandos portrait between the death of Betterton in 1710 and her own demise three years later is no more than that, for Keck clearly indicated to Vertue that he had purchased the portrait at Betterton's death. Unsurprisingly, in her own will Mrs Barry mentions no pictures.* The portrait devolved to a Mr Nicoll, who married the Keck heiress; their daughter became the bride of the Marquis of Caernarvon, afterwards the Duke of Chandos. Thus the painting was passed down. In Boaden's time it belonged to the Duke of Buckingham, who had wed Lady Anne Elizabeth Brydges, Chandos's daughter. At a sale of Buckingham's pictures in 1848, it was purchased by the Earl of Ellesmere. In March 1856 he presented it to the National Portrait Gallery as its first portrait. There it still hangs.

An enthusiast of the Chandos likeness, Malone used a not very faithful engraving of the Humphry copy as the frontispiece to his 1790 edition of Shakespeare; Boswell, sharing his confidence, prefixed a different—and better—engraving of the same portrait in the place of honour alongside the title-page of the first volume of his Variorum Shakespeare. While not explicit on the subject, Boaden evidently accepts the judgement of his hero Malone. But not everyone has championed the picture. Steevens mocked the 'Davenantico-Bettertono-Barryan-Keckian-Nicolsian-Chandosan' canvas; 'our author', he said of a copy, 'exhibits the complexion of a Jew, or rather that of a chimney-sweeper in the jaundice'. For his raillery the Puck of commentators was dubbed by Boaden 'a true Anthropophaginian'. Steevens broke with a tradition going back to the First Folio by issuing his 1793 *Shakespeare* without any prefatory portrait. His scepticism, at least as regards the Chandos

---

* In his 1719 notebook, Vertue indicates that Keck bought the Chandos at Betterton's death in 1710. Mrs Barry died two years later. See Mary Edmond, 'The Chandos Portrait: A Suggested Painter', *Burlington Magazine*, 129 [1982], 146–9.

likeness, has found modern supporters: authorities have noted the discrepancies between the features in that portrait and those of the acknowledged portrayals in the Folio and Holy Trinity Church. The Chandos head is similarly bald, and the line where the hair falls away is sharply defined, as in Droeshout; but the forehead recedes rather than rising perpendicularly, the upper lip is short rather than long, and the chin pointed rather than round.* Some experts still regard the Chandos portrait as genuine, but its status is, and inevitably will remain, debatable.

No such uncertainty surrounds the picture by Federigo Zuccaro (or Zucchero) to which Boaden next turns. An oval, life-sized portrait painted on panel, with the name 'Guglielm: Shakspeare' inscribed on the back, it depicts a dandyish man of about thirty with luxuriant beard, black hair, and bright eyes, 'leaning with his face upon the right hand; the head stooped forward, in earnest meditation, with the evidences of composition lying before him'.[51] Zuccaro came to England in 1574 and stayed until no later than 1580, when Shakespeare was sixteen. Boaden thought the picture might be of Tasso. It is certainly not Shakespeare.

About the genuineness of the Janssen portrait Boaden entertained no doubts. 'Nothing can more distinctly embody our conceptions of Shakspeare', he averred:

It is extremely handsome; the forehead elevated and ample; the eyes clear, mild, and benignant; the nose well formed; the mouth closed, the lips slightly compressed; the hair receding from the forehead, as of one who would become bald; the beard gracefully disposed, and a very neat laced collar thrown over a dress such as the poet, from his circumstances, his character, and his connexions, might be supposed to wear.[52]

When he wrote, Boaden had not yet seen the portrait—only mezzotints of it—but before his monograph appeared he had performed the service, for himself and subsequent bardolaters, of tracking the picture down. He found it at the residence of the Duke of Somerset, by whose name it is to this day sometimes known. There, for Boaden's benefit, it was taken down from its place near the ceiling for examination in the light. In a section of 'Additional Remarks' appended to his book, Boaden reaffirmed his faith in what he described as a tenderly and beautifully executed portrait that had captured for all time the gentleness for which the poet was beloved. His Grace had received the picture as a gift from the Duke of Hamilton, for whom it had been purchased by Samuel Woodburn, a leading picture dealer of the time. This Woodburn had in 1811 published an engraving of the work in his *Portraits of Characters Illustrious in British History*, there stating it to be 'from an original picture formerly in the possession of Prince Rupert'. Charles Jennens, the millionaire

---

* It is fair, however, to record that the most recent authority, Sir Roy Strong (drawing upon the researches of D. T. Piper), finds the similarities greater: 'The main features tally; only the hair, beard and moustache are differently arranged.' He however regards the identity of the sitter as not proven. See Roy Strong, *Tudor and Jacobean Portraits* (London, 1969), i. 279–83.

eccentric of Gopsall Hall in Leicestershire, who owned it in the eighteenth century, is not mentioned. Thus the pedigree is confused. In any event no evidence exists to connect the portrait with the prince.

The picture first came to light in 1770 in an accomplished mezzotint by Richard Earlom for the frontispiece to an edition of *King Lear* prepared by Jennens. The latter fantastic was described by a contemporary as 'a vain fool, crazed by his wealth, who, were he in heaven, would criticize the Lord Almighty; who lives surrounded by all the luxuries of an Eastern potentate'.[53] Yet Jennens achieved his own measure of immortality by writing the libretto for his friend Handel's *Messiah* in 1742, as well as the libretti for two other of the composer's thirty-two oratorios, *Saul* (1735) and *Belshazzar* (1745). Jennens sought to pose as patron rather than editor of the edition of *King Lear*, going so far as to employ the delicious subterfuge of dedicating it to himself, 'with the greatest respect and gratitude': this brilliant stroke satisfied his desire for self-abnegation without sacrifice of his vanity. In the inscription to the frontispiece, however, he boasted his ownership of the portrait, at the same time ascribing it to Janssen.

Immediately above the head in the engraving stands a scroll with the legend 'UT · MAGUS', presumably from Horace's *Epistle to Augustus* (II. i. 213). In this passage Horace evokes the poet–magician who transports his readers one moment to Thebes, the next to Athens. Horace's words, as Boaden observes, fittingly describe the necromancer of the stage who carries his audiences 'over lands and seas, from one kingdom to another, superior to all circumscription or confine'.[54] Alas for speculation! The motto is the conceit of the engraver—or owner—and is absent from the original. In the upper left-hand corner of the painting, as well as of the engraving, there does appear the inscription:

$$\text{Æ}^\text{t}\ 46$$
$$1610$$

This information inevitably points to Shakespeare.

One can appreciate Boaden's enthusiasm. The Janssen is the most elegant of the purported likenesses of the poet, and not unworthy of the easel of a portraitist esteemed in his own day as second only to Van Dyke. Measuring $22\frac{1}{4}$ by $17\frac{1}{4}$ inches and painted in oil on a rough-hewn wood panel now cracked in two places, it depicts a sensitive and aristocratic, almost effeminate, face. The head is a narrow oval. The flesh has an ivory hue; the cheeks glow with red; the mouth—thin and straight—is ruby. The small dark brown eyes have an almond shape; above them the ridges are traced by the delicate curve of the brows. The long nose is finely chiselled. A high bald forehead reinforces one's impression of a sensitive intellectual. The brown hair springing from the temple merges imperceptibly into the dark background. With conscientious solicitude for detail, the artist has delineated the individual hairs of the moustache and beard, which are fair, with touches of auburn in the latter. The subject is richly attired: he wears a large but delicate white lace collar stiffened with wire, surmounting a doublet of silk embroidered with gold. If this is indeed the burgher of Stratford he completes a remarkable triumvirate of portraits of literary geniuses by the same artist; for Janssen also painted Jonson and, in 1618, Milton, then a boy of ten. Those susceptible to myths who can at the same time savour ironies will be pleased to learn that at Charlecote Janssen painted a large picture of the family of Sir Thomas Lucy.

Aware that a Janssen portrait of Elizabeth Wriothesley, daughter of Shakespeare's patron, hung at Sherburn Castle, Boaden inferred as an 'absolute certainty' that the artist was 'Southampton's painter'. It is 'highly probable' that the Earl commissioned him to depict his 'favourite poet' in a likeness which once hung at Titchfield or Beaulieu as 'a shining proof of his own genius, taste, and liberality'.[55] It may have had a place in the collection divided between the Dukes of Portland and Beaufort, either of whom could have presented it to the exquisite of Gopsall Hall. This pleasing fantasy is totally unsupported by evidence.

Jennens did little to further his own cause. When challenged, he failed to produce the picture—was it because he had sophisticated the plate by having the scroll introduced? More important, he offered no pedigree. This despite the cruel hilarity of Steevens, who derided him in the *Critical Review*. The Jennens collection, he sneered in the December 1770 number, 'we are not heartily inclined to treat with much respect, especially as we hear it is filled with the performances of one of the most contemptible daubers of the age'.[56] Jennens made the mistake of protesting, in a letter to the *Gazetteer* (8 January 1771), thus giving Steevens another opening, of which he gleefully took advantage. 'Concerning the print', he declared, 'we will have no controversy; but we still adhere to our former opinion, that the soul of the mezzotinto is not the soul of Shakespeare. It has been the fate of Shakespeare to have had many mistakes committed, both about his soul and body.'[57] In *The Tragedy of King Lear, as Lately Published, Vindicated from the Abuse of the Critical Reviewers* (1772), Jennens defended himself with spirit against the imprudence of a critic who dared cast aspersions on a picture the original of which he had never seen; but he said nothing about its century-and-a-half history or about the circumstances by which he came to own it. We know that a decade earlier he was not the possessor, for no mention of it is made in *London and Its Environs* (1761), which contains an inventory of Jennens's pictures, including a crayon drawing by van der Gucht of the Chandos portrait.

Other obstacles defeat credulity. The tail of the 6 in the 46 of the inscription looks suspiciously like a later addition. The forehead excepted, the elongated visage has little in common with the Stratford bust or the folio engraving. It is not even certain that Janssen is the artist. If he did paint it, he was no more than seventeen at the time, for the parish registers of the Dutch Church in Austin Friars record his baptism on 14 October 1593. Such precocity is not unknown in the annals of art, but it is, to say the least, unusual. A further problem is that Janssen's earliest English picture cannot with certainty be dated until two years after Shakespeare's death—and the portrait in question gives every evidence of having been painted from life. All in all, a doubtful case: much more doubtful than the controversial Chandos.

Devastating in attacks propelled by the twin pistons of vast knowledge and boundless cynicism, the astute Steevens—if taken at his word—was no less vulnerable than his targets when he came to bestow his own allegiance. This he gave to the Felton portrait. Its background is, briefly, as follows: In 1792 the sale catalogue of the European Museum on King Street in St James's Square listed, as item 359, 'A curious portrait of Shakespeare, painted in 1597'. Exhibited for three months or so, it was viewed at the Museum by Lord Leicester and Lord Orford (Horace Walpole), who allowed its authenticity but declined to buy. There too the picture was seen by Samuel Felton of Drayton, Shropshire, and Curzon Street in Mayfair, who asked Mr J. Wilson, proprietor of the Museum, to furnish a pedigree. 'The Head of Shakspeare', Wilson wrote to Felton on 11 September 1792, 'was purchased out of an old house known by the sign of the Boar in Eastcheap, London, where Shakespeare and his friends used to resort,—and report says, was painted by a player of that time, but whose name I have not been able to learn.'[58] An inconvenience of this account is that the Boar's Head, so far as we know, enjoyed the custom not of the poet but of those figments of his imagination, Hal and Falstaff; another awkwardness is that the Great Fire of London in 1666 reduced the whole of Eastcheap to smouldering ruin. Nevertheless, Felton bought the picture, paying the unprincely sum of five guineas.

It is a small portrait, eleven inches high and eight wide, painted on a wood panel split near the right extremity of the ear and cut off just below the flat ruff, which curiously resembles (in Boaden's words) 'a small portable pillory'. The face, representing a man in his early thirties, vaguely resembles the Droeshout rendering, but with notable divergences. The forehead, prominent in the engraving, can only be described as stupendous in the portrait; a 'sugar-loaf skull', one authority calls it. The hairs of the moustache droop down rather than pointing up; the chin is hairless, and the nose flattened. A pensive expression lends a certain attractiveness to the visage. 'There are, indeed,' Steevens enthused, 'just such marks of a placid and amiable disposition in this resemblance of our Poet, as his admirers would have wished to find.' On the back of the panel appears the inscription, seemingly in an Elizabethan hand, 'GUIL. SHAKSPEARE, 1597, R.N.' (Or so it was thought until some time prior to 1827, when Abraham Wivell, the portrait painter and amateur of Shakespearian iconography, discovered, while applying linseed oil to the back of the picture in order to preserve the wood, that the initials were R.B. Not surprisingly, in view of Wilson's letter, he concluded that they stood for Richard Burbage.)

On 9 August 1794 William Richardson, a print-seller in Leicester Square, communicated word of the portrait to Steevens, whose interest was at once aroused. Felton permitted Richardson to carry the picture to the great scholar, who compared it with the Droeshout engraving. Steevens also sought information from Wilson, who on 11 August elaborated on his previous story: He assured Steevens 'that this portrait was found between four and five years ago at a Broker's shop in the Minories, by a man of fashion whose name must be concealed: that it afterwards came (attended by the Eastcheap story, &c.) with a part of that gentleman's collection of paintings, to be sold at the European Museum.'[59] One cannot resist an uneasy suspicion that the secretive gentleman was a cousin of Mr H., Ireland's benefactor.

Still, Steevens professed himself a believer and, in his unsigned article in the October number of the *European Magazine*, declared the Felton 'the *only genuine portrait of Shakspeare*' (italics Steevens's). It served, he maintained, as the model for Droeshout and also for Marshall, who in 1640 modelled his engraved frontispiece to the *Poems* on it rather than, as commonly supposed, the Folio likeness. The name of Shakespeare in the inscription, Steevens insists with unwarranted confidence, is 'set down as he himself has spelt it'. He goes on to conjecture that

this picture was probably the ornament of a club room in Eastcheap, round which other resemblances of contemporary poets and players might have been arranged . . . that when our Author returned over London Bridge from the Globe Theatre, this was a convenient house of entertainment; and that for many years afterwards (as the tradition of the neighbourhood reports) it was understood to have been a place where the wits and wags of a former age were assembled, and their portraits reposited.[60]

Yet, in the same piece, Steevens expresses contempt for the Eastcheap legend that accompanied the majority of fraudulent Shakespeare portraits: it was 'high time that picture-dealers should avail themselves of another story, this being completely worn out, and no longer fit for service'.[61] Never mind, too, that no such painting was recalled by Mr Sloman, who retired as publican of the Boar's Head in 1767, at which time all the furnishings which had devolved on him from the two previous landlords were sold off; he did, however, remember a crude daubing of the Gadshill robbery on the tavern wall. In Crooked Lane, Steevens found the widow of the preceding governor of the Boar's Head now employed as a wire-worker. Her apprentice—not a young man—testified that he had often heard his mistress reminisce about the tavern, and in so doing refer to the Gadshill painting; but she had never so much as hinted at the existence of any other picture.

The pretensions of the Felton head were scorned in *Chalcographimania or the Portrait-Collector and Printseller's Chronicle* (1814), 'a Humorous Poem in Four Books' by William-Henry Ireland, who may be supposed not inexpert in such matters. In a long, gleefully destructive chapter of his monograph, Boaden reports that Fuseli thought the Felton the work of a Flemish hand; for himself, he believes that, to satisfy public demand for likenesses of Shakespeare, an original but mutilated early portrait of an unknown person with a more than usual resemblance to the position of the poet's head and the disposition of his hair in Droeshout was touched up and the ruff superimposed. Several years after the publication of Boaden's book, in the London *Literary Gazette* for 7 July 1827, the editor revealed that 'We have the most conclusive evidence that the Felton is a forgery; for it was altered and painted by John Crauch (Cranch). The story about the Boar's Head, &c., is an auctioneer's trick.' Cranch died at Bath in 1821 in his seventieth year. The notable modern authority on Shakespearian portraiture, M. H. Spielmann, thinks that the Felton head was inspired by Sherwin's line engraving— advertised as 'a striking likeness of Shakespeare'—which adorned the 1790 Ayscough edition of the *Dramatic Works*. A genuine panel, probably Flemish in origin, was altered on this model and then baked in the Westminster ovens to achieve overnight the mellowed look of age.[62]

The Felton portrait has since come into the possession of the Folger Shakespeare Library, where a member of the staff, Dr Giles Dawson, removed the picture from its frame for examination. The inscription on the back, he discovered, had disappeared, nor did it re-emerge under ultraviolet—perhaps heavy varnish had permanently obliterated the legend. At the National Gallery in Washington experts subjected the panel to X-ray, as well as to ultraviolet, and found that the pigment had been heavily restored in the nineteenth century, perhaps 60 to 70 per cent of the surface being new paint. Still, they concluded, nothing prevented the picture from dating back, in its original state, to the beginning of the seventeenth century.

One may yet wonder whether Steevens in the last decade of his life had finally managed to dupe himself. Was not the episode perhaps another manifestation of his saturnine humour at the expense of scholarship and, ultimately, of himself? He had his mischievous quirks: in his 1793 *Shakespeare* Steevens glossed indecent expressions in the text with obscene notes to which he affixed the names of two eminently respectable clergymen, Richard Amner and John Collins, with both of whom he had quarrelled. Steevens is thought to be the acquaintance of whom Lord Mansfield remarked to Johnson, 'Suppose we believe one *half* of what he tells'; to which the Doctor replied, 'Ay; but we don't know *which* half to believe.'[63] Can *we* believe that Steevens accepted the Felton portrait as genuine? In the Advertisement to the Third Variorum, Boswell recalls a revealing anecdote:

There are not, indeed, wanting those who suspect that Mr. Steevens was better acquainted with the history of its manufacture, and that there was a deeper meaning in his words, when he tells us, 'he was instrumental in procuring it', than he would have wished to be generally understood. . . . My venerable friend, the late Mr. Bindley, of the Stamp-office, was reluctantly persuaded, by his importunity, to attest his opinion in favour of this picture, which he did in deference to the judgment of one so well acquainted with Shakespeare; but happening to glance his eye upon Mr. Steevens's face, he instantly perceived, by the triumph depicted in the peculiar expression of his countenance, that he had been deceived.[64]

Whatever Steevens truly felt, his endorsement carried weight. To Richardson the print-seller it meant immediate profits, which he showed no nice conscience in garnering. In his 'Proposals . . . for the Publication of Two Plates', dated 5 November 1794, he advertised Thomas Trotter's engravings of the Felton, 'in the most finished style', at 7s. 6d. the pair. The public, he declares, is entitled to 'scrupulous fidelity' (which these prints conspicuously lack). Richardson deftly makes a virtue of Trotter's crudely delineated costume: 'should any fine ladies and gentlemen of the present age be disgusted at the stiff garb of our author, they may readily turn their eyes aside, and feast them on the more easy and elegant suit of clothes provided for him by his modern tailors Messieurs Zoust, Vertue, Houbraken, and the humble imitators of supposititious drapery'. Anyway, in 'this ancient picture' Shakespeare's dress '*might* have been a theatrical one' (it was not). The actual wood panel, it will be recalled, consists solely of the head and ruff, with no suggestion of a torso. Trotter had clumsily set the Felton head on the Droeshout body, with the shoulder line parallel to the lower part of the nose, creating a curiously stooped figure wearing an impossibly placed wired band. The effect of this deliberate fraud was to encourage public confidence that the Felton portrait had served as Droeshout's model. After Steevens's death, Reed used Neagle's oval-shaped engraving of the picture as the frontispiece to the First Variorum of 1803. Another engraving of the same work, by William Holl, adorned the Second Variorum in 1813. Through the century the portrait, in one version or another, dubiously enhanced study walls and editions of Shakespeare. If the Felton caper was indeed Steevens's last prank, it did as much mischief as he could have wished.

The Felton completes the gallery of major Shakespearian or pseudo-Shakespearian portraits up to the time that Boaden wrote his *Inquiry*. To be sure, there were other cases, still more dubious, which he lumps together in a chapter entitled 'Miscellaneous Heads'. Boaden merely alludes to the elegant engraving by Vertue of King James surmounted by a laurel wreath and a scroll bearing the name William Shakespeare, to which Pope, in a monumental lapse, gave the place of honour as frontispiece to his edition of the dramatist. More space is devoted to the Hilliard (or Somerville) miniature, Agar's engraving of which Boswell affixed 'with great satisfaction' to the second volume of the 1821 Variorum. In 1818 the owner, Sir James Bland Burges, wrote to Boswell that the painting had been commissioned by his ancestor, Mr Somerville of Edstone, near Stratford, who 'lived in habits of intimacy with Shakspeare, particularly after his retirement from the stage'.[65] A fine example of Hilliard's mastery, the miniature is utterly irreconcilable with the authentic likenesses: the subject has a tuft of hair on the forehead and wears a goatee and long horizontal moustache. Around 1725, I. Simon published a mezzotint, purportedly of the poet, which he claimed to have based on a painting by Gerard Zoust (or Soest) 'in the collection of T. Wright, Painter, in Covent Garden'. It shows, in three-quarter view, the face of a man with a copious beard and a full head of hair. Zoust was born twenty-one years after Shakespeare's death: an inconvenient fact. The costume and attitude of the Zoust portrait were imitated in Robert Cooper's engraving of 1811, marketed by Machel Stace, a vendor of puritan books who had (in Boaden's words) 'the audacity to write under it the name of Shakspeare'.

Boaden's unpretentious volume paved the way for other monographs on Shakespearian iconography. In his *Inquiry into the History, Authenticity, & Characteristics of the Shakspeare Portraits, in Which the Criticisms of Malone, Steevens, Boaden, & Others, Are Examined, Confirmed, or Refuted* (1827), Abraham Wivell offers his expertise as a professional portrait painter, and little else: so much the novice was he that he did not know of the existence of the Droeshout engraving until Wheler showed it to him in Stratford two years before he prepared his book. The *Inquiry* is a scrappy affair which reprints the Felton documents (Richardson's proposals, the *European Magazine* articles, etc.), Wivell's own 1825 pamphlet on the Stratford monument, and long extracts—sometimes whole chapters—from Boaden. Despite this exploitation, Boaden receives singularly ungrateful treatment at Wivell's hands. Since the latter has as his main purpose in the book espousal of the Felton cause, his predecessor is the enemy, and he is subjected to repeated attack in Wivell's shambling, semi-literate prose. He adds little that is unfamiliar (apart from correction of the initials on the back of the picture); nor does importance attach to his *Supplement* of the same year or his *Inquiry* of 1840.

Between them Boaden and Wivell do not, of course, mention all the Shakespeare portraits; there are too many, and too many that are patently spurious. But in the bellows portrait they miss a curiosity. It stirred Charles Lamb. In Paris in 1822 Lamb had eaten frogs ('It has been such a treat!') and, shown the portrait by its owner, the great tragedian Talma, he had knelt down and kissed it with bardolatrous reverence, just as Boswell a generation earlier had bussed the Shakespeare relics in Ireland's house in the Strand. Let Lamb describe the picture in his own words:

It is painted on the one half of a pair of bellows—a lovely picture, corresponding with the Folio head. The bellows has old carved *wings* round it, and round the visnomy is inscribed, near as I remember, not divided into rhyme—I found out the rhyme—

> 'Whom have we here,
> Stuck on this bellows,
> But the Prince of good fellows,
> Willy Shakspere?'

At top—

> 'O base and coward luck!
> To be here stuck.—Poins'

At bottom—

> 'Nay! rather a glorious lot is to him assign'd,
> Who like the Almighty, rides upon the *wind*.—PISTOL.'

This is all in old carved wooden letters. The countenance smiling, sweet, and intellectual beyond measure, even as He was immeasurable.[66]

Lamb knew that people laughed at him for being taken in by a forgery, but he believed that in England the portrait (no doubt inspired, as a wag remarked, by a Muse of fire) would convert the sceptics.

He believed in vain; the bellows Shakespeare is of course a fake. The work of one Zincke, it was planted, along with fabricated letters 'proving' that the picture had been painted at Queen Elizabeth's command on the lid of her favourite pair of bellows, in a château near Paris. There a confederate discovered it in the presence (how fortunate) of witnesses. The purchaser had it brightened by the celebrated Parisian picture-cleaner Ribet, whose sponge disclosed a lady in lofty headgear ornamented with blue ribbons beneath the poet's sweetly smiling countenance. Ribet soon restored Shakespeare, and the portrait passed into Talma's possession. Even after he heard of the fraud, he did not cease to treasure his memorial of the Immortal Bard. Today, the head-dress still faintly discernible underneath the repainted surface, it forms part of the splendid collection of Viscountess Eccles (formerly Mary Hyde) at Four Oaks Farm in New Jersey.

# 5

# *Gifford and the Mermaid Club*

SOMETIMES understanding of Shakespeare's career is modified by a source not directly concerned with the dramatist. In 1816 William Gifford, best remembered for his management of the *Quarterly Review*, published an edition of *The Works of Ben Jonson*, with an ambitious biographical memoir in which he deploys all his considerable powers of argumentation and sarcasm to demolish the view—first expressed in Dryden's time and most influentially promulgated by Malone—that Jonson persecuted Shakespeare with unremitting malevolence. To this strange distortion of truth, Gifford offers a diametrically opposing view. 'I am . . . persuaded', he declares, 'that they were friends and associates till the latter [Shakespeare] finally retired—that no feud, no jealousy ever disturbed their connection—that Shakespeare was pleased with Jonson, and that Jonson loved and admired Shakspeare.'[67] If the degree of serenity here envisioned seems excessive, Gifford is nevertheless playing a usefully corrective role. In defending his hero, he clears away ill-founded traditions that perhaps have their fountainhead in the *Worthies* of Fuller, who conjured up in his mind's eye the 'wit-combats' between learned Ben, the Spanish great galleon, and gentle Will, the dextrous English man-of-war.[68]

Yet the same editor bears responsibility for a notable accretion to the Shakespeare-Mythos. He first narrates the legend of the Mermaid Club; a legend as much inspired by Fuller as the discreditable stories which Gifford successfully challenged. Let the charming scene be described in its creator's own words:

About this time [1603] Jonson probably began to acquire that turn for conviviality for which he was afterwards noted. Sir Walter Raleigh, previously to his unfortunate engagement with the wretched Cobham and others, had instituted a meeting of *beaux esprits* at the Mermaid, a celebrated tavern in Friday-street. Of this Club, which combined more talent and genius, perhaps, than ever met together before or since, our author was a member; and here, for many years, he regularly repaired with Shakspeare, Beaumont, Fletcher, Selden, Cotton, Carew, Martin, Donne, and many others, whose names, even at this distant period, call up a mingled feeling of reverence and respect. Here, in the full flow and confidence of friendship, the lively and interesting 'wit-combats' took place between Shakspeare and our author; and hither, in probable allusion to them, Beaumont fondly lets his thoughts wander, in his letter to Jonson, from the country.[69]

But would the proud Ralegh, Captain of the Queen's Guard, have assembled or joined a tavern convocation of poets, wits, and players? A question Gifford does not think to ask. Nor does he remember that from 1603 until after Shakespeare's death, the courtier was imprisoned in the Tower, and thus effectively barred from such literary merriment.

We know from a contemporary Latin poem of unknown authorship that convivial philosophers, including Donne and Inigo Jones but not Jonson, met in some unspecified year, perhaps 1611, at the Mitre in Cheapside or Fleet Street. Perhaps they gathered at other public houses as well. Feastings at the Mermaid Tavern by 'the right worshipfull fraternity of Sirenaical gentlemen' are first referred to by Thomas Coryate, the traveller celebrated for his *Crudities*, in a letter written in 1615 from Ajmere in India. Donne and Jonson are mentioned, but to Shakespeare, living in Stratford retirement the year before his death, there is (unsurprisingly) no reference. Nor does the poet enter into lyric praise of the Mermaid in the verse epistle, of uncertain date and authorship, addressed to Jonson—that epistle, mellow with nostalgic recollection of full Mermaid wine and words nimble and full of subtle wit, which Gifford quotes in his evocation of the Club. In the second decade of the seventeenth century, Jonson, corpulent and rousing, may well have graced a table of wits and witlings at the Mermaid Tavern. But the Club presided over by Ralegh and boasting Shakespeare as a member has no more real existence than any other invention of the willing imagination dreaming of things past.[70]

Yet the world would not willingly let this insubstantial legend die; it would become part of literary folklore. Artists would sketch Shakespeare and Jonson and other wits with meticulous verisimilitude. Poets would furnish the scene with words. Shakespeare appears as dramatic monologist in 'At the "Mermaid",' addressing (as the author of the Browning *Guide-book* puts it) 'his literary friends, especially . . . Ben Jonson, gathered at "The Mermaid" tavern, the favourite resort in London of the Elizabethan wits'. A scholar's sentimental romance has been canonized as fact. Inevitably the supposed fact would be fictionalized as romance. There is a charming scene in a light modern novel in which Shakespeare, Jonson, Ralegh, Sidney, and other worthies are assembled at the Mermaid. Through the open window float the voices of prentice lads singing 'Drink to me only with thine eyes. . . .' 'O rare Ben Jonson', exclaims Stratford's pride, weeping, as the voices die away. 'God,' Jonson nods sadly, 'What genius I had then!'*

# 6

# *Severn and Ward's* Diary

IN the late 1830s Charles Severn, member of the Royal College of Physicians and amateur Shakespearian enthusiast, was appointed Registrar to the Medical Society of London. The duties of this 'unknown individual', as Severn modestly described himself, consisted of looking into the state of the Society's extensive library, and examining the holdings. In setting about this tedious task he was cheered by the hope of lighting upon some valuable memorials that had escaped the notice of his predecessors. He was not disappointed, for his attention was soon arrested by a set of seventeenth-century common-

---

* The association of Shakespeare with the Mermaid persists; thus William Riley Parker writes in his magisterial *Milton: A Biography* (1968), 'Shakespeare almost certainly visited the Mermaid Tavern while little John Milton was playing in the house on Bread Street barely a stone's throw away.' It is fortunate for Milton's biographer that his subject was under-age at the time.

place-books that the Medical Society had acquired late in the preceding century. These duodecimo notebooks, excellently preserved in their original binding, contained a diary in the legible hand of the Revd John Ward, who, a half-century after Shakespeare's death, had ministered as spiritual and bodily physician to his Stratford parishioners.

Severn in 1839 edited a volume of extracts from Ward's *Diary*, prefaced by a desultory series of chapters on the diarist's life, the spelling of Shakespeare's name, the youth of the poet, Shakespeare's property, and his final illness and death—oddly followed by additional chapters on the marriage licence bond and Shakespeare's friends. Severn shows himself no scholar by printing, without scepticism, the complete text of a patently forged letter, but in his section on Shakespeare's last days he produces a choice morsel yielded by the *Diary*. The 'veil of apparently impenetrable obscurity is now removed', he announces triumphantly, 'and we are at length made acquainted with the disease which proved mortal to the great Poet.'[71] Thus came to light Ward's note, jotted down shortly after the Restoration, describing a merry meeting at which Shakespeare, Drayton, and Jonson drank too hard, and as a result of which Shakespeare contracted a fatal fever.

Elsewhere Severn brings together Ward's half-dozen Shakespearian scraps, which include one other item of more than ordinary interest:

I have heard that Mr. Shakspeare was a natural wit, without any art at all; he frequented the plays all his younger time, but in his elder days lived at Stratford, and supplied the stage with two plays every year, and for it had an allowance so large, that he spent at the rate of 1,000[li.] a year, as I have heard.[72]

Surely, the medical registrar reflected, Malone must have underestimated Shakespeare's earnings. Hopeful of confirmation, Severn in 1840 wrote to the man whom he regarded as the most competent authority on such matters, Robert Bell Wheler. The former's letter may now be seen at the Shakespeare Birthplace Record Office, but not, alas, Wheler's reply.

Not everyone would share Severn's confidence in the trustworthiness of his authority. On the other hand, future biographers would scarcely be able to ignore Ward's notes, and particularly his remark on Shakespeare's last illness—the only seventeenth-century record of this event which has ever come to light.

# 7

## *Various Lives*

IT is scarcely surprising, with Shakespeare enthroned as a full-fledged culture hero, that a large and bardolatrous public—not by any means uniformly well educated—should wish to know something about his life and writings. Where could they turn? Certainly not to Boswell's Malone, which, in twenty-one volumes, lay beyond their ken—and pocketbook. They would find daunting as well as expensive Drake's two enormous folios, and most of them would not have access to Wheler's guidebooks, published in a provincial town for locals and visitors. Of similarly restricted circulation would be W. T. Moncrieff's entertaining *New Guide to the Spa of Leamington Priors* (1822), which, although mainly concerned with the waters, nevertheless offers a life of Shakespeare; those consulting it would find some curious evidence on behalf of the deer-poaching tradition: 'if Dr. Gall the craniologist's assertion is to be believed, the organ of robbery (covetiveness) and the organ for forming good dramatic plots, are one and the same; he certainly proved himself a great adept in the latter, and no doubt was so in the former'. But in the earlier nineteenth century more widely disseminated sources of information became available to appease the hunger for knowledge of that middle class which seems forever to be expanding. The curious might turn to popularizations, historical works (such as Hume's *History of England*), biographical compilations, or introductions to editions; above all, to those wondrous new repositories of learning, the encyclopaedias. Those who could afford no better might buy a penny pamphlet, such as the *Life and Times of Shakespeare: Actor and Dramatist* (undated but mid-century), the anonymous author of which promises to sift the few grains of biographical truth from the many bushels of improbabilities. In fact he plagiarizes Rowe's 'Account', his only original contributions to biographical knowledge being the startling information that Shakespeare was born on 14 April and that his elder daughter was christened Judith.

To some present-day readers it may come as a surprise that the author of the *Enquiry Concerning Human Understanding* and the *Dialogue Concerning Natural Religion* was in his own time, and for a long while thereafter, better known as an historian than as a philosopher. Hume's eight-volume *History of England, from the Invasion of Julius Caesar to the Revolution in 1688* (1763) went through numerous editions in the nineteenth century and enjoyed a phenomenal success. Macaulay described him as 'the ablest and most popular' of British historians. Perhaps so, but he is no critic: in the brief space allotted to Shakespeare in *The History of England*, Hume takes a dim view of the poet. In a passage savouring of the neoclassicism of Rymer's day, Hume deplores Shakespeare's 'total ignorance of all theatrical art and conduct' and 'that want of taste that often prevails in his productions'.[73] But what could one expect? He was 'born in a rude age, and educated in the lowest manner, without any instruction, either from the world or from books'.[74] Having thus pronounced on the dramatist's learning, Hume concludes with an egregious blunder of his own: 'He died in 1617, aged 53 years.'[75] It did not go unremarked. In a note to the burial entry in the Stratford parish registers reproduced in Malone's and later editions of Shakespeare, Farmer twitted the learned historian: 'No one hath protracted the life of *Shakspeare* beyond 1616, except Mr. Hume; who is pleased to add a year to it, contrary to all manner of evidence.'[76] The error was corrected in later editions.

Henry Hallam, a later and greater historian and better educated than Hume as a literary critic, also has his blind spots: he can wish the Sonnets, which record the folly of a misplaced and extravagant affection, had never been written—an opinion that would echo through subsequent Victorian criticism. Father of the Arthur Henry Hallam whose premature death Tennyson mourned in *In Memoriam*, he made his reputation with *A View of the State of Europe during the Middle Ages* (1818) and confirmed it with *The Constitutionial History of England from the Accession of Henry VII to the Death of George II* (1827), a work that still commands respect. His remarks on Shakespeare occur in the *Introduction to the Literature of Europe, in the Fifteenth, Sixteenth, and Seventeenth Centuries* (1837–9), which went through five editions. Hallam can only wonder at the unmeasurable chasm separating the indifferent provincial player from the mighty genius who conceived *Macbeth* and *King Lear*: the records of the life, like the Sonnets, yield the unacceptable knowledge that the god was but a man.* And how trivial those records are! 'All that insatiable curiosity and unwearied diligence have hitherto detected about Shakspeare', Hallam laments,

serves rather to disappoint and perplex us than to furnish the slightest illustration of his character. It is not the register of his baptism, or the draft of his will, or the orthography of his name that we seek. No letter of his writing, no record of his conversation, no character of him drawn with any fulness by a contemporary has been produced.[77]

The 'petty circumstances relating to Shakspeare' brought to light by the labours of John Payne Collier (of whom we shall speak at length) did not incline Hallam to change his mind in subsequent editions of the *Introduction*.

By a curious irony this despairing pronouncement was made by a critic in whom we find the first green shoots of a new biographical method that would

---

* Hallam was conscious of what troubled him; in a note written in 1842 he remarks: 'If there was a Shakespeare of earth, as I suspect, there was also one of heaven; and it is of him that we desire to know something' (*Introduction to the Literature of Europe* . . . (4th edn.; London, 1854), ii. 176).

discover in the plays much which the prosaic external records fail to supply. The crucial passage appears in Hallam's discussion of *Timon Of Athens*:

there seems to have been a period of Shakspeare's life when his heart was ill at ease, and ill content with the world or his own conscience; the memory of hours misspent, the pang of affection misplaced or unrequited, the experience of man's worser nature which intercourse with unworthy associates, by choice or circumstance, peculiarly teaches;—these, as they sank down into the depths of his great mind, seem not only to have inspired into it the conception of Lear and Timon, but that of one primary character, the censurer of mankind.[78]

This preoccupation, which belongs to the years 1600 to 1604, first manifests itself in the melancholy Jaques of *As You Like It*, deepens with the Duke of *Measure for Measure*, and is in Hamlet 'mingled with the impulses of a perturbed heart under the pressure of extraordinary circumstances'; with the later plays, *Macbeth* and *The Tempest*, the period is passed. Later in the century Dowden, most influential of the dramatist's subjective biographers, read Hallam and did not forget the passage when he conceived the outlines of Shakespeare's Third Period.

A much fuller and more circumstantial account of the poet appeared in 1824 in Augustine Skottowe's two-volume *Life of Shakspeare* [*etc.*]. As his preface indicates, this enthusiast sought to provide admirers of the dramatist with 'a COMPANION TO SHAKSPEARE': his book, an exercise in *haute vulgarisation*, anticipates modern reader's guides with similar pretensions. It qualifies too as a representative, and hence revealing, middle-brow view of Shakespeare's career. The *Life* itself occupies only some 120 pages (inclusive of notes) preceding chapters on the individual plays. Even so, the biographical section has much extraneous matter, including a persistently disparaging account of the pre-Shakespearian drama, a description of the physical playhouse and stage practices (this portion is heavily indebted to Malone), and a critical survey of the various modern editions of Shakespeare. Skottowe was not an original scholar in the sense that he went back to the records or sought new data, but he knew all the principal secondary sources: Rowe, Aubrey, Oldys, Wheler, above all Malone, whose posthumous Life appeared handily for Skottowe's use. Thus his biography is informed.

One would not, however, describe it as intelligent. Malone's critical attitude towards legends and traditions—one of his chief contributions as a scholar—leaves Skottowe unmoved. He knows his great predecessor's position, but dismisses it in his contentious notes. Himself incapable of discriminating between the relative importance of fact and rumour, or between well- and ill-founded traditions, Skottowe hospitably embraces contradictory reports. Thus, on the occupation of the poet's father: 'John Shakspeare . . . was originally a glover, and, subsequently, a butcher, and also a dealer in wool in the town of Stratford.'[79] (Note the positiveness of statement.) On at least one occasion Skottowe shows some hard-headedness, when he dismisses, as based on insufficient evidence, the common assertion that Shakespeare was born on the 23rd; but even this flash of rigour on a minor matter is only momentary, and is indeed effectually retracted later in the Life when the author forgetfully remarks, 'He died on the 23rd of April, the anniversary of his birth.'[80] Skottowe rejects Malone's dismissal of the deer-poaching tradition, which is presented as fact in the narrative and defended in a note, and he embellishes his account with a stanza from the spurious Lucy ballad, the authenticity of which he stubbornly defends (again in a note). Only once does he reject, outright, inherited lore: he will have no part of the horse-holding story. Instead Skottowe vaguely suggests that Shakespeare began his professional career 'in a very mean capacity', presumably acting minor roles. Convinced that the Elizabethan theatre was an impoverished institution (he knows nothing of how Dulwich College came to be founded), Skottowe has difficulty in understanding how the playwright came into a considerable estate; he is for this reason all the more receptive to the legend that Southampton gave Shakespeare a gift of £1,000. Skottowe believes too in the mulberry tree, in the Combe epitaph as an instance of the poet's extemporaneous wit, even in the tradition of Shakespeare's 'gallantry' at Oxford with Mistress Davenant. Of the poet's marriage he takes a disenchanted worldly view: Shakespeare 'neither bettered his circumstances, nor elevated himself in society by the connection'.[81] The 'cold and brief notice' of the wife in the will leads Skottowe to conclude that the domestic life of the Shakespeares was unblissful.

His audience, one suspects, turned to this Life with confidence. If Skottowe is 'destitute of poetic feeling' (an accusation he levels against Malone), he does give the appearance of being up to date and scholarly; he adopts an authoritative tone, and his book is documented. But readers would come away from it with a largely traditional idea of Shakespeare, one hardly modified by the fruits of Malone's labours. The work would seem to illustrate the lag that often exists between original achievement and the popular dissemination of that achievement.

Skottowe's vapid biography furnished one reviewer with the occasion for a sensational suggestion regarding Shakespeare. It appears in a notice in the *Monthly Review* for December 1824 by an anonymous critic who had earlier demonstrated his ingenuity by conjecturing that Shakespeare began his stage career under the pseudonym of Christopher Marlowe. Now the same reviewer darkly—but unmistakably—implies that the poet was a homosexual. 'The age of James the First', he laments, 'was an age of impurity; and the manners of the sovereign, and the consent of Catholic Europe, had given a license to practices which may not in these happier days be tolerated, but of which then a sonneteer could boast.'[82] The character of the vice, the reviewer declares, makes an honest biography of Shakespeare impossible: one cannot treat it indulgently, nor can one very well rake the National Poet over the coals. No doubt others had worried about the Sonnets: how else to account for such elaborate defences as Drake's of Shakespeare's celebration of male friendship? But nobody had previously made the issue public.

This affront to moral as well as biographical orthodoxy drew a quick response from the Revd Charles Symmons in a conventional life of Shakespeare otherwise distinguished chiefly by its persistent disparagement of Malone. One cannot very well blame the good parson for his outburst of embarrassment and anger. 'I blush with indignation', he fairly splutters, 'when I relate that an offense, of a much more foul and atrocious nature [than adultery], has been suggested against him [Shakespeare] by a critic of the present day, on the pretended testimony of a large number of his sonnets. But his own proud character, which raised him high in the estimation of his contemporaries, sufficiently vindicates him from this abominable imputation.' Nevertheless Symmons undermines his vindication somewhat by admitting that in the language of the Sonnets 'love is too strongly and warmly identified with friendship'.[83] The thought of a homosexual strain in Shakespeare, however angrily rebutted, evidently touched a sensitive spot in the pre-Victorian consciousness. Once made, the suggestion would be made again, but not until the less happy days of the *fin de siècle* would critics discuss, without diffidence or outrage, the Bard as Deviant.

Equally heterodox views about Shakespeare and his family crop up in the strange Life included, in 1837, as part of one of the volumes devoted to 'Eminent Literary and Scientific Men' in the widely distributed *Cabinet Cyclopaedia* conceived and managed by the extraordinary Revd Dionysius Lardner. The scandals in the Dionysian private life of the amorous divine do not here concern us. His *Cyclopaedia*—a reference library rather than an encyclopaedia in the usual sense—came to 133 volumes when completed in 1849, and is an early feat of commercially successful educational publishing; all the more impressive by reason of the eminent contributors he managed to enlist: De Morgan on probabilities, Phillips on geology, Herschell and Baden Powell on the history and study of astronomy and natural philosophy, Lardner himself on geometry, heat, and hydrostatics and pneumatics. It must be allowed, however, that Lardner, who occupied the first chair of astronomy and natural philosophy at what was to become University College, London, knew stars and steam engines better than literature. How else explain his choice of a writer for the vital article on Shakespeare? Charles Armitage Brown, justly incensed by the piece, protested in a letter to Lardner, who mildly replied, '. . . I am not the author of any part of the volume containing the Life of Shakespeare; and . . . I am responsible merely for selecting fit persons to write the different articles.'[84] The identity of this particular fit person he never saw fit to divulge.*

Among the curious features of the *Cabinet Cyclopaedia* Life is the denigration of Shakespeare's father and his father's lineage. Descended from a 'very obscure family', John Shakespeare (we are told) was an illiterate glover, 'very honest, perhaps, but very poor'. The author dwells with relish on this presumed penury, which knew no relief: the Bailiff of Stratford was 'little better than a pauper'. As for the representations in the grant of arms: these are regrettably false. The ancestor mentioned in the instrument as having been honoured by Henry VII belonged not to John's line but to his wife's; hers also was the estate, the value of which was in any event greatly exaggerated. (It is true that John Shakespeare's paternal descent cannot be traced beyond his father.) 'Altogether', the writer severely concludes, 'the affair is discreditable to the father, to our poet himself, and to the two kings at arms.'[85] At Stratford grammar school, which threw open its doors to 'the most indigent boys', the future poet received the rudiments of an education, including the little Latin he came to possess. That the anonymous author has profited from the researches of 'the industrious Malone', as he more than once terms him, we may infer not only from the correct identification of John Shakespeare's occupation but also from the treatment of the deer-poaching, tradition. This the biographer almost rejects, on Malone's grounds, only at the last moment to snatch absurdity from the jaws of reason: Sir Thomas Lucy, he conceives, was the magistrate before whom Shakespeare was haled for stealing deer from *someone else's* park.

What infuriated Brown, however, were not these eccentricities of biographical interpretation, but the discussion of Shakespeare's moral character. The spirit of Bowdler, whose *Family Shakespeare* appeared in 1818, walks abroad in these pages of the Life. Perhaps, we are told, Shakespeare's character deserves the encomiums bestowed on it by modern writers, 'but there is greater probability in supposing that it was not wholly untainted by the vices of the period'.[86] Tradition after all holds that the poet was 'not averse to the bottle, or to pursuits still more criminal' (the *still* is a nice touch). As evidence, the biographer cites the anecdote of Shakespeare taking with lust and *élan* the unsuspecting Burbage's place at an assignation with a feminine patron of the Globe.† But more persuasive testimony comes from the writings themselves. The plays—in which no retributive justice prevails— 'absolutely teem with the grossest impurities,—more gross by far than can be found in any contemporary dramatist'.[87] The Sonnets swell the indictment,

---

* William Jaggard in his *Shakespeare Bibliography* erroneously assigns the *Cabinet Cyclopaedia* Life to the miscellanist Robert Bell. A responsible critic, Bell furnished his *Annotated Edition of the English Poets*, by which he is best known, with a competent memoir of Shakespeare based chiefly on Halliwell's 1848 Life. Presumably Jaggard's misconception arose because Bell contributed two later volumes of literary biography to the *Cabinet Cyclopaedia* under his own name.

† See above, p. 17.

exhibiting as they do the progress of a vicious passion in all the colours of licentiousness. In quoting one of the poems (41: 'These petty wrongs that liberty commits . . .') the biographer silently alters a crucial pronoun, *her* to *she*, to strengthen the accusation: Brown might well complain of deliberate falsification. Other quotations from the Sonnets reveal, astonishingly, that the anonymous author is ignorant of the sex of the person addressed; perhaps it is just as well.

Elsewhere, however, he brings forth information possessed by no one else. It is interesting to learn that Shakespeare while in London never permitted his wife and children to reside with him, and possibly never even to visit him. The 'most corroborative' circumstance for this assertion is that Anne bore him no more children after 1584. 'How account for this', the biographer rhetorically demands, 'except on the hypothesis of a separation?'[88] She stayed on in Stratford, while the dramatist pursued his affairs in a London then more corrupt than at any time since.

The defects of character described by the writer inspire melancholy feelings in his bosom; he sighs, 'His principles, alas! were not equal to his intellectual gifts.' But even the latter leave something to be desired, for in the concluding paragraph we are told that 'the greatest defect of Shakespear is his ignorance or disregard of chronology, geography, the manners, institutions, and opinions of nations'.*[89]

One wonders what the subscribers to the *Cabinet Cyclopaedia*—solid citizens in search of enlightenment—made of this oddly derogatory picture, so different from the customary panegyrics, of the National Poet. Still, some of the biographer's perversities do not entirely lack precedent. Lardner's outrageous hack is, for example, anticipated in his view of Shakespeare's marriage by Thomas Moore in his 'Notices of the Life of Lord Byron'. In extenuating Byron's unfitness for domestic life, Moore cites Shakespeare (along with Dante, Milton, and Dryden) as an instance of the illustrious poet who endured matrimonial infelicity. 'By whatever austerity of temper or habits the poets Dante and Milton may have drawn upon themselves such a fate,' Moore writes,

it might be expected that, at least, the 'gentle Shakspeare' would have stood exempt from the common calamity of his brethren. But, among the very few facts of his life that have been transmitted to us, there is none more clearly proved than the unhappiness of his marriage. The dates of the birth of his children, compared with that of his removal from Stratford,—the total omission of his wife's name in the first draft of

---

* A disparaging tone towards Shakespeare appears again, if fleetingly, in the course of a literary discussion between Lord Cadurcis and Herbert in Disraeli's *Venetia*, published in the same year: '"And who is Shakespeare!" said Cadurcis. "We know of him as much as we do of Homer. Did he write half the plays attributed to him? Did he ever write a single whole play? I doubt it. He appears to me to have been an inspired adaptor for the theatres, which were then not as good as barns. I take him to have been a botcher up of old plays. His popularity is of modern date, and it may not last . . ."' (iii. 231). From such barren soil would Baconianism, that exotic desert blossom, shortly spring.

his will, and the bitter sarcasm of the bequest by which he remembers her afterwards,—all prove beyond a doubt both his separation from the lady early in life, and his unfriendly feeling towards her at the close of it.[90]

But of course this compares neither in scope nor vehemence with the onslaught of the *Cabinet Cyclopaedia* author; and it is at least easy to see what axe Moore is grinding.

These slighting conceptions of Shakespeare the man represent, then, an atypical facet of popular response to a figure of incomprehensible magnitude and elusive personality. An opposing image, in its own way as excessive, emerges from the remarkable anecdote of playwright-player as darling of princes told by Richard Ryan—bookseller, dramatist, and raconteur—in his *Dramatic Table Talk*, published anonymously in 1825. The story is sufficiently brief and entertaining to be reproduced entire:

It is well known that Queen Elizabeth was a great admirer of the immortal Shakspeare, and used frequently (as was the custom with persons of great rank in those days) to appear upon the stage before the audience, or to sit delighted behind the scenes, when the plays of our bard were performed. One evening, when Shakspeare himself was personating the part of a King, the audience knew of her Majesty being in the house. She crossed the stage when he was performing, and, on receiving the accustomed greeting from the audience, moved politely to the poet, but he did not notice it! When behind the scenes, she caught his eye, and moved again, but still he would not throw off his character, to notice her: this made her Majesty think of some means by which she might know, whether he would depart, or not, from the dignity of his character, while on the stage.—Accordingly, as he was about to make his exit, she stepped before him, dropped her glove, and re-crossed the stage, which Shakspeare noticing, took up, with these words, immediately after finishing his speech, and so aptly were they delivered, that they seemed to belong to it:

> 'And though now bent on this high embassy,
> Yet *stoop* we to take up our *Cousin's* glove!'

He then walked off the stage, and presented the glove to the Queen, who was greatly pleased with his behaviour, and complimented him upon the propriety of it.[91]

Ryan cannot be credited with originating this story, for a less elaborate variation on it had appeared in the popular press at the time of the Ireland forgeries in 1796; according to this briefer version, the 'mimic Monarch' had stood near the Queen's box and declaimed (more familiarly than in Ryan), 'But, ere this be done, | Take up our SISTER's handkerchief.' Ryan, however, and not his source, would provide the basis for the frequent retellings of the anecdote in the nineteenth century. Perhaps in an attempt to render it less defiant of credulity the locale of the story would be shifted, late in the century, to the Queen's palace at Richmond. How pleasant, too, to usher Shakespeare into the seat of royalty itself.

A pity that a few trifling circumstances hinder belief in this heart-warming romantic encounter: in the public playhouses performances were not held in the evening, the stage as yet afforded no scenery for eavesdroppers to conceal themselves behind, the Queen is not known to have professed a great admiration for Shakespeare, she did not expose herself to the multitude by visiting the playhouse, and she restrained herself from flirting publicly (or in private) with subjects of inferior station. Otherwise the anecdote is plausible enough. While not enhancing the biographical record, it contributes to the Shakespeare-Mythos.

Had she known it, Ryan's fable would have captivated the ineffably feminine Mrs Anna Jameson, the flutterings of whose heart are enshrined in her *Memoirs of the Loves of the Poets*, which was published in 1829 and went through a number of editions. Perhaps her own unsuccessful marriage prompted her rather specialized literary pursuit. Approaching 'the POET OF WOMANKIND' with reverence and trepidation, she looks, as so many others have looked, to the Sonnets for intimate revelations. She seems to have been the first to suggest that, while some of these poems are directed to Southampton as the Fair Youth, 'others . . . are addressed in Southampton's name, to that beautiful Elizabeth Vernon, to whom the Earl was so long and ardently attached'; a suggestion which would, in later commentaries, bear curious fruit. The poet's mistress, the subject of sonnets 'without doubt inspired by the real object of a real passion', understandably fascinates Mrs Jameson, but she does not press for an identification: of her 'nothing can be discovered, but that she was dark-eyed and dark-haired, that she excelled in music; and that she was one of a class of females who do not always, in losing all right to our respect, lose also their claim to the admiration of the sex who wronged them, or the compassion of the gentler part of their own, who have rejected them'.[92] Having, she says, ransacked everything that had been discovered, said, or surmised relative to Shakespeare's private life, Jameson came away disappointed with the dry results of scholars' labours: 'registers of wills and genealogies, and I know not what'. She is herself above—or beyond—the trivial demands of factual precision; Shakespeare, we are informed, married 'at the age of seventeen, Judith Hathaway, who was eight or ten years older than himself'.[93] Jameson's version of the poet's career is as satisfying as it is simple. The period of the Sonnets was a 'time when Shakspeare was living a wild and irregular life, between the court and the theatre, after his flight from Stratford'; eventually, 'having sounded the depths of life, of nature, of passion', he settled in his native town and there 'ended his days as the respected father of a family, in calm, unostentatious privacy'.[94]

Readers seeking more startling disclosures than Jameson's must have been disappointed with 'The Confessions of William Shakspeare' of 1835, spread over four instalments of the respectably middle-brow *New Monthly Magazine*. The title is the most sensational feature of this essay, written in a tone of sustained hysterical adulation. What the anonymous author, who describes himself as a friend of Leigh Hunt, essentially has to offer is a highly subjective reading of the Sonnets. Little information emerges; the author of the 'Confessions' attempts no identifications, and he seems to be unacquainted with—or indifferent to—Drake's theory about Southampton or Boaden's about Pembroke. He has obviously had a hard time with the poems. The printer, he complains, has flung them into a 'strange and confused jumble of arrangement, or rather non-arrangement'. This confusion the critic makes no attempt to set right. He does, however, beg his readers not to interpret the vocabulary of the Sonnets as reflecting anything strange or unmanly about Shakespeare's character; it was merely the usage of the time. What the friend meant to the poet, as this essayist sees it, is best conveyed in his own breathless dithyramb:

Here was the pillow his spirit reposed on—here was the object to which he clung, as connecting him in actual life with the moral beauty and sweetness of the world. Here was at last some peculiar and captivating medium, through which he could even look out upon the creatures that walked the street before him, and feel one of them not only in sympathy and love, but in the positive scale of being, without remorse or uneasy shame. All that his great heart sought for, he set up here. Here was something that it had thirsted for in vain among his fellow actors and fellow writers.[95]

And so on; but this will suffice to illustrate the style of the 'Confessions'. The panegyric to masculine friendship would carry greater conviction if its author had not found the Dark Lady in a number of sonnets addressed to the Fair Youth. From this error proceeds also the supposition of a liaison between the mistress and the Rival Poet.

A remorselessly sentimental picture of Shakespeare emerges even in the pages not concerned with the Sonnets. Introduced to the stage by his countryman Tom Green, he began his professional life as an amender and rewriter of other men's plays: 'One production was brought to him after another. Fancy the amazement of the poor original authors when their works came back with the touches of that divine hand! It soon fell out that plays altered by him had a surer market than plays written by others.'[96] Hence the calumnies of Greene (misdated 1591). Shakespeare was a fine actor, 'as far in advance, indeed, of his contemporaries and of his audience, as his writing was'. His marriage was compatible: the second-best bed probably had more affectionate associations than the first and may have served on the wedding night. (It is no mean achievement to represent Shakespeare as happily married, passionately involved in an illicit amour, and impeccably moral.) The dramatist was furthered in his career by the chivalrous Earl of Southampton, whose gift of £1,000 was used to purchase theatrical property; hence Shakespeare's position in the patent granted by James (misdated 1602). Thus it goes. What the public made of this farrago one cannot say; but the 'Confessions' did find receptive editors, for it was reprinted the same year in the United States in the *Museum of Foreign Literature and Science*.

Less colourful accounts of the poet preface innumerable nineteenth-century editions of Shakespeare. Inevitably, like the texts they introduce, these are derivative and (on the whole) contain little to excite the historian's interest; but mercifully they are brief. They are also sometimes misleading. Thus W. Harvey, the obscure editor of an 1825 *Works* confuses Birthplace with New Place and presents a curious portrait of the artist as time-server. In compliance with the servile spirit of his day, the poet assiduously courted first Elizabeth, then James. From the former, Shakespeare received only praise—such was her parsimony—but the latter, his ear 'ever open to the blandishments of flattery', humoured the sycophant with that amicable letter which Davenant purportedly treasured.

Representative of the class, although with a difference, are the 'Remarks on the Life and Writings of Shakspeare' contributed by John Britton to the edition published by Whittingham in 1814. Britton brings scholarly, if not literary, credentials to his task. Having commenced his love affair with learning in the hours he could snatch away from bottling wines in his uncle's London tavern, he achieved distinction as an authority on British topography and antiquities, and especially on archaeological art and architecture; he wrote as an expert on Stonehenge and Salisbury Cathedral. In 1816, in 'Remarks' accompanying two woodcuts of the Stratford bust, he argued its superiority, as a likeness of Shakespeare, to portraits he maintained were spurious. His essay on the life and writings, based for its facts mainly on Rowe and Malone (denigrated as a 'laborious commentator'), offers a conventional biographical résumé but takes an independent tack with respect to the generally accepted deer-stealing tradition, which Britton inclines to regard as fictitious. The poet, he feels, was more likely to have left home for London because of natural inclination or estrangement from his wife. Britton entertains grave misgivings about Anne's morals. He points to the burial, on 6 March 1590, of 'Thomas Green *alias* Shakspere', and, supposing without good reason that this Green was a child, adds: 'The inference of which this circumstance is susceptible must be obvious.'[97] To Britton, apparently, belongs the distinction of being the first to suggest that the woman who bore the dramatist three children also mothered a bastard.

Britton's 'Remarks', in their original and expanded versions, went through a succession of editions, but hardly compare in influence or intrinsic interest with the idiosyncratic preface written by the former editor of the *New Monthly Magazine* for one of those single-volume collected editions, all double columns and fine print, that made the National Poet cheaply accessible in this period to a large literate audience.

Those who knew him had reservations about Thomas Campbell—this 'small thin man, with a remarkably cunning and withered face, eyes cold and glassy, like those of a dead haddock'. So he was described by an Irish magazine writer who preferred, understandably, to remain anonymous.

Carlyle took offence at the auctioneer's smirk on Campbell's face, and disparaged his talk as 'small, contemptuous, and shallow'. His drunkenness (so Dyce records in his *Reminiscences*) lowered him in the estimation of the world, nor did his morbid eagerness for praise, the thin skin and soft underbelly of sentimentality, pass unnoticed by his contemporaries. As a poet, however, he was astonishingly overvalued; Byron, surveying his contemporaries, ranked him above Wordsworth. Today Campbell is best remembered, if at all, as a minor versifier caught in transition between an outmoded didacticism, which he could not entirely relinquish, and a romantic lyricism, which he could not wholeheartedly embrace; he was himself inclined to regard as 'the most important event of his life's little history' his role in the founding of London University. This unappetizing man also wrote criticism of a surprising warmth and humanity—qualities that find expression in the preface which he contributed in his later and less productive years to Shakespeare's *Dramatic Works* (1838). The essay also typifies some of the more rational tendencies in early nineteenth-century Shakespearian biography.

If unambitious, Campbell at least knows well enough to consult respectable sources: Drake, Wheler, above all Malone's Life in the Third Variorum. He knows too the scurrilous biography in the *Cabinet Cyclopaedia*, and challenges its insistence on John Shakespeare's penury; paupers, Campbell drily observes, seldom leave their heirs houses and orchards. Farmer's demonstration of Shakespeare's non-learning fails to shake Campbell's equanimity: 'Instead of being surprised that Shakespeare had, to all appearance, great genius with small erudition, I am inclined to ascribe the greatness of his genius to his good fortune in having so small a portion of his youthful powers absorbed in the forced fatigue of acquiring learning. By learning I mean not the knowledge he got from general reading, but the knowledge which he missed acquiring from grammars and dictionaries.'[98] Only once is Campbell notably prejudiced, when he castigates Anne Hathaway for decoying her man into a match which (since he was a stripling and she a mature woman) does her no credit. The deer-robbing tradition he finds, despite Malone, irresistible; but so do most others in this century. Like almost everyone else, too, Campbell supposes that Shakespeare began his playwriting career as an adapter of other men's leavings. Although his mind tells him otherwise, he leans in his heart to belief in Southampton's gift of £1,000; again he has plenty of company. Another characteristic touch is the glowing evocation of wit and hilarity at the Mermaid, where Shakespeare caroused with Jonson, Donne, and the rest. Contrary to Schlegel and other Romantic critics, however, Campbell does not accept the idea that with the Sonnets Shakespeare unlocked his heart; he is amused by Drake's attempts to discredit the existence of the troublesome Dark Lady, while excoriating her 'for faults committed by her during her state of nonentity'. When he turns to the

Shakespeare likenesses, Campbell considers only the Stratford bust and the Chandos portrait; the Droeshout engraving does not even rate a mention. In this way he is again a man of his age.

Yet he makes his own contribution to the tradition of Shakespearian biography. In an eloquent passage he sees the dramatist as representing himself in *The Tempest*:

Shakspeare, as if conscious that it would be his last, and as if inspired to typify himself, has made its hero a natural, a dignified, and benevolent magician, who could conjure up spirits from the vasty deep, and command supernatural agency by the most seemingly natural and simple means. Here Shakspeare himself is Prospero, or rather the superior genius who commands both Prospero and Ariel. But the time was approaching when the potent sorcerer was to break his staff, and to bury it fathoms in the ocean—

> Deeper than did ever plummet sound.

That staff has never been, and never will be, recovered.[99]

No matter that Prospero buries his staff in the earth, not ocean—as an attentive reader triumphantly pointed out in the Shakespeare Society *Papers* for 1844—the effect of the parallel drawn between magician and playwright, each relinquishing his art, would be immediate and enduring. One encounters it in innumerable prefaces to *The Tempest* by critics who (one suspects) never heard of Campbell's edition, and Robert Graves goes so far as to inform us that 'tradition has always identified Prospero with Shakespeare himself'.

Among the earliest to quote him in this connection is the Revd Joseph Hunter, in *A Disquisition on the Scene, Origin, Date, Etc. Etc. of Shakespeare's Tempest*, published the next year. Hunter (who will reappear in this narrative) finds Campbell's view 'beautifully expressed' but rather less than persuasive, for he is himself promoting an early date for the play. His argument is ingenious but, as one might expect, ineffective. Nevertheless it was enough to convince Campbell, who thus demonstrated that he was no scholar. On 12 December 1839 he thanked Hunter for a presentation copy of the *Disquisition* and made an extraordinary recantation: 'You have [he wrote] demolished my comparison between Shakespeare finishing his career with the Tempest & Prospero throwing his magic wand into the sea—with so much truth and urbanity—that I make you welcome to your victory—.'[100] But publicly he kept silent.

The 'Remarks' have all the limitations of amateurism; in an edition which goes under his name, Campbell ignorantly alleges that Heminges and Condell excluded *Troilus and Cressida* from the First Folio. But the preface is far from being totally misleading and it glows with the mellow flavour of Campbell's public personality. Campbell is capable of seeing the mighty Shakespeare as 'a little cherub only four years old, toddling about, and thinking more of sugar-plums than of the Heralds' College'. Sentimental perhaps, and too sweet for the modern palate, which disdains sugar-plums; but nineteenth-century readers must have found Campbell a humane and homely guide to an author too often regarded with awed reverence.

# 8

## *Encyclopaedias*

FOR readily accessible information about Shakespeare's life the public at large, then as now, probably turned more naturally not to isolated editions or scattered magazine pieces, but to the encyclopaedias, general or biographical. Earlier generations, as we have seen, had their assimilative works of reference: Fuller's *Worthies*, Winstanley's *Lives of the Most Famous English Poets*, Collier's *Great Historical, Geographical, Genealogical, and Poetical Dictionary*. But in this period encyclopaedias truly waxed and multiplied, as speculative publishers and booksellers recognized, from the example of the *Britannica* in England and the *Encyclopédie* on the Continent, that such works had a large potential audience and that handsome profits might accrue from their sale.[101] Today scattered and inconvenient to consult, they lack originality of thought and distinction of style; too often the articles are the work of hacks who, secure in their anonymity, cheerfully plagiarized one another. Yet these humbler narratives no doubt had a more widespread influence than the learned (and similarly derivative) tomes of a Drake. The encyclopaedia entries for Shakespeare were read by non-specialists who, puzzled by the apparent incompatibility between the uneducated provincial depicted in the articles and the universal genius revealed by the plays, descended into the bog of Baconianism or other deviations—which too are part of our story. And the encyclopaedia contributors numbered among them not only nameless mediocrities but also at least one writer of enduring interest: Thomas De Quincey, who furnished the article on Shakespeare for the seventh edition of the *Britannica*. As the encyclopaedias were already well established by the nineteenth century, a backward glance may prove helpful.

Credit for the first encyclopaedia article on Shakespeare seems to belong to Thomas Birch, DD (1705–66). The son of a coffee-mill maker, and without the advantages of a university education, Birch between 1734 and 1741 brought out his English edition of Pierre Bayle's *Dictionaire historique et critique*. Under the title *General Dictionary, Historical and Critical*, Birch's volumes were 'interspersed with several thousand lives never before published', of which he himself contributed 618. His article on Shakespeare occupies eleven folio pages in volume IX (1739). Expectedly Birch quotes and paraphrases Rowe, but he also draws upon the editions of Pope and Theobald, and Rymer's *Foedera* (for the 1603 royal licence for the King's Men), and he has the doubtful benefit of being able to offer specimens from Warburton's projected *Shakespeare* 'in several curious remarks, which that excellent Critic has communicated to me'. Birch's article manages to be at once pedestrian and pioneering; the *General Dictionary* represents the first attempt in England to apply the inductive method to biography, and fairly qualifies as the progenitor of the *DNB*.

In 1763 Philip Nicolles contributed the article on Shakespeare to Birch's successor, the even more ambitious *Biographia Britannica . . . Collected from the best Authorities, both Printed and Manuscript, and digested in the Manner of Mr Bayle's Historical and Critical Dictionary*. Already passingly referred to in these pages, the article furnished the first complete text of Shakespeare's will, with the notorious misreading of the brown best bed.[102] Other documents appear in the dozen double-columned folio pages on Shakespeare: the grant of arms (from Pope's edition), James's patent to the King's Men (from Rymer's *Foedera*), the dedication of *The Rape of Lucrece*. Mainly the writer follows Rowe for the facts of Shakespeare's life. Hence the familiar errors (the poet was the eldest of ten children) and the traditions: the deer robbing (pardoned by no less a personage than the Queen!), the first theatrical employment 'in a very mean rank', the £1,000 from Southampton, the epitaph on John a Combe. The deer-stealing incident prompts a learned note on *A Compendious or Brief Relation*, a tract published in 1581 and reprinted in 1751 with a hopeful ascription to Shakespeare. In it the author refers to 'your Majesty's late and singular clemency, in pardoning certain my undutiful misdemeanours'. Shades of Sir Thomas Lucy? Alas, as Nicolles recognizes, the early date of the pamphlet confutes the claim for Shakespeare. (It was, in fact, probably written by Sir Thomas Smith and published by his nephew William Smith, whose initials appear on the title-page; hence the attribution to Shakespeare.[103]) On Shakespeare's learning, Pope is cited as well as Rowe. The limitations of the poet's equipment and circumstances receive stark statement: 'Shakespeare . . . set out without the advantage of education, and without the advice or assistance of the learned; equally without the patronage of the better sort, as without any acquaintance among them.'[104] Thus it was that he directed his playwriting efforts to hitting the taste of the Lower Orders which in those days principally comprised theatre audiences.

The discrepancy of age between Shakespeare and his bride leads Nicolles into one of his few conjectures unsupported by some sort of authority—he suspects a prudential motive on the groom's part 'in respect of fortune'. Precise factual accuracy is not this author's forte. He gives the date of Rowe's edition as 1714, Pope's as 1721. On the number of the poet's offspring he is vague, crediting Shakespeare with 'two, if not three, children'—perhaps an improvement upon Rowe's positively mistaken assertion that there were three daughters. Nicolles erroneously gives Shakespeare's age as seventeen at the time of his marriage: a blunder which others would repeat, and for the same reason: the writer knows the parish-register entry for Susanna Shakespeare's christening and is unwilling to countenance the possibility that the first-born was, as he charmingly puts it, 'the fruits of ante-nuptial fornication'. In sum, the *Biographia Britannica* article offered what was for its time a more than usually informative and misinformative account of the dramatist.

To later memorialists it became a boon second only to Rowe's *Account* for the grand succession of eighteenth-century editors. The anonymous scribbler for the *London Magazine* (as previously noted) pillaged from the *Biographia Britannica* for the Stratford Jubilee number of August 1769. This hack in turn furnished inspiration for David Erskine Baker, who followed him almost verbatim and without acknowledgement in his popular *Biographia Dramatica, or, A Companion to the Playhouse* (1782). In the expanded edition of 1812, the compilation is still shelved, if not consulted, in the reference rooms of libraries; thus were the sins of plagiarism visited on succeeding generations and on those yet unborn. There are some additions (Malone's chronology, for example), and to Baker we are indebted for one improvement on his source: the previously unreported information that Shakespeare first drew breath on 16 April. This intelligence finds a place in *A New and General Biographical Dictionary* (1784), the assemblers of which have the grace to give Baker credit for his plundered wares, although they do not go so far as to confess that they have reprinted him.

A physician and naturalist with a zest for literature, John Berkenhout cites the *Biographia Britannica* as a source (along with Rowe, Oldys, and Pope) for his life of Shakespeare in *Biographia Literaria; or a Biographical History of Literature: Containing the Lives of English, Scotish [sic], and Irish Authors from the Dawn of Letters in These Kingdoms to the Present Time, Chronologically and Classically Arranged* (1777). Berkenhout's enormous undertaking not surprisingly proved abortive; he never got beyond a first volume and the sixteenth century. His article on Shakespeare blends fact with the by now familiar traditions, errors, and misconceptions. Shakespeare, the son of a woolstapler who had nine other children, 'acquired but little classical learning' before marrying the daughter (Christian name unspecified) of John Hatheway [sic] at the age of seventeen. We are told, as solemn truth, that the future poet in the company of hardened deer-stealers more than once robbed the park at Charlecote, and was therefore obliged to seek refuge in London, where he at first supported himself by 'the servile employment of holding the horses of those who rode to the theatre'. But success followed for Shakespeare, first as an actor who reached the top of his bent with the Ghost in *Hamlet*, then as a writer whose dramatic genius delighted Elizabeth. The account ends with the following tribute:

They say Shakespeare was illiterate. The supposition implies more than Panegyric with a hundred tongues could have expressed. If he was unlearned, he was the only instance of a human being to whom learning was unnecessary; the favorite child of Nature, produced and educated entirely by herself; but so educated, that the pedant Art had nothing new to add.[105]

This judgement so favourably impressed some later encyclopaedia contributors that they followed it in letter as well as spirit.

Both Berkenhout and the *London Magazine* Life of 1769 are silently exploited in the three columns given to Shakespeare in the second edition, '*greatly Improved and Enlarged*', of the *Encyclopaedia Britannica* in 1783. The article does not mark a very auspicious start for the treatment of Shakespeare in the most famous of all general reference compilations in English. The Shakespeare described in this memoir received from the free school in Stratford 'the rudiments of grammar-learning', married at seventeen, etc.; all this—about three-quarters of the notice—comes, word for word, from the *London Magazine*. The passage already quoted from the *Biographia Britannica*, about Shakespeare setting forth without the advantage of education or the advice of the learned, reappears; but again the source is the periodical plagiarist of 1769. The concluding paragraph, on Shakespeare as the illiterate darling of Nature, provides the spice of variety, as this is faithfully copied from Berkenhout.

The article in the third edition of the *Britannica* in 1797 retains passages from the second, but has grown to seven columns. The great editions by Johnson and Malone had appeared in the interim and made their impact: Johnson is quoted at length on Shakespeare as a dramatic writer; the chronology of the plays is Malone's, and presented as such. The conception of the poet is also refurbished. Shakespeare, no longer illiterate, knows the Latin poets, especially Terence, and has some acquaintance with French and Italian. But familiar and questionable episodes are retold, with the dubious benefit of melodramatic heightening. The poaching is described thus:

Having the misfortune to fall into bad company, he was seduced into some profligate actions, which drew on him a criminal prosecution, and at length forced him to take refuge in the capital. In concert with his associates, he broke into a park belonging to Sir Thomas Lucy of Charlecote, and carried off some of his deer. Every admirer of Shakespeare will regret that such a blemish should have stained his character. . . .[106]

'One thing at least is certain,' the writer continues, 'that Shakespeare himself thought that the prosecution which Sir Thomas raised against him was carried on with too great severity.'[107] So much for certainty. Nor does this authority entertain any doubts regarding the genuineness of the spurious

Lucy ballad, a stanza of which comprises the only quotation attributed to Shakespeare in the article. Despite its deficiencies, this must have been one of the most frequently consulted of all encyclopaedia accounts of Shakespeare, for the editors of the *Britannica* saw fit to keep it unaltered in the fourth edition (1810), the fifth (1817), and the sixth (1823).

Shakespeare is also the subject of a notice, not remarkable for factual precision, in the ninth volume (1814) of the ambitious *General Biography; or, Lives, Critical and Historical, of the Most Eminent Persons of All Ages, Countries, Conditions and Professions*, put together by John Aikin, MD, and William Johnston. Based on Rowe and the *Biographia Britannica*, the brief memoir informs us that the poet married 'at the age of 17 or 18' a woman 'several years' his senior and sired by her three daughters. This Shakespeare languished for years in a humble condition, 'conversant only with that class of society of which the inferior players make a part', and ultimately, in retirement, 'forgot that he had been any thing in the busy world.' Such a life record raises certain questions. 'Whence', asks Aikin, 'shall we derive that elevation of sentiment, and feeling of genuine dignity of character, which breaks forth with so much lustre in all his capital pieces; whence that beauty and sublimity of imagery which have placed him as high among poets, as his knowledge of nature has done among dramatists?'[108] Whence indeed; the compiler can appeal only to the fire of native genius. The time was not very long to come when some lay readers would reject such an answer—would reject, in fact, Shakespeare's authorship of the plays.

Oddly, the author of the article in Rees's *Cyclopaedia* (1819) gives almost a third of his space to a consideration of music in Shakespeare. The piece is curious in other ways as well. We learn that the date of Shakespeare's birth is 'well ascertained' and that the dramatist purchased New Place three years before his death. Frankly idolatrous, this anonymous contributor withholds full credence in the deer-poaching tradition, which reflects adversely on 'an amiable man and super-eminent author'. He displays, however, no such solicitude for the reputation of Anne Hathaway. The puzzling Stratford register entry (6 March 1589/90) for the burial of Thomas Green *alias* Shakspere leads him to entertain 'some suspicions respecting the fidelity of our bard's wife'—suspicions which, he not quite accurately claims, have 'hitherto escaped the inveterate researches and countless opinions of his [Shakespeare's] biographers and commentators'. The suggestion that Anne was unfaithful would be more than once repeated, and ultimately it would find its way even into the pages of Joyce's *Ulysses*.

In offering greater detail than some predecessors, the 1830 *Edinburgh Encyclopaedia* article at the same time achieves a higher standard of inaccuracy. This memoir quotes Johnson at second hand and prints tables of dates from Malone and Chalmers, but seems to belong to a dimmer biographical past. John Shakespeare was 'a considerable dealer in wool'. After being removed from school to help in his father's business, William married and had three daughters: Susannah, Judith, and Hannah. He fled Stratford after poaching deer, reportedly held horses at the theatre entrance, and received a gift of a thousand pounds from Southampton 'to make some purchases'. To the Earl he dedicated both *Venus and Adonis* and *The Rape of Lucrece* 'as the *first piece of my invention*'. Evidently this anonymous drudge felt no compulsion to check his quotations or consult the latest authorities.

A far more ambitious and original article than these appeared in 1841 in the *Penny Cyclopaedia*, which, as the name suggests, is yet another nineteenth-century venture in the cheap dissemination of knowledge. This encyclopaedia had as its general editor Charles Knight, about whom we shall have more to say hereafter. The notice itself is in the customary fashion unsigned, but elsewhere Knight leaves no doubt that he wrote it. Indeed, he repeats whole passages word for word in his article on Shakespeare for *Knight's Store of Knowledge for All Readers*, also published in 1841, and he elaborates the same ideas in his full-scale biography of the poet. Knight offers a Shakespeare at the other end of the social scale from Carlyle's astonishing peasant or the lowly glover's son of the *Cabinet Cyclopaedia* or the illiterate darling of nature larcenously celebrated in some of the other compilations.

This Shakespeare comes of genteel stock. Never mind that his father signed the corporation book with his mark; the poet was 'originally of the rank which is denominated gentleman at the present day':

... a well-nurtured child, brought up by parents living in comfort if not affluence, and trained in those feelings of honour which were more especially held the possession of those of gentle blood. His father and mother were, we have no doubt, educated persons; not indeed familiar with many books, but knowing some thoroughly; cherishing a kindly love of nature and of rural enjoyments amidst the beautiful English scenery by which they were surrounded; admirers and cultivators of music, as all persons above the lowest rank were in those days; frugal and orderly in all their household arrangements; of habitual benevolence and piety.[109]

Nor did John Shakespeare suffer financial distress in later life. The relevant entries in the corporation books and elsewhere, Knight mysteriously suggests, 'are . . . all capable of another interpretation'. His scepticism serves him better when he rejects (with some concessions) the autobiographical readings of the Sonnets by Charles Armitage Brown and the German critics. He is similarly tough-minded about the poaching legends: ambition, not prosecution, drove Shakespeare to London, there to transform the drama 'from a rude, tasteless, semi-barbarous entertainment, into a high intellectual feast for men of education and refinement'. Through 'honest labours, steadily exerted', Shakespeare amassed his fortune. In the last years of his retirement he renewed his boyhood studies in much the same way that Alfieri took up Greek after he was fifty. Knight's portrayal of Shakespeare the frugal civilizer must have been inspiring to the middle-class readers of the *Penny Cyclopaedia*, although they might well have experienced some confusion if they also consulted the second edition of the *Britannica*.

A different and less bourgeois Shakespeare emerges from the pages of the seventh edition of the *Britannica* in 1842. The editors made amends for the threadbare earlier articles by commissioning De Quincey—who had already contributed the notices of Goethe, Pope, and Schiller—to write on Shakespeare. His piece, spread over forty columns, was the most elaborate encyclopaedia treatment of Shakespeare up to that time. By nature desultory, De Quincey threw himself into the assignment, which came his way just after the death of his wife, to whom he was deeply attached. 'No paper ever cost me so much labour,' he wrote on 16 July 1838 (to Macvey Napier, the editor of the *Britannica*): 'parts of it have been recomposed three times over. And thus far I anticipate your approval of this article, that no one question has been neglected, which I ever heard of in connection with Shakespeare's name; and I fear no rigour of examination, notwithstanding I have had no books to assist me but the two volumes lent me by yourself, (viz. 1st vol. of Alex Chalmers's edit. 1826, and the late popular edit. in one vol. by Mr Campbell.).' He goes on to boast: '. . . if you examine it, you will not complain of want of novelty, which luckily was in this case quite reconcilable with truth,—so deep is the mass of error which has gathered about Shakspeare'.[110]

As a Shakespearian commentator, the Opium-Eater and connoisseur of murder as a fine art is justly remembered for his brief essay inspired, fifteen years previous to the *Britannica* assignment, by the Ratcliffe Highway murders: 'On the Knocking at the Gate in "Macbeth"'. Here De Quincey offers his piercing insight into the dramatic device by which the human makes its reflux upon the fiendish, as with the knocking the pulses of life begin to beat again and the world returns to its everyday round. The danger exists, however, of De Quincey's biography being relegated to oblivion altogether. 'On Shakespeare he wrote only once', the distinguished author of the volume on the Romantic period in the *Oxford History of English Literature* remarks in connection with the *Macbeth* essay.[111] Yet the *Britannica* Life has a stylistic elegance rare in encyclopaedias. It is, too, a stunning example of the unabashed bardolatry of the age; a bardolatry which leads the writer to surrender himself willingly to the mysterious dynamism of genius.

The bardolatry is foreshadowed in the earlier essay, with its concluding apostrophe to the 'mighty poet' whose 'works are not as those of other men, simply and merely great works of art; but are also like the phenomena of nature, . . . which are to be studied with entire submission of our own faculties, and in the perfect faith that in them there can be no too much or too little, nothing useless or inert'. For his biography De Quincey did not set himself the task of synthesizing information about a playwright; rather, he offered homage to a god. He ransacks the vocabulary of adoration to celebrate the 'immortal son' of the glover of Stratford, 'the transcendent poet, the most august amongst created intellects', 'the protagonist on the great arena of modern poetry, and the glory of the human intellect'. Those who choose to stand outside the shrine must suffer the thunderbolt. For the sin of quoting *Macbeth* in the Davenant adaptation, Addison receives a column and a half of abuse; Shaftesbury, who dared complain of Shakespeare's 'rude unpolished style', is denounced as 'the most absolute and undistinguished pedant that perhaps literature has to show'.

Inevitably De Quincey's suppliant posture before the altar of his idol colours his interpretation of the biographical materials. Thus he thinks that Shake-speare was perhaps born on the 22nd rather than (as usually thought) the 23rd, because Lady Bernard married on 22 April 1626, choosing this day in honour of her illustrious grandfather's birthday, which, 'there is good reason for thinking, would be celebrated as a festival in the family for generations'. The mere thought of Shakespeare's mother is sufficient to put the Opium-Eater in an undrugged trance: 'To have been *the mother of Shakspeare,*—how august a title to the reverence of infinite generations, and of centuries beyond the vision of prophecy.'[112] The documentary records that shed their meagre light on the poet's family and his Stratford background are for De Quincey a source of embarrassment: 'the little accidents of birth and social condition are so unspeakably below the grandeur of the theme, are so irrelevant and disproportioned to the real interest at issue, so incommensurable with any of its relations, that a biographer of Shakspeare at once denounces himself as below his subject if he can entertain such a question as seriously affecting the glory of the poet.'[113] Yet even the most antiseptic biographer cannot entirely escape being contaminated with such questions. In De Quincey's hands they cost Shakespeare no glory; he is presented as the son of a gentleman who lost his social standing in Stratford but eventually retrieved it through 'the filial piety of his immortal son'.

As worship replaces disinterested evaluation, De Quincey commits the critical treason of arriving at the right conclusion for the wrong reason. In his anxiety to cleanse the image of his deity of any sullying imputations, he sweeps away with high disdain the doubtful myths and legends that had clung to Shakespeare for two centuries. The persecution of the divine youth by Sir Thomas Lucy is dismissed *in toto* as 'a slanderous and idle tale'; De Quincey repeats Malone's objections and adds that, had Shakespeare suffered the disgrace of a penal whipping, the scandal would have pursued him to London to find a place among the insinuations of the dying Greene. 'This tale is fabulous and rotten to the core,' and its sequel—the bitter ballad—is 'idiot's drivel', probably first set down after the Restoration. Rejected also is the anecdote, retold by Rowe, of the satiric epitaph that so stung old John a Combe: Shakespeare himself took 'ten in the hundred', the customary interest rate at the time. Short shrift too for the legend of Shakespeare holding horses at the stage door, and for the equally absurd fable that he at first served the theatre in the humble capacity of call-boy, or deputy prompter. And then there is the gravestone malediction, so far beneath the poet's intellect and scholarship. Perhaps the parish clerk supplied this specimen of doggerel; perhaps it was an antique formula such as those inscribed on the flyleaves of books to signify ownership. In this way was blasphemy reproved.

Shakespeare's marriage presented a more delicate problem. There is the inescapable fact that Anne Hathaway was eight years her boy-husband's senior; there is the baptism record for the first-born, six months after the marriage bond. 'Oh, fie, Miss Susanna,' De Quincey playfully chides, 'you came rather before you were wanted.' In the end he settles for orthodoxy. Anne, the daughter of a substantial yeoman, had waited too long for a suitor of fortune, and had outlived her market—the gallant phrase is De Quincey's. Now she turned to a handsome youth, tempted his natural frailty with her blandishments, and took advantage of the consequences of precipitate passion. Inevitably there was post-marital *triste*. For confirmatory evidence the Opium-Eater offers his reading of the 'subtile hieroglyphics' hidden in the plays: Prospero warning Ferdinand not to break Miranda's virgin knot before the wedding rites, lest hate and discord strew the marriage bed; Orsino advising the pretended Cesario to take a lover younger than himself, 'Or thy affection cannot hold the bent.' The marriage, according to De Quincey, led to Shakespeare's emigration to London. After four years of misery, chafing under 'the humiliation of domestic feuds', he hit upon the plan of going off alone to the metropolis—a plan which was to have such fateful consequences for literature.

Elsewhere De Quincey sees Lucullus's turning his back on Timon in his hour of need as the dramatist's recollection of Roger Sadler (not Robert, as stated in the article) trampling under foot the financially embarrassed John Shakespeare over a debt of five pounds. (It pleases the author that, after 270 years, Sadler—a type of the selfish and illiberal creditor—should be raked over the coals in the *Britannica*; but the indignation is unrighteous, for Sadler did no more than mention in his will that Edmund Lambert and Edward Cornwall owed him the sum for John Shakespeare's debt.) And naturally De Quincey responds with enthusiasm to Campbell's inspired identification of Prospero with his creator. The plays have now become fully entrenched sources of biographical revelation.

It would be an understatement to say that De Quincey's 'Shakspeare' is notably subjective for an encyclopaedia article. How remarkable, for example, are the omissions of basic facts and information. There is no discussion of the Sonnets (we do not so much as hear of Mr W.H.); not a word about *Venus and Adonis* and *The Rape of Lucrece*, or of their dedications; no mention of the poet's will. De Quincey does not reproduce the text of a single document. His claim that he has neglected no question involving

Shakespeare must be dismissed as a preposterous exaggeration; he has paid a price for having only two books. He does, however, round off his discussion with several pages of sustained rapturous eulogy which, if irrelevant to the present purpose, holds much interest as a manifestation of the excesses of the Romantic movement. All in all, with its eccentricity and passion, the article on Shakespeare in the seventh edition of the *Britannica* is an extraordinary performance; indeed, a *tour de force*.

For all his omissions, De Quincey along the way furnishes a startling piece of information: as early as 1589 (he reports) Shakespeare, then twenty-five, held property in a London playhouse. This intelligence derives from no drug-induced imagining; he gives credit to his authority. Other encyclopaedia contributors, as well as Campbell and Charles Armitage Brown, also describe Shakespeare's astonishing rise to eminence during what were, once supposed, his lost years. The *Penny Cyclopaedia* writer, more precise than De Quincey, names the theatre as Blackfriars and cites Shakespeare's position among the shareholders as fourth. The same author quotes (even more wonderful!) a lengthy letter from Southampton to the Lord Chancellor in which that nobleman recommends the poet as his 'especial friend'. Also cited in the *Penny Cyclopaedia* article is a petition to the Privy Council, presented in 1596, bearing Shakespeare's name; and another document, from the year 1608, with the playwright's evaluation of the wardrobe and properties of Blackfriars. Even more interesting is a letter from the poet Daniel to the Lord Keeper hinting that Shakespeare aspired to the mysterious office (unknown to theatre historians) of Master of the Queen's Revels. Other commentators around this time refer to the identical records (for example, Rose's *New General Biographical Dictionary* of 1848, still occasionally to be found on reference shelves, refreshingly casts aside the deer-poaching legend as 'at once absurd and improbable'—but cites the 'attested fact' that in 1589 Shakespeare obtained a share in the ownership of thè Blackfriars Theatre); so De Quincey's failure to include them all may be regarded as yet another instance of his indifference to the concrete.

But whence came these remarkable additions to Shakespearian knowledge? And why have twentieth-century readers been kept ignorant of them? The explanation is only too easily guessed. Once again a forger had imposed upon credulity, this time a forger more cunning, skilful, and mature than ever Ireland was. Poisoning the wellsprings of learning itself, he achieved a success beyond the pathetic William-Henry's fevered dreams. For this pollution he earned praise from the best critics of the day, and grateful acknowledgement in encyclopaedias, biographies, and editions of Shakespeare. Even the anonymous hack responsible for 'The Confessions of William Shakspeare' knew and valued this authority's 'discoveries'. His name was J. Payne Collier.

# 9

## *The Peele Letter and Fenton's* Tour

FROM the eighteenth century onwards the inventions of forgers had stirred enthusiasm and debate in London literary circles. Macpherson had invented Ossian, and moved Gray to meditate on the imagination that found a habitation, many centuries before, in the cold and barren Scottish Highlands. William-Henry Ireland had found inspiration in the Marvellous Boy with his fifteenth-century Bristol monk. And if the Ireland papers were the most spectacular Shakespearian fabrications, they did not stand quite alone. In 1763 the anonymous author of a life of the actor Alleyn in the first (and only) volume of an obscure journal, the *Theatrical Review*, claimed that a commissioner of the peace for Middlesex, a 'gentleman of honour and veracity', had shown him a curious letter long in the family's possession on the mother's side. Dated 1600, and addressed to 'master Henrie Marle livynge at the sygne of the rose by the palace', it reads:

Friende Marle,
I must desyre that my syster hyr watche, and the cookerie booke you promysed, may be sente by the man.—I never longed for thy companie more than last night; we were all verie merrie at the globe, when Ned Alleyn did not scruple to affyrme pleasauntely to thy friende Will, that he had stolen hys speeche about the excellencie of acting, in Hamlet hys Tragedye, from conversaytions manyfold whych had passed betweene them, and opiniones gyven by Alleyn touchyng that subjecte. Shakespear did not take thys talke in good sorte, but Jonson put an ende to the stryfe with wittielie sayinge, thys affaire needeth no contentione; you stole it from Ned no doubte; do not marvel; have you not seene hym act tymes out of number?—believe me most syncerelie

Harrie
Thyne
G. PEEL.[114]

This entertaining document was reprinted many times. It appears in the *Annual Register* for 1700, in *The Life and Death of David Garrick, Esq.* (1779), by 'an Old Comedian' who attributes it to 'H Peel', and in Berkenhout's article on Shakespeare in the *Biographia Literaria*; on 5 January 1796, at the height of the Ireland uproar, Boaden reproduced it in the *Oracle*. Isaac D'Israeli, in his *Curiosities of Literature*, attributes the letter to Steevens, a likely candidate in this period for any clever hoax. Writing nine years before Steevens's death, D'Israeli moved in literary circles familiar with his subject's pranks. Still, he offers no evidence, and the hoax is in any event not so clever: if Steevens, who after all was a great scholar, did not know that Peele was dead by 1600, he would surely be aware that Marlowe (if he is intended) was, and that this dramatist did not have the Christian name of Henry.*

A more enterprising deception enlivens *A Tour in Quest of Genealogy, through Several Parts of Wales, Somersetshire, and Wiltshire, in a Series of Letters to a Friend in Dublin. . . . Together with Various Anecdotes, and Curious Fragments from a Manuscript Collection Ascribed to Shakespeare*, published in 1811 as by 'A Barrister'—actually Richard Fenton, the poet, topographer, and historian of the Welsh counties. He sorts oddly with the forging rogues— this FSA, intimate of Garrick and Goldsmith, and subject (along with his beautiful wife) of miniatures by Sir Joshua Reynolds.

In one of the letters to his friend Charles, dated 19 October 1807, the barrister tells of purchasing at an auction in Carmarthen a quarto manuscript volume that had belonged to a mysterious and eccentric stranger who passed himself off as an Irishman but actually (people thought) hailed from North Wales. The volume was certainly a lucky find. It contained previously unknown poems by Shakespeare; letters that passed between him and Anne Hathaway, Sir Christopher Hatton, Sidney, Southampton, and others; and (most marvellous of all) memoirs that 'let you into his private and domestic life, and the rudiments of his vast conception'. An account of the provenance, we are told, prefaced the relic, which 'appears to have been copied from an old manuscript in the hand-writing of Mrs. Shakespeare, which was so damaged when discovered at a house of a gentleman in Wales, whose ancestor had married one of the Hatheways, that to rescue it from oblivion a process was made use of, by which the original was sacrificed to the transcript'.[115] Thus were the tracks covered.

To a later letter (dated 8 November) Fenton appends a sample poem of five stanzas, meant to accompany 'a ringe in forme of a serpent, a gift to his belovyd Anna, From W.S.' An improvement upon its obvious progenitor, the ditty to Anna Hatherrewaye that Ireland found in Mr H.'s chest, it concludes with these stanzas:

> The frute that Hymen in our reche
>     By Heven's first commaund hath placed,
> Holy love, without a breche
>     Of anie law maie pluck and taste:
>
> Repeted taste—and yett the joye
>     Of such a taste will neaver cloie,

---

* In the *Biographia Literaria*, which omits the date, Berkenhout appends a note: 'Whence I copied this letter, I do not recollect; but I remember that at the time of transcribing it, I had no doubt of its authenticity.' D'Israeli not very reasonably ascribes this note to Steevens too. On this subject, the late James Osborn has observed that '. . . D'Israeli, who attributed the hoax to Steevens, did so nine years before Steevens's own death; D'Israeli offered no documentation, but he moved in literary circles who knew Steevens's puckish character and pecadilloes' ('English Literature: A Current Bibliography', *Philological Quarterly*, 50 [1971], 397).

So that oure appetits wee bringe
Withinn the cumpass of this ringe.[116]

In this way is the calumny of 'antenuptial fornication' tastefully rebuked. Another excerpt is a letter from Shakespeare to his 'good cozen' Judith Hathaway, written at a country lodging 'adjoyning the paddock of Sir Waulter Rawleigh, at Iselinton'. This homey missive touches upon a number of interesting matters: tobacco; *Romeo and Juliet* (the Nurse is modelled on 'ould Debborah, at Charlecot', and the Apothecary on 'ould Gastrell, neere the churche at Stratford'); a joint of Stratford brawn hugely commended by young Ben; the mulberry tree planted 'after the cuckow had sung on Anna's birth-daie'; and poor Burton the schoolmaster, lately deceased, who every Monday morning would sharpen the birch, 'growen blunted in the service of the forgone week; a practise felt throw the whole schoole, from *top* to *bottome*'.[117] The date of this letter, which contains such an interesting variety of news, is unfortunate: 12 June 155–. Other bits and pieces from the Carmarthen volume crop up later in the *Tour*. Southampton, 'mie noble and trulie liberall patrone', praises the half-finished *Richard III*; Succaro paints the poet's portrait; Francesco Manzine, wool-dyer and alchemist, teaches him Italian in his father's house. A verse alludes to the deer-stealing adventure, as a result of which William was 'rudelye torne | Farre from the muses' haunts I love'. His effusions have their counterpart in two songs, exquisitely modern in feeling and style, addressed to him by Anne, not previously recognized as a female bard.

Although presented with a straight face, these inventions have an engaging lightness of touch. They also constitute an intriguing mystery, to which the manuscript notebooks of Fenton's travels around Wales, preserved in the Cardiff Central Library, provide no answer, for Shakespeare is not so much as mentioned in them. Was Fenton hoaxed by yet more Ireland forgeries innocently purchased in Wales? So Ingleby would suggest late in the nineteenth century.[118] Or is the purported Shakespeare miscellany a playful *jeu d'esprit* executed by Fenton himself? So Chambers, most authoritative of modern Shakespearians, believes, and certainly the Hathaway poems are beyond Ireland's art.[119] More recently an article in the *TLS* (28 October 1955) has endorsed the Chambers hoax theory: the anonymous correspondent notes that the first Shakespearian gleaning appears in the *Tour* immediately after a conversation about literary impostors, that Fenton had forgeries much on his mind, and that one of the Shakespeare letters speaks glowingly of the barrister's namesake and possible ancestor, that 'goodlie ould gentilman' Sir Geoffrey Fenton. Several tantalizing clues, however, seem to point in a different direction.

We know that Ireland's confederate, Montague Talbot, was in Carmarthen in 1795: from there, on 25 November, he wrote to Samuel Ireland the letter reproduced in *Mr. Ireland's Vindication of His Conduct*. The following January, we learn from the *Monthly Mirror* of that date, he was playing Othello at the 'very pretty little Theatre' in Carmarthen; at the same time he was commuting to Dublin to pursue the thespian career that would make him the darling of the Irish stage. He would use Fishguard in South Wales as his port of embarkation. Now, in January 1796, the editor of the London *Morning Herald* printed a letter from one B.B. of Fishguard describing a Christmas visit, some thirty years previous, to 'the house of an old Lady a relative of mine in Wales, one of whose ancestors had married one of the family of the Hathaways of Warwickshire where I saw and was much delighted with a manuscript volume in 4to richly habited in crimson velvet, inscribed on the cover in embroidery of gold "The Swan of Avon" containing poems from Shakspeare to Anna Hathaway before and after his marriage with her, and from her to him (for she too had tasted of Helicon) together with etc.'.[120] Returning to the house, B.B. has retrieved the tattered miscellany, from which he quotes a sample of Anna Hathaway's 'simple, poetical, and impassioned' versifying. Other recovered poems appeared in succeeding numbers of the *Morning Herald*. On 28 February the paper published a lyric addressed 'To her oune Lovynge Willie Shakspeare' and beginning 'From mie throune in Willie's love . . .'. The same poem, under the same title, appears in Fenton's *Tour*. The conclusion seems irresistible that Talbot was B.B. and also the mysterious Irishman of Carmarthen, and that Fenton actually bought the volume at an auction there.

Or does it? According to the *Tour* the Irishman had lately died, whereas Talbot survived until 1831. About the latter's literary gifts we cannot speak, but Fenton was an accomplished versifier, and (so his grandson's memoir informs us) he had settled in Fishguard around 1796, when B.B.'s letters began appearing in the London press. Is Fenton the hoaxer or the hoaxed? The mystery of these relatively innocent fabrications remains.*

---

* Of all the fabrications described in this narrative, those published in the *Tour* have alone found a modern believer, one Arthur Field, MA, of Southampton. In *Recent Discoveries Relating to the Life and Works of William Shakspere* (1954) the happy fruit of years of research among old tomes, Field sets forth some intriguing ideas; he thinks, for example, that the horses purportedly held by Shakespeare at the playhouse door were (by a curious symbolism) actors detained for instruction in their parts by the dramatist. But Field chiefly prides himself on re-presenting the Fenton Shakespeare papers, in the genuineness of which he manifests a touchingly uncritical faith.

# 10

## *J. Payne Collier: A Forger's Progress*

No extenuation can be offered for Collier. He forged in deadly earnest, for glory, and staked his reputation on his 'discoveries'. He deserves the judgement, from De Quincey's 'Secret Societies', passed on him by the man who did as much as anyone to topple him: 'Now, reader, a falsehood is a falsehood, though uttered under circumstances of hurry and sudden trepidation; but certainly it becomes, though not more a falsehood, yet more criminally and hatefully a falsehood, when prepared from afar, and elaborately supported by fraud, and dove-tailing into fraud, and having no palliation from pressure and haste.' The pity of it is that Collier was an excellent scholar whose genuine contributions to Shakespeare studies and English stage history would have assured him the fame and honour to which he aspired. Regarded as a dear man by his family, amiable and by instinct generous, he kept to the high road that avoided petty jealousies and squabbling. 'I hate disputes of all kinds', he wrote to his friend Halliwell (later Halliwell-Phillipps), 'and, if I can help it, I never will mix myself up with any hostility personal or literary.'[121] Yet by his deeds Collier involved himself in hostilities of an intensity and magnitude unprecedented in his day. His case confronts us with a strange manifestation of the criminal mind.

A remote descendant of that celebrated excoriator of the immorality and profaneness of the English stage, the Revd Jeremy Collier, John Payne was born on 11 January 1789, the son of John Dyer Collier, a failed wool-trader who accumulated some money from soap manufacturing, only to lose it in farming. The family knew poverty, but eventually prospered when John Dyer turned to journalism. Eminent writers visited the Collier household: Wordsworth, Coleridge, Lamb, and Hazlitt, the latter two being especially fond of Mrs Collier. The boy was well educated without formal schooling: his father taught him Latin, Greek, and shorthand; from a boarder he learned German. John Payne took up writing before he turned sixteen, and was not yet twenty when John Walter, proprietor of *The Times*, appointed him to join the elder Collier as a reporter on the staff. More than half a century later the forger still recalled Walter with affection as 'the first person who discovered any ability in me, who employed it and rewarded it'.[122] During his leisure hours Collier played billiards in Fleet Street with John Keats and beat him. In 1811 he enrolled in the Middle Temple; five years later he married Mary Louisa Pyecroft, who brought him (in addition to her charms and a modest income) six children.

His career on *The Times* suffered a blow when he misreported a speech in the House of Commons and received a reprimand from the Speaker. Collier injured his own prospects as a lawyer when in 1819 he published—for the money, he says in his manuscript autobiography—*Criticisms on the Bar: Including Strictures on the Principal Counsel Practising in the Courts of King's Bench, Common Pleas, and Exchequer*: a sufficiently comprehensive indictment. Although these sketches denigrating leading advocates of the day were printed anonymously (as by 'Amicus Curiae'), the author soon became known. The pamphlet earned him £200, but the hostility it engendered hindered by a decade his call to the Bar. On the flyleaf of his own copy he later wrote, 'Foolish, flippant, and fatal to my prospects, if I ever had any.' Hindsight enables us to see the two episodes—the misreporting and the libels on authority—as manifestations of the forger's character; in his fabrications he would also misrepresent records, and, by imposing on their credulity, he would assert his superiority over colleagues in his own fraternity, in this case not of attorneys but of scholars.

A disagreement with one of the powers on *The Times* led to Collier's transfer to the *Morning Chronicle*, whose law and parliamentary reporter he became. Long after he had made a name for himself as an antiquary he continued to write for the paper, enlarging his sphere to include literary and dramatic criticism. 'My early employments were irksome and wearisome', he recalled near the end of his long life; 'but stimulated in some degree by my first success and by my love for the best poetry the world has produced, I lightened my labours by the collection and perusal of old English books.'[123] The first old folio he ever possessed was the Shakespeare Third, which he bought in Baldwin Gardens in 1806. Seventy years later the thrill it gave him still retained its glow. 'I fancied it the first Edition and a great prize, and what pleasure I had in making up its deficiencies,' he wrote on the flyleaf of the book, which is now in the Shakespeare Centre at Stratford-upon-Avon. And he added darkly: 'I was then grossly ignorant, and was only beginning what I wish I had never begun.'

Even Collier's juvenile verses reflect, as he recognized himself long afterwards, a passion for the past. Perhaps Lamb first stirred his interest in the early literature of his country. Another encouragement was his introduction, by his father, to the shop in Great Newport Street owned by the bookseller Thomas Rodd, with whose son—a more successful antiquarian bookdealer—he formed a lifelong association. 'I valued him', Collier testified, with qualified enthusiasm, in his nineties, 'but he was always a bookseller, and never parted with a volume out of which he did not derive some profit.'[124] Collier's antiquarian attainments were already considerable when in 1820 he published his *Poetical Decameron*, a series of ten dialogues in which the author ranges about widely in Elizabethan and Jacobean literature, and furnishes along the way the earliest account of Shakespeare's source for *Twelfth Night*,

Barnaby Riche's *Farewell to Military Profession*. As a scholar Collier was now launched.

In 1831 he published the first work to bring him great public recognition: *The History of English Dramatic Poetry to the Time of Shakespeare; and Annals of the Stage to the Restoration*. It deserved its warm welcome, for it is in truth a monument of early nineteenth-century scholarship, superseding Malone (as was evidently the aim) by reason of its greater comprehensiveness, new documentary information, and more sympathetic view of the non-Shakespearian drama. Collier also adds directly to knowledge of the poet. Among the Harleian manuscripts at the British Museum he has found the invaluable Manningham *Diary*, with its reference to a performance of *Twelfth Night* in February 1602. In the same document he came upon the scurrilous anecdote telling how Shakespeare overheard Burbage arranging for an amorous encounter, substituted himself by using the password, 'Richard III' (one of Burbage's parts), and then, when the actor knocked, reminded him that William the Conqueror preceded Richard III.* Collier prudently refrains from committing himself as to whether the story 'be true, or untrue . . . a mere joke, or the invention of "vulgar scandal"', but which way he leans is clear. The historian is not at liberty to suppress a discovery because it represents the Immortal Bard 'as a human being, with human infirmities', and this one has 'tolerably good authority' to recommend it. In any event, if we can trust the evidence of the Sonnets, Shakespeare was 'not immaculate'.[125]

In the *History* Collier's peculiar demon, which was to prove irrepressible, takes control of his pen for the first time. He was not content to let a notable synthesis stand; he must contaminate the flow of factual narrative with his own fabrications. Sometimes he adds only a single word to an authentic document: the insertion 'played', for example, after *Julyus Sesar* in Machyn's account of the masque given on 1 February 1562. Sometimes he is more ambitious and creative, as when he incorporates a long ballad—otherwise unknown—supposedly 'copied from a contemporary MS.' on the Cockpit riot of 1617, containing an allusion to *Troilus and Cressida* ('False Cressid's hood, that was so good | When loving Troylus kept her.'). He comments disingenuously in a note: 'This might be Shakespeare's play, acted surreptitiously at the Cockpit, as it was the property of the King's servants; possibly it was a different play on the same subject.'[126] Ireland was not so cunning. The most arresting invention is a petition ('unfortunately, not the original, but a copy, without any signature') of the Lord Chamberlain's players. They entreat the Privy Council in 1596 for permission to continue their work of renovating the Blackfriars Theatre, and to go on acting there without

---

* Although credit is due to Collier for recovery of the Manningham manuscript, this story—but not its source—was known a half-century earlier; it appears in the *Universal Magazine* for February 1785 (p. 109). The story also appears in W. Harvey's biographical memoir prefixing an undated (?1825) edition of *The Works of Shakespeare*.

inhibition. Shakespeare's name appears fifth in the enumeration of principal actors at the commencement of this 'remarkable paper'. We need not be surprised that the inventor of this 'fact' did not miss its importance: 'It is seven years anterior to the date of any other authentic record, which contains the name of our great dramatist, and it may warrant various conjectures as to the rank he held in the company in 1596, as a poet and as a player.'[127] With what disarming objectivity, with what refreshing lack of dogmatism, does he make the point!

Despite these contributions to knowledge, sales of the *History* disappointed expectation, and the publishers wholesaled the book. No matter, for Collier ensured his success by tactfully dedicating his work to the Lord Chamberlain. That office was then held by the Duke of Devonshire, who had a liberal mind and dearly loved plays, especially old plays. In the diary Collier published forty years later he recalls those heady days in 1832 when the Duke showed his appreciation with a gift of a hundred pounds and then, in February, summoned him to Devonshire House, where he was asked to look after the dramatic library and serve as literary adviser. For these leisure-hour responsibilities he would receive a pension of one hundred pounds annually. He accepted, of course, and further signs of favour followed. The Duke sponsored Collier's membership in the Garrick Club and introduced him at Holland House.

He made friends easily. Around this time he met Lord Francis Egerton, soon to become Earl of Ellesmere—'a poor weak man', Collier uncharitably recalled in his *Autobiography* 'whom anybody turned round their finger'. And in his *Diary* he complained, 'Lord Ellesmere never was my patron: he deserted me when I wanted him most and after encouraging me, & almost promising me, left me to shift for myself.'[128] Egerton generously gave Collier 'instant and unrestrained access' to the Ellesmere papers at Bridgewater House, 'with permission to make use of any literary or historical information' he found therein. Here were reposited the papers of Lord Ellesmere, Elizabeth's Keeper of the Great Seal and James's Lord Chancellor. These documents had been partly catalogued by the Revd H. J. Todd, who preceded Collier at Bridgewater House, but much material had lain unopened since being tied together, perhaps by the Lord Chancellor himself, in the seventeenth century. Now relieved of financial pressures, and given entrée to a vast treasure-house of unexplored records, Collier could freely indulge his twin passions: research and forgery.

His gratifying progress, especially with respect to the latter, is demonstrated in the little volume of some fifty pages published in a limited edition by his friend Thomas Rodd in 1835, *New Facts Regarding the Life of Shakespeare*, which exhibits Collier's principal contributions to biographical misrepresentation. In it he prints and analyses the documents which (he later said) he happened upon one day just after being left alone in the room by Lord

Egerton. At once Collier rushed to his benefactor and read the papers aloud, then deposited the originals and transcripts with His Lordship. Returning the next day eager to follow up his discoveries, he encountered Egerton alighting from his horse at Bridgewater House, and was told by him that Murray the publisher had offered fifty or a hundred pounds for an edition with an introduction. But Collier declined to profit from the liberality of his patron, who, a little taken aback by this display of hyper-squeamishness, 'replied, with his habitual generosity, that the documents were as much mine [Collier's] as his. . . .'[129] But the forger always had a nice sense of propriety. Rodd sold only a few copies of the *New Facts* over the counter to his customers, but the number was sufficient for Collier's purposes. In the book he did not fail to express personal gratitude to His Lordship for having 'laid open the manuscript stores of his noble family with a liberality worthy of his rank and race', and he draws a salutary moral from this display of generosity: '. . . if the example were followed by others possessed of similar relics, literary and historical information of great novelty and of high value might in many cases be obtained.'[130] Before such accomplished duplicity one pauses breathless.

Prospective forgers will look in vain for a more exemplary model than the *New Facts*. The design of the book is chaste, with none of the pretentiousness that contributes to the absurdity of the Ireland folio. The tone is for the most part dryly factual. There are no facsimiles to excite the suspicions of the expert. We are always conscious that the author is an established authority on his subject; he alludes to his *History* in his first sentence. The 'discoveries' themselves are highly interesting and significant but unsensational: no love letters, no professions of faith, no testamentary recollections of rescue from drowning. Little wonder that everyone was for a time taken in.

The records have mostly to do with Collier's *idée fixe*, Shakespeare's connection with the Blackfriars Theatre, which in fact the dramatist's company did not begin to play in until 1609. From Collier we learn that Shakespeare had by November 1589 made such professional headway that his name appears, eleventh out of sixteen, in a list of shareholders in that theatre. The next forgery (already printed in the *History*) shows the dramatist elevated to the fifth rung, and the authentic patent of 1603 places him second; thus does Collier furnish a chart tracing the dramatist's rise to eminence. Another document, entitled 'For Avoiding of the Playhouse in the Precinct of the Blacke Friers', gives a precise evaluation of Shakespeare's holdings in 1608:

> Item   W. Shakespeare asketh for the Wardrobe and
> properties of the same play house 500[li] and for
> his 4 shares, the same as his fellowes Burbidge
> and Fletcher viz 933[li].6[s].8[d] . . . . . . .1433[li].6[s].8[d]   [131]

This paper also shows that Burbage and Shakespeare were entitled to equal shares, four out of a total of twenty, in the profits. On the basis of sober computations, Collier estimates the playwright's yearly income at £300—the same figure that Gildon had offered without authority and that Malone had rejected as too high.

More exciting is the letter signed H.S.—who else but Henry Southampton?—designed to induce Lord Ellesmere to exert himself on behalf of the actors at Blackfriars—where else?—when they were under attack, probably in 1608, from the Corporation of London. This letter, Collier surmises, was placed in the Lord Chancellor's hands by Burbage or Shakespeare himself when they waited upon the nobleman. It reads in part:

These bearers are two of the chiefe of the companie; one of them by name Richard Burbidge, who humblie sueth for your Lordships kinde helpe. . . . The other is a man no whitt lesse deserving favor, and my especiall friende, till of late an actor of good account in the cumpanie, now a sharer in the same. . . . This other hath to name William Shakespeare, and they are both of one countie, and indeed almost of one towne: both are right famous in their qualityes.[132]

With scholarly precision Collier points out a factual error elsewhere in the document, but this lapse does not lessen his appreciation of its significance in establishing that Shakespeare had recently given up acting and that the Burbages originally hailed from Warwickshire. Even a professional antiquary may be allowed to yield for a moment to enthusiasm, and so Collier admits to the delight of discovery. To see such names as Shakespeare and Burbage in conjunction in a contemporary record! And by a happy coincidence to find the paper on the anniversary of the poet's birth and death! It is almost too good to be true.

The Southampton letter is not the last of the magician's wonders. He produces the draft of a patent or privy seal which does 'appoint and authorize the said Robert Daiborne, William Shakespeare, Nathaniel Field and Edward Kirkham from time to time to provide and bring upp a convenient nomber of Children, and them to instruct and exercise in the quality of playing Tragedies Comedies &c. by the name of the Children of the Revells to the Queene, within the Black fryers in our Citie of London or els where within our realme of England'.[133] Collier infers that the destruction of the Blackfriars was contemplated and that Shakespeare therefore considered transferring his interest to another company, but then decided to stay when the King's Men were not expelled after all. It is an especially ingenious fabrication, for it allows Collier to conclude by rejecting the information he claims to have found. The final Shakespearian revelation comes from the hand of Samuel Daniel, who writes to thank the Lord Keeper for being appointed in 1603 to supervise the productions of the Queen's Revels children. Without naming anyone, Daniel refers to a disappointed candidate:

I cannot but knowe that I am lesse deserving then some that sued by other of the nobility vnto her Ma^tie for this roome. . . . it seemeth to myne humble iudgement that one who is the authour of playes now daylie presented on the public stages of London, and the possessor of no small gaines, and moreover him selfe an Actor in the Kings Companie of Comedians, could not with reason pretend to be M^r of the Queenes Ma^ties Revells, for as much as he wold sometimes be asked to approve and allow of his owne writings.[134]

Collier learnedly deduces that Daniel can have meant no one but Shakespeare and that the nobleman who sued on his favourite's behalf was 'most likely' Southampton.

Other documents, having no bearing on Shakespeare, appear in the volume. A few are authentic scraps. Copies of the forged papers remain to this day at Bridgewater House. Lord Egerton could have little suspected, in allowing access to a researcher, that he would be rewarded by an increase in his holdings, but his gratitude for such benefactions, had he known of them, may be doubted.

The *New Facts* is Collier's only book devoted entirely, or almost entirely, to the elucidation of Shakespearian biography. The editing of the master's works now engrossed him, and by a fortunate coincidence succeeding discoveries tended to concern the text. Collier's next pamphlet, *New Particulars Regarding the Works of Shakespeare* (1836) contains much interesting matter, including 'indisputable' evidence of a performance of *Othello* before Elizabeth at Harefield on 2 August 1602—all the more remarkable for (as seems likely) antedating the play's composition. This allusion Collier had inserted in the Egerton Papers at Bridgewater House. In 1839 his *Farther Particulars Regarding Shakespeare and His Works. In a Letter to the Rev. Joseph Hunter, F.S.A.* had as its most entertaining item a ballad, 'The Inchanted Island', which in its story resembles *The Tempest*. It demonstrates that Collier could write a passable ballad. 'Mr. Douce', he remarks with an effrontery that commands awe, 'called it "one of the most beautiful ballads he had ever read", and shook his venerable head (as was his wont) with admiring energy and antiquarian enthusiasm at different passages in it; but I am by no means prepared to give it so high a character.'[135] At the time Collier wrote, Douce was already dead.

In 1840, with a trio of eminent scholars—Dyce, Charles Knight, and J. O. Halliwell (the youthful prime mover)—and the particular encouragement of the antiquary Thomas Wright, he truly advanced the cause of learning by founding the Shakespeare Society, which was modelled on the Percy Society, an organization concerned with the preservation and collection of popular poetry. Collier at once became director of the Shakespeare Society. Among its notable achievements was the first publication, in Dyce's edition, of *Sir Thomas More*. In the space of a decade, 1841–51, before the Society was discredited and ingloriously dissolved, Collier himself prepared no fewer than twenty-one of its publications, including editions of rare works by such neglected authors as Thomas Heywood. But Collier also used the organization as a super-respectable cover for his criminal hobby. He had gained access to the Henslowe and Alleyn papers at the College of God's Gift at Dulwich. There he abused the trust of the Governors and the librarian, the Revd Mr Howe, by tampering in his usual fashion with the holdings. *The Memoirs of Edward Alleyn* (1841), his first Shakespeare Society volume, includes as one of its fabrications a list of King's players. Shakespeare's name appears second, proving 'that up to 9th April, 1604, our great dramatist continued to be numbered among the *actors* of the company'.[136] By the insertion of 'Mr Shakespeare vjd' in an assessment list for the Liberty of the Clink, the forger contrived to show that the dramatist resided in Southwark in 1609 and that, there being no higher levy than sixpence, 'he lived at that time in as good a house as any of his neighbours'.[137] (Always willing to reconsider the implications of his mischief, Collier later revised what he described as a hasty inference, and suggested in his Life that maybe Shakespeare paid the rate not for any dwelling, 'good or bad, large or small', but for his Bankside theatrical property.) A rough memorandum in Alleyn's hand on a miscellaneous scrap of paper which is today no longer to be found at Dulwich shows the actor paying a total of £599 6s. 8d. for Blackfriars property in April 1612—sold to him, Collier guesses, by Shakespeare just prior to his retirement from the stage. This suggestion he would afterwards under pressure withdraw, but he would not renounce the document itself.

The Alleyn *Memoirs* behind him, Collier now gave himself to his gargantuan editorial labours on Shakespeare. In his *Reasons for a New Edition of Shakespeare's Works* (1842) he quite properly attached greatest weight to the settling of the text. In this connection he revealed the existence of a First Folio, belonging to Lord Egerton, with a number of happy marginal emendations in a hand which apparently dated from the time of Charles I. This 'discovery' would later haunt him. Collier also promised subscribers a biography that would form 'an important portion' of the first volume. The large quantity of new material which had come to light since the Third Variorum strengthened Collier in his resolve, and if he himself unearthed most of this information, he modestly refrained from calling attention to his own dazzling success as a literary sleuth. Instead he impersonally summarized the discoveries as though they already formed part of the documentary record and were for all the world as substantial as the monumental bust in Stratford Church. And indeed, when he wrote, who thought otherwise?

With *The Life of William Shakespeare*, included in 1844 in the first volume of the *Works*, dedicated to the Duke of Devonshire 'by his devoted and grateful servant, the Editor', Collier reached the apogee of his career as the poet's biographer. Yet it retains many conventional features. What Malone subjected to searching analysis, Collier accepts. He does not challenge the received interpretation of the allusion to luces in *The Merry Wives of Windsor*, the deer-stealing tradition, or the bitter ballad; this last he views as the principal offence which motivated Sir Thomas's prosecution. About Shakespeare's marriage, on which others held such strong views, Collier writes with moderation, but after judiciously appraising the evidence he concludes with the majority that 'Shakespeare was not a very happy married man.' And, a lawyer himself, he accepts like other lawyers the barrister Malone's theory that the future playwright had employment for a time in an attorney's office after quitting the free school in Stratford.

In the mid-nineteenth century we should perhaps expect Shakespeare to be seen through a sentimental haze of bardolatry: it need not surprise us that Collier stresses (despite the total absence of records) the dramatist's filial solicitude for his aged and unprosperous parents, or that he suggests (as though it were no fantasy) that Shakespeare out of loyalty to his fellows used his influence and money for the reconstruction of the Globe after fire destroyed it in 1613. Bardolatry too may account for a certain recklessness of interpretation; not entirely, however, for recklessness was deeply ingrained in the forger's character. Thus Collier resurrects the theory, demolished for all time (one thought) by Malone, that Shakespeare is the pleasant Willy whose demise Spenser laments in *The Tears of the Muses*; but not content with a single tribute, he finds another allusion in *Colin Clout's Come Home Again*. And, with no departure from his matter-of-fact tone, Collier proposes that Spenser at one time resided in Warwickshire and there befriended Shakespeare, twelve years his junior. Veneration for the culture hero also prompts an extravagant extension of the meaning of Greene's deathbed attack on his upstart rival:*

It . . . establishes . . . that our great poet possessed such variety of talent, that, for the purposes of the company of which he was a member, he could do anything that he might be called upon to perform: he was the *Johannes Factotum* of the association: he was an actor, and he was a writer of original plays, an adapter and improver of those already in existence, (some of them by Greene, Marlowe, Lodge, or Peele) and no doubt he contributed prologues or epilogues, and inserted scenes, speeches, or passages on any temporary emergency. Having his ready assistance, the Lord Chamberlain's servants required few other contributions from rival dramatists.[138]

On Collier's behalf it may be said that he makes no false claim for his reading: 'In our view, therefore, the quotation we have made from the "Groatsworth of Wit" proves more than has been usually collected from it.'[139]

Biography has many uses, and for Collier it included the legitimizing by craft of what he had by criminality perpetrated. His previously announced discoveries fall into place in his magisterial narrative; he need no longer even claim credit for them. In his notes he gives the texts of the forged records as

---

* An added curiosity is Collier's suggestion that 'possibly' Chettle wrote the *Groatsworth of Wit* and foisted it upon Greene because of the latter's 'high popularity'.

well as stanzas from his own forged ballads, including such piquant items as the doleful ditty on how Marlowe came to break his leg while acting on the Curtain stage. At the same time Collier expresses high-minded contempt for unfounded speculation, which, he assures his readers, will not vitiate his lofty undertaking. Nevertheless he speculates constantly, as he weighs what he complacently regards as the 'probabilities'. The pose is in character.

For this Life Collier forged his last biographical record of Shakespeare. Malone, in his inquiry into the Ireland papers, had referred to a document before him which showed the poet to be living in Southwark in 1596. Whether Collier was moved by rivalry with the great man or by an unconscious identification with him, we cannot say; but he produced the document. He says it lists 'Mr. Shaksper' as one of a group of inhabitants of Southwark who had complained (of what is left unspecified) in July 1596. With the condescension which he had mastered, the forger remarks on the ignorant handwriting and peculiar spelling of the paper.

At times one has the eerie feeling that Collier mesmerized even himself. It is difficult otherwise to account for some of his flights of interpretative fancy. The Blackfriars certificate of 1589 is the starting-point for one such imaginative excursion. Like any good bardolater, Collier accepts the tradition that the Earl of Southampton gave his protégé a gift of a thousand pounds. Now, assuming that the Globe opened its doors in 1595 (in fact it did not commence operations until four years later), Collier proposes that it was financed by the sharers in the Blackfriars—those sharers whose existence he had invented. 'Is it, then, too much to believe', he asks, 'that the young and bountiful nobleman . . . presented Shakespeare with 1000*l*., to enable him to make good the money he was to produce, as his proportion, for the completion of the Globe?'[140] The answer is Yes.

Yet one cannot dismiss the Life as a clever but reprehensible literary fraud. For Collier makes genuine contributions to the store of biographical knowledge. Some of the documents described fill in the shadowy family background. One record concerns a Richard Shakespeare, probably the poet's grandfather, and his rental of house and land belonging to Robert Arden, whose daughter Mary married John Shakespeare. In two deeds (dated 7 and 17 July 1550) Arden made over lands and tenements to Adam Palmer and Hugh Porter in trust for six of his daughters, who are named. (Malone believed that Arden had only four daughters, but Collier is aware that he had in fact eight.) Another document shows John Shakespeare employed in 1592 to assist in appraising the goods of Henry Field, tanner of Stratford, after his decease. As evidence of John Shakespeare's illiteracy Collier produces two warrants bearing the former's mark. John Shakespeare's declining fortunes are indicated by a deed of 1579, according to which he disposed of his wife's interest in two Snitterfield tenements for a paltry £4. The most intriguing of these records, recovered at the State Paper Office by the archivist Robert

Lemon and communicated to the author, has Sir Thomas Lucy and other justices naming the poet's father in 1592 as one of those not attending church either because of recusancy or for fear of being served with a process for debt.

Greater intrinsic interest of course attaches to the new information about the dramatist himself. An inventory of corn and malt in the borough of Stratford, dated 4 February 1598, indicates that Shakespeare, a resident of the Chapel Street ward, had ten quarters (equivalent to eighty bushels) of grain. Collier notes that only two persons in the ward held more, and with his scholar's practised eye for the significance of a record, he infers that Shakespeare must by then have owned New Place, which is situated in Chapel Street. Collier also supplements the deed (first noted by Wheler) by which Shakespeare purchased part of a leasehold interest in the tithes of Stratford and adjoining villages for £440: he produces an undated Chancery suit regarding the tithes, in which Shakespeare and others complained that certain lessees refused to pay their annual rent of £27 13s. 4d. In this bill Shakespeare placed the value of his moiety at £60 per annum. Lastly, Collier publishes additional notes of Thomas Greene, Town Clerk of Stratford, on Shakespeare's freehold in the fields of Old Stratford and Welcombe at the time of the projected enclosure of the common lands in 1614. These documents, dryly legalistic though their terminology be, provide welcome information about Shakespeare's economic situation and Stratford ties during his later career.

In a revised edition of his Life (1858) Collier first pointed out a link between Shakespeare and Henry Willobie's *Willobie His Avisa*. In the Argument of this poem of 1594 H.W., suffering from unrequited love for Avisa, confides in 'his familiar friend W.S. who not long before had tried the courtesy of the like passion, and was now newly recovered of the like infection'. Might not W.S. stand for William Shakespeare? The account of him, after all, 'employs so many theatrical expressions, that it seems as if the mention of him had led the author to the use of them'. Future biographers would not ignore the tantalizingly allusive—and elusive—passage. (Two years later, in *Notes and Queries*, W. C. Trevelyan, perceiving no distinction between H.W. and W.H., identified Willobie with the dedicatee of the Sonnets, Mr W.H.) Had Collier satisfied himself with these valid contributions to knowledge, his place would have been secure in the great tradition of English literary scholarship which extends from Malone to Chambers. That he could not so content himself is a measure of his perversity.

He consistently did good as well as mischief. Collier is responsible, for example, for making available Dowdall's letter of 1693 with its gossip, derived from the Stratford Church clerk, of Shakespeare being bound apprentice to a butcher, running off to London to be received as a servitor in the playhouse, and finally being laid to rest in a grave that his wife and daughters earnestly longed to share with him. This document Rodd published in 1838 as *Traditionary Anecdotes of Shakespeare*. If good and evil mysteriously coexist in Collier's Life, so do they elsewhere in his work, as in his character. Of the Alleyn *Memoirs* we have already spoken. Collier was also the first to publish *in extenso* (in 1845) the theatrical material in Henslowe's *Diary*, that most precious of Elizabethan playhouse documents, which the librarian allowed him to remove indefinitely from Dulwich College. The edition reflects Collier's expertise and erudition, but the volume that he returned was defaced with his forged insertions, erasures, and other mutilations. It has not been lent again.

For the services that he performed, praise, honours, and responsibilities came to him. In 1847 he was made secretary to a royal commission on the British Museum that had as its chairman the Earl of Ellesmere, and he played a constructive role in urging the merits of a printed as opposed to a manuscript catalogue. Collier had in 1830 been made a fellow of the Society of Antiquaries; in 1847 he became treasurer, and, in 1849, vice president. In 1850 the Crown rewarded him with a civil list pension of one hundred pounds 'in consideration of his literary merits'. How did the forger feel when, in May 1856, he read a letter from an admirer which concluded, 'Of all the Commentators on S. you are preeminently the most *honest*, and you stand almost alone among modern Editors, in that, while advancing your claims as a Critic, you have not at the same time forgotten that you are a Gentleman'?[141] If he appreciated the irony, this gentleman's pleasure must have been tempered by his awareness of the dangers which now hemmed him in. For in 1853 the first blows had fallen.

# 11
# *J. Payne Collier: Exposure*

THERE had been uneasiness earlier, but it had found only private or qualified expression. 'I observe you quote and rely upon the letter signed "H.S." discovered among Lord Ellesmere's papers by Mr. Collier,' wrote J. W. Croker in 1841 to Charles Knight, then on the Council of the Shakespeare Society. 'If that letter be genuine I must plead guilty to a great want of critical sagacity, for somehow it smacks to me of modern invention, and all my reconsideration of the subject, and some other circumstances which have since struck me, corroborate my doubts. Mr. Collier is, of course, above all suspicion of having any hand in a fabrication.'[142] Not surprisingly, this communication caused Knight unease, and he re-examined the Southampton document closely.

What he now found in it disturbed him even more, for the letter took on a very peculiar appearance. The absence of a superscription, customary in those days, puzzled him. When dealing with Ireland's forged Southampton letter, Malone had observed that the Earl signed himself *H. Southampton*; Chalmers had contended that he used only his surname. But no one was aware of any letter from Southampton subscribed 'H.S.' And the closing— 'Your Lo[rdship's] most bounden at com[mand]'—was not the formula of a great lord in familiar correspondence with an equal: he would say 'affectionate friend' or 'assured friend' or the like. It was odd that the foremost legal officer of the realm should be told, concerning two actors of no special concern to him, that 'they are both of one countie and indeede almost of one towne'. It was odder still that Southampton in his letter should unconsciously quote *Hamlet* ('one who fitteth the action to the word and the word to the action'). Most curious of all was the fact that the letter entirely corroborates a number of familiar biographical points but furnishes so little not already known. In his *William Shakspere* (1843) Knight agonizes publicly over the document. He seems to tremble on the verge of declaring it a forgery, but at the last moment backs away. His faith in Collier's bona fides overcomes the scruples offered by his reason. 'Looking at the decided character of the external evidence as to the discovery', Knight concludes, 'and taking into consideration the improbability of a spurious paper having been smuggled into the company of the Bridgewater documents, we are inclined to confide in it.'[143] Thus Collier survived a close call, but he did not escape entirely unscathed; doubt regarding the genuineness of one of his discoveries had for the first time found public expression.

He faced a severer test at the hands of the Revd Joseph Hunter, who as Assistant Keeper of the Public Records had—as Knight had not—an antiquarian's intimate knowledge of the phraseology of old records and a genealogist's passion for the facts about great or obscure men of the past. Hunter's *New Illustrations of the Life, Studies, and Writings of Shakespeare* (1845) makes a positive contribution to biography which elsewhere in these pages will not pass unnoticed; now it is his usefully destructive analysis of the Ellesmere papers that concerns us. Hunter carries scepticism a long stride beyond Knight. The Southampton letter, he finds, is 'not in the style of the times at all', and moreover betrays inaccuracies that undermine trust. Hunter does not help his case by confusing the Blackfriars playhouse referred to by 'H.S.' with The Theatre, but he rightly questions the statement that Burbage and Shakespeare were 'both of one countie, and indeed almost of one towne'—the former having probably been born in London where his father was known to be residing in 1576. (The urge of Shakespeare-obsessed critics to find a regional connection between their hero and his principal stage interpreter goes back to Malone, who discovered that a John Burbage was bailiff of Stratford in 1555.) The wording of the 1589 document listing the

Blackfriars sharers sounds to Hunter's practised ear 'not like the phrase in which a genuine certificate of that time would be conceived, but very like what fifty years ago would be thought a good imitation of that phrase'.[144] And what shall we think of the names registered in the certificate? It is odd that Richard Burbage, then no more than nineteen, should have proprietor's status; odder still that so too should Nicholas Rowley, who, being Burbage's apprentice, must have been still younger. And what of William Kempe, who belonged at that time to a different company of actors? How strange also that the 1608 paper revealing in detail the value of the Blackfriars property should unite the names of Heminges and Condell, thus anticipating their union as joint-editors of the First Folio! But if these documents are, as Hunter strongly suspects, modern fabrications, who forged them? Perhaps the mischievous Steevens had gained access to the riches of Bridgewater House (somewhere Hunter had a note about this, but alas he has misplaced it) and there tucked away, with perverse humour, his own inventions which at a later day would beguile the innocence of Collier.* After all, 'No one who knows Mr. Collier, can for a moment doubt that they were found by him there.'[145]

At length other scholars than Collier began to obtain permission from the Earl of Ellesmere to inspect the papers at Bridgewater House. On 3 March 1846 Hunter went, and came away with his dark suspicions confirmed. The various documents were suspiciously similar in handwriting and state of preservation, and the paper looked less like writing paper of the period than like the stock used in books. 'My impression', he summed up with nice understatement, 'was not on the whole favourable to their genuineness as a mass.'[146] But these observations he reserved for his notebook, which remains in manuscript; publicly he added nothing to what he had already said. Another visitor was W. H. Black, also Assistant Keeper of Her Majesty's Public Records; but he too kept his peace. Halliwell, who visited Bridgewater House in 1853, was less reticent. The title of the pamphlet he published that year, *Observations on the Shaksperian Forgeries at Bridgewater House*, sufficiently indicates his position: he categorically denounces as modern forgeries all the documents from this source. (Actually Halliwell had not inspected all of them, for he was able to find there only a recent transcript of the Daniel letter.) He does not accuse Collier personally, allowing for the possibility that the researcher had been duped by papers inserted by another. The most important of these records, the certificate of 1589 listing Shakespeare as a shareholder in the Blackfriars Theatre, is 'at best merely a late transcript'. Doubts induced by the appearance of the certificate were strengthened for Halliwell when he examined another paper in the same volume, the warrant purporting to appoint Daiborne, Shakespeare, and others instructors of the

* Later, in his notebook, Hunter refers to his 'uncertain opinion' that Steevens had access to the Ellesmere papers, and acknowledges his confusion on this score (British Library MS Add. 24497, fo. 46ᵛ). Steevens paid a heavy posthumous price for his mischievous streak.

Queen's Revels children; for the information in the warrant cannot be reconciled with that in an authentic patent for the same company issued in 1610. 'Fortunately for the interests of truth', Halliwell justly observes, 'indications of forgery are detected in trifling circumstances that are almost invariably neglected by the inventor, however ingeniously the deception be contrived.'[147] The paper and ink used in the Southampton letter and in the memorandum setting forth the value of Shakespeare's Blackfriars shares appear to be too early, and—a most peculiar circumstance—these records are in the same hand. 'It is clearly', Halliwell concludes, 'Mr. Collier's duty as a lover of truth, to have the originals carefully scrutinised by the best judges of the day.'[148] Being no great lover of the truth, Collier failed to heed the call. His deafness hurt him, for the alarm felt by Halliwell and others on the Council of the Shakespeare Society contributed to the dissolution of that body later in the same year. But the *Observations*, having been printed for private circulation, scarcely caused a ripple. Collier's reputation, so far as the larger public was concerned, remained unsullied, and such was the thickness of his skin that he does not appear to have suffered any loss of equanimity.

It was not the Bridgewater House papers that brought about his ruin but the Perkins Folio. This was his supreme, his overreaching imposture. Although it pertains to the text of Shakespeare rather than to the life records, the episode deserves a place in this narrative, for the exposure led by extension to the discrediting of the biographical discoveries as well. When in 1844 Dyce published his *Remarks on Mr. J. P. Collier's and Mr. C. Knight's Editions of Shakespeare*, he made it clear that his censure applied only to the former's editorial labours. 'Mr. Collier's *Life of Shakespeare*', he went out of his way to declare, 'exhibits the most praiseworthy research, a careful examination of all the particulars which have been discovered concerning the great dramatist, and the most intimate acquaintance with the history of the early stage.'[149] Fifteen years later such an exemption would no longer be possible.

On 31 January 1852 Collier announced in that most respectable of Victorian weeklies, the *Athenaeum*, his purchase of a remarkable book from the late Mr Rodd for the trifling sum of thirty shillings. A shabby and imperfect copy of the Second Folio, it had come to the shop as part of a consignment of books from the country. On the outside of one of the covers appeared the inscription 'Tho. Perkins, his booke'. Sometime later (Collier continues) he discovered that the volume contained numerous marginal and textual annotations—literally thousands and thousands—in an old hand. The Folio would be placed before the Council of the Shakespeare Society, not yet defunct, at the next meeting. On 7 February, in the same periodical, he mused on whether the Old Corrector had garnered his readings from 'purer manuscripts' or whether perhaps they derived from stage recitations. On 27 March Collier made it known, once again in the *Athenaeum*, that the relic would be displayed the following Friday to the fellows of the Society of

Antiquaries for two hours, from 12 until 2, at Somerset House. In the summer of 1853 he published *Notes and Emendations to the Text of Shakespeare's Plays, from Early Manuscript Corrections in a Copy of the Folio, 1632, in the Possession of J. Payne Collier*. He speculates in his Introduction on the possibility that Thomas Perkins may have been descended from the Richard Perkins who had acted in the revival of Marlowe's *Jew of Malta* shortly before 1633. Collier also cunningly covers his tracks by revealing for the first time the damage which the book has suffered: '. . . many stains of wine, beer, and other liquids are observable: here and there, holes have been burned in the paper, either by the falling of the lighted snuff of a candle, or by the ashes of tobacco. In several places it is torn and disfigured by blots and dirt, and every margin bears evidence to frequent and careless perusal.'[150]

None of Collier's previous discoveries had caused such a stir. From the first the merits of the Old Corrector's emendations excited controversy. In 1853 Samuel Weller Singer, a prolific contributor to *Notes and Queries*, entered the lists with *The Text of Shakespeare Vindicated from the Interpolations and Corruptions Advocated by John Payne Collier Esq. in His Notes and Emendations*. In this polemic he heaped scorn on the improvements but accused Collier of no malfeasance; rather he preferred to believe that 'such a staunch defender of the integrity of the old text' had been deluded by a pseudo-antique commentary. A. E. Brae in 1855 was the first to cast public aspersions on Collier's bona fides, in his pseudonymous pamphlet, *Literary Cookery*, although this has more to do with supposed Coleridge forgeries than with the Perkins Folio. The victim responded by suing the publisher for criminal libel. In an affidavit dated 8 January 1856 Collier swore that he had not, 'to the best of my knowledge and belief, inserted [in *Notes and Emendations*] a single word, stop, sign, note, correction, alteration, or emendation of the said original text of Shakespeare, which is not a faithful copy of the said original manuscript, and which I do not believe to have been written, as aforesaid, not long after the publication of the said folio copy of the year 1632'.[151] Thus to the crime of fraud he added that of perjury.

Accompanying the affidavit was a copy of the second edition (also of 1853) of *Notes and Emendations*. Now in the British Library, it is marked exhibit A in the Court of Queen's Bench, where the motion was heard. Sitting in judgement was Lord Campbell, who knew the plaintiff well from his *Morning Chronicle* days (they had worked together as law reporters for the paper). The court properly declined to rule in the case, but the learned justice characterized Collier as 'a most honourable' man and declared the affidavit to be sufficient vindication of the plaintiff's moral character. A couple of years later Campbell, who would end his career in triumph as Lord Chancellor, cast his *Shakespeare's Legal Acquirements Considered* in the form of a letter to Collier, his 'old and valued friend'. At the time the recipient owned himself well enough pleased, and commended the Lord Chancellor's 'easy gossipping style'; but eventually he came to despise his admirer. '[H]is impudence was excessive', the forger writes of Campbell in his *Autobiography*, 'and having drawn up a small tract on the legal learning of Shakespeare, he thought it would answer his purpose to address it to me, and he did so. He afterwards invited me to dine with him, and I did so once or twice; but his parties were really so dull and formal that, at last, purposely I kept away from them, notwithstanding his personal reproaches.'[152] Such are the vicissitudes of friendship.

Meanwhile Collier had found support as welcome as it was unexpected. In April 1853 he received a letter from one John Carrick Moore, informing him that a friend, Mr Parry, had many years past owned a copy of the 1632 Folio with marginal notations; a copy since lost. When shown a facsimile of the Perkins page used by Collier as his frontispiece, Parry had recognized it as belonging to his former copy. Collier hastened over to Parry at his home in St John's Wood, where he lay recuperating from a fall, and obtained from him a pedigree of sorts for the volume. This he published, with suitable embroidery, in the *Athenaeum* for 4 June 1853, after first showing the old man the letter in proof. Rather to the surprise of some, on neither visit to Parry did Collier carry with him the book which alone could settle the question. Years later, Collier, under intense pressure from unbelievers, described in a letter to *The Times* (20 July 1859) a third occasion when he met Parry, this time outside his house, and showed him the Folio. 'That was my book,' the latter had cried out, 'it is the same, but it has been much ill-used since it was in my possession.' So Collier reported. When questioned, Parry recalled the meeting; but he denied seeing the book, which in any event he could hardly have examined because at the time he was lame from his accident, and supporting himself in the road on walking-sticks. (Collier claimed, falsely, that there was only one stick.)

Only expert examination of the Folio could satisfy the sceptics. Such inspection Collier forestalled by presenting his treasure to his benefactor, the Duke of Devonshire, in whose library it eluded prying eyes: the Duke did not know where it was, his librarian refused to exhibit it. But the old Duke died, and his son succeeded him. In May 1859 Sir Frederic Madden, Keeper of Manuscripts at the British Library, asked—and was granted—the loan of the volume. Madden had dabbled in matters Shakespearian in his *Observations on an Autograph of Shakspere and the Orthography of His Name* (1837), and was the greatest of nineteenth-century British palaeographers. He was also Collier's friend, but a man of stern and impeccable rectitude. The doubts about the Perkins Folio being circulated by C. M. Ingleby and others gave him pain. Still he recognized (in a letter to Ingleby dated 8 June 1859) that 'In spite of friendship, *truth* should & must always prevail'; then added, 'but I really cannot bring myself to believe the case to be as you infer.'[153]

A cursory glance at the Perkins Folio showed that the corrections could not be in a genuine hand of the seventeenth century; many of the forms, Madden noted, had been altered to give them the appearance of greater antiquity. On

19 June he examined the Folio leaf by leaf, and discovered to his astonishment thousands of pencil corrections, partly erased, in the margins; pencil
too underlay some of the ink notations. 'These corrections are *most certainly*
in a modern hand', he wrote in his private *Journal*, 'and from the extraordinary resemblance of the writing to Mr. Collier's own hand (which I am well
acquainted with) I am really fearful that I must come to the astounding
conclusion that Mr. C. is himself the fabricator of the notes!'[154] Himself not
doubting that the pencil annotations antedated the ink, Madden put his
assistant, N. E. S. A. Hamilton, on to the investigation, and both men turned
to Professor M. H. N. Maskelyne, Keeper of the British Museum's Mineral
Department, who undertook a series of microscopic and chemical tests. Thus
was Victorian scientism applied to a literary problem. The tests showed that
the pencil notations underlay the ink (and not vice versa), that the fluid
employed was apparently a water-colour rather than any ink ancient or
modern, and that the Old Corrector had attempted to erase the pencil
markings before applying the ink substitute. On 13 July 1859 Parry visited
the Museum and was shown the Folio. He immediately declared that it was
not the volume he had owned over half a century previously—that had been
the 1623 edition. 'I now give up Mr C. altogether', Madden fulminated in his
*Journal* on 27 July, 'and think he deserves to be chased from all literary
Society. It is really too bad in the present day to have one's confidence so
grossly abused by a man, who from his position & access to muniments, had
it in his power so successfully for a time to execute a series of frauds. I feel
great digusted [*sic*] at the dishonesty & falsehood which must have long been
practised.'[155] (In the months that followed, Collier became only too well
aware of Madden's feelings. 'Sir F. Madden is in a fury', he wrote to S. Leigh
Sotheby on 1 April 1860, '& can contain himself so little that he condescends
to vulgar abuse. . . .'[156]) Yet, asked to strip off Collier's mask publicly, Madden
refused: he had known him for too long. That task was entrusted to
Hamilton.

Troubles now closed in remorselessly on Collier. Hamilton went down to
Dulwich College and discovered that in the printed text of the letter from Mrs
Alleyn to her husband, Collier had interpolated the lines mentioning Shakespeare. In the same month, Ingleby, a solicitor who had abandoned the law
for Shakespeare and philosophy, called on Madden and asked him also to
borrow the Bridgewater Folio that Collier had first described in 1841. As he
was leaving the Museum Ingleby (in one of those extraordinary coincidences
which any competent novelist would disdain) met Lord Ellesmere entering
with the very book under his arm. He had brought it in order to obtain
Madden's opinion as to the authenticity of *its* manuscript corrections. Then
and there the Folio was opened to inspection, and once again tell-tale pencil
markings were discerned beneath the ink.

That month the results of the Museum investigations of the Perkins Folio appeared in a series of letters to *The Times*. The next year Hamilton brought out his *Inquiry into the Genuineness of the Manuscript Corrections in Mr. J. Payne Collier's Annotated Shakspere, Folio, 1632; and of Certain Shaksperian Documents Likewise Published by Mr. Collier*. This work extends the indictment, for not only does Hamilton present the case against the Perkins Folio, but he also considers the Bridgewater House papers which Collier had announced more than two decades previously. A single glance was sufficient to reveal the spurious character of these documents. Two (the Daborne warrant and the Daniel letter) are indeed 'such manifest forgeries, that it seems incredible how they could have cheated Mr. Collier's observation, even under the circumstances of excitement described by him as consequent upon their discovery'.[157] Collier fought back as best he could. He denied the allegations, hinting desperately that the Library officers had themselves inserted the pencil markings, charging that Parry had been confused by the rapid passing of Folios back and forth before his eyes. But all to no avail.

The final blow fell in 1861. It was administered by Ingleby, who had become Collier's nemesis. A man of sombre disposition, Ingleby was relentlessly severe on himself. 'I am morally weak in many respects,' he wrote in a manuscript sketch of his life and character. 'In some matters I have been systematically deceptive, & occasionally cowardly & treacherous. I am passionately fond of personal beauty; but on the whole, I dislike my kind, & my natural affections are weak.'[158] Is it any wonder that he would show no mercy to a conscienceless rogue? In *The Shakspeare Fabrications* of 1859 Ingleby had already argued the recent origin of the manuscript notes in the Perkins Folio. Always in precarious health, he now mustered all his energies to produce *A Complete View of the Shakspere Controversy, Concerning the Authenticity and Genuineness of the Manuscript Matter Affecting the Works and Biography of Shakspere, Published by Mr. J. Payne Collier as the Fruits of His Researches*. It is not so much a work of original investigation as an indictment drawn up by a highly intelligent prosecuting attorney who has profited from the expert criminological testing done by the laboratory technicians over at headquarters, in this case the British Museum. Ingleby marshals the evidence against the old scoundrel with overpowering destructive force. The coverage is comprehensive as he treats, in order and with precise detail, the Bridgewater Folio, the Perkins Folio, the Bridgewater documents, the Dulwich manuscripts, and the State Paper with Shakespeare's name that Collier had described in his *History of English Dramatic Poetry*. Periodical correspondence is resurrected, the forgeries are reproduced in facsimile; Ingleby has even interviewed Parry. He misses nothing. (He knew of course that Madden, not Hamilton, deserved credit for first detecting by external evidence the fabrication of the Perkins notes, and on 18 July 1859 Ingleby sounded out Sir

Frederic, in a letter now at the Bodleian, as to whether honour might be bestowed where it was due; but because of the continuing publicity, Madden said no.)

The arraignment is all the more devastating by reason of the author's unimpassioned tone, his reluctance to press a doubtful point, his willingness to make concessions (as when he allows that even a truthful man might, like Collier, vary the details of an event in the retelling). Only extraordinary self-control could purchase such restraint, for Ingleby truly detests Collier. In a supplement to the *Complete View*, which he never published, he is less guarded. Speaking of the Alleyn *Memoirs* and the Henslowe *Diary*, he remarks: 'We know that the great literary slug has crawled over both. What wonder if we shall still be able to trace his slime.'[159] Ingleby's book remains the definitive treatment of this *cause célèbre*, recalling a classic work of the preceding century in which another attorney-turned-Shakespearian scholar unmasked a brasher and less cunning forger.

This time Collier made no reply. What was there for him to say? He could not wriggle out from under the proofs which science and logic had adduced. Although he must have realized that his reputation lay in ruins, he maintained an outward good cheer during the rest of a long life which extended to his ninety-fourth year. But inwardly he rankled over former friends who had turned their backs on him.

The defection of Dyce, his old associate in the Shakespeare Society, hurt the most: Collier alludes spitefully to him more than once in his *Autobiography* and diaries. Why did Dyce do it? he mused, and, unwilling to admit principle as a consideration, settles for envy as the whole answer. Had not Collier beaten him to the book-stalls with an edition of Shakespeare? 'When I told him, about 1842, that I had engaged to prepare a Shakespeare', Collier noted on 30 March 1876, 'he turned as white as a sheet, and our *intimacy* at least such as it had been was at an end from that day.' The outburst follows: 'He was a thoroughly selfish man: Think of his having known me so intimately for 30 years, of his dining at my table scores & scores of times, and at last passing me in the streets without recognition.'[160] Nevertheless, when Dyce lay dying from jaundice, Collier twice attempted to arrange an interview of reconciliation, but Dyce refused to see him—'this, after 30 years of uninter-rupted friendship, & after I had in every way, during the whole of that time, lent him my best aid in every work he produced'. And finally: 'He could not forgive me for stepping before him in publishing an edition of Shakespeare, when he never gave me a hint, even, that he contemplated such a work.'[161] In fairness it may be granted that Collier had some cause for thinking that Dyce was nettled by anticipation: a letter, now in the British Library, addressed to 'My dear Collier' in 1844, reads in its entirety: 'Many thanks for the concluding volume of a work, which I wish most heartily that you never had begun.' It is a matter for regret that Dyce's manuscript *Reminiscences*, now at the Victoria and Albert Museum, contain no sketch of Collier.

He republished his history of the stage in a revised edition with all the forgeries intact; to do otherwise, after all, would have amounted to an admission of guilt.* For private circulation among his friends he printed a diary of those exhilarating days, long since past, when noble lords rewarded him with their favour for his services to scholarship. His *Trilogy. Conversations between Three Friends* (1874) finds him still unrepentant: rejecting his enemies and their evidence, he insists upon the authenticity of the Perkins Folio (even the notorious pencil-markings testify to its genuineness!), and shows how editors have adopted many of the reviled emendations. He was, in truth, incorrigible. In the twilight of his career, when he was eighty-six, he turned, like the uncouth swain in *Lycidas*, to fresh woods. 'I have just discovered a most interesting book,' Collier wrote to J. Parker Norris on 17 November 1875, '—a folio—full of Milton's brief notes and references; 1500 of them.'[162] Touching as is the old rogue's youthful enthusiasm, one need not regret that posterity has been spared the Milton Folio.

In his *Autobiography* intended for presentation to his Glasgow friend Alexander Smith, he tries to keep up appearances, but his mind wanders (he was in his nineties), and betrays him into revelations:

I had made the most extraordinary discovery of a copy of the Second Folio of Shakespeares Works in 1632 containing many admirable and indisputable improvements (in manuscript) of the text as it had descended to us in various quartoes and in the first folio of 1623. Nobody could deny the excellence of many of them, they have been gladly adopted since. . . . if the proposed emendations are not genuine, then I claim them as mine; and there I intend to leave the question without giving myself further trouble: anybody else is welcome to solve the enigma.—Good or bad, mine or not mine, no edition of Shakespeare, while the world stands, can now be published without them: I brought them into life and light, and I am quite ready to be answerable for them.[163]

The megalomania apparent here reasserts itself in another passage, in which the malefactor brings himself to speak the forbidden word, but is cautious enough to employ the third person: 'If the emendations be forgeries how the inventor of them, if alive, must laugh at the ridiculous result of his unrejectable fabrications: they now form an essential part of every new edition of Shakespeare, and never hereafter can be omitted.'[164] Thus, without quite admitting his own responsibility, the forger enjoyed his laugh.

More communicative than the *Autobiography* is his unpublished *Diary*, of which twelve volumes survive. In a note prefacing the first, Collier states that he has destroyed eleven preceding parts. The extant journal covers the period from 7 November 1872 until 11 December 1882, but there are gaps in

---

* In light-hearted moments, however, he could be reckless, if we may trust Dyce, admittedly not a disinterested witness. On 24 March 1860 he wrote to Madden: 'I have *a very decided impression* that Collier once confessed to me, long ago, that he forged as a joke, that Walton paper; but, after the lapse of so many years, *I could not swear to his having so confessed*. I am, however, sure that Thomas Rodd, on the first appearance of the (Ellesmere) Southampton Letter, said to me, "Why should not Collier have invented *it*, since we know that he invented a Walton paper?"' (Bodleian MS Eng. misc. C.96, fo. 252).

continuity; some segments are missing. On the last page of volume xv Collier solemnly enjoins his family never to say a word in his defence after his death: 'If my memory cannot support and defend itself, let it fall.' Again and again, in succeeding instalments, he reverts to this theme with obsessive insistence. At such moments, as he envisages his enemies closing in upon him, his hand, enfeebled by rheumatism, gathers strength: the script becomes larger, the inking heavier; he underlines words, writes NB above or alongside, draws a line in the margin or around an entire passage, affixes his signature. 'I know that Enemies are only lying in wait for me me [*sic*] to assail me,' reads an entry dated 30 November 1880. 'I defy them and charge all my Relations & Friends never to say one word in my defence: if they do, they will incur my heaviest displeasure. I DESPISE ALL MY ENEMIES & spit at them I cannot forgive ALL.'

Along with defiance, remorse and penitence. On 21 November 1877 he admits, 'It is my own fault and folly that I am not now justly considered the first and best emendator of Shakespeare.' This passage too he signs. (He still kept up appearances. Offered—with disguised malice?—a portrait of Ireland in 1878, Collier declined with impressive hauteur: 'I do not like the man', he replied, 'nor his attempt at imposition.'[165]) On 19 February 1881 he is again contrite: 'I have done many base things in my time—some that I knew to be base at the moment, and many that I deeply regretted afterwards and up to this very day.' This is sufficiently vague, but it is suggestive that in his next paragraph he turns to the Perkins Folio. The most forcible expression of his repentance comes near the very end, in the twelfth and last volume of his *Diary*. In a barely legible scrawl Collier writes, on Sunday, 14 May 1882:

I am bitterly sad and most sincerely grieved that in every way I am such a despicable offender I am ashamed of almost every act of my life

J. Payne Collier<br>Nearly blind

My repentance is bitter and sincere[166]

The next year he died at the age of ninety-four.

# 12

# *Joseph Hunter*

THE character of Collier's eminent contemporary Joseph Hunter holds less mystery—and hence less fascination—than that of the forger caught in the toils of his strange compulsion. Indeed, there is little about Hunter's life to stir a biographer's interest. The son of a Sheffield cutler, he was for twenty-four years a Presbyterian minister of Bath. In middle life he found employment in London in the Record Office (Assistant Keeper, first class) calendaring documents, and lived quietly in Torrington Square. There Hunter reared six children, collected notes on sixteenth- and seventeenth-century English poets, and assembled the vast library, the sale of which, after his death in 1861, occupied four days. Never more content than when inhaling the dust of the muniment room, he savoured wills, parish registers, account rolls, and visitation books. His prose style testifies to an urbanity of temper to which genealogists rarely aspire. No fanatic, Hunter realizes that his researches are open to objection as being, by turns, too particular or too speculative. 'It may signify little from whom a great man is sprung,' he sadly admits; 'and Shakespeare, like ten thousand other men, well or ill descended, owes more to himself than to his ancestors.'[167] Nature, after all, cannot choose her origin. Yet something, if perhaps not very much, may be gained from determining the kind or class of persons from whom a great man traced his lineage, and from seeing what went to form the characters of his immediate ancestors. And so, in his 'Prolusions Genealogical and Biographical on the Family of William Shakespeare, and Other Families Connected with Him', comprising the first part of the *New Illustrations* (1845), Hunter gives his account of Shakespeare's paternal and maternal antecedents, as well as of his descendants and representatives—the Halls and Bernards—and of Stratford families that the poet knew: the Combes, Quineys, and Nashes.

Hunter's generalizations, evidence of his desire to escape the narrowing confines of genealogical discipline, nevertheless reflect biases to which genealogists are especially prone. He tends to upgrade Shakespeare's ancestry, in order to show 'that the original prejudices of the poet would be aristocratical, that is, that the influences communicated by his parents would be of that character; and further, that they would be likely to educate him as to them it appeared the heir of a family of some consideration ought to be educated'.[168] On the mother's side, Hunter notes with satisfaction, Shakespeare's family belonged to the ancient gentry of the Midland counties. John Shakespeare may have been for some little time a glover, but after marriage he lived as a gentleman farmer on the proceeds of inherited property; that he ever fell upon hard times Hunter doubts. The Hathaway match brought no valuable connection to the Shakespeares. The antiquary sniffs at the rude marks made by Sandells and Rychardson, the illiterate husbandmen who witnessed the marriage licence bond. Such unseemly persons to grace a poet's bridal! Where were the friends of the family? Hunter demands. 'It seems but too evident, that this was a marriage of evil auspices, and it may have been one principal cause of that unsettled state of mind in which the poet left Stratford, about four years afterwards.'[169] A pity that the last will and testament of the poet's father-in-law had not yet come to light, for it reveals Rychardson and Sandells to have been no such unspeakably low rustics as Hunter believed. The former served as a witness on this solemn occasion, and Hathaway asked the latter, as a trusty 'friend' and neighbour, to be one of two supervisors (trustees, we would say) of his will. In another touching fancy, Hunter conjectures that, had the deceased poet's family neglected to erect a monument in his honour, this duty would have been discharged by the Countess of Pembroke, Dorset, and Montgomery.

Fortunately Hunter leaves aristocratic prejudices behind when he discusses Shakespeare's religion, concluding disinterestedly that 'the truth probably was, that he rested at a point between Rome and Geneva, rejecting what was bad, and receiving what was good from both'.[170] Hunter is enlightening on the growing Puritanism of seventeenth-century Stratford. He suggests that a new spirit of religious melancholy may be responsible for the failure of Shakespeare's heirs to preserve those vanities, a playwright's papers. Elsewhere Hunter is typically a man of his age in accepting unquestioningly the deer-poaching tradition and the hypothesis that the dramatist's early employment consisted of adapting old plays to contemporary taste.

His chief contribution, however, lies on the factual side. More than once Hunter catches the great Malone in error. The latter was confused about the Combes, although one can hardly blame him for having trouble sorting out the facts about a man named John who had two sons with the same Christian name. Then too Malone drew mistaken inferences from the will (abstracted by him and printed in full for the first time by Hunter) of Agnes Arden, the widow of Robert, the poet's grandfather: Robert, it is clear, was her second husband, and she could not have been the mother of Mary Shakespeare. This is a point, however slight, that future biographers would be unable to ignore. Greater interest attaches to Hunter's discovery, from a Court Roll, that John Shakespeare had settled in Stratford as early as 1552, and in Henley Street, where according to tradition William was born. (An attentive reader of the *New Illustrations* followed Hunter up the next year by publishing in the *Historical Register* a record of 1590 that shows John Shakespeare as possessing two tenements in Henley Street.) Hunter scores another first in printing, from the Lansdowne manuscript in the British Library, Lieutenant Hammond's account of his visit in 1634 to the 'neat monument of that famous English poet, Mr. William Shakespeare', with its intriguing reference to the witty and facetious verses fanned up by the poet on 'an old gentleman, a bachelor, Mr. Combe'. These verses, Hunter thinks, Hammond saw displayed in Stratford church. But the most important documentary revelation in the *Prolusions* is an assessment roll of 1598 listing Shakespeare as a resident of the parish of St Helen's, Bishopsgate, and levying a tax on him of 13s. 4d. St Helen's was about a ten-minute walk from the Theatre and the Curtain in Shoreditch, and—in another direction—from the Crosskeys Inn in Grace

Church Street; at these houses in the 1590s played the Lord Chamberlain's Men, Shakespeare's company. Later scholars would explain the significance of the abbreviation *Affid*, for *affidavit*, alongside Shakespeare's name in the document (which Hunter prints in full). This means that the Queen's officers swore that they had tried unsuccessfully to collect the tax.

The genealogical tradition to which Hunter's essay belongs is persistent in Shakespearian scholarship. Later in the century it is represented by the still serviceable pedigrees of the Cloptons and the Combes, the Nashes, Forsters, and later Hathaways, with which J. C. M. Bellew, admittedly a tyro, embellished his *Shakespere's Home at New Place, Stratford-upon-Avon* (1863). George Russell French's *Shakspeareana Genealogica* (1869) mostly recapitulates previous authorities, of whom Hunter is acknowledged chief. French's proudest achievement is the establishment of the exact degree of relationship between Mary Shakespeare's father and Walter Arden, whose son Sir John was Esquire of the King's body in the reign of Henry VII; in the satisfaction derived from such a triumph only another genealogist can share. An altogether less stylish performance than Hunter's, the book none the less amasses a great quantity of detailed information—French records more remote Shakespeares than any of his predecessors—and the thirteen elaborate tables of descent, never superseded, would be consulted with gratitude by later biographers.

# PART V

*Victorians*

# 1

## A Victorian Popularizer

FOR the historian of Shakespearian biography the period ending with Malone represents the high road: a great unimpeded stretch paved by a monumental succession of editions incorporating all the important fruits of research. To be sure, there are some peripheral figures of consequence, such as Oldys and Farmer, who did not edit the plays; but the editors took note of them. There are, too, the diversionary antics of William-Henry Ireland. Still, the way remains clear, and in Malone the pilgrim reaches his first grand objective: a golden apotheosis, the New Jerusalem of eighteenth-century Shakespearian studies. After that the road branches off into various footpaths of articles, monographs, editions, compilations, and full- and small-scale Lives. These become increasingly numerous with the Victorian age: in a single month—May 1844—if we can trust the results of one census, sixty weekly periodicals issued from the London presses; there were 38 quarterlies, 227 monthly magazines, and (throughout the United Kingdom) 447 newspapers. Selectivity now becomes essential. Fortunately, a few striking guide-posts stand clear of the surrounding shrubbery. One of these is Charles Knight, whose achievement as a biographer of Shakespeare is impressive, strange—and finally disappointing.

In this capacity Knight's destiny would be intertwined with Collier's in a way that he could scarcely foresee. Indeed, the outward circumstances of his early life recall the forger's. Born at around the same time, in 1791, Knight was the son of a printer, stationer, and bookseller of Windsor. Knight, unlike Collier, had some formal education, but not much: he spent two years at Dr Nicholas's famous classical school in Ealing, from which he was removed at the age of fourteen to assist in his father's business. Early possession of an imperfect Shakespeare Folio left an indelible impression on Knight, as on his notorious contemporary. Like Collier, Knight had a journalistic phase: while the former covered Parliament for *The Times*, the latter was serving as apprentice reporter in the House for the *Globe* and the *British Press*.

With respect to moral character, they had nothing in common, for Knight was an honourable man. The house in which he grew up stood in the shadow of Henry VIII's Gateway to Windsor Castle; from the sitting-room he could look upon the Round Tower. One day George III visited his father's shop, examined there a copy of Paine's *Rights of Man*, and departed without saying a word. Knight could also see the squalor of the poor of Windsor. As a child he watched a furious and hungry mob gather one autumn evening outside the baker's shop next door to his house to vent their frustration upon the tradesman who sold bread at prices they could not afford. Trembling, he stood behind his father, who exhorted them, in a firm yet kind tone, to go home. They dispersed, and the boy's social conscience was born.

When he entered man's estate, Knight devoted his energies—as publisher, author, and editor—to the cause of Victorian humanitarianism. He resolved to uplift the Lower Orders with wholesome, instructive, and attractive literature that was cheap enough for them to buy. If unscrupulous publishers catered for ignorance with almanacs filled with superstitious rubbish and quack remedies, he would bring out a genuinely informative competitor; thus originated the highly successful *British Almanac and Companion*. The same aim of popular enlightenment lies behind all Knight's varied enterprises through a long career during which he shrugged off Establishment fears that the working classes, emancipated through knowledge, would topple throne and altar.

His profession of independent publisher brought him a wide circle of acquaintances. He befriended De Quincey. Ireland visited him; wretchedly poor and never entirely trustworthy, he looked back joyously on his glorious hoax, 'preserved by his inordinate vanity from any compunctious visitings that might lead him to think that a fraud was not altogether to be justified by its cleverness!'[1] Their meeting serves to remind us of the dramatic pace of change. How close in time to Knight—and how far removed in sensibility— was Ireland's (and hence Malone's) world.

It was natural, given his programme, that Knight should associate himself with the Society for the Diffusion of Useful Knowledge, which had as its guiding spirit Henry Peter Brougham. Today best remembered for the species of carriage to which he donated his name, Brougham was an unamiable man of ferocious energy and liberal ideals; a barrister, he would eventually become Lord Chancellor. Of him it was said, with some acerbity, that if he had known a little law, he would have known a little of everything.[2] But the Society thrived. Under these auspices Knight inaugurated the *Penny Magazine*, an instantaneous commercial hit, and published, in twenty-seven volumes, *The Penny Cyclopaedia*, which brought him prestige and left him thirty thousand pounds out of pocket. As author, he contributed with zest and facility to his own projects. Knight's exercises in the Higher Journalism include *Results of Machinery, Capital and Labour*, even a book on the elephant. He helped make good literature accessible to the many with the lavishly illustrated series he sponsored: the *Gallery of Portraits*, the *Pictorial Bible*, the *Pictorial History of England*, the pictorial *London*. His labour of love was *The Pictorial Edition of the Works of Shakspere*, which he published in instalments from 1838 to 1841. Profusely illustrated with engravings not only of scenes from the plays but also of Elizabethan life and manners, the *Pictorial Shakspere* achieved deserved popularity and went through numerous editions. Knight is probably justified in his claim that he first used the word *pictorial* in this special sense.

His editorial labours on *Twelfth Night* first led Knight to consider Shakespeare's domestic character. He had read De Quincey's article in the *Brittanica*, just off the press, and lingered over the passage on the exchange between the Duke and Viola, interpreted by De Quincey as the playwright's appeal to the lessons of his own experience. 'Shakespeare himself', De Quincey declares, 'looking back on this part of his youthful history from his maturest years, breathes forth pathetic counsels against the errors into which his own inexperience had been ensnared.'[3] This interpretation led Knight to the discovery on which he most prided himself. In his Postscript to *Twelfth Night*, on Shakespeare's will, he remarks:

Shakspere knew the law of England better than his legal commentators. His estates, with the exception of a copyhold tenement, expressly mentioned in his will, were *freehold*. HIS WIFE WAS ENTITLED TO DOWER. . . . She was provided for amply, *by the clear and undeniable operation of the English law.* Of the houses and gardens which Shakspere inherited from his father, she was assured of the life-interest of a third, should she survive her husband, the instant that old John Shakspere died. Of the capital messuage, called New Place, the best house in Stratford, which Shakspere purchased in 1597, she was assured of the same life-interest, from the moment of the conveyance, provided it was a direct conveyance to her husband. That it was so conveyed we may infer from the terms of the conveyance of the lands in Old Stratford, and other places, which were purchased by Shakspere in 1602, and were then conveyed 'to the onlye proper use and behoofe of the saide William Shakespere, his heires and assignes, for ever.' Of a life-interest in a third of these lands also was she assured.[4]

Twenty years later he still recalled the pleasure given him by composition of this paragraph. 'Well do I remember the glee', he writes in his autobiography, 'with which . . . I showed it to my dear friend, Mr. Thomas Clarke, a sound lawyer, who confirmed my opinion, as fully as did Mr. Long and Mr. Hill, with whom I subsequently discussed the matter.'[5] Knight had reason to exult. Others had placed a more benevolent interpretation than Malone or De Quincey on Shakespeare's provision for his widow in the will; for example, Boswell (as we have seen) supposed that the poet had made arrangements for his wife during his lifetime. But nobody previously had grounded an explanation in the law itself.* Knight repeated his point in his *Penny Cyclopaedia* article and in his large-scale biography. His argument won over many eminent Shakespearians, including Collier, who had not yet been discredited.

To accompany his *Pictorial Edition* Knight undertook the work that most concerns us here: *William Shakspere: A Biography*, which comprises the eighth and final volume of the *Pictorial Shakspere*. He settled down to his task in Stratford in the summer of 1842. There he met the antiquarian solicitor, Robert Bell Wheler, who told him of inspecting hundreds of title-deeds and

---

* The issue is more complicated than Knight—or his contemporaries—could have been aware; for a brief summary discussion see Schoenbaum, *William Shakespeare: A Compact Documentary Life* (rev. edn.; Oxford, 1987), 301.

other legal papers from the period 1580–90 in hopes of finding Shakespeare's signature; for, if the future poet had served as a lawyer's clerk, he would on many occasions have been called upon to sign documents as an attesting witness. But Wheler had found nothing. In Stratford, Knight visited the Birthplace on Henley Street, the grammar school, and the Guild Chapel. Especially keen to see Shottery, he wandered over to the neighbouring villages. His steps also took him to Kenilworth and Coventry and Warwick. He followed the Avon where it descends to Bidford and Evesham, and where it winds upward to Charlecote and Hampton Lucy. Traversing the ground with him was his friend, the artist William Harvey, who brought along his pad to complete sketches made the previous summer; for this was to be, after all, a Pictorial Life. 'I wrote a very little', Knight recalled afterwards, 'but my mind was completely filled with the matter upon which I had to write.'[6]

At Oxford, in the solitude of the Bodleian during the Long Vacation, he wrote in earnest. In these surroundings, so rich in association, Knight, who had himself never attended a university, felt called upon to re-create the feelings of his subject as he paused in Oxford on his fateful first journey from Stratford to London:

So noble a place, raised up entirely for the encouragement of learning, would excite in the young poet feelings that were strange and new. He had wept over the ruins of religious houses; but here was something left to give the assurance that there was a real barrier against the desolations of force and ignorance. A deep regret might pass through his mind that he had not availed himself of the opening which was presented to the humblest in the land, here to make himself a ripe and good scholar. Oxford was the patrimony of the people; and he, one of the people, had not claimed his birthright. He was set out upon a doubtful adventure; the persons with whom he was to be associated had no rank in society; they were to a certain extent despised; they were the servants of a luxurious court, and, what was sometimes worse, of a tasteless public. But, on the other hand, as he paused before Baliol [*sic*] College, he must have recollected what a fearful tragedy was there acted some thirty years before. Was he sure that the day of persecution for opinions was altogether past? Men were still disputing everywhere around him; and the slighter the differences between them the more violent their zeal. They were furious for or against certain ceremonial observances. . . . The spirit of love dwelt in the inmost heart of this young man. It was in after-time to diffuse itself over writings which entered the minds of the loftiest and the humblest, as an auxiliary to that higher teaching which is too often forgotten in the turmoil of the world. His intellect would at any rate be free in the course which was before him.[7]

Knight must have worked with extraordinary concentration, for that November he brought out the first instalment of his biography, comprising about half the book.

The following May found him in Edinburgh endeavouring to investigate the 'curious problem' of whether Shakespeare visited Scotland. His discoveries, he admits, were of no great consequence, but they provided the groundwork for speculations in the Life. Knight went on to Glasgow, where he profited from the antiquarian and topographical lore of John Kerr; that knowledge, and the theories founded upon it, would also find expression in the *Shakspere*. The second portion of the book was completed in the same year, and the entire biography published.

# 2

# *Knight's* Shakspere

OPPOSITE his title-page Knight placed two mottoes. The first is Steevens's brutally terse reduction to essentials of biographical knowledge concerning Shakespeare: a man born in Stratford who married and begot children, left for a career as actor and playwright in London, and returned to his place of origin, where he drew up his will, died, and was buried. For his second epigraph Knight selects a passage from Carlyle on Johnson, substituting Shakespeare's name for that of the great lexicographer and critic:

Along with that tombstone information, perhaps even without much of it, we could have liked to gain some answer, in one way or other, to this wide question: What and how was ENGLISH LIFE in *Shakspere's* time; wherein has ours grown to differ therefrom? In other words: What things have we to forget, what to fancy and remember, before we, from such distance, can put ourselves in *Shakspere's* place; and so, in the full sense of the term, understand him, his sayings, and his doings?

This statement affords a clue to Knight's method: he will endeavour 'to associate Shakspere with the circumstances around him, in a manner which may fix them in the mind of the reader by exciting his interest'.[8] Only by so doing can the biographer triumph over the limitations of his data. It is not a novel programme—Drake, a quarter of a century earlier, had attempted to place Shakespeare in the foreground of a picture of the age. But Knight has a different and more sophisticated technique. In the place of separate chapters on the life, the works, and contemporary manners and letters, he weaves these elements together in a sustained narrative of over five hundred pages, rendered seductive by hundreds of facsimiles of documents and pictures of scenes, monuments, and persons. Knight's work is an exercise in popular biography that employs every possible device to ensure the widest readership. Admittedly speculative and fanciful, it is nevertheless based on the facts as he understood them.

The book holds much interest as an experiment in Shakespearian biography. Knight, it is true, cannot lay claim to a distinguished literary style, but he writes with clarity and warmth. His chief strength, however, is constructive. He possesses a sense of geography, and places his subject firmly in the midst of recognizable terrain. Stratford, after all, is a town near other locales: Shottery and Temple Grafton and Snitterfield, Luddington and Fulbrook and Hampton Lucy. The Avon follows a course that Shakespeare presumably knew and which we too can trace. The poet's journey to London must have carried him over the hills dividing Warwickshire from Oxfordshire, through bare downs and the famous park of Woodstock, through Oxford, and on from hamlet to hamlet until he reached the outskirts of London, where the road divided through fields and hedgerows leading to the hills of Hampstead and Highgate on the north and to Westminster on the south. Then we see the great city as it would have appeared to the young Shakespeare. Knight's rambles around Stratford and his intimate experience of London served him well. Previous biographers had mentioned many of the same places, but without Knight's sense of distance and proximity, of the English topography.

Places too have associations which preserve the heritage of the past for the living. Poetic feeling may spring in the beholder of battlefields, castles, and monastic ruins. On the walls of the Guild Chapel in Stratford, Shakespeare as a child may have looked upon an unknown artist's portrayals of saints and martyrdoms, including that of Thomas à Becket. Not far from the town the great thoroughfares of Icknield-way and Foss-way evoked an ancient past. In Worcester Cathedral the boy may have gazed upon the tomb of King John. Within ten miles of Stratford, on the secluded knoll called Blacklow Hill, he perhaps pondered the fate of Gaveston. From this spot Shakespeare could have seen the battlements of Warwick, and, higher than all the rest, Guy's Tower, associated with the Black Dog of Arden. And so on; through these means Knight introduces the historical background without interrupting the story he has to tell.

Every other aspect of the Shakespearian context is worked into the Life in the same way. Someone growing up in Stratford would participate in the festivals of the community: there would be holidays, fairs, wakes, and weddings. Accordingly, the biographer depicts them. Perhaps he was present at Kenilworth in the summer of 1575 when the great Earl of Leicester entertained the Queen with plays and other amusements for nineteen days. He may have seen the Corpus Christi pageants at Coventry. Troupes of actors, we know, visited Stratford in 1574, 1577, 1579, and 1580, and brought with them the fare of the professional London stage. These events permit Knight to introduce some account of the drama prior to Shakespeare. Into all the departments of his biography the author weaves apposite passages from Shakespeare's writings; these serve at once to illustrate the materials and to demonstrate their relevance. Because it has a conscious design adhered to throughout, Knight's *Shakspere* is (except for Drake's grander but less imaginative experiment) the first work on the subject to reflect the art of literary biography.

Knight's decency permeates the biography. Even the loathed Gastrell—he who cut down Shakespeare's mulberry tree and razed New Place—comes in for tolerant if not exactly sympathetic treatment. A prosaic man with strong notions about property, Gastrell merely wanted to have his house and garden to himself, without being pestered by tourists; the wood and stone, after all, were his to do with as he pleased. As for Knight's portrait of the dramatist, we have already encountered the essential elements in the *Penny Cyclopaedia* article published two years previously.[9] Nothing about this Shakespeare and his family would challenge the sensibility of a morally righteous and liberally oriented middle-class reader of the Victorian era.

Knight's Shakespeare, no illiterate and penurious hind, comes of good stock. The father was not a butcher, as Aubrey had jotted down, nor was he (as Malone had contended) a glover. These seemingly contradictory assertions Knight attempted to reconcile by citing a passage from Harrison's *Description of England* (1577), in which the author laments that 'men of great port and countenance are so far from suffering their farmers to have any gain at all, that *they themselves become graziers*, BUTCHERS, *tanners*, SHEEPMASTERS, *woodmen, and denique quid non*, thereby to enrich themselves, and bring all the wealth of the country into their own hands. . . .'[10] Hence the mystery of the butcher is solved and the tradition of the wool-driver explained; if John Shakespeare engaged in the glove trade, it was before he married and inherited an estate. If this man of substance was eventually excused from contributing his share to the poor relief, it was not as a consequence of his own poverty but because he had ceased residing in the borough.

The son was therefore able—*pace* Rowe—to complete his grammar-school training and to acquire that larger instruction in Nature afforded by the fields around Stratford. At home he stored up knowledge for future use by reading the volumes in the library assembled by his well-educated parents; a library that included, as 'the closest companion of the young poet', the costly Chaucer folio of 1542. It was a Protestant household, for the father necessarily took the Oath of Supremacy when he became chief magistrate in 1564. Knight sees a Shakespeare comfortable in his worldly circumstances marrying Anne, 'the prettiest of maidens', raising a family, and experimenting with verse. After the birth of the twins, when he is twenty-one, ambition prompts him to try his luck in London. There he does not hold horses—although, being a practical sort at all times, he may have run a horse-holding concession—nor did he revise the plays of inferior writers. 'The door of the theatre was not a difficult one for him to enter.' He became an actor because he was a dramatist, and not, as others had suggested, the other way round. Finding the stage in a state of chaos and semi-barbarism, he transformed this despised branch of literature, through the force of his genius, into 'a great teacher' of the people. So Knight, the popular educator, interpreted Shakespeare's achievement.

Much in his account runs contrary to the received opinion of his own day, and much would fail to find acceptance. But at a time when the poet's writings were increasingly being ransacked for personal revelations, Knight expresses a minority view that anticipates the modern reluctance to dwell on the mythical sorrows of Shakespeare:

To one who is perfectly familiar with his works, they come more and more to appear as emanations of the pure intellect, totally disconnected from the personal relations of the being which has produced them. Whatever might have been the wordly trials of such a mind, it had within itself the power of rising superior to every calamity.[11]

So too Knight rejects Campbell's biographical interpretation of *The Tempest*: 'Shakspere had to abjure no "rough magic," such as his Prospero abjured.'[12] Knight also has the good sense to dismiss the crab-tree legend, the Lucy doggerel, and the John a Combe epitaph. But in rejecting the deer-poaching tradition, he offers an even more dubious interpretation of the allusion in *The Merry Wives of Windsor*. Shakespeare, he suggests, aims his barb not at the elder Lucy but at his son, who—Knight conjectures wholly without evidence—had attacked the armorial honours granted in 1596 to John Shakespeare. He agrees with Charles Armitage Brown that Shakespeare probably visited Italy, and he devotes a long unpersuasive chapter to showing that the poet also travelled to Scotland.

Other biographies of Shakespeare are not exempt from similar eccentricities; even Malone had his quirks. A more fundamental objection to Knight's method is that he repeatedly crosses the not always distinct boundary between speculation and outright fictionalizing. Sometimes he offers an extenuating 'perhaps' or 'he may have' or 'it is possible', but he does not always make these concessions to normal factual rigour. His Shakespeare shudders; he feels wonder and elation. At his mother's knee he imbibes the cardinal doctrines of the faith. A zealous minister of Stratford lends him a Puritan diatribe against the stage. He exchanges small talk with an ancient minstrel who sings in a tremulous but clear voice. We catch a glimpse of the Shakespeares at their family hearth:

The mother is plying her distaff, or hearing Richard his lesson out of the A B C book. The father and the elder son are each intent upon a book of chronicles, manly reading. Gilbert is teaching his sister Joan *Gamut*, 'the ground of all accord'; whilst the little Anne, a petted child, is wilfully twanging upon the lute which her sister has laid down. A neighbour comes in upon business with the father, who quits the room; and then all the group crowd round their elder brother, who has laid aside his chronicle, to entreat him for a story; the mother even joins in the children's prayer to their gentle brother.[13]

Knight assumes too low a threshold of boredom on the part of his mass audience. His is the first ambitious life of Shakespeare to manifest this uneasy amalgam of the biographer's art with the novelist's. It would not be the last.

Some disapproved. Collier (so we learn from his correspondence) sought to make *his* Life as much as possible the opposite. Nor did the essential weaknesses of Knight's method escape the early reviewers. 'Mr. Knight is not a writer of biography to our liking', decided the *Athenaeum* (2 March 1844); 'he builds hypothesis upon hypothesis, and Towers of Babel that totter and shake, like children's castles made of cards: but he writes agreeably, and if he would confine his fancy within the bounds, he might weave together the disjointed materials of a life, into one continuous and harmonious narrative.' In April 1845 the *Edinburgh Review* described the biography, with fairness, as 'speculative, critical, and not seldom imaginative'. The speculation, in the reviewer's judgement, was essentially sound, but he thought the book too bulky and diffuse.

A reasonable man, Knight did not take the objections to heart—although he could not conceal his displeasure with one critic who summarily dismissed the effort as a 'Burlesque'. Instead he sought to improve the Life when he revised it for a new edition in 1850. He pared down the original 544 pages to 317. He tried to remedy lapses in taste, he reorganized much of his material, and he introduced recently uncovered information. Knight also modified some of his views: he is prepared to admit that John Shakespeare suffered worldly reverses, and that necessity as well as choice led the son to embark on a new course of life. Thus the revised biography has fewer eccentricities than the original, and the excesses of the author's unconventional biographical technique are reduced.

Not a specialist or in any sense a professional scholar, Knight makes mistakes;* yet he brings considerable learning to the cause of popular biography. The editing of Shakespeare lies behind him, and from it he has gained an invaluable intimacy with the entire *œuvre*. He has read fairly widely in Elizabethan literature. At Stratford he has consulted first hand the early records, and found that Malone is not impeccable in his transcriptions. (Knight, however, would have done well to suppress his gloating over the carelessnesses of his great predecessor, for he is not himself a trained palaeographer.) Chalmers, Drake, Wheler, and the rest are all familiar to him. He knows Collier only too well.

How insidiously the forger's poison worked! He had contaminated De Quincey's *Britannica* article; now he would corrupt the lifestream of Knight's *Shakspere*. In his pages Collier is cited more often than any other authority. All the important fabrications receive mention, and thus the documents Collier had described in a volume printed for limited circulation were disseminated far and wide; that, no doubt, was how their creator wished it. But the full extent of his influence on Knight cannot be gauged by the isolated

---

* He thinks that Rowe's Life appeared in 1707 and (along with many others) that the Globe was built in 1593; in text and illustration he furnishes the Blackfriars playhouse with a superfluous curtain, and he shows himself unable to distinguish properly between a public and a private playhouse.

bits of concocted information reproduced in the biography. For this information, naturally enough, furnishes a basis for inferences about Shakespeare's career. The Lost Years vanish because the dramatist had achieved sufficient prominence by the time he was twenty-five to own shares in a major theatrical enterprise; he must have begun his employment some years before then. One thing leads to another. The Blackfriars document makes plausible to Knight the identification of his subject with Spenser's pleasant Willy. The dramatist's professional life is associated largely with the playhouse of Collier's obsession. Around it the entire account of the London years centres; the Globe scarcely matters. So too conjectures about Shakespeare's personal life are affected. From the high assessment on the poet's Clink residence we may assume that it was large enough to accommodate his wife and children, and so we need not entertain uncomfortable thoughts about his separation from his family. Thus it goes; the whole pattern of the narrative is shaped by the Collier forgeries.

It is a sad business. Yet one must forbear to smile condescendingly at the *naïveté* of Knight the amateur: as we have seen, nobody then knew any better. It is true that the Collier materials remain in the last, 'carefully revised' edition of Knight's work, published posthumously as *A Biography of William Shakspere* in the Imperial Shakspere. Knight of course eventually learned the worst—he refers in his memoirs to 'objections of the present day to discoveries of this apocryphal character'—but he was old, his sight had for some time been failing, and in any event he could not expunge Collier without recasting much of the Life. His biography, so imaginatively constructed and written with such zest, is, unlike its subject, for an age rather than for all time; an age which the scientific experts at the British Museum would shortly sweep away. Today Knight's book stands as a curious monument in the landscape of Shakespearian scholarship—in some ways still remarkable, in others mysteriously of another culture.

# 3

# *Halliwell[-Phillipps]: The Cambridge Manuscripts Affair*

By mid-century, relations among the founding quartet of the Shakespeare Society—Collier, Knight, Dyce, and Halliwell—had taken a curious turn. Such harmony as still existed flourished upon deceit: Collier, the guiding spirit, maintained a cordial association with Knight, who did not yet realize that he had been duped. Otherwise there was little good feeling. Friends fell out: Dyce, who in 1840 had dedicated his edition of Middleton to Collier, offered in his *Remarks on Mr. J. P. Collier's and Mr. C. Knight's Editions of Shakespeare*, four years later, a sampling of evidence (the full record, he said, would require volumes) to support his opinion that 'Shakespeare has suffered greatly from both'. Collier had his revenge by assailing his colleague in a later edition of his *Shakespeare*, and Dyce widened the breach with his *Strictures on Mr. Collier's New Edition of Shakespeare, 1858* (1859). Halliwell, who in 1853 would be the first to cast doubt, in print, on Collier's bona fides, referred to Knight in his 1848 *Life of William Shakespeare* as

the only one of late years who has referred to the originals [of Stratford records], but the very slight notice he has taken of them, and the portentous mistakes he has committed in cases where printed copies were not to be found, would appear to show that they were unintelligible to that writer. Malone, with all his errors, possessed some knowledge of palaeography, a science essentially necessary in the investigation of contracted records of the sixteenth century, especially of those written in Latin.[14]

Stung by this public rebuke on the part of a man almost thirty years his junior, and especially by the offensive *portentous*, Knight listed thirty-three of his errors—most of them trifling—in the 1850 revision of his biography. 'Of course I assume that in reading these mouldy and blurred records', he remarks bitterly, 'Mr. Halliwell is infallible in matters of *ys* and *it*. In his case no one can believe in the possibility of a doubt.' The heavy sarcasm of the concluding sentences of his rejoinder reveal how deeply Knight was hurt: 'One has come to enlighten the world, who, by the light of "science," does know that *ibm.* means *ibidem*, and *dnae. dominae*. I am grateful.'[15]

It is Halliwell who now concerns us. The bulk of his work has to do, directly or indirectly, with Shakespeare's life—a subject which eventually crowded out his other interests, and to which he abandoned himself with passionate archaeological zeal. Halliwell investigated every obscure nook of the poet's biography. Of all the nineteenth-century scholars, he made the most enduring contribution to this line of inquiry; one still consults his books. As a man, his character compares strangely with Collier's. Unlike the eminent Victorian, Halliwell did not invent manuscripts. Instead he stole them, and apparently books as well.

The various arts have astonished the world with their child prodigies, and it was perhaps inevitable that one day antiquarianism should produce its Mozart. In this role Halliwell made his meteoric entrance upon the London literary scene in the late 1830s. Born at Sloane Street in Chelsea in 1820, he was the son of a prosperous tradesman who ran a glove-and-hosiery shop in the Haymarket. ('There's a point of resemblance in Shakespeare's personal story and my own personal story,' Halliwell would tell friends in later life. 'His father sold gloves, and so did mine.')* The boy began collecting books and manuscripts, mainly scientific and mathematical, when he was fifteen. At sixteen he was contributing lives of the mathematicians to the *Parthenon*. Before he turned nineteen he had been elected a Fellow of the Society of Antiquaries, and—a more significant honour—a Fellow of the Royal Society. With exquisite presumption Halliwell published, in his nineteenth year, a work entitled *A Few Hints to Novices in Manuscript Literature*. He prepared ten titles for the press in 1840; thirteen in 1841. As a callow youth of twenty he sent an inscribed copy of his two-shilling pamphlet on freemasonry to Collier—more than twice his age—with a brash letter requesting a puff in Collier's newspaper, the *Morning Chronicle*. Collier was taken aback. 'I never puffed myself', he replied, 'nor procured myself to be puffed in my life.'[16] Never daunted, Halliwell was, by the time he achieved his majority, the youngest councillor of the Shakespeare Society, an energetic participant in all the newly formed London literary organizations, and a member of ten antiquarian societies on the Continent and in America.

His university career had surprising repercussions. Halliwell entered Trinity College, Cambridge, as a pensioner in 1837, and during his two terms there the library rarities engaged his attention more than his studies. Recognized as an unusual undergraduate, he was allowed unlimited access to the locked-up manuscripts. In 1838 the Revd Charles Warren, newly appointed as College Librarian, undertook a census of the holdings and discovered to his astonishment that seventeen volumes of catalogued manuscripts had disappeared. When they had vanished could not be determined, but the Library Keeper thought he had seen Halliwell examining the missing volume with the press mark O. 8. 16. Aware of the suspicions, the youth prudently transferred to Jesus College, 'in the hope [he later explained] that he would have a better chance of obtaining a fellowship there than in a large institution like Trinity College'.[17]

He could have used a fellowship. In a letter written in 1839 he complains of being 'overcome with anxiety and deep labyrinths' and being unable to think of anything 'but a way of escape from the Shylock money-lenders of the City of London'.[18] Thus he felt compelled to part with 150 of his valuable manuscripts. These he sold, for the modest sum of fifty pounds, to that great familiar of scholars and collectors, the bookdealer Thomas Rodd. In turn Rodd sold thirty-three of the items, including the lost Cambridge manuscripts, to the British Museum.

Four years later Sir Frederic Madden, who was to take so important a part in the unmasking of Collier, discovered that at least one of the volumes, now in the Egerton Collection, had been abstracted from Trinity College. Informed

---

* He also made this facetious concession: 'The chief difference between us in our respective careers appears to be that he knew how to write plays and I don't.'

by the College authorities of the other missing rarities, Madden broadened his investigation of the Halliwell manuscripts, which bore tell-tale signs of dismemberment and mutilation. Madden's most important discovery concerned volume O. 8. 16: additional leaves had been inserted, the flyleaves at both ends had been removed, and (most suspicious) the pages had been renumbered in Halliwell's hand.

He was informed of the inquiry. 'In the meantime, and until the case has been thoroughly investigated,' the principal librarian wrote to him on 20 January 1845, 'you may perhaps think it proper to abstain from frequenting our Reading Room, or consulting our collections.'[19] Halliwell protested his innocence in letters to the Museum, to Cambridge, and to *The Times*. In a self-exculpatory pamphlet he maintained that the manuscripts had come to him from Denley, a bookseller unfortunately deceased; that this dealer's son recalled seeing one of the suspected items; that he, Halliwell, had offered his collection to Trinity, as he would not have done had he stolen part of it. But Trinity knew nothing of such an offer, Madden frequented Denley's shop and would have noticed the manuscripts had they been there, and the son said that Halliwell had misrepresented him. Others acquainted with the Denleys and the shop challenged Halliwell's defence.[20]

His supporters, however, besieged *The Times*, which printed letters from 'A Lover of Justice', 'A Reader at the Museum', 'A Poor Student', and 'A Hater of Oppression'. 'I have no personal acquaintance with Mr. Halliwell', protested F.S.A., 'but appreciate his constant inquiry and unwearied researches in the field of literary antiquity, as much as I do a fair and honest mode of attack and defence, and despise mere insinuation, so fatal to character and so unlike what we pride ourselves in calling "our national character".' *The Times* ('with its usual *fairness*', sneered Madden) closed its columns to damaging replies, and on 22 November carried an eloquent leader deploring the sacrifice of a scholar caught between two powerful bodies which coldly and civilly contemplated the ruin of his reputation. There could be no doubt of it; Halliwell had the Thunderer on his side.

The Museum now considered instituting criminal proceedings, but was deterred from doing so by Trinity after consultation with its solicitors. Instead they proposed that the College bring an action against the Museum, which would call Halliwell as a witness. He would then be cross-examined, 'a proceeding by which all the facts might be publickly elicited, without that risk which would be attendant on the criminal prosecution'.[21]

But the two parties fell out hopelessly over the strategy of litigation, and had no choice but to drop the whole affair. On 12 June 1846 the Museum authorities informed Halliwell that readmission would be granted him upon application. Soon he was working in the Reading Room again. To the public at large it appeared that his innocence had been vindicated, and he was now free to do the work by virtue of which he would become a great Shake-spearian scholar. The missing Cambridge manuscripts remain in the Egerton Collection of the British Library.*

The Trinity College episode, far from being an isolated misadventure of youth, reflects a deep-seated aberration of character. The evidence points to Halliwell as the offender who stole and disfigured one of the two extant copies of the First Quarto of *Hamlet*. Eventually, like the Cambridge manuscripts, it found its way into the national repository. By then it lacked the title-page, which presumably bore the stamp of the owner, Sir Thomas Phillipps. The Museum had purchased the book from Halliwell, who knew Phillipps's collection and was accused by him of the theft. When he was about sixty Halliwell received a visit from E. V. Lucas and his uncle. Lucas recalls being shown 'many rare books and documents,' by the large man with a white beard, 'but what I chiefly remember of this meeting is his remark that if he ever chanced to see anything in anyone else's house or in a museum that he thought he was more worthy to possess, and (obviously) more able to protect, than its owner, he had no scruples about taking it. This may have been a humorous and idle boast; but he said it.'[22]

Halliwell had a fateful relationship with Phillipps. The latter has already earned a place in this chronicle for his discovery of the most important Shakespearian document to be unearthed during the century: the poet's marriage licence bond, which Phillipps allowed Robert Bell Wheler to publish in the *Gentleman's Magazine*. In the same repository the Baronet came upon the will made by Richard Hathaway's shepherd, Thomas Whittington, in which the testator on 25 March 1601 bequeathed to the poor of Stratford '40s. that is in the hand of Anne Shaxspere, wife unto Mr. Wyllyam Shaxspere or his assigns'. This find Phillipps sat on for a decade before communicating it in April 1847 to the Society of Antiquaries, in whose journal, *Archaeologia*, it appeared the same year. 'His name is now indissolu-bly connected with the biography of our great dramatist,' J. Payne Collier declared appreciatively of Phillipps in the Shakespeare Society's *Papers* in 1847. But time may dissolve even seemingly indissoluble connections: the Shakespearian discoveries receive no mention in the only modern biography of Phillipps: A. N. L. Munby's admirable *Portrait of an Obsession* (1967).[23]

'I wish to have *one Copy of every Book in the World*!!!!!' shrieked the Baronet, a harmful eccentric, and he made a fairly long stride in that direction. It was natural, given their common interests, that he should befriend Halliwell, who dedicated to him the first volume of his *Reliquiae*

---

* Among Halliwell's private papers in the Edinburgh University Library is a note in his hand intriguingly headed, 'The Trinity College MSS. Question'; but it merely points out that College manuscripts were constantly being pillaged. This is very likely true. So far as this writer knows, Halliwell never, either in public or private, disclosed his true role in the affair.

*Antiquae* (edited with Thomas Wright) in 1841. Phillipps invited the youthful bibliophile to his stately manor of Middle Hill, commanding a splendid view of the vale of Evesham. There Halliwell met and fell in love with Phillipps's daughter Henrietta. The father, enchanted with him, endeavoured to promote a match with Henrietta. At first she objected to having a husband chosen for her, but soon melted before the courtship of the tall young man with curly dark hair. Phillipps now turned round and objected to Halliwell, protesting (in a violent fit of temper) that he would not involve his daughter in the misery of being married to a man with so violent a temper.

Friendship between the two collectors soon gave way to coolness, and finally to outright hostility on Phillipps's part. He disapproved of the match not out of sheer perversity (always an element to be reckoned with in his character), nor because of the loss of his *Hamlet*—that was not yet at issue—but for financial reasons: with inherited wealth but heavily in debt as a result of his bibliomanic excesses and other irregularities, Phillipps was disinclined to furnish Henrietta with a dowry, and did not want to have a son-in-law at the mercy of creditors. Moreover, unpleasant rumours about Halliwell's moral character had begun to circulate. Phillipps did what he could to prevent the wedding (he locked up his daughter's clothes), but it nevertheless took place on 9 August 1842. Within the next ten days Phillipps received a poisonous anonymous letter. 'As to his character,' wrote the correspondent, who signed himself Truth, 'ask what character he left at Cambridge. . . . Ask who employed the plea of minority as bar to an action for money lent him? . . . Ask where those valuable books came from that were sold by Sotheby two years ago, to save him from a prison?'[24]

Instead of burning this venomous communication, as a rational man might, Phillipps printed and circulated it. For he now had a new obsessive passion: the destruction of his own son-in-law. He hounded Halliwell relentlessly. He urged the Royal Society to expel its youngest Fellow. When the British Museum purchased a Halliwell collection, Phillipps protested as a Trustee to the Keeper of Printed Books, to the Director, even to the Treasury. If any scraps of information tending to blacken Halliwell's character came into his hands, he printed them. Thus when Hunter accused Halliwell of fraudulently passing off on the Camden Society a transcript ('grossly corrupted') of a purported journal of Sir Henry Wotton, Phillipps gleefully published Hunter's statement.

His daughter made pathetic efforts to soften her father, but he remained implacable; for he was, in truth, a sadistic monster. 'I am sorry for your Children', he concluded, pitilessly, a letter to her during the Trinity College scandal, 'for if he is convicted & transported your Children will not be able to inherit anything.'[25] Twenty-two years later the fierce Lear of Middle Hill thus addressed his own flesh and blood:

It appears that you have no power now Harriet [sic] to effect a reconciliation between us. The consequence will be that you will fall under the Curse destined for all disobedient Children 'unto the 2nd & 3rd Generation.' I understand the Curse has already commenced by your eldest Daughter being halfwitted, & your second is afflicted with a Spinal Complaint. Your husband seems determined that the third shall also incur some misfortune by refusing to make me the Compensation which I understand he once promised.

In such case neither you nor he can expect any blessing of

THOS PHILLIPPS[26]

'My dear Papa,' Henrietta replied with poignant courage and restraint, 'I received your letter last week which amused me not a little.—Some one must have made up their mind to play off a hoax upon you, for it is very certain that none of *my* children show the slightest tendency to being "*half-witted*", and as to the eldest she is now engaged upon a work which her father thinks will be worth printing. . . . They are all good dear children, and a great comfort to us, and though Charlotte has had a weak spine I am very thankful to say the mischief is stopped, and she is better.'[27]

# 4

# *Halliwell[-Phillipps]: Achievements of an Antiquary*

THE situation confronting Halliwell after his marriage might well have daunted a lesser man, but he shrugged off adversity. Then, too, he was in some ways fortunate. He managed to weather the Cambridge manuscripts storm. The Baronet's hatred failed to shake Henrietta's loyalty to her husband: she cut up, pasted, and collated for her 'dear Jamie', made transcripts, read proofs, and prepared indexes. Chronically short of funds during these years, and with a growing family to look after, he yet managed to indulge his collector's passion for books. He would build a library, be forced to sell it off (invariably at a profit, for he was a shrewd buyer), then start all over again. 'For a long time', he recalled two years before his death, 'attempting too much in several directions with insufficient means, and harassed, moreover, by a succession of lawsuits, including two in the Court of Torture,—I mean Chancery,—I was unable to retain my accumulations: and thus it came to pass that bookcase full after bookcase full were disposed of, some by private contract, many under the vibrations of the auctioneer's hammer.'[28] Despite the pressures, his scholarly productivity continued unabated. Halliwell's growing eminence as a man of letters exacerbated the resentment of Sir Thomas, who redoubled his efforts to ruin his son-in-law. The latter carried on undeterred.

Each year saw at least several titles published. Eventually the entries under his name in the British Library Catalogue of Printed Books would occupy twenty-two folio columns. He prepared brief lists, catalogues, and skeleton handlists; he compiled a well-received dictionary of archaic and provincial words, and another of old English plans; he edited romances, poems, dramas, tracts, the letters of British monarchs, popular rhymes and nursery tales; he investigated the history of prices and the antiquities of Islip. From early on, he was drawn to the National Poet. Of Shakespeare, he wrote to Joseph Hunter on 15 January 1842, 'I grow fonder every day'. Shakespearian biography became his master passion, to which he devoted all the resources at his command. When he discovered an old well, choked with refuse, in the Shakespeare country, he had it opened up and the rubbish sifted four times in hopes of unearthing some trifle, maybe a scrap of paper, associated with the poet. 'Up to the point of our visit', a melancholy pilgrim reported in 1875, 'there were no results to show.'[29]

For the historian anxious to maintain some semblance of order and proportion, the sheer mass of Halliwelliana offers an intimidating challenge. The Shakespearian publications alone are of the most diverse and miscellaneous kind. They include facsimiles of documents; extracts from parish registers, subsidy rolls, and other records; a history of New Place; abstracts and copies of indentures illustrative of the fortunes of the Birthplace; memoranda on special topics, such as the Charlecote traditions or the spelling of the poet's name; elaborate editions of the works; and, most important for present purposes, full-scale Lives and outlines of Lives. Halliwell burrowed in every cobwebbed corner which might yield its crumb about the dramatist, his antecedents and descendants, his property, his Stratford associations, and the legends about him that flourished in aftertimes. The Shakespearian relevance of Halliwell's researches is sometimes impressively tangential. Thus he investigated Elizabethan Shakespeares *not* related to the poet; he published the will of Sir Hugh Clopton of New Place and that of John Davenant of The Taverne; he asked, in a pamphlet of 1864, *Was Nicholas ap Roberts That Butcher's Son of Stratford-on-Avon, Who is Recorded by Aubrey as Having Been an Acquaintance of Shakespeare in the Early Days of That Great Poet?* Of all the scholars who so far have entered these pages, Halliwell is the first to show an interest in the history of Shakespearian biographical scholarship: he published Jordan's *Original Memoirs*, Wheler's account of the Birthplace and extracts from his *Collectanea*, Malone's correspondence with Jordan and Davenport. Some of Halliwell's volumes consist merely of random scraps, and he may be driven to apologize (as in the preface to his *New Boke about*

*Shakespeare and Stratford-on-Avon*) for the miscellaneous nature of his offering. This record of ceaseless and not always co-ordinated publication seems to
have been motivated, at least in part, by fear lest he might 'like Malone, die,
leaving the results of much laborious research to be scattered and lost'.[30]

If Halliwell published with discouraging frequency, his individual titles are
too often frustratingly difficult of access; for he would issue them in editions of
one hundred, fifty, thirty, twenty-five, or—not seldom—ten copies only. The
peculiar mode of circulation thus adopted brought protests in Halliwell's own
day. Why not, he was asked, have print runs of five hundred? He defended his
practice by insisting that the collation, transmission, and keeping of accounts
encroached severely upon his time; were he to print greater quantities and
keep large stocks, he would require premises and a staff to carry on the
business. His justification fails to explain why—if he was so eager to
economize on time and labour—he would sometimes print twenty-five copies
and himself take the trouble to destroy all but ten. A letter to W. P. Hunt,
dated 28 March 1868 and now at the University College, London, Library,
suggests that his true motive was a collector's desire to create rarities which
would afterwards command 'marvellous' prices. Such a motive will account
also for the lavish editions he published by subscription at (for then) unheard-
of prices; an enterprise which anticipates the limited editions clubs of modern
times.

For all his eccentricity, and despite the streak of larceny in his character,
Halliwell is the greatest of the nineteenth-century biographers of Shakespeare
in the exacting tradition of factual research which extends from Malone to
Chambers. And it is factual research alone at which he excels; when he
ventures forth from the record office, he is likely to find himself outside his
*métier*: ignorant of the uses of figurative language in verse, Halliwell takes a
metaphorical reference to Tarquin as an allusion to *The Rape of Lucrece*, and
hence to Shakespeare. On his own ground, however, Halliwell has no peer in
the nineteenth century. The achievements of scholarship are notoriously
ephemeral, consigned as they are to oblivion by new information and
improved methods of research. Yet one still turns to Halliwell's books for their
antiquarian riches.

His first major experiment in Shakespearian biography is the maturely
scholarly Life he published when he was only twenty-eight. An astonishing
amount of investigation lies behind the book. In the Stratford archives he has
stirred the dust settled there since the time of Malone. He has pursued the
Shakespeares of Rowington in the Chapter House in Westminster, and at
Longbridge House he has combed the notices of the Shakespeares of Warwickshire in Mr Staunton's manuscripts. The parish registers of Snitterfield,
St Nicholas, and other hamlets have been opened to his trained and
inquisitive eye. The Black Book of Warwick and the King's Silver Book at
Carlton Ride have not eluded him. Originally Halliwell's ambition did not

extend beyond publishing his discoveries separately, but his publisher (so the preface informs us) refused to accept the material unless presented in the form of a consecutive narrative. Thus he was forced into the 'bold and arduous' task of Shakespearian biography.

In practice, however, Halliwell makes only token concessions to the demands of narrative, which he uses primarily to stitch together the texts and summations of records. His style—spare, dry, graceless—rarely rises to the level of banality; the musty air of the muniment room flavours his pages. For Halliwell the biographical quest served mainly to appease his voracious appetite for minute, buried, or out-of-the-way facts. Shakespeare's genealogy stirred his imagination more profoundly than the verse; the kind of challenge to which he rises superbly is the 'very difficult task' of attempting 'to identify the exact position of the room in which Shakespeare was educated'. Halliwell is antiquarianism incarnate, and in Shakespeare he has found a single object for his unrelenting energies. One cannot imagine a more striking contrast than that between his parched and factual Life and the ebullient, sometimes extravagant, reconstruction of the poet's career essayed by Knight five years earlier.

The considerable value of Halliwell's 1848 Life resides in his presentation of the records. He is the first biographer of Shakespeare to appreciate fully the significance of the Stratford documents, and to exploit them systematically. For documents already known he did not trust to printed texts, as almost everyone else was content to do; instead, wherever possible, he made fresh collations with the originals. Frequently he improves upon the transcriptions of his predecessors, and, not least, upon the great Malone. Halliwell raises an eyebrow over the nearly forty errors in Malone's copy of a list of contributors in 1564 to Stratford poor relief. Halliwell's own transcriptions, regrettably, are not impeccable, and he would have done well to keep in check the relish with which he exposes the lapses of others. ('A gentleman who is very sharp on the blunders of other people', a reviewer chided him on a later occasion, 'should be a little more accurate himself.') Sometimes, too, he imparts misinformation, as when he says wrongly that the Blackfriars Gate-house mortgage deed has been lost or misplaced. Nevertheless there can be little doubt that, on the whole, Halliwell raised the scholarly standard of his day.

In many instances his texts are the fullest ever published, before or since. Recognizing that 'for all ordinary purposes brief abstracts would have been sufficient', he none the less devotes a page and a half to a 1592 inventory of the goods and chattels of Henry Field. Halliwell is prompted to do so by 'the desire of rendering my collection of documentary evidence as complete as possible'. The same desire—a scholarly counterpart to the collector's avidity for ownership—leads him to allot four pages to the record of corn and malt in Stratford in February 1598, when Shakespeare is listed as holding ten quarters. Collier was satisfied to quote a few lines from the document; others since have been content with the four words of the pertinent entry ('Wm. Shackespere. x quarters.'). It is perhaps difficult to foresee many occasions when one would wish to peruse the entire text of this unsensational document, but the biographer will not find it without interest, and if he wants it he has it to hand in Halliwell's pages. Having greater general significance is Halliwell's wide knowledge of legal instruments, which enables him to avoid interpretative pitfalls and to cast new light on substantive biographical issues. Thus he sees nothing peculiar about Shakespeare's marriage licence bond, and is able to refute authoritatively Collier's opinion 'that the whole proceeding seems to indicate haste and secrecy'. On the bequest of the second-best bed, Halliwell remarks that 'it was the usual mode of expressing a mark of great affection'—a view he supports tellingly with analogous extracts from other contemporary wills.

In the title of his book Halliwell promises *Many Particulars Respecting the Poet and His Family Never before Published*, and with this Life the documentary record of the Shakespeares is enhanced. John Shakespeare and his fortunes occupy a quarter of the work. The Snitterfield registers, Halliwell reports, show that John had a brother Henry there; this fact supports the likelihood that Richard Shakespeare of Snitterfield was the poet's grandfather. All the entries concerning John Shakespeare in the registry of the Court of Record of Stratford are presented. An item from this source lists him as securing in 1587 a writ of habeas corpus to remove a suit for debt to the Queen's Bench. Halliwell gives the text of the conveyance, recently found in a Birmingham solicitor's office, by which the dramatist's father parted in 1597 with a portion of the Henley Street property for 50s., and he prints for the first time the deed, announced by Collier, showing John and Mary Shakespeare selling property in Snitterfield. Also produced is a hitherto unknown inventory of Ralph Shaw's goods, which the dramatist's father helped to make. As a whole, these documents tend to confirm John Shakespeare's illiteracy* and his declining prosperity in later life. Halliwell also furnishes the text of the will of Richard Hathaway, Anne's father, which he discovered in the Prerogative Office.

Respecting the poet himself, Halliwell establishes the exact date of the purchase of New Place—Easter Term, 1597—from the fine levied on that occasion; a fact that Collier had correctly inferred from other evidence. In the Chamberlain's accounts in Stratford, Halliwell has found the entry of payment to Shakespeare of tenpence for a load of stone in 1598. The same source yields an entry of twenty pence for 'one quart of sack and one quart of claret wine' to slake the thirst of a preacher at New Place in 1614. Halliwell supplies the text (known but not previously printed) of the poet's action

---

* Invariably the documents are signed either with his mark or with a pictogram—a gracefully drawn pair of compasses, the instrument used for making ornamental cuttings in the backs of gloves. The fully literate—even those who had become infirm or senile—tended to make a simple scrawl for their signatures rather than crosses.

against Philip Rogers in 1604 with respect to malt sold and delivered to him on several occasions. A document not before noticed is the fine levied in 1610 on the Old Stratford freehold bought by Shakespeare from the Combes (this is erroneously dated 1611 by Halliwell). He also prints two letters alluding to the dramatist in connection with financial dealings: one, dated 4 November 1598, was written by Abraham Sturley to Richard Quiney; the other, probably from the same year, was by Quiney. Lastly, one should not omit mention of Richard James's letter of around 1625 referring to Shakespeare as having to change the name of Oldcastle to Falstaff; this document, reproduced by Halliwell, had already been printed by him in 1841 in a pamphlet *On the Character of Sir John Falstaff*.

The records unearthed by Halliwell's diligence are uniformly unspectacular, concerned as they are with the family tree and with property and money. This last, Halliwell grants, is 'not a very poetical theme, but one in which the dramatist evidently took a lively interest, having seen, perhaps, that "if money go before, all ways do lie open", and that it is "a good soldier, and will on"'.[31] This unpoetical theme the unpoetical Halliwell would expatiate on in subsequent biographical forays. Not surprising, in the mean time, that his audience should be disappointed with the revelations promised by the title and also by a publisher's prospectus trumpeting IMPORTANT SHAKESPEARIAN DISCOVERIES. No letters here from noble lords, no love poems, no lock of hair. 'We have read and re-read the whole of Mr. Halliwell's book', the *Athenaeum* reviewer complained, not quite fairly, on 8 January 1848, '—and cannot, for the life of us, find more than three new facts in his thick octavo volume.' (One of the three cited Halliwell does not claim to have unearthed.) What the multitude craved only an Ireland or Collier could supply.

'No more is attempted', the biographer informs his readers at the outset, 'beyond placing before the reader an unprejudiced and complete view of every known fact respecting the poet.'[32] Still, from these documentary pages a conception of Shakespeare does emerge; one best summed up in Halliwell's own words: '. . . he was prudent and active in the business of life, judicious and honest, possessing great conversational talent, universally esteemed as gentle and amiable; yet more desirous of accumulating property than increasing his reputation, and occasionally indulging in courses "irregular and wild", but not incompatible with this generic summary.'[33] The 'irregular and wild' course to which Halliwell alludes is, as one would expect, the deer-poaching escapade. Halliwell looks tolerantly upon traditions, applying to them the motto (to which Malone subscribed) *omnis fabula fundatur historia*. Even the absurd anecdote about Queen Elizabeth dropping her glove in the theatre and having it picked up on stage by Shakespeare finds a place in Halliwell's narrative.[34] 'I cannot say who invented this story, but there is no good authority for it', he recognizes; yet he goes on to add a proviso:

'however possible it may be that it is founded on an earlier and less circumstantial tradition.'[35]

He is capable of straining the evidence to favour a wrong-headed notion. Keen to show that the bride's father, Richard Hathaway, stood by when the marriage licence bond was executed, Halliwell points triumphantly to the initials R.H. on the seal. But the initials are probably R.K., and in any event Hathaway by then lay in his grave; his will, which Halliwell himself recovered, was proved on 9 July 1582, four months previous to the bond. Only once, however, does he indulge in pure biographical fancy unsustained by fact (however misinterpreted) or tradition—when he refers to 'the dark eyebrow of Anne Hathaway, a lovely maiden of the picturesque hamlet of Shottery'. The records vouchsafe no information regarding the shade of the lovely maiden's eyebrows, as the *Athenaeum* critic triumphantly observes. Truth is an elusive commodity, and even the biographer who restricts his faith to tangible evidences may find himself cheated. So it was with Halliwell. His readers are given assurance that this Life will contain no notice of the many Shakespearian forgeries which had abused the public, but the promise is unwittingly violated. Occupying pride of place alongside the first page of Halliwell's text is a facsimile of a portion of the Southampton letter. Halliwell knows that the document is suspect—Knight had publicly voiced his doubts, as had Hunter—and he expresses the wish that Collier would publish facsimiles of all the manuscripts relating to Shakespeare among the Ellesmere papers. But on the basis of the palaeographical evidence, derived not from personal inspection of the document but from a reproduction, he vouches for the genuineness of the Southampton epistle. So too Halliwell admits all the rest. The Collier mischief had worked well.

It continues its operation in Halliwell's *New Boke about Shakespeare and Stratford-on-Avon* (1850), the first of the biographical miscellanies he would from time to time issue. This 'boke' has as its chief object the initiation of a 'gradual collection of fac-similes of every document of any real importance respecting Shakespeare'. The documents thus exhibited number among them the genuine marriage licence bond but also, alas, the Dulwich College papers noticing the dramatist, every one of which is a Collier forgery. Facsimiles, Halliwell claims in his preface, 'prove, or invalidate, the authenticity of the documents'. That he was unable to evaluate correctly the evidence thus staring him in the face shows the limited usefulness of facsimiles and also gives the measure of Collier's skill. The other bits and pieces in the collection are more reliable: documents relating to Shakespeare's maternal ancestors, a list of John Shakespeare's attendances at meetings of the Stratford Corporation, a transcript of the poet's will clearly representing the corrections and interlineations. The most interesting item is a writ of 1596, not previously known, citing 'Mr. Shaxpere' (probably John rather than William, although the editor thinks otherwise) in connection with 'one boke' which he had apparently purchased.

Collier is partially exorcized from *The Works of William Shakespeare, the Text Formed from a New Collation of the Early Editions: to Which Are Added All the Original Novels and Tales on Which the Plays Are Founded; Copious Archaeological Annotations on Each Play; an Essay on the Formation of the Text; and a Life of the Poet*, published in sixteen volumes from 1853 to 1865. In this sumptuous edition, lavishly illustrated with woodcuts by F. W. Fairholt, Halliwell has set himself no less a task than that of superseding the Third Variorum of 1821. The set was limited to 150 copies, and sold for £63 (with the plates on ordinary paper) or £84 (with plates on India paper). In addition to a sprinkling of dukes and earls, Halliwell's subscription list included the King of Prussia and Mr Zelotes Hosmer of Boston. For the biography, which alone concerns us here, Halliwell vowed to examine all documents of the slightest biographical importance, and he enlisted the expert assistance of the accomplished palaeographer W. H. Black. While the volume was in press, Halliwell gained access to the Ellesmere papers. The same evening he cancelled at the printers the portion of the Life containing these documents, and also a costly lithographic facsimile of the Southampton letter. 'The consummate skill with which one of these papers is executed,' he alluded ruefully to the latter in his preface, 'the fac-simile exhibiting a fluency of the old character that might deceive the most practised, demonstrates the necessity that existed for scrutinizing the originals.'[36] Halliwell had thus learned his lesson.

Yet if Collier lost one of his subjects, he did not suffer deposition; that remained years off and required the strange eventful history we have already recounted. In a hostile notice of Halliwell's volume, the *Athenaeum* reviewer (2 July 1853) owned to not finding the biographer's views in opposition to the Collier papers much more valuable than his earlier arguments in their favour. His thin skin pierced, Halliwell struck back the same year in an intemperate pamphlet, *Curiosities of Shaksperian Criticism*, hinting broadly that he suspected his anonymous antagonist to be Collier himself—a suggestion disdainfully rebuked in the *Athenaeum* for 13 August. It should not go unremarked that Halliwell makes no attempt to deprive Collier of his entire kingdom. Other forgeries—not at Bridgewater House but in the State Paper Office and Dulwich College—mock his expertise by intruding upon his sober documentary pages.

As the author should have informed his readers but did not, the 1853 biography is not a fresh undertaking but merely a revision of the Life he published five years previously. Although the amount of revision is fairly considerable, for the most part he reproduces the earlier work verbatim. Still, a special interest attaches to some of the new material. There are morsels of fresh information: Halliwell has found a subpoena (dated May 1582) calling John Shakespeare as a witness in a suit over property involving his cousin Robert Webbe of Snitterfield. The document itself is less intriguing than the circumstances of its discovery, which illustrate Halliwell's intuitive grasp of record-office sleuthing. He has found the subpoena where few others would have sought—in a tiny wedge of vellum folded as a knot at the bottom of a string holding a bundle of writs of the Court of Record of Stratford.

Other passages in this Life show a new sensitivity on Halliwell's part to his role of antiquarian biographer. He acknowledges a fashionable tendency to deride labours yielding only recondite facts of no use to 'the philosophical biographer'. But these individually insignificant facts have, he insists, great value in the aggregate. They fill in the portrait of the dramatist suggested in the 1848 Life; Halliwell's unpoetical theme returns with fuller orchestration. Viewed together, the Shakespeare records show that his ambition, like Sir Walter Scott's, moved him to accumulate a fortune and establish a family line, rather than to preserve for posterity the writings which have immortalized him. Halliwell's Shakespeare is the exemplar of thrifty bourgeois virtue. 'No doubt . . . can exist in the mind of any impartial critic, that the great dramatist most carefully attended to his worldly interests; and confirmations of this opinion may be produced from numerous early sources.'[37] Among Halliwell's sources is the familiar bust in Holy Trinity Church. The unphilosophical biographer has paused before this likeness of a plump, well-fed, middle-aged citizen, and he has come away persuaded that his subject was much as other men.[38]

This Shakespeare of the middle-class imagination presents an opposing image to that of the supreme being created by a century of bardolatry; one need only look upon Coleridge's concept, and then on Halliwell's, in which the poet's eye is no longer represented as rolling with a fine frenzy. The image of Shakespeare as stout burgher would reappear in other Victorian biographies. It was fittingly produced by an England in the heyday of a material splendour largely achieved (it was felt) by those virtues of prudence and industry which Shakespeare's career exemplified.*

Astonishingly, the enterprise of editing the complete *Works* on a monumental scale left Halliwell with energy for other activities. In the columns of periodicals he announced his documentary finds. The *Athenaeum* for 30 April 1864 carried word of his discovery, in a large manuscript volume, of the allotment to Shakespeare of $4\frac{1}{2}$ yards of 'skarlet red cloth' on the occasion of the procession of James I through London in 1604. The word 'skarlet' does not in fact appear in the manuscript; to Halliwell we presumably owe this

---

* This view reaches a *reductio ad absurdum* in the pages of that gentleman of engagingly Dickensian name, Samuel Smiles, who collected for the benefit of the Lower Orders biographical facts about poor men who made good by dint of their own efforts. Of Shakespeare, the pygmy doctor of Edinburgh writes in *Self-Help; with Illustrations of Character and Conduct* (1859): '. . . he must have been a close student, and a hard worker', who prided himself more on his practical qualities than on his literary attainments. 'It is certain . . .', he concludes, with genial complacence, 'that he prospered in his business, and realized sufficient to enable him to retire upon a competency to his native town of Stratford-upon-Avon' (p. 191). *Self-Help* went through a staggering number of editions and translations. Benthamite utilitarianism was then in full bloom.

improvement. His discovery that Shakespeare and his fellows attended on the Spanish ambassador at Somerset House, also in 1604, was reported in the Literary Gossip columns of the *Athenaeum* for 8 July 1871. In *Notes and Queries* (26 August 1871) Walter Thornbury urged caution in accepting Shakespeare's name in connection with the occasion, because of the interpolations of forgers such as Ireland in authentic records. Hesitation was uncalled for, Halliwell assured readers of *Notes and Queries* on 2 September; his find was 'beyond suspicion'. The episode is peculiar in a minor way, for although Shakespeare probably took part in the ceremony at Somerset House, his name does not actually appear in the Declared Accounts of the Treasurer of the King's Chamber. One can never *entirely* trust in Halliwell's bona fides.

During these years he published other editions, as well as facsimiles and catalogues. His most useful work, however, was accomplished in Stratford. In 1863 he initiated the successful movement for the purchase of the sites of Shakespeare's residence and gardens at New Place with a view to making them a public trust in municipal hands. In the same year, at the invitation of the Corporation, he completed arranging and calendaring almost five thousand separate documents dating from the thirteenth century until 1750; documents which had been thrown confusedly together in boxes in a room of the Birthplace on Henley Street. This work he did without fee, in the same spirit that he supported out of his own pocket the campaign to purchase New Place. The results of his labours were made available to the outside world in a series of volumes that appeared in rapid succession. The most important is *A Descriptive Calendar of the Ancient Manuscripts and Records in the Possession of the Corporation of Stratford-upon-Avon; Including Notices of Shakespeare and His Family, and of Several Persons Connected with the Poet*, published by Halliwell at his own expense in 1863 in a folio volume that came to 467 pages. Between 1864 and 1867 Halliwell published an enormous quantity of material (seven titles in all) which makes available a mine of local records of much value to the historian or biographer—especially the Shakespearian biographer, whose needs Halliwell has most in mind. In the next century an edition of *The Minutes and Accounts of the Corporation of Stratford-upon-Avon and Other Records*, prepared by Richard Savage and E. I. Fripp for the Dugdale Society, would partly supersede Halliwell. He offers extracts; the new transcription would make available the entire municipal record without omission or abbreviation, but would cover a briefer period. Few in the whole history of English literary scholarship can have matched Halliwell's energy. 'I wonder at your enthusiasm and physical power', an admirer wrote to him. 'I should have thought you would have been heartily sick of the subject.'[39]

Work must have dominated all other considerations in a life of such plentiful accomplishment. Even his vast accumulations of books, pictures, documents, and relics, the acquisition and possession of which gave Halliwell such evident pleasure, served chiefly as practical libraries; hence his willingness to part with his collections once he had completed the project for which he had purchased them. The antiquary's compulsion to record minutiae prompted Halliwell in 1870 to issue a curious (to use his favourite word) leaflet, without title, describing the furnishings of his flat at 11 Tregunter Road in South Kensington, where he had moved in 1867 from West Brompton. Halliwell tells how, to economize on time, he has separated and distributed objects among 328 drawers in his three rooms. On the west side of the Biography Room he stored his papers on various subjects: 'The Crabtree', 'Portraits—Personal Appearance', 'The Will.—Illness and Death.—Temperance', etc. The drawers on the east side held paper, other supplies, and his numerous prospectuses. Halliwell spares us no details: the upper case of drawers in the ante-room contained such fascinating items as study-candle ends, new pins, old pins, and wafers.

In the same year that he published this leaflet, Halliwell speaks of having 'abandoned the critical study of the text of Shakespeare'. Of the treasures which in his youth he had picked up from the stalls outside bookshops Halliwell would say, 'Plentiful as blackberries in the blackberry season, they were almost as cheap.' They were no longer so, and from February 1872 he restricted his competitive resources to assembling materials illustrating the life of Shakespeare.

In that month Sir Thomas Phillipps died. His daughter recoiled in horror at reports of the Baronet's conduct during his last days—'Papa never spoke to me', she lamented in the privacy of her diary, 'and carried his revenge beyond the grave.' His decease was commemorated by an obituary of 'the greatest book collector of modern times' in the *Athenaeum* (10 February) that did not so much as hint at what might euphemistically be described as the eccentricities of the Baronet. That this handsome tribute should have been written by the object of Sir Thomas's unrelenting hatred, James Orchard Halliwell, is one of the mysteries of the human heart. '*De mortuis . . .*', perhaps he thought. In his will, Phillipps remembered his son-in-law characteristically by expressly forbidding him and his wife, as well as any Roman Catholic, from entering Thirlestaine House, Cheltenham, the great mansion literally filled with the books and manuscripts brought together by an overpowering compulsion. There he passed the last decade of his long life (Sir Thomas was born in 1772). The scandalous clause (which became public knowledge) puzzled Henrietta. 'I must say that neither James or myself were ever the slightest way inclined to Romanism & I never will be,' she wrote in her diary. 'All our children are brought up in the principles of the Christian religion.'[40]

Sir Thomas could not, however, prevent the Halliwells from inheriting Middle Hill. Justly apprehensive about his son's extravagance, the Baronet's father had stipulated in his will that, while Thomas Phillipps Junior should enjoy the family estates and their income during his lifetime, they must thereafter devolve upon his children and their issue. To ensure preservation of his name, the husbands of the respective daughters who might come into the inheritance were obliged to 'use assume and take upon themselves respectively the Surname of Phillipps'. Sir Thomas endeavoured to circumvent the provisions of his father's will, but to do so required Halliwell's consent, and this (not surprisingly) was unforthcoming. On the very day that word of the collector's death reached his detested son-in-law at Weston-super-Mare, the latter rushed up to town, and before evening fell he had given his solicitor instructions for the preparation of the disentailing deed. Now by royal letters-patent he adopted the additional surname Phillipps, and took over the management of the ancestral properties.

Middle Hill stood desolate and dilapidated. The banisters were gone, the windows were without a single pane of glass, the boards on the library floor were swollen with water; cattle wandered aimlessly through the rooms on the ground floor. But, with characteristic energy, Halliwell threw himself into the work of repair. A few months after Sir Thomas's death he could command £80,000 on his interest in the Middle Hill estates. 'Well done, Mr Halliwell Phillipps!' Sir Frederic Madden, who knew the Baronet only too well, exulted in his private journal. 'This looks like business indeed! Oh, ghost of Sir Thomas, how you must gnash your teeth, to see what is taking place!'[41]

That summer tragedy blighted the triumph. The accident that befell Henrietta as she rode her high chestnut horse she described in her diary shortly after the event: '. . . we rode by Offington & home by Tarring but unfortunately just as we got to Tarring a bird flew out of the hedge & frightened my horse and it bolted with me & in the village as I was turning a corner a large cart with brushes was in the way & not having room to pass at the rate the horse was going, it slipped down on its hind legs & threw me into the road.'[42] She recovered, or seemed to recover; she walked, and bathed in the sea again; she humbly thanked God for His mercy. But inexorably she disintegrated—softening of the brain, the doctors called it—and her last years were consigned to invalidism and mental dissolution. The ghastliness of her situation is conveyed by the terse observation of an acquaintance. 'Her screams are . . . more feeble', he wrote, near the end, 'tho' still very loud at times.'

For a time her husband soldiered on. In 1874 he published his *Illustrations of the Life of Shakespeare*. Seven years previously he had invited the public to subscribe to a series of folio volumes which would 'accumulate a collection of materials illustrative of the details of Shakespeare's Life and Works, . . . interpreted by the aid of contemporary documents and books, and by an elaborate system of truthful artistic illustration'.[43] The *Illustrations* comprised the first instalment in this encompassing programme; there would never be a second. Arranged according to the author's whim, without regard to chronology or any other rational scheme, these essays represent (he says in his preface) 'merely one of the amusements of the declining years of life'.

In the same preface Halliwell-Phillipps (as he became known to posterity and as we shall hereafter refer to him) clings to his view that Shakespeare wrote not for immortality but 'as a matter of business', that he undertook authorship 'as the readiest path to material advancement'. The papers themselves, however, relate not to such broad interpretative considerations but, rather, to more specific issues—the commencement of Shakespeare's professional career, New Place, the mulberry tree—amenable to documentary treatment. A passage on Shakespeare's manuscripts rekindles the hope, voiced by Malone, that autograph remains, perhaps awaiting discovery in some obscure corner behind a wainscot, might one day come to light. Has anybody, Halliwell-Phillipps wonders, thought of inspecting Abington Hall, where Shakespeare's granddaughter resided, and searching behind the elaborate panelling for treasures that the wealth of the Indies would purchase cheaply? Shakespeare's literary correspondence perhaps, or holographs of his published (or even unpublished) plays. Finds of such an order always eluded Halliwell-Phillipps, but he has much to offer in these essays written for his private amusement. His discourse on the playhouses and companies is the soundest up to that time, and in it the author establishes a fact of prime significance: the Globe, he proves, was erected in 1599—not 1594, as Malone had conjectured, and as subsequent students had uninquisitively accepted.

In the *Illustrations* too Halliwell-Phillipps makes public his most important Shakespearian discoveries. A summer spent in the murky record room of the Lord Chamberlain's office has yielded a memorandum, in the accounts of the Treasurer of the Chamber, registering payment 'to William Kempe, William Shakespeare, & Richarde Burbage, servants to the Lord Chamberlayne', for two comedies performed before the Queen at Greenwich in December 1594. This scrap furnishes the earliest official record of the dramatist's name; it identifies his acting company and establishes his prominence in the early 1590s. In the same repository Halliwell-Phillipps came upon a thin manuscript volume bearing the title, *Presentations and Warrants in the Years 1631, 1632, &c.* Here, among other contemporary transcripts of papers passing through the Chamberlain's office, he found the *Answer* of Cuthbert Burbage, his sister-in-law Winifred, and her son William to a petition regarding shares in the Globe and Blackfriars theatres. The affidavit recounts how, with borrowed money, Cuthbert's father James built The Theatre on leased ground, 'by which means the landlord and he had a great suit in law, and by his death the like troubles fell on us, his sons; we then bethought us of altering from thence, and at like expense built the Globe with more sums of money taken up at interest, which lay heavy on us many years, and to ourselves we joined those deserving men, Shakspere, Hemings, Condall,

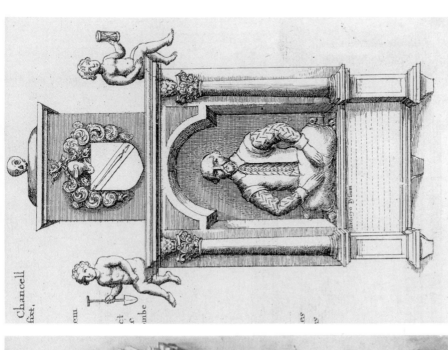

1. Shakespeare monument,
Holy Trinity Church, Stratford-upon-Avon.

2. The Dugdale engraving of the Shakespeare monument.
   *Folger Shakespeare Library.*

3. 'A Sculptor's Workshop, Stratford-upon-Avon, 1617' by H. Wallis, 1857.
*From the RSC Collection, with the permission of the Governors of the Royal Shakespeare Theatre* ©.

28. Delia Bacon.
*Folger Shakespeare Library.*

Very gratefully yours
Delia Bacon.

29. Dr and Mrs Wallace in the Public Record Office.
*Henry E. Huntington Library and Art Gallery.*

30. Sir Edmund Chambers.
*National Portrait Gallery, London.*

Philips and others, partners in the profits of that they call the House.' Little wonder that Halliwell-Phillipps could hardly believe his eyes. For the first time a document had been recovered expressly mentioning Shakespeare in relation to the management of the playhouses of the King's Men. 'These important evidences', the Literary Gossip column of the *Athenaeum* enthused, 'contradict all recent theories and opinions respecting Shakespeare's business connexion with the theatres.' Had Halliwell-Phillipps retrieved nothing else, these finds would have assured him a place of honour in this narrative.

He tried to carry on with a second instalment of the *Illustrations*, but had to give up—his powers of mental labour, he discovered to his horror, were failing. 'The hopeless state of my Wife's malady', he admitted, piteously, 'has changed all my plans & all my aspirations.'[44] His publishers were asked to withdraw from circulation part one of the *Illustrations*. To bookdealers he sent a printed notice requesting them to remove his name from their mailing lists: he was no longer collecting. His friends he informed individually that he was abandoning his scholarly activities altogether, even to the extent of avoiding literary correspondence. 'Henceforth', he wrote to Professor Blackie in Edinburgh, 'for the remainder of my life my tastes & pursuits are rural and agricultural'; and he added, with bitterness, 'There is no real encouragement for the literature of research in this country & I fear never will be.'[45] Given his frame of mind, it is not surprising that he should now turn in revulsion against himself and all his works. A request for some of his books drew this reply:

How *can* you think at my age that I am such a conceited ass as to keep a lot of my rubbishy books, every one nearly of which I am heartily ashamed of & only wonder at my gooseosity & insolence in printing them. . . . I thought you were wiser than to collect such trash.[46]

No longer able to cope with his situation, he sought escape by day through drink; for sleep he came to depend upon narcotics. These anodynes in combination wrecked his vigorous constitution. Reduced almost to nervous prostration, he was told by his physicians that he must free himself from drugs and spend at least nine months of the year breathing sea air, if he was to survive.

# 5

# *Halliwell-Phillipps: The Last Years*

THUS it came to pass that Halliwell-Phillipps forsook London for the Sussex Downs. In 1877–8 he bought thirteen acres on one of the highest elevations of the Downs, overlooking Brighton and Preston, intending there to build a large house of brick and stone. As a stopgap he put up a timber bungalow; then, abandoning his more ambitious architectural design, he gradually added a number of rooms, galleries, and outhouses, also of wood but with an outer casting of vulcanized sheet-iron. Morbidly fearful of dogs and fires, Halliwell-Phillipps bounded his property with a stout oaken fence and installed sheet metal between floors and carpets to prevent the possible spread of a conflagration in his tinder-box. This low, shambling structure with its many tall iron chimneys he described as his 'quaint wigwam'; neighbours referred to it as Phillipps's Folly.

In 1879 his wife died, and shortly thereafter he remarried. The bride, a Stratford solicitor's daughter, had a pale face and lustrous black eyes; she was less than half her husband's age, but the match was (so far as we can tell) successful. Halliwell-Phillipps now returned to his literary pursuits. He set up a system of numbered drawers similar to his arrangements at Tregunter Road. From his many-windowed study, with its fireproof safes and glazed bookcases, he could look out upon the long line of the English Channel. Here, at Hollingbury Copse near the sea, Halliwell-Phillipps regained his health and devoted the last years of his life to the subject now of consuming interest to him: Shakespeare's life. His recovery testifies eloquently to the recuperative powers of the human spirit.

He collected again. A printed notice he took to attaching to his letters begins, 'Few favours are greater than those which are conferred by so many of my friends in giving me the earliest intimation that reaches them of purchasable articles of Shakespearean interest; and they will, I feel sure, kindly excuse my drawing their attention to the following lists. . . .' As Halliwell-Phillipps claimed with greater accuracy than diffidence, his strange bungalow came to house '*more record and artistic evidences connected with the personal history of the Great Dramatist than are to be found in any other of the World's libraries*'.[47] In addition to books, the collection included the Droeshout engraving in its exceedingly rare first state, six title-deeds respecting New Place in the poet's lifetime, and a vast assemblage of drawings and engravings illustrative of Shakespearian biography. In the declining hours of the afternoon Halliwell-Phillipps would walk the Downs in his tattered grey coat and scarecrow hat, or dispense lavish hospitality to any stranger with like interests who visited his wigwam. But the early part of his day—from five in the morning until 2 p.m., with a short interval for breakfast—was sacred to the work he pursued surrounded by his rarities.

In his study Halliwell-Phillipps assembled a collection of large scrap-books in which he pasted notes, letters, newspaper and magazine clippings, original documents, transcripts, and excerpts from books. The curious student turning the pages of these fifty-six scrap-books, now deposited at the Folger Shakespeare Library in Washington, DC, starts in involuntary horror at finding bits and pieces from rare old folios and quartos: a single miscellaneous volume (no. 7) contains clippings from the blackletter *Book of the Use of Silk*, *The Spanish Tragedy*, the anonymous *King Leir*, Brome's *The Antipodes* and *A Jovial Crew*, the 1641 Jonson Folio, and many more rarities. Another instance of Halliwell-Phillipps's vandalism? Perhaps, but in those days one could still obtain defective copies of old books for pennies at the bookstalls.* The scrap-books nevertheless testify to Halliwell-Phillipps's willingness to sacrifice all things, old or new, to his Shakespearian passion.

The title-page of the *Illustrations* bears an engraved sketch of the remains of New Place: a few foundation stones stand on the level ground as the only evidence of the ancient edifice. To Halliwell-Phillipps this image had the force of metaphor; it was symbolic of the fragments of the poet's history that had escaped devouring time. The same picture had adorned the title-page of Halliwell-Phillipps's historical account of the house; it would be reproduced in his most imposing monument, the *Outlines of the Life of Shakespeare*, which went through seven editions in his lifetime. In the *Outlines* he would also reuse materials, and even the phraseology, from the *Illustrations*. In its first incarnation the *Outlines*, published in 1881 'for presents only', ran to 192 pages; before proceeding further he hoped to elicit opinions from his literary acquaintances and correspondents. The response must have gratified him, for the *Outlines* grew until in 1887 it required two oversized volumes of 848 pages.

His industry puts lesser mortals to shame. Hopeful of finding traces of Shakespeare's footsteps, Halliwell-Phillipps personally examined the records of Banbury, Barnstaple, Bewdley, Bridport, Bristol, Cambridge, Canterbury, Coventry, Dorchester, Dover, Faversham, Folkestone, Hythe, Kingston-on-Thames, Leicester, Leominster, Lewes, Ludlow, Lyme Regis, Maidstone, New Romney, Newcastle-upon-Tyne, Newport, Queenborough, Rye, Sandwich, Shrewsbury, Southampton, Warwick, Weymouth, Winchelsea, and York. Alas for patient effort; he awakened no drowsy hamlet with a notice of the National Poet. But, ever the optimist, he expresses satisfaction with having uncovered 'curious material of an unsuspected nature respecting his company and theatrical surroundings'.

Viewed against the background of these heroic labours, Halliwell-Phillipps's narrative of Shakespeare's life for the first edition of his *Outlines* seems disappointingly slight. Coming to under a hundred pages, it dwells mostly on factual particulars, matters of chronological order and the like, concerning the plays. Notable is the author's championship of the integrity of the *Henry VI* trilogy and *Titus Andronicus*; later scholarship would vindicate

---

* I understand, however, from Giles E. Dawson, formerly Curator of Books and Manuscripts at the Folger Shakespeare Library, that an interior leaf from the first edition of Ralegh's *History of the World*, pasted by Halliwell-Phillipps in his scrap-books, derives from an otherwise apparently perfect copy that he sold to the Earl of Warwick, whose collection now belongs to the Folger.

these unfashionable views. The only new record of interest, found by Halliwell-Phillipps in the register of the consistory court of Worcester, is an action for defamation taken in 1613 by Shakespeare's elder daughter, Susanna Hall, against John Lane Junior of Stratford who had reported that the plaintiff 'had the running of the reins and had been naught with Rafe Smith at John Palmer'. (To have 'the running of the reins' meant to suffer from gonorrhoea, 'reins' signifying kidneys or loins; but Halliwell-Phillipps offers no explanation for the phrase.) It is proper, however, to judge the *Outlines* not on the basis of the first edition—after all, avowedly a trial balloon—but on the seventh, his ultimate statement on the subject of the poet's biography.

In the preface, written five years before his death, a note of sadness enters into Halliwell-Phillipps's reflections on what a lifetime of research had accomplished. 'No matter what pains a biographer may take to furnish his store', he sighs, 'the result will not present a more brilliant appearance than did the needy shop of Romeo's apothecary. He is baffled in every quarter by the want of graphical documents, and little more can be accomplished beyond a very imperfect sketch or outline of the material features of the poet's career.'[48] The plain and unobtrusive style at which the author aims, although inelegant, is at any rate serviceable, and an improvement upon the ungainly, sometimes unsyntactical, prose of his earlier Lives. Collier has at last been totally expunged, but one notes with interest that Halliwell-Phillipps could not entirely forbear romantic elaboration in his last biography, the principal intention of which is 'to furnish the reader with an authentic collection of all the known facts respecting the personal and literary history of the great dramatist'. The documentary narrator sees Shakespeare's mother as an 'unlettered damsel' who enchanted her son with recitals of romances of knights and fairies, and provided him with an image of 'perfect womanhood'. If in Wincot, we are told in a later passage, the poet had been observed 'to pass Mrs. Hacket's door without taking a sip of ale with the vigorous landlady, he might perhaps no longer have been enrolled amongst the members of good-fellowship'.[49] No muniment room could have furnished Halliwell-Phillipps with this intelligence.

The biography as a whole shows that its compiler's ideas had changed little over the years. It presents a view of Shakespeare still current in the late nineteenth century: a view haunted by traditions which the next era would exile. This Shakespeare was untimely ripped from the Stratford free school to dwell 'with illiterate relatives in a bookless neighbourhood'. While still a boy he was apprenticed to a butcher. For recreation he poached deer and other small game, was prosecuted by an irate justice of the peace, found himself— the penniless son of an impoverished father—without influential friends, wrote a bitter lampoon, fled to the metropolis, and years later vented his spleen by caricaturing the knight of Charlecote in *The Merry Wives of Windsor*. In London he effected his transition from the rural stable to the urban playhouse via the intermediate phase of holding the horses of patrons. Commencing his professional career as an actor, Shakespeare achieved fame and considerable fortune by writing plays 'under the domination of a commercial spirit'. At Bidford (it seems) he participated in a drunken revelry remembered in a merry jest, and at Stratford he planted the mulberry tree which would provide an endless stream of souvenirs for pilgrims. There, in his twilight years, he enjoyed the estate secured by his practical wisdom, and made a will that left his wife, as an affectionate token, his second-best bed. This summary is no doubt somewhat unfair, ignoring as it does the mass of verifiable data in the Life; but these extra-factual materials are certainly present. They serve to make Halliwell-Phillipps's account seem so foreign, in many essentials, to modern biographical treatments, and at the same time not too far removed from the sketches of the early encyclopaedists. Were the narrative itself the only—or even the primary—feature of the last *Outlines*, we would consult the work chiefly for its authoritative restatement of a conception of Shakespeare rendered quaint by time.

But Halliwell-Phillipps put much more into his final large-scale work, making of it, indeed, a distillation of all his researches. Mere enumeration must serve to indicate the range, if not the informational riches, of the series of appendices which comprise the greater part of the *Outlines*. Unaccountably there is no table of contents, but the compiler furnishes an analytical index— a feature conspicuously absent from his earlier biographies. Halliwell-Phillipps provides the texts of documents relating to the Blackfriars and Globe playhouses, an account of The Theatre and the Curtain, notes on the Birthplace, estate records relating to the properties in which Shakespeare had interests, various early versions of the scandalous Davenant legend, extracts from the Stratford parish registers, wills and inventories, the prefatory matter to the First Folio, early biographical notices of the poet, documents pertaining to Shakespeare's litigation on money matters, early records of performances of the dramatist's plays, notes on Stratford neighbours, an illustrated history of New Place, an account of John Shakespeare the shoemaker, contemporary allusions naming the poet, the text of his will, documents relating to the Snitterfield properties of the Ardens and (later) the Shakespeares, an excursus on both ancestral families (with the pertinent documents), and all the numerous references to John Shakespeare. A section of voluminous illustrative notes follows the appendices.

Here the most interesting entry (vol. ii, p. 396) concerns the marriage licence bond, for Halliwell-Phillipps makes public for the first time an important and perplexing record that the Revd T. P. Wadley had drawn to his attention some years previously. The Episcopal Register of Worcester for 27 November 1582—the day before the signing of the bond—records the issuance of a licence permitting the wedlock of William 'Shaxpere' and Anne

Whateley of Temple Grafton, a little village five miles west of Stratford. This Shakespeare is presumably the poet, but Anne Whateley enters oddly into the arrangements. Halliwell-Phillipps remains unruffled: the scribe, he suggests, miswrote the Register entry, and anyway the bond has incomparably higher authority than the Register. He is probably right about the source of confusion, and he is certainly correct in allowing greater authority to the bond, which is supported by the tradition concerning the name of Shakespeare's wife reported by Rowe in his 1709 *Life*. Later scholars, however, would not view the problem with so much equanimity.

The *Outlines* is Halliwell-Phillipps's *summa*, a species of biographical variorum that became the primary source-book for all subsequent Lives. It does for the nineteenth century what Sir Edmund Chambers would accomplish for the twentieth with his *William Shakespeare: A Study of Facts and Problems*. In Chambers's pages no previous work of scholarship is cited more often than the *Outlines*, which the later effort does not entirely supersede. That is perhaps the highest compliment one can pay it.

Throughout his long professional career Halliwell-Phillipps tried to live peaceably and get on with his work; unlike some other eminent Shakespearians, he did not court controversy. But again and again events conspired, with or without his complicity, to rob him of his quiet. In youth there was the Cambridge manuscripts affair which he brought upon himself. His middle years were disturbed by the prolonged defensive struggle with his father-in-law. And in old age he found himself involved in two acrimonious quarrels not of his own choosing.

The first is the more trivial. In 1880 Swinburne dedicated his *Study of Shakespeare* to Halliwell-Phillipps in recognition of the latter's services to the poet. The *Study* included as an appendix a 'Report of the Proceedings in the First Anniversary Session of the Newest Shakespeare Society', a gleeful burlesque of the minutes of the New Shakspere Society, with the founder of which—Frederick James Furnivall—Swinburne had been feuding. In a letter (dated 23 September 1879) he informed Halliwell-Phillipps in advance of publication that the book attacked the Society 'with some vigour of expression'; but, although he had corresponded cordially with Furnivall, the dedicatee raised no objection. Furnivall, a man whose violence of temper matched in magnitude his prodigious scholarly productivity, reacted predictably. In a foreword to a facsimile edition of *Hamlet* (1880), he alludes to Halliwell-Phillipps as a leading member of the firm of Pigsbrook (i.e., Swinburne) and Co.; he refers moreover to the venerable biographer's 'porcine vagaries' and his observations held up 'on the prongs of a dung-fork'. Understandably infuriated by these discourtesies, their recipient protested to the president of the New Shakspere Society, Robert Browning, who replied mildly that his position was purely honorary, that he had never attended one of the Society's meetings, and that anyway he failed altogether to understand the nature of the controversy. All this no doubt is true.

Dissatisfied, Halliwell-Phillipps wrote to Browning again, then published the entire correspondence. Resignations from the Society followed. Furnivall sent the departing members a printed letter that illustrates not only the quality of his rage but also the paranoiac streak in his make-up. 'I regard as an impertinence', he fumed, 'your intrusion of yourselves into a dispute declared by me to be private between Mr. Hl.-Phillipps and myself, and I am now glad to be rid of you, whose return for the faithful work I have given you (and others), is this present censorious caballing against me.' The same year (1881) the 'incomparable blackguard', as Swinburne termed him, thundered against Swinburne and Halliwell-Phillipps in a pamphlet, *The "Co." of Pigsbrook & Co.*, as ineffective as it was vituperative. Dowden tried to serve as peacemaker, then withdrew from the whole sorry fray; 'I mean to stand aloof', he wrote sadly to Halliwell-Phillipps on 22 February 1881, 'having failed in any degree to do good.' Anyway Halliwell-Phillipps now felt easier on the subject of Furnivall, for a stay in London persuaded him of (as he wrote to Justin Winsor on 16 March) 'the universal contempt & ridicule which surround him and the New Shakspere Society'.[50]

The other episode had more fateful repercussions. In Stratford in the early 1880s ill will towards Halliwell-Phillipps arose over what was regarded as his proprietary attitude respecting the documents, some of which he was seeking to autotype at his own expense; it was even suggested that he had left records in a decaying and dangerous condition. His pamphlet, *The Stratford Records and the Shakespeare Autotypes* (1883), exacerbated the strain by seeming to charge the Shakespeare Trustees with dereliction of duty in not competing at a recent sale for a set of documents relating to Stratford; documents which Halliwell-Phillipps bid for successfully. This imputation he denied making, and characteristically he printed the correspondence. Persuading himself that a leading townsman, C. E. Flower, had maligned him, he took to addressing letters to 'Mr. Flower, Common Brewer, Stratford-on-Avon', and on one occasion replied to a question by sending him a photograph of himself with thumb to nose and fingers outstretched.* This did not help matters. No doubt there was over-sensitivity on both sides, but the local authorities would have done well to humour their erstwhile benefactor, for in his will he had made the Corporation heir to his priceless Shakespearian collections. George Boyden, the proprietor of the *Stratford-on-Avon Herald*, indeed tried to arrange for a meeting of reconciliation, but it never came off. Halliwell-Phillipps failed to make the retraction that Flower demanded, and the Corporation failed to placate the scholar by conferring upon him the honorary freedom of the

---

* He had used the same photograph previously when urged to resume his scholarly pursuits.

Borough which he coveted. The aftermath was melancholy. With accusations of rudeness and ingratitude, Halliwell-Phillipps revoked the bequest and turned his back forever on the town that he had loved and that had furnished him with the materials for his most notable contributions to scholarship.

Towards the close of a full life directed overwhelmingly to Shakespearian research, he thought of those scholars, the eminent dead, with whom he had been associated. In a letter written on 10 December 1884 he recalled 'that dear old room' in which, a mere youth, he had been welcomed by Dyce a half-century earlier. 'I was good friends all their lives', he goes on, 'with Dyce, Collier, Harness, Singer & Knight, & I believe with now every English & American editor save only F. J. Furnivall.'[51] Elsewhere, glancing back at his work, he realized (in his own metaphor) that he had dredged only a few corners of the great river; plenty of fish remained. Was there a chance of recovering the book or roll of the inhabitants of Stratford-upon-Avon ordered by the Corporation in 1603 for a benevolence towards the city of Geneva? Did records exist noticing Shakespeare as a Groom of the Chamber to James I? Such questions passed through Halliwell-Phillipps's mind. Much still remained to be done with the Indices Finium, the uncalendared State Papers, and with the records of litigation belonging to the various courts. But the most awesome challenge was the great repository of documents in Chancery Lane. If only the government were considerate enough to transplant the Public Record Office to the Brighton Downs—notices of Shakespeare must lie buried among the millions of papers there. So he guessed correctly in 1884 in *Memoranda, Intended for the Use of Amateurs, Who Are Sufficiently Interested in the Pursuit, to Make Searches in the Public Record Office*. It would remain for a remarkable American in the next century to make the major discoveries that Halliwell-Phillipps knew still lay ahead. At the British Museum, in October 1915, Charles W. Wallace of Nebraska patiently transcribed the *Memoranda* in order to have by him, for his own research, the suggestions that his predecessor tried to work out but could not.

Halliwell-Phillipps died on 3 January 1889, in his sixty-eighth year, and was buried in Patcham churchyard, near Hollingbury Copse. By the terms of his will he left his principal acquisitions, including the Blackfriars conveyance with Shakespeare's signature, to the Birmingham Corporation for £7,000: a bargain at the price. In the event that the Corporation refused the offer, the collections were to be sold undivided for £10,000, and, if no buyer came forward within twelve years, they were to be sold at auction as a single lot. Birmingham declined, against the advice of the Libraries Committee, mainly because of legal difficulties. In 1899 the Halliwell-Phillipps Shakespeariana—accurately described as a 'really national collection'—was bought by Marsden J. Perry of Providence in Rhode Island, USA. It has since come into the possession of the Folger Shakespeare Library.

# 6
## *Alexander Dyce*

ALL four founding members of the Shakespeare Society at one time or another tried their hands at biographies of their hero. Of these men Alexander Dyce was the most balanced scholar. A clergyman who had taken orders to escape the service in the East India Company planned for him by his father, Dyce gave up his provincial curacy and, settled in bachelor comfort at London's Gray's Inn Square, found fulfilment in his literary pursuits. He was the ablest of the early Victorian editors of the Elizabethan drama: his pioneering editions of Peele, Middleton, and others proved indispensable later in the century to Bullen; his Shirley has not yet been superseded, although it is time it was. No streak of perversity soils Dyce's character, as it does Collier's and Halliwell-Phillipps's; unlike Knight, he was no enthusiastic popularizer but a deep student of his subject. 'So long as the best traditions of English scholarship survive', Bullen eulogized him in the *Dictionary of National Biography*, 'his name will be respected.' Yet Dyce's Life is the least considerable of any produced by his group.

An editor averse to writing literary biography, he tried to escape a task he regarded as ungrateful: his understanding with his publisher, Moxon, expressly released him from having to supply his edition of Shakespeare's complete works with an introductory Life. Circumstances, however, made him relent, and prepare the 'slight memoir' (as with characteristic deprecation he refers to it) which runs to 124 pages. This is an elaboration, with many changes, of the slight but admired Life he composed twenty-five years previously for the Aldine edition of *The Poems of Shakespeare*.

Dyce does not dwell on genealogies and dry legal records. When it comes time to describe the poet's action against Philip Rogers for recovery of a trivial debt, he shakes his head over the petty details that Shakespeare's biographers must enter upon. Does he have Halliwell-Phillipps in mind when he refers disdainfully to 'professed antiquaries, with whom the mere mention of a name in whatever kind of document assumes the character of an important fact'? The reference does not appear in his 1832 'Memoir of Shakespeare'. Dyce has no discoveries of his own to offer, nor any startling insights, such as Knight's point about dower. He sticks to the known facts and quotes the more important life records.

For Dyce, as for all his contemporaries, the Collier heritage proved troublesome. The Ellesmere papers were under attack, but the British Museum specialists had not yet demonstrated their spuriousness. To this awkward problem Dyce provides an awkward solution. He includes the Collier materials in his narrative, but within square brackets, and, informing the reader that their genuineness has been 'violently assailed', leaves him to determine their reliability. It is difficult to see how the general reader (or, for that matter, the expert) could reasonably be expected to arrive at an intelligent decision on the basis of mere quotation of the documents. The petition in the State Paper Office and the Dulwich College 'discoveries', not yet suspect, Dyce presents as part of the biographical record.*

The intrusion of Collier apart, this Life is a sober nineteenth-century performance. Farmer still holds sway as *the* authority on Shakespeare's education. Along with almost everyone else, Dyce interprets Greene's attack upon Shakespeare as evidence that the poet began his playwriting career as an adapter. While more severe on traditions than Halliwell-Phillipps, Dyce accepts several popular ones: the deer-stealing frolic (*pace* Malone), the munificent gift from Southampton, the sociable club at the Mermaid Tavern. He rejects the identification of Mr W.H. with the Earl of Pembroke, he denies that Shakespeare figures as pleasant Willy in Spenser's *Tears of the Muses* but accepts the debated allusion in the same poet's *Colin Clout's Come Home Again*, and he sees no reason to conclude, from passages in the plays, that the dramatist languished in domestic unhappiness. Not presuming to give an overview of Shakespeare the man, Dyce nevertheless reveals, in an occasional aside, that he belongs to Halliwell-Phillipps's party: Shakespeare's 'grand object', he remarks in a footnote, 'from his earliest days . . . [was] the acquisition of a fortune, which was to enable him eventually to settle himself as a gentleman at Stratford'.[52]

About the merits of his biographical sketch Dyce speaks with attractive diffidence. 'Owing to the scantiness of materials for his history', he writes, 'and to our ignorance of what we most wish to know concerning him, a Life of Shakespeare, in spite of its subject, is generally among the least readable efforts of the biographer: and I cannot but feel that, if my own memoir of the poet has any claim to another character, it is solely on account of its comparative shortness.'[53] It is fitting that his essay should begin with the famous summary statement of what we know about the poet by Steevens, to Dyce the most acute of Shakespeare's commentators. Yet this Life, so modest in scope, from the first commanded a respect withheld from numerous similar efforts.

A glance at one of the other prefatory biographies that were a standard feature of nineteenth-century editions of Shakespeare will help us to appreciate why. William Watkiss Lloyd's *Life of William Shakespeare* for Samuel

---

* He relegates to a footnote, however, Mrs Alleyn's letter to her husband with its reference to Shakespeare, and quotes Halliwell-Phillipps's conclusion in the 1853 Life that Collier's 'interesting' *Memoirs of Edward Alleyn* 'conveys an inaccurate reading of the original manuscript preserved at Dulwich College, and cannot, therefore, be received as evidence'.

Weller Singer's edition of the *Dramatic Works* appeared one year before Dyce's, and is of equivalent length. It offers some curious information and opinions. Archdeacon Richard Davies's memoranda concerning Shakespeare are unceremoniously dismissed as 'garbage', although Lloyd tends to accept the deer-poaching story that is Davies's principal contribution. 'The curtains in front of the stage ran upon a rod, and opened in the centre,' the author informs us positively about the curtainless Globe playhouse. He is 'strongly confirmed' in his view that Nashe is lashing out at Shakespeare as the shifting companion who wrote an early version of *Hamlet*. The Collier papers, Lloyd knows, have been impugned, but he is unaware that Daniel's letter concerning the mythical mastership of the Queen's Revels belongs to the challenged series. In the stolid representation of the poet in Holy Trinity Church, the memorialist finds, as have few others, an 'expression of sportiveness or wit', and he infers 'geniality of temperament at least, if not joviality'.[54] Lloyd is here being influenced by his persuasion that poets should be convivial. Not surprising, given the competition, that Dyce's Life should gain acceptance as an authoritative epitome of the current state of Shakespearian knowledge— just as the edition it introduces was received for its time as (in Bullen's words) 'the most readable and satisfactory of all the editions of the dramatist'.

# 7

# *American and Continental Biographies*

WHILE Dyce, Halliwell-Phillipps, and others slaked the thirst of Victorians for information about Shakespeare's life, readers in the USA and on the Continent sampled the produce of local vineyards. Inevitably such wines had a less fragrant bouquet, and they require only passing and selective sampling here.

Although Shakespearian studies had long flourished in the USA (the first collected edition outside Great Britain was published in Philadelphia in 1795), biographical scholarship in the New World was unsurprisingly derivative. Of the two most notable nineteenth-century American authorities one, the Revd Henry Norman Hudson, never ventured abroad, and the other, Richard Grant White, visited England only once, in middle life. A gentleman of aristocratic mien unjustly calumniated as a snob Anglomaniac, White was compelled to earn a living from journalism which he detested, and to pass his days in a city, New York, that he loathed; no wonder a note of asperity creeps into his writing. Not intended for antiquaries or Shakespearian scholars,

White's *Memoirs of the Life of William Shakespeare* (1865) shows a command of the facts as they were then understood, but isolated oddities fleetingly brighten the faded purple of his prose. At one point White describes Shakespeare's daughters as having married 'simple village rustics', a phrase not especially appropriate to the eminent Dr Hall. In a strange passage he bemoans the discovery of Shakespeare's pursuit of two hapless debtors, Rogers and Addenbrooke, in the Stratford Court of Record: 'what have these to do with the life of him whom his friends delighted to call sweet and gentle? Could not these, at least, have been allowed to rest?'[55] This despite the fact that White's Shakespeare, like Halliwell-Phillipps's, is the offspring of Victorian materialism: the thrifty, prudential poet who cherished his pecuniary independence above his literary reputation, and wrote plays solely in order to advance his social position. Why should not such a man have sued those who owed him money? Anne Hathaway evokes more extreme anguish. Poor Shakespeare: married at eighteen to an older woman 'who came to him with a stain upon her name' but without a brain in her head—'could we but think of their life and leave out the world's interest in him should we not wish that one of them, even if it were he, had died before that ill-starred marriage?' This wish is not shared by Hudson, an Episcopalian minister with a vocation for literary evangelism. In the rudimentary biography comprising the first chapter of his huge *Shakespeare: His Life, Art, and Characters* (2 vols., 1872), the divine takes issue with White over the dramatist's conjugal relations ('all sheer conjecture, and rather lame at that') but forfeits confidence when he professes moral certainty that eighteen of the Sonnets to the Fair Youth were in fact addressed to Anne Hathaway.

Most students—whether in the USA, France, or Germany—share White's bleak opinion of the dramatist's domestic situation. Guizot, in *Shakespeare and His Times* (1821), believes that Shakespeare developed a positive aversion to Anne, his attraction to her being 'one of the most fleeting fancies of his youth'. In his *William Shakespeare* (1864) Victor Hugo sees the poet as having merely sipped at matrimony: the man who left his wife a second-best bed must have experienced keener delights on the best one. According to the celebrated *Shakespeare Commentaries* (1849) of Gervinus, for whom the life of Shakespeare is a variation on the Prince Hal story, the playwright's wife, good *hausfrau*, looked after 'economical circumstances' at home while her husband debauched himself in the metropolis and found sinful solace in the arms of a Dark Lady. In his Life (translated in 1876) Karl Elze demonstrates with patient Teutonic logic that the marriage was foredoomed to decline and collapse:

An older woman—nay, even one of the husband's own age—will always be one stage in advance of the man in her physical development, and even more so intellectually and in her ideas about life; the man never overtakes her in this, and any true or enduring sympathy between them becomes an impossibility. She does not look up to him as she ought; on the contrary, is inclined to look down upon him; she comes to find herself unable to live in his interest, and expects him to adapt himself to hers, which is contrary to nature. She possesses, or believes she possesses, more knowledge of life than her husband, and thus instinctively considers it her right—nay, her duty— to direct and guide him, which, in a lively or energetic woman, is bound to lead to a love of domineering. . . . In most cases also, when a woman is married to a man younger than herself, and the period of first and ideal love is passed, it is the older woman who is more concerned about worldly matters, about providing for the future; or it pleases her feminine vanity that, in spite of her age, she can attract another lover. Sensuality may bridge over the gap for a time, but not permanently. . . . These are phenomena based upon the unalterable laws of nature.[56]

The writer of such a passage qualifies as a Spengler of domestic life.

If Shakespeare's marriage is a touchstone of the biographer's spirit, the calf-killing legend recorded by Aubrey tests his powers of critical discernment. Take Guizot, for example, whose work ushered in a new era of historical study of the poet in France. 'Who cannot catch a glimpse, in this story', Guizot asks, with a rhetorical flourish,

of the tragic poet inspired by the sight of death, even in an animal, and striving to render it imposing or pathetic? Who cannot picture to himself the scholar of thirteen or fourteen years of age, with his head full of his first literary attainments, and his mind impressed, perhaps, by some theatrical performance, elevating, in poetic transport, the animal about to fall beneath his axe to the dignity of a victim, or, perhaps even, to that of a tyrant?[57]

Hugo's account of the same episode has the advantages of terseness and muscularity but the same demerit of credulity: 'William Shakespeare made his *début* in a slaughter-house. At the age of fifteen he entered his father's stables, bared his arm, and killed sheep and calves,—"in a high style", says Aubrey.'[58] The Germans reacted to the same legend with predictable sobriety: Gervinus barely alludes to it, and Elze dismisses the story as 'utterly absurd and ridiculous'.

Of these studies Elze's is the most considerable work of scholarship—the achievement of a man conversant with all the relevant material on his subject, conscious of the excessive freedom of conjecture indulged in by his predecessors, and openly sceptical about traditions to which even such distinguished contemporaries as Halliwell-Phillipps were susceptible. Hugo's book is precisely the opposite: the most factually careless account of the dramatist produced by the age. Written during his exile at Hauteville House in Guernsey, away from libraries and reference works, it regrettably does not benefit from Hugo's direct communication (via the good offices of a medium) with the shade of Shakespeare, who accommodatingly preferred to converse with him in French. The author garbles his facts throughout: dates are inexact, names misspelled, events misrepresented. Incompatible theories cheerfully coexist, sometimes within the framework of a single sentence. 'His [Shakespeare's] wife being put aside,' we learn, 'he was a schoolmaster, then clerk to an attorney, then a poacher.'[59] Much of the book has little connection with Shakespeare (in a splendid passage Hugo describes the atrocities of Caligula). Yet, of all the miscellaneous endeavours described in this section, Hugo's alone proceeds from the pen of a genius. Hopelessly uncritical as biography, it triumphs as a document of the Romantic revolt: an incandescently eloquent statement of Hugo's credo with respect to philosophy, science, and (above all) art. In after years the book would have little effect on the biographical mainstream; it would, however, stir Yeats.

# 8

## *The Sonnets: Divers Theories*

'I HAVE long felt convinced, after repeated perusals of the *Sonnets*,' Dyce declared in 1832, 'that the greater number of them was composed in an assumed character, on different subjects, and at different times, for the amusement, and probably at the suggestion, of the author's intimate associates.'[60] This considered judgement he saw no reason to revise when he extensively altered his memoir a quarter of a century later. By then, however, the Sonnets had become the critical battleground which they have remained to the present day. Across the sea, White declared his support for Dyce in *Shakespeare's Scholar* (1854), and Hudson, that judicious digester of commentators, seems to echo the earlier editor when he writes that these poems 'were intended mainly as flights or exercises of fancy, thrown into the form of a personal address, and written, it may be, in some cases at the instance or in compliment of the Poet's personal friends'.[61] In England, Bolton Corney expressed a popular view when, in a privately printed pamphlet on *The Sonnets of William Shakspere* (1862), he concluded that 'they are, with very slight exceptions, mere *poetical exercises*'. But Emerson, sublimely out of touch with criticism, rhetorically asked in his *Representative Men*: 'Who ever read the volume of the Sonnets without finding that the poet had there revealed, under masks that are no masks to the intelligent, the lore of friendship and of love; the confusion of sentiments in the most susceptible, and, at the same time, the most intellectual of men?'[62]

A professional scholar has less excuse for ignorance of scholarship than an inspired transcendentalist, and so we encounter with some surprise the one-sided pronouncement of the distinguished future biographer of Milton and

Professor of English Literature at University College, London, and (after-wards) Edinburgh University. 'Criticism seems now pretty conclusively to have determined, what it ought to have determined long ago,' David Masson writes in his 'Shakespeare and Goethe' (1852), 'that the *Sonnets* of Shake-speare are, and can possibly be, nothing else than a poetical record of his own feelings and experience—a connected series of entries, as it were, in his own diary—during a certain period of his London life.' It is a 'fact' that 'these Sonnets of Shakespeare *are* autobiographic—distinctly, intensely, painfully autobiographic'.[63] Masson's own exegesis of their content is preserved in two unpublished manuscript volumes in the Folger Shakespeare Library bearing the title *Autobiography of Shakespeare from his Thirty-fourth to his Thirty-ninth Year, Derived from His Sonnets: together with the Sonnets Themselves Arranged & Elucidated*. The *Autobiography*, composed when Masson was only twenty-four or so, holds few surprises,* but is enlivened by romantic elaboration. Thus Sonnet 20 ('A woman's face with nature's own hand painted') inspired a detailed picture of the Sweet Boy and the poet who feels an 'almost unmanly' love for him: 'Reclining on a chair, tearing his spaniel's neck with his ringed & ruffled hand from which he has just dropped a volume, is a youth in the first bloom of years, his forehead fair as a girls' [*sic*], although with manlier locks clustering round it, his eye-lids downcast so that their orbs are fringed, & the soft peach of his cheek first dimpling where it curls towards the small proud lip, & then rounding itself away in the white chin & throat. Standing by a window near & looking with a smile of pleasure & affection on the youth, is a man in his early prime, full-browed, clear-eyed, & with a short-close beard.'[64] It is such a portrait as Oscar Wilde might have appreciated.

A different interpretation of the critical consensus appears in Thomas Kenny's *Life and Genius of Shakespeare* (1864): 'The prevailing opinion among the most recent commentators seems to be that those strange compositions [the Sonnets] were, for the most part, produced by the poet in a purely fanciful and fictitious character.'[65] Had the climate changed so much in a dozen years? That is hardly likely. Himself a member of the school of Masson, Kenny prefers to see himself as belonging to an enlightened critical minority rather than as the voice of the status quo, which at this time could be identified as the commentator's pleasure dictated. In the following decades the debate would go on with wearisome iteration until, near the close of the century, it would flare into new life in the writings of Sidney Lee.

Whatever merits the views of the anti-personalists may have, it must be granted that their posture derives, at least in Victorian times, not from the disinterested pursuit of truth, but from an anxiety to redeem the god of their

* One novelty, however, is his reading of 'A man in hue all hues in his controlling'. To Masson this much interpreted line recalls that the Earl of Pembroke was the son of the Lord President in Wales; therefore the pun on the 'name *Hugh*, as a synonym for *Welchman*'. This pun had escaped previous commentators.

idolatry from self-confessed impurities of the flesh. These critics may embrace the notion of Shakespeare the Burgher of Stratford, but never of Shakespeare the Fornicator of Stageland. The position is stated with refreshing directness by Halliwell-Phillipps in the numerous editions of his *Outlines*:

. . . the victim of spiritual emotions that involve criminatory reflections does not usually protrude them voluntarily on the consideration of society; and, if the personal theory be accepted, we must concede the possibility of our national dramatist gratuitously confessing his sins and revealing those of others, proclaiming his disgrace and avowing his repentance, in poetical circulars distributed by the delinquent himself amongst his most intimate friends.[66]

Swallow the autobiographical reading and we must digest a Shakespeare who is, in Forsyth's words, 'a sycophant, a flatterer, a breaker of marriage vows, a whining and inconstant person'.[67] Worse yet, could he reveal himself as (most horrible of imaginings!) the practitioner of unnatural vice: Shakespeare the Invert? One can only sympathize with the bewilderment and distress experienced by Léon de Wailly. 'Mais, grand Dieu!' he cries; 'qu'aperçois-je en relisant quelques-uns des premiers sonnets? *Lui* au lieu d'*elle!* . . . est-ce que ces sonnets seraient adressés à un homme? Shakspeare! grand Shakspeare!'[68]

Another response is possible, of course, to a poet who lays bare in impassioned language the torments of unhallowed desire (whether hetero- or homosexual), and in Nietzsche, writing around 1881, it finds powerful expression:

The passions become evil and malignant when regarded with evil and malignant eyes. It is in this way that Christianity has succeeded in transforming Eros and Aphrodite—sublime powers, capable of idealisation—into hellish genii and phantom goblins, by means of the pangs which every sexual impulse was made to raise in the conscience of believers. Is it not a dreadful thing to transform necessary and regular sensations into a source of inward misery, and thus arbitrarily to render interior misery necessary and regular *in the case of every man!* Furthermore, this misery remains secret, with the result that it is all the more deeply rooted; for it is not all men who have the courage, which Shakespeare shows in his sonnets, of making public their Christian gloom on this point.[69]

But few in an England complacent amid its public reticences would recognize either the courage or the Christian gloom.

In his popular and delightful *Journal of Summer Time in the Country* (1849), the Revd R. A. Willmott, enthusiast of nature and literature, sees the Sonnets as 'a chapter of autobiography, although remaining in cypher till criticism finds the key'. In the latter half of the nineteenth century, many adherents of the autobiographical principle applied their ingenuity to finding the key—or keys—that would settle, once and forever, the identities of Mr W.H. and the Sweet Boy, the Dark Lady and the Rival Poet. The Pembroke legions waged dubious battle with the Southamptonites on their darkling plain. On other fields skirmishes between supporters of new pretenders to the roles of dedicatee, friend, rival, and mistress drew ink. Familiar questions were raised over and over again. Does the *begetter* of the dedication signify *procurer* or, rather, *inspirer*? Is only one friend addressed in the Sonnets, or are several? (Samuel Neil, in *Shakespeare: A Critical Biography*, 1861, uncritically argues the presence of 'noble and beloved friends', as well as young Hamnet Shakespeare, the poet's wife, a daughter, and Queen Elizabeth.) These burning issues enlivened—if that is the word—countless newspaper and magazine columns. The literature produced by the fantastic quest for identities achieved a volume out of all proportion to its significance. Here a brief sampling must suffice.

From 'Hints for the Elucidation of Shakespeare's Sonnets', diffidently submitted from Paris by the Mazarine Librarian Philarète Chasles, readers of the *Athenaeum* for 25 January 1862 would learn that the dedication to the Sonnets is not properly a dedication at all but a monumental inscription comprising two discrete sentences. The first is addressed *by* (not *to*) Mr W.H., the unnamed 'onlie begetter' being the recipient. In the second sentence, beginning 'The well-wishing . . .', the publisher expresses his good wishes for the commercial success of his book. Mr W.H. is William Herbert, and the inspiration for the poems—their begetter—is Southampton. Thus are reconciled the rival claims of the two chief critical factions. This tortured reading was applauded by the editor of the *Athenaeum*: the discovery that W.H. is not dedicatee but dedicator, 'a point not hitherto suspected by the wise' (this is true enough), will lead 'towards an understanding of the personal history unquestionably involved in these poems'.*

Letters raising objections appeared in the 1 February number, and the discussion also found its way into *Notes and Queries* (1 February and 1 March 1862), where Chasles found an ardent supporter in Corney, who praised 'a Frenchman who has routed a whole army of English editors, annotators, pamphleteers, etc.'. Corney improved upon the hypothesis by judiciously emending the punctuation of the inscription. But Gerald Massey rejected the theory in his 1866 edition of the Sonnets, and Chasles returned to the lists in the *Athenaeum* for 16 February 1867 with a mock dedication to Massey illustrative of the argument. At the same time he withdrew Herbert in favour of William Hathaway, Shakespeare's brother-in-law. As for Southampton, '*He* figures at the head of the inscription, crowned with immortality, while T.T. remains humbly crouching at the base, and W.H. kneels in an obscure corner.'[70] It might almost be an emblem. On 23 February Samuel Neil

* For all his ingenuity at explication, M. Chasles appears to have been a singularly gullible gentleman—how else explain his purchase, reported in the *Pall Mall Budget*, of some hundred letters written by Shakespeare to Galileo, Montaigne, etc.? The fact that all these epistles were in French might, one imagines, have given the Mazarine Librarian pause.

protested that he had already put Hathaway forward as W.H. The Mazarine Librarian brushed aside the priority seeker on 9 March with the dry observation that Neil had misunderstood his theory. A week later, Massey replied to Chasles, triumphantly citing Neil's letter as proof that the Frenchman's syntax would never be understood across the Channel. On 13 April Chasles defended himself with another restatement of his reading. Massey answered again (27 April), and in the last sentence of his letter proposes, without fanfare, Sir William Hervey (or Harvey), the Countess of Southampton's third husband, as Mr W.H. The nomination at the time would go unseconded; then, years later, be reintroduced more than once and eagerly embraced by the Southamptonites, most notably Dr Rowse, who—noting that Sir William outlived his wife and (in 1609) took a much younger bride, and might reasonably hope for posterity—argued that Harvey might well be the subject of a dedication wishing him 'all happiness and that eternity promised by our ever-living poet': the procreative eternity which Shakespeare promised the Fair Youth in the eternizing sonnets.

The *Athenaeum* correspondence reached its climax with Neil's letter of 27 April, in which the Rector of Moffat invites us to read *begetter* as '"suggestor or suggestress" . . . the literal adviser of the production of the book'—in other words, the poet's wife, *begetter* being a pun on Anne's matrimonial function. This interesting speculation was received with the silence not of surrender but (one guesses) of stupefaction. Dyce, in a revised edition of his *Shakespeare* (1866), dismisses the Chasles theory as a wild dream. If so it would, as we shall see, prove a recurrent one.

Other interpretations of the mysterious dedication remained in store for faithful readers of the *Athenaeum*. On 22 November 1873, Charles Edmonds, dismissing Chasles's views as preposterous, suggests 'the true interpretation of the inscription to be that "T.T." the publisher (or as he quaintly calls himself "the well-wishing adventurer in setting forth") feeling deeply indebted to "Mr. W.H." for having obtained for him the privilege of publishing such a popular work as Shakespeare's Sonnets were likely to be, wishes him all happiness, and that eternity promised by the great bard to those who are instrumental in preserving things which the world "would not willingly let die"'.[71] According to Edmonds, the thought was probably suggested to Thorpe, that close student of early Shakespeare, by the opening lines of *Love's Labour's Lost*, 'Let fame, that all hunt after in their lives, | Live regist'red upon our brazen tombs.'

This reading seems prosaic alongside some others. A Forsyth—either Ebenezer, author of *Shakspere: Some Notes on His Character and Writings* (1867) or the J. Forsyth, editor of the *Inverness Advertiser*, referred to in the same year by Neil—seems to have first suggested that the full stop after the H in 'W.H. ALL' was inserted by the printer in error. Thus W. Hall becomes the dedicatee. Who was this individual? Perhaps (as Underhill proposes in *Notes* *and Queries* in 1890) William Hall of the Halls at Hallow, near Worcester; but more likely the stationers' assistant William Hall, who piratically offered the manuscript of the Sonnets to Thorpe for surreptitious publication. This dubious theory, to which Sidney Lee subscribes, has had many followers. Why Hall, who in 1609 had his own press, should have procured the manuscript for another printer is a question to which these enthusiasts do not address themselves. A still more novel interpretation of the initials is offered, merely as a guess, by D. Barnstorff in *Schlüssel zu Shakspeare's Sonnetten* (1860; translated 1862). Could not W.H. stand for *Shakespeare Himself?* The artist, who in his poems reveals his innermost self, thus modestly dedicates the sequence to his own genius. This odd *Schlüssel* provoked howls of derision from scholars in England and on the Continent; scholars whose own views were often not much less far-fetched. Even M. Chasles allows himself a Gallic sneer. 'What a recondite conclusion!' he exclaims with heavy sarcasm. 'And if not very german to the matter, at least eminently German!'[72] But F. G. Fleay, about whom more hereafter, refers in his *Biographical Chronicle of the English Drama* (1891) to Barnstorff's 'great Teutonic discovery', and he points to the dedication of *Abuses Stript, and Whipt* (1613)—'To himself, *G.W.* wisheth *all happiness*'—as a parody of the inscription for the Sonnets. Barnstorff's theory would not want supporters;* nor, for that matter, would almost any other, however eccentric.

On 30 August 1873 C. Elliot Browne published in (where else?) the *Athenaeum* a letter headed 'Shakespeare's Sonnets: An Old Theory', that revived, with variations, a guess of the previous century. Tyrwhitt, it will be recalled, had detected a punning allusion to W. Hughes in Sonnet 20 ('A man in hue, all hues in his controlling'). Browne does not, like his predecessor, concern himself with the possible identification of Hughes with the W.H. of the dedication; he focuses only on the personal reference in a particular line. He notes that a John Hughes is mentioned by Drayton as having made the company at Polesworth gravely merry with his lyre, and that Polesworth lies just a few miles from Stratford; so this musician probably lived within the circle of Shakespeare's acquaintance. There may have dwelt in the same circle another Hughes, to whom more interest attaches because of his Christian name. In Captain Walter Bourchier Devereux's history of the Earls of Essex, Browne had come upon the account by Edward Waterhouse (in time Chancellor of the Exchequer in Ireland) of the death of the first Earl, whose faithful retainer he was, on 22 September 1576; an account which, Browne knows, also appears prefixed to Hearne's edition of Camden's *Annales*. 'The night following', Waterhouse reports, 'the Friday night, which was the night before he died, he called William Hewes, which was his musician, to play

* One variant holds that W.H. is a misprint for W.S. or W.SH., the printer in this way acknowledging the author (J. M. Nosworthy, 'Shakespeare and Mr. W.H.', *Library*, 5th ser., 18 [1963], 294–8).

upon the virginal and to sing. "Play", said he, "my song, Will Hewes, and I will sing it myself". So he did it most joyfully. . . '.[73] Were Browne more of a scholar, he would have gone back to the *Annales* and found that the reference, as given there, is not to William Hewes the musician but to William Howes the physician. He could not have known, of course, that Devereux, amateur bungler that he was, had consulted the Broughton papers for the passage, and misread the name as it appears there: Hayes. From such confusions do myths grow. Will Hughes would now attract his sect of believers.

# 9

## *Oscar Wilde*

THE description of William Hewes singing for the Earl of Essex on his deathbed is quoted by Oscar Wilde in his *Portrait of Mr W.H.*, in which the creative imagination vivifies the bleached bones of pseudo-scholarly controversy. Through a minor triumph of literary art, homo-erotic fantasy successfully masquerades as fiction and criticism, the whole sufficiently sublimated so as not to disturb the staid sensibilities of readers of *Blackwood's Edinburgh Magazine*, which carried the *Portrait* in the July 1889 issue. (The *Fortnightly Review* had turned it down.)

It need not surprise us to find a portrait the central device of a Wilde piece almost exactly contemporaneous with *The Picture of Dorian Gray*. The full-length painting of Mr W.H. hangs in the London house of a man named Erskine, who is a good deal older than the unnamed narrator. It depicts a youth, about seventeen years old, of marvellous girlish beauty, with short golden hair, wistful, deep-sunken eyes, and white lily hands. He wears a black velvet doublet of the sixteenth century, with fantastically gilded points. Masks of comedy and tragedy hang from a marble pedestal; the background is peacock-blue. The youth's hand rests on a book on which is written, in a crabbed early hand, 'To the only begetter of these ensuing sonnets. . . .' The portrait, Erskine reveals, had belonged to his friend Cyril Graham, an effeminate aristocrat of exquisite grace and comeliness, who had always taken the female leads in Shakespeare's school plays.

This Graham had evolved a theory of the Sonnets. Mr W.H. is the Hews-Hughes of the twentieth poem, and this Hughes is Willie—not the musician of Browne's conjecture but a boy actor in Shakespeare's company. This youth, not of noble birth (a lord would never have been styled Mr), the poet addressed in terms of passionate adoration; for him were created Viola, Rosalind, Juliet, Desdemona, and the rest, even Cleopatra. Unfortunately the list of actors prefacing the First Folio names no Willie Hughes, so demonstration must depend on spiritual and artistic responsiveness to the Sonnets themselves. Such evidence, Erskine advises his young friend, will not convince the world. But Graham produces tangible proof of the existence of Mr W.H. in the form of a portrait found (like the Ireland manuscripts) in an old chest. Like the Ireland papers too, the picture—as Erskine by chance discovers—is a fraud. The friends quarrel, and Graham commits suicide: a martyr to literature.

'I will take up the theory where Cyril Graham left it', vows the narrator, 'and I will prove to the world that he was right.' And so he studies the cycle, separating the lines in which the poet speaks of his quatorzains from those in which he alludes to his dramatic writings. Everywhere Graham's disciple sees the visage of the actor: the lovely boy, master-mistress of Shakespeare's passion. He discovers puns on *use* and *usury*, and notes that *shadows* has a technical sense associated with the stage: points that later commentators would make again without acknowledgement, or indeed awareness, of their source. But why should the poet urge Willie Hughes to take a wife when his own early marriage had been (as we well know) miserable? This question leads the narrator to his great discovery: the conjunction proposed is with the Muse, and the children the actor is begged to procreate are the *personae* of his art. 'The whole cycle of the early sonnets is simply Shakespeare's invitation to Willie Hughes to go upon the stage and become a player.'[74] Marlowe is the rival poet who lures the golden youth to his theatre to take the part (in a delicious touch) of Edward II's homosexual favourite Gaveston; on Marlowe's death, the actor returns to Shakespeare and is forgiven. Who was this Willie Hughes? Perhaps the son of the musician who played the virginals for Lord Essex and may have been the relation of Margaret Hews, the Restoration actress beloved of Prince Rupert. Perhaps too he went across the sea with a troupe of English comedians, to meet with them his end in an uprising in Nuremberg and be secretly buried by young men in a little vineyard outside the city.

The narrator draws together all the passages corroborative of his theory which he describes with all his faith and enthusiasm in a letter to Erskine. Then, having given expression to his passionate belief, the narrator loses it: Willie Hughes becomes myth, a boyish dream. 'The one flaw in the theory', he tells the sceptic he has persuaded, 'is that it presupposes the existence of the person whose existence is the subject of dispute.'[75] But Erskine will not be shaken; his last act, before death, is to protest his faith. As for the narrator, he now keeps in his study the portrait of Mr W.H., which wins the admiration of his artistic friends. 'I have never cared to tell them its true history,' he concludes. 'But sometimes, when I look at it, I think that there is really a great deal to be said for the Willie Hughes theory of Shakespeare's Sonnets.'[76]

So Wilde himself thought; he found his theory, like himself, irresistibly fascinating. In a letter to W. E. Henley (July 1889) he referred to the 'wonder and beauty' of his discovery of 'the real Mr W.H.'. Others too were convinced, including the philosopher and statesman A. J. Balfour and, less surprisingly, Wilde's destructive minion Lord Alfred Douglas, who long afterwards wrote a volume in advocacy, *The True History of Shakespeare's Sonnets* (1933). Yet Wilde draws back from total commitment. In *The Portrait of Mr W.H.* he chooses a species of fiction rather than the essay form, and escapes the responsibilities of scholarship through the licence of art. His narrative structure permits him to state the theory and to set forth the evidence with impassioned argument; but his words are uttered not in his own person but by a figment of his imagination. The edifice of argument is undermined by Erskine's scepticism and by the faked portrait, then is patiently rebuilt, only to be rejected and, at the close, doubtfully affirmed. ('You *must* believe in Willie Hughes,' Wilde once said after a recital that left his audience deeply impressed; then added, 'I almost do myself.' It was a favourite notion of his that in converting another to an idea, one lost one's own faith in it.[77]) That the whole exercise meant much to Wilde may be seen from the fact that he returned to *The Portrait of Mr W.H.* after publication, and wrote a new version more than twice the length of the original. The story came uncomfortably close to home. As the late Richard Ellmann observes in his biography, Wilde imagined Shakespeare, like himself a married man with children, captivated by a boy as he had been captivated by Robert Ross: the knowing youth who had first seduced him to homosexual love, and who would go on to hold a permanent place in his life.[78]

The revised *Portrait* has a strange history. 'Our English homes will totter to their base when my book appears,' Wilde prophesied with characteristic modesty. It was announced for publication in 1893 by Mathews and Lane of the Bodley Head, but the partnership broke up, the volume did not appear (despite Wilde's protestations), and the manuscript vanished. Rumour had it that it had been stolen from Wilde's Tite Street lodgings. Charles Ricketts, who painted a portrait of W.H. for the frontispiece to the projected book, met Wilde for the last time in 1897. 'I must return to literature', the disgraced but still irrepressible Wilde told the Royal Academician and part-owner of the Vale Press, 'and you must print "A Portrait of Mr. W.H." I know it needs retouching, though one of my early masterpieces.' Ricketts indicated a preference for another Wilde piece, and the latter replied: 'Yes, perhaps you are right . . . "Mr. W.H." might be imprudent . . . the English public would have to read Shakespeare's sonnets.'[79] Apparently he did not know that the manuscript had been lost. It turned up many years later among the effects of Lane's office manager and was sold to the American dealer Rosenbach. In

1921 the revised *Portrait* appeared in print for the first time in a limited edition, without explanation.

In the enlarged version Wilde does not tamper with his original fictional framework, but instead expands the essay side of the work. The homo-erotic strain becomes more open and pronounced. Wilde includes an apostrophe to Elizabethan boy actors; he rhapsodizes over Neoplatonic concepts of the Renaissance that gave warrant for the permanence and ardour of friendship—his Willie Hughes is to Shakespeare what Cavalieri was to Michelangelo. To the question he raises in the 1889 version—'Yet as Shakespeare forgave him [Hughes], should not we forgive him also?'—Wilde adds to the reply significantly if ambiguously. In the original it reads: 'I did not care to pry into the mystery of his sin'; in the revision: 'I did not care to pry into the mystery of his sin or of the sin, if such it was, of the great poet who had so dearly loved him. "I am that I am"', said Shakespeare in a sonnet of noble scorn.'[80] The fantasist thus joins those many who have refashioned the master in their own image. But mainly Wilde elaborates and reinforces his theory of the Sonnets. He multiplies supporting passages from the sequence. He points out that the Sweet Boy is fair, whereas the preferred candidate of the scholars, Pembroke, was swarthy and dark-haired. He argues that 'Nash with his venomous tongue had railed against Shakespeare for "reposing eternity in the mouth of a player"', the reference being obviously to the Sonnets.'[81] He finds Willie Hughes in the shepherd of *A Lover's Complaint* and in the rose-cheeked god of *Venus and Adonis*. The Dark Lady enters his discussion for the first time, and is identified with the citizen's wife of Manningham's scurrilous anecdote; in a novel twist we learn that Burbage, Shakespeare's rival according to the jest, has through the vagaries of tavern gossip 'no doubt' replaced the true actor-rival: none but Willie Hughes. She also figures as the heroine of Thomas Cranley's *Amanda: or, The Reformed Whore*. What matter that the romance did not see print until 1635? Wilde is above chronology. To round off his case, Wilde presents a whole scheme for the Sonnets, which, with some rearrangements necessitated by the theory, are now seen to constitute a perfectly unified whole. Reordering always seems to be the price paid for a demonstration of unity.

The scholar may read *The Portrait of Mr W.H.*, and be moved by the power of Shakespearian debate to arouse the enthusiasm and (more important) engage the imaginative faculties of a writer of some genius. But no specialist, it goes without saying, could take seriously the case for the actor Willie Hughes, whose very existence may owe its origin to the aberration of a compositor who capriciously italicized and capitalized a single word, *Hews*. Wilde did not foresee this objection, but he grasped better than many of his followers the element of presupposition in his argument.

Other artists have responded differently. 'The most brilliant of all is that story of Wilde's,' says Mr Best during a memorable discussion of Shakespeare at the National Library in Dublin during the afternoon of 16 June 1904, and he goes on to say, not quite accurately: 'That *Portrait of Mr W.H.* where he proves that the sonnets were written by Willie Hughes, a man of all hues. . . . Of course it's all paradox, don't you know. . . . The light touch.' But Best is a dilettante enthusiast of Wilde; a character in a novel, neither Joyce nor his spokesman. We come closer to the creator's view in Mr Lyster's rejoinder: 'Do you think it is only a paradox. . . . The mocker is never taken seriously when he is most serious.'[82] Wilde's theory, Joyce's contemporary André Gide told Julien Green in 1931, 'is the only, not merely plausible, but possible interpretation of the sonnets addressed to the unknown man who is the counterpart of the *Dark Lady*'.[83] And in George Sylvester Viereck's *My First Two Thousand Years*, the Wandering Jew meets Willie Hughes, the delicately fashioned actor who is enchanting as Juliet. To him Shakespeare has dedicated his Sonnets, which may not shock the classical scholar but which have the power to set ribald tongues in London wagging; the scandal of the affair sears the performer's character. Willie, however, is not Wilde's Sweet Boy, nor the one known to most of us from the cycle, but, rather, a delectable girl. In his heart of hearts the poet suspected the secret from the first. 'Do you like to play the boy,' she is asked, 'to be his play-boy?' 'Yes,' she replies shyly after a moment's hesitation, and then, with a blush: 'He called me master-mistress of his passion. . . .' The critics too have done their share to keep alive this epicene figment of an invert's imagination. That redoubtable high priestess of serious Sunday journalism, Brigid Brophy, has described Wilde's improbable hypothesis as having 'the merits of social and psychological plausibility' (*Queen*, October 1963). In this story Wilde 'dazzlingly offered a theory, withdrew it, and half offered it again, in fiction within fiction which anticipates Borges.' Thus Wilde's most recent and greatest biographer.[84]

# 10
## *Samuel Butler*

SAMUEL BUTLER fails to mention *The Portrait of Mr W.H.* in his *Shakespeare's Sonnets Reconsidered* (1899), an edition with elaborate introduction. His own theories respecting Willie Hughes and the sexuality embodied in the poems make it unlikely, however, that he missed *Blackwood's* for July 1889.

Butler achieved his minor but secure literary niche posthumously with *The Way of All Flesh*, but he wrote incessantly, sometimes at the rate of almost a book a year. Part of this work is amateur scholarship of considerable erudition and maddening eccentricity—precisely what one would expect of a man who was in his private life (as F. W. Dupee puts it) 'by turn, or in perpetual combination, a hero and an ass'. His asininity comes to the fore in *The Authoress of the Odyssey, Where and When She Wrote, Who She Was, the Use She Made of the Iliad, and How the Poem Grew Under Her Hands* (1897), which displays much learning. Debunker and satirist, artist and photographer, Butler was also an eccentric classicist. His study of the Sonnets is no less perverse than that of the *Odyssey*, and the quality of his scholarship, as regards accuracy of detail, is ominously suggested by the title he gave one of his chapters: 'The Rev. Edmund Malone'. Butler published the book at his own expense under the Longmans imprint, although Charles Longman protested—not because the argument was hopelessly misguided but because it took up 'an unpleasant subject'.

Two articles in the *Fortnightly Review* prompted Butler to try to solve the riddle of the Sonnets. In the first (December 1897) William Archer, a Pembrokist, showed the baselessness of the contention that the poems were addressed to Henry Wriothesley; in the other (February 1898) Sidney Lee, sympathetic to Southampton, performed a like service for Pembroke. Amid the ruins of this mutual destruction, Butler determined to adapt to the Sonnets the method that with Homer had brought the results which so satisfied him. He memorized the poems. He cut them out (excising the numbers) and rearranged them until he arrived at 'their original order'. He brushed up on the commentators, consulted some historical sources and the *DNB*, and worked out his own ideas. Butler never questions the autobiographical purport of the sequence, which scarcely exists for him as poetry. The veil that always cloaks Shakespeare the dramatist is lifted, and 'we look upon him face to face'.

What we see is a very youthful writer infatuated with Willie Hughes. The latter is Mr W.H.: the inspiration for the poems and also the procurer who marketed them to Thorpe for ready cash. This Hughes is not, however, Wilde's boy actor, but (despite the reality of his existence) an equally unlikely candidate. In the State Papers Domestic, Butler found a Willie Hughes, or Hewes, who after many years in the navy as a steward, applied for the post of sea-cook in 1634 and died three years later.

Butler's mariner would seduce even fewer students than Wilde's adorable thespian, but such interest as *Shakespeare's Sonnets Reconsidered* holds resides not in the identification of the persons but in the story which the author pieces together from the poems. Belonging to an earlier period in the poet's life than was previously realized—the 1580s—the Sonnets are really love letters that tell a 'very squalid' tale of youthful folly.* 'He was vain, heartless,

* Butler's dating of the Sonnets, one of the more striking features of his study, anticipates Leslie Hotson's *Shakespeare's Sonnets Dated* (1949). Hotson apparently did not know Butler's edition.

and I cannot think ever cared two straws for Shakespeare, who no doubt bored him,' Butler writes of W.H.; 'but he dearly loved flattery, and it flattered him to bring Shakespeare to heel; moreover, he had just sense enough to know that Shakespeare laid the praise on thicker and more delectably than any one else did, therefore he would not let him go.' Instead, playing as a cat would with a mouse, W.H. sets a trap. Sonnet 23, with its reference (in Butler's misreading) to Shakespeare's desire for the 'perfect ceremony of love's rite', testifies to the poet's readiness to fall victim to 'a cruel and most disgusting practical joke'. At this point the story becomes murky because of Butler's prissy reticences. (Despite his libidinal preoccupations, the word *sexuality*, or any variant, appears not once in his essay.) Apparently Shakespeare is enticed to a rendezvous with the sailor boy, and a pederastic interlude takes place; later we are told that 'the love of the English poet for Mr W.H. was, though only for a short time, more Greek than English'.[85] In the course of the Greek ceremonies, W.H.'s confederates burst in by prearrangement, catch the poet *in flagrante delicto*, beat and temporarily lame him ('Speak of my lameness and I straight will halt'), and force him to flee leaving his cloak behind ('And make me travel forth without my cloak'). An element of literalism in these readings of Sonnet 34 is difficult to deny.

The 'hideous offence', we learn with some relief, was not repeated, and the remainder of Shakespeare's writings over a period of twenty-five years display reassuringly heterosexual tendencies. Butler, himself homosexually oriented, inclines towards forgiveness:

Considering, then, Shakespeare's extreme youth . . . his ardent poetic temperament, and Alas! it is just the poetic temperament which by reason of its very catholicity is least likely to pass scatheless through what he so touchingly describes as 'the ambush of young days'; considering also the license of the times, Shakespeare's bitter punishment, and still more bitter remorse—is it likely that there was ever afterwards a day in his life in which the remembrance of that 'night of woe' did not at some time or another rise up before him and stab him? nay, is it not quite likely that this great shock may in the end have brought him prematurely to the grave? . . . I believe that those whose judgement we should respect will refuse to take Shakespeare's grave indiscretion more to heart than they do the story of Noah's drunkenness; they will neither blink it nor yet look at it more closely than is necessary in order to prevent men's rank thoughts from taking it to have been more grievous than it was.[86]

Butler's interpretation of the Sonnets finds curious parallels, as his biographers have noted, in the events of his own life. He too was lured into a liaison with a handsome male, Charles Paine Pauli, who recognized Butler's weaknesses and exploited them ruthlessly. Pauli sponged off him, at the same time making no effort to conceal his boredom with his benefactor, whose literary and philosophical interests he did not share; he played to the hilt the role of the vain and heartless mistress. In 1897 he died. At the funeral Butler

discovered to his horror that Pauli, who over the years had fleeced him of
£6,000 or £7,000, had another keeper, made a comfortable income from his
legal practice, and left a substantial fortune in which his patron did not share.
'Did he ever borrow money from *you*?' Butler asked Pauli's friend Lascelles
that traumatic day. 'No, never', was the reply, 'not so much as a five pound
note.' Butler would not forget Pauli when he studied the Sonnets. The poems
would become the receptacle for his own frustrations and the vehicle of
unconscious self-analysis. On to them he projected himself as the victimized
poet. He attributed to him his guilt-stricken response to his own inversion
and, in pardoning Shakespeare, came to terms with himself. In Mr W.H. he
depicted the cruel lineaments of the man whom he had loved and who had
spurned him.

So much is private history. Readers in 1899 were confronted with a book
ostensibly on Shakespeare. The theory it expounded was unpalatable and
therefore best ignored; Butler complained petulantly about the lack of
reviews. But the Bishop of London wrote to commend him on his common
sense—one wonders what *he* made of the book. At the British Museum Butler
was introduced to Furnivall, whom he describes with characteristic mis-
judgement as 'a most amiable, kindly old gentleman'. At a nearby A.B.C.
cafeteria they took tea together and discussed the Sonnets. Furnivall argued
against Butler's opinions and 'denied that there was anything in the Sonnets
to indicate that Shakespeare had been fonder of Mr. W.H. than he should
have been'.[87] The kindly old gentleman confessed that he had not read
Butler's book through, only skimmed it, and he had never studied the
sequence closely. When they parted, Butler felt understandably depressed.
Robert Bridges did read *Shakespeare's Sonnets Reconsidered*, and found himself
agitated. 'The more I read, the stronger your position seemed to be', the
future laureate wrote to Butler on 25 January 1900, '—but yet I was not
convinced. It only seemed to me that the refutation was more difficult.'[88]
Butler's critics have treated the study respectfully; considerably more so than
have Shakespearians. 'It must be admitted', Phillip Henderson assumes
without warrant, 'that with Butler's interpretation in mind everything does,
as he claims, fall into place'.[89] Anyway Shakespeare, not least in his Sonnets,
has a way of bringing out the bizarre in critics, and of the bizarre Butler had a
sufficiency. On behalf of the book, it may be granted that it did help to bring
into the open the question of the sexuality revealed in the poems. And the
theories it argued have found at least one notable modern partisan in the
person of the writer who, more than any other, may be given doubtful credit
for having inherited the mantle of Butler's eccentric scholarship: Robert
Graves. In 'The Sources of *The Tempest*' (1925), Graves improves upon Butler
by suggesting that the cruel joke was devised by W.H. with the collaboration
of the Dark Lady and a rival playwright.

# 11
## Dark Lady and Rival Poet

OUR concern with Mr W.H. and the Fair Youth is not intended to suggest an indifference, during this period, to the identity of the unfashionably beautiful black-haired enchantress of the later Sonnets. Far from it: speculation, as the phrase goes, was rife. If the Dark Lady was Mrs Shakespeare, then (Heaven be praised!) a cloud disappears from the poet's good name and his married life— or so a few believed, not realizing that they were merely removing one cloud for another. Or perhaps these poems have to do with Sidney's Stella, Lady Rich, who aroused the jealousy of Elizabeth Vernon, Southampton's bride, and inspired a deep and desperate devotion on the part of Pembroke. Is her name not punned upon in Sonnet 146: 'Within be fed, without be rich no more'? (The answer is no.) Perhaps some of the Sonnets were written to express Mistress Vernon's emotions; perhaps others are concerned with Pembroke's passion, and were in fact written by him. This complicated explanation is offered and defended by Gerald Massey, working-class poet and untutored mystic (often the worst kind), in his six-hundred-page *Shakspeare's Sonnets Never before Interpreted: His Private Friends Identified: together with A Recovered Likeness of Himself* (1866), which he cruelly enlarged as *The Secret Drama of Shakspeare's Sonnets* (1888). Of his theory Massey diffidently claims, 'It is the only one that surmounts the obstacles, disentangles the complications, resolves the discords, and out of various voices draws the one harmony. It ignores no difficulty, violates no fact, strains no point for the sake of making extremes meet; it gathers up every possibility, and is consistent from beginning to end.'[90] Still, there are a few difficulties. Lady Rich was old enough to be the mother of Southampton or Pembroke, had half a dozen grown children, and was mistress (afterwards wife) of Lord Mountjoy. The two Earls, moreover, are not known to have loved her. These points were made in an *Athenaeum* review of Massey (28 April 1866) but did not discourage him from his oracular persuasion. Others too have found Lady Rich in the Sonnets. In an article in the *Nineteenth Century* (1884) Charles Mackay repeats Massey's point about Pembroke writing the Dark Lady poems, but identifies her as Christian Bruce, wife of the second Earl of Devonshire. She was born in 1595—some seven years later than Butler's date for the cycle. The identification nevertheless appeals to one R., who, in *Shakespeariana* (1884), speaks of the precocious infant as 'beautiful, dark-haired, and accomplished'.[91]

That way madness lies. The Mary Fitton theory tirelessly espoused by Thomas Tyler at least comes within the bounds of sanity. In his daily visits to the British Museum during the 1880s this neckless, waistless, rectangular man, wearing a frock-coat and tall hat and disfigured by a monstrous goitre hanging from his left ear to the point of his chin, met the young Bernard Shaw. Shaw recalls him as 'a gentleman of such astonishing and crushing ugliness that no one who had once seen him could ever thereafter forget him'. Certainly Shaw did not: 'He remains a vivid spot of memory in the void of my forgetfulness, a quite considerable and dignified soul in a grotesquely disfigured body.'[92]

The Sonnets seem to have given point to Tyler's lonely life. In speeches, articles, pamphlets, and editions, he argues that the dark mistress is Mary Fitton, a Maid of Honour to Queen Elizabeth. To believe in Mistress Fitton presupposes faith in the Pembroke cause, for she was the Earl's paramour and bore him a short-lived son. She also had two bastard daughters by Sir Richard Leveson before settling down to domesticity in 1607 as the wife of Captain William Polwhele. Unfortunately there are certain obstacles in the way of accepting Mary Fitton as the Dark Lady. One cannot prove a personal acquaintance between Shakespeare and the promiscuous Maid, although Tyler received valiant aid from the Revd W. A. Harrison, who pointed out that in 1600 William Kempe, the clown in the Lord Chamberlain's company, dedicated his *Nine Days' Wonder* to 'Mistress Anne Fitton', mistaking her Christian name; presumably the poet knew her more intimately. Tyler strains his ingenuity to find a reference to Fitton as the 'fit one' in Sonnet 151, a pun so deeply buried that it has eluded all other readers. Because the dark woman of the Sonnets is married, and the cycle by general consent antedates Mary Fitton's 1607 wedding, Tyler is driven to conjecture that she had a prior husband in early youth and that the contract had been annulled 'either on the ground that the previous consent of parents had not been obtained, or from some other cause'; but for this wishful thought not a shred of supporting evidence exists.[93] Then there is the Sonnet mistress's 'exceptionally dark complexion, a matter of great importance', in Tyler's mind. And so he journeyed to Mary Fitton's tomb in Gawsworth Church in Cheshire and returned in triumph with word that traces of dark paint were still visible on her effigy there.

He won notable converts: Furnivall, Georg Brandes, William Archer, and for a time Sidney Lee. Then the roof fell in on the whole edifice of speculation. In 1897 Lady Newdigate-Newdegate reproduced two previously unknown portraits of Mary Fitton in her *Gossip from a Muniment-Room*. They reveal that—unlike Shakespeare's swarthy, black-eyed, and raven-haired anti-heroine—the Queen's Maid of Honour had a fair complexion, grey eyes, and brown locks. Tyler protested that the pictures were fakes, but Lady Newdigate-Newdegate, in a second edition of her book the following year, furnished indisputable proof of their authenticity. Tyler's British Museum friend Shaw was convinced, and so was everyone else. The Tyler theory was finished, at least in reputable scholarly circles, although it would flourish again disreputably in Frank Harris's *The Man Shakespeare and His Tragic Life-story*.

The identity of the Rival Poet, the least prominent personage of the Sonnets, has mercifully occasioned the least speculation, most of the contenders for the honour being named before the century ended. Malone, as we have seen, thought of Spenser, and Boaden argued for Daniel. In his 1859 edition of the sequence, Robert Cartwright first proposed that Marlowe was 'just the splendid and dissipated character to dazzle and lead the young lord [Southampton] astray'.[94] Marlowe would have supporters: he is, we recall, the seducer of Willie Hughes in Wilde's *Portrait*. Mackay, keen to unravel what he regarded as the tangled skein of the Sonnets, went so far as to suggest, with some confidence, that Marlowe actually wrote a number of them; as supporting evidence, he naïvely cites a Collier forgery. In *Notes and Queries* (12 February 1859), J.G.R. nominated Jonson, perhaps an inevitable although never a popular candidate. William Minto, in *Characteristics of English Poets from Chaucer to Shirley* (1874), first put forward Chapman. The proud full sail of the rival's great verse, referred to in Sonnet 86, is displayed in Chapman's translation of Homer, just as the supernatural inspiration alluded to in the same sonnet ('his spirit, by spirits taught to write | Above a mortal pitch') is indicated in his *Shadow of Night*. Minto's guess quickly established itself as the favourite, and remains so. In concluding this survey we should not, however, overlook G. A. Leigh's discovery, announced in the *Westminster Review* for 1897, that the rival poet was Torquato Tasso. After all, does not the mention in Sonnet 78 of an 'alien pen' point unmistakably to a foreigner?

# 12

## *Faith and Works*

IN this period some bardolatrous amateurs studied Shakespeare's writings not for clues to the loves and rivals of his private life but for evidences of his doctrinal persuasion, or of his occupation before he turned to the stage. In *An Inquiry into the Philosophy and Religion of Shakspere* (1848), William John Birch, editor of the *Oracle of Reason*, proved the National Poet an atheist; the Presbyterian minister Thomas Carter offers a *Shakespeare Puritan and Recusant* (1897); Richard Simpson and Henry G. Bowden, the latter of whom was called Father Sebastian, demonstrate Shakespeare's Catholicism in *The Religion of Shakspere* (1899). It was even suggested, in the *Franciscan Annals* for August 1898, that Shakespeare clandestinely joined the Third Order, and was indeed buried in Franciscan garb—hence the gravestone malediction, for discovery of his religious affiliation would have resulted in the confiscation of his estate. Thus does each man convert Shakespeare to his own belief or infidelity. More interest attaches to speculations about his professional (albeit non-theatrical) skills.

William J. Thoms in 1859 proposed that Shakespeare had for a time seen service as a soldier in the Low Countries; perhaps in some encounter there he acquired the lameness to which he alludes in his Sonnets. Thoms points to passages in the plays indicative of military expertise. He attaches weight to Sir Philip Sidney's reference, in a letter dispatched from Utrecht in 1586, to 'my Lord of Leicester's jesting player'. For his trump card Thoms cites the William Shakespeare listed in a muster roll of trained soldiers within the hundred of Barlichway (in the town of Rowington) in 1605.* 'I think few of my readers will deny', Thoms concludes, 'that I have succeeded in my endeavour to establish the fact that SHAKESPEARE WAS A SOLDIER.'[95] The confidence of the theorizer is in this case, as in so many others, built upon shifting sands. William Shakespeare of Barlichway was of course not the poet but his namesake.

No martial type, Thoms was a gentle antiquary who founded *Notes and Queries*; the one occasion, in 1848, when he was called upon to shoulder a brown bess filled him with apprehension. But in the speculations of other students on Shakespeare's second profession we encounter the familiar pattern of self-projection. By the middle of the eighteenth century a tradition in the Royal College of Surgeons maintained that Shakespeare had stuck up his pole next door to the Boar's Head tavern in Eastcheap and there employed himself as a barber-surgeon. (The suggestion that Shakespeare flourished saw and scalpel as well as goose quill appears in print in 'A Medico-Chirurgical Commentary on Shakespeare', carried in 1829 by the *Quarterly Journal of Science of the Royal Institution*.) On the basis of the ruffian billows in *Henry IV* and the terminology of nautical manœuvring in *The Tempest*, old salts concluded that the poet must have been bred to the sea.† Physicians determined, on the basis of such evidence as the stuffed alligator, musty seeds, and other paraphernalia of the apothecary shop in *Romeo and Juliet*,

* Thoms is not the first to notice this record. The actual discovery may now, after the passage of over a century, be credited to Robert Lemon of the State Paper Office, who communicated his find to Collier in a letter dated 19 January 1854 and preserved in a copy of the latter's 1842–53 *Shakespeare* (vol. i) in the British Library. Collier discussed the record in the second edition of his Life (1858) without mention of Lemon.

† One anonymous mariner, the author of 'Shakespeare, a Seaman' in the *Saint James's Magazine* (July 1862), issues the following call: 'Messmates and shipmates! Let us clasp hands round the capstan, and take our solemn ship's oath, by the mainmast and by the rudder, and by the sheet anchor, that "Shakspeare was once a captain of the forecastle!"'

that Shakespeare had practised medicine during the Lost Years. An article in the *Gardener's Chronicle* (29 May 1841) argues, from the multitudinous horticultural references in the plays, that Shakespeare's second livelihood derived from professional gardening. Another piece, in the *Scottish Typographical Circular* for 2 August 1862, bears the title 'Shakspeare a Printer', a thesis independently developed by William Blades in *Shakspere and Typography; Being an Attempt to Show Shakspere's Personal Connection with, and Technical Knowledge of, the Art of Printing* (1872). One must add, in fairness to Blades, that his tone is not sober but facetious; his humour, however, was lost on the sober commentators who reported his argument.

The suggestion made in the preceding century by the barrister Malone, that the dramatist spent some years as a lawyer's clerk, received warm support from succeeding generations of attorneys. *Shakespeare a Lawyer* (1858), by William L. Rushton, adopts a less positive stance than the title would indicate. He sets out to show that Shakespeare had a technical knowledge not only of the law of real property—the principles and practice of which non-professional Elizabethans might be capable of understanding—but also of the common law, criminal law, and statute law. Rushton supports his case with numerous illustrations from the works. His advocacy stops short, however, of insistence that Shakespeare belonged to the legal profession; Rushton merely says that the poet displays 'a general knowledge of the laws of England'—a conclusion which few today would care to deny.

The same hypothesis occupies Lord Campbell in *Shakespeare's Legal Acquirements Considered*, published the following year. In his *Shakespeare Bibliography* Jaggard denounces this little book as an 'amazing plagiarism' of Rushton. The charge, however, is on the face of it unlikely, for Campbell dates his preface 1 January 1859: too close on the heels of *Shakespeare a Lawyer*. Collation of the two works reveals that Campbell cites a number of the same passages which Rushton had drawn attention to in the plays and poems; but such duplication is inescapable, given the undertaking. Otherwise the two treatises have little in common; Campbell's is fuller, better organized, and more urbane—just what one would expect when the Chief Justice enters the lists with a Liverpudlian law student.

Because of Campbell's special authority, some interest attaches to his interpretation of the will, which Rushton does not discuss. He sees, surely correctly, Shakespeare's aim as the perpetuation of the estate in a male heir descended from one of his daughters. When it comes to the second-best bed, Campbell resurrects Malone's opinion: it is a 'sorry bequest', appropriate to one '"no better than she should be"', whatever imaginary personal charms may be imputed to her'. The terseness of the document leads Campbell to believe that the dramatist himself composed it, and the precision of the language suggests professional legal experience on the part of the testator.

Thus the will strengthens Campbell's evident inclination to claim Shakespeare for his own fraternity. To do so he is not above tampering with the rules of evidence. Nashe's epistle to Greene's *Menaphon*, with its famous references to the trade of *noverint* and an early version of *Hamlet*, is taken to allude not to Kyd but to Shakespeare. Others had made Campbell's application, but he goes further and posits collusion between Shakespeare's enemies. 'This evidence', he concludes, '. . . seems amply sufficient to prove that there was a conspiracy between the two libellers, Nash and Robert Greene, and that Shakespeare was the object of it.'[96] Sufficient for the Chief Justice perhaps, but not for the impartial layman; for Campbell's 'evidence' consists of no more than a summary description of the passage in Greene.

In a pleasing fantasy Campbell sees Shakespeare as occupying a desk in the office of a prosperous country lawyer 'attending sessions and assizes,— keeping leets and law days,—and perhaps being sent up to the metropolis in term time to conduct suits before the Lord Chancellor or the superior courts of common law at Westminster. . . .'[97] Yet reality must at last intrude upon this idyll: Campbell has read Knight's popular Life, and he cannot ignore the biographer's devastating point about the absence from Warwickshire records of the poet's signature as a witness to deeds and wills. Shakespeare could have obtained his legal knowledge (which is anyway less impressively technical than the author believes) in other ways. Reason in the end prevails over desire, as Campbell settles for an agnostic position. (Others evinced less caution—for example, 'H—— T——', who answers in the affirmative the question posed by his title, *Was Shakespeare a Lawyer?* (1871). Unmoved, the *Athenaeum* reviewer took the author's initials to stand for Horribly Tiresome.)

The question of what Shakespeare did during the Lost Years would engage others who came after the Chief Justice; but the answer, despite no shortage of ingenious suggestions, would remain as elusive as ever.

# 13

## *Masks, Bones, and Portraits*

WHILE some probed the *œuvre* for clues to Shakespeare's personal and professional activities, others bestirred themselves over the portraits, purportedly of the poet, that were announced with increasing frequency to a briefly delighted world; perhaps, to these idolaters, Shakespeare's fancied physiognomy conveyed outward and visible signs of his inner and invisible genius. Of the portraits recovered in the nineteenth century, most famous is the Flower.

Around 1840 H. C. Clements, an artistic gentleman of Peckham Rye, obtained a portrait of Shakespeare—the identity of the subject would in this instance not be questioned—from an unnamed dealer. (Later it would be alleged that the source was a descendant of the dramatist's family; so do legends grow.) In the upper left-hand corner of the painting is written, in cursive script, '*Willm Shakespeare* 1609'. At the time of the purchase Clements pasted on the box in which he kept his prize a note with these words:

The original portrait of Shakespeare, from which the now famous Droeshout engraving was taken and inserted in the first collected edition of his works, published in 1623, being seven years after his death. The picture was painted nine years before his death, and consequently sixteen years before it was published. . . . The picture was publicly exhibited in London seven years ago, and many thousands went to see it.

Curiously, of the many thousands who viewed the portrait none left a record of the public exhibition. Clearly, however, the painting bears a relationship to Droeshout: stance, features, expression, and costume coincide.

The artist's panel, now worm-eaten and marred by other surface imperfections, is of ancient English elm on which (as radiography has shown) a quattrocentro artist had previously painted a Madonna and Child, with St John. The outline of her shoulder may still be discerned under Shakespeare's white wired band. His costume is otherwise black, with gold embroidery less elaborate than in the engraving. In 1895 Clements died, and the picture came into the possession of Mrs Charles Flower, who donated it to the Shakespeare Memorial Gallery in Stratford. Thenceforth known as the Flower Portrait, it has stirred much enthusiasm and debate.*

Connoisseurs—among them Sir Sidney Colvin, Sir Edward Poynter, and Sir Lionel Cust—have argued that it was indeed Droeshout's original. No doubt a good portrait, and an early one (perhaps *the* earliest authentic painting of the dramatist), it could not, however, have been made from life. One associates cursive script with a later period, and other evidence points inevitably to the conclusion that the artist worked from Droeshout, rather than vice versa. The painting lacks the animating touches, 'the accidents of a face' (in Spielmann's phrase), characteristic of a portrait made from a living subject. It improves upon the engraving, correcting the lighting and reducing exaggerations: surely Droeshout would not have deliberately introduced infelicities into his copy. But one telling circumstance clinches the argument. The painting

---

* The 'Droeshout Original' betrays tell-tale signs of cropping and sophistication: the Flower head and ruff now surmount a tunic which gives the appearance of having been, for the occasion, dry-cleaned. The picture is reproduced in an essay reconsidering it by Paul Bertram and Frank Cossa. '"Willm Shakespeare 1609": The Flower Portrait Revisited', in *Shakespeare Quarterly*, 37 [1986], 83–96. I discuss the Flower as well as other Shakespeare portraits in my paper, 'Artists' Images of Shakespeare', given at the Third Congress of the International Shakespeare Association, 1986, and included in its published proceedings, *Images of Shakespeare*, ed. Werner Habicht, D. J. Palmer, and Roger Pringle (Newark, Del., 1988), 19–39.

follows not the first state of the engraving, a copy of which was first brought to light by Halliwell-Phillipps, but a later version. Hence the long arched eyebrows, the heavier moustache and accentuated tuft under the lower lip, the shadow on the wired band.

Yet the derivative character of the Flower Portrait has failed to diminish the potency of its appeal. Sidney Lee, who knew the evidence, chose it as the frontispiece to his *Life of William Shakespeare*, for so long regarded as standard. More recently the 'Droeshout original' also has served as the frontispiece to A. L. Rowse's *William Shakespeare* (1963).*

Another icon dependent upon Droeshout is the Ely Palace portrait, found in 1845 in an obscure broker's shop, where it was bought for a few shillings by or for Thomas Turton, Bishop of Ely; cleaning revealed the inscription 'AE39. + 1603'. An announcement of the discovery appeared on 21 November 1846 in an architectural weekly, the *Builder*. Painted in oils on an oak panel, the picture has artistic merit, despite the general likeness of the head to the Droeshout engraving (moustache and right eyebrow are in fact identical with it). After Turton's death, the Ely Palace portrait came into the possession of the Birthplace, where for years it hung unnoticed in a peak of one of the upper chambers and gathered so much dust as to become almost invisible. But eventually it found an American champion, John Corbin, who in 1903 published a monograph entitled, not quite accurately, *A New Portrait of Shakespeare*. One cannot dismiss the painting out of hand as an imposture, but the absence of a verifiable pedigree challenges faith. So too any picture found in an obscure broker's shop must arouse instinctive suspicion. Inquiries revealed that the broker obtained the portrait from among the effects of an old London family established in Little Britain before the poet's time; Shakespeare knew and visited the family, and gave them the painting as a memento. Such a tale is not calculated to disarm scepticism. The Ely Palace is nevertheless the only painting of Shakespeare today hanging in the Birthplace.

So many other imputed portraits of Shakespeare came to light in the nineteenth century that to describe them all would be an impossible task, and anyway hardly worth the effort. The Lumley, Hampton Court, Stace, Dunford, and Auriol are still remembered; the Ashbourne, Bath, and Hunt deserve brief notice.

In 1847 the Revd Clement Usill Kingston of Ashbourne published an engraving of a portrait acquired for him by a friend from a London shop. 'I really believe it to be the best, and certainly the most interesting portrait of

---

* Shakespeare likenesses, genuine and dubious (the Stratford monument, the Droeshout engraving, and the Flower, Chandos, Janssen, Felton, Ashbourne, and Grafton portraits) are discussed and reproduced in my *William Shakespeare: Records and Images*, 155–200. The Flower portrait, which hangs in what is now called the Royal Shakespeare Theatre Gallery in Stratford, was dramatically restored in 1979, resulting in a painting with a dual identity: a half-length portrait of Shakespeare superimposed upon a slightly more than half-length Madonna and Child with Saint John (464–5). See the essay with illustrations by Bertram and Cossa, '"Willm Shakespeare 1609": The Flower Portrait Revisited'.

the immortal bard in existence,' Kingston wrote to the leading authority of the day, Abraham Wivell, on 8 March 1847. Wivell too expressed enthusiasm. And it *is* an interesting portrait: a three-quarter length figure, dignified in bearing, with fair skin, auburn hair, light beard, and an almost Spanish sobriety. The forehead is high, the expression contemplative, the attire rich yet sober. In his right hand the subject clutches a small book, perhaps a missal; a skull supports the arm (perhaps alluding to *Hamlet?*). In the top left-hand corner is inscribed in gilt:

AETATIS SVAE 47

A$^{\circ}$ 1611

It is a pity that the sitter—a physician? a philosopher? Shakespeare?—cannot be traced.

Underneath the present paint, as revealed by X-ray photography, lies another portrait of an unknown subject (could it be, as was suggested, Edward de Vere, seventeenth Earl of Oxford: a possibility calculated to make an anti-Stratfordian's heart skip a beat?), belonging—expert testimony in time concluded—to the second half of the seventeenth century. The earl gave up the ghost in 1604, and the present overpainting took place not earlier than the eighteenth century. Most probably the Ashbourne belongs with the other 'supply-portraits' manufactured in the eighteenth century from existing portraits to satisfy the burgeoning market for new portraits of Shakespeare. Today the Ashbourne hangs in the Folger Shakespeare Library. Cleaning of the painting by a conservator has revealed that the date 1611 of the inscription has been altered from 1612, the outline of the original '2' now being visible.

Several paintings during this period were attributed to Zuccaro, among them the portrait to which the Bath Librarian Archer first drew attention in 1859. It depicts an Italianate gentleman of around thirty, a jewel-string threading his ear and a lace ruff setting off his head; on the back is written, in an Italian hand of later date, 'W. Shakespear'. But Zuccaro, who left London not long after 1574, cannot be responsible for this portrait, nor for the others whose paternity has been laid to his fertile brush.

The third portrait, the Hunt, turned up in Stratford in 1860 in the home of the town clerk, William Oakes Hunt. He had employed Simon Collins, a well-regarded restorer, to remove the white paint which since Malone's time had defaced the waist-length figure of Shakespeare in Holy Trinity. This task done, Collins was commissioned to clean some old pictures in Hunt's possession. Under one, which had served admirably for target practice when the town clerk as a boy disported himself with bow and arrow, the restorer found a plump figure clad in red and black—the colours of the bust!—that bore an unmistakable resemblance to the church effigy. The similarity became even more uncanny when Collins finished his job of retouching. Had the seventeenth-century sculptor used the painting as the basis for his bust? More likely the unknown artist had followed the monument for his portrait, which may have been occasioned by the Garrick jubilee. This picture, Hunt's gift to Stratford, for long hung in the Birthplace encased in oak taken from the old structure of Shakespeare's house. The display case, the most authentic part of the portrait, has since been discarded, and the picture itself relegated to the Reserve Collection in New Place.

Not only paintings, but newly recovered pieces of statuary as well, rewarded nineteenth-century bardolaters. In 1848 workmen razing a wall in Spode and Copeland's china warehouse in Lincoln's Inn Fields, in order to make possible the enlargement of the museum of the College of Surgeons, came upon a red terracotta bust, overlaid with black paint, that was instantly recognized as a *vera imago* of Shakespeare. The warehouse stood on the site of the old Duke's Theatre of which Davenant became the first manager in 1660. Might not this *objet d'art* have adorned the proscenium of the playhouse governed by the poet who claimed Shakespeare as his godfather, and sometimes (when in his cups) even as his spiritual—or maybe natural— parent? The bust was acquired from the College by the surgeon William Clift, who passed it on to his son-in-law, the celebrated anatomist Professor (afterwards Sir Richard) Owen. He in turn sold the bust to the Duke of Devonshire, who presented it to the Garrick Club in 1855. Owen, described by a friend as 'a man not to be contradicted', believed firmly in its genuineness, and for a long time others did too. But the artist's source is almost certainly the Chandos portrait, and the style is eighteenth century. The skilled sculptor who carved the Garrick Club bust was probably the same L. F. Roubiliac who made the famous statue of Shakespeare for the great actor. The Duke had unwittingly bestowed his gift appropriately.

In the same year, 1849, that Owen was exhibiting his find in London, a foreigner brought to the British Museum a relic that made a more dramatic and lasting commotion. The stranger's name was Ludwig Becker, a portrait painter from Darmstadt who had settled in Mainz. His work had brought him recognition, and he had indeed been appointed Court Painter to the Grand Duke of Hesse. In Mainz in 1845 or 1846 he had purchased a tiny oil-on-parchment miniature, measuring little over three inches by two, that depicted a dead man crowned with bays lying in state. In the background a burning taper is dimly discernible and between the candle and the corpse appears, in bright gilt, the inscription 'Aō 1637'. Becker identified the body with Shakespeare, despite the fact that the likeness (especially the nose), the wreath, and the date all point to Jonson, the laureate who died in 1637. The awkward presence of the date Becker rationalized away by supposing it the year when the picture was copied from another portrait, or possibly from a statue or cast, although artists do not customarily date their copies so prominently. The miniature had belonged to the Count and Canon Francis von Kesselstadt, and was believed to have hung in his own bedroom with the label 'Traditionen nach Shakespeare'. In 1841 the Count died in Mainz, and the following year the portrait was sold at auction to an antiquary of the town, S. Jourdan; from him Becker bought it. There had been in the Kesselstadt collection, he learned, a plaster of Paris cast, and this item—perhaps the model for the painting—he now sought. After two years of fruitless inquiries Becker found it in a broker's rag shop in Mainz. He did not doubt that picture and cast were of the same person; others did. It was on the cast that interest would focus.

This object has come to be known as the Kesselstadt Death Mask. In the nineteenth century it was not doubted that the Mask had been made from a dead face: traces of auburn-coloured hair clung on the inside to the moustache, the beard, the eyebrows and lids. The foremost modern authority on Shakespearian iconography, however, regards the relic as not properly a death mask at all but a cast from one, and probably not even a direct cast; the facial hairs perhaps being transferred from the original during casting. There can be no question that the narrow oval face is pleasing: to many it would not seem incompatible with the ideal image of a philosophic poet. The receding forehead is full, yet delicate; the nose is long and finely arched; the narrow chin is pointed; the eyes are sunken, the cheeks drawn. It is a dignified and melancholy visage—Hamlet's creator might have had such an aspect. The effect of the Mask, which brings the viewer as it were directly into the presence of the great and memorable dead, is ineffably moving. In 1880 Ronald Gower expressed what no doubt many had felt when looking at the Kesselstadt Death Mask: 'Sentimentally speaking, I am convinced that this is indeed no other but Shakespeare's face; that none but the great immortal looked thus in death, and bore so grandly stamped on his high brow and serene features the promise of an immortality not of this earth alone.'[98] On the back of the Mask is cut, in three places, the legend '+ A$^{\underline{o}D}$m̄ 1616'. It is the one tangible link with Shakespeare—and may have been introduced later.

How did the Mask, if genuine, come into the possession of the German nobility? Had some ancestor of the Count visited England and there purchased a memento of Shakespeare? Unfortunately no record exists of a Kesselstadt visit to Britain. Perhaps Gheerart Janssen's sons brought back with them to their native Amsterdam the 'flying-mould' from which their father had sculpted the Holy Trinity bust. Possibly thence it passed to Cologne, where the Kesselstadts had formerly resided. But that the Mask had ever belonged to the Count is not verifiable. Conceivably the rag broker, having heard of Becker's quest, hoodwinked him; possibly Becker himself was a rogue. Such thoughts did not occur to the English enthusiasts in 1849, nor did they concern themselves with the irreconcilability of the Mask with the two authentic likenesses of Shakespeare. Experts examined the cast and were impressed, and the multitude flocked to the British Museum to gape at it in its glass case. Sir Richard Owen professed his faith. So too did the American painter William Page, for whom the Mask became a passion. He took many measurements and compared them with the statistics for the Stratford monument. Almost half, he asserted, corresponded; the majority that did not failed to trouble him. Becker went off to Australia and later perished in a British expedition across the continent. The Mask was exhibited in Stratford during the Shakespeare tercentenary, and until at least the end of the century pilgrims could there purchase stereoscopic photographs of it as souvenirs.

Eventually the relic returned to the Becker family and came into the custody of Ludwig's brother, Dr Ernest Becker, curator of the Grand Ducal Museum at Darmstadt. Enthusiasm for the Mask, as for so many purported portraits of Shakespeare, gradually waned, and little was heard about it until in 1961 it was reported as up for auction, with a catalogue price of 40,000 DM. By an odd coincidence an article in advocacy appeared the same year. The author, Frederick J. Pohl, an enthusiast of the Vikings and Amerigo Vespucci, concluded on the basis of comparative measurements that the Mask represented the same man as the Stratford bust.[99] Pohl put the odds in his favour at a trillion to one; some bookmakers, one suspects, would offer rather less. The Kesselstadt Mask resides to this day—appropriately enough—in the Darmstadt Museum.

Earlier controversy over the authenticity of the Kesselstadt Mask largely inspired the strange movement, in the name of Victorian scientism, to exhume Shakespeare's bones. 'If he had had a Boswell,' Dickens said of Shakespeare, 'society wouldn't have respected his grave, but would calmly have had his skull in the phrenological shopwindows.'[100] That was in 1847; beginning in the 1860s we hear serious talk of graves, of worms, and epitaphs. The tercentenary number of *Chambers's Journal* (23 April 1864) asked that 'in the interests of science physical and moral, the relics of the great Shakespeare . . . be subjected to a thorough examination'. But it was J. Parker Norris, a US student of the Shakespeare portraits, who first aroused emotion when he proposed disinterment in the 'Shaksperian Gossip' section of the *American Bibliopolist* for April 1876. 'If we could even get a photograph of Shakespeare's skull', he owned, 'it would be a great thing, and would help us to make a better portrait of him than we now possess. . . . Open the grave reverently, have the photographers ready, and the moment the coffin lid is removed (if there be any) expose the plates, and see what will be the result.'[101]

But had the grave been, in fact, already opened? After Garrick's jubilee, it was reported, Horace Walpole had promised George Selwyn three hundred guineas for Shakespeare's skull, if he could come by it. Years later, in 1794, a young doctor named Frank Chambers took up this sporting offer; breaking into Stratford Church by night, he procured the skull with the aid of local ruffians. But the master of Strawberry Hill declined to buy, and the *memento mori* found its way to the family burial vaults of the Sheldons, presumably hard by to the Beoley parish church in Worcestershire, where so many Sheldons were interred. So the story goes, but it was first published (in part) almost a century later, in the *Argosy* for October 1879, and five years later expanded into a pamphlet, *How Shakespeare's Skull Was Stolen and Found*, by 'A Warwickshire Man'. The author, C. L. Langston, happened to be the incumbent of Beoley Vicarage. His narrative reads like what it must surely be: a lurid fiction.*

There is better evidence that the grave had actually yielded its secret. During his memorable visit to Stratford-upon-Avon early in the nineteenth century, Washington Irving was told how the earth had caved in as labourers dug a vault adjoining Shakespeare's. The old sexton stood guard to prevent vandals from entering the vacant space that reached into the dramatist's grave, but could not resist peering in himself; there he saw neither coffin nor bones, only dust. 'It was something, I thought,' Irving muses, 'to have seen the dust of Shakspeare.'[102] A somewhat different version of the same episode appeared in the *Monthly Magazine* for February 1818. 'Mr. ——— told the writer, that he was excited by curiosity to push his head and shoulders through the cavity', the correspondent reports, 'that he saw the remains of the Bard, and that he could easily have brought away his skull, but was deterred by the curse which the poet invoked on any one who disturbed his remains.'[103] Another witness who peered through the aperture, however, could discern no coffin, skull, or undecomposed bones; only a small quantity of mouldering dust on the level floor.[104]

In any event, angry outcries greeted the desecration proposed by Norris. The *Birmingham Daily Post* of 29 September 1877 reported the reaction of the vicar of Holy Trinity: 'Having dilated upon the cool presumption of the author of the letter, Dr. Collis continued, that persons proposing such an experiment would have to walk over his prostrate body before they did it; adding that the writer even forgot to say "if you please"'. But the vicar died, thus obligingly providing a prostrate body for his opponents to walk over, and Norris, despite his crude American manners, found a valuable ally in C. M. Ingleby, Collier's nemesis.

---

* Langston received enquiries from everywhere. One reader, Samuel Timmins, he favoured with a mischievous doggerel, now pasted into the copy of his *Argosy* article in the Birmingham Shakespeare Library:

> Should you enquire 'Pray sir what particle
> Of truth pertains to this strange article?'
> It needs perhaps some little tact
> To trace the fiction in the fact:
> For what is fact and what is fiction
> I must refer you to its diction.

The review that appeared with a version of this rhyme is, I suspect, by Timmins.

Ingleby argued eloquently against the twin grounds for abstention—sentiment and the gravestone malediction—in his curious little tract, *Shakespeare's Bones. The Proposal to Disinter Them, Considered in Relation to Their Possible Bearing on His Portraiture: Illustrated by Instances of Visits of the Living to the Dead* (1883). He spoke with a special authority, not only as a respected Shakespearian, but also as a Life Trustee of the Birthplace, Museum, and New Place. With perhaps excessive tact, Ingleby dedicated his book to the Mayor and Corporation of Stratford and the vicar of Holy Trinity. He presented gift copies of the book to influential persons, who responded in character. A. E. Brae looked forward to the exhumation with touching antiquarian optimism: soon it might be possible to measure Shakespeare's leg bone, and decide once and for all the question of his lameness. Dowden was moved to a flight of humorous fancy. 'My own notion', he wrote frivolously to Ingleby, 'is that Bacon after his decease made away with Shakspere's skull, & then having forgotten which was his own & which Shakspere's & being ashamed of what he had done, he never returned to his grave; which accounts for no remains having remained in St. Michaels Church.'[105] Cardinal Newman, on the other hand, displayed a pious gravity appropriate to a prince of the church: 'As to your proposal itself,' he confided, 'I should rather reserve such veneration of relics for the instance of canonized Saints, and I have sometimes fancied there was something of idolatry in our treatment of William Shakespeare.'[106] But support came to Ingleby from Samuel Timmins, the Birthplace Librarian who had written a history of Warwickshire, and (more important) from the Revd George Arbuthnot, custodian of Shakespeare's grave, who yielded with misgivings to the imperious necessities of phrenological science.

They had not reckoned on the intensity of public indignation. Ingleby was crucified in the correspondence columns of *The Times*. Halliwell-Phillipps, still Stratford's friend, wrote to the Corporation in vehement protest: if the skull matched the Stratford bust it would only confirm the already known; if it did not, maybe it was someone else's skull. When the town council met on 4 September 1884 the mayor informed the Corporation that he had received indignant letters from around the world. Not surprisingly, sentiment and the malediction prevailed. The diggers put aside their spades; Shakespeare's mortal remains—if they still existed—were left undisturbed.*

---

* It should not be assumed from this fiasco that the grave has since remained immune from attempted assault. *The Times* (3 September 1969) reports an unidentified gentleman as preparing to apply to open the poet's grave on behalf of the Shakespeare Action Committee.

# 14
## *Arnold and Bagehot*

WE need not wonder that some wished to finger the hard bone of the master's skull, to confront his material presence; for he appeared so remote, so inscrutable. 'I keep saying, Shakspeare, Shakspeare, you are as obscure as life is,' wrote Matthew Arnold to Arthur Hugh Clough in December 1847. To Arnold, as to Schiller, Shakespeare was the objective poet hiding himself behind his work like Deity behind the universe. In his sonnet addressed to the self-schooled, self-scanned, self-honoured, self-secure artist, Arnold made the point most memorably for his own age and for all time:

> Others abide our question. Thou art free.
> We ask and ask—Thou smilest and art still,
> Out-topping knowledge.

Yet others found Shakespeare the myriad-faceted accessible.

Among these was Walter Bagehot, no literary scholar but banker, politician, editor of *The Economist*, and a most attractive humanist. At the age of twenty-seven he published in the *Prospective Review* for July 1853 his essay 'Shakespeare', which he later slightly revised as 'Shakespeare—The Individual'.* In it the author reveals quite nakedly the limitations of his biographical knowledge, which has approximately the precision of an encyclopaedia article vaguely remembered. Not very promising; yet Bagehot has read the plays and poems with warm responsiveness, and they serve him well. He knows the arguments against deducing a writer's personality from his work, deflects them gracefully, and goes on to handle his elusive subject with such civilized grace and gentle humour as to silence protest. This despite our awareness of an element of self-portraiture—idealized self-portraiture—in his image of Shakespeare.

The poet is a hunter of hares, a judge of dogs, an outdoorsman. From boyhood Bagehot was, as his friend Hutton observes, 'excessively fond of hunting, vaulting, and almost all muscular effort': he kept a pack of hounds and chased the rabbit across the moors. From the pain and suffering of his grammar-school years, Shakespeare derived 'not exactly an acquaintance with Greek and Latin, but like Eton boys, a firm conviction that there are such languages'. (Bagehot knew the classical tongues rather better than that.) Shakespeare's great familiarity with the common people gave him a

---

* Since Hutton's edition of Bagehot's literary essays in 1879, it has been customarily entitled 'Shakespeare the Man', but this designation has no authority.

sympathy with the narrow intelligence induced by narrow circumstances; he recognized but was unangered by the stupidities of the 'illogical classes'. Still, he disbelieved in the 'pecuniary classes'. Politically he was loyal to the ancient traditions of the British Crown. (Bagehot, who advocated moderate reform of the franchise, opposed allowing the ignorant to govern the educated: he appreciated too the limitations of the business world.) Like the essayist, Shakespeare was a cheerful and humorous man, quick of mind, whose gaiety was in solitude softened by a musing sadness. His religion—like Bagehot's—was of weekdays rather than of the Sunday pulpit; he saw the unity of Nature and penetrated, in Hamlet and other creations, to the secret principles of the inner life.

Agreeable sentiment colours the concluding picture of Shakespeare in his last years, after retirement to Stratford as a moneyed man:

We seem to see him eyeing the burgesses with good-humoured fellowship and genial (though suppressed and half unconscious) contempt, drawing out their old stories, and acquiescing in their foolish notions, with everything in his head and easy sayings upon his tongue,—a full mind and a deep dark eye, that played upon an easy scene— now in fanciful solitude, now, in cheerful society; now occupied with deep thoughts, now and equally so, with trivial recreations, forgetting the dramatist in the man of substance, and the poet in the happy companion; beloved and even respected, with a hope for every one, and a smile for all.[107]

The dominant hues of Bagehot's palette are too roseate for the modern taste; Shakespeare our contemporary wears many masks, but not a smiling one to greet all comers. Yet with its cunningly chosen quotations, its felicity of phrase and warmth of feeling, this portrait so curiously resembling the cultivated literary amateur who limned it, retains an unfashionable attractiveness. For this, credit must go to Bagehot the Individual.

# 15

# *Other Amateurs*

BAGEHOT was merely one among many Victorian amateur enthusiasts of Shakespeare who sought to characterize their idol in print. As the century wears on, their numbers multiply and become a source of weariness to the historian, who, even if he has no intention of recording all, must nevertheless read them. Occasionally, it is true, the work of an obscure devotee arrests us by its originality and soundness of judgement. Such is the case with *The Life and Genius of Shakespeare*, published in the tercentenary year by the almost altogether unknown Thomas Kenny.

The British Library Catalogue of Printed Books lists no other title by Kenny; he is not memorialized by the *Dictionary of National Biography*. Neglected for almost a century, the *Life and Genius* (which never achieved a second impression) has recently been accorded due honour for anticipating the modern position with respect to the authorship of the *Henry VI* plays. Kenny argues that Shakespeare wrote the trilogy unaided, and that *The First Part of the Contention of York and Lancaster* and *The True Tragedy of Richard Duke of York*, formerly regarded as source plays, are in fact imperfect versions of Shakespearian originals. This thesis has a crucial biographical implication that does not escape its promulgator: it topples the traditional view that Shakespeare began his stage connection as a play-adapter. Kenny writes:

We can hardly imagine this large, negligent workman engaged in the literary drudgery of omitting, enlarging, transposing, and amending the thoughts of another writer, and proceeding, at each step of his progress, with a constant and minute reference to his model. We believe that his rapid and airy fancy would have wholly failed him in such a task; and this, surely, was not the kind of work by which he was to astonish and to overshadow all the dramatists of his age.[108]

Here Kenny aligns himself with the wave of the future in Shakespearian biography.

Recognizing that he is no Collier or Halliwell-Phillipps, he disclaims any pretence to profound scholarship. His book is disproportioned and at times haphazard: the *Henry VI* sequence usurps almost a third of Kenny's pages, whereas other plays go undiscussed in the critical survey. He can be eccentric in the true fashion of the amateur, as when he insists that the Sonnets are 'feeble, diffuse, indistinct, without any concentrated interest of thought and feeling'. In his chapter on the life Kenny too readily concludes that the poet married unhappily, and he exaggerates John Shakespeare's financial misfortunes. But by and large Kenny presents the facts in unadorned fashion, without that admixture of traditionary lore which is a standard feature of most biographies in this period. His sensible handling of the deer-poaching story is characteristic: he refuses to believe that Shakespeare, when he came to write *The Merry Wives of Windsor* in the maturity of his powers, bitterly remembered a youthful escapade. In a chapter on Shakespeare's character, Kenny avoids the usual pieties, attributing the scantiness of the evidence to 'the absence of any very marked incidents in the poet's career, and of any very imposing personality in the poet himself'.[109]

In a sensitive passage, Kenny seeks to reconcile those opposing *personae*, the practical businessman of the documentary record and the poet, impassioned and profound, of the dramatic record:

The direct controlling influence in his daily life, the special incentive to all his labours, was the desire to accumulate a fortune, and to secure those social advantages by which the possession of wealth is naturally accompanied. This was the counterpoise to the extravagant emotional and meditative tendencies of his nature. It was by this practical instinct that he held on to the realities of human existence—that, in its agitations and its struggles he was a steadfast actor, and not a mere amazed observer and a passionate dreamer—that he resisted the ceaseless pressure of a restless imagination—that he offered a determined front to the ever-rushing invasion of the wonder and the mystery of this changeful world of time and place. It was the familiar landmark that fixed for him his own little home in the infinite ocean of life.[110]

Later critics would consciously or unconsciously (more likely the latter) echo this view.

*The Life and Genius of Shakespeare* is, in sum, a remarkable book, a unique product of non-professional scholarship. If, however, the historian reads Kenny with delight, he must pay a heavy price for such isolated pleasures by contending with the host of offerings—boring or incompetent or both—that inspired amateurism has placed on the shrine of bardolatry. Take, for example, James Walter's *Shakespeare's True Life* (1890), which boasts as its chief attraction several hundred insipid sketches by Gerald E. Moira. Regrettably these illustrate a text. One must trudge with Walter through the poet's Warwickshire and London, suffer the biographer's denigration of 'Mr Haliwell Phillips', and suppress a yawn as he proclaims his single 'revelation'—the non-fact that Shakespeare and Anne Hathaway underwent a ceremonial 'whatever it was' (in Walter's poignant phrase), in the Roof Chapel or the Oratory of the Old Manor House in Shottery. But the historian must endure severer trials. Walter merely challenges his patience; with Heraud and Fullom he must look to his sanity.

John Abraham Heraud, author of *Shakspere: His Inner Life as Intimated in His Works* (1865), would seem to bring the right qualifications to his task. A friend of Carlyle and the acquaintance of Coleridge and Wordsworth, he made a career for himself as poet, playwright, and dramatic critic for the *Athenaeum*. But he also quaffed deep draughts from the stein of German philosophy, and the resultant intoxication proved disastrous. 'I had to settle what Life was', Heraud ominously announces early on in his five-hundred-page 'reverential inquiry'; 'and to trace it from mere individuation to individuality.'[111] A metaphysical quest requires chronological guidelines, and so Heraud establishes to his own satisfaction the succession of the plays—a succession in which *Hamlet* precedes *Henry VI*, and *Julius Caesar* follows *The Winter's Tale*. Heraud's arrangement of the canon into four periods perhaps murkily anticipates Dowden's famous division, but his interpretative readings belong to this critic alone. 'The Man has become a Messiah', Heraud writes of the Sonnets, 'and the Woman the Church. The former retains his fairness, but the latter is depicted as black. She is, in fact, the black but comely bride of

Solomon. . . . Both claim the allegiance of Shakspere,—his "female evil" and his "better angel:" . . . Thus it is that Shakspere parabolically opposed the Mariolatry of his time to the purer devotion of the Word of God, which it was the mission of his age to inaugurate.'[112] From such commentary the reader starts, as from a waking dream.

In his *History of William Shakespeare* (1862), S. W. Fullom subjects his audience to an ordeal of another kind. The popularizing author of *The Human Mind* and *The Miracles of Science* has no metaphysical pretensions: the credentials he brings to Shakespearian biography differ from Heraud's, and are (one would expect) more relevant. Fullom bears a superficial resemblance to the professional investigator. He has immersed himself totally in his subject. In Warwickshire he has tirelessly sought out new information. From the Revd Donald Cameron, Rector of Snitterfield, who was present at the demolition of Richard Shakespeare's cottage, he has obtained a description of the structure that stood on the green, over a muddy brook. Robert Arden's property in Wilmcote had for long eluded discovery, but there the local patriarch, one John Price, aged eighty-four, led Fullom to the site of the old homestead, a dilapidated edifice bearing the appropriate name of Starve Hall. At Luddington he has had even better luck.

In this village on the north bank of the Avon, some three miles west of Stratford, Fullom pursued the record of Shakespeare's marriage. Fire had long since consumed the chapel, but the old parsonage still stood:

The house is occupied by a family named Dyke, respected for miles round, and here the report of the marriage can be traced back directly for a hundred and fifty years. Mrs. Dyke received it from Martha Casebrooke, who died at the age of ninety, after residing her whole life in the village, and not only declared that she was told in her childhood that the marriage was solemnized at Luddington, but had seen the ancient tome in which it was registered. This, indeed, we found, on visiting the neighbouring cottages, was remembered by persons still living, when it was in the possession of a Mrs. Pickering, who had been housekeeper to Mr. Coles, the last curate; and one cold day burnt the register to boil her kettle![113]

This account is sufficiently circumstantial to suggest at least the possibility of a true report.

Whatever else, Fullom has obtained from his rambles a nice sense of the terrain. He has followed the Avon's meandering course, he has seen the forest of Arden and the countryside that, viewed from the top of Bowden Hill, stretches away to the horizon beyond Snitterfield Bush. Yet, with the various tools of biography to hand, Fullom has written what can only be regarded as a bewildering disaster.

Facts, the poet's works, and the traditions—above all, the traditions—strangely blend. Shakespeare's grandfather furnished the model for the Corin of *As You Like It*; the dramatist represented Anne Hathaway as Perdita, and himself as Prince Hal. Through the character of Cranmer and the events leading to the fall of the favourite Wolsey, Shakespeare appealed in *Henry VIII* to Elizabeth for the life of Essex (executed a dozen years before the composition of the play). He addressed his sonnets of separation to Anne Hathaway during their engagement; as for the 'sweet boy', he is no lordling but the poet's little son. Such biographical gems does the corpus yield. Somehow the recorded facts of Shakespeare's career arrange themselves according to the pattern of Victorian melodrama—John Shakespeare in his time of legal troubles over debt plays his part in a curious scene involving 'the distressed father; the dejected mother; the avaricious, plotting uncle; himself [William] the rightful heir'. Fullom rejects the apocryphal stories which represent the dramatist as enjoying the pleasures of bottle or bed, but otherwise this biographer hugs to his bosom any traditional tale, however dubious, for in traditions he finds the flesh and blood that the documents so frustratingly lack.

Fullom's Shakespeare is the butcher's apprentice of Stratford, who killed a calf in high style and made a speech. 'Could we but recover one of those orations!' he sighs. 'Crude it would be, no doubt, but we should see Mind sparkling through it—the precious metal veining the quartz.'[114] The legend of the Seven Champions of Stratford in their mighty contest with the Sippers of Bidford, followed by the sobering aftermath under the crab-tree, receives its most lovingly detailed rendering: 'The tradition now appears as it was related to us on the spot by Mr. Bagshawe, of Bidford, to whom it was told in his childhood by the old people of the village, which carries it back for at least a hundred years; and it was confirmed to us by old people still living at Bidford.'[115] Epic proportions are assumed by the deer-stealing story, which stretches out across four chapters. Here Fullom makes his special late contribution to the Shakespeare-Mythos.

At Charlecote Park he has extracted from Mrs Lucy a manuscript pedigree, made about ninety years previously by an old man named Ward with help from family papers available to him; this pedigree (Fullom reports) contained a note on the poaching affair. Infuriatingly, he neglects to give the text of his prize documentary recovery, but one can fit together the essential pieces of the story. After cutting up with his butcher's knife the stag intended to grace his nuptial feast, Shakespeare was captured and imprisoned, perhaps at Warwick. While he languished in jail, 'his friends', Ward reveals, 'interested in his behalf the most important man in Warwickshire, no less a person than Robert, Earl of Leicester; and this great magnate now interceded with Sir Thomas Lucy, and prevailed on him to abandon the prosecution'.[116] After release, the impenitent scapegrace (in Ward, as in Rowe) nailed his satirical ballad to the park gate at Charlecote, and as a result found himself obliged to leave Stratford, 'at least for a time'. Years afterward Leicester demanded of Lucy that he 'adopt his badge of the bear and ragged staff, which he wished to see worn by the servants of all the gentry of Warwickshire, and the Charlecote Knight refused to comply, at the same time calling him an upstart'. Recollecting Justice Shallow in *2 Henry IV*, the Earl asked Shake-speare to take up his quarrel by representing Sir Thomas 'again' in *The Merry Wives of Windsor*. Thus did Shakespeare repay his debt to the noble lord, and thus is explained the armorial allusiveness of the comedy—'this was natur-ally made the point of retaliation by Leicester, as a sneer at his own arms was the ground of offence'.[117]

The rest of the narrative is of a piece with these revelations. Of course this Shakespeare, after his arrival indigent in London, organized a horse-holding service outside the theatre: '. . . the horsemen rode up, calling for Shake-speare, and being met by a cry of "I am Shakespeare's boy, sir! I am for Shakespeare," unhesitatingly resigned their steeds, knowing that, at the close of the performance, they would find them at the door. The boys held the horses, and Shakespeare took the charge. He did it in high style!'[118] The story gets better with each retelling. As a strolling player still in his teens Shakespeare toured every part of England and Wales, and naturally he noted each detail of the scene for future reference when in Dover he stood atop Shakespeare's cliff. By his twentieth year he had established his reputation in London and was making copious additions to *The Merry Devil of Edmonton* and the other apocryphal dramas of which every sober scholar acquits him. Sidney and Spenser befriended Shakespeare, and he enjoyed the protection of Leicester, a subsidy from whom 'beyond doubt' enabled the dramatist to relieve his father's wants. Successful and the companion of many powerful nobles, Shakespeare responded with bitter grief to the death of Hamnet in 1596, and began to prepare for his retirement. These preparations must have occupied him for some time, for (so Fullom informs us) he did not write his last play until 1605. The dramatist's later years were brightened by a reconciliation with Sir Thomas, who threw open to the erstwhile poacher the splendid library at Charlecote.

Written with a seeming rationality, Fullom's *History of William Shakespeare* is not really biography as it is usually understood, but, rather, phantasma-gorical romance. It illustrates what can happen when a totally uncritical mind operates on the materials for a Life and mimics the processes of scholarship. This latter nineteenth-century *History* testifies also to the abiding power of legend. Against the background of scientism and the methodical researches of a Halliwell-Phillipps, Fullom stands aloof: a trea-surer of the tradition, a maker of the mythos.

# 16

## The New Shakspere Society

IT would be hard to imagine a more striking contrast than that between the miscellaneous outpourings of these amateurs and the vast organized programme, professional in orientation, which came into being with the establishment of the New Shakspere Society. The Society's first meeting came to order at 8.00 p.m. on Friday, 13 March 1874, at University College in Gower Street, London.[119]

In the chair sat F. J. Furnivall: Butler's amiable, kindly greybeard; Swinburne's ferocious blackguard. 'Devoid of tact or discretion in almost every relation of life', observes his long-time associate Sir Sidney Lee, 'he cherished throughout his career a boyish frankness of speech which offended many and led him into unedifying controversies. He cannot be absolved of a tendency to make mischief and stir up strife.'[120] Furnivall's temper may perhaps be most charitably viewed as a manifestation of titanic energies that produced antagonisms, failures, and splendid achievements. All serious students of English language and literature owe him an everlasting debt of gratitude for his contribution to the *New English Dictionary* as originator, organizer, editor, and contributor. An abstainer from flesh, alcohol, and tobacco, Furnivall obtained solace from spelling reform—a Shavian cause irresistibly alluring to teetotalling vegetarians. In 1854 Furnivall helped to found the Working Men's College in Red Lion Square. Later, adapting his principles of group enterprise to literary studies, he set up the Chaucer Society, the Ballad Society, the Wiclif Society, the Shelley Society, and the Browning Society. They have all perished, but not before making their contribution. The Early English Text Society, which Furnivall formed in 1864, survives to carry on its valuable work of providing sound editions of medieval texts.

By 1874 the first generation of Victorian scholars—the school of Dyce, Collier, and Halliwell-Phillipps—had already made their impact, and a new company, armed with new techniques, was arriving on the scene. These scholars would find a forum in the meetings and publications of the New Shakspere Society. The Society was founded in the hope that advances in other fields would find a correlative in Shakespeare studies—'. . . in this Victorian time,' Furnivall wrote in his prospectus, 'when our geniuses of Science are so wresting her secrets from Nature as to make our days memorable for ever, the faithful student of SHAKSPERE need not fear that he will be unable to pierce through the crowds of forms that exhibit SHAKSPERE'S mind, to the mind itself, the man himself, and see him as he was.' To see him as he was; that, after all, had been Bagehot's endeavour. Like Bagehot, too, the Society would focus on the writings, not the documentary record, which Halliwell-Phillipps had explored so thoroughly (it was felt) as to leave little for others to do.

The method of the New Shaksperians would be to establish by objective tests, mainly metrical, the chronological succession of the dramatist's works, to verify the results by the Higher Criticism,

and then to use that revised order for the purpose of studying the progress and meaning of Shakspere's mind, the passage of it from the fun and word-play, the lightness, the passion, of the Comedies of Youth, through the patriotism (still with comedy of more meaning) of the Histories of Middle Age, to the great tragedies dealing with the deepest questions of man in Later Life; and then at last to the poet's peaceful and quiet home-life again in Stratford, where he ends with his Prospero and Miranda, his Leontes finding again his wife and daughter in Hermione and Perdita; in whom we may fancy that the Stratford both of his early and late days lives again, and that the daughters he saw there, the sweet English maidens, the pleasant country scenes around him, passt [spelling reform] as it were again into his plays. So that, starting with him in London, . . . we get him at last down quietly in his country home again, with the beauty of that country, wife and girl, and friends around him; with sheep-shearings to be talkt of, and Perdita with the Spring flowers to be lovd, and everything else serenely enjoyd; and so he ends his life.[121]

This sentimental and hallucinatory sketch, in which the creator and his creations hopelessly merge, shows that Furnivall was no critic.

Later he expanded his précis into a long introductory memoir for the popular *Leopold Shakspere* (1877). As biography it holds little interest. Furnivall vigorously (he is never anything but vigorous) affirms his faith in the autobiographical validity of the *œuvre*: 'I refuse to separate Shakspere the man from Shakspere the artist,' he declares. The problem, acknowledged in characteristically insensitive phrasing, is to pick out 'the extra-dramatic bits from the plays, and . . . [combine] them with the like bits in the poems'. His own reconstruction of the life-in-the-work is scrappy, disproportioned, and crudely written—all of which Furnivall cheerfully admits, explaining that much of it was dictated to an amanuensis, amid the pressure of other work, from old notes and recollections. But Furnivall concerns us less as biographer than as founder of the New Shakspere Society, of which he was primarily the compère. Other performers did the turns.

An entertainment must have a stellar attraction. During the first months of the Society's life the centre of the stage was held by Frederick Gard Fleay. His preparation augured well. At Trinity College, Cambridge, he had applied himself so diligently to his studies that his fellow undergraduates termed him 'the industrious flea'. His training in the natural sciences ideally prepared him, he felt, for applying quantitative tests to literature. 'The great need for any critic who attempts to use these tests', Fleay insisted, 'is to have had a thorough training in the Natural Sciences, especially in Mineralogy, classificatory Botany, and above all, in Chemical Analysis. The methods of all these sciences are applicable to this kind of criticism, which, indeed, can scarcely be understood without them.'[122] Increasingly alienated from the Established Church in which he had taken orders, and burdened with uncongenial duties as headmaster of Skipton Grammar School in the North, he welcomed the advent of the New Shakspere Society, which would listen to his novel theories, and discuss and publish them.

In common with Furnivall and other eminent Victorians, Fleay had seemingly limitless reserves of energy: he read prodigiously, mastered the available information on the Elizabethan stage, and subjected scores of plays to statistical analysis. His intuition, moreover, led him to make a number of lucky guesses. But demons of eccentricity and error plagued the industrious Fleay. The scientist of literature would state dogmatic conclusions without evidence, and reverse himself with equal positiveness; then firmly take still another incompatible position. He is forever discovering the truth, and the truth is always different. He is also forever making mistakes. In his *Shakespeare Manual* (1878) Fleay found, or had pointed out to him, over one hundred errors. His method is mathematical, yet the mysteries of simple arithmetic seem always to elude his grasp. To these oddities must be added a streak of childish capriciousness. Bullen tells a delightful anecdote: '. . . on one occasion, when I objected to some peculiarly far-fetched theory of his, he protested that it was not to be taken seriously but was "intended as a skit on the New Shakspere Society"; whereupon I reminded him of the fact (which he had forgotten) that he originally announced this theory in a school-edition of *King John* (when he was headmaster of Skipton Grammar School); and I mildly expostulated with him for mystifying schoolboys.'[123]

That Fleay's habits yielded eccentric results need hardly surprise us. In his inaugural paper for the New Shakspere Society he concluded, on the basis of metrical data, that *Macbeth* preceded *Hamlet*, *Othello*, and *Lear*; he assigned *The Taming of the Shrew* to 1600, *Cymbeline* doubtfully—with good cause—to 1604, and *Julius Caesar* (with a query) to 1607. The minutes of the discussions following the presentation show that from the first gratitude for Fleay's toilsome labour was mingled with scepticism regarding his methods and findings. Alarmed, Halliwell-Phillipps, who kept away from the Society's meetings, wrote to Furnivall urging him to turn the attention of members to the necessity of searching for facts and documents before theorizing about the dates of Shakespeare's plays.

Fleay nevertheless persevered. Armed with his infallible tests he attacked the Shakespeare canon, which was promptly defeated. He found numerous spurious passages in the plays: he detected signs of Jonson's hand in *Julius Caesar*; he took the Porter scene away from Shakespeare, then returned it when Furnivall exploded. But Furnivall remained unappeased, and Fleay's wild speculations became too much for his other early supporters in the Society. Eventually he dropped out, and he resigned also from his Skipton headmastership and (in time) from the ministry. In the seclusion of his home in Upper Tooting he devoted himself to his grand projects: *A Chronicle History of the Life and Work of William Shakespeare* (1886), *A Chronicle History of the London Stage, 1559–1642* (1890), and *A Biographical Chronicle of the English Drama, 1559–1642* (2 vols.; 1891). At some unspecified time he also wrote—but never published—*Shakespeare in London*, a biographical excursus giving special prominence to the theatrical environment; the 419-page manuscript languishes unconsulted at the Folger Library.

To the printed volume of Shakespeare, which chiefly concerns us, Fleay devoted ten years of his life. It makes for heavy going. The style is extremely terse and unseductive; as Fleay acknowledges, there is much repetition of matter from one section to another. A long, barely readable, segment is devoted to 'Annals'; another offers a rehash of Fleay's latest thinking on the chronology of the plays. The single narrative chapter, 'The Public Career of Shakespeare', seems to be organized on a principle of non sequitur—in three consecutive sentences of the same paragraph (for example), Fleay deals with the authorship of *The Revenger's Tragedy*, the marriage of Susanna Shakespeare, and Court performances by the King's Men.

His intentions, it must be granted, are commendable. Fleay disassociates himself from what he describes as the 'peeping school' of biography. Instead he emphasizes Shakespeare's professional life: his companies and theatres, the plays besides his own in which he might have acted, the authors with whom he came in contact, the rival operations of other troupes. This whole province, Fleay correctly maintains, the biographers have neglected. To his task he brings his rich knowledge of the stage records and comprehensive familiarity with the old drama. He also brings to it his peerless eccentricity. Much of the *Chronicle History*, which is presented as a dry factual treatise, is an exercise in uninhibited fantasy, with improvisations throughout masquerading as likelihoods and facts. Shakespeare 'probably' joined Leicester's Men during or after their visit to Stratford in 1587: a supposition for which no evidence whatever exists. He 'most likely' acted in *Fair Em*, an anonymous romance of the 1590s; again there is no evidence. *1 Henry VI*, Fleay remarks parenthetically, is 'a refashioning by Shakespeare of an old Queen's play, into which he introduced the Talbot scenes, celebrated by Nash, which drew such crowded audiences'.[124] His dating of the plays is rendered generally untrustworthy by his unfounded assumption that authors continually revised their own work and that of others. Greene's deathbed attack on the only Shakescene was answered in *A Midsummer Night's Dream*, in which Bottom and his players represent Greene and Sussex's company. '*Edward III.*, by Marlowe, was, with alterations by Shakespeare, acted about the city in

1594.'[125] (There is no conclusive proof of Shakespeare's hand, and no proof at all of Marlowe's; the date of *Edward III* is uncertain.) As regards the Sonnets, Gervase Markham is nominated for the role of Rival Poet, the candidacy being offered with a contagious lack of enthusiasm; previously Fleay had suggested Thomas Nashe, and he would go on to offer others. Fleay achieves an apotheosis of his own with the suggestion that Edmund Shakespeare acted for the King's Men and wrote the anonymous *Yorkshire Tragedy* under his brother's supervision—a circumstance that accounts for its being published as the work of William Shakespeare.

It is characteristic of Fleay that in the *Biographical Chronicle of the English Drama* Shakespeare should appear out of alphabetical sequence. The author does not attempt to cover the same ground as in his Life, but instead gives a list of the plays, with theorizings as usual labelled as certainties. A long 'Excursus on Shakespeare's Sonnets' enters rather oddly into a volume concerned with the stage. Without withdrawing Markham as the Rival Poet, Fleay proposes Michael Drayton for the same honour, although he acknowledges a 'missing link' in the evidence. On another page of the same work he suggests Gabriel Harvey. That Southampton is the Sweet Boy, Fleay regards as 'beyond doubt'. Unaware that Massey had anticipated him, he elaborates arguments, made earlier in his *Life and Chronicle History*, that William Hervey is Mr W.H.: unwilling to publish the Sonnets himself because of the Dark Lady episode, Southampton persuaded his stepfather to issue them without mention of the Earl's name. Only Hervey would have access to the manuscript, be willing to accept the dedication, and yet demand preservation of his own incognito. Elsewhere Fleay deplores wild guesses and castigates another scholar, Karl Elze, for presenting statements without proof.

Fleay's extravagances manifest themselves elsewhere in his chronicle of the stage, but are sometimes offset by inspired insights and genuine contributions to knowledge. The same cannot be said of his writings on Shakespeare. A windmill of scholarship, Fleay uncovered no new documents (he sought none), and the interpretations he placed upon existing records must always be taken *cum grano*. He fancied himself in the vanguard of a brave new movement, but the results of his literary experiments were seldom constructive. With his pretensions to scientific method, he fostered dangerous illusions of certitude as regards Shakespearian chronology, and his disintegration of the canon initiated a wasteful misdirection of energies that went on for decades. No doubt Fleay meant well, but as a Shakespearian he can be regarded only as a mischief-maker.

His colleagues in the New Shakspere Society paid even less attention than he to the outward facts of the dramatist's career. The directors of the first Shakespeare Society had after all produced notable Lives, and well-wishers had brought before the committee—and in particular its leader Collier—interesting new documents pertaining to Shakespeare's career. The later organization fulfilled no such function. Members diligently gathered literary references which formed the basis of the valuable Shakespeare Allusion-Books published by the Society; but the strictly biographical side was left to Halliwell-Phillipps, who by himself constituted a Shakespeare society.

Of all the papers read at the Gower Street meetings over a period of two decades, only Thomas Tyler's have more than a tangential bearing on Shakespeare's life. On 30 May 1884 he argued the identification of Mr W.H. with William Herbert. Tyler's paper was described by the chairman, Furnivall, 'as the most important contribution to the difficult subject of the *Sonnets*'. Bernard Shaw sat in the audience that night. So too did Charlotte Carmichael Stopes. As she was leaving she said to Tyler, 'I hope I may live long enough to be able to contradict you!' 'No, you won't,' he replied complacently, 'for my theory is going down Time!' And she rejoined, 'Not if I live long enough.'[126] We shall encounter the formidable Mrs Stopes again in this narrative. At the next meeting, on 13 June 1884, Tyler urged the merits of his pet notion that Mary Fitton was the Dark Lady. This time Furnivall objected: the Queen's Maid of Honour could only be Shakespeare's unfaithful mistress if she were married and thus in a position to break her 'bed-vow'. Tyler's views were discussed again five years later, at the meeting on 11 October 1889, and won enthusiastic applause from everyone present except the lone Southamptonite, Mrs Stopes. Were one to judge only from these sessions and the printed *Transactions*, the New Shakspere Society made the slenderest of contributions to our subject.

But there is more to the story. Furnivall had been moved to found the group out of profound dissatisfaction with the state of Shakespeare scholarship in Britain in the last quarter of the nineteenth century. 'It is a disgrace . . . to England', he thundered in his prospectus, 'that even now, 258 years after SHAKSPERE's death, the study of him has been so narrow, and his criticism, however good, so devoted to the mere text and its illustration, and to studies of single plays, that no book by an Englishman exists which deals in any worthy manner with SHAKSPERE as a whole.' The reproach did not fall on deaf ears. Furnivall's words reached a quiet and urbane young Irishman of picturesque appearance who is included in the original list of vice-presidents of the New Shakspere Society. His name was Edward Dowden.

# 17

# *Edward Dowden*

A GRADUATE of Trinity College, Dublin, he had been appointed at the age of twenty-four to the newly established chair there in English literature—a post he held until his death. Dowden never attended a meeting of the New Shaksperians; he did not know a single member of its working committee. But he respected Furnivall and followed the *Transactions* closely; above all, his sensibility was attuned to what the age craved. In 1875, one year after the Society's formation, Dowden published *Shakspere: A Critical Study of His Mind and Art*, a work that introduced a new species of Shakespearian biography and decisively influenced future approaches. The title-page lists Dowden's vice-presidency of the New Shakspere Society. In his preface the author reproduces, from Furnivall's introduction to a new edition of Gervinus's *Shakespeare Commentaries* a trial table of the order of the plays; here Dowden found the organization of Shakespeare's writings fell into four periods—an organization (deriving from Coleridge) of rich critical suggestiveness. That Dowden was responding specifically to Furnivall's prospectus is illustrated by the fact that he takes from it, with due acknowledgement, the word 'youngmanishness' as a descriptive term appropriate to the First Period. A link between Furnivall's New Shaksperians and Dowden's *Shakspere* therefore clearly exists. With this book the elder scholar's plea was answered, and Dowden's reputation made.

About half the *Shakspere* was delivered in the form of Saturday lectures, in the spring of 1874, in the Museum Buildings of Trinity College. That first audience must have found the performance dazzling. The study throughout testifies to a ripe wisdom beyond its author's years. To Dowden, Shakespeare's plays constitute an incomparable gallery of character portraits, and he offers many subtle insights reaching into the heart of Juliet, Brutus, Hamlet, and the rest. Dowden, moreover, develops his argument with such eloquence, such unaffected grace of style, that some at least of his auditors must have wondered whether the spirit of the poet had not entered the critic. In Catholic Dublin the lecturer described a Shakespeare moved by a spirit of Protestantism, considered not as theological dogma but as part of a great human movement—'It may be asserted, without hesitation, that the Protestant type of character, and the Protestant polity in state and nation, is that which has received impulse and vigour from the mind of the greatest of English poets.'[127] Was Dowden challenging what his audience held dear? Far from it; for Trinity College had been founded as a Protestant institution which (among other functions) trained clerics for the Established Church in Ireland. Nothing that Dowden said would stir dissent; much would move appreciative applause.

His attempt, which distinguishes this study from preceding criticism, is 'to connect the study of Shakspere's works with an inquiry after the personality of the writer, and to observe, as far as is possible, in its several stages the growth of his intellect and character from youth to full maturity'.[128] The intense biographical hunger of the period manifests itself in Dowden's priorities. Whatever splendour resides in the artist's creations, even more splendid is the mind of their creator:

There is something higher and more wonderful than St. Peter's, or the Last Judgment—namely, the *mind* which flung these creations into the world. And yet, it is when we make the effort which demands our most concentrated and most sustained energy,—it is when we strive to come into the presence of the living mind of the creator, that the sense of struggle and effort is relieved. . . . There is something in this invigorating struggle with a nature greater than one's own which unavoidably puts on in one's imagination, the shape of the Hebrew story of Peniel. We wrestle with an unknown man until the breaking of the day.[129]

So Dowden wrestles with Shakespeare, and for him the day breaks.

In confronting the English history plays, Dowden poses his 'main question' thus: '—What was Shakspere gaining for himself of wisdom or of strength while these were the organs through which his faculties of thought and imagination nourished themselves, inhaling and exhaling their breath of life?'[130] Few today would see this as the main question; indeed, few would even think it an appropriate one to ask. But such is the tendency of Dowden's interrogation. When Shakespeare wrote *As You Like It*, he was himself in the Forest of Arden, whither he had repaired to find refreshment after Court and camp. So too he undertook *The Winter's Tale* because he needed relief from the overstrained imagination of which *Timon of Athens* bears evidence. The *œuvre* is an autobiographical poem, and the autobiography holds more importance than the poem.

Dowden of course knows the essential facts of the documentary record, and he restates the Halliwell-Phillipps view of Shakespeare the self-possessed worldling. Only such a view, after all, will account for the suit brought by the poet in 1604 against Philip Rogers for malt sold and delivered to him— 'Shakspere evidently could estimate the precise value for this temporal life (though possibly not for eternity), of £1, 15s. 10d.; and in addition to this he bore down with unfaltering insistence on the positive fact that the right place out of all the universe for the said £1, 15s. 10d. to occupy, lay in the pocket of William Shakspere.'[131] But at around the same time, in 1602, that the dramatist paid down £320 of current English money for 107 acres of arable land in the parish of Old Stratford, he has Hamlet describe the courtier Osric, holder of 'much land and fertile', as 'spacious in the possession of dirt'. And the same Prince ponders the fate of another property owner. 'This fellow', Hamlet says, regarding the skull in his hand, 'might be in's time a great buyer of land, with his statutes, his recognizances, his fines, his double vouchers, his recoveries. Is this the fine of his fines, and the recovery of his recoveries, to have his fine pate full of fine dirt?' No one had previously brought into conjunction in this way the facts of Shakespeare's mundane life with those of his life in art. Dowden sees the poet as moving in two spheres, the one limited and practical, the other opening into the infinites of passion and thought.

The two worlds interact. The artist struggles to achieve sanity and self-control; his plays constitute an intimate diary of that quest. He feared that he might become a Romeo abandoning himself to the life of feeling, or a Hamlet succumbing to nerve-sapping thought; having suffered grievous injury, he must steel himself not to become a Timon. By buying up houses and land, Shakespeare vanquished the destructive selves within him; but acquisition did not destroy his idealism, which told him that we are such stuff as dreams are made on. To Stratford, as to his dukedom, he returned in his latter days, and there, living somewhat aloof from his fellow Warwickshire magnificoes, he gazed serenely down on the spectacle of human life. In the bust in Stratford Church—that bust which has presented so many different aspects to its viewers—we see him as he last was. The lips and the massive animation of the face reveal to Dowden a Shakespeare who was rewarded at the close with the mastery of life he had sought in the decades of tempestuous creation.

The critic's double vision thus furnishes an antidote to the Victorian-philistine reconstruction of Shakespeare's life, without denying the facts on which that reconstruction is based. At the same time it serves Victorian optimism by glorying in the poet's ultimately purged and purified soul. Privately, however, Dowden remained uneasy. While his book was still in press he wrote to a friend: 'This "Study of Shakespeare" I only partly like myself, and I expect you will only partly like it. One who loves Wordsworth and Browning and Newman can never be content to wholly abandon desires and fears and affinities which are extra-mundane, even for the sake of the rich and ample life of mundane passion and action which Shakespeare reveals.'[132] But like the hero of his own imagination Dowden managed to overcome his self-doubts.

In his chapter on the dramatist's humour Dowden arranges the grand design of Shakespearian development into four periods, carrying the dramatist from his years of tentative experimentation to the final quietude of a Warwickshire field where, one breezy morning when daffodils began to peep, he conceived Autolycus. In *Shakspere* (1877), an unpretentious primer that went through numerous editions and had greater influence than its ambitious predecessor, Dowden repeats his classification. But, for the benefit of students, he imposes a stricter (and inevitably reductive) pattern on the several periods, each of which receives a descriptive tag. In the first, to which Dowden gives the name *In the workshop*, Shakespeare was learning his craft as a playwright. The second, *In the world*, shows him beginning to lay hold of real life in the political arena of the history plays; in his practical affairs Shakespeare was building the fortune that would allow him eventually to retire as a country gentleman. Before the close of the third period his son and father had died, and he had been wronged by the Fair Youth of the Sonnets;

during this unhappy time the poet penetrated to the heart of darkness and probed the mystery of evil, without however losing faith in human virtue. This phase Dowden calls *Out of the depths*. In the last period the clouds dissipate, the artist casts off gloom and suffering, and emerges 'wise, large-hearted, calm-souled'. The final romances dwell on the necessity of repentance and the duty to forgive; Shakespeare is *On the heights*. Thus in Dowden we see the flowering of that subjective approach to Shakespearian biography of which the first intimations appeared two generations earlier in the criticism of Hallam.

With the passage of time Dowden grew in learning and mellowness; his style became, if anything, more elegant. But his reading of the poet's mind and heart hardened into a fixed mould, altering if at all only in the direction of increased subjectivity. Dowden's introductions to the Oxford Shakespeare present us once again with the sinned-against playwright of the Third Period, now disillusioned with the Dark Lady and expressing his revulsion in *Troilus and Cressida*, which Dowden had left largely undiscussed in his first book. In his edition of the Sonnets (1881) Dowden argues, as we would expect, against the view held by Dyce and others that the sequence was a mere exercise in the poet's fancy, that his passion was a painted fire. To the contrary, in the sonnet, 'real feelings and real experience, submitting to the poetical fashions of the day, were raised to an ideal expression'.[133] Shakespeare in these poems speaks in his own person. What he says, happily for the critic, fits neatly into the scheme for the Four Periods:

Shakspere of the Sonnets is not the Shakspere serenely victorious, infinitely charitable, wise with all wisdom of the intellect and the heart, whom we know through *The Tempest* and *King Henry VIII*. He is the Shakspere of *Venus and Adonis* and *Romeo and Juliet*, on his way to acquire some of the dark experience of *Measure for Measure*, and the bitter learning of *Troilus and Cressida*.[134]

About the historical personages behind Mr W.H., the Fair Youth, and the Dark Lady, Dowden maintains a sensible agnosticism. Although he leans towards Chapman as the Rival Poet, such questions do not really engage him. The true identifications must be with the personages of the plays. And so Dowden cannot resist speculating on whether the courtesan queen of *Antony and Cleopatra*, black from Phoebus' amorous pinches, has not some imaginative kinship with the swarthy temptress, wily in the ways of love, who took the measure of Shakespeare the man.

To the Henry Irving Shakespeare in 1888 Dowden contributed two biographies. One, factual and straightforward, takes full advantage—with generous acknowledgement—of the labours of Halliwell-Phillipps; the other, charting the inner life of the poet's spirit, shows Shakespeare passing through a moral crisis around 1600, but weathering betrayals of friendship and love to attain 'at the close a high table-land where the light is clear and steadfast and the finest airs of heaven are breathed by man'.

No biographical pattern imposed on Shakespeare before or since has made so profound an impact as Dowden's. The Dublin professor is the only academic critic of Shakespeare whose work would remain uninterruptedly in print for almost a century.[135] (His first *Shakspere* is still in print.) Dowden's theory of the poet's four periods would be promulgated in a succession of textbooks and in innumerable classrooms, to the great satisfaction of pedagogues and pupils. If to a modern scholar, Hardin Craig, the Dowden groupings of Shakespeare's plays 'suggest somewhat fancifully the phases of experience through which he was passing', he nevertheless patiently summarizes the phases in the introduction to the *Complete Works* (1951), and in his text he arranges the plays according to the Four Periods. Thus was the Dowden formulation passed down to multitudes of twentieth-century college students. And how tidy and unambiguous that formulation is, and how memorably phrased! Heart-warming too: a pleasant glow suffuses us as we consider that the greatest of writers had to learn his craft in much the same manner as humbler mortals, that he achieved mastery not only in his art but also in the market-place, and that in his personal life he passed from the darkness of disillusionment to the tender daylight of inner peace. Gone is Arnold's remote Everest of a Shakespeare.

One man at least preferred Everest. In the contemporaneous 'At the "Mermaid"' (1876) Robert Browning, who loathed biography, may glance obliquely at Dowden when he has Shakespeare protest:

> Here's my work: does work discover—
>     What was rest from work—my life?
>         .     .     .     .
>
> Blank of such a record, truly,
>     Here's the work I hand, this scroll,
> Yours to take or leave; as duly,
>     Mine remains the unproffered soul.
>         .     .     .     .
>
> Which of you did I enable
>     Once to slip inside my breast,
> There to catalogue and label
>     What I like least, what love best,
> Hope and fear, believe and doubt of,
>     Seek and shun, respect—deride?
> Who has right to make a rout of
>     Rarities he found inside?

It is not one of Browning's best poems.

If others thought it naïve to make a simplistic equation of the creator with his creation—to assume that the joyous mood must be on the comic dramatist, or that the tragedian himself suffers the tragic emotion—they held their peace. Did everyone forget that the author of *Oedipus the King* was known as the happy playwright? So it seems. And did no one see the danger of divorcing Shakespeare from his professional milieu? In a book of over four hundred pages Dowden not once mentions the Globe Theatre, as though Shakespeare had no relation to his company or audience or playhouse. Might he not have been in London surveying the facilities at the Blackfriars Theatre (also unnoticed by Dowden) rather than lying outstretched in a Warwick-shire meadow when he conceived Autolycus? Nor does Dowden ever think in terms of dramatic forms subject to their own laws; it does not occur to him that *Hamlet*, if supreme in its class, is yet a specimen of the revenge play, a genre with its own special conventions. He must adjust the canon to the requirements of his scheme. *Titus Andronicus* sorts oddly with Shakespeare's 'years of bright and tender play of fancy and feeling', so this brutal tragedy must be cast out—despite admittedly strong external evidence of its genuine-ness. On the other hand, 'an entirely comic subject somewhat disconcerts the poet', and therefore Shakespeare cannot be entirely responsible for *The Taming of the Shrew*.

Of course eventually the reaction came. In *Shakespeare in Fact and in Criticism* (1888) Appleton Morgan, president of the New-York Shakespeare Society, glanced reprovingly at Dowden for supposing that the plays evolved out of the dramatist's moods, despondent or joyous, rather than out of a hundred theatrical contingencies: the preferences of managers and audiences, the advice of colleagues, the experience of rehearsals. But the echo of this distant and muted drum failed to reach England, let alone Dublin. The first forcible opposition was not expressed until 1909, when on 11 June Sidney Lee addressed members of the English Association on *The Impersonal Aspect of Shakespeare's Art*.

The poet's peculiar thoughts and emotions, Lee grants, may find expres-sion in the plays: 'But what is the critical test whereby we can distinguish Shakespeare's private utterances and opinions from the private utterances and opinions of his dramatic creations? Where is the critical chemistry which will disentangle, precipitate, isolate his personal views and sentiments?'[136] Critical science yields no such chemistry, and if we try to get at Shakespeare's opinions by arbitrarily tearing passages from their context, we court hopeless perplexity; in this way we can prove Shakespeare either a collectivist or an individualist. Are we not on safer ground in seeking larger correspondences between the dominant mood or tone of a work and the author's internal state when he wrote? But the burden falls on the subjective biographer of furnishing corroborative documentation for, say, the internal upheaval the poet suffered while composing *Othello* or *King Lear*. Unfortunately, no such

documentation is forthcoming, and the unsurpassed excellence of the tragedies may in fact point to an opposite conclusion:

I would contend that had Shakespeare been in this 'third period' the victim of a private
calamity or the prey of searing anxiety, he could never have approached the highest
pitch of artistic perfection, and could never have written the work assigned to him. No
such unfaltering equilibrium, in treatment of plot and character, as distinguished, for
example, *Othello* and *Coriolanus*, would have been within his power, had he sought
expression in tragedy for agonies of his own heart, for moral and mental catastrophe
within the scope of his own conscience.[137]

The personalists ignore Shakespeare's dependence on written sources, rather
than private experiences, for the material of his plays, and 'his well
ascertained disposition to reconcile his dramatic work with the accidental
calls of public taste, and the requirements of his theatrical managers'. The
*œuvre* discovers not the artist but his supreme dedication to his art. Browning
was right.

Lee's paper did not escape the notice of the principal object of his attack.
Within six months of the lecture Dowden had replied not only to his
antagonist but also to Browning, whom Lee had quoted, in 'Is Shakespeare
Self-Revealed?', in the *Contemporary Review* for November 1909. The Victorian Shakespeare of 'At the "Mermaid"' is its author responding to critics who
had pestered him and clearly betraying the 'inscrutable' Browning's own
passions, convictions, and prejudices; the poem thus stands as another
instance of the autobiographical character of literature. To the challenge of
Lee, Dowden brings undiminished eloquence and urbanity. Cunningly he
catches his opponent quoting critics out of context or forgetting his thesis by
citing passages from Shakespeare to support a point about the dramatist.
Lee's psychology of poets is too simple: they do not fashion their worlds out of
nothing, or out of mere reading or observation, nor do they know passion
only vicariously. The imagination may veil but it does not hide the personality of the creator. Dowden's examples of writers who reveal themselves in
their works—Grillparzer, Flaubert, Scott, Balzac, Goethe—show the splendid
breadth of his reading.

But if he is more sophisticated and (in the highest sense of the word) literate
than Lee, he does not quite carry the day. Lee's points about the dramatist's
need to fulfill himself artistically and to please his public wring significant
concessions from Dowden: Shakespeare 'may have turned about 1600 from
comedy to tragedy not because some recent experiences had saddened or
darkened his view of life, not because he needed the aid of art to overmaster a
private grief or mood of trouble, but because he had for the time exhausted his
comic vein, or because, as Dr. Lee suggests, he now aspired to the highest
form of dramatic poetry, or because tragedy had become the fashion of the
day'.[138] From the plays we do learn, as Dowden maintains, that their author

had wit, irony, and pity; so much Lee surely would grant. But when, near the end of the essay, Dowden gives us yet again the familiar formula ('a season came when a pure and serene light—not without a touch of pathetic beauty in it—was again shed over Shakespeare's art') we withhold assent. The arguments from analogy—the examples of Goethe and the rest—have not demonstrated the truth of the personal allegory Dowden derives from the plays; nor can the magician's felicity of phrase overcome our scruples.

In the years to come those scruples (as we shall see) would be more frequently expressed. The subjective argument for a finally serene Shakespeare would be opposed by another argument, equally subjective, for a finally bored Shakespeare. In an influential lecture the dramatist's sorrows would be dismissed, without tears, as mythical. One suspects, however, that Dowden's Shakespeare, ascending in four gigantic strides from the workshop to the heights, has after almost a century not entirely taken leave of the academy; from time to time he still returns to haunt the classroom, like the ghost of Hamlet's father stalking the battlements of Elsinore.

# 18

## *Georg Brandes*

Across the North Sea he haunted late nineteenth-century Denmark itself. The influence of Dowden lies heavily upon the three large volumes of Georg Brandes's *William Shakespeare: A Critical Study*, published in Copenhagen in 1895–6, and (in a translation by William Archer and others) in London two years later. This work introduced the Dowden approach to a vast foreign audience to whom the English critic was inaccessible.

A friend of Ibsen, who commissioned him a leader in the revolution of the spirit, Brandes stirred the torpid waters of provincial Danish intellectual life as profoundly as Ibsen did those of Norway. Jew, political radical, and literary modernist, Brandes stood outside a society which he passionately loved and abhorred. He felt with especial acuteness the undisguised hostility to his Judaism. In *Reminiscences of My Childhood and Youth*, written in middle age, he still recalls the horror of an episode from his childhood when a grinning boy taunted him with the epithet Jew and, having asked his mother what a Jew was, she held him up to the large oval mirror above the sofa. 'How these Christians hate me!' he cried to Edmund Gosse after a social rebuff from a clergyman pillar of Danish respectability.

He first won renown and stimulated controversy with a celebrated series of lectures begun in the autumn of 1871, *Hovedstrømninger i det 19de Aarhundredes Litteratur* translated as *Main Currents in Nineteenth Century Literature*. Here Brandes attempted to assess scientifically (in France, England, and Germany) the early nineteenth-century reaction against the neoclassicism of the preceding epoch, and the overcoming of that reaction. Published in six volumes, the *Main Currents* quickly established itself as a classic. In it, however, Brandes attacked the literature and politics of his own country; he had already outraged religious orthodoxy by a dispute with Bishop Martensen. And so the chair in aesthetics to which he aspired at the University of Copenhagen was denied the 'atheist Jew'. In 1877 Brandes exiled himself to Berlin, where he assimilated German *Kultur*, to which he was already sympathetic. When, five years later, he returned to Denmark, the climate of opinion was more favourable, partly as the result of his teachings. Brandes himself had changed, however: he declared himself an aristocrat and, as a disciple of Nietzsche, the upholder of a radical new ethic. In this phase, at the age of fifty-four, he published his *Shakespeare*. It enhanced his fame and helped to bring him, in 1902, a professorship in aesthetics at Copenhagen. But the victory rang hollow: he was already sixty, and the appointment conferred none of the privileges and responsibilities of a university chair.

A synthesizer, Brandes in his *Shakespeare* attempts to furnish his Continental readership with that species of comprehensive biography which Drake had essayed on a much different plan. Brandes threw all his astonishing energy into preparation for the task. He studied the plays and poems with intense personal involvement; he mastered the best scholarship and criticism—British, German, French—of his own day: he knows Halliwell-Phillipps, the New Shaksperians, Elze, Taine, and the rest. He does not, however, share the scepticism of the best English scholars with respect to myths. The discredited ballad maligning Sir Thomas Lucy is 'possibly . . . genuine', and every probability favours the tradition that the enraged knight had the poacher whipped and imprisoned. After escaping to London, Shakespeare held horses at the playhouse door, but soon rose above such menial duties to become a revamper of old plays. His personal amiability and lofty forbearance discouraged his disparagers; he made no retort to Jonson's 'ill-will and cutting allusions'. At the Mermaid Tavern he rubbed shoulders with Ralegh, Burbage, Lyly, Chapman, and Florio, as well as Jonson—a motley assortment. So far this Shakespeare has much in common with the poet of the early encyclopaedists; the spirit of Malone sees far distant. But Brandes knows and accepts the latest views on the Sonnets expressed by Tyler: Mr W.H. is William Herbert, and Mary Fitton is the dark temptress. He also endorses the careerist interpretation of Shakespeare's motivation, as propounded by Halliwell-Phillipps. The dramatist was 'preoccupied with the ideas of acquisition, property, money-making, wealth. . . . like the genuine country-born

Englishman he was, he longed for land and houses, meadows and gardens, money that yielded sound yearly interest, and, finally, a corresponding advancement in rank and position'.[139] But, in a distinctive variation on this theme, Brandes links the preoccupation with the composition of *The Merchant of Venice*.

For he is writing the most speculative and subjective kind of biography. Brandes's natural tendency, he admits in his introduction, is to search out the human spirit half concealed, half revealed, in the canon; at the end, in the last chapter of his third book, he triumphantly proclaims his success: he has repudiated the notion of Shakespeare's impersonality, and discovered the artist's 'whole individuality' in his writings. That individuality mirrors, sometimes bizarrely, the mind of the biographer. In his thinking, Shakespeare (we learn with some alarm) more than once anticipates Schopenhauer. With Nietzsche he held, in his inmost heart, that 'there were no such things as unconditional duties and absolute prohibitions. . . . the ethics of intention are the only true, the only possible ethics'.[140] It is as though not only Hamlet but also his creator had attended a German university. Perhaps inevitably, the aristocratic spokesman for a cultural élite sees Shakespeare as a fellow aristocrat. From hob-nobbing with young patricians he imbibed his royalist view of history. For the common herd, the unwashed groundlings who crowded into the Globe, he felt unalloyed scorn:

Their struggles are ridiculous to him, and their rights a fiction; their true characteristics are accessibility to flattery and ingratitude towards their benefactors; and their only real passion is an innate, deep, and concentrated hatred of their superiors; but all these qualities are merged in their chief crime: they *stink*.[141]

Only the illustrious few made life worthwhile. Shakespeare is a Coriolanus.

It follows then that Volumnia is Shakespeare's mother, buried 9 September 1608—significantly, only a little while before he composed *Coriolanus*. Representative of the 'haughty patrician' strain in the Shakespeares, she found a fitting incarnation in the imperious matron of Rome. Is it, one wonders, too fanciful to suppose that she is also Brandes's mother? Gosse has described the fierce old Jewess, contemner of Danish public life, sitting in biblical majesty, like Deborah or Jael; she bequeathed to her son the iconoclasm and vital genius which mark his character. So far as Shakespeare himself is concerned, Brandes states 'He had lived through all of Hamlet's experience—all.' He had lost his benefactor Southampton to imprisonment; he had lost his father to the grave. He had discovered the unworthiness of all he reverenced:

The woman he loved, and to whom he had looked up as to a being of a rarer, loftier order, had all of a sudden proved to be a heartless, faithless wanton. The friend he loved, worshipped, and adored had conspired against him with this woman, laughed at him in her arms, betrayed his confidence, and treated him with coldness and distance. Even the prospect of winning the poet's wreath had been overcast for him. Truly he too had seen his illusions vanish and his vision of the world fall to ruins.[142]

Shakespeare, who had thus suffered from others' baseness, is also Timon, 'whose wild rhetoric is like a dark essence of blood and gall drawn off to relieve suffering'. Lastly, and expectedly, he is Prospero burying his magic staff forever.

In its larger outlines, Brandes's reconstruction of Shakespeare's career has heavy meteorological overtones. The middle years, when he wrote *Much Ado About Nothing* and *Twelfth Night*, were bathed in sunshine. Then black clouds massed over his mental horizon; he endured the dark night of the soul. Instead of comedies he wrote gloomy tragedies, and hence must have been gloomy. The essential Brandes paradigm of the life finds exquisite expression in a single long paragraph:

From this point, for a certain period, all his impressions of life and humanity become ever more and more painful. We can see in his Sonnets how even in earlier and happier years a restless passionateness had been constantly at war with the serenity of his soul, and we can note how, at this time also, he was subject to accesses of stormy and vehement unrest. As time goes on, we can discern in the series of his dramas how not only what he saw in public and political life, but also his private experience, began to inspire him, partly with a burning compassion for humanity, partly with a horror of mankind as a breed of noxious wild animals, partly, too, with loathing for the stupidity, falsity, and baseness of his fellow creatures. These feelings gradually crystallise into a large and lofty contempt for humanity, until, after a space of eight years, another revolution occurs in his prevailing mood. The extinguished sun glows forth afresh, the black heaven has become blue again, and the kindly interest in everything human has returned. He attains peace at last in a sublime and melancholy clearness of vision. Bright moods, sunny dreams from the days of youth, return upon him, bringing with them, if not laughter, at least smiles. High-spirited gaiety has for ever vanished; but his imagination, feeling itself less contrained than of old by the laws of reality, moves lightly and at ease, though a 'deep earnestness now underlies it, and much experience of life.[143]

Why does the sky turn blue and the fever ebb away? Nowhere in his enormous study does Brandes express more bafflement than over the springs of the poet's convalescence. External circumstances had altered little; Shakespeare could have derived little uplift from keeping as close as possible (as Brandes informs us he did) to the unappetizing King and his Court. Perhaps in Shakespeare's harvest time some innocent young girl—the prototype of Marina or Perdita or Miranda—came into his life and gradually reconciled him to existence. It must be reckoned a disadvantage of Brandes's vision of the life that it calls for supporting biographical evidence totally absent from the documentary record.

In any event, Shakespeare was restored. Amid the autumnal scenery of *The Tempest* he took leave of his art. Unattached to London although he had spent the greater part of his lifetime there, he went back to Stratford, perhaps stopping along the way in Oxford, where he received a specially cordial greeting from the pretty hostess of the inn, Mistress Davenant, and saw her son, little William, who bore a curious resemblance to the guest. At home in the town of his birth, he fell into the humdrum domestic routine. His wife had turned fanatical Puritan and during his absence had entertained a travelling preacher; his elder daughter Susanna and her husband were desperately pious; Judith, twenty-eight and unmarried, had the mind of an ignorant child. None of them had seen or read any of Shakespeare's plays. He exchanged greetings with the neighbours—the ploughman, the fishmonger, the butcher whose trade he knew well—but was among them, not of them. In nature he found his chief companionship. The poet, Candide-like, cultivated his garden, in which grew the famous mulberry tree planted by his own hand.

'Many and various emotions crowded upon Shakespeare's mind in the year 1601', begins one chapter, and for a moment we feel ourselves in the midst of a fictional romance. Generally, we learn, the dramatist wrote in the morning, but he conceived *King Lear* 'on a night of storm and terror, one of those nights when a man, sitting at his desk at home, thinks of the wretches who are wandering in houseless poverty through the darkness, the blustering wind, and the soaking rain—when the rushing of the storm over the house-tops and its howling in the chimneys sound in his ears like shrieks of agony, the wail of all the misery of earth'.[144] How fortunate for literature that Shakespeare knew to come in from the rain. The study has many such imaginative touches which better agree with the novelist's than with the biographer's art.

Although Brandes does not speak of Four Periods, his Life reads like a melodramatized version of Dowden. The untrammelled romanticism is the Dane's, but his method derives from the Englishman, as does the essential pattern of Shakespeare's career. Brandes knew Dowden: he refers to him eight times in eight hundred pages. But always it is with reference to specific interpretative points unrelated to the grand biographic design. When Brandes makes Dowden's connection between the dramatist's purchases of property in Stratford and Hamlet's graveyard musings on a great buyer of land, we look in vain for a citation. He might well have been more generous in acknowledging an indebtedness which can only be described as substantial.

Today this *Shakespeare* seems almost as remote as Knight's Life. In its own time, however, the Brandes trilogy evoked world-wide admiration. It was printed in America as well as in England. Not only Denmark but also Germany hailed it as epoch-making. In Italy the work received praise as 'la maggior monografia su Shakespeare che sia mai stata scritta da un autore di lingua non inglese'. Translated into several languages and published in numerous editions, the book furnished countless readers with their conception of Shakespeare; a Shakespeare they could respond to as human.

Brandes's *Shakespeare* also made an impact on minds more profound and creative than his own. Two great makers of the modern sensibility reacted to it inimitably in the next century. Freud in a famous footnote to his autobiography (1925) remembers that he found in Brandes the intelligence that Shakespeare composed *Hamlet* immediately after his father's death in 1601; that is, when the dramatist was still mourning his loss and during a resurgence of his own childish feelings with respect to a parent. Thus did the master herald psychobiography. Joyce read Brandes closely. Through the transforming power of his genius, a number of Brandes's ideas, and several of his phrases, continue to live in the pages of *Ulysses*. 'He drew Shylock out of his own long pocket', Stephen Dedalus says of Shakespeare. For Stephen, as for Brandes, *King Lear* and *Troilus and Cressida* show events casting their shadow over the hell of time; then the shadow lifts; the heart of Shakespeare, shipwrecked and tried like Pericles, is softened by 'A child, a girl placed in his arms, Marina'. Of Anne Hathaway, Stephen says: 'In old age she takes up with gospellers (one stayed at New Place and drank a quart of sack the town paid for but in which bed he slept it skills not to ask) and heard she had a soul.' Once, at least, repetition is almost literal. Of *Venus and Adonis* Brandes had written, 'contemporaries aver that it lay on the table of every light woman in London'. In Joyce this appears as: 'That memory, *Venus and Adonis*, lay in the bechamber [*sic*] of every light-of-love in London.' Such is the tribute that art pays to scholarship.[145]

# 19
## *Sidney Lee:* DNB

'MR. Brandes accepts it as the first play of the closing period,' Stephen says of *Pericles*. 'Does he?' comes the challenge. 'What does Mr. Sidney Lee, or Mr. Simon Lazarus, as some aver his name is, say of it?' Like Brandes a Jew, Sidney Lee was born in Keppel Street, in the shadow of the British Museum, on 5 December 1859, the son of Lazarus Lee, a merchant, who named him Solomon Lazarus. At the City of London School in Cheapside, Dr Edwin Abbott first aroused his interest in Elizabethan literature. As a classical scholar Lee showed sufficient promise to obtain, in 1878, an exhibition at Balliol College, Oxford. There (it was widely believed although never confirmed) he came to the notice of the master, the great translator of Plato, Benjamin Jowett, who—foreseeing a brilliant future for the youth—counselled him to change his name. Such advice would have been in character for the worldly Jowett. In any event Lee altered his name to Sidney L[auncelot]; eventually he dropped the middle initial.

In his later career recognition and honours would come to him. Lee would become Professor of English Language and Literature at the East London College in the University of London. He would be elected a member of the Royal Commission on Public Records, a Trustee of the National Portrait Gallery, Registrar of the Royal Literary Fund, president of the English Association, and (the appointment he most cherished) chairman of the Trustees of Shakespeare's Birthplace. He would receive honorary degrees from Manchester, Glasgow, and Oxford. He would be knighted. But, as the passage from *Ulysses* testifies, Lee would not be permitted to forget that he was a Jew. In that standard authority, the 1911 *Shakespeare Bibliography*, William Jaggard lists him for all time under the heading:

Levi *Sidney L. Lee* (Solomon Lazarus)

When forced by bibliographical imperatives to cite the adopted name, Jaggard customarily puts it in inverted commas; in the entry for Jews he considerately furnishes the cross reference: *See* Levi. The predicament of Lee (Jaggard's malicious Levi has no warrant) and Brandes can be understood only in the context of European anti-Semitism before the First World War. But Jaggard's way with Lee may also owe something to the latter's less than flattering treatment of Jaggard's namesake and remote ancestor—the printer of the First Folio—when it came time for Lee to compose a book-length Life; the larger currents of world events sometimes take a back seat to personal pique. Still, anti-Semitism determined the form of Jaggard's antipathy. And when at last it was Lee's turn to cite Jaggard in an appendix on 'The Sources of Biographical Knowledge', the bibliographer to his lasting credit did so with complete dispassion.

At Oxford, Lee's academic performance must have disappointed expectation. In 1880 he took third-class honours in classical moderations; he was awarded a second class in modern history in 1882. But while still an undergraduate he published two precocious articles in the *Gentleman's Magazine*, on Dr Lopez as the original of Shylock and on the topical allusiveness of *Love's Labour's Lost*, that brought him to the attention of the influential Furnivall. Lee was on the verge of accepting a newly established professorship of English at the University of Groningen when Furnivall strongly recommended him to Leslie Stephen, who had just embarked upon a momentous editorial undertaking, the *Dictionary of National Biography*, and was scouting about for an assistant. Stephen needed a man with zeal, knowledge, skill at research and abstracting information, and the pertinacity to keep contributors and printers on schedule. Lee filled the bill, and in addition brought to his task that indispensable qualification for a sub-editor: clear handwriting.

He assumed his duties in March 1883, at a salary of £300 a year, and remained with the enterprise until 1917. During this period Lee wrote 870 articles for the *DNB*, most of them (such is an editor's lot) on secondary figures. With the passage of time, as the Shakespearians of his generation departed the scene, he furnished sketches of their lives: Lee is responsible for the entries for Fleay, Furnivall, and Halliwell-Phillipps, and he commissioned others, many of them by A. H. Bullen, who had been his schoolmate at the City of London School. Thus Lee contributed to the history of literary scholarship.

His relations with his chief gradually blossomed into friendship. 'My greatest piece of good fortune perhaps', Stephen recalled appreciatively in 1903, 'was that from the first I had the co-operation of Mr. Sidney Lee as my sub-editor. Always calm and confident when I was tearing my hair over the delay of some article urgently required for the timely production of our next volume, always ready to undertake any amount of thankless drudgery, and most thoroughly conscientious in his work, he was an invaluable help-mate.'[146] Eventually the strain of the drudgery began to tell on Stephen— 'That damned thing goes on like a diabolical piece of machinery, always gaping for copy', he complained, 'and I fancy at times that I shall be dragged into it and crushed out into slips'—and he began to disengage himself. In 1890 Lee became joint-editor; with the twenty-seventh volume, in 1891, his name appears alone on the title-page.

Surprisingly, his work on the *DNB* allowed him sufficient leisure to publish, in 1885, an attractive history of *Stratford-on-Avon from the Earliest Times to the Death of Shakespeare*, illustrated with numerous ink sketches by Edward Hull. 'It is possible', Lee writes in his Introduction, 'that an account of the town that shall treat it as a municipality not unworthy of study for its own sake . . . will be richer in suggestiveness, besides being more in harmony with the perspective of history, than a mere panegyric on the parochial relics as souvenirs of the poet's birthplace, home, or sepulchre.'[147] His book he regarded as an experiment in the direction of the former. Lee traces the history of the town back to 781, when Offa the King of Mercia confirmed the right of the Bishop of Worcester to 'Stretforde'; he closes with Sir John Bernard's death and the transfer of New Place to the Clopton family. Along the way, in addition to chapters on Holy Trinity Church and other monuments, on the municipal government, on gardens, plagues, and indoor and outdoor amusements, there are sections on the Shakespeares, father and son. Part of a chapter is given to poaching in Charlecote Park, but Lee treats the supposed episode in Shakespeare's career as an unproved allegation; somewhat inconsistently he takes Justice Shallow to be a satirical portrait of Sir Thomas Lucy. The inconsistency disappears from the enlarged 1890 edition of the book. So too, unfortunately, does Lee's scepticism. The theft of the deer took place, the poacher was prosecuted, and 'Shakespeare undoubtedly took a subtle revenge' by immortalizing Charlecote and its master in *The Merry Wives of Windsor*. It is not the only time that Lee changed his mind.

Volume fifty-one of the *DNB*, published in 1897, contains Lee's first biography of Shakespeare, which the author expanded into the *Life of William Shakespeare* published the next year. He had, however, previously entered the lists in connection with a vexed problem facing the poet's biographers. In his articles on Mary Fitton (1889), Lady Pembroke (1891), and William Herbert (1891), Lee had aligned himself with the Pembroke party. He announces his position in no uncertain terms: 'Shakespeare's young friend was doubtless [that dangerous word!] Pembroke himself, and "the dark lady" in all probability was Pembroke's mistress, Mary Fitton. Nothing in the Sonnets directly contradicts the identification of W.H., their hero and "onlie begetter", with William Herbert, and many minute internal details confirm it.' If any readers of this passage from the sketch of William Herbert turned to the same author's article on Shakespeare, they must have learned with some puzzlement that the identification of Mr W.H. with Pembroke is confuted by the form of address, that the Earl's liaison with Mary Fitton belongs to a period *after* the composition of the Sonnets, and that the whole theory is (in a word) 'inadmissable'. The Pembrokist had turned Southamptonite with a vengeance. In an article in the *Fortnightly Review* for February 1898, Lee denounced the Herbert argument as flimsy and, by implication, of doubtful sanity. That Southampton was the patron of these poems, Lee proclaims in an essay in the *Cornhill Magazine* for May 1898, 'can be proved with almost mathematical certainty'. No mathematical demonstration, alas, follows; nor did Lee ever explain his change of views, or for that matter even hint that he had changed his mind—omissions for which he was criticized. He did, however, carefully erase traces of his Pembroke partisanship from later editions of the *DNB*.

This stunning volte-face on the personages of the Sonnets is not Lee's most abrupt reversal of opinion. In the London edition of the *DNB* he wrote as follows on an issue of more general biographical significance:

Shakespeare's personal relations with men and women of the court involved him at the outset in emotional conflicts, which form the subject-matter of his 'Sonnets.' The 'Sonnets' consequently bear to his biography a relation wholly different from that borne by the rest of his literary work. Attempts have been made to represent them as purely literary exercises, mainly on the ground that a personal interpretation seriously reflects on Shakespeare's moral character. . . . But only the two concluding sonnets (cliii. and cliv.) can be regarded by the unbiassed reader as the artificial product of a poet's fancy. . . . In the rest of the 'Sonnets' Shakespeare avows, although in phraseology that is often cryptic, the experiences of his own heart.[148]

Between the time that the London printers set up the article in type and the appearance of the New York edition later in the same year, Lee underwent an extraordinary change of heart. For in the American edition he insists: 'While

Shakespeare's poems bear traces of personal emotion and are coloured by personal experience, they seem to have been to a large extent undertaken as literary exercises. His ever-present dramatic instinct may be held to account for most of the illusion of personal confession which they call up in many minds.' Again Lee fails to explain his sudden conversion, and nowhere does he alert readers that two different versions of his article exist. The perform-ance is worthy of Fleay.

Lee's idea of biography was formed by his association with Stephen, his acknowledged literary mentor, and with the *DNB*. In lectures on 'Principles of Biography' (1911) and 'The Perspectives of Biography' (1918), composed with the advantages of hindsight, he sets forth his theories. To Lee, biography serves a national, an institutional, function. A Life satisfies the commemora-tive instinct: it protects from oblivion the memory of buried humanity by transmitting in enduring form the character and achievement of a great man. Right biographic method is distinguished by a discriminating brevity; Lee's model is not Boswell but Plutarch. The life-writer must not be prejudiced by any ethical or official bias, and he must not yield to hero-worship, which produces panegyric rather than biography. Nor should he wander from the straight path of his subject's personality into the byways of the historical context; such works as Drake's, combining the life with the times, leave Lee cold. He assigns biography no very high place among the literary genres, and would limit the creative aspect to the attempt, on the biographer's part, to bring his protagonist alive—'His purpose', Lee says, 'is discovery, not invention. Fundamentally his work is a compilation, an industriously elabor-ated composition, a mosaic.'[149] If it all sounds cut and dried, how could it be otherwise for Lee, to whom the chief virtues of biographical art were soundness and utility?

His 'Shakespeare', which occupying almost one hundred columns of print was the longest life to appear in the *DNB*, exemplifies the advantages and weaknesses of his concept of biography. The article is undoubtedly the most solid condensed treatment of the subject yet published—unrivalled in ampli-tude and in richness of documentation. Lee strikes a rare balance between presentation of the facts—the dry dust of genealogy, of marriage, christening, and burial records, of litigation, mortgages, and the like—and discussion of the achievement: the dates, genesis, and individual character of the works. Attempting no original contribution, he misses little of consequence in previous scholarship. The whole is supplemented by valuable appendices dealing with portraits and memorials, bibliography, and Shakespeare's reputation. Lee's style, never brilliant, has the virtues of lucidity and conciseness; the discipline of the *DNB* had served him well. Readers presum-ably turn to reference works for edification rather than rapture, and in Lee's article they were not distracted by the heavy burden of eulogy that sinks De Quincey's contribution to the *Britannica*. Yet this Life, however masterly in some ways, has peculiar deficiencies that reflect the author's limitations. Chief among these is his essential mediocrity of mind.

One cannot always rely on Lee in matters of precise factual detail. Of John Shakespeare he writes, 'Contemporary documents often describe him as a glover'—but only two contemporary documents, dated 1556 and 1586, so describe him. Richard Shakespeare of Snitterfield was most probably but not certainly (as claimed by Lee) the poet's grandfather. Of Shakespeare's uncle Henry, Lee says that 'he remained at Snitterfield all his life, and died a prosperous farmer in December 1596'. But Henry Shakespeare farmed land at Ingon, in the parish of Hampton Lucy, as well as in Snitterfield, and he was constantly in trouble, at one time being imprisoned for debt. The state of our knowledge hardly gives us leave to describe John Shakespeare as 'a trader in all manner of agricultural produce', and specifically in corn, malt, meat, wool, skins, and leather. Nor is it quite fair to credit Aubrey, on the basis of his random jottings, with being Shakespeare's first biographer, a distinction more properly earned by Rowe. Lee asserts that the 'Gilbert Shakspeare, adolescens' buried in Stratford in 1612 was 'doubtless' the poet's nephew rather than the brother who, by Oldys's testament, 'survived to a patriarchal age'. But Lee is doubtless wrong. The term *adolescens*, which he could not see as being applicable to a man of forty-five, may simply have signified the bachelor status of the deceased;* Capell in his version of the tradition reported by Oldys speaks merely of a relation, not specifically of the poet's brother; Shakespeare in his will mentions no brothers, all of them presumably being dead. There is no evidence that he ever had a nephew Gilbert.

Lee sometimes falls into inconsistencies. In one place he cites Plautus as the original of *The Comedy of Errors*; in another, the lost *History of Error*, with Plautus a 'possible' subsidiary source. Lee insists upon the objectivity of Shakespeare's dramatic art, thus early parting company with the subjective school of Dowden; but, with well-justified diffidence, he quotes the plays to substantiate his belief in Shakespeare's marital unhappiness. Moreover, in a single astonishing sentence he simultaneously rejects and takes advantage of the personalist approach. 'In Prospero', he writes, 'the guiding providence of the romance, who resigns his magic power in the closing scene, traces have been sought without much reason of the lineaments of the dramatist himself, who in this play probably bade farewell to the enchanted work of his life.'[150]

Elsewhere Lee displays confidence in circumstances one would think conducive to hesitation. 'There is little doubt that Spenser referred to Shakespeare in "Colin Clouts come home againe" (completed in 1594), under the name of "Aetion" . . .' and 'Criticism has proved beyond doubt that in these plays [the *Henry VI* trilogy] Shakespeare did not more than add, revise, and correct other men's work.' (Of course criticism had proved no

---

* In the parish register *adolocentulus* and *adolocentula* are used more than once to describe unmarried adults.

such thing.) But Lee is most reckless on the subject of the poet's early marriage. The sureties of the Worcester bond, 'representing the lady's family, doubtless secured the deed on their own initiative, so that Shakespeare might have small opportunity of evading a step which his intimacy with their friends' daughter had rendered essential to her reputation'.[151] The shotgun wedding here envisaged belongs to Lee's imagination, not to the documentary record. His statement, late in the article, that Jonson and Drayton seem to have been the retired dramatist's chief literary companions in his last years, has only the doubtful warrant of Ward's apocryphal anecdote about Shakespeare dying after a drinking bout with the two worthies.

Lee's article comes fittingly at the close of the century. It sums up an era of scholarship rather than pointing ahead to the future. In this Life Shakespeare is still the Stratford lad, injudiciously mated, who robbed the park at Charlecote and was therefore obliged to flee his native haunts. Shallow, with the dozen white luces on his coat, is still Lucy. The story that Shakespeare held the horses of playhouse patrons is yet again retold; soon it would be forgotten. We see once more the dramatist consuming his leisure hours in witty conviviality at the Mermaid Tavern. But if Lee does not purge the life of the traditions, it must in fairness be said of him that he is sound enough a scholar not to make a lawyer or a soldier of Shakespeare or to send him packing on an unchartered journey to Italy.

We need feel no surprise that the poet who emerges from Lee's pages has the prosaic mentality of his chronicler. 'Shakespeare, in middle life, brought to practical affairs a singularly sane and sober temperament,' we are assured. The author devotes several pages to analysis of his hero's financial position, and he repeats with approval Pope's couplet that Shakespeare 'For gain not glory winged his roving flight, | And grew immortal in his own despight.' One senses the satisfaction underlying Lee's final paragraph of appreciative evaluation:

With his literary power and sociability there clearly went the shrewd capacity of a man of business. His literary attainments and successes were chiefly valued as serving the prosaic end of providing permanently for himself and his children. His highest ambition was to restore among his fellow-townsmen the family repute which his father's misfortunes had imperilled. Ideals so homely are reckoned rare among poets, but Chaucer and Sir Walter Scott, among writers of exalted genius, vie with Shakespeare in the sobriety of their personal aims and the sanity of their mental attitude towards life's ordinary incidents.[152]

It is the highest praise Lee can bestow upon the supreme artificer.

The mention of Scott in the same breath with Chaucer and Shakespeare marks Lee as a man of his century. If today his view of the author of *Hamlet* and *King Lear* seems oddly limiting and philistine, Lee (as we have seen) does not stand alone: he has the great example of Halliwell-Phillipps before him.

Nor did the biographer ever see fit to qualify his opinion. The words just quoted appear, with only minor stylistic alterations, in the fourteen successive editions (including one major revision) of Lee's *Life of William Shakespeare*. They outraged readers sensitive to any suggestion that the National Poet might have had the character of a careerist. Hence the spluttering response of Logan Pearsall Smith, for whom Lee—'one of the most matter-of-fact of all the hard-boiled Shakespeare critics'—raves, 'though in more prosaic accents, as much as the maddest sentimentalist and blatherskite of them all'.[153]

# 20
## *Lee's* Shakespeare

THE first edition of Lee's full-scale Life (1898) followed hard upon the heels of the *DNB* essay. Although swollen to almost five hundred pages, the expanded version reproduces the general outlines and often the phraseology of its predecessor, and avowedly adheres to the implicit principles of the *DNB*. Lee aims to set before his readers, as concisely as clarity and completeness allow, 'a plain and practical narrative of the great dramatist's personal history'. He steers clear of strictly aesthetic criticism, elsewhere superabundantly available, and concentrates instead on providing 'an exhaustive and well-arranged statement of the facts of Shakespeare's career, achievement, and reputation, that shall reduce conjecture to the smallest dimensions consistent with coherence, and shall give verifiable references to all the original sources of information'.[154] The statement of purpose has a becoming unpretentiousness; and it is precisely because he is not given to trumpeting the merits of his own efforts that a special weight attaches to Lee's statement that no one has previously attempted such a work. He is right. In this biography Lee avoids the pseudo-fictional excesses of popular reconstructions such as Knight's; unlike Halliwell-Phillipps he comes to grips with the *œuvre* as well as with the often trivial outward records. Within the confines of a single volume Lee furnishes the essentials of a library. Small wonder that this Life should receive an enthusiastic welcome, and indeed excite awe. 'A definitive biography', pronounced the *Pall Mall Gazette*. 'A marvel of research, and, on the whole, remarkably temperate, judicious, and convincing,' declared the magisterial *Times*: 'Never before has learning been brought to bear on Shakespeare's biography with anything like the same force.' Oddly enough, however, Lee's former chief responded with more qualified enthusiasm. In 'Shakespeare as a Man', an urbane but light-weight essay in the Bagehot mode, Stephen maintains that 'Self-revelation is not the less clear because involuntary or quite incidental to the main purpose of a book.'[155] *Hamlet* is grist to the biographer's mill after all, the ideal approach lying somewhere between the extremes of objectivity and subjectivity represented by Lee and Brandes.

To his credit, Lee catches errors in the *DNB* article. He is no longer so positive about Richard Shakespeare of Snitterfield being the poet's grandfather. Uncle Henry, we are now told, followed his farming with 'gradually diminishing success' and gave up the ghost 'in embarrassed circumstances'. (It was not yet known that Henry Shakespeare, although much indebted, was described as having died in his own house, with money in his coffers and a mare and plentiful corn and hay in his barn.) The sentence about John Shakespeare's occupation undergoes revision—'Documents of a somewhat later date [than 1551] often describe him as a glover'—but it cannot be said that the revision straightens out this part of the record. Other doubtful statements, such as those about Gilbert, the poet's brother, remain unaltered.

Patent inaccuracies, which if usually minor are perhaps symptomatic, stir unease. As in the *DNB* article, *The Winter's Tale* is referred to as *A Winter's Tale*. New peccadilloes enter the narrative; the Arthur Mainwaring who figured in the affair of the Welcombe enclosures is rechristened Arthur Mannering. Lee gives the text of Anne Shakespeare's epitaph, absent from the *DNB*; but in the line 'Quam mallem, amoueat lapidem, bonus angel[us] ore[m]', he reads the last word as *ore*, although in the inscription a circumflex abbreviation plainly appears. One is not surprised to find Lee misspelling Malone's Christian name *Edmund*. More consequential are the lapses in the account of Shakespeare's finances. The *DNB* passages on the poet as astute businessman clearly made an impression on their author (as they did on readers), for he expanded the material into a chapter, fittingly entitled 'The Practical Affair of Life'. For this section Lee reviewed his calculations (based on estimated playhouse receipts, payments for plays, and the like) and came up with a preposterously exaggerated annual income for Shakespeare of over £600 (earlier he had estimated the dramatist's professional earnings at 'near' £600). When he overhauled his Life almost twenty years later, Lee went over the same terrain again, and, as we shall see, arrived at an even more outlandish set of figures.

As we might expect, he subjects to no searching critical scrutiny the hoary biographical traditions he had rehearsed in the *DNB*. If anything he is less critical. The story about Southampton's gift to the poet of a thousand pounds, regarded by many previous biographers with well-justified incredulity, is reported as a 'trustworthy tradition'; in the earlier article Lee merely described it as an 'independent tradition'. He accepts (despite Gifford's impatient efforts at disproof, which Lee knows) the legend that Shakespeare intervened to reverse his company's decision not to mount *Every Man in his*

*Humour*. Lee admits as genuine the dramatist's improbable witticism about Latin spoons when he purportedly stood godfather to an offspring of Ben. 'Shakespeare is known to have been a welcome guest at John D'Avenant's house,' we are informed with unwarranted assurance. So it goes.

Elsewhere too Lee presents uncertainties as facts, and sometimes he mars a plain tale in the telling. John Shakespeare's Spiritual Testament (misleadingly termed a will) is dismissed as a clumsy fabrication perpetrated by John Jordan, this being his 'most important achievement'. In Lee's version, the scurrilous Manningham story about a tryst involving Shakespeare, Burbage, and a citizen's wife submits to a fine Victorian reticence:

Burbage, when playing Richard III, agreed with a lady in the audience to visit her after the performance; Shakespeare, overhearing the conversation, anticipated the actor's visit, and met Burbage on his arrival with the quip that 'William the Conqueror was before Richard the Third.'[156]

Were it not for a discreet allusion in the next paragraph, one might infer that the lady was entertaining her guests for tea. Manningham tells us that the 'visit' took place at night, and he reports Shakespeare 'at his game'.

On other occasions Lee falls into the inconsistencies to which he was peculiarly prone. From one sentence we learn that 'Jonson was of a difficult and jealous temper, and subsequently he gave vent to an occasional expression of scorn at Shakespeare's expense'; from the next, that 'amicable relations' between the two men 'habitually subsisted'.[157] But the most serious confusions occur in the several chapters given over to the Sonnets.

Amounting almost to a fifth of the entire Life, this section represents Lee's most original contribution to Shakespearian biography. He had already performed his remarkable turnabout in the pages of the New York edition of the *DNB*. Now in a chapter on 'The Borrowed Conceits of the Sonnets', he argues at length on behalf of his conviction (announced in the preface) 'that Shakespeare's collection of sonnets has no reasonable title to be regarded as a personal or autobiographical narrative'. The novelty of Lee's approach resides in his application of historical and comparative criteria. He maintains that, 'As soon as the collection is studied comparatively with the many thousand sonnets that the printing presses of England, France, and Italy poured forth during the last years of the sixteenth century, a vast number of Shakespeare's performances prove to be little more than professional trials of skill, often of superlative merit, to which he deemed himself challenged by the efforts of contemporary practitioners.'[158] The poet assimilated the thought, and sometimes the words, of the English sonneteers, especially Drayton. The recurring theme, so movingly expressed, of Shakespeare's claim of immortality for his verses is a borrowed one handled by Ronsard and Desportes, and unblushingly iterated by Drayton and Daniel. Why literary imitation should be incompatible with personal emotion Lee makes no attempt to explain.

He moreover undermines the elaborate edifice of his argument by characteristically muddled and imprecise thinking. Directly following his chapter on the borrowed conceits, another presses for acceptance of Lee's identifications of the personages of the Sonnets. Southampton is the Lovely Boy; the Rival Poet is Barnabe Barnes (a suggestion that would win little favour). Although Lee questions the sincerity with which the poet yields up his mistress to his friend, he sees the liaison with the Dark Lady as actual, not feigned: 'The character of the innovation and its treatment seem only capable of explanation by regarding it as a reflection of Shakespeare's personal experience.'[159] Thus Lee in practice endorses the private-revelation school that in theory he rejects. Elsewhere in the book, it may be granted, he does not deny a residue of autobiography in the sequence, but of course this admission goes against the plain statement in his introduction. How much is a small proportion? In his article published in the same year in the *Cornhill Magazine*, Lee detects 'genuine autobiography' only in the dozen or more poems addressed to the patron. The restriction of the personal to a small group of sonnets does separate Lee from others; but again he has shifted his ground, and because he failed to revise later printings of his *Shakespeare* in the light of this change, it is difficult to take him seriously.

A more concrete, if not necessarily more valuable, contribution is made by Lee in his discussion of Mr W.H. On independent grounds he determines that this gentleman was William Hall, a stationer's assistant 'professionally engaged, like Thorpe, in procuring copy'. The same Hall, there is 'little doubt', had in 1606 obtained for publication the Jesuit Robert Southwell's *Fourfold Meditation*, which he provided with a dedication, signed 'W.H.', attesting his religious satisfaction in the recovery. 'When Thorpe dubbed "Mr. W.H.", with characteristic magniloquence, "the onlie begetter [*i.e.* obtainer or procurer] of these ensuing sonnets"', Lee writes, 'he merely indicated that that personage was the first of the pirate-publisher fraternity to procure a manuscript of Shakespeare's sonnets and recommend its surreptitious issue.'[160] In Lee's view, only Hall, of those known familiarly to Thorpe, had the right initials. But how can he be so certain of Thorpe's circle? The name Hall is common, and so too are the initials: William Holme, a bookseller known to Lee, belonged to the Stationers' Company and followed his trade in London in the first decade of the seventeenth century. And why assume that the W.H. of the Catholic *Fourfold Meditation* is the printer Hall, who after all had Puritan leanings? Would the pious W.H. of the Southwell dedication engage in such hanky-panky? Why did not Hall himself publish the precious sonnets? And why should Thorpe dedicate his book to a printer he had not employed to set it up in type? To these awkward questions Lee does not address himself. Some years later, in the 1905 edition of the *Life*, he expresses a fleeting misgiving when he admits in a footnote that Mr W.H. eludes positive identification. In the final enlarged version of the biography, however, the old, unsupportable certainty again manifests itself. This backing and filling, it should by now be clear, is characteristic.

On the basis of such contributions to knowledge Lee won fame as Shakespeare's biographer—he would be knighted in 1911. Later editions of the *Life*, it is true, do contain a new biographical record derived from previously unexamined contemporary manuscripts: the payment to the dramatist of 44s. for devising an impresa to be worn by the Earl of Rutland on his shield for the tilt upon the King's Accession day in 1613. Lee announced the discovery of this document in *The Times* for 27 December 1905. But he had not found it himself. Sir Henry Maxwell-Lyte, deputy-keeper of the Public Record Office, and W. H. Stevenson, the historical scholar, had come upon it in Belvoir Castle while scrutinizing (on behalf of the Historical Manuscripts Commission) the sixteenth- and seventeenth-century household books of the Duke of Rutland. These investigators communicated their find to the age's foremost authority on Shakespearian biography, and Lee made it the basis of his article. His *Life* had already received the most coveted of accolades for a scholarly work; it had come to be regarded as standard. When the sixth edition, larger than the original by more than half, appeared on the eve of the tercentenary of the poet's death, it obtained from an England engulfed in war as much respectful attention as any distinguished original work published in times more propitious to humanistic endeavour.

Into this major recension Lee incorporated the documentary discoveries, some of them exciting, which the new century brought; these our narrative will in due course chronicle. He made a few minor finds of his own. At Somerset House he transcribed the wills of William Combe the elder, and of his nephews Thomas and John Combe. Lee was thus able to correct errors in earlier accounts of Shakespeare's relations with the Combe family. Among the testaments in Somerset House too he came upon some new facts concerning Shakespeare's tomb-maker, Gheerart Janssen, which he supplemented with information from the papers of the Duke of Rutland, for whose family the Janssens had on two occasions carved monuments.

Lee retains the general arrangement of the 1898 edition, but he introduces much new detail and fleshes out the notes to make of them a rich reference source. Everywhere we see the author attempting to improve upon his earlier effort. The discussion of the Stratford grammar school now includes information about the masters in Shakespeare's day: Roche, Hunt, Cottom. Expanded also is the material on the London stage. The chapter on 'Survivors and Descendants' now boasts an elaborate and useful genealogical chart; another section, 'Autographs, Portraits, and Memorials', has grown into a succinct monograph providing information unavailable in other biographies. Similarly Lee elaborates his discussion of the plays: he enters into greater detail on the dramatist's use of his sources, he suggests more amply the distinctive nature of each of the works. Thus Lee maintains his equivalence between commentary on the *œuvre* and strictly narrative biography. Reviewers, it is true, missed in these sober pages the visualizing power of a Macaulay and the exultation of a Swinburne. But awed by Lee's encyclopaedic design, they bowed down before (in the phrase of the *Sunday Times* critic) a 'monument of erudition'. 'New sources will have to be yielded from the buried past', intoned the *Daily Telegraph* (15 December 1915), 'before the authority and completeness of this work can be called in question.' While not calling into question his own authority and completeness, Lee did make minor corrections and additions for a new edition in 1922 and another (one year before his death) in 1925. This last edition, then, represents his final statement on the subject.

The century was already one-quarter over. The First World War had, in four cataclysmic years, swept away the Edwardian sensibility. A single year, 1922, brought two works which transformed consciousness: *The Waste Land* and *Ulysses*. But the air that circulates through Lee's last Life is not of the first freshness; rather, an aroma of the nineteenth century clings to it. This Shakespeare of the legends and the counting house—in youth a deer-stealer and in middle age a prosperous burgher with a burgher's values—is such a Shakespeare as Halliwell-Phillipps had earlier envisaged. Lee attempts to embrace modernity but does not know how to go about it: so he underestimates the importance of E. K. Chambers's decisive lecture, 'The Disintegration of Shakespeare', and fails to attach sufficient weight to the evidence, adduced in 1923, for Shakespeare's share in the play of *Sir Thomas More*.

The Lee of the last edition has changed little over the years. He is as disinclined as ever to make essential distinctions between fact and speculation. In 1592, we are told without a hint of qualification, 'Marlowe was already working with Shakespeare, and showed readiness to continue the partnership.'[161] Since no evidence of such collaboration exists, none is forthcoming; but Lee elsewhere cites as the fruits of partnership two plays on Henry VI, the *Contention* and *True Tragedy*, originally produced by Greene and Peele. These historical dramas are in fact corrupt texts of Shakespeare's *Henry VI*, Parts II and III; so Peter Alexander would demonstrate in 1929, and help consign to oblivion the long-standing legend of Shakespeare the playmender.* In Sonnet 107, the ambiguously allusive poem in which the mortal moon endures her eclipse and 'peace proclaims olives of endless age', Lee casually informs us that 'reference is made to Queen Elizabeth's death'. Quite likely so, but other scholars have variously assigned the sonnet to 1594, 1595, 1596, 1598, 1599 or 1600, 1602, and 1609. So much for certainties.

---

* Lee's statement about the authorship of the trilogy nicely illustrates the fallibility of confidence: 'criticism has proved beyond doubt that in the three parts of "Henry VI" Shakespeare with varying energy revised and expanded other men's work' (p. 117).

Lee typically finds himself at a loss in coming to grips with difficult or complicated issues. He gives a muddled account of Shakespeare's marriage, finding irregularities in the bond where none exists, and groundlessly slurring the clergyman performing the ceremony—whoever he was—as 'obviously' of an 'easy temper'. Lee is also unreliable on the property taxes assessed against Shakespeare as a resident of St Helen's parish, Bishopsgate, in 1596. The waters become muddier still in Lee's lengthy account of the threatened Welcombe enclosure. He gets off on the wrong tack by tracing the enclosure ferment to the death of John Combe—John's small landholdings lay in a different part of Stratford, and his heir was not the William Combe active in the enclosure manœuvrings but William's brother Thomas, who took no part in the proceedings. Lee states that Lord Chancellor Ellesmere was 'ex-officio lord of the manor of Stratford, in behalf of the Crown', but Stratford was not then a Crown manor, and, even had it been, administration of it would not have devolved upon the Lord Chancellor. The estate of a Thomas Barber of Stratford also enters into the tangled skein of narrative. Here Lee does not help matters by endorsing the sentimental view that 'Shakespeare would seem to have been benevolently desirous of relieving Barber's estate from the pressure which Combe was placing upon it.' This appealing supposition is undermined by the fact that the relevant record concerns not Shakespeare but the maligned Combe.[162]

Lee may be pardoned for having his troubles with so complex and at the same time comparatively minor an issue as the Welcombe enclosure. It is less easy to excuse his difficulties, brought upon himself, with the more crucial matter of the dramatist's income. Lee estimates the Globe's receipts during its first years at £3,000 per annum, and the operating expenses at half that figure; Shakespeare's share would come to no more than £150 annually. In later years profits increased, but so did the number of shareholders. Black-friars brought Shakespeare another £150. For his acting he received a maximum of £180 in fixed salary; then, for his plays, an average of £40 yearly. To these sums add £20 for '"benefits" and other supplementary dues of authorship', £15 for Court performances, and £2 or £3 for services as an officer of the royal household (groom of the chamber). Hence Lee's grand total of over £700 annually for Shakespeare during the last fourteen or fifteen years of his professional life.[163] Lee's statistics, it goes without saying, are arbitrary and unsound, and he misconceives the financial structure of the King's Men: as actor-sharer Shakespeare drew no salary, only a percentage of the profits. Malone in the eighteenth century came closer to the truth when, on the basis of less elaborate calculations, he estimated the play-wright's earnings at £200. But Lee, not Malone, was 'standard' during the first decades of the present century, and his figures, arrived at with a show of arithmetic and set forth with matter-of-fact assurance, would mislead the unwary.

What was Lee really like? In the Shakespeare Birthplace Record Office in Stratford-upon-Avon there is a vast hoard of Lee's lecture notes, letters, and professional records, until recently uncalendared and encrusted with alluvial deposits of muniment-room sediment. To those viewing the otherwise dull excrementa of a busy life one fact emerges powerfully, and that is Lee's domination of Shakespeare studies in his later years, a domination amounting almost to monopoly. Overseas correspondents interviewed him; he was invited to make speeches, contribute articles, join committees, grace occasions. Copies of his Life, bound in leather, were awarded as school prizes. Owners of purported portraits of Shakespeare sought his opinon. He received such letters as the following:

I saw in the times recently that you are the top dog in England about Shakespearean matters, and I consequently want your advice.

My Father, who left all his Books, papers, M.S., etc., to me—wrote a book on Shakespeare's knowledge of *mad folk*.

Such are the penalties one pays for fame.

The name of Sidney Lee, Augustus Ralli predicts in his *History of Shakespearian Criticism* (1932) 'will be honoured while Shakespearian literature endures'. Alas for prophecy: Lee's monument of erudition has survived less well than that other, shattered, monument which Ozymandias commissioned for his own vainglory. Scholarship is perishable stuff. Yet Lee too does live on, if anonymously, in a work of creative genius: Joyce read the Life, and in *Ulysses* drew upon it for at least a dozen passages, most of them concerned with Shakespeare's family.

In its final incarnation Lee's *Life of William Shakespeare* follows by slightly over a century the posthumous publication of Malone's Life, the first attempt at a biographical synthesis. Lee, of course, had more to synthesize and more formidable decisions to make. Wheler, Halliwell-Phillipps, and other devoted antiquaries had augmented documentary knowledge of Shakespeare's career; a school of Romantic critics had made the discovery, fraught with excitement and danger, that the plays and poems contain a key to the secret heart of their creator. Unlike Malone, Lee would have the perseverance and organizing power to bring together scattered materials hostile to biographical art, and he would be granted over two decades in which to revise his work for a succession of editions. He would assemble much useful information in a unified and readable narrative. He would avoid the pitfalls of impressionism, and he would succeed in striking a balance between chronicling the life and characterizing the achievement. He would demonstrate the feasibility of Shakespearian biography on the large scale. Lee's deficiencies—the muddleheadedness, the imprecisions, the unwarranted assumptions and certainties—belong to his endowment.

Great creative epochs take place, Arnold says in a famous passage of criticism, when the power of the man concurs with the power of the moment, the supreme instance being Shakespeare in the England of Elizabeth. If even the more splendid feats of scholarship are of a less than Shakespearian magnitude, they too require the confluence of the man and the moment. At the close of the nineteenth century—a century of (among other things) biography—the moment would seem to have been at hand for a grand summary achievement of Shakespearian biography, just as in the eighteenth century the stage had been set for Malone. But Lee is not Malone; the man was wanting. Thus it is that the era ends on a note of decrescendo. Yet Lee speaks, if not as well as we could wish, for his time; he captures in lucid but earth-bound prose a transitory biographical consensus. The next age would bring more brilliant and also more fragmentary accomplishment.

# PART VI

## *Deviations*

# 1
## *Delia Bacon*

DURING the palmy days of mid-century bardolatry a rough beast slouched towards Stratford to be born. Outwardly the Shakespearian scene appeared calm enough. Pilgrims thronged to the Birthplace in ever-increasing numbers; editions, critiques, and Lives multiplied; the archives yielded new bits and pieces of biographical information to scholar-sleuths; a Shakespeare Society garnered the fruits of co-operative inquiry. Then, suddenly, in 1857 a streak of crazed lightning flashed across the spring sky. That year an eccentric American spinster published a very large book setting forth a very strange thesis. Will the Jester, that illiterate peasant from Warwickshire, did not write the plays that go by his name. Only a secret society of master wits could have produced such profound philosophical dramas, and no wit attained greater mastery than Francis Bacon. Doubts concerning the true authorship of the Shakespeare canon had (as we shall see) occurred to others, and in any event the vast traditionalist majority remained unruffled by the hopelessly dense prose of Delia Bacon's *The Philosophy of the Plays of Shakspere Unfolded*. Nevertheless, the lady from Connecticut had started a movement; other voices—eventually, many voices—would join to form a shrill dissenting chorus. As the century waned, even such sober biographers as Sidney Lee would feel themselves obliged to devote some of their space to refuting the claims of the anti-Stratfordians. The new century did not put an end to the proselytizing of Delia Bacon's followers, who hailed her as their patron saint and revered her book (which most of them did not read) as holy writ. They faced competition, however, from rival sects championing their own unorthodoxies. The historian may lament the necessity of having to make his way through thousands of pages of rubbish, some of it lunatic rubbish. He must, however, reckon the heretical movements as part of his story, for anti-biography is, after all, an aspect of biography.*

* While the development of legitimate Shakespearian biography has received little attention from students, they have, by a curious irony, abundantly chronicled the anti-Stratfordian movement; Clio is a fickle muse. In addition to numerous articles, chapters of larger studies, and one casebook for undergraduate edification (*Shakespeare and His Rivals*, ed. George McMichael and E. M. Glenn, 1962), the heretics as a literary phenomenon have inspired three volumes during the past few decades: Frank W. Wadsworth, *The Poacher from Stratford* (1958), R. C. Churchill, *Shakespeare and His Betters* (1958), and H. N. Gibson, *The Shakespeare Claimants* (1962). To these helpful surveys, and especially to Wadsworth, I am indebted for insights, leads, and information. All three, however, are not without their drawbacks. The generally excellent Wadsworth book lacks documentation, and has faults of emphasis; thus the author gives perfunctory notice to Elizabeth Gallup, who created more of a stir and had greater influence than some figures whom he treats at length. While agreeably modest and a cheerful read, Churchill's book suffers from the limitations of amateurism. His summaries (e.g. of Abel Lefranc's views) are not always to be trusted, and he displays a disconcerting indifference to factual

Born in 1811 in a log cabin in Ohio, Delia Bacon was the daughter of a failed pioneer, who died when she was six, bequeathing to his widow (in addition to his poverty) six children. Family friends reared Delia in Hartford, Connecticut; she would never know real home life. Her serious education was limited to one year at Catherine Beecher's well-regarded school for the young ladies of Hartford. Delia had other disabilities to overcome: malarial fever wrecked her originally sturdy constitution, headaches and neuralgic pains prostrated her, destructive demons lurked in her unconscious. The cause with which Delia Bacon's name is linked has exposed her to the mirth of non-believing posterity. But the mirth is unseemly. In her own time her passionate sincerity elicited selfless efforts on her behalf on the part of distinguished men: Emerson, Hawthorne, Carlyle; today, when the absurdity of her views is fully appreciated, only the most unfeeling can fail to be moved by the desolation of her last years.

Delia Bacon early resolved to be somebody and accomplish something. For a time, in partnership with a sister, she ran private schools on the Beecher model; but they all failed. She dabbled in fiction, besting Edgar Allan Poe in a short-story competition, and published a closet drama. These experiments brought Delia a modest literary reputation but little income; she always remained financially (as well as emotionally) dependent on her brilliant elder brother Leonard, a New Haven pastor. At the age of thirty-four she slipped into an unhappy love affair with a candidate for the ministry ten years her junior. He dallied with her strong emotions and proposed matrimony—so she thought; was it merely a wish-fulfillment?—then mocked her by showing her ardent letters to his companions. Later he claimed she was deluded, and had five or six times asked him to marry her. The scandal aroused Leonard, and resulted in an ecclesiastical trial. The verdict, a mild rebuke for the unwilling suitor, failed to vindicate Delia. Her cause received unsolicited support from Catherine Beecher, who told the story in *Truth Stranger Than Fiction*; one fancies that Catherine's sister, Harriet Beecher Stowe, would have managed a more effective literary defence. Delia never fell in love again.

Instead she took to the lecture platform, addressing fashionable assemblages in Boston and Cambridge, Manhattan and Brooklyn. Her audiences responded with enthusiasm to the fragile lady in a black gown which set off her pale yet radiant face; she was hailed as a 'seer' and 'prophetess'. But the

precision. (Churchill thinks that *The Learned Pig*, clearly the work of a naval officer, was possibly written by a country parson—a small lapse, but the critic dealing with these controversies should be careful about authorship attribution.) Gibson, the most recent authority, mentions neither of his predecessors but naïvely overvalues 'that very great Elizabethan scholar', J. M. Robertson—a repudiated disintegrator. In a more specialized contribution, *The Shakespearian Ciphers Examined* (1957), William F. and Elizabeth S. Friedman analyse with professional expertise the cryptographic systems employed to deny Shakespeare his writings. This wholly persuasive book is indispensable for anyone concerned with the ciphers, but it too lacks documentation, which was however present in the original typescript now on deposit in the Folger Shakespeare Library.

lecturing brought insufficient money or renown. Already Delia had embarked on the great work of her life, a revolutionary study of the writings ascribed to Shakespeare.

She compared Shakespeare with Bacon, and everywhere found parallels; a single conclusion, challenging centuries of orthodoxy, seemed inescapable— a certainty to which the identity of surname, it goes without saying, contributed. Delia expected her discovery to confer incalculable benefits on humanity, and to reap, for herself, commensurate rewards. In this spirit she sent Emerson a preliminary outline of her ideas (had he not written, 'Other admirable men have led lives in some sort of keeping with their thought; but this man, in wide contrast'?), and piqued his curiosity by referring to a cipher: not the mathematical cipher devised by the later Baconians but a hidden current of political meaning in the plays. Perhaps for a time sharing her vision, Emerson encouraged Delia, and offered to act as her literary agent. Now she needed to cross the ocean and seek in England the external historical evidence that would confirm the revelations from the text. With a subsidy from Charles Butler, the New York banker and philanthropist, she sailed for Liverpool on 13 May 1853. With her she carried a letter of introduction to Emerson's friend Thomas Carlyle.

The Carlyles received Delia at Cheyne Row on 10 June. They had also invited James Spedding, the great editor of Bacon. Over tea and scones the American described her theory. Spedding sat in speechless astonishment, but Carlyle could not contain himself. 'Do you mean to say', he thundered, 'that Ben Jonson, and Heminges and Condell, and all the Shakespeare scholars since, are wrong about the authorship, and *you* are going to set them right?' 'I am,' Delia rejoined, unintimated. 'And much as I respect you, Mr. Carlyle, I must tell you that you do not know what is really in the *Plays* if you believe that that booby wrote them.' A shattering explosion of laughter followed. 'Once or twice', Delia wrote to her sister several weeks later, 'I thought he would have taken the roof of the house off.'[1] Still unshaken, Delia insisted that Truth, not opinion, was the ware she offered, and at length Carlyle, wearied or amused, agreed to receive what his strange visitor had further to deliver on the subject. 'For the present', he wrote to his brother John on 13 June, 'we have (occasionally) a Yankee lady, sent by Emerson, who has discovered that the "*Man* Shakespeare" is a *Myth*, and did *not* write those Plays which bear his name . . . she has actually come to England for the purpose of examining that, and if possible, proving it, from the British Museum and other sources of evidence. *Ach Gott!*'[2] Yet he too helped her, arranging a letter of introduction to Collier (of which nothing came), taking up her cause (also unsuccessfully) with publishers, even offering her a vacant room in his house (which she did not accept).

She remained aloof from the British Museum, for, as Carlyle eventually perceived, Delia was 'working out her Shakspere Problem, from the depths of her own mind, disdainful apparently, or desperate and careless, of all *evidence* from Museums or Archives'.[3] In truth, she was devoting herself to what her revered namesake would describe as vain matter, and adopting the method of the scholastic philosophers as derided in *The Advancement of Learning*. She had come to resemble one of those spiders which, 'knowing little history, either of nature or time, did out of no great quantity of matter, and infinite agitation of wit, spin out unto us those laborious webs of learning which are extant in their books'. It is not to be expected that she should heed Bacon's warning: 'For the wit and mind of man, if it work upon matter . . . worketh according to the stuff and is limited thereby; but if it work upon itself, as the spider worketh his web, then it is endless, and brings forth indeed cobwebs of learning, admirable for the fineness of thread and work, but of no substance or profit.'[4] When she died, letters of introduction to Sir Henry Ellis and Anthony Panizzi, high officials of the British Museum, were found unopened among her effects.

St Albans beckoned her. There Delia revelled in the soft delightful air, flitted wraith-like about the ruins of Gorhambury, and stood awestruck before Bacon's tomb. Regrettably the beadle declined to open the sepulchre, which perhaps contained evidence in support of her theory; nor could she later prevail upon Lord Verulam to do so. Nevertheless, Delia took lodgings a quarter of a mile from the tomb. Now the shade of her great hero hovered about her and haunted all her steps.\* How close Bacon and the ghosts of other Elizabethan wits were to Delia is revealed by a letter she wrote in March 1854:

Let us hope that these people will begin to take care of themselves and pay their house debts at least. . . . Nobody has trusted them to such an extent as I have. It is more than ten years since I have had their whole business thrust upon me, and I never had a penny from them yet. But perhaps I do wrong to say that. Indirectly, I have had help from them. It is through their means that I am here.[5]

Apart from these spirits, which she struggled to exorcise, Delia had little company: during her ten months at St Albans, she wrote to Emerson in exaggerated terms, she did not speak to one of the natives. From St Albans Delia migrated to Hatfield, and thence back to London. Now a recluse, and suffering from poverty and malnutrition, she wrote exultantly to her sister Julia from Sussex Gardens on 12 January 1855: 'Think, if you can, what it is to feel that I am delivering myself from it at last, that here in this land of my fathers God has at last given me the utterance that I have all my life lacked, and that this great secret, in which the welfare of mankind is concerned, will not perish with me for want of the means of telling it.'[6] But Delia found the book trade surprisingly indifferent to the welfare of mankind, despite her cautious testimonial from Carlyle ('I can freely bear witness in general that

---

\* So she expressed herself to Butler. To Emerson, a few months later, she reported that the shade 'seems satisfied at last'. She was wrong.

she writes in a clear, elegant, ingenious and highly readable manner; that she is a person of definite ideas, of conscientious veracity in thought as well as in word, and that probably no book written among us during these two years has been more seriously elaborated, and in all ways made the best of, than this of hers'). With Emerson's help, however, she placed an article in a New York magazine, *Putnam's Monthly*, 'William Shakespeare and His Plays; an Inquiry Concerning Them', which took pride of place in the January 1856 number, at last made her hypothesis public.

Her discourse is obscure, allusive; great tides of rhetoric—incantatory, brilliant, inchoate—swirl over the seaweed and scattered pebbles of thought. Delia espouses Bacon's claim to the plays but never names him, preferring to speak of 'ONE, with learning broad enough, and deep enough, and subtle enough, and comprehensive enough, one with nobility of aim and philosophic and poetic genius enough, to be able to claim his own, his own immortal progeny.'[7] But elsewhere the one becomes several, for Sidney is quoted and Ralegh unmistakably hinted at. What emerges most clearly from this hysterical exposition, prodigal of question marks and exclamation points, is the authoress's conviction that Shakespeare could not have written the plays. Delia heaps all the scorn at her command on the Stratford poacher, the poor peasant leagued with a 'dirty, doggish group of players', the 'stupid, ignorant, third-rate play-actor', 'the pet horse-boy at Blackfriars—the wit and good fellow of the London link-holders, the menial *attaché* and *elevé* [sic] of the play-house'. (In common with later Baconians, the daughter of an improverished preacher must have her aristocratic gibe at the expense of the Lower Orders.)

If Delia Bacon's theory is mad, the nineteenth century furnished her—as it did others—with the matter for the madness. Phrases from the Collier forgeries infiltrate her account of Shakespeare. She does not mention the Third Variorum, with Malone's Life, but she knows that doubts have been cast on the Charlecote escapade. 'We cannot spare the deer-stealing,' she insists. 'As the case now stands, this one, rich, sparkling point in the tradition, can by no means be dispensed with. Take this away, and what becomes of our traditional Shakespeare? He goes! The whole fabric tumbles to pieces, or settles at once into a hopeless stolidity.'[8] For Delia Bacon, as for others in her time, whole continents divide the sublime Bard of idolatrous criticism from the provincial thespian of contemporary scholarship: the youth from a bookless neighbourhood; the representative of the Commercial Spirit who wrote plays in order to make real-estate investments, and regarded with inexplicable indifference the fate of an *Othello*, a *Macbeth*, a *King Lear*. Surely

it is not this old actor of Elizabeth's time, who exhibited these plays at his theatre in the way of his trade, and cared for them precisely as a tradesman would—cared for them as he would have cared for tin kettles, or earthen pans and pots, if they had been in his line, instead; it is not this old tradesman; it is not this old showman and hawker of plays; it is not this old lackey, whose hand is on all our heart-strings, whose name is, of mortal names, the most awe-inspiring.[9]

The man Shakespeare belongs with the strolling players at Elsinore; for the true creator of *Hamlet* we must look to the Prince. In the Players' scene we have a mystic allegory of the true origin of the plays. They could have originated only among 'the selectest instrumentalities of the ages', that circle possessing 'a culture which required not the best acquisitions of the university merely, but acquaintance with life, practical knowledge of affairs, foreign travel and accomplishments, and, above all, the last refinements of the highest Parisian breeding'.[10] Having such a culture, Bacon and his fellow wits chose as the forum for their revolutionary New Philosophy the theatre rather than the book-stalls, the Parliament, or the pulpit; for in so doing they escaped an oppressive censorship. It does not occur to Delia that government control applied no less oppressively to the drama. Had not her idol Bacon referred in a celebrated sonnet to art made tongue-tied by authority?

With timorous good judgement, *Putnam's* withheld endorsement: 'In commencing the publication of these bold, original, and most ingenious and interesting speculations upon the real authorship of Shakespeare's plays', went the prefatory note, 'it is proper for the Editor of *Putnam's Monthly*, in disclaiming all responsibility for their startling view of the question, to say that they are the result of long and conscientious investigation on the part of the learned and eloquent scholar, their author; and that the Editor has reason to hope that they will be continued through some future numbers of the Magazine.'[11] But no sequel followed, although Delia submitted three articles more. Like the first, they consisted of generalities, whereas the editors expected a demonstration. Meanwhile readers, including the well-known Shakespearian scholar Richard Grant White, protested. The second article, already set up in type for the February number, was scrapped, and Delia was advised to publish her findings in book form.

Nothing would have pleased her better, for a great mass of material awaited the printers. Because of her imperious demands, however, the leads she received from publishers slipped through her fingers. How could she part on the usual terms with a work of such momentous import for humanity, a work to which she had consecrated eleven years of her life? Let the publishers take 10 per cent of the proceeds—she would settle for the remaining ninety. Given Delia's attitude, Emerson could do nothing for her. In desperation she turned, in May 1856, to the American consul in Liverpool.

'This is not *Consular* business exactly, I suppose. . . .', she admitted in her letter to Nathaniel Hawthorne. 'But I think when President Pierce appointed one so eminent as yourself in the world of letters, to represent him in this country, he deserved the return which he will have if through your aid this discovery should be secured to the country to which it properly belongs instead of being appropriated here—or instead of being lost rather as it is more likely to be, unless someone will help me.'[12] In a reply dated 12 May, Hawthorne stood aloof from her theory but, like other notable men, he responded to her passionate faith in her eccentric ideas; a faith which, he felt, transmuted them into gold, and conferred upon her the privileges of one inspired. So he helped her. He gave her money, and she invited him, ever so diffidently, to call upon her:

The reason I shrink from seeing anyone now is, that I used to be somebody, and whenever I meet a stranger I am troubled with a dim reminiscence of the fact, whereas now I am nothing but this work, and don't wish to be. I would rather be this than anything else. I have lived for three years as much alone with God and the dead as if I had been a departed spirit. And I don't wish to return to the world. I shrink with horror from the thought of it.[13]

Suddenly, for a piercing moment, the pitiful recluse saw herself as she was: 'this is an abnormal state, you see, but I am perfectly harmless.' They met. Hawthorne would long remember the striking, expressive face and the dark hair, still untinged with grey, of the slender tall woman who was once a lovely girl. He gave generously of himself to her. He asked a friend to edit her manuscript, he dealt with the printers, he read proofs; most important, he footed the publication bills, of which she was kept ignorant. 'The woman is mad', Hawthorne said of these arrangements, 'but the book is a good one; and as she threw herself on me, I will stand by her in spite of her nonsense.'[14]

She *was* mad. Delia had moved to Stratford, where, in September, took place one of the strangest events in her strange life. The hieroglyphics of Lord Bacon's letters, she imagined, supplied the key to the whole mystery; for they contained minute instructions for locating a will and other relics secreted in a hollow space beneath Shakespeare's gravestone. Hence Will the Jester's malediction, equivalent to the curse by which pirates protected their buried treasure. Now Delia hovered ghoul-like about Holy Trinity. Before attempting an actual assault upon the grave, she consulted the sexton, and was referred by him to the vicar, who granted her (she thought) permission. One evening, late in the month, after the last tourist had departed, Delia entered the church, candle in one hand, lantern in the other. On the altar steps, beneath the blackness of the vaulted roof, she paused to light the lantern. Its feeble ray failed to illuminate the bust of the Old Player, although she knew he stared down at her. Delia walked to the rear of the nave, where her landlady—an uneasy accomplice—passed her a small shovel and then fled. Delia now proceeded to measure the gravestone. Was it thick enough to hold (in her phrase) the 'archives of this secret philosophical society'? Maybe it served merely as a lid, and another slab rested beneath it (the inscription, after all, spoke of *stones*). Delia carefully examined the crevices on either side. 'If I only had the proper tools', she complained to herself. 'I could lift the stone myself, weak as I am, with no one to help.' The hours passed. Now and again she heard a stealthy footstep amid the dusty tombs: it was the sexton, who studied her no less intently than did the impostor Bard. Delia did not dig, for her faith had deserted her. In her mind she reviewed the clues. Bacon's letters referred to a tomb—no question of that—but whose? Perhaps Ralegh's or Spenser's; perhaps that of the Master Wit himself, which she had sought in vain to open. A strange weariness overcame her. She left, her mission unaccomplished.

For a time the book sustained her. Hawthorne supplied a tepid preface consisting mainly of quotations from one of Delia's unpublished manuscripts. He praised her sincerity and heroic devotion, but concluded that 'It is for the public to say whether my countrywoman has proved her theory.' Her benefactor insisted that Delia not dedicate the work to him; even the appearance of his name on the title-page disheartened him—he would just as soon, he confessed, stand in the pillory. Delia understandably protested—she had expected more—and the quarrel between author and sponsor led to the defection of the publisher. 'This shall be the last of my benevolent follies', Hawthorne vowed, 'and I never will be kind to anybody again as long as [I] live.'[15] But another, less well-known publisher took the volume, and in 1857 *The Philosophy of the Plays of Shakspere Unfolded* was finally made available to mankind.

Its 675 closely printed pages do little to develop or support the thesis Delia had advanced the previous year in *Putnam's*. How could it? 'The question of the authorship of the great philosophic poems which are the legacy of the Elizabethan age to us', she warns in her introduction, 'is an incidental question in this inquiry, and is incidentally treated here.'[16] Correct attribution of the plays is no more than a preliminary to a consideration of their nature and design; great sections of the book deal with *King Lear*, *Julius Caesar*, and *Coriolanus*. In others she takes up Ralegh and Montaigne, the Baconian Rhetoric and the Fables of the New Learning. Delia's incantatory prose, abristle with dashes, makes for heavy going, as the following extraordinary early sentence will illustrate:

The proposition to be demonstrated in the ensuing pages is this: That the new philosophy which strikes out from the Court—from *the Court* of that despotism that names and gives form to the Modern Learning,—which comes to us from the Court of the last of the Tudors and the first of the Stuarts,—that new philosophy which we have received, and accepted, and adopted as a practical philosophy, not merely in that grave department of learning in which it comes to us professionally as philosophy, but in that not less important department of learning in which it comes to us in the disguise of amusement,—in the form of fable and allegory and parable,—the proposition is, that this Elizabethan philosophy is, in these two forms of it,—not two philosophies, not two Elizabethan philosophies, not two new and wondrous philosophies of nature and practice, not two new Inductive philosophies, but one,—one and the same: that it is philosophy in both these forms, with its veil of allegory and parable, and without it; that it is philosophy applied to much more important subjects in the disguise of the parable, than it is in the open statement; that it is philosophy in both these cases, and not philosophy in one of them, and a brutish, low-lived, illiterate, unconscious spontaneity in the other.[17]

And so on. She resembles, for all the world, Jonson's rare scholar Doll Common in *The Alchemist*, gone made with studying Broughton's works, and belabouring Sir Epicure Mammon when the fit is on her.

None the less, from the nightmare scaffolding of Delia's paranoia an argument hangs which may be dimly apprehended. The authorship question, however incidental, must enter in. She must counter Jonson's avowed admiration for Shakespeare. ('Ben Jonson must be answered, first,' Emerson had written to her. 'Of course we instantly require your proofs.' 'I know all about Ben Jonson,' Delia sneered.) Why not assume that Jonson had three patrons—Ralegh, Bacon, and Shakespeare—that he had deep personal obligations to Bacon (as well as to Shakespeare), and that the two men met and liked one another? Such a doubtful supposition explains little, but for Delia it settles the matter; until her opponent addresses himself to this point, 'any evidence which he may have to produce in opposition to the conclusions here stated will not be of the least value'. She goes on to envision a coterie of high-born wits and poets, having Ralegh at their head, which originated an esoteric mode of expression for their revolutionary ideas, and thus avoided the Star Chamber. The Shakespeare canon implicitly—if enigmatically—sets forth their doctrines, and specifically embodies the missing fourth part of the Great Instauration described by Bacon in the *Novum Organum*. The plays represent Bacons 'examples of inquiry and investigation, according to our own method, *in certain subjects of the noblest kind*'. A note of exultation creeps into Delia's concluding pages:

There is no room here for details; but this is the account of this so irreconcileable difference between the Man of these Works and the Man in the Mask, in which he triumphantly achieved them;—this is the account, in the general, which will be found to be, upon investigation, the true one. And the more the subject is studied, even by the light which this work brings to bear upon it, the more the truth of this statement will become apparent.[18]

Regrettably Delia could not find place for an HISTORICAL KEY containing 'irresistible, external, historical proof'; that would take another book, and anyway she needed to do more research. Perhaps the unsatisfactory night in Holy Trinity had something to do with her decision.

*The Philosophy of the Plays of Shakspere Unfolded* was not received as Delia expected. The reviewers hooted in derision, no one wrote Delia an appreciative letter, the volume failed to sell. 'I believe that it has been the fate of this remarkable book', allowed Hawthorne in his 'Recollections of a Gifted Woman', 'never to have had more than a single reader'; he confessed that he had himself only dipped into it. Whether neglect and vituperation crushed her spirit we cannot say, perhaps she would have broken down anyway once she no longer had a focus for her monomania. But break down she did. The fragments of her disordered thoughts—her divided loyalties to the Old World and the New, her megalomaniac persuasion that it had been vouchsafed to her to discover a great truth—find expression in a memorandum scrawled on 14 June 1857 to the mayor of Stratford, who had shown her sympathy:

it was a mmrble mmt when you at the Avn. The signs of it are in the strt whether you understand them or not. There reason in those reasons whether you observe them or not. You have forded the Avn. You have untied the *spell*. Theres reason in the whole. You have crossed the Avon twice. You crossed it in the E. & you crossed in the West. *You* crossed the old in the N *Haven*. The E in the West and you crossed the New, the *West* in the East. *History* rest in me a clue & run a right spirit millim me.[19]

A few months later she claimed, for the first time, her own descent from Francis Bacon. There remained for her incarceration, dissolution, and, in 1859, death.

She would not be forgotten. A sect of many unbelievers would revere her name. On 24 May 1911, when the great New York Public Library opened its doors for the first time, and eighty thousand visitors filed through them, the first book requested was *The Philosophy of the Plays of Shakspere Unfolded*. The Library unfortunately did not have a copy, but Charles Alexander Montgomery, an enthusiast of Shakespearian anagrams, quickly donated one. On the verso of the title-page he extolled the authoress as a martyr to the cause of Baconianism, and underneath he copied the following quatrain:

> Ohio! Greetings! Gifted Daughter, thine,
> Has e'en thy Brilliant sons, surpassed: enshrine,
> In rev'rent Tribute, Delia Bacon's name—
> Out-live, 't will, e'en our marbled '*Hall of Fame*'!

Of such stuff are saints made.

# 2
## *The First Unbelievers*

MIGHTY causes from obscure beginnings spring. Sometimes, indeed, the beginnings may be so obscure as to prove, upon examination, non-existent. The Revd James Townley's *High Life below Stairs*, first produced in 1759 at the Theatre Royal in Drury Lane, is an instance. This two-act play points an exemplary moral—if persons of rank lived up to their standard they would cease furnishing ridiculous models for those low creatures, their servants—and was, accordingly, booed by the footmen of London and Edinburgh. 'I never read but one Book,' Lady Bab's maid boasts in a scene showing the menials aping their betters. 'What is your Ladyship so fond of,' inquires another domestic, Kitty.

> LADY BAB'S MAID. *Shikspur*. Did you never read *Shikspur*?
> KITTY. *Shikspur*? *Shikspur*?—Who wrote it?—No, I never read *Shikspur*.[20]

This exchange, a variation on a hoary wheeze, illustrates the ignorance of Kitty, a former Chelsea half-boarder with pretensions to gentility. But the Baconians, who discern in Townley's harmless farce an early manifesto of the anti-Stratfordian creed, have never been remarkable for their sense of humour.

Nor do they fare better with the dark conceit of *The Life and Adventures Of Common Sense: An Historical Allegory* (1769), by Herbert Lawrence, apothecary and surgeon. Offspring, in ancient Athens, of Truth and Wit, the hero-narrator makes his way to England in 1588 (passing the Spanish Armada *en route*) to join his mother, her affectionate friend Wisdom, Genius (the latter's first cousin), and Humour (half-brother to the narrator). In London he meets 'a Person belonging to the Playhouse; this Man was a profligate in his Youth, and, as some say, had been a Deer-stealer, others deny it; but be that as it will, he certainly was a Thief from the Time he was first capable of distinguishing any Thing; and therefore it is immaterial what Articles he dealt in'.[21] On a shabby pretext this rogue hustles off to Holland his new friends, with whom he has entered on 'a sudden and violent Intimacy'. He then seizes from Wit's baggage a commonplace-book containing 'an infinite variety of Modes and Forms, to express all the different Sentiments of the human Mind, together with Rules for their Combinations and Connections upon every Subject or Occasion that might Occur in Dramatic Writing'. A cabinet yields the glass of Genius, capable of penetrating into the deep recesses of man's soul; from a hat-box the thief purloins Humour's mask, which has the power of making the wearer's every sentence appear pleasant and entertaining. 'with these Materials, and with good Parts of his own, he commenced Play-Writer, how he succeeded is needless to say, when I tell the Reader that his name was *Shakspear*.'[22]

This parable neatly exemplifies a popular eighteenth-century notion of the early manhood of the National Poet, but how it serves the anti-Stratfordian cause is not immediately apparent. For the allegory does not deny Shakespeare his own works, but, rather, shows the dramatist equipping himself with the attributes of his art; elsewhere, in analogous vein, the narrator refers to Vergil, Horace, and Ovid as mere amanuenses whose literary productions were dictated by Wit and Genius. Such considerations carry little weight with the Baconian apologists, who seem not to understand Lawrence or even, as Wadsworth observes, to have read him through. These sectarians point to Wit, Genius, and Humour as characteristics of Bacon; they note with triumph that he kept a commonplace-book. Somewhat inconsistently, they identify Wisdom with their hero. Thus, contrary to the dictates of common sense, did they interpret *The Life and Adventures of Common Sense* as anticipating their heresy.

Some have genuflected before the Learned Pig as an early prophet of the anti-Stratfordian revelation. The main theme of *The Story of the Learned Pig* (1786), by 'an Officer of the Royal Navy', is the sad neglect of naval officers by an ungrateful nation, but it takes the odd and entertaining form of a chronicle of the metempsychoses of a soul which, after a strange medley of lodgings (Romulus, Brutus, a black cat, a fly), has come to inhabit a performing pig at Sadler's Wells. In Elizabethan times the soul found its incarnation in Pimping Billy, son of a fishwife and Cob, the immortal water-carrier of Jonson's *Every Man in his Humour*. At the playhouse, where he serves as horse-holder and panders for country folk 'unacquainted with the ways of the town', the swaggerer makes friends with the Immortal Shakespeare. This privileged relationship enables Billy, in his inimitable fashion, to absolve the poet of the calumny of poaching:

... the prevailing opinion of his having run his country for deer-stealing ... is as false as it is disgracing. The fact is, Sir, that he had contracted an intimacy with the wife of a country Justice near Stratford, from his having extolled her beauty in a common ballad; and was unfortunately, by his worship himself, detected in a very aukward situation with her. Shakspeare, to avoid the consequences of this discovery, thought it most prudent to decamp. This I had from his own mouth.[23]

One revelation leads to another: 'With equal falshood has he been father'd with many spurious dramatic pieces. "Hamlet, Othello, As you like it, the Tempest, and Midsummer's Night Dream", for five; of all which I confess myself to be the author.'[24] For a time Pimping Billy lives happily on the proceeds of his plays, but then (such is the transience of felicity) preference is shown to Shakespeare by 'the *first* crowned head in the world, and all people of taste and quality'. A violent fit terminates the pander's existence, and we next see him as a bear baited in every town in England. Only the remorselessly literal-minded could detect the seeds of heresy in this quaint narrative, but such literal-mindedness is not alien to the anti-Stratfordian sensibility.

Still, Baconianism does indeed trace its origin to the eighteenth century: the thesis was unequivocally formulated, without the disguises of allegory or fiction, by the Revd James Wilmot, to whom *The Story of the Learned Pig* has sometimes been ignorantly ascribed. By a curious irony, however, Wilmot's epoch-making hypothesis would remain buried for over a century before being discovered and announced to the world by someone in the ranks of the orthodox.

Born at Warwick in 1726, Wilmot was a Fellow of Trinity College, Oxford, before cutting a figure in the glittering political and literary society of London. Noted Parliamentarians became his friends; he knew Johnson and Sterne and Warton. Wilmot retired, about 1781, to the Warwickshire of his origins, becoming Rector of Barton-on-the-Heath, a little village on the Avon some six or seven miles north of Stratford. There he amused himself with his friends and books, especially the writings of his great favourites, Shakespeare and Bacon. At the invitation of a London bookseller, Wilmot undertook a life of the former, and in the Stratford environs set out in quest of information respecting the poet. What he found disconcerted him, for he learned that Shakespeare, the son of a butcher who could neither read nor write, was

at best a Country clown at the time he went to seek his fortune in London, that he co<sup>d</sup> never have had any school learning, and that that fact would render it impossible that he could be received as a friend and equal by those of culture and breeding who alone could by their intercourse make up for the deficiencies of his youth.[25]

Yet Shakespeare's writings testify to the training of a scholar and a traveller, and to an intimate acquaintance with the great and learned. It was all very puzzling. Wilmot examined many private collections of letters and documents, and discovered that none of the local gentry of Shakespeare's day seemed to have heard of the great man. Thus the rector's consternation grew.

He had better luck in finding traditional lore concerning the dramatist's Stratford contemporaries. Wilmot was told of an extremely tall and ugly man who had, with the connivance of the justices, blackmailed the local farmers under threat of bewitching their cattle. Why, the rector wondered, had not Shakespeare—if he wrote the plays—introduced this fascinating character into one of his comedies, and thus exposed the corrupt magistrates? Wilmot also recovered curious legends: that the devil had removed a church tower, and that, one Shrove-tide, cakes had hailed down and crippled unfortunate pedestrians. How could Shakespeare have ignored this material, which fairly shrieked for dramatization? What a thrilling stage spectacle the descent of the pancakes would have made: an effect (although Wilmot does not go so far as to suggest so) worthy of Inigo Jones.

Wilmot also searched diligently for Shakespeare's books; surely his next of kin, poor illiterates that they were, would have sold them to the neighbourhood gentry who alone assembled libraries? In vain, however, Wilmot covered himself with the dust of every private bookcase within a radius of fifty miles. And why could not even a page or two of the many sheets in the poet's hand (over a quarter of a million, he calculated) be produced? Had the plays perhaps been the work of someone who wished, for good reason, to conceal his connection with them? The pieces began to fit together.

In the plays Wilmot tracked down additional clues. They revealed an acute knowledge of the law. *Coriolanus* contains an allusion to the circulation of the blood, although Harvey did not publish his discovery until 1619; unaware that the latter was Bacon's physician, Wilmot assumed that a scientist would be more likely to incorporate such a reference than a mere actor. Furthermore, the designations of three characters in *Love's Labour's Lost*—Biron, Dumain, and Longaville—coincide with the names of three ministers at the Court of Navarre, where Anthony Bacon resided and whence he sent his famous brother letters. These hints, no less than the missed opportunities in the *œuvre* and the disappearance without trace of the dramatist's books and manuscripts, led Wilmot around the year 1785 to the astonishing conclusion that Bacon wrote Shakespeare. The former had destroyed his manuscripts in order to conceal the fact that so exalted a personage had descended to the base art of playwriting. For this hypothesis, arrived at diffidently but firmly held, Wilmot found additional confirmation in the numerous extraordinary likenesses of style between the two Elizabethans.

Favoured visitors enjoying the bucolic hospitality of Barton-on-the-Heath heard the novel theory, but Wilmot published nothing: he did not wish to offend his Stratford neighbours, who prided themselves fiercely on Shakespeare's association with the town. When he was almost eighty Wilmot summoned before him his housekeeper and the schoolmaster of Long Compton. 'Take, then, my keys', he commanded, 'and burn on the platform before the house all the bags and boxes of writings you can discover, in the cabinets in my bedroom.' Thus were Wilmot's papers, with their records of local traditions and the Baconian heresy, committed to the flames.

His claim to notice might have gone up in smoke with them had not James Corton Cowell, a Quaker of Ipswich, visited Warwickshire in 1805 in quest of data on Shakespeare's life for a paper to be read before his local Philosophic Society. 'Everywhere', he sadly recalls, 'was I met by a strange and perplexing silence.' His expedition might have ended in total frustration had he not been entertained in Wilmot's house, where, after the last of the guests—local gentry—had departed near 4 o'clock, he was informed of the

Baconian theory. 'D$^r$. Wilmot does not venture so far as to say definitively that Sir Francis Bacon was the Author', Cowell would afterwards report, 'but through his great knowledge of the works of that writer he is able to prepare a cap that fits him amazingly.' On 7 February 1805 Cowell appeared before the Ipswich Philosophic Society and declared himself 'a Pervert, nay a renegade to the Faith I have proclaimed and avowed before you all'. Without naming the Rector of Barton-on-the-Heath, he then proceeded to unfold his 'strange and surprising story'. Cowell expected it to be greeted with shouts of disapproval and execration, and he was not disappointed. The following April he again faced the Society, and this time sought to bolster his position by identifying his informant, after first extracting a solemn vow of secrecy from the gathering. The secret was well guarded. It did not come to light until the manuscript of Cowell's two addresses before the Ipswich Philosophic Society passed into the possession of the University of London Library, and was there discovered by the late Professor Allardyce Nicoll. In 'The First Baconian', an article published in 1932 in the *Times Literary Supplement*, he gave Wilmot his due.

Prior to the recovery of the Ipswich file, the anti-Stratfordians could hail Colonel Joseph C. Hart as their first indisputable—although not Baconian—standard-bearer. Hart was a New York lawyer, an officer in the National Guard, and United States consul at Santa Cruz in the Canary Islands; he was also a priceless eccentric. In that year of revolution, 1848, Hart did his share by publishing *The Romance of Yachting*. Before getting launched he prophesies, with no more accuracy than diffidence, that his work will be taken up by most intelligent persons going to sea, and will indeed 'form an addition to nautical libraries generally'.[26] The book contains tables of longitude and the like, and a glossary of sea terms; but the author's maritime theme does not prevent him from airing strong opinions on a variety of topics. Nowhere is he more idiosyncratic than on the subject of gentle Shakespeare, who reduces Hart to a state of apoplexy. In *The Romance of Yachting* Hart describes a voyage aboard the *J. Doolittle Smith* from Sandy Hook, in eastern New Jersey, to Cadiz. Through a complicated process of free association he arrives at a saying of Purchas, who, being an Elizabethan, in turn recalls Shakespeare.

It is one of the misfortunes of literary history that, in his thirst for biographical information, Hart should have seized upon the notorious Life in Lardner's *Cabinet Cyclopaedia*. The effect on an untutored and volatile mind is predictably mischievous: from the *Cyclopaedia* derives Hart's scurrilous portrait of the Shakespeares. The poet's father was extremely poor and hopelessly illiterate—'So says Lardner, and he proves it beyond dispute.' The family procured its coat of arms by fraud. As for William: 'he grew up in ignorance and viciousness, and became a common poacher—and the latter title, in literary matters, he carried to his grave'. Thus deer-stealing furnishes excellent training for a career of plagiarism, the 'proof' of the latter being Greene's deathbed invective against an upstart crow beautified with University Wit feathers. (No novelty attaches to the concept of Shakespeare the Thief; as far back as 1738, in *Réflexions historiques et critiques sur les différens théâtres de l'Europe*, Luigi Riccobini had declared: 'Guillaume *Shaskpear* ayant consomé son patrimoine, entreprit le métier de voleur.') A worthless person on whom too much good Christian ink has been spilled, Shakespeare left his wife to languish in Stratford while he disported himself with a mistress in the metropolis. Not surprisingly, given his dissolute ways, Shakespeare's principal literary occupation (plagiarism excepted) was to stuff other men's plays with his obscenities.

It is the obscenities that kindle Hart's fury; he fairly explodes with indignation over the smut which, in his innocence, he finds everywhere. Once again Lardner has given Hart his lead, for the latter quotes with relish the *Cabinet Cyclopaedia* judgement that Shakespeare's plays *'absolutely teem with the grossest impurities'* (emphasis Hart's). The secret of the dramatist's success with playgoers was, indeed, his skill as pornographer:

The plays he purchased or obtained surreptitiously, which became his 'property,' and which are now called his, were never set upon the stage in their original state. They were first spiced with obscenity, blackguardism and impurities, before they were produced; and this business he voluntarily assumed, and faithfully did he perform his share of the management in that respect. It brought *money* to the house.[27]

Of all the plays in the canon *The Merry Wives of Windsor* comes nearest to being Shakespeare's own, for 'this revolting piece of trash' bears on every page the imprint of his 'vulgar and impure mind'.

Hart envisages a curious scene in the early eighteenth century. Shakespeare has been forgotten, along with the ignorant theatre manager who owned the playbooks, and the wits ('educated men, men of mind, graduates of universities') who wrote and delivered them and then went off to starve. In a garret an actor—Betterton—and a writer—Rowe—have come upon a great mass of unattributed plays from which they pick and choose. '"I want an author for this selection of plays!" said Rowe. "I have it!" said Betterton; "call them Shakespeare's!" And Rowe, the "commentator", commenced to puff them as "the bard's" and to write a history of his hero in which there was scarcely a word that had the foundation of truth to rest upon.'[28] The identity of the true author (of the clean bits, that is) Hart does not attempt to guess. But his great and oddly pure hero is Jonson, who alone had the 'poetical power' and 'deep philosophy' for the more elevated passages of *Hamlet* (the 'To be, or not to be' soliloquy being, after all, a literal translation from Plato). The view that Hart is the first Baconian, sometimes still expressed by responsible scholars, thus has no basis.

Apart from his public repudiation of Shakespeare, which ensures for Hart everlasting anti-Stratfordian fame, his principal contribution lies (as Wadsworth has remarked) in his hectoring tone, so unlike the mild apostasy of Wilmot, and in the ferocious application of snob values to the authorship question: what is of value in the plays must be denied the Stratford barbarian. Both the tone and the snobbery would characterize later Baconian polemics, but it is difficult to see Hart as furnishing the lead. *The Romance of Yachting* apparently sold poorly (it is today scarce) and in England seems to have been unknown even to literate seafaring types. For a while longer the still waters of orthodoxy would remain untroubled.

An article contributed anonymously, apparently by Robert Jamieson, to *Chambers's Edinburgh Journal* in 1852 is another pebble that caused few ripples, but the author of 'Who Wrote Shakespeare?' expresses doubts which would soon assail Delia Bacon and others. He is unable to reconcile the plays, those most wondrous achievements of human intellect, with the Shakespeare who bought and sold land and theatrical shares and drew up a last will and testament 'as plain and prosaic as if it had been the production of a pigheaded prerogative lawyer'. Perhaps, the essayist suggests, this Shakespeare intent only on profit hired himself a poet: 'some pale, wasted student . . . who, with eyes of genius gleaming through despair, was about, like Chatterton, to spend his last copper coin upon some cheap and speedy means of death'.[29] The inventor of this improbable hypothesis candidly admits that he does not much fancy it himself, but he finds it no less plausible than most of what has been written to connect Shakespeare the man with the poet of *Hamlet*.

None of these published expressions of discontent with Stratfordian orthodoxy argues that Shakespeare is Bacon. The glory of Delia Bacon's martyrdom is not dimmed; she remains the first to express the great heresy, however obscurely, in print. But she achieved her priority by the narrowest of margins. In September 1856, eight months after Delia's *Putnam's* article, a Harley Street recluse named William Henry Smith circulated privately a printed letter to Lord Ellesmere (then president of the Shakespeare Society) asking, *Was Lord Bacon the Author of Shakespeare's Plays?* His imagination (such as it was) nourished by a handful of books—*The Anatomy of Melancholy* and *The Pilgrim's Progress* for theology, Bacon for solid reading, and Shakespeare for pastime—Smith's interest in the authorship question emanated from a youthful quest for a debating topic. By the following year Smith had expanded his pamphlet from fifteen pages to over a hundred in *Bacon and Shakespeare. An Inquiry Touching Players, Playhouses, and Play-writers in the Days of Elizabeth*.

Unlike Delia, Smith haunted no graves, shovel in hand. A man of simple and untormented mind, he first sketched in drab prose some of the essential features of the Baconian argument. Unlike Delia, too, he is not a Groupist imagining a syndicate behind the Folio; to Bacon alone belong the plays, poems, and Sonnets (in which may be discerned 'many of the phases of Bacon's early life'). Smith's point of departure is the meagreness of the Shakespearian biographical record, which he summarizes in a chapter occupying less than a page:

WILLIAM SHAKESPEARE'S is indeed a negative history.
   Of his life, all that we positively know is the period of his death.
   We do not know when he was born, nor when, nor where, he was educated.
   We do not know when, or where, he was married, nor when he came to London.
   We do not know when, where, or in what order, his plays were written or performed; nor when he left London.
   He died April 23rd, 1616.[30]

But it is mainly the refinement of tone of the *canon*, so happily free from uproariousness, that leads Smith to doubt the traditional attribution: surely a courtly presence rather than a licensed vagabond lurks behind the plays. (Such an argument, one suspects, would have puzzled the author of *The Romance of Yachting*, gnashing his teeth over filth and blackguardism, but Hart and Smith agree in making social class a criterion of authorship.) Bacon lived at the right moment, and he had the equipment—the learning, wit, and poetical facility—for the task. In offering parallels of diction and phraseology as evidence, Smith introduces a new strategy which would prove popular. The words he cites, however, are not very unusual—*knee, eager, inkling*—and the phrases not very parallel: 'Nor are those empty hearted, whose low sound | Reverbs no hollowness' is compared with 'For the sound will be greater or lesser, as the barrel is more empty or more full'. But then Smith does not pretend to be a literary student.

A stumbling block to heretics is, of course, Jonson's eulogy of Shakespeare, and this impediment enables Smith to show off the limitations of his ingenuity. The phrase 'Soul of the age', he decides, better describes Bacon than Shakespeare, while (a brilliant stroke this) the lines

> Thou art a monument, without a tomb,
> And art alive still, while thy book doth live,

are more appropriate to the quick than to the dead. But what of the small Latin and less Greek: hardly a respectful way to speak of the sage of St Albans. But maybe Bacon did not understand Latin *critically*, and anyway it is just the sort of nastiness that Jonson would stick into a panegyric. (Smith forgets that Jonson wrote, 'thou *hadst* small Latin and less Greek', which is a pity, for it rather suggests that the party eulogized is no longer among the living.)

The Droeshout engraving in the Folio does not resemble the Stratford bust, and may 'alike shadow forth Bacon, or Shakespeare, indifferently', a supposition strengthened by the fact that Bacon, when eighteen, posed for a portrait that resembles in form the Shakespeare engraving. On the question of why Bacon failed to publish his plays in his own name, Smith's thought processes are less quaint: the noble lord wished to avoid the approbium that attached to hireling writers for paltry playhouses.

Smith grants that his reasoning is inferential and hypothetical, and he does not claim to have proved anything; he merely hopes that his evidence will spur further inquiry. It did. The letter to Lord Ellesmere and its elaboration as *Bacon and Shakespeare* evoked a broad spectrum of public response. Smith won a distinguished, if non-literary, convert in Lord Palmerston, who, towards the close of a long life, rejoiced in three triumphs: the reunification of Italy, the opening up of China and Japan, and 'the explosion of the Shakespearian illusions'. But Smith was drubbed by the *Athenaeum* ('Mr. Smith has scarcely made a semblance of a case'), and generally ignored by the scholarly Establishment. There were popular outcries. 'I won't have Bacon,' protested 'John Bull' in a letter to the *Illustrated London News* for 10 January 1857:

I will have my own chershed 'Will'. I have borne a great deal, and never changed my faith. . . . I know the pestilent vapour will pass away, and the steady glories of Will. Shakspeare blaze forth again; but in the mean time we shiver under the passing cloud . . . . Why not Sir Walter Raleigh? Why not Queen Elizabeth herself? But, as I began, we won't have 'Bacon'!

Mock candidates are a useless weapon against the anti-Stratfordians, for whom no possibility is too extravagant: Delia Bacon, as we have seen, suggested Ralegh, and in the ripeness of time the plays would be claimed for the Virgin Queen, thought by some heretics to be the mother of Bacon.

In *William Shakespeare Not an Impostor* (1857), by 'an English Critic', the infant hypothesis provoked the first serious attempt at refutation. 'Wherever Englishmen go', gravely intones the critic, George Henry Townsend, 'they carry with them their English Bible and their English Shakespeare; and neither of these can we suffer to be lightly spoken of or undervalued.' Townsend decries Smith's prejudiced treatment of Shakespeare's life, the inexactness of the phraseological parallels, the false premises and erroneous conclusions. Solemnly he pronounces judgement on the malefactor:

The new reading-room of the British Museum seems to be the proper arena for the punishment of those who offer violence to our great literary heroes. Let a large black board be erected in this new temple of learning, on which the names of all those condemned by a fairly-constituted jury, of wanton and wicked assaults upon the reputations of the illustrious dead, and other literary misdemeanours, may be inscribed. We doubt not that first and foremost upon the list will appear the name of William Henry Smith, found guilty of traducing the characters of Bacon and Shakespeare.[31]

The patriotic defender of the national monument fittingly dedicates his book 'To the English People'.

Undismayed, Smith bided his time. Indeed time was on his side, for the pestilent vapour did not dissipate; rather, it swelled into a great blanket of smog that hung oppressively over Shakespeare studies. By 1884 the authorship controversy had stirred France, Germany, and India, as well as England and the United States, and it had produced over 250 books, pamphlets, and articles. In that year Smith, by now a bright-eyed old man so enthusiastic about the cause 'that he can hardly allow himself to speak upon it, it excites him too much', had a last fling with another tract entitled *Bacon and Shakespeare*. A rambling affair with little new to offer, apart from the suggestion that Shakespeare could neither read nor write, it is notable for Smith's complacent faith that the war has almost been won; the military metaphor is perhaps inevitable, 'Now that the triumph seems so near at hand', he addresses his troops, 'we cannot resist coming to the front to congratulate those that have fought the battle upon their success, and we candidly own to show ourselves as a veteran who has survived the campaign....'[32] Alas, the ink had only begun to flow.

# 3

# *Representative Baconians*

OF anti-Stratfordianism's numerous nineteenth-century converts, none made a bigger splash than Ignatius Donnelly (1835–1901). Maybe he was not mad, although some of his contemporaries denounced him as the 'prince of crackpots'; but certainly he was larger than life. 'America, the land of "big things"', the *Pall Mall Gazette* wrote of him with awe, 'has, in Mr. Donnelly, a son worthy of her immensity.' As a young man he dabbled in verse; in later life he dallied with spiritualism. He read law, and was admitted to the bar. This training would afterwards incline Donnelly to the belief that the author of the Shakespeare plays must have been a distinguished attorney. (It is yet another manifestation of the projection mechanism that lawyers especially should be drawn to the Baconian heresy.) He dreamed a grandoise dream: at Nininger City, on the west bank of the Mississippi, Donnelly sought to create the metropolis of the future, a flowering of urban culture out of frontier toil; but the panic of 1857 defeated him, and the houses already erected were hauled off to neighbouring farms. He entered politics, serving as lieutenant governor of Minnesota and then, for three terms, as congressman; but this career soured when the local party chieftains, disturbed by his ambitions, turned against him. Undismayed, Donnelly edited a liberal weekly, the *Anti-*

*Monopolist*, ran for office and lost, and enthralled great throngs with his golden oratory on behalf of the People's Party—the 'Populists'—whose 'Great Apostle of Protest' he became.

Eventually the Sage of Nininger retired to his study, where he produced works espousing theories more radical, if anything, than his political and social views. In *Atlantis: The Antediluvian World* (1882) he accepted literally the strange fable of Critias in Plato's *Timaeus*, and postulated an ancient island-continent, long since submerged in 'a terrible convulsion of nature'. With this work Donnelly became the founder of Atlantology; his name to this day inspires reverence among the faithful. In *Ragnarök: The Age of Fire and Gravel* (1883), he argues that, aeons past, a mighty comet pounded and crushed the earth. These outpourings of eccentric erudition provided a fitting preparation for Donnelly's magnum opus, *The Great Cryptogram: Francis Bacon's Cipher in the So-Called Shakespeare Plays* (1888).

This enormous tome, running to a thousand pages, affords a good deal less entertainment than either of its predecessors. Donnelly scores already familiar points about the illiteracy of Shakespeare and the profound learning (especially legal learning) displayed in the plays. Two hundred pages are required for parallelisms of word, phrase, and thought between Bacon and Shakespeare. Donnelly cites the 'curious word' *eager*, the 'strange word' *thicken*, the 'rare word' *phantasm*, and other such novelties. The sceptical reader, uneasy in his awareness that these words occur elsewhere, must remember that Donnelly's hero wears many masks. This busy scribbler penned Montaigne's *Essays*, Burton's *Anatomy*, the numerous plays of the Shakespeare apocrypha, a bit of Peele, and the whole Marlowe corpus. Donnelly, moreover, countenances the suggestion made by Mrs Constance Mary Pott, who had not yet startled the world with her discovery that Bacon founded the Rosicrucian society) that the philosopher had a thumb in the plays of Greene, Marston, Massinger, Middleton, Shirley, and Webster—after all, Donnelly calculates, if Bacon took time out from his public life and private studies to dash off a play every fortnight from 1581 to 1611 (why not?) he would have written '*seven hundred and eighty plays!*' Such substantial production materially reduces the chances of finding curious words in writers other than Bacon.

But Donnelly's essential contribution lies in his book ii, 'The Demonstration'. He has found the Bacon cipher—not in the *De Augmentis*, as one would expect of a dedicated student of Lord Verulam, but in the chapter on cryptography in *Every Boy's Book*, which one day he 'chanced to open'. This experience led Donnelly, in the winter of 1878–9, to reread the Shakespeare canon with the sole aim of finding a cipher there. He looked (he tells us) for a brief message, something on the order of 'I, Francis Bacon, of St. Albans, son of Nicholas Bacon, Lord Keeper of the Great Seal of England, wrote these

Plays, which go by the name of William Shakespeare'; but instead, after false starts (one of his chapters is entitled, with delicious innocence, 'Lost in the Wilderness'), he discovered more than he bargained for imbedded in *1* and *2 Henry IV*: 'A long, continuous narrative, running through many pages, detailing historical events in a perfectly symmetrical, rhetorical, grammatical manner.' Such a narrative, *'always growing out of the same numbers, employed in the same way, and counting from the same, or similar starting-points, cannot be otherwise than a prearranged arithmetical cipher'*.[33] But surely, as Gibson suggests, it would have been far simpler for Bacon to set down his narrative and deliver it, sealed, to some trusted associate, with instructions that the packet not be opened until some specified future date; and how much safer than making elaborate arrangements, inimical to secrecy, with authors, editors, and printers.[34] Why should a super-subtle mind have resorted instead to the juvenile device of incorporating a hidden cipher message in the plays? 'Why, I answer, have men in all ages performed great intellectual feats?' Donnelly rejoins. 'What is poetry but fine thoughts invested in a sort of cipher-work of words?'[35] Next question.

Donnelly misunderstands Bacon's biliteral cipher, which conveys messages by means of alternative founts of type rather than by an arithmetical sequence of words. Nor does he accurately describe his own computations, for he promises a strict system—the same numbers and the same starting-points (although weakened by the qualifying *or similar*)—but in fact liberates himself from mathematical restraint by the use of 'multipliers', 'modifiers', and 'subordinated root numbers'. That is not all. Sometimes Donnelly counts hyphenated words as two, sometimes as one; sometimes he passes over bracketed passages, and sometimes not. He varies the points at which he begins his counts, and also their direction, up or down.[36] In this way he deceived himself into believing he had unexpectedly discovered a cipher narrative, whereas he had in fact selected words from the plays, arranged them in a semiconsciously willed pattern of nineteenth-century Elizabethan phraseology, and rationalized the results by an appeal to impersonal mathematical method. When he comes to describe the death of Marlowe, Donnelly inevitably gives the version accepted in his own day. How was he to know that in the next century Leslie Hotson would discover the true facts of Kit Marlowe's demise in Dame Eleanor Bull's at Deptford?

Greater fascination, if not reliability, attaches to Donnelly's account of Shakespeare. We receive the anticipated revelation about the authorship of the plays ('Shak'st spur never writ a word of them'), and also some unexpected titbits about the personality and adventures of the poet; even about his appearance. The son of a poor peasant who followed the glover's trade 'in the hole where he was born and bred, one of the peasant-towns of the west', this Shakespeare (Lord Cecil informs the Queen) is 'a poor, dull, ill-spirited, greedy creature'. The deer-poaching story, which had survived through the century despite the battering-ram of Malone's scholarship, is confirmed by the cipher and puffed up to melodramatic proportions. Anne Hathaway, 'far gone in pregnancy', is his match; the widow of one Whatley, she is gross and vulgar, 'with a good heart, 'tis true, but with a loud tongue and rough manners'. She served as the model for Mistress Quickly. All this and more does the wonderful cipher, imitating art, reveal.

While allowing that he had not achieved mathematical perfection in setting forth the rules for the cipher, Donnelly did not doubt the persuasiveness of his demonstration. 'The proofs are *cumulative*,' he exulted. 'I have shown a thousand of them.' But an ungrateful world responded with disbelief, indifference, or laughter. 'Mr. Donnelly seems but imperfectly acquainted with the properties of numbers, a fact that may partially account for his delusion,' wrote one reviewer, John T. Doyle, in the *Overland Monthly* (July 1888). 'He announces (page 568) as a notable discovery of his own the "curious fact" that if you take away the last ten words of a sentence, the tenth word from the end has thereby been removed, and he puts this discovery forward in a way to leave it doubtful whether he does not regard it as a peculiarity of Shakespeare, or even of this particular edition of his plays!'[37] Of the hundreds of notices of *The Great Cryptogram* only a handful expressed approval, or even tolerance. Financially the book, unlike Donnelly's two predecessors, was a disaster, and eventually his publisher, who had expected it to be the crowning work of his business career, sued the author for four thousand dollars of unearned advance royalties.

Donnelly's challenge to sceptics—'Let them try to create a similar sentence, in the same way, with numbers not cipher numbers'—did not go unheeded. The year that produced *The Great Cryptogram* also brought forth *The Little Cryptogram*, by a fellow Minnesotan, Joseph Gilpin Pyle. This merciless parody bears the subtitle, *A Literal Application to the Plan of 'Hamlet' of the Cipher System of Mr. Ignatius Donnelly*. With computations very similar to those of his model, Pyle extricated the following cipher message: 'Don nill he (= Donnelly), the author, politician and mountebanke, will worke out the secret of this play. The Sage is a daysie.'

A cold wind wafted across ocean and prairie from St Albans. The incumbent rector of Bacon's seat, the Revd A. Nicholson, applied Donnelly's principles to pages 73-4 of *2 Henry IV* (Donnelly had launched his cipher hunt with these pages), and presented his findings in a scornful pamphlet, *No Cipher in Shakespeare*, published in the same year as *The Great Cryptogram*. Nicholson's methods do not merely resemble Donnelly's; they are identical. Using a Donnelly 'root number', 523, and applying it to Donnelly's specially selected modifiers, Nicholson came up with this communication: 'Master Will i a Jack Spur (William Shakespeare) writ this Play and was engaged at the

Curtain.' Donnelly's four other roots yielded almost precisely the same message.

One might guess, from its reception, that *The Great Cryptogram* was a great fiasco. So Donnelly thought. To settle his obligation to his publisher, he traded real estate for the plates, then wrote morosely in his *Diary*: 'I will . . . put them in my garden, and build a little house to cover them. . . . The little building will be my monument of colossal failure.'[38] But the Donnelly Proposition rose, phoenix-like, from the ashes of its demolition. Despite a disastrously slow start, the book eventually sold (he claimed) 20,000 copies—the Sage of Nininger had latched on to something that appealed to the inherent paranoia of the anti-Stratfordian mentality, preoccupied as it was with secrecy, plots, and mystic codes. If the inventor of the biliteral cipher in his own time concealed the authorship of his dramatic masterpieces, he had no doubt planted somewhere a claim to them which posterity would one day disclose to a true believer. That Donnelly's methods were loose and vulnerable only spurred on others to find the key that would break the code. Their efforts—to which we shall return—comprise what David Kahn in his massive history of secret writing, *The Codebreakers* (1967), fittingly describes as 'The Pathology of Cryptology'.

That the representative Baconians (and there were many) were Americans rather than Britons is not mere coincidence. By the latter decades of the nineteenth century the heresey had become an international phenomenon. In 1878 J. Villeman argued 'un procès littéraire' in the pages of a Paris weekly. 'Tout ce qu'il y a de bon dans les drames de Shakespeare, est de Bacon,' he concludes with Gallic clarity; 'tout ce qu'il y a de mauvais dans les drames de Bacon, est de Shakespeare.' In Germany Edwin Bormann, the librettist, wrote voluminously on behalf of the cause, his contribution reaching its apogee with *Francis Bacons Reim-Geheimschrift* (1906), translated in the same year as *Francis Bacon's Cryptic Rhymes and the Truth They Reveal*. According to Bormann, the Viscount reveals his secret by means of 'curiously rhymed' words in the *Essays*; as these rhymes include such pairings as *Wheels* and *Tails*, the argument is most effective when read aloud in a German accent. Bormann strengthens his case with ingenious explications. In the famous opening sentence of the essay 'Of Truth'—'*What is truth?* said jesting Pilate; and would not stay for an answer'—Pilate represents Shakespeare ('"Pilum" in Latin, is a "spear"; "Platus" is one armed with a "Pilum", a "spear", i.e., a "spear-hurler", a "Lancer", "*Shakespeare*."'), and the phrase 'and would not stay for an answer' signifies 'that the actor, feeling the London pavement growing hot under his feet, had, in the very prime of his manhood, left the City, and retired to the quiet of his native town, Stratford-on-Avon, where he died a few years after'.[39] Bormann is equally persuasive on the essay 'Of Vicissitude of Things', in which the expression 'a Circle of Tales' can refer only to the First Folio.

Other Baconians propagated the word in Holland, Italy, and India. However, the most plentiful support continued to come from the United States, where, after all, the schism had first made significant public headway.

# 4
# *Repercussions: Dr Owen*

EVERYWHERE the controversy, now a *cause célèbre*, caught the popular imagination. Newspapers, magazines, and lecture halls reverberated with echoes of the great debate; an article by Richard Grant White in the *Atlantic Monthly* (April 1883) bears the apt title, 'The Bacon–Shakespeare Craze'. By 1885 support for the movement warranted the establishment of The Bacon Society, which seven years later began regular publication of a journal, *Baconiana*. In 1888 the Society's Honorable Secretary, R. M. Theobald, was able to garner enough letters from the correspondence columns of the *Daily Telegraph* over a period of six weeks to fill a volume, *Dethroning Shakspere*; and these constituted only a sampling. The ferment caused by the Shakespeare Question touched distinguished men of letters as widely separated in sensibility and style as Henry James and Mark Twain. Each responded in character.

Troubled by doubt, James stood aloof from public declarations and a final commitment. 'I am "a sort of" haunted by the conviction that the divine William is the biggest and most successful fraud ever practised on a patient world,' he confided to Violet Hunt on 26 August 1903. 'The more I turn him round and round the more he so affects me. But that is all—I am not pretending to treat the question or to carry it any further. It bristles with difficulties, and I can only express my general sense by saying that I find it *almost* as impossible to conceive that Bacon wrote the plays as to conceive that the man from Stratford, as we know the man from Stratford, did.'[40] In his novella 'The Birthplace' James never specifically mentions Shakespeare, but whose birthplace he intends is never in doubt. Morris Gedge, formerly in charge of a town Library ('all granite, fog and female fiction') becomes custodian of *the* Birthplace. Abetted by his wife Isabel, he exhibits to pilgrims—many of them American—'the sublime chamber of Birth', and ladles out with mounting disbelief the legend of the shrine and its idol. Was He (the upper case is James's) really born in the Birthchamber? 'There's very little *to* know,' Gedge perforce must admit. 'He wasn't there in the Birthplace, in the same sense that Goethe was at Weimar.' The evidence, particularly about that upstairs room where He was supposedly born, is nil—'It was so awfully long ago.' In the end, the play's the thing; leave the author be. If there is no author, only all the immortal people in the work, nobody else, why in the world should there be a house? '"There shouldn't," said Morris Gedge.' So ends 'The Birthplace', and thus did the tide of nineteenth-century anti-Stratfordianism affect the nuances of Jamesian literary art.

Twain too concluded that Shakespeare was (along with Satan and Mary Baker Eddy) a fraud, and he proclaimed his conviction in the free-wheeling prose of *Is Shakespeare Dead?* (1909). Twain's interest in the controversy went back to his apprentice-pilot days on a Mississippi steamboat, when he argued about Delia Bacon with a Shakespeare-spouting master as they glided down the great river. Now, late in life, he received from a friend George Greenwood's *The Shakespeare Problem Restated* (1908), in which the talented attorney showed the plays to be the work of a talented attorney. On the half-title page Twain wrote, 'this book reduces him [Shakespeare] to a skeleton & scrapes the bones';[41] his copy, now in the Berg Collection of the New York Public Library, is heavily and approvingly annotated. Twain was no scholar (he describes the Folio as a quarto and mistakes the £ sign as an abbreviation for *number*), and in Greenwood he found sufficient ammunition for his purposes. The barrister's urbanities, however, give way to frontier invective, as Twain refers fondly to his opponents as troglodytes, thugs, Stratfordolators, Shakespearoids, banga-lores, herumfrodites, blatherskites, buccaneers, bandoleers, and Muscovites. Still the troglodytes at least knew where they stood with Twain.*

The most sensational exploitation of popular curiosity occurred in the pages of a Boston monthly, the *Arena*, whose editors in 1892 set up a solemn 'Tribunal of Literary Criticism' to try the case. For over a year, in a series of articles, the *Arena* reported the deliberations of the tribunal. The heretics scarcely received a fair hearing. They found themselves with only one unequivocal attorney, Donnelly, whereas a whole phalanx of the faithful served as defence counsel: the Revd A. Nicholson, W. J. Rolfe, F. J. Furnivall, and Felix Schelling—solid scholars all. Edwin Reed, it is true, argued at great length for Bacon, but at the last moment he switched sides and, in a flush of belated enthusiasm, concluded with a peroration on the virtues of the defendant—'A man born, where nearly all the benefactors of the human race have been born, in a cottage; descended from a line of husbandmen to whom the soil they tilled gave a silent strength; educated in a school where the mind unfolds as naturally as a flower; brought into contact with the world's literature at a time of life when curiosity and ambition have their keenest edge; a man beloved for the gentleness of his spirit, and revered for his genius.'[42]

* A brief, cogent psychological analysis of Twain's anti-Stratfordianism is to be found in Alfred Harbage's *Conceptions of Shakespeare* (1966). In the transplanted aristocratic waifs of *Pudd'nhead Wilson* and *The Prince and the Pauper*, Harbage discerns Twain's 'sublimation of his own "cloudy sense of having been a prince." . . . He never claimed that he, the boy from Hannibal, was really a king or messiah. Instead he claimed that Shakespeare, the boy from Stratford, was really a lord. And yet he professed to despise lords' (p. 116).

Donnelly might well fume: 'After talking for three hours in behalf of his illustrious client, Mr. Reed whirls round and employs the last half-hour of his speech in telling the jury they should give their verdict for the other side.'[43]

Perhaps not surprisingly under the circumstances, the trial failed to produce any new evidence. We hear yet again about Shakespeare's ignorance and his parents' illiteracy, about the non-existent manuscripts, about the Donnelly cipher, and the undated letter (familiar enough to Bacon partisans) addressed to Bacon by his friend Sir Tobie Matthew, with its ambiguously teasing postscript, 'The most prodigious wit that ever I know of my nation, and of this side of the sea, is of your Lordship's name, though he be known by another.' The jury of eminent personages assembled for the occasion, including among them Edmund Gosse, the economist Henry George, the Honorable William E. Russell (governor of Massachusetts), and the actor Henry Irving, decided overwhelmingly in favour of Shakespeare. Only one juror, Mr G. Kruell, found for the plaintiff. Mr Kruell was a wood engraver. The most original opinion came from A. B. Brown, who concluded that occult forces helped Shakespeare to write his plays, which are evidently examples of automatic writing.

The verdict did not satisfy Dr Orville Ward Owen, a Detroit physician, who published a pamphlet of rebuttal, *A Celebrated Case. Bacon vs. Shakespeare. In the Court of "The Arena." Request to Reopen Brief for Plaintiff* (1893). The request was denied—or, rather, ignored—but Dr Owen did not lose heart. Indeed, for the rest of his life he devoted himself heroically to the Baconian cause. Between 1893 and 1895 he published five volumes of *Sir Francis Bacon's Cipher Story*; the sixth and final instalment, alas, still languishes in manuscript. By deciphering the writings he attributed to Bacon, Owen produced 'Sir Francis Bacon's Letter to the Decipherer', a long verse epistle explaining (superfluously, one would think) how to decipher the writings of Bacon. Sir Francis mentions the four 'key' or 'guide' words, FORTUNE, HONOUR, NATURE, and REPUTATION. The cryptographer's task is to

> Match the syllogisms duly and orderly,
> And put together systematically and minutely
> The chain or coupling, links of the argument.
> That is to say, the connaturals, concurrences,
> Correspondents, concatenations, collocations, analogies,
> Similitudes, relatives, parallels, conjugates and sequences
> Of everything relating to the combination, composition,
> Renovation, arrangement, and unity revolving
> In succession, part by part, throughout the whole,
> Ascending and descending, leaving no tract behind,
> And sifting it as faithful secretaries and clerks
> In the courts of kings, set to work, with diligence and
> Judgement, and sort into different boxes, connaturals

> Concerning matter of state, and when he has
> Attentively sorted it, from the beginning to the end,
> And united and collected the dispersed and distributed
> Matter, which is mingled up and down in combination,
> It will be easy to make a translation of it.[44]

Thus spake Bacon, in accents strange.

The various 'connaturals', etc., of the four key words yield (according to one estimate) some 10,650 words for Owen to conjure with, and as he is flexible about where to locate the quotation suggested by the guide word—in one case it lies forty-seven lines away—he can manipulate his sources at will. Moreover, Owen and his co-workers subtly altered the quotations; for example, the lines, 'Yea, mock the lion when he roars for prey | To win the lady', from *The Merchant of Venice*, become 'Yea, mock the lion when he roars for prey | To win the cipher'. True, only one word is substituted, but it is a fairly important word. The extent of the tampering is difficult to determine, for Dr Owen does not furnish his readers with the sources of his quotations (or, for that matter, the guide words he has used to find them); but we have the testimony of his friend and critic, Dr Frederick Mann, that 'it is doubtful if a single page is made up of extracts quoted fairly'.[45]

In an important passage Sir Francis gives the decipherer some practical advice:

> Take your knife and cut all our books asunder,
> And set the leaves on a great firm wheel
> Which rolls and rolls.[46]

From this hint Owen constructed his wonderful Wheel. The machine consisted of two huge reels mounted on horizontal axles. Seated on a stool on a raised platform, an operator rotated the drums by means of a crank. Stretched on to them was a canvas, a thousand feet long and over two feet wide, on which Owen had mounted in four neat rows the printed pages of Bacon's works: his acknowledged writings, plus the plays of Shakespeare, Marlowe, Greene, and Peele, *The Anatomy of Melancholy*, *The Faerie Queene* and Spenser's other poems; even Dr Rawley's English translation of *In Feliciam Memoriam Elizabethae*, prepared by Rawley (who was Bacon's secretary) twenty-two years after his employer's death.

The decipherer—either Owen himself or one of the three women who assisted him—turned the reels on their axles, located (by means of the guide words) the relevant extracts as they passed along on the belt, and dictated to a typist the execrable verses of Sir Francis Bacon's cipher story. This narrative, which fills a thousand printed pages, treats of the Spanish Armada, the St Bartholomew massacre, the execution of Mary Queen of Scots, Queen Elizabeth's secret marriage to the Earl of Leicester, the banishment of Bacon to France, and the strangulation of Elizabeth by Robert Cecil. Not all of this material is otherwise familiar to historians. One of the most startling revelations concerns Bacon's parentage. 'Slave!' Elizabeth shrieks at him in a dramatic encounter:

> I am thy mother.
> Thou mightst be an emperor but that I will not
> Bewray whose son thou art;
> Nor though with honourable parts
> Thou art adorned, will I make thee great
> For fear thyself should prove
> My competitor and govern England and me.[47]

After recovering from his swoon Bacon runs to his foster mother, Lady Ann Bacon, and learns from her that the noble Earl of Leicester is his father. Unable to assert his lawful title to the throne, and burning with the sense of his humiliation, Bacon (as Owen helpfully explains) 'sought to right the great wrong heaped upon him, through the aid of masks and cipher which should eventually be discovered and rescue his name and fame from the cloud under which he was compelled to remain, during the life of Queen Elizabeth and her immediate successors'.[48] Hence he turned to Spenser, Burton, and other pliant hacks, and paid them off for the use of their names. Shakespeare presented a special case: 'Anthony Bacon, the foster-brother of Francis, was the unknown owner of the Globe Theatre. Shakespeare, while uneducated, possessed a shrewd wit and some talent as an actor. He received, as a bribe, a share in the proceeds of the theatre, and was the reputed manager.'[49]

As he decoded the cipher story, abetted in the later stages by visitations of Bacon's shade, a strange fancy took hold of Owen: he became persuaded of the existence, within a set of iron boxes, of Bacon manuscripts which alone could settle the authorship question incontrovertibly. But where had Lord Verulam secreted the chests? To locate their whereabouts Owen used a new cipher system. Starting from a key word, he selected letters this time, rather than words, by moving (like the king in a chess game) up, down, horizontally, or vertically; this method Owen called 'The King's Move Cipher'. A new repository of clues presented itself in the form of *The Arcadia*, for centuries mistakenly attributed to Sidney.

Finally, in 1909, the cipher revealed that the iron boxes lay buried near a castle two and a half miles above the point where the Wye River joins the Severn. With the financial support of Dr William Prescott, a Boston physician, and others, Owen journeyed to England. He planned a stay of six weeks, but was to remain for six years. Luring him with silent force was the castle, a vine-covered ruin splendidly situated on the summit of a cliff; below, through the castle estate, flowed the River Wye. From the owner, the Duke of Beaufort, Owen obtained permission to excavate, on the understanding that the relics, once recovered, would be presented to the nation in the Duke's name. Dr and Mrs Prescott accompanied him. Fred S. Hammond, an engineer prudently hired by the Duke to keep an eye on Owen, joined the party. A horde of reporters trailed after. Secret government agents, Owen darkly fancied, were set to spy on him. In nearby rock formations, he searched for a cave sheltering the boxes, then gave up; 'the cipher', he sighed, 'was incomplete and left much to unravel'. Some tinkering with his method furnished Owen with an explanation for his false start: 'Bacon feared [according to Mrs Prescott's *Reminiscences*] the cliff might fall away or be cracked by the winter frosts, thus disclosing the hiding place of the manuscripts.'[50]

The cipher now led Owen to the river bed. Fresh financial support was forthcoming from the Chicago Baconian, Colonel George Fabyan, director of the Riverbank Laboratories at Geneva, Illinois. Encouragement of a less material nature came to Owen in the form of an anonymous communication furnishing an anagrammatic reading of Jonson's poem 'To the Reader' in the First Folio, which begins, 'This figure, that thou here seest put, | It was for gentle Shakespeare cut.' When anagrammed, the second line can be made to read: 'Seek, sir, a true angle at Chepstow—F'. So Owen sought the true angle by sinking shafts—eight or ten all told—in the river bed. He did not find the boxes, supposedly placed in a raft, 'like eels in an eel trap', although the digging did turn up an abandoned cistern and the remains of a Roman bridge. Years later someone reported that the excavations also yielded, beneath a dozen feet of mud, 'a small, gray, stone structure . . . marked with inscriptions of Francis Bacon', which was regrettably empty. The exploreres kept a stiff upper lip. 'Our readers may feel that so far the story spells only defeat and failure', Mrs Prescott declared, 'but we never lost faith or hope.' Maybe, she mused sadly, the 1638 *Arcadia* had led them astray.

In 1920 Prescott mounted another assault with money furnished by Mr Harold Shafter Howard. This time Owen stayed behind—his health had collapsed—but the expedition acted on his clues. Prescott excavated the cellar at Chepstow, thus endangering the castle walls, and found the handle of an iron chest, but neither the chest itself nor the manuscripts. Four years later Howard returned to look for steps 'which the cipher said were there, but which had not been found due to Owen's miscalculations'. A flight of steps was duly uncovered on the nearby Hastings Clay estate, but Howard lost interest in this lead: the manuscripts—sixty-six boxes of them—lay hidden, he now felt, in a grotto in Piercefield Park. Hammond, a convert to the lunatic enterprise, disagreed—the cipher, he insisted, pointed to a chamber in the wall of the castle tower. In any event, Howard did not dig, for the owner of Piercefield Park, suspecting his visitor of eccentricity (Howard arrived flying the American flag from poles which he stuck in high trees facing the house), refused permission.

Meanwhile Owen lay paralysed in Detroit. Shortly before his death in 1924 he poured out his feelings to a young admirer. 'Never go into this Baconian controversy', he warned,

for you will only reap disappointment. When I discovered the Word Cipher, I had the largest practice of any physician in Detroit. I could have been the greatest surgeon there, if I had staid [*sic*] by it; but I thought that the world would be eager to hear what I had found. Instead, what did they give me? I have had my name dragged in the mud,—had more calumny heaped upon my character than many people can imagine,—lost my fortune,—ruined my health,—and today am a bedridden almost penniless invalid.[51]

Thus did the noble endeavour claim another martyr.

# 5

# *Elizabeth Gallup*

YET, despite the bitterness of his disappointment, Owen's sacrifice was not entirely in vain, for he had attracted disciples no less committed than he to the cause. Of these the most remarkable was Mrs Elizabeth Wells Gallup, whose fame would outstrip her master's.[52] A high school principal educated at the State Normal College in Michigan and at the Sorbonne in Paris, she assisted Owen in preparing the later books of *Sir Francis Bacon's Cipher Story*. While Gallup was engaged in this task, it dawned upon her that Bacon's explanation of the biliteral cipher in the *De Augmentis* might serve a covert end. Experimentally she applied the method, which unlike her predecessors she understood, to a photographic facsimile of the First Folio. In the italic words and passages scattered prodigally through the text, Gallup discovered two founts of type, which may be labelled *a* and *b*: the one (*a*) normal, the other (*b*) slightly divergent, with the capitals easily discerned and the lower case letters accessible to the educated eye. From the permutations of the *b* fount emerged the cipher message. Surprised and gratified, Gallup extended her coverage to the enlarged Bacon canon, as determined by Owen; the canon that included Spenser and Rawley. For good measure she added Jonson's *Sejanus* and masques, nor was she so fastidious as to exclude the plays of the Shakespeare apocrypha.

Her quest produced such cipher messages as the following:

When th' Masques—in my friend Ben Jonson's name—with Part o' th' King's Coronall Entertaynment have been entir'ly decipher'd, take Greene's and Peele's workes in th' order giv'n in th' Faerie Queene. My plaies are not yet finisht, but I intend to put forth severall soone. However, bi-literall work requiring so much time, it will readily be seene that there is much to doe after a booke doth seeme to bee ready for the presse, and I could not well saye when other plays will come out. The next volume will be under W. Shakespeare's name. . . . When I have assum'd men's names, th' next step is to create for each a stile naturall to th' man that yet should [let] my owne bee seene, as a thrid o' warpe in my entire fabricke soe that it may be[53]

The disclosures relating to Bacon's personal history confirmed those of Owen's word cipher; Gallup's hero was indeed of royal birth. Other revelations, she later confessed, were repugnant to her very soul, but with stoic forbearance she pressed on. 'As a decipherer I had no choice', she declares, 'and I am in no way responsible for the disclosures, except as to the correctness of the transcription.'[54] *The Anatomy of Melancholy* yielded an unexpected dividend in the form of the summary and partial translation of *The Iliad* and *Odyssey*. These discoveries enhanced Gallup's admiration for Owen: surely Bacon had expected his future unraveller to find the biliteral cipher first and, thus rewarded, then to proceed to the word cipher; Owen, with the inspiration of genius, had reversed logical sequence.

In 1899 she published, for private circulation, *The Bi-literal Cypher of Sir Francis Bacon Discovered in His Works and Deciphered by Mrs. Elizabeth Wells Gallup*. The furore created by it ensured a wide audience for the enlarged edition of 1900, which was followed by a third edition the next year. Newspapers around the world reported her story. The book received some temperate reviews; W. H. Mallock acknowledged in the *Nineteenth Century* (December 1901) that Gallup had succeeded in shifting the authorship controversy to wholly novel territory, and that prima-facie grounds now existed for entertaining the possibility that the Baconian theory was true. But elsewhere Stratfordians leaped to the attack with unedifying virulence, some even going so far as to accuse the pious lady from Detroit of insanity or plagiarism (the translation of Homer suspiciously resembled Pope's). Although everywhere the hearts of Baconians skipped a beat, the Society was not yet ready to accept Gallup's revolutionary contribution. In December 1900 the Council solemnly resolved: 'That in view of the failure to produce satisfactory key-alphabet for the cipher narratives, declared by Gallup to have been inserted by Francis Bacon in various books, and the inconclusive nature of her demonstrations, the Society is unable to give any support or countenance to the alleged discovery.'[55] There the matter did not rest.

Gallup answered her detractors in articles and pamphlets, and went about her business of decipherment. At the British Library the dim light taxed her eyes and forced her to desist for a time. Later she worked in American collections. In the library of Dr John Dane of Boston she examined a copy of the rare 1623 London edition of the *De Augmentis*, and near the end of the book found the reference that so many had for so long sought. The cipher revealed the whereabouts of the original Bacon manuscripts:

Our task is often shared . . . by one most devoted always, the constant and faithful friend William Rawley. He it is which must fulfil our plann of placing certain MSS . . . to insure their preservation, in tombes, graves, or in monuments, intending to give unto every man his owne, i.e., it is our design to put MSS. (of playes, poems, histories, prose— . . . etc., in a marble monument and in tombes wherein the cinders of our masques may lie. With much care we shall carve upon the stones placed to mark their lowly or lofty sepulchres . . . such cypher instruction as must leade unto true knowledge of all we shall hide within, . . . All are in time to bee plac'd in the graves or in memoriall marble tables or monuments. . . . There cannot be founde a better device than that of the stone of the Stratford Tablet . . . to preserve a large part of the playes. . . . A boxe shall thereby appear after much quest. Thence the plays mayst thou take, if the century shal be pass'd; if it be ere long, touch none. . . . So whilst these tombes do stand shall hope for this our work live.[56]

Aflame with enthusiasm, in July 1907 Gallup sailed to England to search out the graves and monuments.

Frustration awaited her at every turn. The grave of Robert Greene lay out of reach beneath the Liverpool Street railroad station in London. Marlowe rested fifteen feet below the tower of St Nicholas's Church in Deptford. Of Peele's burial spot no trace remained. The original monument to Edmund Spenser in Westminster Abbey had crumbled during the previous century; still, in the engraving of it in the 1679 edition of *The Faerie Queene*, she discerned a hidden message inscribed on the stones: 'A small inner space at the west end contains the MS. named.' (Knowledge of Gallup's labours might have spared Roderick Eagle his own excavations in Westminster Abbey, where in 1938 he dug up someone's skull and bones before learning that the location of Spenser's grave was no longer known.) From the Latin inscription beneath the bust of Burton in Christ Church, Oxford, Gallup exhumed another directive: 'Take heed: In a box is MS. Fr. B.' She tapped on the metal plate bearing the inscription, and it resounded hollowly—surely a box might lie within. The authorities, however, unreasonably refused to let her excavate. Stratford she bypassed, for a message, just deciphered, indicated that Rawley and Burton had altered the plans made by Bacon to hide the manuscripts there. At St Michael's Church in picturesque St Albans she discovered that the inscription on the pedestal of Bacon's monument had been recut upon an earlier text. 'This makes it impossible', she lamented, 'to translate the Cipher message which it undoubtedly contained.'[57]

That year in October, Gallup took up residence in Oxford, and at the Bodleian Library deciphered Rawley's translation of the *Resuscitatio*, published thirty years after Bacon's death. The cipher discourse turned out to be an old man's rambling soliloquy, but it contained a message so important that Gallup reproduces it in italic:

*Certain old panels in the double work of Canonbury Tower, and at our countrie manor, Gorha'bury, alone sav'd most valu'd MSS. Thus co'ceal'd, more closely watched, more suited to escape sublest* [sic] *inquiry, you shall find th' dramas hee wisht to hide in th' stone he proposed should bee sett up in the Ch. of Stratf'd.*[58]

But Canonbury Tower, once Bacon's residence, had since been largely reconstructed; it was no secret that the concealed chamber behind 'panel five' had been shut up as unsafe. Gorhambury manor lay in ruins. The expedition ended in failure.

Still Gallup did not despair; faith in the cipher sustained her. In 1910 she wrote:

That the Cipher message is enclosed in the works I have deciphered I *know*, from years of hard and exhaustive study. There is no more doubt of the existence of the Cipher and its message than there is of the Morse alphabet and its use at the present day. The study has been of thousands of pages, comparison and classification of hundreds of thousands of the Italic letters, and I have the right to claim, and insist, that I *know*.[59]

Over the years support came to her. Colonel Fabyan, the eccentric millionaire who had financed Owen's excavations, became her Maecenas; on his spacious estate in Geneva, Illinois, maintained by a pension and assisted by a research staff, she passed her latter days in tranquillity. She had the satisfaction of receiving the endorsement of General Cartier, head of the cryptological service of the Deuxième Bureau of the French Army General Staff during the Great War. The Bacon Society, cowed by the General, retreated from its earlier position; in 1936 the President, after quoting Cartier, concluded 'that no case has been made out for distrusting Mrs Gallup's work as a whole'.[60] Later numbers of *Baconiana* confirmed the rehabilitation.

But of course she pursued a chimera. Gallup's methods have been subjected to exhaustive and devastating analysis by Colonel William F. Friedman and Mrs Elizabeth S. Friedman.[61] The arguments of these professional cryptographers are too complicated for detailed summary in these pages, but a few cardinal points may be briefly noted: type analysis reveals not two italic forms, as required by the cipher; but a host of different forms in the books decoded by Gallup. No peculiarities exist to support her classification of *a* and *b* founts, nor can the experts detect evidence of the purposeful distribution of the multitudinous variety of founts that are in fact employed. The method, moreover, requires that all copies of the work containing the cipher be identical, so that the cryptographer may obtain the same (correct) message no matter which copy he happens to use. But, as the bibliographers—Charlton Hinman and company—have shown, no two copies of the First Folio are identical; Gallup simply did not understand the conditions of Elizabethan printing. Lastly, examination of sample Gallup decipherments reveals a disturbing degree of error and arbitrariness. What she had disco-vered was not a biliteral cipher but a biliteral Rorschach blot. The Friedmans recall Gallup as an intelligent, sincere, and upright matron. She was not mentally unbalanced, nor did she perpetrate a deliberate fraud. The *agents provocateurs* of her unconscious dictated the enormously long Gallup cipher story, amounting in all to perhaps 150,000 words; even the items that struck her very soul with repugnance came from this mysteriously unbidden source. Such a phenomenon is perhaps almost as remarkable as the disclosures she claimed to have found.

# 6

## *Other Cryptanalysts*

THE endeavour to strengthen the Baconian case took ever more extravagant forms. Clues from the master's *Alphabet of Nature* and *History of the Winds* supplied Mrs Natalie Rice Clark, wife of a professor of Greek at Miami University in Ohio, with the inspiration for the device she describes in *Bacon's Dial in Shakespeare* (1922). Clark drew a clock face with twelve hours and superimposed it on a mariner's compass with thirty-two points. At the intersections of lines and circles on the compass she placed letters of the alphabet. A pointer, like a clock hand, showed the correspondences between the Dial and Shakespeare's plays. Various phenomena in the text—Gates and Keys, time references, compass points, Bacon acrostics, etc.—were tallied with the Dial. She put capital letters on the contrivance, and from lines traced between them created 'Maze Pictures', wonderful cobweb designs forming Baconian signatures, which she called 'Blazons'. A maze of the constellation Dipper, found in the Epilogue to *The Tempest* (which 'seems to have been written partly as an allegory of the cipher') reads 'I, W.S. Am F. Bacon.' The capitals of the prefatory poem to the First Folio by 'I.M.'—the translator James Mabbe, 1572–?1642, of Magdalen College, Oxford—form a curtained room, or stage, in the rear of which lies the pointed shadow of a grave; within this small but sufficient space is to be found, 'plainly . . . for all to see', the message, 'Exit W.S.—Re-enter F. Bacon.' These are the only coherent statements produced by the Dial: precious little ore for so elaborate a mining operation.

However, Clark did not become cross with her cipher, which she cherished as 'a most sane and human and worthy cipher, both comrade and critic at once'. She went on, in *Hamlet on the Dial stage* (1931), to postulate that the dramatist 'gave reality to the people of his imagination by using, as he wrote, an actual miniature stage or board, upon which he moved figures that represented the characters in the play'.[62] The very existence of this stage, marked with a series of circles and lines, reinforced her conviction that Bacon was the author, but she now began to suspect that he worked in collaboration: 'Those of us who are impressed by the friendly and genial quality of Bacon's mind will rather enjoy the fancy that he could not help taking one or two other brilliant wits into the Dial circle with him.'[63]

Clark's Dial, the operations of which she described with a mistiness which thickens into pea soup, is an exotic offshoot of anti-Stratfordian inquiry which would attract few proponents. More and more, in the earlier twentieth century, a rage for acrostics and anagrams swept the Baconian legions. Inevitably they were drawn to the Clown's nonsense word in *Love's Labour's Lost*, 'Honorificabilitudinitatibus'—an unusually meaty morsel for the hungry anagrammist to sink his teeth into. (An informative brief genealogy, tracing the word back to Papias, *c*.1055, and noting its use by Dante and Rabelais, is given by Douglas Hamer in his review of the original edition of *Shakespeare's Lives* in the *Review of English Studies*, 22 (1971), 484.) A piquant sauce was the fact that the word appears among scribbled notes in the Northumberland Manuscripts (found in 1867 in Northumberland House in the Strand), which contain copies of some Baconian writings; it also occurs, diagrammed, in the Bacon papers in the British Library. (That it should crop up here and there is in itself hardly surprising, for the facetious coinage was popular; it did not originate with Shakespeare but first entered print in the *Catholicon* of Joannes Balbus as far back as 1460.*) Of the various messages produced by rearranging the letters of the word, none was more positively announced than that of Sir Edwin Durning-Lawrence. In *Bacon is Shakespeare* (1910)—a characteristically tentative title—he writes

the true solution of the meaning of the long word 'Honorificabilitudinitatibus', about which so much nonsense has been written, is without possibility of doubt or question to be found by arranging the letters to form the Latin hexameter.

HI LUDI F. BACONIS NATI TUITI ORBI
These plays F. Bacon's offspring are preserved
for the world.
It is not possible to afford a clearer mechanical proof that
THE SHAKESPEARE PLAYS ARE BACON'S OFFSPRING. . . .

It is not possible that any doubt can any longer be entertained respecting the manifest fact that

BACON IS SHAKESPEARE.[64]

---

* It appears as a familiar long word in Marston's *Dutch Courtesan* (*c*.1604).

Little wonder that Sir George Greenwood—barrister, MP, and eminent schismatic—scrawled on the flyleaf of his copy, 'This book is the work of a conceited lunatic. The folly of it is unspeakable!'*

To his anagrammatic expertise Durning-Lawrence adds numerological wizardry. The twenty-seven letter word falls on the twenty-seventh line of page 136 of the First Folio, and is the 151st word in roman type on the page. The numerical values of the initial and terminal letters of the revealed sentence, as determined by their positions in the alphabet, total 136; the sum of the equivalents of the intermediate letters is 151. Put these figures together and we have 287, which is the sum of all the letters of 'Honorificabilitudinitatibus'. It does not occur to Sir Edwin that these totals must needs be identical, having been produced by the same set of letters; nor does he explain why he omits italic words from his calculations. Would the profound scholar have written such lame Latin even when constrained by anagram? Durning-Lawrence, moreover, does not know that Bacon Latinized his name as *Baconus*, with genitive *Baconi*: never *Baconis* (from *Baco*). And why does the word appear as the 150th, on the twenty-fifth line, in the 1598 Quarto? Was Bacon anticipating typographical contingencies a quarter of a century hence? Silence. So sure was Durning-Lawrence that he had found the only correct solution that on 11 November 1910 he offered a reward to any reader of the *Pall Mall Gazette* who could invent 'another sensible anagram' giving the numbers 136 and 151. This sporting challenge was accepted by one Ralph J. Beevor of St Albans, who offered 'Abi invit F. Bacon Histrio ludit' (Be off, F. Bacon, the actor, has entered and is playing). Sportingly Sir Edwin produced his cheque for one hundred guineas.

'In printing this book I wish to present, as concisely as I can, some acrostics which have come to my notice.' Thus modestly does William Stone Booth begin his volume of 631 enormous pages, *Some Acrostic Signatures of Francis Bacon* (1909). He urges the merits of the string cipher, by means of which he locates, in Shakespeare's text, the name, title, and armorial motto of Bacon, as well as a few other names: Jonson, Milton, Anthony Bacon. According to Booth this cipher operates as follows:

Suppose that each letter is the initial letter of a word; then in order to keep them in a string all that was necessary was to fall back on the zig-zag method of writing used by the early Greeks. . . . This string or zig-zag order will give an acrostic on initials, terminals, capitals, or all letters in the text, and running alternately with and against the *sense* of the text or composition, and absolutely independent of its meaning.[65]

But the true string cipher, known as far back as the seventeenth century, is a far cry from Booth's loose system, which he extends well beyond the limits of valid cryptography.[66]

---

* Afterwards, perhaps embarrassed by this lapse from his customary urbanity, Sir George scored the words through. The copy is now in the Folger Shakespeare Library.

Eventually disillusionment overtook him, despite his substantial following, which included a converted Harvard University professor. Looking back at his earlier work in a book quaintly entitled *Subtle Shining Secrecies Writ in the Margents of Books* (1925), Booth confesses, 'I made the great mistake of printing much which a logical examination will obviously discard as accidental, while leaving the reader to make his own decision as to which spelling trick was likely to be intentional.' *Some Acrostic Signatures*, as well as its sequel, *Marginal Acrostics, a Catalogue* (1920), was merely 'a collection of laboratory notes, a record for the examination of students'.[67] Unfortunately, the shining secrecies that are the fruits of Booth's newly acquired wisdom have little more glitter: the revised method, which he calls 'devices', yields not phrases or sentences but words, and many of these are abbreviated or Latinized. Chance, just as well as design, might have produced them.

Booth's contribution found little favour with Walter Conrad Arensberg. Millionaire occultist and art collector, Arensberg flourished in the southern California sunshine which has nurtured so many idiosyncratic visionaries, and he left, as his most enduring monument, the assemblage of twentieth-century and pre-Columbian art now housed by the Philadelphia Museum. Harvard instilled in him a love of Italian literature which found expression, along with his passion for anagrams, in *The Cryptography of Dante* (1921). Dante led him to Shakespeare, whose language he found similarly rich in anagrams. In *The Cryptography of Shakespeare* (1922), Arensberg deployed such heavy artillery as the anagrammatic acrotelestic, the compound anagrammatic acrostic and acrotelestic, and the cross-gartered acrostic. Later, in *The Baconian Keys* (1928), he brought reinforcements in the form of an Autonomous Master-Key and a Pseudonymous Master-Key. Despite their imposing names, these methods hardly bear analysis; essentially they permit Arensberg to select the letters he wishes and to arrange them in patterns which suit his fancy. As he says, 'In its general character, the Baconian system of cryptography is flexible.'[68]

Meanwhile Arensberg read Mrs Pott, author of *From Francis Bacon and His Secret Society*. From it he learned that at the age of fifteen the Magus (as she refers to him) began to assemble 'his "*Invisible Brotherhood*"' so a Rosicrucian, '*the last of his circle*', revealed to her. That is not the only startling intelligence Pott shares with her readers. In the preface to her revised edition she writes with conspirational urgency:

Long research, collation of books and records, and finally corroborative and emphatic assurance from two authorities as important and indisputable as any, have independently testified to the truth of conclusions arrived at by the present writer as to the death of Francis St. Alban—that *he did not really die in 1626*. The witnesses agree not together; yet neither do they check or correct each other. We say, then, *they are in league*; they are of that Fraternity which is bound '*to Conceal as well as Reveal*' the Secrets of their Great Master. *In 1626* he died to the world—retired, and by the help of many friends, under many Names and Disguises, passed to many Places. As Recluse, he lived a life of study; revising a mass of works published under his 'Pen-names'— enlarging and adding to their number. They form the Standard Literature of the 17th Century.[69]

A very learned German correspondent privately informed her that the Miracle of Men died in 1668 at the fairly advanced age of 107. But where did his secret associates bury him? 'Was St. Paul's his monument?' asks Pott. Arensberg sought the answer.

Cryptic expressions in the plays furnished it. They pointed not to St Paul's or to St Albans, where Bacon asked in his will to be placed, but instead to the Chapter House of Lichfield Cathedral in Staffordshire; there members of the clandestine fraternity of which he was founder or head buried his mortal remains, along with evidence (including the original manuscripts) of his authorship of the Shakespeare canon. This society, Arensberg believed, still existed, and maybe its members had knowledge of the hidden grave with its treasure. Four features of the Chapter House identified the location precisely: the chimney in the southern wall, the rose in the stained glass window to the west of the chimney, the cross in the window to the east, and a curious stain on the pavement. (It did not matter that all four were of comparatively recent origin.) According to Bacon's master plan, the truth, when ultimately discovered, would constitute the crowning proof of his inductive method:

This false tradition . . . that the author of the Shakespearean works was the actor William Shakespere, was deliberately established by Bacon as his supreme symbol of all the falsehood which he saw enshrined as truth in all departments of human knowledge; and as a tradition that is established by the essentially fallible method of deduction from "authority," it was intended, in its eventual overthrow, to illustrate, in a manner which would excite the imagination of the world, the possibility of the existence of error in all the traditions in all departments of human knowledge which are based on deductions from authority and not on the Baconian method of induction.[70]

Religion no less than philosophy dictated the arrangements, for Arensberg's hero had come to take on the attributes of deity—'he must be understood to have intended his secret grave at Lichfield to be the symbol of his death and rebirth, just as the Holy Sepulchre is the symbol of the death and rebirth, or resurrection, of Christ . . .'.[71] On 19 April 1923 Arensberg petitioned the Dean and Chapter of Lichfield Cathedral for permission to excavate. Five days later they refused.

The next year, in another pamphlet, Arensberg addressed the Dean and Chapter again, persuaded now that they were associated with the covert society and guarded the grave as a religious shrine. Had they not cunningly introduced cryptograms, expressing the Baconian secrets, in their official publications, such as the twelfth edition of the *Hand Guide to Lichfield*

*Cathedral*? In the same tract Arensberg announced that the remains of Bacon's mother also rested at Lichfield. He pointed to the presence of four loose stones in the Chapter House: 'a reference to the location of the Holy Grail as symbolised in the secret grave'. Bacon, we moreover learn, ascribes to himself—in connection with 'his divine identity as incarnation and as reincarnation'—the following names:

WILLIAM SHAKESPEARE, JEHOVAH, DEVIL, JESUS CHRIST, SAINT AMPHIBALUS, SAINT ALBAN, LADY ANN BACON, SIR FRANCIS BACON KNIGHT, BARON VERULAM OF VERULAM, VISCOUNT SAINT ALBAN.[72]

That the surreptitious burial had indeed taken place in accordance with Bacon's intentions is revealed by cryptograms in various later texts, including the then-sixpenny ticket of admission to Holy Trinity Church. Arensberg sent a draft of his pamphlet to the long-suffering Dean and Chapter, but they chose to ignore it. They refused his request to take photographs in the Chapter House. They even removed the loose stones. But Arensberg did not despair; the truth, he knew, would one day prevail.

In 1928 he published a huge, impenetrable interpretative tome, *The Shakespearean Mystery*, which contains a chapter dealing exhaustively with the cryptic references in the plays to Lichfield as the locale 'of the secret grave of the divine mother and the divine son in the Shakespearean mystery'. In *Francis Bacon, William Butts, and the Pagets of Beaudesert* (1929) and *The Magic Ring of Francis Bacon* (1930), Arensberg fills in gaps in knowledge about Bacon's ancestry and the conditions he required for the ultimate disclosure of the Shakespearian secret. Baron Verulam was the offspring not of Elizabeth and Leicester but of Lady Ann Bacon and Sir William Butts, eldest son of the physician to Henry VIII. By right of this descent Bacon claimed the English throne, for Betts was the heir of the house of Lancaster. In founding his secret society the Viscount contracted for the services of a Paget: probably Lord Thomas Paget, third Baron of Beaudesert, whose seat became Bacon's secret residence. To the custody of his heirs Bacon entrusted a ring (alluded to punningly in *All's Well That Ends Well* and explicitly in *Henry VIII*), which possessed magical properties the interpreter must understand in order to penetrate the mysteries guarded by the custodians of the grave. Hence Arensberg's new appeal, this time to the Marchioness of Anglesey and the Marquess of Anglesey, Baron of Beaudesert; unless Bacon's plan had miscarried, one of them must have the magic ring. But they failed to respond, and in his home in Hollywood, city of insubstantial dreams, Arensberg consoled himself with a cryptographic equivalent in Shakespeare's text. He died in 1954, having by the terms of his will endowed the Francis Bacon Foundation, which, on the campus of Pomona College in Claremont, California, houses the largest Bacon collection in the world. To the end the secrets of Bacon's grave eluded Arensberg.

Viewed as a whole, his career assumes the lineaments of archetype. Arensberg was the last of the major Baconians. Beginning, like so many of the others, with pseudo-intellectual constructs, he moved rapidly beyond the horizon of rationality to an outer darkness of occult speculation. Those who knew him recall Arensberg as a rational man, but on this subject he was mad, and mad from the outset. Surely it is madness—if madness in reason— to believe that the hilarity of Falstaff, the agony of Othello, and the rage of Lear serve merely the puerile requirements of a game of words or numbers: telling an inpossible tale of courtly intrigue, conveying signatures or broken fragments of thought. For this the poet's vision, the playwright's craft? For this the tangled skeins of metaphor, one figure melting before full realization into the next; the even-handed justice metamorphosing into heaven's cherubin horsed upon the air's sightless couriers? For this the lyricism of *Romeo and Juliet*, the ripeness of *Antony and Cleopatra*? Accept so much, and we may swallow the rest: that Bacon was the legitimate offspring of a Queen or the bastard son of a Butts, that he was the incarnation of the living God, that he founded Rosicrucianism, that his remains await resurrection in the Chapter House of Lichfield Cathedral. Spirits of unreason always seem to lurk, like Furies, in the Baconian shadows. Delia Bacon was prophet to the rest.

# 7

# *Groupists*

DELIA was of course not a proper Baconian but a Groupist envisioning a secret association of high-born wits, Ralegh and Bacon chief among them, as the true progenitors of the Shakespeare canon. Others—many others, alas— followed her example and argued for multiple, or at least dual, authorship. In William D. O'Connor—friend of Walt Whitman, author of *The Good Gray Poet* (the title gave a phrase to the language), and librarian of the United States Treasury—she found her first convert. To him belongs the distinction of being the lone reader, alluded to by Hawthorne, of the whole of 'Delia Bacon's splendid sybyllic book on Shakespeare'. O'Connor discerned in her that species of madness to which great wits are near allied; more profitable, he thought, to be insane with Miss Bacon than rational with Dr Johnson. Her views find a spokesman in the hero of O'Connor's novel of the Fugitive Slave Law era, *Harrington; a Story of True Love* (1860). Long years afterwards he returned to the authorship controversy in a polemical tract in defence of Pott, *Hamlet's Note-book* (1886). Here, dismissing Shakespeare as 'a grotesque anomaly', he identified Mr W.H. as Walter RalegH, and the T.T. of the

Sonnets dedication as Ralegh's companion, Thomas HarioT (O'Connor has forgotten that the poems were published by Thomas Thorpe). This evidence had eluded Delia Bacon.

The Groupists attracted a more learned advocate, although (like all the rest) an amateur, in Appleton Morgan. A lawyer who had written with authority on the principles of evidence, he was one of the founders of the Shakespeare Society of New York (odd springboard for deviationist expression) and the publisher of the Bankside Shakespeare. In *The Shakespearean Myth: William Shakespeare and Circumstantial Evidence* (1881) Morgan disposes, with insinuating dispassion, of the 'legal presumption' of Shakespearian authorship. Having distinguished between the 'junta' theory of Delia Bacon and the 'unitary' theory of Smith, Holmes, *et al.*, he offers the New Theory, which is admittedly less novel than the designation suggests. According to this hypothesis, Shakespeare, stage-manager or stage-editor, touched up the plays of others to make them palatable to the groundlings. The natural wit, as the dramatist's contemporaries described him, becomes in Morgan's translation a Warwickshire clown. How cosy and plausible to suppose that

this funny Mr. Shakespeare—who happened to be employed in the theater where certain masterpieces were taken to be cut up into plays to copy out of them each actor's parts—that this waggish penman, as he wrote out the parts in big, round hand, improved on or interpolated a palpable hit, a merry speech, the last popular song, or sketched entire a role with a name familiar to his boyish ear—the village butt, or sot, or justice of the peace, may be; or, why not some fellow scapegrace of olden times by Avon banks? He did it with a swift touch and a mellow humor that relieved and refreshed the stately speeches, making the play all the more available and the copyist all the more valuable to the management. But, all the same, how this witty Mr. Shakespeare would have roared at a suggestion that the centuries after him should christen by his—the copyist's—name all the might and majesty and splendor, all the philosophy and pathos and poetry, every word that he wrote out, unblotting a line, for the players![73]

But who then contributed the philosophy and pathos and poetry? Confronted with the overwhelming question, Morgan retreats into vague conjecture. In an imagined scene Shakespeare is approached at the theatre by certain noblemen of the Court—maybe Southampton, Ralegh, Essex, Rutland, and Montgomery; perhaps also that 'needy and ambitious scholar named Bacon, who, with an eye to preferment, maintained their society by secret recourse to the Jews or to any thing that would put gold for the day in his purse'.[74] The name of a living man, their hireling, would protect their incognito better than a pseudonym. Shakespeare agreed and became rich. Morgan did not do too badly either, his book achieving sufficient celebrity to be tranlsated into German in 1885 as *Der Shakespeare-Mythus*.

Similar theories are expounded, in much the same judicial tone, in John H.

Stotsenburg's depressingly long *An Impartial Study of the Shakespeare Title* (1904). We encounter such sentences as, 'The true rule, both of law and reason, is that when direct evidence of facts can not be supplied, reasonable minds will necessarily form their judgment on circumstances and act upon the probabilities of the proposition under consideration.'[75] Such wearisome legalistic formulations need not surprise us, for Judge Stotsenburg sat on the Indiana bench, and the *Impartial Study* is the medlar fruit produced by his stolen hours. A sampling of the chapter headings will sufficiently indicate the operative assumptions that guide the learned jurist's argument:

> Doubts Raised as to Shaksper's Ability and Learning
> William Shaksper Has No Place in Henslowe's Diary
> Shaksper Commended No Contemporary
> Shaksper Left No Letters and Had No Library
> Shaksper Gave His Children No Education
> Shaksper's Utter Indifference to Literary Proprieties
> Shaksper Not the Shakescene of Robert Greene
> Lies Fabricated in Aid of the Shaksper Pretension
> Shaksper's Real Name and Traditional Life
> The Learning of the Author or Authors of the Poems and Plays.

Since the Groupists rarely agree on the constituents of the group, it is not surprising that the judge comes up with his own set of candidates. He concludes that Bacon wrote *Venus and Adonis* and *The Rape of Lucrece*, that Sidney produced the Sonnets, that a consortium of professional dramatists—Drayton, Dekker, Munday, Chettle, Heywood, Webster, Middleton, and Porter—participated in the original composition of the plays, and that these plays (or some of them) were 'polished and reconstructed' by Bacon. Court adjourned.

Usually associated with the Baconian stalwarts, Sir George Greenwood (in his dauntingly voluminous writings) more than once denies membership in the club. 'It is no part of my plan or intention to defend that theory,' he insists in the preface to *The Shakespeare Problem Restated*, and indeed he remains comfortably agnostic, contenting himself with negative onslaughts against the orthodox citadel. These he manages with a curious mixture of suavity and abuse that makes him unique among the heretics. The illiterate peasant from squalid Stratford and his partisans, especially Lee and Churton Collins (Professor of English Literature at the University of Birmingham), elicit from Sir George the full gamut of patrician denigration. Serpents of cultivated malice lurk in the fine print of the footnotes; in one, Sir George, a Cambridge graduate, savours the fact that his two principal foes took second-class Oxford degrees, and he goes on to dispose of Lee thus:

In the Calendar of 1880 he is mentioned for the first time as Minor Exhibitioner of Balliol College. For the benefit of the puzzled investigator (and such, at first, was I) it may be mentioned that he there appears under a slightly different form of appelation to that by which he is now familiar to us, not having at that date discarded two Biblical *praenomina* in order to assume the more Saxon name of Sidney. I cannot help thinking, by the way, that Mr. Sidney Lee might be rather more tolerant of those who imagine that some great man in Elizabethan times might have seen advantages in the assumption of a pseudonym.[76]

At first it might seem curious that Sir George's circle of friends included his fellow MP, J. M. Robertson, author of the prolix *Baconian Heresy*, to which Sir George replied with his own massive book. But the true irony in the association of the believer with the arch-heretic lies in the actual closeness of their stances. Prince of disintegrators, Robertson in *Did Shakespeare Write Titus Andronicus?*, *The Shakespeare Canon*, and his other books, doles out great slabs of Shakespeare to Peele, Greene, Chapman, Marlowe, and other dramatists, the Stratfordian indeed stood, as Sir George observes, on 'the slippery slope of Infidelity'. Disintegration of Shakespeare's text furnished the Groupist with a starting-point, and conferred respectability upon pseudo-scholarly endeavour. 'That Shakespeare, whoever he was, did not write a very large portion of the thirty-six dramas which were published as his in the Folio of 1623 is now generally admitted,' Sir George observes. And so, at the time, it was.

If Sir George can pride himself on refusing to indulge in wild theorizing, other heretics did not submit to similar inhibitions. There is, for example, J.G.B., who answered his self-inflicted question, *Who Wrote Shakespeare's Plays?* (1887), by postulating that the author was Cardinal Wolsey. The manuscripts of this prelate, deceased in 1530, came into the possession of Bacon; he prepared them for exhibition and capped the *œuvre* by writing his own play of *Henry VIII* with Wolsey a principal figure. An equally beguiling suggestion was made by Harold Johnson in *Did the Jesuits Write 'Shakespeare'?* (1910). Noting that the only English Pope, Adrian IV (1154–9) bore the name of Nicolas Breakspear, Johnson proposes that the pontiff inspired the pseudonym adopted by members of the Society of Jesus as they varied their devotions by busying themselves with *Romeo and Juliet* and *Antony and Cleopatra*.

Sceptics unpersuaded of the existence of this popish plot may find other hypotheses more seductive. In *The Five Authors of 'Shakespeares Sonnets'* (1923) H. T. S. Forrest of the Indian Civil Service fancies a sonnet tournament, with Shakespeare, Barnes, Warner, Donne, and Daniel fighting it out for a prize offered by the Earl of Southampton. Gilbert Slater, by profession an economist and social historian, does Forrest two better by offering *Seven Shakespeares* (1931). These claimants include Bacon, Ralegh, the Earls of Derby, Rutland, and Oxford; also Marlowe (returned from the dead in 1594 as Shakespeare). Imbibing *Julius Caesar* as connoisseurs sample wines to determine their origin, Slater detected a female bouquet. In *Antony and*

*Cleopatra* too he found feminine rather than masculine intuition. The seventh
Shakespeare is a woman, who in *As You Like It* (a markedly feminine play)
portrays herself as Rosalind! No work inspires Slater to more piercing insight
than *King Lear*; to the authoress 'it is due that we are made to see that there
was something to be said for Goneril, and that Lear was a most undesirable
visitor in the house, sure to upset any hostess's nerves'.[77] This Woman
Shakespeare is Mary, Countess of Pembroke, and sister to Sir Philip Sidney
(in an odd slip she is described as Sidney's brother). Surely the peerless
lady of Wilton, rather than 'the money-lending maltster of Stratford', is
addressed by the manly and gallant Jonson as 'My beloved' and 'sweet
swan of Avon'.

This limited (but, one trusts, sufficient) sampling of Groupist heterodoxy
may fittingly conclude with a more recent contribution. Like the others, it
calls for no rebuttal; gossamer fancies, insubstantial as air, need not be
broken upon the critic's wheel. Alden Brooks clears the ground for his own
theory by deposing the poet in *Will Shakspere Factotum and Agent* (1937). 'In
no sense was this fellow a man of literary genius,' he sums up. 'He was,
instead, country wit, business man, theatrical factotum, play-broker, figure-
head, agent.'[78] Behind the broker, behind the Southampton of the Sonnets
and dedications, looms *the* Poet, whom Brooks does not yet care to name. In
the seven hundred closely printed pages of *Will Shakespere and the Dyer's
Hand*, published by a reputable house (Charles Scribner's Sons) in 1943,
Brooks has another, more violent, fling at Shakespeare. Everywhere in
Elizabethan literature he sees unflattering allusions to the National Poet—
this 'despicable trencher-slave, parasite, blood-sucker, pandar, and corrupter
of young noblemen'; for so Marston had described 'fat-paunch'd Milo'
(obviously Shakespeare) in his satirical *Scourge of Villainy*. Will sabotaged his
company by selling plays on the sly to the printers. The wealth he hoarded
was swollen from the proceeds of the Blackfriars Gate-house, converted by
the whore-master poet into a brothel. Finally, outwardly honoured but
gnawed by inward shame, he succumbed to the excesses with which he had
taxed his obese body. This was Shakespeare. We learn without wonder that
Brooks is a minor novelist.

But who is *the* Poet? For this post Brooks sets up fifty-four requirements,
some conventional ('The Poet was a courtier'), others peculiar ('The Poet
died before the winter of 1608'). Only one candidate meets all the desiderata:
Sir Edward Dyer. Patron of letters and close friend of Sidney, he is today best
remembered for his elegy for Sidney, beginning 'Silence augmenteth grief',
from *The Phoenix Nest* (1593). 'Shakespeare' glances at him in the 111th
Sonnet:

> Thence comes it that my name received a brand,
> And almost thence my nature is subdu'd
> To what it works in, like the dyer's hand.

The recovery of the punning allusion—a recovery made possible by the wedding, for an instant, of fantastic speculation and prosaic literalism— marks the high point of *Will Shakspere and the Dyer's Hand*. Dyer, incidentally, did not write all the Sonnets, but had help from Nashe, Daniel, Barnes, Southampton, and other poets. The Earl also commissioned, supervised, and perhaps partly wrote *Lucrece*, a topical poem in which Tarquin the Ravisher is a skit on Ralegh the Proud.

The Brooks thesis seduced few readers; no Dyer Society followed upon the identification of the Poet's hand. Ignored by Gibson in *The Shakespeare Claimants* and savaged by Wadsworth in *The Poacher from Stratford*, Brooks had the resilience to return to the fray in 1964 with *This Side of Shakespeare*. He stands firm for Dyer, and if Shakespeare inspires a mellow condescension rather than the old rage, he is factotum still:

He chose writers and plays, offered advice, acted as general supervisor. It was undoubtedly his natural wit, his 'facetious grace,' and showman's instinct, that gave to many of the Shakespearean plays that leaven few other plays of the time possessed. If his major role was the trafficking for their composition and sale, he became none the less their bondsman and sponsor.[79]

Thus, in sentiments echoing those of Appleton Morgan almost a century earlier, does the Groupist pass judgement on Shakespeare on the four hundredth anniversary of his birth.

# 8

## *Looney and the Oxfordians*

THE month that saw an armistice bring to an end the Great War witnessed another event hardly less momentous, at least for members of the Shake- speare Fellowship. In November 1918 J. Thomas Looney, a Gateshead schoolmaster, deposited with the Librarian of the British Museum a sealed envelope containing an announcement of his discovery that the plays and poems of Shakespeare issued from the pen of Edward de Vere, the seventeenth Earl of Oxford. Before taking this unusual step the schoolmaster had submitted his work, the result of years of patient investigation, to a publisher; but the latter rejected it when Looney refused to adopt a *nom de plume* to forestall the hilarity of reviewers. Now, covetous of priority, he resorted to the device of the sealed letter with its overtones of mysterious significance so congenial to the anti-Stratfordian mentality.

The book, *'Shakespeare' Identified*, appeared in 1920, and initiated the Oxford movement, which has given the Baconians a run for their madness. In his introduction Looney disclaims an expert's knowledge of literature (when he began he had read only Shakespeare, Spenser, and Sidney among the Elizabethans), nor does he pride himself on a critic's soundness of literary judgement. Instead he makes a virtue of amateurism. 'This is probably why the problem has not been solved before now,' Looney asserts. 'It has been left mainly in the hands of literary men.'[80] For years, however, he had been putting his young charges through their paces with *The Merchant of Venice*, prolonged intimacy with which persuaded Looney that the author knew Italy at first hand, and—more important—had an aristocrat's indifference to business methods and an aristocrat's casual regard for material possessions. It is difficult to escape the conclusion that snobbery led Looney, a gentle retiring soul, to seek a Shakespeare with blue blood in his veins. His own family, the pedagogue boasted, was descended from the Earl of Derby, once kings of the Isle of Man, whence came Looney's immediate forebears. He expresses the heretic's customary disdain for the 'coarse and illiterate circumstances' of Shakespeare's early life, and in an unconsciously revealing passage implies that a great writer must have lords and ladies in coaches driving up to his door.[81]

'My preparation for the work lay', Looney reflected in old age, '. . . in a life spent in facing definite problems, attempting the solution by the methods of science, and accepting the necessary logical conclusions, however unpalatable & inconvenient these might prove.'[82] His impartial science, derived from the Positivism of Comte, led Looney to seek nine 'general features' in the author of Shakespeare's works:

1. A matured man of recognized genius.
2. Apparently eccentric and mysterious.
3. Of intense sensibility—a man apart.
4. Unconventional.
5. Not adequately appreciated.
6. Of pronounced and known literary tastes.
7. An enthusiast in the world of drama.
8. A lyric poet of recognized talent.
9. Of superior education—classical—the habitual associate of educated people.[83]

To these Looney added nine 'special characteristics':

1. A man with Feudal connections.
2. A member of the higher aristocracy.
3. Connected with Lancastrian supporters.
4. An enthusiast for Italy.
5. A follower of sport (including falconry)

6. A lover of music.
7. Loose and improvident in money matters.
8. Doubtful and somewhat conflicting in his attitude to woman.
9. Of probable Catholic leanings, but touched with scepticism.[84]

Without the advantages of historical or literary training, Looney had now to find the candidate who met all the general and special requirements.

Plunging in, he selected *Venus and Adonis* and began to look for a poem with similar stanza and cadence in Palgrave's *Golden Treasury*, which alone constituted Looney's reference library of sixteenth-century verse. In 'Women', by Edward de Vere, he found the poem. He next had to learn something about the poet. After several false starts in history textbooks, Looney discovered with delight from the *DNB* that Oxford '*evinced a genuine taste in music and wrote verses of much lyric beauty*'; also that 'Puttenham and Meres reckon him *among the best for comedy* in his day; but though he was a patron of players *no specimens of his dramatic productions survive.*' (The italics in these misquoted passages are supplied by Looney.)

The de Veres traced their descent in an unbroken line from the Norman Conquest: higher aristocracy, there can be no question. Evidence of Lancastrian sympathies (Looney's third special criterion) may be found in the fact that the twelfth Earl lost his head in 1461 for loyalty to the Red Rose. Sidney Lee's *DNB* sketch describes Oxford as having had a thorough grounding in Latin and French, great prowess at the tilt, and an ambition for foreign travel gratified by a journey to Italy in 1575. As a youth, however, he also manifested 'a waywardness of temper which led him into every form of extravagance, and into violent quarrels with other members of his guardian's household'. At the age of seventeen he fatally wounded an under-cook at Cecil House. Report held that he threatened the ruin of his first wife in order to avenge himself on the father-in-law who had incurred his displeasure. There were other indications of a volatile temper: Oxford grossly insulted Sidney on the tennis court at Whitehall—addressing him as a 'puppy', according to Sir Fulke Greville (Sidney's biographer)—and afterwards plotted his assassination; in 1586 he quarrelled with Thomas Knyvet, duelled with him, and entered into a subsequent vendetta. Irresponsible, he hired lodgings and left others, of humbler station, to foot the bill. The Earl's improvidence brought him into financial straits from which he tried to extricate himself by selling his ancestral estates at perversely low prices. Lee does not dwell on the Earl's seduction of one of the Queen's Maids of Honour, nor does he report Aubrey's presumably apocryphal anecdote: 'This Earl of Oxford, making of his low obeisance to Queen Elizabeth, happened to let a fart, at which he was so ashamed that he went to travel, 7 years.'[85] In any event, Looney does not include flatulence as another of his hero's special attributes. Nor does he list cruelty, perversity, and profligacy as features of the author evident from a perusal of his work.

Looney properly relishes the contemporary evidence that Oxford wrote plays (after all it cannot be demonstrated that Bacon or most of the other chief claimants performed this necessary activity), and he attempts to bolster the testimony of Puttenham and Meres by the familiar tactic of converting Shakespeare's dramas into *pièces à clef*. Indeed the Earl can scarcely restrain himself from putting in appearances everywhere in the canon. In *Love's Labour's Lost* he is Berowne mocking Holofernes—Gabriel Harvey, the 'kissing traitor' who had circulated satirical verses about Oxford behind his back. Elizabeth's royal ward is Bertram in *All's Well That Ends Well*, the jealous husband is Othello, the Patron of Oxford's boys is the master of the revels at Elsinore. It follows that the rest of the dramatis personae must have historical identities; and so Laertes is Thomas Cecil; Polonius, Burleigh (to reappear in Venice as Brabantio); Ophelia, Lady Oxford (reincarnated after drowning only to be strangled as Desdemona); Horatio, the Earl's cousin Sir Horace de Vere—principally, it would seem, because of the partial congruence of Christian names. Such a view of drama implies that plays are secondarily intended for theatrical representation, being pre-eminently literary artifacts. To this reversal of priorities Looney freely subscribes: '. . . if we must choose between the theory of their being literature converted into plays, or plays converted into literature, on a review of the work no competent judge would hesitate to pronounce in favour of the latter supposition',[86] Looney, one suspects, has not polled all the competent judges.

His subjective ruminations do little to strengthen an hypothesis which has certain inherent limitations. The attestation of Puttenham and Meres to Oxford's playwriting activities cuts two ways. Meres after all lists Shakespeare separately in *Palladis Tamia* and names twelve plays, as well as *Venus and Adonis*, *The Rape of Lucrece*, and the Sonnets: clearly he did not believe that the Earl wrote *The Comedy of Errors*, *Romeo and Juliet*, and the rest. And if people knew that Oxford graced the stage with plays, why had he need of employing Shakespeare as a mask? The only motive that Looney can suggest is self-effacement. 'We may, if we wish', he adds, 'question the sufficiency or reasonableness of the motive. That, however, is his business, not ours.'[87] But of course the man who sets out to convince the public of the validity of an eccentric theory *must* make the motivation his business. These considerations, however, pale into triviality alongside the principal drawback of the entire argument: Oxford, born in 1550, died in 1604. Thus he was forty-three when he offered the first heir of his invention to Southampton, and was buried before *King Lear*, *Macbeth*, *Antony and Cleopatra*, *Timon of Athens*, *Coriolanus*, *Pericles*, *Cymbeline*, *Winter's Tale*, *The Tempest*, and *Henry VIII* appeared on the stage.

To get round this perplexity Looney must urge that the authorities have misdated *King Lear* and *Macbeth*, and that Oxford at the time of his death left unfinished manuscripts which inferior dramatists then completed. 'The people who were "finishing off" these later plays took straightforward prose, either from the works of others, or from rough notes collected by "Shakespeare" in preparing his dramas, and chopped it up, along with a little dressing, to make it look in print something like blank verse.'[88] Such a considered judgement emanates naturally from a sensibility to which the music of Shakespeare's final period is ragtime. *The Tempest* presents Looney with his greatest challenge, for topical references and other internal considerations lead him to accept the late date to which the commentators assign it. So he must deny it altogether to his candidate—at the same time admitting that 'but for the theory that Edward de Vere was the writer of Shakespeare's plays we might never have been led to suspect the authenticity of "The Tempest"'.[89] The task of denigration proceeds apace. Prospero's speech on the cloud-capped towers and gorgeous palaces becomes 'simple cosmic philosophy, and, as such, it is the most dreary negativism that was ever put into high-sounding words'.[90] (The disciple of Comte insists upon the positivism of his idol.) Elsewhere in the play Looney finds stolen thunder, muddled metaphysics, witlessness, and coarse fun. Above all, the verse is bad, which by Looney's standard merely means that it has irregular scansion syllables. This evaluation of *The Tempest*, needless to say, has met with a cool reception—even from fellow Oxfordians. Looney had at the outset confessed his lack of critical equipment; in the end, having constructed his elaborately rationalized fantasy, he becomes a casualty of that handicap.

Despite its intellectual *naïveté*—perhaps because of it— *'Shakespeare' Identified* impressed the impressionable. In his introduction to the 1948 reprint (which drew some respectful journalistic notices) the maritime novelist William McFee compared the Looney book, for revolutionary significance, with *The Origin of Species*. He also described the Gateshead pedagogue as a sleuth 'methodically and relentlessly closing in on the author, not of a crime, but of a mystery'. The mantle of Conan Doyle sits more comfortably on Looney than that of Darwin; Galsworthy pronounced *'Shakespeare' Identified* 'the best detective story' which ever came his way. Herein must lie much of the fundamental appeal of the work and of anti-Stratfordian demonstrations generally. Sober literary history is metamorphosed into a game of detection, in much the same manner as James Thurber's American lady in the Lake Country transformed *Macbeth* into a Hercule Poirot thriller ( ' "Oh Macduff did it all right," said the murder specialist.' ). To such a game the cultivated amateur can give his leisure hours in hopes of toppling the supreme literary idol and confounding the professionals. Little wonder that one heretic, Claud W. Sykes, casts his investigation as an exercise in detection, with Sherlock

Holmes tracking down the true perpetrator of the plays by means of Baker Street deduction!

Be that as it may, Looney founded a school. A tangible result of *'Shake-speare' Identified* was the formation in 1922 of the Shakespeare Fellowship, a society hospitable to all heretics but chiefly devoted to perpetuating the claims of Oxford. *The Shakespeare Fellowship News-letter*, issued by the association, performed a service analogous to that of *Baconiana*. In addition to schoolmasters and attorneys the group attracted military and naval types, the novelist Marjorie Bowen, and Christmas Humphreys, QC, an authority on Buddhism. It appealed to the young at heart: Canon Gerald H. Rendall, sometime Gladstone Professor of Greek at University College, Liverpool, read Looney and, at the age of eighty, experienced a conversion. He proceeded to advance the cause with a series of volumes: *Shakespeare Sonnets and Edward De Vere* (1930), *Shake-speare: Handwriting and Spelling* (1931), *Personal Clues in Shakespeare Poems & Sonnets* (1934), and *Ben Jonson and the First Folio Edition of Shakespeare's Plays* (1939). So prodigious was the display of energy that one admirer was prompted to exclaim in 1944 that Canon Rendall 'represents one of the biological reasons why the Germans, despite all their sound and fury, will never overcome the British'. After the outbreak of the Second World War the continuity of the Fellowship's work was assured by the formation of an American branch presided over by Dr Louis P. Bénézet of Dartmouth College. This true believer's own contributions include the suggestion, made in *Shakspere, Shakespeare and De Vere* (1937), that in the Sonnets the Earl addressed his elligitimate son, who acted in his father's company of players under the name of William Shakespeare.

The publications of the de Vere sect are too numerous for listing, much less evaluation, in these pages, but a few of the principal items may be mentioned. A member of the Fellowship, Captain Bernard Mordaunt Ward, produced in 1928 a massive biography from contemporary documents, *The Seventeenth Earl of Oxford, 1550–1604*, aimed at rehabilitating the nobleman's somewhat tarnished reputation. While not overtly concerned with the authorship debate, Ward gives tacit support to the theory (suggested by Lefranc) that Oxford and the Earl of Derby were in some way connected with the composition of Shakespeare's plays. Others too favoured the idea of mixed authorship, for example, Slater's *Seven Shakespeares* mentioned above. In *Lord Oxford and the Shakespeare Group* (1952) Lieutenant-Colonel Montagu W. Douglas ingeniously proposed that the Queen charged the Earl with the control of a Propaganda Department for the issuance of patriotic pamphlets and plays, and that he satisfied this commission by putting together a syndicate which included Bacon, the Earl of Derby, Marlowe, Lyly, and Greene: a motley assortment by any standard. Still others sought to adjust the Shakespeare chronology to the facts of Shakespeare's life and thus get round the embarrassment of denying him *The Tempest*. Mrs Eva Turner Clark, in the 693 pages of *Hidden Allusions in Shakespeare's Plays* (1931), arranges the works in a sequence beginning with *Henry V* and culminating with *King Lear* in 1590; for *The Tempest* she finds a snug niche half-way between. This novel arrangement is made possible by identifying Shakespeare's plays with the titles of lost Tudor interludes, and by correctly interpreting internal historical references which had escaped all other scholars. In *King Lear*, for example, the banishment of Kent parallels the banishment of Drake in 1589, while the play as a whole reflects Oxford's bitterness over 'the failure of the Queen to back him up in his patriotic endeavour to support the throne and country against the factions that were, as he saw them, disintegrating forces in the government, if not actively seditious'.[91] Into such tracts for the times do the plays dwindle in Oxfordian hands.

In a note appended to the last page of *'Shakespeare' Identified* Looney had admitted to a belief that the Grafton portrait of Shakespeare really depicts the Earl. The Shakespeare iconography fascinates the Oxfordians. In the pages of *Scientific American* for January 1940, Charles Wisner Barrell, one of the brethren, revealed that X-ray and infra-red photography had detected underneath the Ashbourne portrait the pigment of another painting representing de Vere. This discovery was greeted with hoots of delight by the Fellowship, but how it materially aids the cause (even if we accept the doubtful findings of a partisan) is not clear, for the Ashbourne picture, like the Grafton, has no standing as a genuine likeness of Shakespeare.

Among those who applauded Barrell were Dorothy and Charlton Ogburn in *This Star of England* (1952), the most monumental contribution ever made to the literature of heresy. As one would expect from a volume running to 1297 pages, all the familiar Oxfordian arguments appear, and also some new ones. The quality of the Ogburn reasoning may be illustrated by a single example. They reproduce Touchstone's interrogation of William in *As You Like It*, with certain words and clauses italicized: 'Art thou *learned? . . . all your writers do consent that ipse is he*: now, *you are not ipse, for I am he. . . . He, sir, that must marry this woman. Therefore, you clown, abandon.*' This straight-faced commentary follows:

*How can a man speak more plainly than this?* Oxford—or William Shakespeare—tells Shaksper, another William, to abandon all pretensions to the plays and clear out, forthwith. 'You are not ipse, for I am he.' All the 'writers'—Jonson, Marston, Dekker, Peele, *et al.*—know this, 'do consent' to it. What other possible interpretation can be put upon these candid lines?[92]

The aggrandizing tendencies of the heretic surface: Oxford must be credited not only with all of Shakespeare, but also with the apocryphal plays, Marlowe's *Edward II*, Kyd's *Spanish Tragedy*, and Lyly's *Endymion* and other comedies. In such a context we learn without astonishment that the Earl of Southampton sprang from the loins of Oxford and the womb of Elizabeth, somehow legitimately mated; the Sonnets to the Fair Youth (pun: Vere Youth) therefore become a touching poetical testament of a father to his son. Without once referring to *This Star of England* the Ogburns—this time Dorothy and her son, Charlton Junior—warmed over their stew as *Shakespeare: The Man behind the Name* (1962), which has at least the merit of comparative brevity.

With *The Mysterious William Shakespeare: The Myth and the Reality* (New York, 1984) Charlton Ogburn goes once more unto the breach, to do battle for his own brand of Oxfordian reality, this time with a volume of almost 900 large pages—not the longest such exercise but very long indeed—which surely qualifies as one of the seven wonders of anti-Stratfordianism, although I would be hard pressed to name the other six. Most of the terrain Ogburn traverses will be familiar to initiates. He argues that de Vere is the Will Moxon of Thomas Nashe's *Strange News*; the same Will who partook of Rhenish wine and herrings with Nashe and Robert Greene a month before Greene's death: this Will is to be identified with another Will, the celebrated—if supposititious—playwright of the English stage. Elizabeth's grant of £1,000 a year to Oxford facilitated the writing and production of plays supportive of the throne. The author dwells upon parallels between Shakespeare's plays and Oxford's life, unmindful of the discommoding truth that literature and life are full of cunning parallels. Ogburn also ruefully recounts one unbeliever's encounters with the Shakespeare Establishment. *The Mysterious William Shakespeare* inspired Richmond Crinkley's sympathetic review article, 'New Perspectives on the Authorship Question', which mysteriously appeared in that Establishment bastion, *Shakespeare Quarterly* the next year (36, 515–22). 'Shakespeare scholarship', Crinkley concludes, 'owes an enormous debt to Charlton Ogburn.' Not everyone would agree.

Washington, DC, attorney, business executive, connoisseur of the arts, philanthropist, and Oxfordian enthusiast, David Lloyd Kreeger was the master spirit behind the moot-court debate sponsored by the American University in the nation's capital, and argued by two members of that university's law-school faculty (Peter Jaszi for the Oxfordian position and James Boyle for the man from Stratford) before a trio of Supreme Court justices in appropriate juridical garb: Harry A. Blackmun, William Brennan, and John Paul Stevens. The event took place on 25 September 1987, at the Metropolitan Memorial United Methodist Church in the presence of bus-loads of high-school students, contingents of Oxfordian and Stratfordian partisans, white-collar Washingtonians, and the youthful Charles Francis Topham de Vere Beauclerk, a collateral descendant of the seventeenth Earl of Oxford. All told, roughly a thousand—maybe more—jammed into the church that autumn morning. The lawyers presented their arguments, with occasional interjections from the bench, and the court recessed until afternoon when the justices returned to their seats to deliver their opinions. Justice Brennan, the acting chief, spoke first, concluding that the case for the Oxford side remained unproven. 'What business have I to be judging this?', Justice Blackmun could not help asking himself. He thought of Isabel in *Measure for Measure* ('Oh, it is excellent | To have a giant's strength but it is tyrannous | To use if like a giant'). He agreed, however, that Justice Brennan's conclusion was 'the legal answer'. Justice Stevens similarly arrived at a legal verdict for the Stratford man, although qualified by a degree of personal uncertainty. The event was chronicled in the *Washington Post* and—more conspicuously, as might be expected—in the *New York Times*. Months passed. The Authorship Question became the subject of a long essay by James Lardner in the 'Onward and Upward with the Arts' department of the *New Yorker* (11 April 1988), 87–106. In a retrospective contribution to the *de Vere Society Newsletter*, a new periodical (1988), Ogburn denounced the moot 'trial' as a 'miscarriage of justice' in which Justice Brennan acted for all practical purposes as a witness for the Stratfordian side.

A permanent record of the great Washington Shakespeare debate was eventually published in the American University Law Review, 37 [1988], 609–826. Included was a verbatim transcript of the Justices' opinion, as well as prefatory remarks by Kreeger and essays by Jaszi ('Who Cares Who Wrote "Shakespeare"?') and Boule ('The Search for an Author: Shakespeare and the Framers'). There the matter did not rest: a reprise with a different dramatis personae (Kreeger, Ogburn, and Shakespeare excepted), took place on 26 November 1988 at the Middle Temple—in the same (then new) 'large and stately' hall where a young lawyer, John Manningham, had the good fortune to see a special production of *Twelfth Night* performed by the Lord Chamberlain's Men on 2 February 1602, and jotted down his impressions. On the occasion of the Middle Temple Moot this author was invited to testify as an expert witness, but, as circumstances worked out, the sponsors were unable to accommodate the expense of my journey. Nor was Ogburn able to take part, so Kreeger, Shakespeare, and the Earl of Oxford had to manage without us. The Oxfordians were represented by L. L. Ware, a founding member of the Mensa Society, and Gordon C. Cyr, former director of the Shakespeare Oxford Society; the Shakespearian side by Stanley Wells, director of the Shakespeare Institute, and Professor Honigmann. The presiding judge, Lord Archer, won applause by delivering the day's closing comments in blank verse. The three Law-Lords judging the Shakespeare Moot, as the mock trial was called, all found for the man from Stratford. Court adjourned.

To the Baconians it was not given to glory alone in a cipher. In *Edward De Vere: A Great Elizabethan* (1931) George Frisbee prints a multitude of ciphers based on the six letters of de Vere's name. Not surprisingly, he finds these characters everywhere: in Gascoigne's *Supposes*, in Marlowe, in Harington, Puttenham, Ralegh, Spenser, James I, above all in Shakespeare (most curiously in the contents page and dedication of the 1623 Folio). Even Canon Rendall gratifies us with a cipher:

> Why write I still all one, E.VER the same,
> And keep invention in a noted weed,
> That E.VERY word doth almost tell my name,
> Showing their birth and where they did proceed?

The Canon takes a special pride in this bit of inanity, which, he modestly allows, rescues Sonnet 76 from inanity.[93]

For the parallelism of the Oxfordians with the Baconians to be complete we need only the spirit from the grave and clues to the whereabouts of the Earl's lost manuscripts. No disappointment, alas, awaits us. In the autumn of 1942 Percy Allen, author of several Oxfordian treatises, consulted a London medium, Mrs Hester Dowden, daughter of the celebrated Dublin authority on Shakespeare.* The seances continued over an extended period, with one *Johannes* serving as control, and Hester Dowden herself taking down conversations in automatic writing, of which she was a most gifted practitioner. At these sessions Allen (through the good offices of his deceased brother) met Oxford, Bacon, and Shakespeare. They described their mode of collaboration with alacrity. 'I was quick at knowing what would be effective on the stage,' Shakespeare owned. 'I would find a plot (*Hamlet* was one), consult with Oxford, and form a skeleton edifice, which he would furnish and people, as befitted the subject.'[94] Often they took their efforts to Bacon, whose advice was requested but (the Viscount sadly reports) seldom accepted. All this Allen found extremely fascinating, as well he might, but a small difficulty troubled him. In 1943 one Alfred Dodd published a book, *The Immortal Master*, containing scripts by Hester Dowden reporting direct communication with Bacon, in the course of which the latter claimed for himself Shakespeare's writings. 'My friend, I *can* help you,' Bacon reassured Allen. 'I was acting through a Deputy in the case of Dodd—a Deputy who has never been personally in touch with me, and who questions nothing; for he is firmly convinced that I wrote the plays and sonnets, and took no trouble to have a direct message from me.'[95] Some spooks, it seems, are unreliable.

Where three centuries of scholarship had failed, Dowden's gatherings succeeded, clearing up disputed points in Shakespearian biography and producing fresh details. The poet indeed entered the world on St George's day, his mother invariably having her infants baptized three days after birth. The parents were Protestant (so much for John Shakespeare's Spiritual Last Will and Testament!). At the free school Will was considered a dull scholar. Although the deer-poaching legend had some basis in fact, the youth ran off to London not because of Lucy's wrath, but rather to escape becoming a butcher, the occupation selected for him by his father. At the as yet non-existent Globe in 1581 there was no stage, only a courtyard. 'My first duties', the shade recalled, 'were connected with preparation, cleaning the yard and

---

* Allen, Dowden's biographer informs us, was selected by Spirit People to be the final unraveller of the Shakespeare Mystery (Edmund Bentley, *Far Horizon* [London, 1951], 148).

seats, and putting them in order. . . . I was receiving so little from an unwilling father, that I had to increase my earnings; and so, being accustomed to horses, I held them while the spectators came.'[96] In 1583 Shakespeare met Oxford, who advised the young actor (as he then was) to set down on paper some of the stories rattling around in his brain. From these beginnings ensued the collaboration of the nobleman and the rustic. Will contributed the villains—Shylock and Iago and Edmund—and the scenes of great passion and simple English humour. To Oxford we owe the more lovable characters and most of the poetry.

All this and much more the *séances* brought to light. Perhaps the most exciting of the disclosures was the location of the priceless play manuscripts. They were buried in Shakespeare's tomb. (Surely the shade is confused—he must mean the grave; it happened so long ago.) One bundle served as the pillow for the corpse, another lay between the hands, a third at the feet; *Hamlet* reposed on the breast. Delia Bacon's intuition had been right after all.

# 9

# *Freud*

IN certain recurring features of anti-Stratfordian behaviour we may discern a pattern of psychopathology. The heretic's revulsion against the provincial and lowly; his exaltation of his hero (and, through identification, himself) by furnishing him with an aristocratic, even royal, pedigree; his paranoid structures of thought, embracing the classic paraphernalia of persecution: secrets, curses, conspiracies; the compulsion to dig in churches, castles, river beds, and tombs; the auto-hypnosis, spirit visitations, and other hallucinatory phenoma; the descent, in a few cases, into actual madness—all these manifestations of the uneasy psyche suggest that the movement calls not so much for the expertise of the literary historian as for the insight of the psychiatrist. Dr Freud beckons us.

Of his abiding interest in Shakespeare we have abundant evidence. Freud began reading the plays at the age of eight, and was always ready with a quotation from them. Shakespeare's pre-Freudian understanding of human nature filled the doctor with admiration; Shakespeare became, with Moses and Leonardo, one of the three extraordinary personalities in whom the founder of psychoanalysis took pre-eminent interest. That the authorship controversy stirred his analytical curiosity need not surprise us. It is, however, both surprising and sad that the schismatics were able to claim Freud as one of their own.

The instructor of his youth, the brain anatomist Meynert—revered by his celebrated pupil as the greatest genius he had ever encountered—was a professed Baconian. But Freud resisted this influence, although on disconcertingly Groupist grounds: if Bacon wrote Shakespeare, he 'would have been the most powerful brain the world has ever borne', 'whereas . . . there is more need to share Shakespeare's achievement among several rivals than to burden another important man with it'.[97] So Freud wrote in 1883; later his scepticism deepened when he discovered that the cult's founder bore the name of Bacon, so suspiciously suggestive of a personal motive. Nevertheless, the Baconians fascinated him, and, prior to the First World War, he urged his disciple Ernest Jones (the Shakespearian of the circle by virtue of his work on *Hamlet*) to study their methods and contrast them with the psychoanalytic approach. That way the theory would be disproved, and Freud's mind set at rest. But Jones shied away from the assignment.

Freud continued to toy with faddist ideas about Shakespeare. Was the National Poet, he wondered, really an Englishman? An Italian, Professor Gentilli of Nervi, had proposed that the name was a corruption of Jacques Pierre; indeed, Shakespeare's features looked more Latin than English. (One suspects that Freud accepted unquestioningly the genuineness of the Chandos portrait.) There the matter rested until around 1923, when Freud read the Looney book. It converted him to the Oxfordian faith. His intuition had found confirmation—if 'Shakespeare' was not actually a Frenchman, at least he had Norman forebears.

At a convivial gathering in celebration of his seventieth birthday in 1926, Freud expounded the de Vere theory at length. 'I remember my astonishment', Jones writes, 'at the enthusiasm he could display in the subject at two in the morning.'[98] The next year Freud reread Looney with no accessions of doubt. In 1928 he turned to Jones again, this time asking him to investigate what new psychoanalytic conclusions would follow from assigning the plays to Oxford. Again Jones prudently remained aloof.

His coolness did not, however, dampen his master's enthusiasm for the theory. 'I no longer believe in the man from Stratford,' he confided to Theodore Reik in 1930. That year, in a speech accepting the Goethe Prize, Freud made his views public: 'It is undeniably painful to all of us that even now we do not know who was the author of the Comedies, Tragedies and Sonnets of Shakespeare; whether it was in fact the untutored son of the provincial citizen of Stratford, who attained a modest position as an actor in London, or whether it was, rather, the nobly-born and highly cultivated, passionately wayward, to some extent *déclassé* aristocrat, Edward de Vere.'[99] He proceeded to revise his earlier pronouncements. In *The Interpretation of Dreams* Freud had likened, tentatively, the repressed Oedipal strivings of Hamlet to the death of Shakespeare's father and of the playwright's son Hamnet; now, in a footnote to the eighth German edition, he included a disclaimer remarkable for the casualness of its phrasing: 'Incidentally, I have in the meantime ceased to believe that the author of Shakespeare's works was the man from Stratford.' Canon Rendall's *Shakespeare Sonnets and Edward De Vere* shored up Freud's conviction; not doubting the significance of the Sonnets as self-confession, he found that Oxfordian authorship made them more intelligible.

As regards the plays, the Oxford heresy opened new vistas of psychoanalytic speculation. Lear had three daughters; so too had de Vere. 'If Shakespeare was Lord Oxford', Freud wrote on 25 March 1934 to James S. H. Bransom,

the figure of the father who gave all he had to his children must have had for him a special compensatory attraction, since Edward de Vere was the exact opposite, an inadequate father who never did his duty by his children. A squanderer of his inheritance and a miserable manager of his affairs, oppressed by debts, he could not maintain his family, did not live with them, and left the education and care of his three daughters to their grandfather, Lord Burleigh. His marriage with Ann Cecil turned out very unhappily. If he was Shakespeare he had himself experienced Othello's torments.[100]

Elsewhere Freud accepts the Oxfordian identification of Lord Derby, the Earl's first son-in-law, with Horatio in *Hamlet* and Albany in *King Lear*.

English-speaking readers did not yet know of Freud's conversion. For their benefit he composed a note for insertion in the 1935 edition of *An Autobiographical Study*. The translator, James Strachey—only too well aware of the contempt felt by orthodox scholars for the Oxfordians—threw up his hands. Patiently he explained to Freud the English connotation of the name Looney; a connotation which (in Jones's apt phrase) 'could only have the effect of adding risibility to derision'. Freud yielded—but in the American edition he stuck to his guns. 'The same sort of narcissistic defense need not be feared over there,' he snapped at Strachey.*

In London in 1938, a refugee from the Nazi occupation of Vienna, Freud received a letter of welcome from Looney. 'Dear Mr. Looney,' the great man replied, 'I have known you as the author of a remarkable book, to which I owe my conviction about Shakespeare's identity, as far as my judgment in

* The correspondence of Freud and Arnold Zweig, first published as recently as 1970 in an edition assembled by Ernst L. Freud, the master's son, reveals the strength of Freud's commitment to his heterodoxy. 'I do not know what still attracts you to the man of Stratford,' he expressed himself petulantly to Zweig on 2 April 1937. 'He seems to have nothing at all to justify his claim, whereas Oxford has almost everything. It is quite inconceivable to me that Shakespeare should have got everything secondhand—Hamlet's neurosis, Lear's madness, Macbeth's defiance and the character of Lady Macbeth, Othello's jealousy, etc. It almost irritates me that you should support the notion' (p. 140). As Peter Gay notes in *Freud: A Life for our Time* (New York, 1988), Freud twice read *'Shakespeare' Identified*, for some years pursuing this 'chimera', and discussing it with—among others—Ernest Jones, who tried in vain to dissuade him. 'Jones', Gay observes, 'shrewdly connects this harmless mania with Freud's fascination with telepathy. Both, he suggests, support the view that things are not what they seem to be' (p. 643 n.).

this matter goes.'[101] The next year he died, but not before making a final attestation of his faith. For his last revision of *An Outline of Psychoanalysis*, published in 1940, he added this note to his original remarks on Hamlet's Oedipus wish: 'The name "William Shakespeare" is very probably a pseudo-nym behind which a great unknown lies concealed. Edward de Vere, Earl of Oxford, a man who has been thought to be identifiable with the author of Shakespeare's works, lost a beloved and admired father while he was still a boy and completely repudiated his mother, who contracted a new marriage very soon after her husband's death.'[102] Thus he transferred from Shake-speare to Oxford his original insight into the psychogenesis of *Hamlet*. Long after Freud's passing, an English psychoanalyst, Victor Kanter, went through the library of the master's house at 20 Maresfield Gardens in Hampstead, and found there fourteen anti-Stratfordian works, most of them by Oxfordians, and only eight by orthodox scholars.

Inevitably some of Freud's followers stumbled after him into the Oxfordian bog; most notably Dr A. Bronson Feldman, who in a series of articles has explored the workings of de Vere's unconscious in *The Comedy of Errors*, *Othello*, and the Sonnets.* But mainly the Freudians have tried to account for their leader's aberration, a quest that may shed psychoanalytic light on the entire anti-Stratfordian syndrome. Certainly Freud's position cannot be understood on purely rational grounds: he knew from the example of Leonardo what a supremely creative mind could accomplish without formal training; he knew from his own triumphs the irrelevance of aristocratic blood to great endeavour. Something in Freud's mentality, Jones suggests, pro-duced a fascination with the idea of men not being what they seemed. Thus Moses, universally reckoned a Jew, must have been a noble Egyptian.

Such obsessions reflect the operation of the Family Romance fantasy. The child, reacting against disappointment with the imperfections of his parents, compensates by replacing them with others of higher birth; he must be a stepchild, or adopted. In later life such fantasies of parental idealization are transposed to a Moses—or Oxford. To the psychoanalyst Dr Harry Trosman, 'the imputation of Shakespearean authorship to a historical figure is another example of the formation of a transference allowed to continue unresolved and continually buttressed by the discovery of "new historical evidence"'. It is not surprising that the Family Romance should flourish in Freud's psyche. The household of his childhood included two half-brothers twenty and twenty-three years his senior; his mother—his father's second wife—was their contemporary. In *The Psychopathology of Everyday Life* Freud relates a slip of the tongue, involving the substitution of a name, to the fantasy of 'how

* Freud's early disciple Wilhelm Stekel embraced a more orthodox unorthodoxy. From his *Autobiography* we learn that Stekel 'shared the belief of many that Bacon actually wrote certain of Shakespeare's plays' (p. 223), an opinion which led to his break with his friend Samuel Tannenbaum, psychiatrist and eccentric Shakespearian. Stekel had already seceded from Freud's circle, which was clearly schism-prone.

different things would have been if I had been born the son not of my father but of my brother'.

Fantasies of this kind testify to feelings of ambivalence towards the father. According to Norman Holland, Freud's urge to dethrone Shakespeare stems from his view of 'the artist as a kind of totem whom he both resented and emulated'.[103] The psychoanalytically oriented will see manifestations of this filial ambivalence throughout the dreary pages of anti-Stratfordian discourse: on the one hand, denigration of the drunken, illiterate, usurious poacher from the provinces; on the other, ecstatic veneration of the substitute claimant, aristocrat and deity. The heretic's selection of de Vere, courtly amateur rather than professional man of letters, confirms his identification with his idealized choice, for the Oxfordians are, almost to a man, dilettante scholars. In Looney's case the tendency towards idealization finds early expression in his gift of Carlyle's *Heroes* to a youthful friend, with the advice that he read it before turning twenty. The British Library deposition shows Looney imagining in his own life a situation parallel to that in which (he believed) Oxford found himself:

Through some untoward event Looney's identity as the discoverer might not be revealed, while someone else was to be acknowledged as having solved the puzzle. By entrusting his deposition with the Librarian of the British Museum, Looney could well imagine that eventually *his* identity would be revealed as the original instigator of the Oxfordian position. In the same way that he states credit must be given 'to the great Englishman' who actually authored the plays, credit would then be given to him who had actually made the Oxfordian discovery first.[104]

Looney's deliverance of his idol from depreciation and obscurity exemplifies the rescue fantasy, interpreted by Freud as the son's defiant wish to settle his account with his father for the gift of life. ('I want nothing from my father,' the boy in effect says. 'I will give him back all I have cost him.') In the rescue fantasy one sees again the operation of the Family Romance, dually functioning 'to mask the hostile impulses and preserve the lost omnipotent object'. In such a way does psychoanalytic theory explain the unconscious origins of anti-Stratfordian polemics.

# 10
## *Other Claimants*

THOSE unmoved by Looney, Freud, and company can choose from other aristocrats. Sir Walter Ralegh, Sir Anthony Sherley (a favourite of Essex), Anthony Bacon (qualified for the Sonnets by reason of lameness), and Robert

Cecil, first Earl of Salisbury, have all found champions. The ingenuity of the arguments in favour of William Stanley, sixth Earl of Derby (1561–1641), devised by Abel Lefranc (1863–1952), a professor of French at the Collège de France, won him a distinguished convert in the author of *Aphrodite*. In an unpublished letter dated 6 April 1919 (preserved in the Folger Shakespeare Library) Pierre Louys observed that 'le stratfordhomme' signed himself 'Wm' or 'William', whereas Derby preferred 'Will', a name or word which occurs often in the Sonnets and plays mistakenly attributed to Shakespeare. More-over, just as in time the composer Wagner would call his children Eva and Siegfried, Will Derby christened *his* son Jacques, and married him to a Frenchwoman (Charlotte de la Trémoille) to ensure that he would not be called James but Jacques by his wife; and we all know that *As You Like It* features a Jacques. *Voilà.*

These candidacies have inherently no less rationality than that of Christo-pher Marlowe, the most recent to achieve wide notoriety. By now only the most innocent will suppose the Siberian expanses separating the literary personalities of the gentle Shakespeare and iconoclastic Marlowe (described by a contemporary as of a cruel and intemperate nature) sufficient to discourage heretical speculation. One fact, however, might: Marlowe's sudden death at the age of twenty-nine on 20 May 1593 at widow Eleanor Bull's place of public refreshment at Deptford, where he may have had lodgings, not far from the plague-stricken capital. The circumstances of the slaying are set forth in detail in the legal records; a jury of sixteen accepted the coroner's findings. But of such impossibilities the anti-Stratfordians make instruments to plague us.

In 1955 Calvin Hoffman of Long Island, described by *Time* magazine as a Broadway press agent and by a disciple as a poet and playwright, published the results of nineteen years of hard labour, *The Murder of the Man Who Was 'Shakespeare'*. The long quest began when he took to jotting down phraseo-logical correspondences between Shakespeare and Marlowe. Then one night, as he tossed restlessly on his mattress counting parallelisms instead of sheep, a dark thought occurred to Hoffman: what if the report of Marlowe's assassination was a hoax? As time passed, the possibilities of a monstrous imposture (how Marlowe must have suffered!) became oppressive; like Delia Bacon, Dr Owen, and other schismatics, Hoffman found his pursuit occupy-ing most of his waking hours and forcing him to sacrifice mundane interests. His researches carried him to England, Denmark, and Germany; he prowled in graveyards, inhaled the dust of tombs, wearied himself in ancient archives. At last a theory took shape.

Less than a fortnight before Marlowe's death, the Privy Council had issued a warrant for the poet's arrest on suspicion of blasphemy and atheism, but allowed him to remain at large when he posted bond and consented to attend daily upon the Council. The high-born Thomas Walsingham, involved with

Marlowe in a homosexual liaison, feared for his protégé's life, and with the latter's assistance concocted a plot to save him: three sinister characters in Walsingham's service—Skeres, Poley, and Frizer—would pass off a corpse as Marlowe's. In some narrow alley of Deptford the trio selected their victim—a foreign sailor, maybe Italian or Spanish—lured him to Dame Eleanor Bull's house, plied him with drink, then stabbed him to death. Meanwhile, Marlowe, who had lain low at his lover's Scadbury estate, hoisted sail for France. ('The figure on the Channel ship watches the tender outlines of the French coast as they emerge out of the morning mist, purple and gold in the rising sun.'[105]) Afterwards he lived in Italy, but eventually he returned in disguise to Scadbury and there dwelt in seclusion, roaming the woods that furnished him with the nature allusions for his plays. These and the poems were submitted to Marlowe's benefactor. Walsingham hired a professional scribe to copy them, hid the original manuscripts, and employed an obscure actor—an unimaginative but steady fellow—to lend these masterpieces his name. Hence it was that, four months after Marlowe's supposed death, Shakespeare made his literary début with *Venus and Adonis*. Later the Sonnets were published with their decication to Mr W.H.—Walsing-Ham, of course. The true date of Marlowe's demise has eluded even Hoffman's patient researches, but the poet must have died before 1623, when Walsingham sought to keep alive the memory of his beloved by publishing a folio edition of the plays. How Shakespeare's name on the title-page would have abetted this end Hoffman does not explain.

Needless to say, he produces not a single record to support this preposterous romance: rather a sad showing for nineteen years of steady work. Nor can he even properly claim priority for his theory. William Gleason Zeigler, a San Francisco lawyer, had put forward the Muses' darling in 1895 in his novel, *It Was Marlowe: A Story of the Secret of Three Centuries*. Not until the project had haunted him for twelve years did Hoffman discover this anticipation, on which he tries to put as brave a face as possible, dismissing the Zeigler performance in a footnote as a 'cinematic "thriller"', and 'the purest fiction and fantasy'. Hoffman does not mention Zeigler's preface and notes, in which the suggestion that Marlowe lived on until 1598 is made with evident seriousness and in straightforward expository prose; nor does he allude to the crucial point that Zeigler credits Marlowe with all of Shakespeare's 'stronger plays'. Slater's *Seven Shakespeares*, cited in the same Hoffman note merely as a work proposing Marlowe as one of the seven, offers the theory that the death was faked, that Marlowe left England at the beginning of June 1593, and that he later contributed substantially 'to the main volume of the Shakespeare plays'. And Hoffman has missed altogether Mackay's theory (1884) about Marlowe's pen in Shakespeare's Sonnets. He does, however, offer ample homage to the Ohio professor Thomas Corwin Mendenhall who, employing a team of put-upon women to count millions of words and letters of words, was able to plot graphs of the vocabulary curves of various writers. The results, published in the hospitable pages of *Popular Science Monthly* (December 1901), demonstrated that the characteristic curve of Marlowe's plays 'agrees with Shakespeare about as well as Shakespeare agrees with himself'.

This 'evidence' Hoffman supplements with his parallels, the value of which he modestly describes as 'enormous'. Many of these enormously valuable parallels are not parallel; thus Hoffman compares 'Some swore he was a maid in man's attire' (from *Hero and Leander*) with this line from *Venus and Adonis*: 'Stain to all nymphs, more lovely than a man'. Some of the correspondences are commonplace phrases that any writer might have used; others may be accounted for by the acknowledged influence of Marlowe on Shakespeare. The remainder involve direct quotation, such as Pistol's 'And hollow pampered jades of Asia, | Which cannot go but thirty mile a day'.

Shakespeare's famous tribute to Marlowe—

> Dead shepherd, now I find thy saw of might:
> 'Who ever lov'd that lov'd not at first sight?'

—presents a special problem, for to ordinary readers it would seem clear that the author placed Marlowe among the departed. Hoffman gets round this awkwardness by proposing that the dead shepherd is Sidney; later he includes the quotation in *As You Like It*, along with the line from *Hero and Leander*, among his parallels. Such procedures are not calculated to satisfy the fastidious. Any confidence in Hoffman's scholarship is further undermined by his indifference to factual precision: names are misspelt, dates rendered inaccurately, and words silently omitted from inaccurate quotations.

These deficiencies did not escape responsible reviewers. The *TLS* (27 January 1956) described *The Murder of the Man Who Was 'Shakespeare'* with accuracy and commendable restraint as 'a tissue of twaddle'. But in the popular press Hoffman created a 'storm of controversy' (*Time*), no doubt in part provoked by the sweaty journalese of his prose style. This is one of those books that introduce readers to the brawling Elizabethan world, teeming with swarthy folk convulsed with Rabelaisian humour. Courts are corrupt, women fecund, and men sensual. When the plague strikes, Elizabethan London becomes 'a den of horror'. Gallants chase their ladies on soft summer nights illuminated by lascivious firefly lanterns. When Hoffman arrives at his high point, the Marlowe murder plot, he resorts to a 'fictionalized approxima-tion'. Under the circumstances, one is not inclined to fault this strategy.

Hoffman's mission did not end with publication of his book. The compul-sion to dig, endemic among anti-Stratfordians, had taken hold of him, and he summoned together his energies for an assault upon the Walsingham tomb in the Scadbury Chapel of St Nicholas's Church in Chislehurst. Manuscripts deposited there would, Hoffman hoped, prove to the world the existence of what he terms, rather unfortunately, 'the Marlowe–Shakespeare fraud'. The

Walsingham family having died out early in the eighteenth century, Hoffman enlisted the co-operation of Major John Marsham-Townsend, who, as lord of the manor, had rights of exclusive use of the chapel. In January 1956 a consistory court granted consent for the opening of the monument; four months later stonemasons pried open the top of the tomb. Within lay no manuscripts, nor even human remains; only hard-packed yellow sand, serving as a damp course, from the Normandy shore. Workmen removed the sand, then cut a small hole in the floor, through which they could discern, about two feet below, a leaden coffin. Alas, the master of Scadbury Park resolutely refused to permit the dismantling of any Walsingham coffins.

The reader of this narrative will be prepared to learn that failure did not discourage Hoffman. 'It has not proved or disproved my theory,' he told a *New York Times* reporter: 'It was a 1,000 to one chance that we would find any manuscripts. I have other clues to follow up while I am in England.'

Inevitably the Marlowe cause, like the others, gained adherents. In 1956 a Scarsdale attorney, Sherwood E. Silliman—how innocently appropriate are some anti-Stratfordian names!—privately printed 'a fanciful play', *The Laurel Bough*, re-creating the slaying and substitution for the poet of a down-and-out actor, after which Marlowe is comforted by the love of Walsingham's wife and continues his playwriting career as Shakespeare's collaborator. This theory, the author announces, was formulated independently of Hoffman. In his preface, Silliman makes some telling points about Marlowe and Shakespeare: 'Each used poetic blank verse', 'Both dote on pomp and ceremony', etc. Surely, Silliman triumphantly concludes, 'no two human minds could have such striking similarities'. In the text to his sumptuous pictorial biography, *In Search of Christopher Marlowe* (1965), A. D. Wraight does not go so far as to endorse Hoffman's thesis, but the influence surfaces and the name Hoffman appears ominously among the credits. An avowed disciple is David Rhys Williams, author of *Faith beyond Humanism*, who in *Shakespeare Thy Name is Marlowe* (1966) thanks Hoffman for generous permission to write on the subject. While in England, Williams reports, he addressed the London chapter of the Marlowe Society at Hoffman's suggestion; mercifully there is as yet no Marlowe Society Newsletter. At Canterbury to visit the shrines sacred to Marlowe's memory, Williams glimpsed a carton containing over five hundred newly discovered documents, many of them identifying the poet as Shakespeare. As Dr Urry, the City archivist, has not yet got round to publishing his report on these intriguing papers, the public must endure suspense for a while longer before the final dispensation of the Marlowe claim.

Why not King James?, asks R. C. Churchill, and he backs up his proposal with a sporting offer: 'In fact, for ten guineas per thousand words, payable in advance, I will undertake to prove to all but the hopelessly prejudiced that King James—or, alternatively, Fulke Greville or Sir Thomas North—was the real author of the works erroneously attributed to William Shakespeare of Stratford-on-Avon.'[106] The suggestion of James as potential author follows the concluding item in Churchill's historical survey, George Elliot Sweet's *Shake-speare the Mystery* (1956). Recommended for its 'persuasive logic' by Erle Stanley Gardner, who entertained his fellow sleuth on his ranch, Sweet's diligently researched book (it cites among other authorities the *World Book Encyclopedia*) presents the case for Queen Elizabeth, who alone possessed the Negative Capability which Keats so admired in Shakespeare; after all, she survived plenty of political uncertainty—a nice gloss on Keats. Sweet's trump card is the Epilogue to *Henry VIII*, which he interprets in novel fashion. ''Tis ten to one this play can never please,' the Epilogue begins; Sweets reads: *There are ten kings in Europe, I am the one queen.* He continues in like manner with the rest of the speech, no doubt allaying the unease of some who might otherwise be troubled by a reshuffling of chronology that results in the assignment of *The Tempest* to 1582. The chief mystery about *Shake-speare the Mystery* is its printing (although not publication) by the Stanford University Press.

The effect of Churchill's whimsical proposal evaporates when one realizes that he is not the first to think of the wisest fool in Christendom in this context:

In the . . . debates I argued for the theory that King James himself was the real poet who used the *nom de plume* Shakespeare. King James was brilliant. He was the greatest king who ever sat on the British throne. Who else among royalty, in his time, would have had the giant talent to write Shakespeare's works?[107]

The debates alluded to did not take place on the platform of some genteel anti-Stratfordian meeting, but, rather, in the Norfolk Prison Colony in Massachusetts. In such a setting do we find expressed presumptions of royalty with respect to the author of Shakespeare's works. The claim of the first Stuart was urged by Malcolm Little, who is unusual among heretics in denying altogether the existence of the historical Shakespeare. Later Little would die violently on another platform. By then he had achieved notoriety as Malcolm X.

With the Black Muslim candidate our own survey comes to an end. Perhaps at this pause in the narrative the writer may be permitted to drop for a moment the historian's mask of impersonality and give vent to private emotion. This section has been the cruellest endeavour I have ever confronted. The sheer volume of heretical publication appals. In the 1840s Joseph S. Galland, a professor of Romance Languages at Northwestern University, compiled a typescript bibliography, *Digesta Anti-Shakespeareana*, that fills six large volumes and describes 4,509 items. A number of these are enormous, and many more have of course appeared since. The voluminousness of output is matched only by the intrinsic insubstantiality of most of it:

two characteristics which together produce an overpowering effect. The lawyers were back at their game in a series of articles in the *Journal of the American Bar Association* in 1959 and 1960, afterwards reprinted as *Shakespeare Cross-Examined*; but just as one despairs of the legal profession, Milward W. Martin replies to his colleagues rationally in *Was Shakespeare Shakespeare? A Lawyer Reviews the Evidence* (1965). Thus the bibliography swells.

Many curious theories lie outside the ken of my selective history. I have not found space to deal with those who claim for Bacon *Don Quixote* (the English translation as well as the original Spanish version), the plays of Lope de Vega (all 2,200; why not?), the dramas of Calderón, Gray's *Elegy*, *Gulliver's Travels*, and Poe's *Raven*. Nor have I been able to savour the contribution of George M. Battey—another endearing name—whose application of 'the alphabetical numerical clock count' led to the inescapable conclusion that the plays were written by Daniel Defoe—or rather, as Battey prefers, Daniel Foe. I regret, however, not being able to consider the ingenious speculation of James Freeman Clarke, who contributed to the *North American Review* (February 1881) an article entitled 'Did Shakespeare Write Bacon's Works?'

A great many of the schismatics are (as we have seen) distinguished in fields other than literary scholarship, and their ignorance of fact and method is as dismaying as their non-specialist love of Shakespeare's plays is touching. One feels oppressed, moreover, by the presence of an irresistible passion in these men and women: the inexorable compulsion that usurps thought, courts ridicule, even (at times) unseats reason. Vanity presses have published some of these anti-Stratfordian diatribes at their authors' expense; others have been sponsored by well-esteemed commercial houses which would refuse, as a poor business risk, the scholar's sober monograph. It would be a nice question to determine which phenomenon has the more depressing implications.

If the well one day should run dry, it might be argued, we would be deprived of the harmless mirth occasionally provoked by heretical extravagance; but it would be an exaggeration to suggest that the gaiety of nations would be thereby eclipsed. In 1969 there appeared a novel, *The Philosopher's Stone*, by that voluble autodidact Colin Wilson, whose hero Newman, travelling in time through an effort of the will, discovers that Bacon not Shakespeare—both 'second rate minds'—wrote the plays. The continuous flow of publication, and the publicity given sensational theories by newspapers throughout the Western world, have understandably induced in laymen—even educated laymen—lingering doubts about the reality of Shakespeare and the true authorship of the canon. Away from the academy, whether in the lounge bar of a cruise ship or in the shadow of the Moorish wall in Gibraltar or on an Intourist bus on the road to Sevastapol, the professor of English (once his identity has been guessed by fellow-holiday-makers) will be asked, as certainly as day follows night, 'Did Shakespeare *really* write those plays?' He will do well to nod assent and avoid explanation, for nothing he says will erase suspicions fostered for over a century by amateurs who have yielded to the dark power of the anti-Stratfordian obsession. One thought perhaps offers a crumb of redeeming comfort: the energy absorbed by the mania might otherwise have gone into politics.

# PART VII

## *The Twentieth Century*

# 1
## Elton and Masson

MODERN times did not arrive simultaneously with the new century. The editors of *The Nineteenth Century*, unprepared to honour with excessive attention an univited guest, rechristened their periodical *The Nineteenth Century and After*. The year 1901, it is true, witnessed the passing of Victoria, who over the years had waxed increasingly Victorian, and to many Britons the event gave an uneasy sense of imperial decline. 'I felt', Elinor Glyn recalls, 'that I was witnessing the funeral procession of England's greatness and glory.' But Edward VII and his Queen Alexandra who followed—he, stout, with top hat, frock coat and boutonnière, spats, and walking-stick; she, beautiful, in mauve velvet, with parasol and feather boa—did not by their vacuous examples stir placid waters. Viewed in nostalgic retrospect, this was not the dawn of a new era but the long afternoon of the Edwardian garden party.

Yet a scent of change flavoured the afternoon air. To think of the arts in the first years of the twentieth century is, after all, to think of Duchamp, Stravinsky, and Diaghilev. It seemed to Yeats that everybody got down off his stilts in 1900: no more absinthe with the black coffee. Henry James had his final phase. *Sons and Lovers* appeared in 1913, *Dubliners* in 1914; Eliot wrote 'Prufrock' in 1910. Edward, as the late Richard Ellmann so eloquently reminded us, has two faces.[1]

In scholarship, where the present builds consciously upon the past, continuity exercises greater sway than in other endeavours. Halliwell-Phillipps's *Outlines* remained the standard resource for facts and documents. Expanded through a series of revisions, Sidney Lee's Life went unchallenged as the authoritative biography along objective lines. The archival investigator is unlikely to find his yellowed manuscripts rustled by the winds of fashion; but one expects the creative biographer to be more sensitively attuned to alterations in the prevailing *Zeitgeist*. Change, however, came slowly. For the internal drama of Shakespeare's life as mirrored in his art, Edwardians looked for guidance to Dowden and Brandes, who lived on as respected worthies in the new age. Even Joyce, whose *Ulysses* is synonymous with the modern sensibility, turned (as we have seen) to nineteenth-century Shakespearian's—Lee and Brandes—for his data. and even for some of his phrases.

The biographical tradition of the previous century found its popular epitome in the humane pages of Walter Raleigh's *Shakespeare* (1907) for the English Men of Letters Series. He affirms the validity of the *mythos*—to do otherwise being 'the very vanity of skepticism'—and gives us a Shakespeare (son of a glover, wool-driver, and butcher) who killed a calf in high style, robbed a park, and lampooned his prosecutor. Nor does Raleigh doubt the presence in the plays of a central drama taking place within the creator's mind, the battalions of imagination arrayed against those of will. Of new documentary finds he offers little hope: just possibly some new facts might swell the meagre store, but that was not likely after generations of antiquarian toil. So Raleigh writes on the eve of a stunning biographical recovery in the Public Record Office.

The impingement of past upon present is illustrated by the posthumous publication, in the early years of the new century, of the Shakespearian writings of two nineteenth-century scholars. In 1904 A. Hamilton Thompson edited a series of papers that Charles Isaac Elton had intended as the nucleus of an exhaustive study of Shakespeare; David Masson's daughter Rosaline performed in 1914 a similar act of piety for her father when she assembled the lectures that he had delivered for thirty years—from 1865 until 1895—at Edinburgh University. These volumes, Elton's *William Shakespeare: His Family and Friends* and Masson's *Shakespeare Personally*, represent in minor key two popular and divergent approaches to Shakespeare in this period.

Elton belongs to that familiar breed of antiquaries who never allow the quiet tenor of their lives to be ruffled by an event. Educated at Cheltenham College and at Queen's College, Oxford, he successfully resisted any stirrings of ambition: called to the Bar in 1865, he devoted himself less to his legal specialty, the laws of real property, than to investigating early English land laws and institutions, especially the tenures of Kent. After inheriting his uncle's estate of Whitestaunton four years later, Elton passed many indolent hours in the surrounding green hills of Somerset. He excavated a small Roman villa on his property, and described the relics in a learned paper. He collected pottery, rare books, and illuminated manuscripts, and he published erudite discourses on *The Origins of English History* (1882) and *The Career of Columbus* (1892).

It is in keeping with such a life that at his death Elton should leave his principal work unfinished, and bequeath to his editor the fragments of an opus rich in learning but without shaping form. From which branch of the Hathaways did Anne derive?—such questions fascinate Elton. He doubts that she could have been the same person as Agnes of Shottery, Richard Hathaway's daughter, for Elton has found evidence that Anne and Agnes were separate names ('As early as the thirty-third of Henry VI, it was decided that Anne and Agnes are distinct baptismal names and not convertible, so that if an action was brought against John and his wife Agnes, and the wife's name was Anne, the variance was essential and could not be amended.').[2]

But the distinction is without a difference, for in Shakespeare's day many people were in fact interchangeably called Anne or Agnes—the theatrical entrepreneur Philip Henslowe, for example, referred to his wife as Agnes in his will, but she appears as Anne in the funeral records. Elton's conclusion, arrived at with such a parade of learning, is wrong.

More often he avoids conclusions. Witness his discussion of the deer-poaching. Elton knows that the Lucys had no park at either Charlecote or Fulbrook (Malone had shown that), but the removal of deer from any enclosed ground was illegal within the Act of Elizabeth. 'After the lapse of centuries, the offence, if it happened, may fairly be condoned.'[3] *The offence, if it happened*; with such qualifying clauses does Elton hedge his argument, and it is little wonder that he never mastered historical narrative.

Irresolution is a weakness, and antiquarianism confines biography within unimaginative bounds. Elton is an irresolute antiquary; yet, paradoxically, therein lies his strength: the virtue of incertitude is dispassion, and antiquarianism requires learning and scrupulosity. Because he has these qualities, Elton's niche is secure. He devotes amply documented pages to the precise date of Shakespeare's birth, concluding, after a display of indecisiveness, that the poet came crying into the world on Friday, 21 April 1564, or maybe the following Saturday or Sunday. Intelligence of this kind will quicken the heartbeat of few readers; nevertheless we wish to know, as exactly as possible, when the National Poet was born, and on this subject Elton remains authoritative. So too he performs a service by sorting out the confusion over John Shakespeare's occupation. Was he a farmer, butcher, or glover—or a combination, as some thought, of all three? Did his eldest son kill calves in high style? As Elton points out, John Shakespeare could scarcely have tilled in the common field at Asbies, for he leased his farm there to a tenant. Nor would he have been allowed to unite the trades of butcher and glover: such a combination would have been incongruous. (Yet two recently uncovered cases in the Exchequer court reveal him as being accused, in 1572, of illegally purchasing quantities of wool; see Part VII, Chapter 21.)

He could not keep a regular meat-shop while trading in skins, and no one has seriously suggested that he worked about as a slaughterman, though such people were classed among butchers. The meat trade was stringently regulated by statute, and nothing was allowed to interfere with the regular official inspection. The killing of calves was the subject of constant restrictions, and it is certain that the inspectors would put a stop to anything that might injure the veal; it is almost inconceivable, indeed, that a boy would be allowed to play such pranks in the shambles as the gossips described. A butcher's business was to sell wholesome meat and suet at a profit not exceeding a penny in the shilling, not taking his veal too young, nor keeping the calf so long that its meat might encroach on the steer-beef, and not selling any lean meat as if he had got it from the fat stock. He was bound, moreover, to keep the horns and hide of every beast till all the beef was sold, so that in the case of theft the owner might identify his property. The Tanners' Act was passed in 1530, and continually renewed.[4]

This passage reveals Elton at his best; the force of the argument is irresistible. A pity that one of Shakespeare's more recent biographers, A. L. Rowse, has overlooked this analysis and, moved by irrelevant recollection of his own Cornish boyhood, resurrected untenable notions about the occupation of the household in which Shakespeare grew up.

Elsewhere Elton gives chapters to Shakespeare's descendants, to historical Stratford and its monuments (with much out-of-the-way lore), to Snitterfield, Wilmcote, and the manor of Rowington, to Midland agriculture, and to landmarks on the Stratford road and in London. Throughout Elton refers to the distinguished students who preceded him: Oldys, Greene, Halliwell-Phillipps, and the rest; for he has a lively sense of the scholarly tradition. More important, he devotes chapters to the Revd John Ward, who late in the seventeenth century made notes on Shakespeare's death; to Mr Dowdall, who visited Stratford in 1693; and to William Hall, who passed through the town the next year and (like Dowdall) recorded his impressions. Thus Elton anticipates to some degree the subject of this book.

Unlike his antiquarian predecessors, Wheler and Hunter, he is not a major figure, even in his own limited sphere, for he brought to light no important new facts. But the serious reader and the professional scholar may still consult Elton for his specialized contributions and rich illustrative detail. Not many scholarly books over half a century old retain a similar usefulness.

Masson's *Shakespeare Personally* offers a study in contrast, just as do Elton and Masson as men. The latter experienced ambition and moved confidently in a larger world. Before occupying the chair at Edinburgh, he was professor of English at University College, London. He helped bring higher education to Scottish women. He founded and edited *Macmillan's Magazine*, and carried to a triumphant conclusion his six-volume *Life of Milton*; he numbered among his familiar acquaintances Thackeray, Charles Knight, and the Carlyles. Not surprisingly, Masson held strong views on Shakespearian biography.

He was a subjectivist: the Sonnets, *pace* Lee, are 'expressly and thoroughly autobiographical'; the plays, with their recurrences and fervours, reveal the interior life of their creator. Shakespeare passes through 'a mood compounded of the passionate impetuosity, the all-for-love recklessness, of Romeo in *Romeo and Juliet*, the instability and variability of Proteus in *The Two Gentlemen of Verona*, and the wit, irony, and sharp-sightedness of Biron in *Love's Labour's Lost*'.[5] This *Romeo-Proteus-Biron mood* (the italics are Masson's) is succeeded by a *Jaques-Hamlet mood*—meditative, inquisitive, sceptical—which is in turn swept away by the hurricane of the great tragedies. There follows a *Coriolanus-Timon mood* of universal detestation and loathing;

but the *Prospero mood* ushers in a final serenity. When the island magician apologizes to Ferdinand—

>              Sir, I am vexed.
> Bear with my weakness. My old brain is troubled.
> Be not disturb'd with my infirmity.
> If you be pleased, retire into my cell,
> And there repose. A turn or two I'll walk
> To still my beating mind.

—the occasion is not dramatic—Caliban's conspiracy—but personal: 'None can read that passage', Masson insists, 'without seeing that Shakespeare likewise, when he had written it, was overcome with agitation, and rose to pace his chamber and so still his beating mind.'[6] Such are the insights afforded by subjective biography. During Masson's thirty years at Edinburgh University, five thousand students heard him lecture on Shakespeare personally, and were thus instilled with the Dowden-Brandes view.

# 2

# *Charlotte Carmichael Stopes*

ONE of those who listened was Charlotte Carmichael. That was before women could take degrees at Scottish universities, but she received tuition from Masson and other sympathetic professors, and was awarded a Diploma in the customary eight subjects. In 1879 she married Henry Stopes, an architect with a passion for palaeontology, and with him journeyed up the Nile as far as the Cataracts. In London Stopes founded a Discussion Society for ladies, did needlework, and championed the historical privileges of British freewomen. She also reared two daughters, one of whom, Marie, would gain notoriety as the liberated rhapsodist of married love. In this eccentric household the family cat, seated in state in a high chair, was served at table from its own dinner-service; not until she was eight or nine did Marie Stopes realize that pussy was not her brother. With her husband away on frequent expeditions, Stopes found time for the scholarly pursuits on which her reputation rests.

Shakespeare had stirred her imagination from early childhood; before she could read, she pored over the huge folio volumes of the family's illustrated edition, and formed characteristically strong views about Prince Arthur and his wicked Uncle John. Now a Hampstead matron, Stopes attended the meetings of the New Shakspere Society and was thus introduced to the most advanced scholarly circles. She did not herself, however, take the New Shakespearian path, although she contributed to one of the Society's publications. Nor did she become a disciple of Masson, despite the fact that his terms 'fervours' and 'recurrences', as well as some Masson *aperçus*, appear, although unascribed, in her introduction to an edition of the Sonnets.

Her own bent was towards the biographical, but objective biography, of the Halliwell-Phillipps variety. A rage for discovery possessed Stopes. With all her feminist ardour, she determined to excel in a field overwhelmingly dominated by men. There were other obstacles to surmount. The death of her husband, who always spent two pounds in earning one, left her hopelessly in debt. Not having the strength to be a general servant, she applied—however incredible it seems—for daily work, presumably as a char, at the Record Office, but was turned away: they had enough girls, they told her. Nevertheless she wrote indefatigably; an unending stream of articles, many of them trifling, issued from her pen and appeared in outlets as divergent in their appeal as the *Shakespeare Jahrbuch* and the *Stratford-upon-Avon Herald*. They did not solve her financial problems. 'The difficulties of my class of work', she complained to Marie on 28 February 1908, 'are, that it requires to be severe & uninterrupted that it is poorly paid, when it is paid, & that payment is always *deferred*, while all domestic needs are urgent and must be paid at once.'\* Still she achieved a journalistic coup in ingratiating herself with the management of the *Athenaeum*, to whose columns she regularly contributed. Later she collected her articles into books.

One such assemblage is *Shakespeare's Warwickshire Contemporaries* (1897; revised and enlarged 1907), which she reprinted from the Stratford *Herald*. This work places Shakespeare in his Warwickshire context, then often misunderstood, by presenting brief sketches of neighbours and relations whose lives touched his own. Stope's *dramatis personae* include Richard Field, who printed Shakespeare's first book; Sir Thomas Lucy, master of Charlecote; Dr John Hall, the dramatist's son-in-law; the poet Drayton, who took physic from Hall; the great Clopton family, owners of Clopton Hall from the time of Henry III; the Combes, from whom Shakespeare bought 107 acres of land near Stratford; and William Underhill, the late owner of New Place poisoned by his young son Fulke. Less familiar figures also receive chapters: John Somerville of Edreston, for example, who with a fever in his brain set out for London to shoot the Queen and put her head upon a pole; although clearly mad, he was executed. There are sections too on the clergy of Stratford, and on the local schoolmasters. It is a novel collection which shreds indirect light on Shakespeare himself. The effect is to counter Halliwell-Phillipps's

---

\* Her money troubles were chronic. On 22 September 1922 she informed the Secretary of the Royal Society of Literature (of which she was a Fellow) that she was too poor to publish her work, and asked the Society to take it up and thus to encourage 'real research'. The letter, in which Stopes describes herself as a pioneer, is in the Society's archives.

unwittingly mischievous description of Stratford as 'a bookless neighbour-hood', so greedily seized upon by the heretics—Stopes points out that one man alone, the Revd John Marshall of Bishopton, left 187 books in 1607.

She is hard-headed about traditions. The impromptu epitaph on John a Combe is dismissed as very improbably Shakespeare's. Of the deer-stealing legend she writes, 'I am sure that "Shallow" was not intended to represent Sir Thomas Lucy; that there was no foundation for the tradition, and that the whole story was built upon a misreading of Shakespeare's plays, and a misunderstanding of his art.'[7] Stopes, it is true, does waver for a moment when confronted with the reference in *The Merry Wives of Windsor* to a dozen white luces—maybe the poet is after all revenging himself upon the knight for supposedly being at the back of those who opposed the Shakespeare grant of arms—but the allusion irresistibly invites such speculation. One is still tempted to cry, 'Bravo, Mrs Stopes; well done!' In these pages of *Shakespeare's Warwickshire Contemporaries* a new note, tough-minded and critical, makes itself felt. That note would distinguish much of twentieth-century biography.

The older approach, fanciful and at times sentimental, subsists alongside in Stopes's work. In *Shakespeare's Family* (1901) the dramatist is treated—refreshingly—not as a subject for literary biography, but as 'an interesting Warwickshire gentleman'. Nevertheless the author cannot resist quasi-fictional embroidery, as when she explains the mysterious marriage licence entry for Shakespeare and Anne Whateley of Temple Grafton:

Travelling was inconvenient on November roads; Will set out for the license alone, as bridegrooms were often wont to do, when they could afford the expense of a special license. He might give his own name, and that of his intended wife, at a temporary address. The clerk made an error in the spelling, which might have been corrected; but meanwhile discovered that Shakespeare was under age, was acting without his parents—that the bride was not in her own home, and that no marriage settlement was in the air. No risk might be run by an official in such a case; the license was stayed; sureties must be found for a penalty in case of error. So poor Will would have to find, in post-haste, the nearest friends he could find to trust him and his story.[8]

Conceivably this is the way it happened; but what is missing from Stope's account is the *conceivably*, and her patronizing air towards Will does not help matters.

Elsewhere in *Shakespeare's Family*, an expansion of articles from the *Genealogical Magazine*, she takes up the name Shakespeare, the localities of early Shakespeares, the cousins, connections, collaterals, and the descendants of the dramatist, and unrelated contemporary Shakespeares in Warwickshire, other counties, and London; part II treats, with similar genealogical assurance, the various Ardens. Stopes attempts a minor correction in the factual record: Halliwell-Phillipps had said that Gilbert Shakespeare, the poet's brother, was described in the *Coram Rege* roll of 1597 as a haberdasher of St Bridget's Parish, London, but she could find no trace of this tradesman,

only a Gilbert Shepheard, apparently misread as Shakespeare by her predecessor. So she thought, but the carelessness is hers, not Halliwell-Phillipps's:
the record, citing Gilbert Shackspere, indeed existed and would resurface.
Failure is compounded by absurdity when Stopes suggests that the word
'adolescens' in the burial entry for Gilbert may be the clerk's malapropism for
'deeply regretted'. Chambers hoped that she was jesting, but Stopes never
jests.

Her next book, *Shakespeare's Environment* (1914), reprints scattered papers
of unequal merit. Stopes has discovered that in 1595 'Mr Shaxpere' (whether
John or William is not evident) was indebted to 'Jone Perat' for 'one book';
but the eager researcher must be denied her small triumph, for Halliwell-
Phillipps produced the same record in his *New Boke* of 1850. In another
article she adds a new detail to knowledge of John Shakespeare's declining
fortunes: the *Coram Rege* rolls for Trinity Term, 1580, show him being fined
£20 for not appearing in Queen's Bench, and another £20 as pledge for John
Awdelay, hatmaker of Nottingham, who also missed his day in court. These
discoveries stir Stopes less than the Dugdale engraving of the Stratford
monument. Although this was the first engraving of the bust, it had been
ignored (except by Halliwell-Phillipps, whom Stopes overlooks), and so her
enthusiasm is understandable. But she goes to absurd lengths to elevate
Dugdale. Maybe he had seen his countryman in habit as he lived (Dugdale
was not quite eleven when Shakespeare died). The bust must be discountenanced: it has been extensively reconstructed, Stopes dubiously claims, the
face 'more than likely' being restored with the aid of cement. In Dugdale's
rendering of a grim-visaged, droopy-moustached Shakespeare, she manages
to discern 'something biographical, something suggestive'; indeed, a
'thoughtful poetic soul' (Chambers sees the lineaments of a tailor). In a
notably fatuous essay on the Sonnets Stopes proposes a new Dark Lady,
Jacquinetta Field, the wife of Shakespeare's printer: 'she was a Frenchwoman, therefore likely to have dark eyes, a sallow complexion, and that
indefinable *charm* so much alluded to'.[9] On such foundations are hypotheses
confidently erected (Field's Christian name was in fact Jacqueline.). An
equally ineffective article, 'Mr. Shakspeare about My Lorde's Impreso', seeks
to cast doubt on the identification of the dramatist with the 'Mr Shakspeare'
who was rewarded, along with his fellow Burbage, for devising an *impresa* for
Lord Rutland on the occasion of a tilt in 1613. Stopes offers her own
candidate: John Shakespeare, royal bitmaker. In none of the records,
however, is this artisan referred to by the honorific 'Mr', and *imprese* were not
horse-trappings but painted shields embellished with emblems and mottoes.
(Elsewhere, while acknowledging that this record of Shakespeare and Burbage was discovered by Stevenson, she claims to have discovered it herself
among the Belvoir MSS. What significance, one wonders, does she attach to
the word *discovery?*) Stopes's tendency to denigrate or ignore the accomplish-

ments of co-workers did not go unnoticed in her own time. In an icy review of *Shakespeare's Environment* for the *Dial* (14 October 1915), Samuel Tannenbaum, a psychiatrist, remarked, 'A psycho-analyst [*sic*] would find it difficult not to conclude that Mrs Stopes's defective logic is due to an overwhelming desire to belittle the discoveries of others and to magnify the importance of her own.'

Stopes published other volumes on Shakespeare and his period. She is the first scholar of any distinction to attempt a reasoned refutation of the anti-Stratfordians; her *Bacon-Shakespeare Answered*, in the enlarged edition of 1889, of course left the believers unmoved, as have all subsequent appeals to reason. Later Stopes complained maliciously that J. M. Robertson had ignored her in *The Baconian Heresy*. She published a book on *Burbage and Shakespeare's Stage* (1913). Her *Life of Henry, Third Earl of Southampton, Shakespeare's Patron* (1922), received as standard, assumes that Southampton was the Lovely Boy of the Sonnets, in which are to be found the 'twined threads of biography and autobiography'. Twenty-eight years in the making, this is still the fullest life of the Earl. (Southampton's most recent biographer, G. P. V. Akrigg, pays tribute to his predecessor as 'that assiduous researcher'.) A miscellany, *Shakespeare's Industry* (1916), is mere book-making occasioned by the Tercentenary.

It is difficult to evaluate Stopes's contribution objectively. By her own lights she probably failed. She spent long years among the public records but found (at least as regards Shakespeare) no major document. A straining after significant achievement is sometimes painfully evident in her work. One would expect from the title of an article she published in the *Jahrbuch* for 1896, 'The Earliest Official Record of Shakespeare's Name', that she had found something new or at least had extracted a hitherto unsuspected significance from a known record; but the document she discusses had been turned up by Halliwell-Phillipps, and what she has to say about it need not detain the historian. Her slovenliness, the vice of amateurism, disheartens: Stopes confuses names and dates (she cannot even get straight the name of the New Shakspere Society with which she was closely associated); she is capable of making two gross blunders in quoting two excessively familiar lines from *Hamlet*. The records Stopes quotes are reproduced with appalling carelessness. When these peccadilloes were pointed out by Watson Nicholson of the British Museum in a lengthy review of *Burbage and Shakespeare's Stage*, Stopes stubbornly attempted to answer the unanswerable—she had injured her knee and, forced to lie on the sofa, was unable to check her transcriptions at the Museum or Record Office. (Then why did she not defer publication until she could?) Anyway, she adds, 'I never pretend to be an "archivist"'.

But if not, what then was she? Lacking the discipline of the professional scholar, Stopes has her own eccentric strength. She did an heroic amount of archaeological burrowing, and found bits and pieces of new information which enhance the record. She drew attention to such neglected items as the Dugdale engraving. She throve on genealogy, but her true forte was describing Warwickshire folk of Shakespeare's day; men and women whose unmomentous lives she could narrate without inflation or romance. In these pages of her too-numerous books she furnishes biographical contexts of lasting interest. Shakespeare himself brought out the worst in Stopes, for he roused in her a lust for scholarly glory which constantly eluded her.

That larger fame would be won by an American couple from far-off Nebraska, who sat a few places away from her under the domed roof of the Public Record Office. The ageing Scottish lady and the brash young Yankee first met, however, not in the Round Room but at the British Museum, in the autumn of 1905. To Charlotte Stopes it was 'a little romantic episode', this encounter with 'a tall handsome dreamy looking, proud, touchy American, who only gradually melted'. Soon he would solidify again.

# 3

# *The Wallaces*

'PRIOR to his researches, which in number of documents and value on Shakespeare and his theatrical compeers exceed the results of the previous three centuries, it was believed and taught for nearly 50 years that everything was known about Shakespeare that ever would be known. His remarkable discoveries have changed all this, given a world-wide impetus to a fresh study of Shakespeare and the drama on the historical side, and brought lasting honor to American scholarship.'[10] So writes Charles William Wallace in his manuscript papers now housed in the Huntington Library. He knows whereof he speaks, for his subject is himself. If he exaggerates—he found no single document as centrally important as the will or marriage licence bond—he does so only slightly: Wallace made the most important contribution of the twentieth century (thus far) to our factual knowledge of Shakespeare.

He came from the American West, unlikely nursery of archive sleuths. The son of a county judge and farmer, Wallace was born in Hopkins, Missouri, on 6 February 1865. Educated after a fashion at Western Normal College in Shenandoah, Iowa, he then became a professor of Latin and English at his Alma Mater. Later he taught at a normal school in Nebraska, and founded a preparatory school to the state university. Meanwhile Wallace took a bachelor's degree at Nebraska and went on to graduate study there. In 1901 he joined the faculty as an assistant instructor in English literature.

The University of Nebraska would remain his academic home for the rest of his career.

In 1893 Wallace married a daughter of the frontier, Hulda Alfreda Berggren of Wahoo, Nebraska. In the autumn of 1904, on an extended leave of absence from his university, Wallace crossed the ocean with Hulda, the first of many such pilgrimages. At the University of Heidelberg he studied as a *hospitant*, then took a Ph.D. at the University of Freiburg im Breisgau. That was in 1906. By this time he and his wife had gained expertise in reading the secretary hand of Elizabethan legal records, and he had made the first of his Shakespearian discoveries at the Public Record Office in London.

In pursuing the children's troupes at the Whitefriars and Blackfriars, Wallace was thorough enough not to exclude from search such unpromising index subheadings as 'lands', 'messuages', etc. Hence it was that he came upon a suit concerning the Blackfriars Gate-house purchased by Shakespeare in 1613. A couple of years after this transaction the widow Anne Bacon, a former owner of the estate, died, leaving her son Mathias sole executor. He now required authorization from the Court of Chancery to surrender the title-deeds to the present occupants. On 26 April 1615 'Willyam Shakespere gent' with six other petitioners jointly prayed the Lord Chancellor to have the 'letters patents, deeds, evidences, charters, muniments and writings' pertaining to the 'messuages, tenements and premises' handed over to the complainants. In his answer of 5 March, Bacon did not contest the claim—he was merely detaining the deeds 'until such time as he may be lawfully and orderly discharged thereof upon his delivery of the same'. On 22 May the court concluded the amicable litigation by furnishing the order. As Wallace recognized, these documents hold little excitement; nor did he entertain much hope of locating the deeds that occasioned the action. But the case supplies its biographical crumb—it shows Shakespeare looking after his property rights a year before his death—and adds a few names to the record. To discover anything of direct biographical pertinence would inevitably command interest—and also stir expectation of larger discoveries to follow.

In 1905 Wallace transcribed the Bill, Answer, and Decree, and prepared an article for the *Athenaeum*. When that journal kept him on the string for nearly a month, he began to fear anticipation, and so he sent the piece to the London daily *Standard*, which printed it on 18 October. To this episode Wallace would later trace his bad relations with the *Athenaeum*. 'Since then', he complained in an unposted letter dated 7 January 1913, 'the *Athenaeum* has used every possible occasion to speak ill of my work. There is a small circle of narrow souls here who are eaten up with envy at my success in research. . . .'[11]

The success to which he refers was achieved not through the gifts of Fortune but as the result of monomaniacal devotion to the cause. Wallace gave up his university salary to stay on in London; a letter to Professor Stuart P. Sherman (28 February 1914) speaks of 'my willingness year after year to make the terrible sacrifices necessary to find and publish the materials'. He broke with the genteel tradition and brought to his work the training, persistence, and systematic methods of the professional.

At all times his wife assisted. On every possible public occasion Wallace expressed his gratitude, and his personal papers include an exquisitely wretched verse tribute of which it will be sufficient to quote one stanza:

> Wide was the world and our spirits' ambitions!
>   Close to my side has she worked,
> Delved in researches of science and letters,
> Circled the realms of known knowledge,
>   Passed me to reach nobles[t] heights,
> Sealing with Shakespeare & Handel & Raphael.[12]

For this team tedium hardly existed. They worked fifteen to eighteen hours a day. They looked at multitudes of documents which held nothing for them; eventually, Wallace claimed, they examined over five million records. Discoveries that less ambitious searchers would cherish as breath-taking rewards they transcribed and referred to in passing, or filed and forgot. They never published the 1640 Salisbury Court Theatre proceedings against Richard Brome, for example, which yields unique insight into the conditions of a Caroline playwright's employment. They unearthed a law suit concerning the lost Jacobean play, *Keep the Widow Waking*, and did nothing with it; a generation later C. J. Sisson would rediscover the same litigation and make his reputation with that find.

But why should they bother when they had so much which was immeasurably more important? In October 1909 Wallace announced in *The Times* his exhumation of the Ostler-Heminges suit from the rolls of the Court of King's Bench. The next year, in popular articles in the *Century Magazine* (August and September) and scholarly papers in *Nebraska University Studies* and *Shakespeare Jahrbuch*, he made public the documents he had uncovered in the Court of Requests. Three theatrical suits are involved, of which two concern the unfortunate widows of actors. In *Ostler vs. Heminges* (29 October 1615) the daughter of Shakespeare's old friend Heminges sued her father when he withheld the shares she expected to inherit after her husband's death. The widow of Augustine Phillips (the actor who had bequeathed Shakespeare a thirty shilling gold piece) made a bad second marriage with John Witter, a wastrel; after running through her capital he went to law for her Globe interest (*Witter vs. Heminges and Condell*, 28 April 1619). In the third action, *Keysar vs. Burbadge and Others* (8–12 February 1610), Robert Keysar, London goldsmith and shareholder in the defunct Children of the Queen's Revels, sued the Burbages and three other King's Men for a one-sixth interest in the Blackfriars Theatre and its profits. These suits enabled scholars to determine for the first time Shakespeare's financial stake in the Globe and

Blackfriars; an interest which fluctuated from one-eighth to one-fourteenth in the former, and from one-sixth to one-seventh in the latter, depending upon the number of shareholders. The actual cash value of the interests is more problematical, although Thomasina Ostler estimated hers at £300, and numerous attempts at evaluation have been made since Wallace published the records. To the layman this may sound like arid arithmetic, but for the biographer it is vital to know the economic organization of Shakespeare's company and the dramatist's role in that organization. Never prone to understatement, Wallace claimed of his documents that, 'On the side of Shakespearian biography, they are the most valuable records discovered since the Rev. Joseph Green, of Stratford-on-Avon, in 1747 found the poet's will.'[13] in his various publications Wallace summarized fairly the sometimes complicated litigation, and he made available full transcripts of all the relevant papers. If he showed an inexpert understanding of the seventeenth-century law of co-ownership, this lapse scarcely detracts from the importance of Wallace's contribution.

His ideology moulded by a capitalistic democracy, he took special delight in presenting records which revealed the operation of a business enterprise founded (in his view) on ideals of democratic brotherhood. 'I have never apotheosized Shakespeare,' he wrote in his personal papers. 'I have always taken him as a man, read him as a man speaking to men, and have searched with hope of finding him wholly a man among men.'[14] His hopes were abundantly realized in 1909—his *annus mirabilis*—during his search of the great bundles of documents which comprise the Court of Requests holdings at the Public Record Office. Among these papers—uncalendared, unindexed, and for three hundred years unperused—Wallace found the twenty-six depositions, witness lists, summonses, orders, and decrees that make up the Belott-Mountjoy case. One of these documents is Shakespeare's signed deposition: the sixth authenticated signature of the poet. Taken together these records, unlike all others, tell a humanly interesting story of Shakespeare in London, as he entered the lives of a French Huguenot family that lived at the corner of Silver and Monkwell streets: an address not previously associated with him. The deposition, moreover, would have a value for scholarship which Wallace did not envisage, for it bears the most natural of all Shakespeare's signatures. For once the hand is not cramped by the small space available on the labels of conveyancing documents, or quavering from mortal illness, as when he revised his will. This signature would be the principal exhibit used by a great palaeographer, Sir Edward Maunde Thompson, when he set out, in *Shakespeare's Handwriting* (1916), to show that the three pages of Addition D in the manuscript play of *Sir Thomas More* are in Shakespeare's autograph.

One may pardon Wallace the trite subtitle he gave to his article in the March 1910 issue of *Harper's Magazine*, which offered to the world 'Shakespeare as a

Man among Men'. One may forgive him also his sentimental eulogizing and the *naïvetés* of his popularizing prose:

The story is of the simplest and most ordinary sort, arising out of the life of the most ordinary people. . . . That Shakespeare lived with a hard working family, shared in their daily life, and even lent his help with the hope of making two young people happy makes him as the world would gladly know him, an unpretentious, sympathetic, thoroughly human Man.[15]

After all, Wallace had made the Shakespearian discovery of the century.

What sensations did the moment of triumph bring? Fortunately he left a record on three roughly scrawled sheets now at the Huntington Library. Headed 'Feelings', they contain the following account, in which the original emotions have been (one suspects) somewhat modified by recollection in tranquillity.

We were working in a room full of other searchers, My wife stood at the opposite side of the table examining another of the great bundles of miscellaneous old skins and papers. I asked her to come over and look at a document. We sat down together to read it, as we had done in hundreds of cases. We looked it through with about as much excitement as we do the morning paper. I saw by the look on her face that she felt as I did,—glad, but disappointed in a measure. We were aware of the bigness of what we had. But we were looking for bigger. We had searched at various periods for some years, always confident as we had long before announced to our friends, that we should get Shakespeare's signature and a personal expression from him. We had fixed our minds on the topic he should speak on,—one that would be finally thereby settled to all the world—and this was so much less than we had wished! We exchanged a few words over the document, but no one in the room might have guessed that we had before us anything more important or juicy than a court-docket. But we saw that we had only a part of the documents in the case. We must find the rest.[16]

They did.

But the bigger game they sought forever eluded them. They had, however, already found much on the child actors, the subject which originally brought them to London, and they would bring to light a good deal more. The results of this research Wallace embodied in two monographs, *The Children of the Chapel at Blackfriars, 1597–1603* (1908) and *The Evolution of the English Drama up to Shakespeare* (1912), which hold more direct interest for the historian of the stage than for the biographer of Shakespeare.

For his contributions to documentary knowledge of the early theatre and the National Poet, honour and universal recognition came his way. No scholar of the period had a more distinguished reputation for archival expertise. *The Times* offered him homage in leaders. Newspapers throughout the world publicized his discoveries, the Hearst chain giving him a two-page spread ('Prof. Wallace's Remarkable Analysis of 3,000,000 Documents Which Prove the Immortal Bard Never to Have Been a Roistering, Reckless

Profligate'). Naïvely Wallace revelled in the attention, collecting newspaper and magazine clippings, copying out—or having his wife copy out—choice morsels of praise, composing panegyrics to himself in the third person. Little wonder, under the circumstances, that a sense of perspective abandoned him. He came to believe that, as a result of his achievements, the whole dramatic literature of the Elizabethan age would require re-editing, and the history of the stage would have to be wholly rewritten.

Inevitably he aroused envy as well as gratitude. How could it be otherwise, given the nature of the man and his accomplishment? A brash stranger had come to England from Nebraska—it might just as soon have been Outer Mongolia—and uncovered records which the greatest British scholars, from Malone to Halliwell-Phillipps, died not knowing. Wallace's self-promotion— as one contemporary sneered, he 'boomed' his finds 'in true Transatlantic style'—aggravated the wound to national pride. And so his awesome labours were deprecatingly praised for their Teutonic thoroughness. His prose style gave rise to cruel mirth: Sir George Greenwood ironically extolled Wallace's 'beautiful and picturesque language'; an *Athenaeum* reviewer snidely observed, 'The style of the writing does not always "economize the reader's attention", owing to occasional peculiarities in construction, odd uses of words, and superabundant "tropes".'

Wallace's megalomania combined with the scholar's psychosis, paranoia, to produce alarming symptoms. The Royal Commission of Public Records established in 1910 had as its primary object (he was persuaded) the installing of students in the Record Office where they could hunt down all the Shakespeare materials before Wallace had a chance to publish them; thus the Commission would 'prevent any more such American successes'. Although constrained to allow in private that the scope of the Commission had been broadened to take in all records, not only Elizabethan ones, he nevertheless indiscreetly voiced his complaint in public, and in an American magazine, the *Literary Digest*. The episode intensified hostility and gave pain to well-wishers in England.

He had other troubles. In the sort of pursuit in which Wallace was engaged, priority counts for all, and so he had to keep a step ahead of competitors no less avid for fame than he. Some wretches, he protested, tried to steal his finds; others accused *him* of theft. Two scholars in particular plagued him: Stopes and her 'little friend' (so she termed him), the French academic Albert Feuillerat. Wallace had started out on amicable terms with both—his pocket diary for 1904–5 contains Feuillerat's calling card as well as the Stopes address—but eventually they made his life miserable. Stopes (he charged) haunted him at the Record Office, cajoling the officials into telling her which records he was searching, then making application for them herself. In the summer of 1908 Wallace had to send back to the Repair Department papers containing new information about Burbage and The

Theatre, and, before their return, Stopes had put in for them. He went over to her in the Round Room, told her they were still in his name, and—responding to her curiosity—went on to describe their contents. Within the year Stopes had published the discovery as her own in the *Fortnightly Review*. On another occasion Wallace told her about the Ostler-Heminges suit; word that sent her scurrying to the *Coram Rege* rolls, which she ransacked without luck. Nevertheless she reported success and repeated Wallace's description of the material.

So at any rate he insisted, but did his own imagination transform legitimate competition into illusory persecution? In the *Jahrbuch* for 1910, he complained, Stopes allowed him priority in transcribing the Ostler deposition, thus implying that the actual discovery was hers; but that volume of the *Jahrbuch* contains no article by Stopes, only the editor's survey of recent work. She was not silent about her dealings with Wallace. 'When I was going consecutively through the Coram Rege Rolls of James I (the most *uncomfortable* of records to read), Dr Wallace plunged into the middle of my work, in 1905.'[17] Such impudence! Moreover, he obtained unclassified papers, with Shakespearian items, of the Court of Requests—papers to which she had been refused access—by simple virtue of the fact that he was not a resident of London. When Wallace repeated in print his version of the Burbage affair, Stopes—in print—flatly denied his story: 'If he ever told anybody it must have been somebody else. Neither then, nor at any time, did he ever tell me anything that I wished to know. I had all my papers before he began his work, which I can prove.'[18] She never produced her proof; clearly one of them was lying. Stope's behaviour on other occasions inclines one to accept her testimony *cum grano*, but, on the other hand, Wallace was foolish to tell her about his discoveries—if, indeed, he did tell her. 'Nor do these things', he lamented, 'show the worst of what I have suffered from her.' The *éminence grise* of the *Athenaeum* had stirred up that journal and Feuillerat against him.

The *Athenaeum* had always treated him scurvily. Expressions of gratitude for his 'laborious researches' are invariably qualified by denigration: he lacks expertise with Latin records, he wants the modesty which should distinguish the scholar, he makes unworthy references to his predecessors. Even the Belott-Mountjoy sensation becomes somehow not too important. But the feud with the *Athenaeum* began in earnest with the notice there, on 2 November 1912, of *The Evolution of the English Drama*. Wallace's subtitle, *A Survey Based upon Original Records Now for the First Time Collected and Published*, seemed to the reviewer 'a little misleading', for Wallace in fact owed debts to others most notably Professor Feuillerat, who had already published some of the chief documents on the Blackfriars Theatre in the *Daily Chronicle* for 22 December 1911 and, the next year, in the *Shakespeare Jahrbuch*. 'We make no attempt to decide who had the real priority in the date of discovery,' the reviewer remarked, choosing his words carefully. 'The public are not particularly interested in close reckonings as to days or weeks. But scholars like to know, and the author would have done well to acknowledge the work of a student in the same domain.'

The next number carried a communication from Feuillerat with unpleasant charges. They concern the Loseley MSS, which contain the Blackfriars material. Feuillerat noted that he had obtained permission to print these manuscripts in 1906, that he had brought out a first instalment in 1908, and that he had two years later announced his intention of publishing others. 'On September 28th, 1911,' he recalls, 'I spoke with Dr. Wallace about my discovery, and I confidentially told him that the documents were to be found among the Loseley MSS. Dr. Wallace answered that *he had not seen the Loseley*, but that he had other documents confirming the thing.' Accordingly Feuillerat went ahead with publication of the articles already cited. 'Meanwhile',

at a date which I can ascertain if necessary, but which cannot be earlier than *October 10th*, 1911, since at that time I was still searching the Loseley MSS., Dr. Wallace obtained permission to examine the Loseley MSS.

In October, 1912, Dr. Wallace has published the book described above, in which he feigns to ignore my discovery, but reproduces all the documents already published by me, and boldly affirms that before 1907 he had in his possession documents to which, to the best of my belief, he had access only in 1912.[19]

Feuillerat also claimed that he had sent Wallace an offprint of his *Jahrbuch* article, and this statement is borne out by the presence among the Wallace papers of the 'Separat-Abdruck' in its original envelope. But the Frenchman had addressed it to Nebraska, and the packet did not reach Wallace's flat in Torrington Square until late August, as the postmark shows. By then his book was already at the printers.

The Feuillerat allegations were the most serious that Wallace ever had to face. In the *Athenaeum* for 23 November he described as 'absolutely untrue' Feuillerat's version of the conversation; the latter had told him that there was nothing on the Blackfriars in the Loseley MSS. Beneath Wallace's letter the *Athenaeum* printed, damagingly, another from W. H. Grattan Flood, who declared that he had made the Loseley references known to Sidney Lee in 1906, and anyway the greater part of the material had already been published in 1879 in the Seventh Report of the Historical Manuscripts Commission; Wallace's claim to novelty is 'surely monstrous'. Another exchange between the principals followed. Wallace crowed that he had demolished Feuillerat, but it is clear that the episode rattled him badly. He sought an interview with the editor of the *Athenaeum*, who refused to see him but maliciously reported the request in his gossip column. The Wallace papers at the Huntington contain numerous drafts of protesting letters to editors, and rejoinders to Feuillerat. Hearing that his now controversial book was to be reviewed in the *Jahrbuch*, Wallace even went so far as to urge the editor to strike from the notice any reference, direct or indirect, to the Frenchman, and to see to it that he, Wallace, received unqualified credit for his discoveries. But Feuillerat would have the last word. In his edition of *Documents Relating to the Revels at Court in the Time of King Edward VI and Queen Mary* in 1914, he rehearsed the quarrel and repeated his most telling point, which Wallace, despite the loquacity of his replies, had not attempted to refute: Feuillerat's competitor had not gained access to the Loseley MSS until October of 1911.

With the cloud of *l'affaire Feuillerat* hanging over him, Wallace returned to the Record Office. In the quiet of the search room, surrounded by his beloved rolls and bundles of documents, he failed to hear the guns of August. The Wallaces stayed on in London during the first years of the Great War, then, short of funds, returned to the United States. There he threw himself into a grandiose scheme for a Shakespeare Foundation that would collect and publish documentary evidences on the dramatist and his theatre, and otherwise support the kind of research with which Wallace was identified. To help launch the Foundation, he went on an extensive lecture tour during the 1916–17 season; before schools, colleges, clubs, and societies, he told 'for the First Time the Consecutive Story of his Discoveries' (so the publicity hand-out reads). The goal was to enlist a million members before the next anniversary of Shakespeare's birth in 1917. But for some reason the million members did not materialize—the nation must have been preoccupied with other matters—and the plan never got off the ground.

He would have to raise money some other way. Geology had long been his hobby, and he decided that he would make a fortune in Texas by striking oil. Wallace moved to Wichita Falls, and in the autumn of 1922 he bought the oil and mineral rights to a 160-acre field, in nearby Archer County, to which he later added two sections of adjoining land. He personally directed the drilling operations. The professionals told Wallace that the area was dry, but he thought he knew better. He was right: his Archer tract would develop into one of the most valuable oil properties in America. His wife now helped in the management of the fields, just as earlier she had shared his work at the Record Office. Whatever he sought, whether old papers or oil, it was Wallace's destiny to find. He became rich and hoped, with his new wealth, to publish a great collection of records. But cancer foiled this dream, and on 7 August 1932 he died in Wichita Falls.

# 4
## Joseph Gray

FOR objective biography the first decade of the twentieth century was an exhilarating time. To the contributions, erratic but relevant, of Stopes, and the major discoveries of Wallace, the finds published by Professor John W. Hales and Dr Andrew Clark serve as modest pendants. In the *Athenaeum* for 26 March 1904 Hales announced that in Pipe Roll 41 Eliz., on membrane 'Residuum Sussex', the name William Shakespeare occurs in the margin with the words 'Episcopo Wintonensi'—indicating that the dramatist then lived in the Bishop of Winchester's liberty of the Clink on the Bankside. Another document (Hales goes on), dated 15 November 1597, shows William Shackspere being assessed 5s. as a resident of St Helen's parish, Bishopsgate Clark's discovery pertains more to the Shakespeare-Mythos than to biography proper. In the library founded by Archdeacon Thomas Plume at Maldon in Essex, among the notes compiled by the Vicar of Greenwich at various intervals from 1657 to 1680, Clark found the anecdote about Jonson offering a spoon of latten for translation at the christening of Shakespeare's child; also the story of John Shakespeare as a merry-cheeked old man who durst have cracked a jest with Will at any time. The Plume jottings were first published, not entirely accurately, in the *Westminster Gazette* for 31 October 1904.

They quickly found a place in the appendix to Joseph Gray's *Shakespeare's Marriage* (1905), the most considerable biographical monograph of the period. Although less specialized than the title suggests—Gray includes chapters on 'The Departure from Stratford' and on 'Facts and Conjectures'— the main substance of Gray's book concerns the marriage. No episode of Shakespeare's career more required sober consideration, for around the two nuptial records, the licence and the bond, had sprung up an unruly thicket of unfounded surmise, irresponsible inference, and moralistic prejudice. Did the dispensing with the full publication of the banns imply that the ceremony took place clandestinely? Was it reasonable to assume that there had been a pre-contract, with a promise *per verba de futuro*, to sanction cohabitation? Or did Anne Hathaway, eight years her husband's senior, force the marriage upon a passionate youth? Did jealousy, loss of affection, and estrangement follow upon wedlock? And what of Anne Whateley of Temple Grafton? Was Whateley, not Hathaway, the bride's true name? Or had it been assumed as a decoy because the Shakespeare family found the match offensive? Or had another Warwickshire Will taken a different bride at precisely the same

moment in history? Why does Anne's name not appear in Richard Hathaway's will—or is she represented therein as Agnes? To these questions and
others Gray addresses himself.

Not enough evidence exists, of course, for him to dispose of them all. But at
least in Gray's pages opinion does not masquerade as fact, and the inferences
have a solid basis in knowledge: he has analysed 166 bonds executed during
the years 1582 and 1583, and he has studied minutely the policies and
procedures which obtained in the diocesan Consistory Court of Worcester
under Bishop Whitgift, a reforming prelate who ruled sternly from 1577 to
1583. For some details, to be sure, Gray must resort to analogies from
Canterbury and London practice, and some of his illustrations necessarily
date from a period somewhat later than that of Shakespeare's marriage. But
never before had the evidence been so scrupulously marshalled and evalu-
ated.

The result is not so much a new view of Shakespeare's marriage as a more
informed choice from among incompatible possibilities. Gray accepts the
traditional belief that Anne was the daughter, otherwise known as Agnes, of
Richard Hathaway of Shottery. So far as the insertion of Temple Grafton in
the Register is concerned, '. . . it may be inferred that the terminal place-
name in the lists of marriage licences was in all cases intended as the
residence of the bride, and that Temple Grafton was probably copied in error
from the allegation, in which it may have appeared as the residence of one of
the persons concerned in the application for Shakespeare's licence, or as the
church named for his marriage.'[20] Gray demolishes the assumption that the
union took place without John Shakespeare's consent, such consent being
'regarded as one of the most important of the precautions then taken against
carelessness, collusion, or fraud on the part of any one concerned in a
marriage by licence'.[21] There is nothing remarkable about the omission of
John's name from the bond; in only twenty-four such documents did Gray
find a surety of the same name as the bridegroom. An unrealistic notion of
Shakespeare's position in 1582 lies behind the abuse heaped by some
commentators on the two friends of the Hathaways who bound themselves in
a surety of £40.

So there was no conspiracy to obtain a licence by fraud. Nor does the
evidence sustain the view that the marriage took place under circumstances
morally discreditable to the contracting parties. Haste there was, as Anne's
condition necessitated and the single asking of the banns attests, but the
behaviour of the couple was not determined by Victorian norms. If they
incurred contemporary censure, it is odd that legend has not memorialized
the scandal, along with the deer-poaching, the Bidford drinking bout, and the
tryst with an innkeeper's pretty wife in Oxford.

Written in a prose identical in hue to the author's name, Gray's
pages convey no sensational revelations; *Shakespeare's Marriage* makes for

profoundly unexciting reading. Therein, curiously enough, lies its strength. The drama and scandal which previous writers had found in this episode derive not from the known facts but from the imagination—sometimes prurient, sometimes florid, sometimes both—of the biographer. Gray's disinterested monograph would not put an end to the excesses of the subjective approach, but in future such excesses would be less pardonable. His book remains the authoritative treatment of the subject.

# 5

## *The Grafton Portrait*

ALONG with the sorting out of facts, the quest for a *vera imago* less hostile to the romantic ideal than the two authorized likenesses. In the early years of the century, in the village of Winston between Darlington and Barnard Castle, a small picture painted on oak panel hung on the walls of a picturesque inn called The Bridgewater Arms. The subject is a youth with oval face, gentle expression, and a shock of curly dark brown, almost black, hair reaching nearly to the base of the neck. He wears a maroon slashed doublet surmounted by a greenish gauze collar. In the upper left- and right-hand corners raised yellow letters (probably formed by a thick application of pigment) read:

AE SVAE · 24                    1·5·8·8

On the back of the stretcher, or frame, are branded the initials W + S.

For over two hundred years the forebears of the Ludgate sisters, proprietresses of the inn, had farmed for successive Dukes of Grafton. The portrait, it was claimed, had been owned by one of these noblemen, who presented it—for reasons left unexplained—to a sturdy yeoman in his service. For five or six generations the picture had lain in an old farmhouse, belonging to the line and tenanted by the Ludgates, in the village of Grafton; eventually it was transferred to The Bridgewater Arms. When early in 1907 the distinguished expert on Shakespeare iconography M. H. Spielmann began to make inquiries concerning the portrait, word leaked to the press. On 18 February the *Manchester Guardian* carried a photograph of 'the supposed portrait of Shakespeare which had been found in a village inn near Darlington'. Soon newspapers throughout England, as well as in America and Germany, were publishing sensationalized stories about the discovery of the Grafton Shakespeare.

It is a genuinely old picture, probably by some Dutch or English follower of Holbein or Bettes, and the raised letters of the inscription appear to be contemporary. The face is attractive, reminding Dover Wilson of Shelley; the wonderfully soulful eyes rivet attention. In its proportions the head corresponds with the Droeshout engraving. The curve of the lips too reminds us of Droeshout, and there is the same impressive expanse of forehead. Shakespeare of course was twenty-four in 1588. The initials, however, do not help, for Miss Ludgate cheerfully admitted that her father had branded them on, remarking as he did so 'that inasmuch as the portrait evidently represented Shakespeare he might as well set it upon record for the guidance of future owners'.[22] And the nose raises doubts: this organ, if not quite bulbous, thickens uncharacteristically at the end, and the exposed nostril is shaped unlike those of the engraving and sculpture. The absence of a reliable pedigree compounds uncertainty; for all we know, the painting may be of Continental origin, and the youth Netherlandish. All told, the evidence is painfully insufficient to connect this portrait, known traditionally as 'Old Mat' in the owners' family, with the poet of Stratford-upon-Avon.

Yet the urge to do so would prove irresistible. A Stockport connoisseur, Thomas Kay, purchased the picture and devoted the last months of his life, on the eve of the First World War, to an amateurish monograph recounting *The Story of the "Grafton" Portrait of William Shakespeare*. When he died in the autumn of 1914, Kay bequeathed his treasure to the John Rylands Library in Manchester. There it still hangs. One enthusiast, wisely making 'no pretensions to Shakespearean scholarship', has suggested in the Library's *Bulletin* (July 1945) that the poet's relations with the Stanleys brought him to Lancashire, there to sit for his portrait, which later passed to Grafton, where the Stanleys had connections. The argument is a tissue of improbable conjecture (we have no verifiable evidence of Shakespeare's intimacy with the Stanleys), but advocacy has overcome reasonable scruples: the writer suppresses the information—surely known to him, for it occurs in two of the sources he cites—that the sunken initials W. S. are a modern improvement. The Grafton has also impressed viewers with considerable pretensions to Shakespearian scholarship. J. S. Smart found in the portrait his own idea of the young Shakespeare, and wished it genuine.[23] The picture appears as frontispiece to Smart's posthumous *Shakespeare: Truth and Tradition*, as it does to Dover Wilson's *The Essential Shakespeare*. Despite awareness that its claims were not very strong, Wilson made Old Mat the banner of his crusade against Janssen's 'self-satisfied pork-butcher' and the 'pudding-faced effigy' of Droeshout. 'I do not ask the reader to believe in it or even to wish to believe in it,' Wilson writes. 'All I suggest is that he may find it useful in trying to frame his own image of Shakespeare. It will at any rate help him to forget the Stratford bust. Let him take it, if he will, as a painted cloth or arras, drawn in front of that monstrosity, and symbolising the Essential Poet.'[24]

The Grafton portrait of the Essential Poet is the last of the major supposititious Shakespeare icons. Other likenesses have since turned up with metronomic regularity, being submitted at the rate of one a year to the National Portrait Gallery for authentication, which is invariably withheld. But none has created a similar stir.

# 6

# *Lytton Strachey*

THE Edwardian afternoon was Bloomsbury's morning. In February 1900 five freshmen at Trinity College, Cambridge, banded together to form a Reading Club that met in the rooms of Clive Bell. This precocious quintet comprised, in Bell's opinion, the nucleus of the legendary Bloomsbury Group. Five years later one member, now in his middle twenties but still hovering about Trinity in hopes of a fellowship, reviewed *Shakespeare's Marriage* for the *Spectator*. One might think that Gray's dry treatise would appeal but indifferently to the temperament, at once sceptical and romantic, of the young Lytton Strachey. And indeed the reviewer begins by asking, 'Who could possibly want to know how, or when, or where he happened to be married?' It is, after all, so personal a proceeding; Shakespeare himself would experience a shock at the sight of Gray's volume. 'One can imagine his amused shade smiling over such speculations, and replying to the anxious searcher after further information very much in the manner of Pontius Pilate in M. Anatole France's story:— "Special license? Bond? Prohibited seasons? . . . I don't recollect." '25

But Strachey comes to praise Gray, not to bury him: Shakespeare's importance justifies historical dissertations, and the marriage is one of the most disputed incidents in his career. Strachey has only admiration for the thoroughness of Gray's research, the closeness of his reasoning, and the lucidity of his style. Contemplating the faded extracts from the Elizabethan records, the reviewer sees, in his mind's eye, a vanished age take shape and substance. 'An atmosphere of mysterious antiquity arises from them [the extracts] like the fumes from an alchemist's alembic; the reality of the present disappears; its place is taken by the phantasma of the past. Society lives again for us as it lived in the England of Elizabeth.'26 Surely none but an intensely romantic sensibility could respond in this way?

But students of Shakespeare recall Strachey chiefly for his iconoclastic paper on 'Shakespeare's Final Period': a realist's onslaught against the sentimental subjectivism of Dowden and kindred Victorian spirits who charted the dramatist's interior life in the plays. Is memory playing tricks upon us?

Shakespeare always elicited from Strachey responses of rapture and identification. 'It is only by holding our breath', he writes in the same notice of Gray, 'that we begin to understand how necessary breathing is; and the best way of bringing before our minds the true magnitude of our debt to Shakespeare is to imagine for a moment or two that he never existed. To suppose ourselves deprived of all the gifts which his writings bring to us . . . must not that be ranked among those speculations "too sad to insist on"?'[27] Given such a theme, the magnificent hyperbole of a Swinburne loses hyperbolic force. In a youthful diary Strachey asked whether Shakespeare had any character of his own, and answered yes, the poet was 'a cynic in his inmost heart of hearts'. But the cynicism, that self-indulgence of a superior intellect, was a projection of the shy diarist on to his subject, as Michael Holroyd, Strachey's biographer, has recognized. 'Unable to contemplate directly his own hated image', Holroyd observes, 'he uses Shakespeare as a convenient looking-glass.'[28] Although 'Shakespeare's Final Period' did not appear between the covers of *Books and Characters* until 1922, it belongs to Strachey's student days. He read it before the Sunday Essay Society at Trinity on 24 November 1903, and it first saw print in the August 1906 number of the *Independent Review*.

Strachey's rhetoric—witty, eloquent, entirely alive—has as its target the 'ordinary doctrine' of his day, as formulated by Dowden and accepted by the leading critics. This is the Shakespeare of the Four Periods, who descended into the depths for his tragic phase but, after the storms and stresses of middle life, found the serenity of meditative romance 'On the Heights'. The most important stage, as Strachey rightly discerned, is the last; in order for biography to achieve its consolatory end, Shakespeare must die happy. To substantiate this personal history, Dowden and the others had pointed to the pastoral lovers and philosophic poise, the atmosphere of forgiveness, and the happy *dénouements* amid spring flowers and blue skies, of *Cymbeline*, *The Winter's Tale*, and *The Tempest*. But such a barometric reading of the poet's inner weather, Strachey points out, requires a selective interpretation of the text. Cordelia too is serene, and *Measure for Measure* ends fortunately; but the Shakespeare of these plays does not look down upon us from the heights. Fastening upon other elements on the last romances, Strachey creates another Shakespeare. Perdita and Miranda retire; the brute Cloten, fiendish Queen, and poisonous Iachimo advance to the footlights. The critic reminds us of the tragic first two acts of *The Winter's Tale* and the cruelties and coarsenesses in all these plays; he confronts the long drawn-out unravelling of *Cymbeline*, and the tiresome conspiracies of the shipwrecked noblemen on the Enchanted Isle. As for Prospero, he is less the embodiment of wise benevolence than 'an unpleasantly crusty personage, in whom a twelve years' monopoly of the conversation had developed an inordinate propensity for talking'.[29]

The Shakespeare of the Final Period was not serene but apathetic: 'Bored with people, bored with real life, bored with drama, bored, in fact, with everything except poetry and poetical dreams.' So Strachey imagines the dramatist in his Stratford retirement—'Half enchanted by visions of beauty and loveliness, and half bored to death; on the one side inspired by a soaring fancy to the singing of ethereal songs, and on the other urged by a general disgust to burst occasionally through his torpor into bitter and violent speech.'[30] Strachey set a high value on boredom, a condition from which he frequently suffered himself.

Now such a view of Shakespeare is certainly at odds with Dowden's, deliciously so—but in much the same way as one side of a coin differs from the other. Is not Strachey's purpose, no less than that of the Victorian bardolaters, to exalt what is sublime in Shakespeare and dismiss what is offensive to fastidious tastes: the grossness, the clowning, the machinery? Strachey's Shakespeare is as much a product of selective reading of the plays as Dowden's, and the iconoclastic author of *Eminent Victorians* is no less sentimental and eulogizing than his eminent Victorian predecessor. Even the boredom may be a cynic's version of serenity. Both portraits require us to assume a one-to-one correspondence between what an author writes and his mood at the time; the creator of *Romeo and Juliet* must be in love (bad luck for Anne back home in Stratford). Such a view is of course simplistic and fallacious. If a work of art exploits the dark forces of the soul, it is (in Camus's words) 'not without channelling them, surrounding them with dikes, so that the water in them rises'. Subjectivists like Dowden and Strachey overlook the dikes, and they ignore utterly the professional aspect of dramaturgy: the truism that the drama's laws the drama's patrons give.

Perhaps the most interesting feature of Strachey's essay is that briefly, tentatively, he stands aloof from his own construct, and regards objectively the tacit assumption on which it rests, 'that the character of any given drama is, in fact, a true index to the state of mind of the dramatist composing it'. The validity of this crucial assumption has never been proved, Strachey grants, but he goes on to turn his back on the whole problem: 'It is not, however, the purpose of this essay to consider the question of what are the relations between the artist and his art; for it will assume the truth of the generally accepted view, that the character of the one can be inferred from that of the other.'[31] Thus peremptorily is the overwhelming question dismissed. It is as though for a moment in the first years of the new century, Strachey trembled on the brink of modernity, only to draw back. Critical theory interested him only passingly, and anyway he was not prepared to scuttle his whole conception of Shakespeare the man. That it should emerge as an inversion of Dowden's, despite the fact that both critics drew their data from the same plays, of course implies a refutation of their method. But this implication was not immediately apparent, and for a time longer the subjective school would dominate Shakespearian biography.*

# 7

# *Frank Harris*

IT was, in truth, the golden age of subjectivism. This was the time when *The Tempest* as personal allegory had become a *donnée* of criticism. Thus, in a notice of J. H. Leigh's production of the play at the Court Theatre in 1903, the incomparable Max Beerbohm remarked, '*Obviously*, Shakespeare, at the close of his career, wished to write an epilogue to his work, an autobiography, in allegorical form.' [Emphasis supplied.] Nowhere are the excesses of this biographical approach more flagrant than in the fantasies of Frank Harris.

Liar, libertine, and blackmailer, Harris (né James Thomas) was a scoundrel. Despite his below-average height—5′ 5″ without the augmentation of Cuban heels—he loomed larger than life; a ruffian to be sure, but (in Shaw's words) 'simply the most impossible ruffian on the face of the earth'. Such a man, as one would expect, was profoundly unacademic; the worlds of action and of power, not cloistered halls, engaged his fierce energy. Born in Galway, probably in 1856, he early emigrated to America. There he shined shoes and toiled on the Brooklyn Bridge before heading west to the frontier, where he met Wild Bill Hickock and rustled cattle from across the Mexican border. After practising law with dubious credentials in Kansas, he covered the Turko–Russian War in 1876 as a newspaper correspondent; when General Skobeleff seized and abandoned the Plevna redoubt at terrible cost, Harris was there. Golden years followed for him in London. The sheer thrust of his personality carried him upward; one did not meet Harris, an acquaintance remarked, one collided with him.

He gained fame as the brilliant editor of the *Fortnightly Review* and, afterwards, of the *Saturday Review*, which he made the vehicle for the literary and artistic ferment of the *fin de siècle*. His circle included Wilde, Beardsley, Wells, Beerbohm, and Shaw. He wived it wealthily with a dowager unresponsive to his sexual demands, which, needless to say, were inordinate. In their funny little house squeezed in between the big ones on Park Lane, they

---

* In his critique of Shakespeare's Final Period' Strachey, Holroyd suggests, 'almost certainly' confused boredom with 'complete exhaustion' (*Lytton Strachey*, i. 144). Thus the fallacy that Strachey dimly apprehended persists in the late 1960s, and in a work of distinguished biographical scholarship.

threw glittering parties. At the Café Royal in his heyday Harris presided at luncheons to which his guests, fifty or so, had been invited by telegram, and held them spellbound with the ceaseless eloquence of his conversation, couched in a voice which seemed like the rustle of the leaves of a brass artichoke. 'To survive you', Oscar Wilde wrote to him, 'one must have a strong brain, an assertive ego, a dynamic character. In your luncheon-parties, in old days, the remains of the guests were taken away with the *débris* of the feast.'[32]

*In old days*. Worse and worse times still succeeded the former. The polite society at whose gates he clamoured, rather too strenuously, for admission never accepted the Bohemian upstart with the loud voice. H. G. Wells found Harris 'too loud and vain . . . to be a proper scoundrel', but others would view him less charitably. His political career collapsed (he stood for Parliament as Conservative candidate for Hackney), and so did his marriage. He drank too much. To appease his sexual vanity he practised on a grand scale the noble art of seduction, although his conquests were probably rather less numerous than he would later recall in his salacious memoir, *My Life and Loves*. (Still they did include the young and still-virginal Enid Bagnold (later Lady Jones), who would go on to achieve celebrity for her novels and plays, among them *National Velvet*, the complexly wrought *Chalk Garden*, and *The Chinese Prime Minister*. Harris's editorial genius now served second- and third-rate papers, *Modern Society* and the *Candid Friend*, to which he contributed puffs, plagiarisms, and libels. One libel landed him in jail. For solace he turned to the sun-drenched south. In his Mediterranean villa he disported himself with olive-skinned Riviera virgins bought for him by his Italian gardener. Ousted from the corridors of power, he meditated the great book he would one day produce on Shakespeare. The virgins gave him pleasure; from Shakespeare he expected immortality.

His three companions, Harris would proclaim—loudly of course—were Christ, Shakespeare, and Wilde. To the awe inspired in Harris by Shakespeare a famous incident at the Café Royal testifies. There, during a lull in conversation at one of his luncheon parties, the conversation shifted to homosexuality. A great hush descended upon the room at the mention of a subject that, in those days, was taboo. Harris, however, thundered on in his powerful basso: 'Homosexuality? No, I know nothing of the joys of homosexuality. My friend Oscar can no doubt tell you all about that.' Further silence, even more profound. Harris continued: 'But I must say that if *Shakespeare* asked me, I would have to submit.'[33] Only for Shakespeare.

Harris discovered the plays in boyhood, when he read them chiefly for the stories. Every few years he read them again. In 1878, at the University of Heidelberg (where did he not turn up?), he heard the celebrated Kuno Fischer lecture on Shakespeare; a performance which aroused sufficient opposition in Harris to send him back to the *œuvre* in a state of feverish excitement. Long passages he committed to memory, and recited aloud in the streets of

Heidelberg to the bemusement of passers-by. At Fluelen he devoted a holiday to the study of the writer who had now become his god. Among the myriad voices of the plays he began to hear, with ever greater insistence, the accents of one voice; in the crowd of personages one face came to stand out, that of the poet, 'for all the world like some lovelorn girl, who, gazing with her soul in her eyes, finds in the witch's cauldron the face of the beloved'.[34] He had discovered—he fancied—the man behind the masks. Harris's first article on 'The True Shakespeare' appeared in the *Saturday Review* for March 1898. Others followed, and Harris's friends urged him to make a book of them. Heinemann announced the volume for the spring of 1899, but spring came and went without the book. According to Harris the publisher dared not bring out a work that exonerated Shakespeare from being Greek sexually. On such occasions one recalls that Harris was also a liar.

'Frank Harris is upstairs', wrote Wilde from Napoule in February 1899, 'thinking about Shakespeare at the top of his voice.' The subject haunted Harris. He revised and expanded his articles. On 17 December 1908 he complained to Arnold Bennett, 'I wish to God I had never begun it! I could have written a dozen novels in the time it has taken me.'[35] But he persevered; for all the critics were wrong, fatally wrong, and a higher power had chosen Harris to set them right. 'I am sending Dowden to-day', he informed Bennett the following September, 'to show you the best of what was known about Shakespeare before I began my work, you will see from that the incredible stupidity of the commentators.'[36] At last, in the same year, 1909, *The Man Shakespeare and His Tragic Life-Story* was unveiled.

The *Athenaeum* and the *Times Literary Supplement* ignored the book, and the more specialized serials—*Shakespeare Jahrbuch*, *Modern Language Review*, and the like—maintained a similarly discreet silence. In the popular press of England and America, however, *The Man Shakespeare* created a sensation. 'This is the book for which we have waited a lifetime,' the *New York Times* proclaimed: 'We know this now it is come, and we mark the day of its publication as a red-letter day in the history of literature.'[37] Assiduously buttered up by the author, Arnold Bennett reviewed the work with an enthusiasm which cannot be accounted for merely on the basis of Harris's flattery. 'A masterpiece on Shakespeare has at last been written,' he declared. 'It has destroyed nearly all previous Shakespearean criticism, and it will be the parent of nearly all Shakespearean criticism of the future.'[38] Bernard Shaw praised Harris's *Shakespeare*. To Upton Sinclair it was quite as wonderful a creation as any character in the plays, a veritable Hamlet of criticism.

These encomiums did not strike Harris as excessive; the work was his dearest creation, to be fondled in all his references to it. When a new book on Shakespeare appeared without mentioning his masterpiece, a look of pained incomprehension would pass over his face. Silence disturbed Harris more than dissent—'A man may be judged by his disciples,' he told a London editor; 'Jesus had Paul, *I* have Arnold Bennett.' Ordinarily he was content to see himself as Paul, declaring unto the ignorant THE UNKNOWN GOD.

Harris's own estimate of his achievement, as well as his identification with the Apostle, has been echoed by his admiring biographers, helpless in the face of Harris's 'power of assertion', which won the admiration of Bernard Shaw, an old friend himself not deficient in such power. Of *The Man Shakespeare* Tobin and Gertz write: 'It is his masterwork, a carved casket filled with wondrous things. . . . It builds up a remarkable hypothesis with such power of language that it takes on unprecedented significance.'[39] And, according to Root, 'To read *The Man Shakespeare* is an experience so tremendous that one may liken it to birth or a conversion like that on the road to Damascus, which changed Saul into Paul. It is the blinding Light. . . .'[40] This marvellous casket we must now open, and, having accustomed our eyes to the Light, examine its riches.

They disappoint the expectations engendered by panegyric; we find not Light but darkness visible, not the jewels and gold of wisdom but the rhinestones and tinsel alloy of a second-rate mind. *The Man Shakespeare* is slapdash in organization, without continuity or progression, but from it one can reconstruct that 'tragedy of tragedies', the poet's life story. He planned, it seems, to marry Anne Whateley of Temple Grafton, but a friend of the Hathaways, the masterful Fulke Sandells, heard what loose Will was up to, and induced the Bishop of Worcester—easy mark!—to grant a licence for the high-spirited youth to espouse his other Anne, far gone with child, without his father's consent. She turned out to be a jealous scold who poisoned Shakespeare's life, so to escape her nagging he fled to London. She, poor creature, was left to suffer the extremes of poverty; for ten years her husband did not set foot in Stratford. In the capital Shakespeare wrote plays, but his energies went mainly into his idolatrous passion for one of the Queen's Maids of Honour, Mary Fitton. To woo her he sent as his messenger the young Lord Herbert (odd that a common player could command the services of a noble lord), but she seduced him. Thus the poet lost both friend and mistress. 'It was her falseness that brought him to self-knowledge and knowledge of life, and turned him from a light-hearted writer of comedies and histories into the author of the greatest tragedies that have ever been conceived. Shakespeare owes the greater part of his renown to Mary Fitton.'[41] He boiled with jealous rage (*Othello*) and raved with erotic mania (*King Lear*). He went mad (*Timon of Athens*). Broken in health, prematurely old and enfeebled, the dramatist returned to Stratford, where his beloved Judith retrieved him from death's door. In gratitude he idealized her as Marina, Perdita, Miranda. His loathing for his wife he carried with him to the grave, for he composed the doggerel malediction that is his epitaph in order to frustrate Anne's desire to be buried with him.

To Harris the plays scarcely exist as objective works of theatrical art. All literature is autobiography, disclosing the creator's inner nature with as much scientific fidelity as the thumb-print reveals the identity of its possessor. Again and again Shakespeare put the one love of his life into his writings. Mary Fitton is the Dark Lady of the Sonnets. She is Rosaline, the hard-hearted black-eyed wench who gives Romeo sleepless nights; also that other Rosaline, the whitely wanton with the velvet brow in *Love's Labour's Lost*. She is the lecherous Gertrude, the sluttish Cressida—above all, she is Cleopatra. Twenty times Shakespeare portrayed himself at different phases of his career—'. . . as a sensuous youth given over to love and poetry in Romeo; a few years later as a melancholy onlooker at life's pageant in Jaques; in middle age as the passionate, melancholy, aesthete–philosopher of kindliest nature in Hamlet and Macbeth; as the fitful Duke incapable of severity in "Measure for Measure", and finally, when standing within the shadow, as Posthumus, an idealized yet feebler *replica* of Hamlet.'[42]

To some readers it may come as a surprise that Macbeth is bracketed with Hamlet as a kindly aesthete, but Harris has little sensitivity as an interpreter of Shakespeare. He fails to see the gentle mockery directed at Duke Orsino in *Twelfth Night* (another 'snapshot'—a favourite Harris word—of the playwright); he does not notice that Jaques stands outside the circle of sympathy in *As You Like It*, scorned as he is by Rosalind and Orlando. The simplistic and reductive tendencies of Harris's criticism everywhere oppress us. What can be said on behalf of a reader for whom *Measure for Measure* is a mere tract for the times, and for whom 'Shakespeare was not a good playwright and took little or no interest in the external incidents of his dramas'? How do we contrive to suppress exasperation with a critic who confidently asserts that 'Brutus was no murderer, no conspirator, no narrow republican fanatic, but simply gentle Shakespeare discovering to us his own sad heart and the sweetness which suffering had called forth in him'? The fact that Brutus murders and conspires will make little headway with an interpreter for whom Shakespeare's plays— *Hamlet*, *King Lear*, and all the rest—are instalments of a personal confession, and nothing more. It is the absence of diffidence, of any inclination to consider alternative possibilities, which makes Harris's criticism ultimately so obnoxious. Throughout he uses the vocabulary of confidence; such words and phrases as 'unmistakable', 'surely', 'manifestly', 'established beyond dispute', appear *ad nauseam*, along with thunderous amplifications of discoveries which (missed by all previous commentators) are 'astounding' or 'astonishing'. Again and again the author attempts to overcome inadequacy of demonstration by stridency of assertion.

The list of Shakespearian self-portraits, it will be noticed, excludes Hotspur, Henry V, and the Bastard in *King John*. Because they go against the grain of Harris's conception of Shakespeare, these characters are wooden marionettes, mere copies of the historical sources, or they lack conviction, betraying their creator's want of manliness by tell-tale inconsistencies and falsities. For Harris would have us believe that Shakespeare had a soft melting nature; he was Richard the dreamer, not Bolingbroke the politician; 'a gentle yet impulsive nature, sensuous at once and meditative; half-poet, half-philosopher, preferring nature and his own reveries to action and the life of courts; a man physically fastidious to disgust, as is a delicate woman, with dirt and smells and common things; an idealist daintily sensitive to all courtesies, chivalries, and distinctions'.[43] *Gentle, impulsive, physically fastidious, a delicate woman*—did Wilde, one wonders, sit for this portrait of the artist? But Harris's Shakespeare also revels in his passion for the opposite sex, giving himself to all the subtle games of love until his health collapses under the strain. Here the biographer, the amorist who lusted after dark-haired beauties, fashions his subject in his own image. This impression finds confirmation—if confirmation were needed—in Harris's account of the dramatist's London reception:

From the moment young Will came to London, he was treated as an upstart, without gentle birth or college training; to Greene he was 'Maister of Artes in Neither University'. He won through, and did his work; but he never could take root in life. . . . He was in high company on sufferance.[44]

So it was with Harris. In his days of wealth and power, with a socially correct wife, he found himself an outsider, looked down upon with condescension by simpletons of high station; Edwardian society accepted him on sufferance.

The Shakespearian composite—Wilde and Harris, the delicate plant and the raging beast—is incongruous, and Harris's other details do not add plausibility to the picture. Like Cassio, Shakespeare could not hold his liquor. Insomnia plagued him, and in general he suffered from the disabilities of the neuropath. Inordinately vain and egocentric, he was an exquisite snob who concentrated all his loathing for the ordinary Englishman into his subhuman Caliban. Yet, like his *alter ego* Antonio in *The Merchant of Venice*, this self-centred snob was a generous friend, extravagantly careless of money.

So much for the personality of the man. What of his physical exterior? Incongruity assumes a new dimension when Harris visualizes the writer described by Aubrey as 'a handsome, well-shaped man':

I picture him to myself very like Swinburne—of middle height or below it, inclined to be stout; the face well-featured, with forehead domed to reverence and quick, pointed chin; a face lighted with hazel-clear vivid eyes and charming with sensuous-full mobile lips that curve easily to kisses or gay ironic laughter; an exceedingly sensitive, eager speaking face that mirrors every fleeting change of emotion. . . .
    I can see him talking, talking with extreme fluency in a high tenor voice, the reddish hair flung back from the high forehead, the eyes now dancing, now aflame, every feature quick with the 'beating mind'.[45]

In this passage the desire to extol the Divine Bard mingles with an impulse, subtle and perhaps unconscious, to reduce him to size: everybody knew that Swinburne was ugly—just as Harris was ugly. In his nature, no less ambivalent than that of the Shakespeare of his portrayal, adulation always co-existed with envy. We are not surprised when ultimately he patronizes his subject: 'Poor, broken Shakespeare!'

For other biographers Harris feels only a superior being's disdain. The host of commentators, whether academic or amateur, are derisively referred to as Professors (in his vocabulary a pejorative) or Dryasdusts. A critic of such prejudices is unlikely to set for himself rigorous standards of scholarly precision: Harris confuses Cassio with Cassius, persistently misspells Furnivall's name, and (more serious) distorts the views of the authorities he cites. They all wrote rubbish anyway; he has, he claims, waded through tons of the stuff to no avail. Actually Harris has only a nodding acquaintance with the scholarly tradition. He knows Coleridge, Halliwell-Phillipps, Goethe on *Hamlet*; Lee he alludes to indirectly. Tyler, whose work on the Sonnets Harris uncritically absorbs, alone moves him to generous acknowledgement. Needless to say, he has missed Charlotte Stopes's demolition of Tyler in the *Jahrbuch*, and he is happily unaware of Lady Newdigate-Newdegate's demonstration that Mary Fitton possessed a fair complexion, grey eyes, and brown hair: not quite ideal qualifications for a Dark Lady. Even less excusable is Harris's ignorance of Gray's well-known monograph exploding that theory of the marriage set down with all the certitude of fact in *The Man Shakespeare*. And how unfortunate that Harris has overlooked Elton's book on Shakespeare's family and friends, which would have spared him his dreary confusion about the occupation of John Shakespeare, who appears in Harris's pages as butcher and dealer in skins, corn, wool, and malt; a jack-of-all-trades—anything but the glover that in fact he was.

Mere details, Harris might shrug, worthy of notice only by a Dryasdust. He exults in the originality of his overall thesis. 'Without a single exception', he crows, 'the commentators have all missed the man and the story; they have turned the poet into a tradesman, and the unimaginable tragedy of his life into the commonplace record of a successful tradesman's career.'[46] Thus he rebels against the Victorian–Philistine conception of Shakespeare; the Shakespeare of Halliwell-Phillipps and Lee. But so too do all the subjective biographies, whose great quest is to recover the soul of the poet. Harris knows Dowden and Brandes, and refers to them some twenty times in his book, almost invariably to take exception on some minor point. But clearly he has profited from Dowden and shamelessly plagiarized from Brandes: 'Its purpose', Brandes said of his study, 'was to declare and prove that Shakespeare is not thirty-six plays and a few poems jumbled together and read *pêle-mêle*, but a man who felt and thought, rejoiced and suffered, brooded, dreamed, and created.'[47] Harris might have described his own opus in exactly the same

terms. Like him, Brandes fixed upon Mary Fitton as the tawny beauty who
enslaved the dramatist, and identified her with Rosaline in *Love's Labour's
Lost* and *Romeo and Juliet*, and with Cleopatra—just as he discovered
Shakespeare himself in Biron, Hamlet, and the other protagonists. (Being,
unlike Harris, a scholar, Brandes read Lady Newdigate-Newdegate, and
revised his views in later editions of his *Shakespeare*.) Brandes's hero, like
Harris's, experiences jealousy and rage; with Timon he breaks down, and
with the Romances he convalesces. Even Harris's strictures on Shakespeare's
snobbery, which seem to be grafted on to his discourse, probably derive from
Brandes, who, more than any other critic, insists on his subject's aristocratic
sympathies. Brandes's work appeared two years before Harris contributed his
first article on Shakespeare to the *Saturday Review*; there he praises his rival
as 'the ablest of Shakespeare's commentators'. When Harris came to write his
book, he was no longer capable of a gesture of magnanimity which would in
any way lessen his own claim to a hearing.

His larcenous propensities did not deter him from complaining bitterly that
others had filched his wares, one of the chief culprits being the famous Oxford
don, A. C. Bradley. The author of the classic *Shakespearean Tragedy* stimulated
Harris's bile with his piece on 'Shakespeare the Man' (*Oxford Lectures on
Poetry*, 1909), a sensitive essay in personalist biography. Against Lee and the
objective school, whose arguments he well knows, Bradley maintains that
Shakespeare in his plays and poems betrayed himself—although how much,
Bradley admits, is a question. But he does not doubt that the character who
reveals most fully the author's personality is Hamlet: like his creator honest
and open in nature, sweet-tempered, the lover of his friend, disillusioned, sad,
fierce, and cynical, a dreamer and a philosopher. That Harris too had seen
Shakespeare in Hamlet, as well as in a score of other figments of the
dramatist's imagination, Bradley was aware, and he gives his predecessor
generous credit. Not enough, however, to soothe a plagiarist's vanity.

*The Man Shakespeare* reveals the man Harris as a literary charlatan. Is
there, nevertheless, a positive side to his accomplishment? In portraying the
National Poet as an unabashed sensualist, he saw himself as striking a blow
at the dominant Puritanism of the times, the same Puritanism that had
crucified Wilde. A certain novelty attaches to the role of passion in the study;
with some justice Harris would later claim that his conception of sexuality as
a forcing-house for genius was foreign to the English mentality. By vigorously
rejecting the view, no doubt held if not expressed by many readers of the
Sonnets, that Shakespeare had homosexual tendencies, he cleared the air of
some fumes. (One is reminded of the delicate proprieties of that vanished era
by the fact that not once in his twenty-page chapter on Shakespeare's
sexuality does Harris employ the forbidden word.) But mainly he furnished a
vulgarized subjective biography, in the Dowden–Brandes mode, for the

edification of a large popular audience disinclined to turn directly to the pages of donnish authorities.

*The Man Shakespeare* found a ready market in Britain and the United States, and achieved a German translation. Harris changed the ideas of people with his book. Acutely aware that he did not know enough to offer expert criticism, Bennett expressed forcibly the impact of the work upon an intelligent lay reader; 'To me', he wrote to Harris,

> ... your portrait was at first most disconcerting. I had an image of Shakespeare as a successful, hustling, jolly playwright of immense artistic power, *but somewhat disdaining that power*, and keen on material ends; always thinking of an easy old-age at Stratford. . . . You smash my image to atoms, but it keeps reconstructing itself again in spite of you—from mere habit. I shall have to get used to it. . . . For the general public your book is at least 30 years before its time. And in 30 years (or so) people will be beginning to admit that in the way of constructive criticism it marked an era.[48]

In thirty years *The Man Shakespeare* would be forgotten. But for Bennett and others enjoying the tranquillity of the Edwardian afternoon, Harris brought Shakespeare vibrantly and coarsely alive. Perhaps that was his aim all along.

Two years after *The Man Shakespeare* Harris entered the ring for another bout with the Bard, and again found himself badly outclassed. Originally he thought of calling his sequel *The Woman Shakespeare* ('for the woman a man loves is the ideal in himself'), but, suspecting that some readers would misunderstand this title, he changed it to *The Women of Shakespeare*. When the invitation to write the work reached him, he informs us, he was on holiday in the south of Italy, away from his books and notes, and so he perforce depended on the texts of the plays for snapshots of the four ladies who conspicuously influenced Shakespeare's life and art: his mother, his wife, his mistress Mary Fitton (a sketch of whom serves as frontispiece to the volume), and his daughter Judith. Separation from his books brought no diminution in the author's confidence. Harris offers the interesting theory that Helena in *All's Well That Ends Well* represents not only Mary Fitton but also the eternally feminine artist. That conceited, shallow cad Bertram is Lord Herbert, and when he vouchsafes that Helena's 'infinite cunning with her modern grace | Subdued me to her rate,' he is describing the seduction which shattered his luckless friend. These lines, avers Harris with the assertiveness that is his trademark, comprise 'Herbert's confession as if taken down from his own lips, with the "i's" dotted and the "t's" crossed'.[49] Although better organized than its predecessor, *The Women Of Shakespeare* has little fresh to add—and it is tiresome to encounter once again at the outset Paul and the unknown God.

# 8
## Harris versus Shaw

MEANWHILE Harris had written a play, *Shakespeare and His Love*, dramatizing—if that is the word for so theatrically unrealized a piece—some of his
biographical obsessions. Shakespeare falls in love with the tall, dark, and
proud Mary Fitton, sends Lord Herbert to woo her for him, then discovers that
they have become lovers. Stock romantic improvisations follow: the Queen,
having got wind of the affair, ensconces Herbert in the Tower, and the
distraught Maid of Honour begs Shakespeare to try somehow to secure his
release. Within his grasp is the mastership of the Revels, which would enrich
him and his colleagues, but instead he petitions Elizabeth to pardon the friend
who, with patrician callousness, has betrayed him. In an epilogue Shakespeare lies dying, haunted by memories of the Dark Lady ('A great woman')
and his mother: 'The gentlest sweetest—the noblest mother in the world.'*
Harris's play reveals, more nakedly his critical works, the cloying sentimentality of his conception of Shakespeare. Lovelorn and wallowing in self-pity,
he mopes about the stage, mouthing dialogue which consists of trite modern
phrases interlarded with snatches and paraphrases from (among others)
*Othello*, *Troilus and Cressida*, and the Sonnets. At one point Harris has the
National Poet throwing '*herself in a seat*'—an unconscious admission,
perhaps, that the author has effeminated his subject? As if aware that
Shakespeare's part is too pallid to sustain interest over four acts, Harris fills
the stage with supporting players. Jonson acts the manly friend; Chettle
(viewed as the model for Falstaff) furnishes what is intended as robust
humour. Southampton, Ralegh, and Essex have bit parts, and now and then
Chapman, Dekker, Marston, and Fletcher swell a scene. Even young Willie
Hughes—touching gesture in the direction of Wilde's memory!—sings a song
or two.

Among the Folger holdings is a copy of *Shakespeare and His Love*, as
privately printed by Harris, with his inscription (dated July 1904) to
Beerbohm Tree. Harris hoped that Tree would produce the play, and even
altered it to suit his presumed taste, excising a scene in which Mary Fitton
declares that she can simultaneously love two men, Shakespeare and
Herbert, in different ways: an avowal too daring for audiences not yet
exposed to Genet and Pinter. But the great actor–manager procrastinated,
and Harris turned next to the Vedreene–Barker management, with no better

* This display of filial enthusiasm was too much for Shaw: 'Englishmen mostly quarrel with their
families', he snarled, 'especially with their mother.'

luck; for Granville-Barker, displaying good judgement, objected to the part of Shakespeare. Finally, in 1910, Harris published a revised version of the drama with an admission that it was second-rate: too loosely constructed, too literary in tone, too redolent of piety. But the weeds of humility sit awkwardly on Harris's shoulders, and elsewhere in his introduction he lashes out at his old friend Bernard Shaw. The latter had seen *Shakespeare and His Love* and was now about to produce a play entitled *The Dark Lady of the Sonnets* into which he had introduced Mary Fitton—'my Mary Fitton', cried Harris with proprietary passion in a letter to Bennett. How dare he! 'It amused me years ago', Harris remarks without amusement, 'to see Mr. Shaw using scraps of my garments to cover his nakedness; he now struts about wearing my livery unashamed.'[50] And to a reporter he complained: 'What does Shaw know about Mary Fitton? What does he understand about women? What does he understand about passion?'[51] The rival was un-deterred; *The Dark Lady of the Sonnets* opened at the Haymarket on 24 November 1910 with (bitter pill for Harris!) Granville-Barker as Shake-speare.

The conceit that the master who already had to his credit *Pygmalion*, *Major Barbara*, and *Man and Superman* would find unconquerable the urge to loot the rejected offering of a writer with no discernible bent for the stage is so extravagantly absurd that only someone as humourless and vindictive as Harris could have seriously proposed it. ('Have you really stolen his Shake-speare?' Shaw was asked during an interview carried on the front page of the *Daily News* the day of the première of the *Dark Lady*. 'Why should I?' Shaw retorted. 'Not that his Shakespeare is not worth stealing, but Shakespeare is common property; and I can dramatise him for myself in half the time it would take me to steal Frank's dramatization. Besides, he will never really understand Shakespeare.') To be sure, the *Dark Lady* is a slight invention, a mere skit of fourteen pages designed to solicit funds for the establishment of a National Theatre; but it is pure Shaw none the less. His Shakespeare jots down for future use unexpectedly Shakespearian lines uttered by the Dark Lady, Queen Elizabeth, a beefeater—anyone who wanders on to the premises. No self-effacing lover, he is brash, proud of his gentility (never for a moment, like Wallace's Shakespeare, a man among men), and impudently cheerful. 'It is no fault of mine that you are a virgin, madam,' he bluntly tells the Queen, 'albeit tis my misfortune.' When he embraces his sovereign and tries to kiss her on the mouth, the Dark Lady enters from behind and, not recognizing the Other Woman, separates them with a right uppercut. (The historicity of this scene is doubtful.) Then, realizing what she has done, Mistress Fitton falls to her knees in terror: 'Will', she shrieks: 'I am lost: I have struck the Queen.' To which he replies, 'Woman: you have struck WILLIAM SHAKESPEAR.'[52] Her Majesty registers stupefaction.

If, in this incarnation, the poet's egoism reminds us more of Shaw than of gentle Shakespeare, the kinship is not lost on his creator. 'I am convinced', he admits in his preface, 'that he was very like myself: in fact, if I had been born in 1556 instead of in 1856, I should have taken to blank verse and given Shakespear a harder run for his money than all the other Elizabethans put together.'[53] The preface as a whole is a remarkable performance, all the more effective for the geniality—even affection—with which he treats the adversary who vilified him; the success of Harris's book, he says, gave him great delight. Unlike his foe he does not believe in the Mary Fitton theory, but merely adopts it as a convenient fiction which allows him to bring Elizabeth on stage. Anyway the hypothesis is not Harris's but Thomas Tyler's, and the former knew of it only because Shaw had reviewed Tyler's edition of the Sonnets in the *Pall Mall Gazette*.

The achievement of *The Man Shakespeare*, in Shaw's eyes, is that Harris has given readers a credible human being rather than a god. The vindication of Shakespeare from the whispered charge of homosexuality (a word which Shaw, like Harris, cannot bring himself to use) brings cheers, but the sage of Ayot St Lawrence takes shrewd exception to Shakespeare's alleged sycophancy—the kings and courtiers of the plays fare not much better than the Lower Orders, and 'A sycophant does not tell his patron that his fame will survive, not in the renown of his own actions, but in the sonnets of his sycophant.'[54] Nor did Shakespeare fall victim to idolatrous passion and have a nervous breakdown; his irrepressible gaiety, which is the distinctive badge of genius, would have prevented that. 'Men have died from time to time, and worms have eaten them; but not for love.' Here Shavian and Shakespearian laughter join, and the sanity of art is reaffirmed.

# 9

# *Studies Mad and Bad*

STRACHEY, Harris, and Shaw make no pretence of being Shakespearian scholars; they promulgated views about their subject that the sober historian must regard as frivolous or perverse or (at the very least) unfounded. But the sober historian himself does not always merit trust, and sometimes, unlike these gentlemen, he bores us. A case in point is W. Carew Hazlitt, the editor, bibliophile, and grandson of the essayist. His *Shakespear: Himself and His Work* (1912) represents the fourth and final version of a study first published in 1902. A very substantial book (it comes to five hundred pages), it is written in unspeakably bad prose. Perhaps just as well, for the uncritical reader who struggles through it will find himself hopelessly misled. That Hazlitt is an unreliable as well as dull scholar his selection of the Ashbourne portrait of Shakespeare for his frontispiece sufficiently forewarns us. From 'unmistakeable allusions', mainly in the Sonnets, he concludes that the young Shakespeare made his way to London on horseback; to such questions does this biographer address himself, and on such evidence does he rely. The same literalness leads Hazlitt to give credence to the old canard about the poet's lameness (again on the testimony of the Sonnets), but we are reassured to learn that the disability was a temporary infirmity rather than a permanent affliction. On larger issues, such as Shakespeare's conjugal relations, this antiquarian bumbler has decided views. He doubts that Shakespeare even saw his wife, much less bedded her, during his visits to Stratford between 1587 and 1611. Living *en garçon* in London during these years, he was not humanly likely, given his voluptuous blood, to have escaped the gaieties and temptations of the metropolis. 'Who shall say', Hazlitt asks in one of his more inspired moments, 'that he never proved a Tarquin to some unchronicled Lucrece?'[55] Thus does he furnish bardolaters with a new icon to contemplate: Shakespeare the Rapist.

We need not be surprised that a student of Hazlitt's calibre flirts with heresy: he thinks that Bacon may have written the first sketches of the English history plays, which Shakespeare afterwards retouched for the stage. In sum, *Shakespear: Himself and His Work* may be taken as representative of the thoroughly bad books, without interest or influence, that the new century spawned. The reader need not be wearied by a survey of the others.*

Along with dullards, lunatics. To one, more privileged than most, Shakespeare spoke directly and in his own person. The recipient of the Word was, not surprisingly, a medium: Mrs Sarah Taylor Shatford of New York City, who impressed many of the people she met in hotel parlours or the park or in cafeterias with her ability to lead them into the light and solace of spirit communion; they gave her testimonials (some she printed) of spook visits. Her first contact with Shakespeare came in December 1916, when she took down in the space of an hour before breakfast three poems written through the Ouija board on themes of war, peace, and God's love. A few days later she heard Shakespeare's voice, and dispensed with the board. On 22 May 1917 he revealed his purposes: 'I was told by the One who speaks for Him, that if I came back and undid my wrong, helped men to rise from their wicked impassioned selves to look to Him instead, to incite nobility of aim and the

---

* William J. Rolfe's long, workmanlike *Life of William Shakespeare* (1904), designed mainly for US consumption, perhaps deserves mention for its mysterious preface. His manuscript, Rolfe writes, was put in a vault for safekeeping, but disappeared from his library. 'Though I had little doubt by whom it was taken, the evidence was purely circumstantial; and for that and other reasons it was impossible for me to make any effort to regain possession of it. The person who took it intended after reading it, to return it without betraying himself, but he was afterwards tempted to put it into other hands with a false statement of its history, possibly with a view to its being utilized, in part if not as a whole, in print. . . . I have therefore been compelled to undertake the depressing task of rewriting it. . . .' This is the most fascinating passage in the book.

love of God instead of enflaming the lusts of the craven for the flesh, that when I had fulfilled this errand, and came again to His presence, my opportunity to rise would be bestowed, and I should rise and be forgiven at last.'[56] In the interminable pages of *Shakespeare's Revelations* ('Through the medium of his pen SARAH TAYLOR SHATFORD Dictated exactly as herein found. No illiteracies, no obliterations, chargable [*sic*] to the Medium. My hand and seal hereon. W.S. In spirit.'), the long-departed dramatist expresses himself on death, sex, the Vast Beyond, reincarnation, and many other interesting topics.

But this communication did not exhaust his message, for he went on dictating to his 'treasured humble clairaudient, Sarah', matter for three more books. In *My Proof of Immortality* (1924), 'By Shakespeare's Spirit', he furnishes the text of his only spirit play. 'A tomb holds my guts', he explains in a preface subscribed Old Bill, 'my brain survives. Not the same methods. A ribald jester for a king's amuse, was I. A light for God would I be, *become*, Sarah, make it.'[57] It must be regretfully confessed that this volume of 517 pages offers little biographical sustenance. There is even less in Shatford's *Jesus' Teachings* (1922), but *For Jesus' Sake* (1920), a collection of prayers, contains an important declaration: 'I am Catholic, have been since my birth. Raised by a devout woman of this faith, whose principles, had I but followed with profit to my soul, I should be making sonnets for my Maker's lute instead of mine own.'[58] Although the information would appear to derive from an unimpeachable source, scholars (perhaps unaware of the scarce publication) would go on debating Shakespeare's faith.

# 10

# *Dark Ladies*

THE sonnets Shakespeare composed for his own lute rather than his Maker's went on inspiring hypotheses, less bizarre perhaps than Shatford's hallucinations, but not necessarily more persuasive. Hopeful explicators continued to pursue the real-life counterparts of Shakespeare's *dramatis personae*. (Did Shakespeare satirize William Cecil, Lord Burghley, a member of the Queen's Privy Council and, throughout most of his career, Elizabeth's most trusted adviser—as Polonius in *Hamlet*? Maybe Shakespeare saw Burghley's *Certain Precepts, or Directions* (1616), written for his son Robert Cecil, in manuscript; after all, the few precepts which Polonius gives *his* son parallel Burghley's *Precepts*. Of course, many cunning parallels obtain between literature and life, and parental maxims, savouring of worldly prudence, were traditional in this period; but these have given comfort to those who envision Shakespeare at home in the corridors of power, as well as to Oxfordian schismatics.)

Meanwhile, the Pembrokists beat a slow retreat, with later counter-thrusts under the capable generalship of Chambers and Dover Wilson, before the better-equipped battalions of the Southamptonites, who lacked only the ultimate weapon of conclusive demonstration. The lesser personages of the drama had their champions too. Chapman remained a heavy favourite for the role of Rival Poet, although to Hubert Ord we owe the interesting suggestion that the competing versifier was Thomas Speght, the bombastic editor of Chaucer, who is eulogized in the Sonnets.[59] This possibility, one must in all fairness to Ord admit, had not previously occurred to anybody. But it was now mainly the Dark Lady's turn to seduce investigators, just as three centuries earlier she had seduced the poet and his friend. The Mary Fitton theory received a new twist from Charles Creighton (*Shakespeare's Story of His Life*, 1904), who proposed that Sonnets 100–54 represent 'the successive stages of a strange attempt by the poet to persuade Lord Pembroke to father an impending infant [Mary's] which was not his. The alleged real father . . . is darkly indicated in S. 124 as a certain courtier or statesman; and in the upshot it appears that Shakespeare himself was impeached of the paternity, falsely as he alleges, and upon suborned information.'[60] Perhaps, as occurs to the Variorum editor of the Sonnets, Creighton's strange attempt helped, by its very extravagance, to discredit the pretensions of Mary Fitton; but of course her fortunes were tied up with those of Pembroke. She fades from controversy, supplanted by new temptresses.

Of these Mistress Davenant seems to have won the most hearts. In its original form Arthur Acheson's theory, first made public in 1913, is remarkable chiefly for coming so late. As far back as the seventeenth century Aubrey noted that Shakespeare, while on the road between London and Stratford, had bedded down at the Crown Tavern in Oxford; that the innkeeper's wife was a very beautiful and witty woman; and that her poet son boasted of being Shakespeare's by-blow. The scandal was repeated by Anthony Wood and others. Not, however, until the appearance of Acheson's *Mistress Davenant the Dark Lady of Shakespeare's Sonnets* was the name of this amiable hostess entered formally into the competition. A decade previously, in *Shakespeare and the Rival Poet*, Acheson had convinced himself that the enigmatic *Willobie His Avisa* referred to Shakespeare and Southampton, and that Avisa represented the Dark Lady. From obscure references in the poem Acheson deduced that she kept an inn, the George or St George and the Dragon. Later the fact that *Willobie His Avisa* is dated from Oxford led him to seek the model for Avisa in a tavern there, and Aubrey and Wood furnished the missing link in the chain of evidence.

Acheson sees Southampton's affair with Mistress Davenant as having its inception in the autumn of 1592. That September he accompanied the Queen and Court on a progress to Oxford where, the town being crowded and accommodation at a premium, the Earl put up at the George Inn on Cornmarket Street, then run by the Davenants. (Not until 1604, in fact, did John procure a city licence to vend wine, and he was never landlord of an inn called the George; but such details escape Acheson's notice.) In this fashion, according to Acheson's romance, the Dark Lady entered Southampton's life. Eventually she would enter literature: not only as the enamorata of the Sonnets, but also as 'My hostess of the tavern' in *1 Henry IV*, Hermia in *A Midsummer Night's Dream*, 'black-eyed Rosaline' in *Romeo and Juliet*, the faithless whore of *Troilus and Cressida*, and the Serpent of old Nile in *Antony and Cleopatra*. These free-wheeling identifications remind us that Acheson belonged to the generation of Frank Harris, and, like him, stood apart from the professors.

These at once raised objections. There is no certainty that the W.S. of *Willobie His Avisa* really stands for Shakespeare, or that in the poem he is in love with Avisa, or that this odd production furnishes a commentary on the Sonnets. Several reviewers tore Acheson's book to shreds. Still the author won converts. He is named by Shaw in his celebrated preface to his Dark Lady skit; had he wished to be with the times, Shaw admits, he would have made Jane Davenant, rather than Mary Fitton, his heroine, but the playlet called for a jealous scene between the poet's mistress and Elizabeth, and the Queen was not known to have frequented Oxford taverns.

Acheson did not rest on his dubious laurels. His *Shakespeare's Lost Years in London 1586–1592* (1920) mainly sets out to prove that the dramatist came to the capital as the bonded servant to James Burbage for a term of years, and was thus early connected with the company that would become the Lord Chamberlain's Men. But obscurely, in a footnote, Acheson announces his startling discovery that John Davenant married twice, his first wife, the *femme fatale* of the Sonnets, being the daughter of the mayor of Bristol. More than sufficient amplification followed, two years later, in the 676 pages of *Shakespeare's Sonnet Story 1592–1598*.

Acheson relates how he sought, from clues in *Willobie His Avisa*, a woman whose name had the initials 'A.D.', the daughter of a tradesman who was also the mayor of a western town legendarily associated with St Augustine. Now, since *Avis* in Latin signifies 'bird' (then, as now, it was also a proper Christian name, but that would make matters too easy), Acheson looked for a tradesman with the surname of Bird. His investigations—or, to be more precise, the investigations he commissioned—drew him on to Bristol, linked by tradition with Augustine's Welsh campaign. Here he found Shakespeare's Bird: a draper named William Bird had held office as a mayor of the town, and in his will, preserved at Somerset House, he left bequests to five 'natural' children. A daughter named Anne received £500, but she will not do for

Acheson, whose satisfaction with an hypothesis is dependent upon its degree of complication. Taking 'natural' to mean 'illegitimate' rather than 'lawfully begotten' (its more usual Elizabethan sense), he furnishes Bird with a bastard daughter Anne, whose nominal father was William Sachfielde, a mercer of Bristol. But we know that Davenant's wife had the Christian name of Jane; therefore the Oxford vintner must have married twice, 'first in 1591–2 to Anne—daughter of Mayor Bird—who succumbed to the temptations of her calling as the hostess of a popular Elizabethan tavern or inn . . . and who died sometime before 1600; Davenant marrying again at about the latter date; his second wife being named Jane, and, according to contemporary report, a highly moral and virtuous woman.'[61] Thus does Acheson preserve unbesmirched the reputation of good Mistress Jane by creating, from whole cloth, an imaginary first wife, miraculously transferred from the west country to Oxford. Such pyrotechnics of rationalization must be a Baconian's envy.

The impact of Acheson's reconstruction is blunted by the heavy dense fog of his prose—torpid, shapeless—through which drizzle such expressions, redolent of certitude, as 'clearly shown', 'demonstrated', 'palpable', 'in all probability'. Perhaps the complaisant prolixities lulled Sidney Lee, never the most acute of critics, into allowing, in the *Year's Work*, that Acheson had 'good ground' for his lunatic marriage tale. But few other reputable scholars suspended disbelief, and the vintner's non-existent first wife soon joined other discarded shades.

She enjoyed another incarnation, however, in the more congenial surroundings of overtly fictionalized romance. Anne figures passionately in the Comtesse de Chambrun's *Two Loves I Have* (1934). 'Mistress Davenant thinks as I do', husband John harangues his tavern cronies, '—not agreeing however that the Lady of whom they [the Sonnets] speak must be Mistress Fytton nor yet French Jackline, the printer Field's wife. Nan sayeth that the dark lady will never be known, and doubts whether she ever really existed outside of Will's imagination. . . .'[62] So the cuckold persuades himself; but we, as readers, have seen Nan—her smouldering black eyes ringed with red—implore Will to take her with him to the world's end: 'For behold! I am no ale-house wife, nor yeoman's daughter of the sleek midland shires, but Queen of all Egypt.'

In the pages of Elizabeth Goudge's 'The Dark Lady' (subtitled 'A story Based on the Truth'), the imaginary daughter of Bristol's mayor exists with a particularity that the biographer, slave to evidence, can only admire:

She was a glowing brunette with dark sparkling eyes, and black hair with copper lights in it that curled most wickedly round her enchanting face. Her skin was a clear golden colour, flushed with pink like an apricot on a sunny wall, and her red lips were richly curved and deeply indented at the corners. She had dark winged eyebrows and long silky lashes that were like curling fans. She had a dimple in her chin, and though she was light and airy on her feet as thistle-down she had at the same time a delicious robin-like plumpness. She was quick-tempered, warm-hearted, gay and witty, yet now and then thoughtful. She was a wild creature, but yet there were times when she could be most exquisitely gentle. She was made of fire and dew, cold snow and hot sunshine. She was all sorts of women rolled into one, one of those in whom every type and condition of man seems to find his ideal.[63]

Little wonder that in her presence Shakespeare's visage takes on the look 'sometimes seen on people's faces when they were saying their prayers in church, or when they were looking at a sunset or a harvest moon'. But the romantic idyll does not last. A storm breaks in the respectable, cliché-haunted Davenant household, Anne is turned out of doors and dies young, and her lover is left to console himself 'with the thought that we and our sorrows are, after all, only such stuff as dreams are made of'. He would find a use for that reflection.

Along with fantasy-spinners, relentless literalists for whom the Dark Lady was dark indeed; their Shakespeare might have said, with the master of Locksley Hall, 'I will take some savage woman, she shall rear my dusky race!' In his Aryan retreat an obscure German translator, Wilhelm Jordan, apparently first introduced a racial aspect to the already complicated relationships of the Sonnets. In *Shakespeares Gedichte* (1861) he asks whether the seductress whose breasts were dun (dark brown) and has wires (hence twisting, curling) was not a mulatto or quadroon from the West Indies, with African blood coursing through her veins, and the musical sense common to her rhythmic race. Few, if any, seem to have taken any notice of Jordan, and not until 1933 did G. B. Harrison independently broach a similar theory in *Shakespeare under Elizabeth*, a biography offered (the preface forewarns) without scholarly diffidence as an imaginary portrait. 'She was a courtesan', Harrison says of the Black Woman, 'notorious to fashionable young gentlemen of the Inns of Court who took their pleasures in Clerkenwell; and for a time Shakespeare became her lover.'[64] A note draws attention to the Gray's Inn Revels of the 1590s, in which mock homage was paid to the Prince of Purpoole by (among others) '*Lucy Negro, Abbess de Clerkenwell*'.

Harrison's suggestion is taken up, without reference to him, by Edgar I. Fripp in his monumental posthumous biography, *Shakespeare, Man and Artist* (1938). To Fripp the Dark Lady is an amalgam of half a dozen 'ladies' of the tavern, kitchen, and drawing room. It was the poet's lot, as an actor, to be cast among the whores of Shoreditch (not, rest assured, as a client); he knew the drinking-bars and brothels of the theatre district, and in writing *Venus and Adonis*, no less than the Sonnets, drew his inspiration in part from Lucy Negro, the infamous head of the sisterhood of Black Nuns of Clerkenwell, ever ready with the burning lamps of venereal disease to chant *placebo*—'I shall gratify you'—to randy gentleman law students of the Inns.

This identification, like the others, is of course pure fancy. Better perhaps to abandon the pursuit and conclude, with Simone Arnaud, 'Point n'est besoin de rechercher le nom de la femme. Shakespeare l'a écrit d'un mot: c'est Perfidie!'\*⁶⁵

# 11
## *Edgar I. Fripp*

NOTWITHSTANDING his indiscretion with Lucy Negro, which however is symptomatic, the Revd Edgar I. Fripp is a distinguished scholar in the great Stratford antiquarian tradition; the last in line of lineal descent from Robert Bell Wheler. From boyhood Fripp loved and ardently studied Shakespeare, but his earliest publications after graduating from London University in 1893 were two unremarkable tracts on the origin and composition of the Book of Genesis. Later he lectured in University extensions in London, Liverpool, and Belfast. In 1925 he became a Life Trustee of Shakespeare's Birthplace, and in 1930 William Noble Research Fellow at Liverpool University. Like Wheler's, his was an uneventful life which allowed ample leisure for rambles in the Shakespeare country and uninterrupted hours in muniment rooms. To strangers visiting the Cotswolds he must have presented an odd sight as, in full clerical garb, he roared by on his motorcycle, his long white hair flying in the wind.

Such an enthusiast we might expect to see represented in *Notes and Queries* by the occasional brief article rich in precise information about some cousin, twice removed, of the poet's great aunt. He does not disappoint us: between October 1920 and April 1921 Fripp contributed to that estimable repository of antiquarian lore seventeen articles under the apt general heading, 'Among the Shakespeare Archives', and concerned with such subjects as 'Widow Townsend of the Wold', 'Sir Thomas Hargreave, Vicar of Snitterfield', and 'The Expulsion of Master William Bott from the Stratford Chamber'. But

---

\* The pursuit is not abandoned by David Scott, author of an article entitled, with poignant confidence, 'Shakespeare, Essex, and the Dark Lady: Solutions to the Problems', which appeared in the *Dalhousie Review* (1969). In it, Scott puts forward that 'extraordinarily interesting person', Jean de la Kethulle, wife of John Daniel (or Daniell) of Daresbury. It seems that the poet seduced her in The Hague while serving Leicester as junior aide, player, and spy—activities regrettably unknown to previous biographers. If Mrs Daniel describes her liaison with Shakespeare in her 'partial autobiography', now a Public Record Office manuscript, Scott neglects to mention this presumably relevant fact. He does, however, inform us that Shakespeare witnessed the death of Sidney, left Leicester for service with Essex, counselled the 'love-crazed' James VI of Scotland, and helped bear the canopy during that monarch's wedding ceremonials in Oslo (a stone's throw from Elsinore). This account—seasoned with references to James Bond, Vyshinski, Profumo, Mata Hari, and the Cuban missile crisis—might pass as a tolerable parody of the wilder excesses of Sonnet pseudo-scholarship, but it is apparently not written with tongue in cheek.

fortunately Fripp, who knew Warwickshire and its records better than anyone of his own day or since, set a more ambitious scholarly programme for himself. With his friend Richard Savage, Deputy Keeper of the town archives, he edited four volumes of *Minutes and Accounts of the Corporation of Stratford-upon-Avon and Other Records, 1553–1620* (1921–9), a work highly serviceable to biographers of Shakespeare; on his own he produced four slender volumes: *Master Richard Quyny* (1924), *Shakespeare's Stratford* (1928), *Shakespeare's Haunts Near Stratford* (1929), and *Shakespeare Studies Biographical and Literary* (1930). These writings he intended as prelude to a *magnum opus* that would sum up a lifetime of devoted research. Death terminated his hopes in November 1931, but by then his great project was almost completed. Seven years later F. C. Wellstood, the Birthplace librarian and Fripp's 'old friend and comrade in scholarship' (so the minister's wife described him), saw the work through the press.

Although he chronicled the London years, for Fripp the dramatist was always Shakespeare of Stratford. Throughout the plays and poems this local partisan saw reflections, conscious or unconscious, of the Warwickshire setting, which, he devoutly believed, was revealed no less intimately in the Shakespeare canon than medieval Florence in the *Divine Comedy*. The artist who leads the life of the imagination runs the risk of forfeiting his sanity, and Shakespeare—oppressed by rehearsals and performances and by the pestilence and summer heat of the metropolis—would flee to Stratford as to sanctuary, there to be soothed by the quietude of Avon and Arden, and the fellowship not of players but of ordinary honest folk, friends and neighbours. So Fripp persuaded himself, although he attempted (never very strenuously) to steer clear of Stratfordolatry.

Stratford is context; its history, a chapter in the history of Warwickshire, the Midlands, even of England. The very name Stratford-upon-Avon takes on the glamour of a remote past, conjuring up the English settlement at the *ford* where the Roman *street* crossed the river called by the Welsh *afon*. Fripp savours the Welsh place names of rivers, hills, hollows, and camps converted by the Romans into strongholds; also the Welsh names of Stratford families of Shakespeare's time, including the Fluellens, humble townsmen mainly, but numbering among them an alderman, Lewis ap Williams. The history at Shakespeare's door also stood outside Fripp's—a history evoked for him, as for Charles Knight in the previous century, by the associations of locales: Warwick Castle, seat of the Kingmaker; the battlefield of Evesham, where Simon de Montfort fell; Bosworth Field, where Henry Tudor defeated Richard Crookback and brought to an end the Wars of the Roses. But mainly Fripp finds himself drawn to settings with biographical rather than political connections. The footpath from Back Lane leads him, as it did the young poet, through tilled fields and past the grounds and dovecotes of the Manor Farm to the hamlet by the stream where stands the solid double messuage, or dwelling, of Anne Hathaway. So too Fripp's steps take him to Bidford-upon-Avon, to Luddington and Bishopton, to Wilmcote, Rowington, Snitterfield, and Henley-in-Arden. But people fascinate him more than topography.

They are for the most part folk overlooked or but slightly noticed by history. In examining such lives Fripp follows the example of Stopes, whose 'fearless researches' he generously acknowledges; but he is the more reliable scholar. Some of his subjects are relatives of Shakespeare (Uncles Harry of Ingon, and Edmund of Barton-on-the-Heath, and Cousin Robert Webbe of Snitterfield) or of the Hathaways: Richard, Anne, and Bartholomew, as well as their cousins—Frances, for instance, who married David Jones of Stratford, fondly memorialized by Fripp for supervising 'a Whitsun Pastime' in 1583. Then there were the Hathaway neighbours, among them Roger Burman, with plentiful corn and hay in his barns, lands sown with wheat and peas, and a hall richly furnished with brass and copper and pewter; maybe he knew young Shakespeare when he came a-courting Anne. In Stratford William Greenaway the carrier lived nearby to the Shakespeares on Henley Street. A draper, he owned two small shops in Middle Row, and augmented his income by conveying goods—cheese, brawn, conyskins, woollen shirts, and hose—on pack horses to London. On one occasion he delivered a letter and 'two cheeses, a loaf of bread and five shillings in money' to John Debdale's son Robert, imprisoned in the Gatehouse at Westminster. Perhaps the latter came under the influence of the Catholic master Simon Hunt when he attended Stratford grammar school a few years before Shakespeare's enrollment, for Debdale left England and was admitted to the Society of Jesus in Rome. Eventually he was tried and executed in London as a seminary priest who had practised exorcism. Robert Southwell, the Jesuit poet, reported that Debdale met his martyrdom 'most steadfastly', as in time he would himself. Such are the individuals that Fripp rescues from oblivion by means of wills, conveyances, court records, an occasional letter. 'An inventory', he once said, 'is a dead list or a living picture according to the artist in the reader.'

These lives help to establish a matrix for biography. A few have more obvious relevance to Shakespeare even if they do not always impinge directly on his career. In the days when his father served as constable and helped to keep the peace, John Bretchgirdle came to Stratford as vicar. A man of Calvinist leanings who had been schoolmaster at Witton, he shaped the religious life of the town as, with the whitewashing of frescoes and removal of stained glass it made the transition from Catholicism to Protestantism. Bretchgirdle baptized and buried Margaret Shakespeare, and on 26 April 1564 he officiated at the christening of her brother William. One of the parson's devoted pupils from the Witton days, John Brownsword, followed him into Warwickshire and on Sunday, 1 April 1565, signed a contract with the bailiff and burgesses of Stratford to 'serve in their Free School, as a good and diligent schoolmaster ought to do for the term of two years from the date above said . . . in consideration of the sum of £20 yearly and his dwelling-house'. Brownsword had a flair for Latin verse (which Fripp reproduces and translates), but he did not instruct Shakespeare, for the schoolmaster kept his post for little over two years before moving on. A sadder case is that of Katharine Hamlet, spinster, who drowned in the Avon at Tiddington, about a mile east of Stratford, in 1579. At the inquest it was concluded that she did not wittingly seek her own damnation, but had accidentally slipped and fallen into the water. Did Shakespeare (Fripp asks) think of Katharine Hamlet long years after, when his Ophelia sank to muddy death? An intriguing speculation, but no more.

The Shakespearian pertinence of the Quineys is another matter. Adrian Quiney was John Shakespeare's neighbour in Henley Street for half a century; both were yeomen and tradesmen who ascended the gradations of municipal preferment together. Adrian's son Richard wrote the only surviving letter, a familiar one, to the dramatist, and his son Thomas married Shakespeare's younger daughter Judith. The union of the two families was cemented in the name they bestowed on their first infant, Shakespeare Quiney. Such a family merits a book (the records are ample) and in *Master Richard Quyny* Fripp, the ideal biographer for the occasion, furnishes it.

These studies are subsumed by the posthumous Life. 'My endeavour', Fripp introduces his book, 'has been (and it is but an endeavour) to see Shakespeare in his context—to study and interpret him in the light of his environment, geographical, domestic, social, religious, dramatic, literary.' So he had always endeavoured: not surprising that he should reproduce much of the matter, often word for word, of his earlier publications. He goes on:

'I have approached him from the end of his antecedents, not of his sequel, from the end, that is, of the first half of the sixteenth century, not of the second half of the seventeenth: on the side of Latimer's Warwickshire and the Reformation in Stratford, the old morality Drama still going strong, hand in hand with the preaching of "God's Word", and the reading for the first time in English of the Bible as a new and sacred discovery—not on the side of the Restorationist gossip, a degenerate and discredited Stage (from which Ben Jonson had turned in disgust), and an exhausted Puritanism.'[66]

The last phrases reflect the temper of the pious oralist, and betray a bias which warps the fabric of the enterprise; this critic will be offended by the indecencies of the Sonnets, that 'mass of fictitious nonsense'. There are other disabilities. By occupation and habit of mind the antiquary attunes himself to detail, the small currency of local history; but an ambitious biography running to almost a thousand pages requires proportion, a sense of the whole, and steady narrative thrust. These skills elude Fripp, and his book breaks down into bits and pieces—163 chapters in all. It is rather as though a gifted miniaturist found himself confronted with the task of decorating the ceiling of the Sistine Chapel.

The antiquarian tradition, moreover, is essentially amateur, made possible by leisure and sustained by enthusiasm and local pride. Such amateurism has its glories, but the modern biographer of Shakespeare is confronted by fierce methodological problems which call for professional expertise: he must sift fact from legend, distinguish among possibility, probability, and certainty, and tread delicately in relating the corpus to the life. Too often in Fripp's pages scholarly discretion yields to sentimental fancy or unwarranted certainty. The young Shakespeare 'surely' witnessed the Princely Pleasures at Kenilworth in July 1575. His father, of course literate, emerges as an obstinate Protestant recusant against episcopacy, symbolized by that little black martinet Archbishop Whitgift; the documented decline in John Shakespeare's fortunes in his later years was a dodge, abetted by his fellow burgesses, to mislead the authorities about his failure to attend church services. For this improbable reconstruction there is no more evidence than for Fripp's identification of the godparents at every christening in the Shakespeare family. The future poet was 'most manifestly' trained in an attorney's office, probably that of Henry Rogers, the town clerk. Anne Hathaway is described thus:

Godly, quiet, clinging, frail, is our thought of the great Poet's helpmate, pious like her daughter, silent like Virgilia ('my gracious silence'), wifely as his best-loved heroines are (Perdita and Imogen above the rest), delicate (the mother of only three children that lived, and only two who survived her), infirm in March 1616 when he added to his will the affectionate little bequest of the 'second-best bed with its [sic] furniture', to ensure her possession of the fourposter and chamber which they had shared in New Place.[67]

Hard to recognize in this portrait of a saintly semi-invalid the passionate (or calculating) village lass who took to her bosom and her bed, without benefit of clergy, a youth eight years her junior—and who outlived him by another seven.

Restless in the face of dry technical arguments, Fripp could never proceed beyond a romantic and uncritical love for his theme. To him municipal records became the *madeleine* which retrieved the golden treasury of the past. Even the penmanship of the Elizabethans evokes a nostalgic tribute: 'They wrote the old English (Gothic) hand, and for the most part wrote it beautifully; with a quill, as pliable and skilful an instrument as the modern fountain pen is stiff and mechanical [what would Fripp think of the ball-point?]; in ink that was ink ('incke' as Shakespeare spelt it), and on paper that was paper, worthy to hold a good thought for centuries.'[68] *Shakespeare, Man and Artist* Fripp calls his volume, but in his pages the man and the artist sometimes coalesce with surprising results. Celia's description of her 'sheep-cote fenced about with olive trees' in *As You Like It*—

> West of this place, down in the neighbour bottom.
> The rank of osiers by the murmuring stream
> Left on your right hand brings you to the place.   [IV. iii. 79–81]

—is 'unmistakably' the Anne Hathaway homestead, a photograph of which duly appears in the text with the simple legend, 'Celia's Cottage'.

Politically a Liberal and in religion a devout Unitarian, Fripp in his last paragraph sums up Shakespeare as 'supreme among English laymen for his Reverent Liberalism'. Not everyone will recognize the poet in these words, although Fripp himself is clearly present. If the portrait is biased, it is also, despite the massive scale, incomplete (the unedifying Manningham anecdote about the dramatist's assignation with a citizen's wife escapes mention). Judged by the highest biographical standard, *Shakespeare, Man and Artist* amounts to a grand failure. Due emphasis, however, should fall on the grand. On some matters—local customs, persons, places—pre-eminently requiring the antiquary's gifts, the two volumes comprise a unique treasury of information. In this work, taken along with his others, Fripp demolishes as was his purpose Halliwell-Phillipps's canard about rude Stratford: the bookless neighbourhood inhabited by illiterate peasants. For ordinary readers this Life makes for heavy going; probably few have struggled through to the end without skimming. But subsequent biographers would find *Shakespeare, Man and Artist* an invaluable source-book, despite the grossly inadequate index, and would help themselves (with or without acknowledgement) to its antiquarian riches.

# 12
## *Joseph Quincy Adams*

THE tercentenary of the First Folio in 1923 did not pass unnoticed. A somewhat anachronistic survival from the previous century, Sir Sidney Lee graced many a celebration as the *doyen* of Shakespearians. The Worshipful Company of Stationers and the British Museum mounted suitable exhibitions. But the nobler tributes were the published fruits of scholarship. A quintet of distinguished students—A. W. Pollard, W. W. Greg, Sir Edward Maunde Thompson, J. Dover Wilson, and R. W. Chambers—pooled their expertise and critical powers to make an impressive case for *Shakespeare's Hand in the Play of*

*Sir Thomas More*. The three pages of Addition D reveal Shakespeare in a new professional relation: the collaborator called in by four fellow dramatists (Munday, Dekker, Chettle, and possibly Heywood) to salvage a play they were having difficulty in getting past the censor. The tercentenary also elicited a major American biography. The first such work produced in the former colonies to win respectful attention in England, J. Q. Adams's *Life of William Shakespeare* was also the first serious rival to Lee's biography.

Adams's early years seemed hardly calculated to go to form a distinguished savant. The son of a Baptist preacher, he attended schools in various Southern states as his father moved from one pastorate to another. But after graduation from Wake Forest College in 1900 he received a first-class professional training on the modern Teutonic model. At the University of Chicago he enjoyed a period of study under the great medieval and Elizabethan scholar John M. Manly. Adams's preparation also included a year of advanced work in London before he took a Ph.D. in 1906 at Cornell University. This was followed by a summer at the University of Berlin, where the eminent Dr Brandl held sway. Adams taught for many years at Cornell and rose rapidly through the ranks from Instructor to Professor. He collected books, fished and shot pheasant and quail, and (being of orderly scholarly habits) amassed an exhaustive card-index file for the Elizabethan period. In 1917 he edited *The Dramatic Records of Sir Henry Herbert, Master of the Revels, 1623–1673*, and in the same year published *Shakespearean Playhouses: A History of English Theatres from the Beginnings to the Restoration*—a survey which, although outdated, has not yet been altogether superseded. Thus the student still turns with profit to Adams's discussion of the decades of litigation—comprising bills, answers and depositions, and decrees and orders—respecting The Theatre, its construction, tearing down, and carting away, and the land on which it stood, and the lawsuits between the Burbages and their Shoreditch landlord, Giles Allen: all this complicated and theatrically momentous history, which the Shakespearian biographer will scarcely find irrelevant. These efforts prepared him for the large-scale biography. Honour came to him when in 1934 he was appointed the first Director of the Folger Shakespeare Library. There he once demonstrated his unflappability by calming an hysterical lady suffering from Baconianism, who screamed to the visitors in the Exhibition Gallery that the Library, being a monument to a lie, should be pulled down. Adams disregarded her advice and went on to help create the greatest specialized collection of its kind in the world.

If Fripp is the representative antiquary, Adams belongs firmly in the academic tradition at a time (although recent, already remote) when the university was a place removed, neither knowing nor caring that the world passed, and professors, often unmarried, endured stoically their genteel poverty. Adams's scholarship typifies both the strengths and limitations of such an ambience. In the peaceful isolation of Ithaca, by the quiet waters of

Lake Cayuga, he wrote a biography which—understandably enough, given the circumstances of composition—breaks no new ground on the documentary side. Anyway Adams felt (somewhat prematurely, as it turned out) that the end of the line had already been reached in record-office research and that a period of stagnation had set in.

If he resists the allure of primary resources in England, he remains in other respects too the outsider. Sometimes he betrays his alien status by a trifling slip, as when, in a phrase that falls oddly on the British ear, he refers to a 'Fellow of Oxford'. More serious is his misconception of the historical importance of Stratford. He describes the corporate borough as a 'little village', and sees it as peopled by yokels who supplied a pattern for Dogberry, Sly, and Bottom; even the splendid Collegiate Church of the Holy Trinity is patronizingly dismissed as 'pretentious'. An eminent theatrical historian should have known better. How, one wonders, did Adams account for the fact that the outstanding professional troupes of London included this nondescript hamlet in their provincial itineraries?

But he displays solid learning too. He has at his fingertips over two centuries of Shakespearian scholarship,* and, to illustrate an historical point, he can pluck at will an apt passage from a rare sixteenth- or seventeenth-century printed text: Ferne's *The Glory of Generosity* (1586) or Davies's *Wit's Bedlam* (1617) or Hoole's *New Discovery of the Old Art of Teaching School* (1659). Although no technical bibliographer, Adams grasps (as Fripp did not) the significance of recent work by Pollard, Greg, and Dover Wilson on the transmission of Shakespeare's text. A novel feature of this *Life* is the concluding quartet of chapters on the making of playhouse manuscripts and the printing of the Quartos and Folio. These sections, in which incidentally Adams accepts as Shakespeare's the *More* fragment, constitute the first popular treatment of a complex and specialized subject.

The task of Shakespeare's biographer, as Adams sees it, is to consolidate and interpret; he must somehow fit together disconnected pieces of knowledge into a meaningful pattern and present the assembled picture with clarity. This means placing the dramatist in his context, which for the author of *Shakespearean Playhouses* is London and its theatres:

However much we may think of him as a genius apart, to himself and to his age he appeared primarily as a busy actor associated with the leading stock-company of his time; as a hired playwright—often, indeed, a mere cobbler of old plays—writing that his troupe might successfully compete with rival organizations; and, finally, as a theatrical proprietor, owning shares in two of the most flourishing playhouses in London. Thus his whole life was centred in the stage, and his interests were essentially

---

* Despite his professionalism, however, Adams does not always get his facts straight: he misdates *The Malcontent* and *The Atheist's Tragedy*, and confuses the two versions, one perhaps a Bad Quarto, of *The Taming of the Shrew*.

those of his 'friends and fellows', the actors, who affectionately called him 'our Shakespeare'.[69]

If the play-cobbler reference causes unease, who will gainsay the propriety of the general approach? Yet Adams is the first major biographer to think of Shakespeare mainly in terms of his theatre. Throughout Adams discusses the companies and their fortunes, and he includes whole chapters on the playhouses and on the rise of professionalism in the drama. Prior experience has equipped him well for this part of his task, but it is Adams's misfortune that his work appeared in the same year as that most monumental of all treatments of the subject: the four volumes of E. K. Chambers's *The Elizabethan Stage*.

In his preface Adams wrinkles up his nose at the 'fanciful speculation' and 'bizarre' hypotheses of the subjective school. Further on he condemns, in the most forcible terms, 'that specious type of scholarship which seeks to disclose the life of so practical a man and objective a poet as Shakespeare by a closet examination of his plays'.[70] Thus he casts his lot with the impersonal school. The identity of Mr W.H., that great ink blotter, interests Adams not a whit, and he refuses to waste a paragraph on the subject (the Sonnets, it goes without saying, are literary exercises).

When Shakespeare embarked upon his tragic phase, he was not wallowing 'in the depths' but was, rather, the mature artist—ambitious, confident— stretching his wings in the most demanding of dramatic forms. And if, with *The Tempest*, he bade the stage farewell through the vehicle of Prospero, this biographer is not one to remind us. Aesthetic criticism, indeed, he avoids as much as possible; a wise decision, for Adams's gifts that way are at best inconsiderable. What can we think of the sensibility of a reader for whom the impassioned Dark Lady poems represent 'a laughing satire on the amorous cycles of the day'?

The Lost Years provide a test of mettle. Adams is the first biographer of note to build upon Aubrey's remark, derived from the actor Beeston, that Shakespeare 'had been in his younger years a schoolmaster in the country'. Acceptance of this possibility frees one from the painful necessity of separating a young husband from his wife and small children. Moreover, *The Comedy of Errors* is based upon Plautus, who had a place in the grammar-school curriculum; both this play and *Love's Labour's Lost* include pedants in their *dramatis personae*, both contain echoes of the classroom and tags from school textbooks. But Adams clearly best relishes his final, triumphal argument: 'Lastly, not to exhaust the reasons that might be cited, a career as schoolteacher would splendidly equip the non-university trained Shakespeare for his subsequent career as a man of letters.'[71] Why such should be the case is not self-evident, at least to this writer, and one suspects that the soul of the pedagogue in Adams has surfaced.

His Life, it must be allowed, is studded with presumptions, put forward with some such formula as 'probably' or 'we may suppose' or 'it is more than likely' to remove the stigma of fictionalization, and intended to show the hero and his kin in an exemplary light. Upon Will, the 'fond young parents' lavished all their affection. At his mother's knee the child—auburn-haired, with big hazel eyes, rosy cheeks, high forehead, and gentle disposition—acquired that noble idea of womanhood (Tamora? Goneril? Lady Macbeth?) which informs his plays. When school let out he ran whooping and hallooing home—that is, when he did not skip across the fields to visit the house of Uncle Harry and Aunt Margaret, on whose farm he picked up some of his extensive knowledge of rustic types and country living. The poet married for love, not for money or out of compulsion. On the extremely doubtful basis of the tax assessment on Shakespeare's London residence, Adams concludes that he had his wife and children with him in Bishopsgate. In the summer of 1596, however, he packed them off to Stratford for the fresh air, and at a prematurely early date he deprived himself of the gratifications of a brilliant career in order to return to the bosom of his family and breathe the fresh air with them (elsewhere, however, Adams suggests that Shakespeare retired because of Bright's disease). While still in London, he 'no doubt' enjoyed the tavern hospitality of noble lords—Southampton, Rutland, Pembroke—or suitably distinguished knights (Ralegh or Salisbury); that is, when he was not being entertained at the Mermaid by Jonson, whose release from prison he helped secure. How thankful must we be that Shakespeare absented himself from conviviality long enough to write his plays. The success of *Venus and Adonis* was equalled by that of *The Rape of Lucrece* (it was not), and the author achieved recognition in his own lifetime as England's pre-eminent man of letters (he did not). The Essex episode distressed Shakespeare greatly, and he never forgave Elizabeth her callous treatment of her erstwhile favourite (if this is so, she seems not to have taken notice).

Thus it goes, and the sentimentality is aggravated by the author's spinsterish pudency. (He aligns himself with the 'wiser sort' who objected to the excessive lasciviousness of *Venus and Adonis*.) Occasionally one longs—terrible admission!—for a countervailing dose of Frank Harris. Scoundrel and wild man, Harris at least portrayed a Shakespeare with blood flowing in his veins, rather than an aqueous infusion of chalk and ink. We recall that a Baptist preacher's son wrote this Life, and are inclined to sympathize with the reviewer who thought that in Bunyan, Adams might have found a more congenial subject.

Perhaps *A Life of William Shakespeare* is best regarded as, in serious biography, the last gasp of an expiring bardolatry. It is now badly dated, and even its clichés of style—'true love', 'golden opportunity', 'chorus of praise'—today appear calculated to produce an unbidden shudder. Yet, the idolatry excepted, it is a sane work, reasonable in tone and generally sound in the

evaluation of documentary evidence not susceptible of sentimental coloration. From fragmentary materials Adams has had the skill to fashion a
continuous and proportioned narrative which is readable, if not absorbing,
and at all times lucid despite the weight of detail. For two generations this
work would stand without rival as the biographical reading in the undergraduate Shakespeare course in US colleges and universities. From Adams
innumerable students would derive their sum of knowledge not only about
the dramatist but also about his theatrical environment. On a visit a while
ago to the Henry Suzzallo Memorial Library of the University of Washington
in Seattle, this writer noticed seventeen copies of Adams's *Shakespeare* in the
stacks. No subsequent Life had, at least so far, achieved a similar status.

# 13

## *Smart and Alexander*

EVEN during the palmy days of subjective biography, when the poet and his
Dark Lady were executing their *pas de deux* through the canon, the documentary record and its interpretation consistently engaged interest. As early as
1904 D. H. Lambert, solicitor and amateur Shakespearian, furnished in
convenient handbook form a chronological sequence of the texts of the extant
life records. A sort of poor man's Halliwell-Phillipps, *Cartae Shakespeareanae*
(not impeccably proof-read by Lambert) paved the way for other such
compilations: C. F. Tucker Brooke's *Shakespeare of Stratford* (1926), which
has had wider appeal, perhaps in part because of its modern spelling; also
Pierce Butler's altogether less notable *Materials for the Life of Shakespeare*
(1930). G. E. Bentley's *Shakespeare: A Biographical Handbook* (1961),
although arranged in narrative chapters, continues the tradition. During the
early years of the century Stopes and Wallace conducted their archival
researches; specialized monographs were produced by Gray and others—
Ernest Law, for example, who studied in detail *Shakespeare as a Groom of the
Chamber* (1910) and *Shakespeare's Garden* (1922), which has sections on such
burning issues as 'The Position of the Mulberry Tree'. In the second quarter
the critical examination of legends and traditions, the pursuit of new
information, and the work of synthesis proceeded apace. Three figures stand
out above the rest: Smart, Alexander, and Chambers.

'Among the scholars of his own generation', writes his memorialist, 'John
Semple Smart attracted little attention.' Such a summing up, discouraging to
most men, would have pleased Smart, for he studiously avoided attention; his
life was a placid stream into which no pebble was ever thrown. Appointed

Queen Margaret College Lecturer at the University of Glasgow in 1907, he remained in that post until his death, and rarely strayed, except for a Continental holiday, from his Scottish domicile. A bachelor and a solitary, he acquired a fastidious love of the best produced by Greece and Italy, but no mere aesthete, he improved his learning amid the austere surroundings of the record office. In his pursuit of exactness Smart would patiently wait years for a fact to turn up that would justify a conclusion of which he was already certain. His scholarly output, accordingly, was severely limited: a book on the Ossian controversy, an edition of Milton's Sonnets, the occasional article. A brief note on 'Shakespeare's Italian Names' (*Modern Language Review*, 1916) shows Smart's temper; resisting the temptation to re-create Shakespeare in his own image as a traveller to Italy, he contents himself with the suggestion that perhaps the poet picked up his knowledge of the language from Italian acquaintances residing in London. When he died in 1925 Smart left incomplete a draft of a study written in pencil, on stray sheets, in a difficult hand. His successor at Glasgow, Peter Alexander, transcribed the work for the press. Published in 1928 as *Shakespeare: Truth and Tradition*, this modest volume is responsible for Smart's considerable posthumous reputation.

He says much that later biographers would heed, and says it gracefully and sometimes with wit. With Fripp (who writes not nearly so well) he shares the desire to restore the reputation of Stratford. Shakespeare's town, Smart reminds us, was no mere rustic village but 'a small metropolis for the district round, and stood higher in relative size and rank among English towns than it does at the present day'.[72] Adams might have read this section with profit. In the great central chapter of the book, 'Things That Never Were', the principal legends comprising the Shakespeare–Mythos receive a cold, hard look, and more often than not crumble beneath Smart's unromantic gaze. He begins with the Ward anecdote about the dramatist's merry—and mortal— meeting with Drayton and Jonson, notes Ward's qualifying 'it seems', and wonders what we should make of an authority who elsewhere informs us that Milton was a Papist in disguise. So too Smart rejects Aubrey's improbable tale of the butcher's boy who slaughtered calves to the accompaniment of poetical declamations. The horse-holding episode fares no better (did Shakespeare, Smart asks with fine scorn, remit to his wife and three children in Stratford the coppers he picked up in his menial London occupation?). It is joined in Smart's limbo of rejected myths by the drinking bout with its aftermath under the Bidford crab-tree, and by Archdeacon Davies's poaching story. In connection with the poaching Smart produces a new record hostile to romance: the licence, from the Patent Rolls for James I, permitting Sir Thomas Lucy the third to maintain a deer park at Charlecote; the document, dated 1618, had eluded Malone, whose search of the Patent Rolls did not extend beyond Elizabeth's reign. Only one traditionary report receives respectful, brief notice. Of Aubrey's jotting about Shakespeare as a school-master in the country, Smart writes, 'No tradition can ever be implicitly relied on; but this is the best supported and most creditable of all.'[73]

But Smart's chief contribution appears near the end, in two fragments of an unfinished chapter. Here he takes up the notorious diatribe against Shakespeare in Greene's *Groatsworth of Wit*, and argues that the crucial term *Johannes Factotum* signifies not a jack-of-all-trades—and hence a refurbisher of old plays—but an individual of boundless conceit, i.e., an actor who had the audacity to set himself up as a play-maker. In this sense is the dramatist an upstart crow. A second fragment follows up this *aperçu* with the suggestion that *The True Tragedy of Richard, Duke of York* is not a Marlowe or Greene play revised by Shakespeare as *3 Henry VI*, but a mangled version of the latter.

The great bibliographer Pollard hailed *Shakespeare: Truth and Tradition* as 'a new landmark in Shakespeare scholarship'. Indeed it was. If Smart's pages no longer hold the excitement of novelty, that is because his views have met with so high a degree of acceptance. Few have had the temerity to offer dissent, the most notable of these being Dover Wilson, who puts forward ingenious but ultimately unavailing arguments in his essay on 'Malone and the Upstart Crow' in *Shakespeare Survey* 4 (1951) and in his introduction to *2 Henry VI* (1952) for the New Shakespeare. Smart has effectively carried the day. Not many scholarly books have had comparable impact.

But, then, few dons have been so fortunate in their disciples. At Glasgow, Smart taught and inspired Peter Alexander, and on his deathbed exhorted his brilliant pupil to follow in his steps. In particular he urged the young scholar to examine the text of *Henry VI*. He did so, and in 1929 published *Shakespeare's Henry VI and Richard III* with an introduction by Pollard, whose endorsement carried weight. Alexander begins with Greene's *Groats-worth of Wit*, restating with additions of his own his master's argument. Smart had dwelt on the *Johannes Factotum*; Alexander pounces on the 'upstart crow', the phrase harbouring (he suggests) an accusation not of literary theft but of the theft of a literary reputation.

This conclusion in turn becomes the first link in an elaborate chain of evidence. On the basis of irrefutable textual data Alexander demonstrates that the *First Part of the Contention of the Two Famous Houses of York and Lancaster* and *The True Tragedy of Richard Duke of York and the Good King Henry the Sixth* are Shakespearian Bad Quartos (actually the latter is an octavo) and not plays in which Greene had a hand. This finding leads to an affirmation of the integrity of *2* and *3 Henry VI*. Heminges and Condell, who had included these works in the First Folio, are thereby vindicated. It now becomes necessary to re-examine the grounds for denying Shakespeare other Folio plays, princi-pally the much maligned *Titus Andronicus*. The evidence is reviewed, and *Titus Andronicus* restored to the canon. This Senecan and Ovidian tragedy, along with *The Comedy of Errors* and *Venus and Adonis*, testifies to the young Shakespeare's absorption in Latin literature, and lends credence to Aubrey's story of a schoolteaching phase. Alexander's radical revision of the canon implies, moreover, that the date agreed upon since the time of Malone for the dramatist's stage début, 1591, is too late: Shakespeare may have been in London as early as 1586 (he could not then have instructed his pupils for very long). Such is the train of cause and effect which the upstart crow initiates.

In 1935, Alexander was appointed Professor of English Language and Literature at the University of Glasgow; in time he became Regius Professor at Glasgow. Alexander is gratefully remembered for his one-volume edition of the Complete Works for Collins (1951), which has often been reprinted. His later books, *Shakespeare's Life and Art* (1939) and *A Shakespeare Primer* (1951) contain many good things but reinforce our sense that, as a biographer, his importance rests chiefly upon his demonstration and extension of Smart's ideas about Shakespeare's early professional career. Alexander's last biographical excursion, *Shakespeare* (1964), for the Home University Library, reveals him as having lost none of his zest for scholarly controversy. The book is curiously lopsided—the dramatist's later life receives only perfunctory notice—and Alexander must have bemused his popular readership by crossing swords for several pages with Dover Wilson over the notorious crow and his borrowed plumes. But after the dust and feathers have settled, there is no doubt (at least in this reader's mind) that Alexander has carried the day. The edifice of argument in his 1929 monograph remains intact.

# 14

## *E. K. Chambers*

NOTABLE contributors though they be to biographical scholarship, Smart and Alexander are overshadowed—along with Adams, Fripp, and almost everybody else—by a single massive presence. The achievement of Sir Edmund Kerchever Chambers exceeds one's reasonable expectations of what may be accomplished in several lifetimes. After editing *Richard II* for school use and taking part in the Warwick Shakespeare, he prepared (from 1904 to 1908) single-volume editions of all the plays for the Red Letter Shakespeare; the introductions to these he later collected in *Shakespeare: A Survey*. For a time he wrote articles and reviews for the *Academy* and the *Athenaeum*, and during several months in 1904–5 was dramatic critic of the *Outlook* and the

*Academy*. For the Malone Society, which he helped to form and which he served as President, Chambers edited dramatic records from the Patent Rolls, the Privy Council Register, the Lansdowne Manuscripts, and the Remembrancia of the City of London. Over the years he published numerous scholarly and popular editions of a wide range of major and minor writers—Arnold, Donne, Beaumont and Fletcher, Milton, Vaughan, Aurelian Townshend, Landor—as well as collections of English pastorals, English lyrics, and fifteenth-century carols, not to mention *The Oxford Book of Sixteenth Century Verse*, and standard books on *The English Folk Play* and *Arthur of Britain*. To *Shakespeare's England* he contributed the chapter on 'The Court'. His fourteen-page distillation of available knowledge of Shakespeare for the eleventh edition of the *Britannica* in 1911 withstood changing fashion so well that before finally expiring it was retained, as revised by John Crow, in the 1968 *Encyclopaedia Britannica*; it must be the most widely consulted of all encyclopaedia articles on Shakespeare. Chambers's miscellaneous lectures have more than casual interest. One, 'The Disintegration of Shakespeare', read before the British Academy in 1924, effectively turned the tide against those—the disintegrators—who assigned passages and sometimes whole plays of Shakespeare to other hands. In its aridity Chambers's volume, *English Literature at the Close of the Middle Ages*, for the Oxford History of English Literature, betrays the author's weariness—but, then, he was almost eighty when he completed it. These accomplishments are but the pendants to his major works on dramatic history: *The Mediaeval Stage* (2 vols., 1903) and *The Elizabethan Stage* (4 vols., 1923). And these in turn are but preparation for *William Shakespeare: A Study of Facts and Problems* (2 vols., 1930). Apart from this tangible achievement, there is his silent contribution over the years as a frequent reader and adviser for his publishers, the Clarendon Press.

What staggers credulity, however, is less the sheer quantity of output, or even its overall excellence, than the fact that most of Chambers's essays, lives, editions, and syntheses were the fruit of his spare hours—the work, as he once said, 'not of a professed student, but of one who only plays at scholarship in the rare intervals of a busy administrative life'. For Chambers held a responsible full-time position, non-literary in character, as a public servant.

He was born at West Ilseley, Berkshire, on 16 March 1866. After being educated at Marlborough, he went up to Corpus Christi College, Oxford, where he read classics and (like Alexander) came under the influence of an inspiring teacher, Arthur Sidgwick, with whom he chased butterflies on the Cumnor Hills. Although Chambers took firsts in Honour Moderations and in Greats, and was awarded the Chancellor's English Essay Prize, he lost out in his bid for a fellowship; not a tragic disappointment, for his interests had already shifted from classical studies to English. A university post being unavailable—Oxford had no English School until 1894—he turned to government service. In 1892 Chambers entered the Education Department (soon to be reorganized as the Board of Education) as a junior examiner entrusted with day-to-day correspondence in the elementary branch. Despite an off-putting manner interpreted by friends as shyness and by provincial deputations as hauteur, he moved up through the hierarchy during a stirring period of reform; by 1921 he had reached the lofty position of Second Secretary. He was responsible, according to a former colleague, for the creation of the administrative machinery of the central office; it was Chambers who drew up the special regulations for university tutorial classes. His chief Morant wrote to him in 1912: 'I shall *never* forget your steadfast persistence and hard work and loyal devotion and splendid brainwork.'[74] When Chambers received his knighthood in 1925 it was as much in recognition of his services to education as to literature.

How did he manage to lead a double life so successfully? In his early days as a civil servant he would check in at the office in the morning, read his mail, dictate a few replies, and then adjourn to the British Library, returning to the office to sign the letters before going home; a long lunch break enabled him to compose most of an introduction to one of his Shakespeare editions. But in later years, as his responsibilities increased, his burdens became onerous. Having put in a depleting day at the office, Chambers would retire to his study after dinner, leaving any company behind, to push on with *The Elizabethan Stage*. This project occupied him for twenty years. Small wonder that he should sometimes feel tired and played out, or long for those placid academic surroundings where scholars sit 'from morn to eve, disturbed in the pleasant ways of research only by the green flicker of leaves in the Exeter garden, or by the statutory inconvenience of a terminal lecture'. But he did not often allow himself the luxury of such reflections. Chambers must have had a will of iron. His mind was quick, and his pen fluent; he possessed an administrator's capacity to organize his work and keep his desk clear. His physical constitution too must have been superb. A photograph reveals him as a man of handsome, somewhat rugged features, with unlined face and a full head of black hair; he might pass for thirty-five but was fifty when it was taken.*

It may be said that the virtues of a high-grade civil servant—the organizational and synthesizing powers—distinguish his literary labours. Speculation he regarded as a manifestation of the scholar's melancholy: facts absorbed him, and minutiae did not arouse his impatience. Chambers is an assimilator

---

* In a notice of the first edition of this book in the *Chicago Sun-Times* (18 October 1970) G. B. Harrison, who avouches that he knew well most of the great Shakespeare scholars of the 1920s and 1930s when he was a junior lecturer at the University of London, remarks that Chambers neither knew nor was known by younger scholars: 'He was an unfriendly man and not loved in the Board of Education where he was one of the senior officials, as was shown by the departmental riddle: "What is the difference between Madame Tussaud's Waxworks and the Board of Education? One has a Chamber of Horrors, the other a horror of Chambers"'). Harrison also reports that a story current in the 1920s held that when Sidney Lee, writing his biography of Queen Victoria, interviewed King Edward VII, the latter advised, 'Stick to Shakespeare, Mr Lee; stick to Shakespeare. There's money in it.'

rather than an explorer of unfrequented paths, an encyclopaedist rather than a restricted specialist. His prose style exquisitely matches his needs: at its best concise, disdaining floridity, without eloquence but enlivened by occasional flashes of caustic wit. At its worst, when he grew old, it becomes neutral and dry, yielding to bureaucratic greyness.

Of such a man we do not expect a highly developed sensibility. The modern temper seems hardly to have touched Chambers: for him the poetry of the Augustans is versified prose; Eliot makes him squirm. Prudently, he devoted little of himself to criticism. Thus he wrote a prize-winning life of Coleridge remarkable mainly for never coming to grips with the subject as poet or critic. If we learn with surprise that this child of the nineteenth century himself wrote verse, the verse itself is what we might expect. Pallid Edwardian romanticism perfumes the poems he gathered together in *Carmina Argentea*, a pamphlet printed privately in Oxford in 1918. But Chambers's sonnet on Shakespeare deserves quotation for its revelation of a vein of fancy and sentiment which as a biographer he endeavoured ruthlessly to suppress:

> I like to think of Shakespeare, not as when
> In our old London of the spacious time
> He took all amorous hearts with honeyed rhyme;
> Or flung his jest at Burbage and at Ben;
> Or speared the flying follies with his pen;
> Or, in deep hour, made Juliet's love sublime;
> Or from Lear's kindness and Iago's crime
> Caught tragic hint of heaven's dark way with men.
> These were great memories, but he laid them down.
> And when, with brow composed and friendly tread,
> He sought the little streets of Stratford town,
> That knew his dreams and soon must hold him dead,
>> I like to think how Shakespeare pruned his rose,
>> And ate his pippin in his orchard close.[75]

Any discussion of Chambers, however, must focus on him less as a sonneteer than as the author of three great works on Shakespeare and the English drama.

While still a young man, in the waning years of the nineteenth century, he began to think of writing 'a little book . . . about Shakespeare and the conditions, literary and dramatic, under which Shakespeare wrote'. Such a volume would naturally begin with the mid-sixteenth century, but Chambers thought it appropriate to include a preliminary account of the origins of *mimesis* in England and its evolution in the Middle Ages. Soon it became apparent to him that a history of the medieval drama had never been undertaken from the English point of view. Thus he came to write *The Mediaeval Stage*. Analysis of this work, which established Chambers's scholarly reputation, lies outside the scope of the present volume. Suffice it to say here that for over half a century his assumptions and methods went unchallenged, until O. B. Hardison raised large questions about the bond between drama and religious ritual in *Christian Rite and Christian Drama* (1965). Few scholarly works have held on for so long.

Two decades passed after the appearance of *The Mediaeval Stage*, and the little book on Shakespeare remained unwritten. 'Perhaps it was only a mirage', Chambers remarked in 1923, 'since working days have their term, and all that I can now offer, after an interval of twenty years, is another instalment of *prolegomena*.' The *prolegomena* to which he refers, with characteristic understatement, is *The Elizabethan Stage*. Arranged in five divisions, the work organizes the available facts about the Court; the settlement of the players in London, and the struggle of Court and city over control of the stage; the acting companies; the playhouses; plays and playwrights. Appendices furnish a Court calendar, records of Court payments, documents of criticism and control, and (among other materials) lists of printed, manuscript, and lost plays. Here Chambers surveys more familiar territory than in *The Mediaeval Stage*, and his monumental undertaking is which he lists, often with brief evaluative comments, his vast array of authorities.

The work remains standard—nobody before or since has attempted a synthesis on such a scale—but is marred by an awesome miscalculation which the author came to realize after he could no longer do anything about it. The natural terminus for a history of the Elizabethan stage is 1642, when the playhouses were shut down by official fiat, but Chambers looking forward to the little book that would crown his labours, made his cut-off point the year of Shakespeare's death. But 1616 as a date has no theatrical significance; it does not even coincide with Shakespeare's retirement from the stage. Players and playwrights went on functioning after Shakespeare's demise—it cannot be said that all nature felt this wound. The practical disadvantages for Chambers are great. When, for example, he arrives at Shakespeare's friend and arch-rival, he discusses no play after *The Devil is an Ass* in 1616; yet Jonson lived on until 1637. Fortunately, a superbly equipped continuator was found in G. E. Bentley, although he cannot duplicate (who could?) his predecessor's extraordinary spareness of exposition. Chambers requires four volumes to cover half a century of unprecedented theatrical ferment, whereas *The Jacobean and Caroline Stage* takes seven volumes to chronicle, with inevitable overlap, a period of twenty-six years.

To Shakespeare, Chambers allots a meagre eleven pages, limited to the principal bibliographical data and a conjectural chronology; this was not the occasion for a proper biography. The pre-Alexander vintage of his opinion is indicated by a reference to the *The First Part of the Contention of the Two Noble Houses of York and Lancaster* as the anonymous source play for *2 and 3 Henry VI*. But Chambers had a chance to revise his views in 1930, when at last the little book was written.

Contemplating the two volumes, which run to over a thousand pages, he confessed, 'I have not found it possible to use quite that brevity of words which the confident surmise of youth anticipated.' Chambers would later refer to his *magnum opus* as a Life, but his subtitle describes it more accurately as *A Study of Facts and Problems*. Of continuous narrative there is little: by page 91 Shakespeare is dead and his line extinct. But let the author himself describe his threefold purpose:

The present volumes complete the design of *The Elizabethan Stage* by a treatment of its central figure, for which in that book I had no space proportionate to his significance. I collect the scanty biographical data from records and tradition, and endeavour to submit them to the tests of a reasonable analysis. And thirdly, I attempt to evaluate the results of bibliographical and historical study in relation to the canon of the plays, and to form a considered opinion upon the nature of the texts in which Shakespeare's work is preserved to us.[76]

Thus does Chambers's *William Shakespeare* form the central panel of his great triptych.

The account of the life does not reach beyond the first three chapters— 'Shakespeare's Origin', 'The Stage in 1592', and 'Shakespeare and his Company'—and of these the second is given over to a retrospect of stage conditions when Shakespeare made his début early in the last decade of Elizabeth's reign. So compressed is Chambers's style, moreover, that he finds space in these chapters for relevant digressions on the Stratford environment, grammar-school education in the sixteenth century, and the theatres and personnel of the Chamberlain-King's Men. The approach is what we would expect: dispassionate, aloof from bardolatry, meticulous, totally informed. Chambers resists romantic extenuation. 'A kindly sentiment has pleaded the possible existence of a pre-contract amounting to a civil marriage', he reports matter-of-factly; but later, in the second volume, he returns more sceptically to the same theme: 'No doubt canon law recognized a contract *per verba de praesenti*, or a contract *per verba de futuro* followed by cohabitation, as constituting a marriage. But this does not seem to have been accepted by English secular law. Nor does contemporary moral sentiment appear to have approved the anticipation of the fuller ceremony.'[77]

So it goes. Being a blank, the Lost Years invite speculation and allow for the operation of that mechanism of self-projection we have noted so many times: the barrister Malone putting the young Shakespeare in an attorney's office, the donnish Adams making him a pedagogue stuffing his brains with Latin texts. Will Chambers furnish him with a minor government secretaryship? He meets the challenge with wonted austerity:

Whatever imprint Shakespeare's Warwickshire contemporaries may have left upon his imagination inevitably eludes us. The main fact in his earlier career is still that unexplored hiatus, and who shall say what adventures, material and spiritual, six or eight crowded Elizabethan years may have brought him. It is no use guessing. As in so many other historical investigations, after all the careful scrutiny of clues and all the patient balancing of possibilities, the last word for a self-respecting scholarship can only be that of nescience.

> 'Ah, what a dusty answer gets the soul
> When hot for certainties in this our life!'[78]

*It is no use guessing*—the words might do as Chambers's motto.

The rest of the first volume mostly takes up matters which, although not strictly biographical, no biographer can ignore. The most notable advances in Shakespearian scholarship in this century surely lie on the textual side. Chambers's survey and analysis are beyond praise, but he wrote during a period of ferment in bibliographical study, and consequently time has to some extent passed him by. Today's student is better advised to turn to W. W. Greg's *The Shakespeare First Folio* (1955) and to more recent work by Charlton Hinman, Fredson Bowers, Alice Walker, Stanley Wells, and Gary Taylor than to the equivalent discussion in Chambers's huge section of over 220 pages on the Folio plays.* His chronology has fared better. A refinement of the one set forth in the 1911 *Britannica* article, it is the most rational treatment ever of a subject that has inspired a good deal of irrationality. All the questions are rethought: Chambers, for example, discards Fleay's metrical tables, reproduced as late as 1926 in Tucker Brooke's *Shakespeare of Stratford*, with the devastating observation that the sum totals fail to correspond with the number of lines in the plays, divergences sometimes amounting to the hundreds. Fleay could not add, a deficiency in a scholar whose method was arithmetical. Chambers's own attitude towards metrical evidence reflects his customary caution. His findings with respect to the chronology have worn so well that J. G. McManaway, in 'Recent Studies in Shakespeare's Chronology', *Shakespeare Survey* 3 (1950), could offer only a few modifications.

'So far as my researches enable me to judge', wrote Father Herbert Thurston† in 1930 apropos of the identification of Pembroke with the Fair Youth, 'this solution has now been completely abandoned.'[79] Not quite, for that year Chambers revived it. In his early days a Pembrokist, then a hard-line critic who rejected (in the *Academy*, 1897) 'the specious structure of Mr. Tyler's theory' as an 'unsubstantial pageant faded into nothingness', he came round again to his first opinion in his *William Shakespeare*. Tyler, he now feels, mishandled a potentially strong case because of his obsession with Mary Fitton, which led him to date the bulk of the Sonnets 1598–1601, after

---

* It is only fair to add that in this section Chambers does not limit himself to the Folio texts, but also discusses the Quartos, as well as dates and sources.

† The same Thurston who here appears in rather a foolish light deserves credit for recovering in 1923 a Spanish text, published in Mexico City in 1661, of the Spiritual Last Will and Testament purportedly subscribed by John Shakespeare. This, the first such version to turn up, confirmed the authenticity of the document except for the first 2½ articles.

the abortive attempt in 1597 to match the Earl with Lady Bridget Vere. If, as Chambers believes, the Sonnets were written earlier—between 1593 and 1596—Herbert, then in his early teens, might more naturally be called a 'boy' than Southampton, who was born in 1573. Moreover, Tyler and other Pembrokists had overlooked the effort made to betroth Herbert to Sir George Carey's daughter when he was barely fifteen, the negotiations collapsing because of the Earl's 'not liking'. Chambers notices that Carey was son to Lord Hunsdon, the patron of the Chamberlain's Men, and that—although Southampton was still living at the time—Shakespeare's fellows dedicated the Folio to the Pembrokes. Admittedly some other youth of good quality and breeding, but of less degree than an Earl, might satisfy the conditions for an identification; but clearly Chambers favours Herbert. (Later his position hardened; in *Sources for a Biography of Shakespeare*, in 1946, he owned to little doubt that Pembroke was the man.) No previous scholar had made so cogent a case for this candidate. Hyder Rollins, who as editor of the Variorum *Sonnets* had to read masses of trivia on the subject, inclined to Herbert, in part (I suspect) as a result of Chambers's influence; but Rollins preserved his agnosticism. Perhaps Chambers would have done well to keep his also, for his argument, like the others, rests heavily on circumstance and conjecture. It is reassuring, though, to realize that he was human after all.

But at the journey's end the rigid self-discipline prevails. What emotion did he experience upon completion of his life's work? Did he feel exhilaration or regret—or merely exhaustion? What, in the last resort, was his conception of Shakespeare the man? A personal postscript might have followed the last chapter, but none is forthcoming. Nor does Chambers permit himself a final burst of eloquence, a terminal peroration. The last paragraph concerns The Taverne in Oxford, and the final sentence reads: 'There is no reference to the inn in Davenant's will.' Thus ends the first volume of Chamber's *William Shakespeare*.

The rest is appendices. After a prefatory table of principal dates and a pedigreee of Shakespeare and Arden, volume two proper begins with the records, set forth *in extenso*. The section commences with christenings, marriages, and burials, carries on through the Arden inheritance, the Belott–Mountjoy suit, and the Stratford tithes, and concludes with Shakespeare's will and the epitaphs. Chambers's transcripts are not, alas, impeccable, nor are his extracts from printed sources, although most of the errors are trivial, involving mere accidentals. Oddly Chambers (quoting Fripp's *Richard Quyny*) modernizes spelling for a brief extract from Daniel Baker's letter to Quiney (ii. 103), although the original is readily accessible in the Stratford Records Office. Nor does Chambers treat consistently capitalization, abbreviations, and the like. The reader wishing to compare the transcriptions with the originals for himself may place Chambers alongside the facsimiles (preferably the more legible ones) in B. Roland Lewis's sumptuous but sometimes inept

*The Shakespeare Documents* (1940); Lewis's own transcriptions compare disadvantageously with those of Chambers. Almost as valuable as his texts of records are Chambers's commentaries. Thirteen pages of discussion in small type follow a brief extract from John Combe's will: 'Item, I give . . . to Mr William Shackspere five pounds.' The section on the dramatist's financial interests in the Globe and Blackfriars rescues this important subject from the confusions perpetrated by Lee, whose wit on such matters was unequal to the task. Chambers exercises similar magisterial authority with the other much-discussed issues of the life: the grants of arms, the marriage, the Welcombe enclosure, the poet's bequests. A second appendix supplies texts of contemporary personal references to Shakespeare, including a few (for example, *The Tears of the Muses* and *Colin Clout's Come Home Again*) wherein Chambers himself fails to see veiled allusions. The third substantial appendix, 'The Shakespeare–Mythos', brings together the principal legends and traditions, commencing with Richard James, *c.*1625, on the transformation of Oldcastle into Falstaff, and concluding with Fullom's *History of William Shakespeare* in 1862. Not all the myths are traced back to their source in this admirable compilation. Briefer appendices follow on performances of plays before the closing of the theatres, the name Shakespeare, and the forgeries. Lastly, Chambers gives a table of quartos, metrical tables, and a list of books. It may be safely claimed that no single volume has proved more indispensable to twentieth-century students, serving them as Malone and Halliwell-Phillipps served previous generations. For this writer it has been a constant companion.

The reviewers immediately hailed Chambers's *Shakespeare* as a classic. At the same time the work has limitations, some of which were noticed from the first. The author's scepticism at times borders upon insensitivity, as when he can only attribute to a 'super-subtle criticism' discernment of a progression in intimacy in Shakespeare's two dedications to Southampton. Misprints are few, but one, the careless substitution of *b* for *d* (the letters being used for enumeration) muddies a complicated argument about the authorship of *Pericles*. Chambers attributes to Malone in 1790 a passage concerning the corruption of Shakespeare's text (I. 144), but Malone is quoting Dr Johnson's *Proposals* of 1756. Much more serious is Chambers's failure to facilitate use of what is after all a reference work intended for consultation rather than steady reading. Paragraphs tend to be overlong by modern standards, sometimes running to half a dozen or more pages (for example I. 196–204). Chambers furnishes no analytical table of contents. The index, unhelpfully arranged in three sections (Plays, Persons, and General), is painfully inadequate—it occupies a mere twenty-three pages—and not impeccably accurate. Malone, whose Christian name is here as in the text misspelled, appears in the index of Persons, but not Furnivall, Lee, Stopes, and Wallace, although Chambers cites these worthies many times. Dissatisfaction was to a degree assuaged by publication in 1934 of Beatrice White's *Index to 'The Elizabeth Stage' and 'William Shakespeare'*, but here too Furnivall, Lee, Stopes, and Wallace remain unlisted. Ireland and Collier, however, find places; presumably 'The good that men do. . . .' Meanwhile, in 1933, Charles Williams had brought out a convenient abridgment of Chambers, *A Short Life of Shakespeare, with the Sources*.

Inevitably this grandest of monuments to modern Shakespearian scholarship has roots in the past. Chambers after all reached maturity in the *fin de siècle*, and all his life he remained (in his own phrase) 'an impenitent Victorian'. Despite the flat timelessness of his style and the objectivity of his subject-matter, his predispositions sometimes surface. His way with legend, what he calls 'the Shakespeare–Mythos', is in this regard illuminating. He of course recognizes the superiority of fact to tradition, and seeks to maintain a critical balance; but he is more indulgent than some (most notably Smart) to the picturesque stories and details that began to circulate in the late seventeenth century. 'I think that perhaps the current attitude to tradition is rather inclined to err on the side of excessive skepticism'; so Chambers wrote in 1946. He rejects the Bidford crab-tree (too late in sprouting) but is drawn to the deer-poaching legend in its original form, before modern sophistications, because it derives from no fewer than four early sources: Davies, Rowe, Jones (as reported by Oldys and Capell), and Joshua Barnes. Chambers does not entirely discard the possibility that John Shakespeare, as Aubrey reports, exercised the butcher's trade, but he receives more cautiously the same authority's suggestion about William teaching school in the country. 'The course at Stratford, even if not curtailed', Chambers observes, 'would hardly have qualified him to take charge of a grammar school; but his post may have been no more than that of an usher or *abecedarius*.'[80] Of course the young Shakespeare might have served as an assistant teacher not in Stratford but in some provincial household in another county; Lancashire, as we shall see, has been suggested.

Really arresting is the brief emergence, at one crucial point, of unabashed subjectivism. Chambers perhaps forewarns us in his preface when he pays tribute to Dowden as one of the earlier writers, along with Malone and Halliwell-Phillipps, to whom he owes most. In his chapter on chronology Chambers draws upon Dowden's 'admirable treatment' of Shakespeare's four phases. But this is by the by; the shades of Dowden and Brandes stalk in earnest when he turns to *Timon of Athens*. We have the spectacle of this most restrained and objective of biographers offering a speculative reconstruction as romantic as it is undemonstrable:

Shakespeare's spirit must have been nearly submerged in *Lear*, and although the wave passed, and he rose to his height of poetic expression in *Antony and Cleopatra*, I think that he went under in the unfinished *Timon of Athens*. The chronology of the plays becomes difficult at this point, and it is therefore frankly a conjecture that an attempt at *Timon of Athens* early in 1608 was followed by a serious illness, which may have been a nervous breakdown, and on the other hand may have been merely the plague. Later in the year Shakespeare came to his part of *Pericles* with a new outlook. In any case the transition from the tragedies to the romances is not an evolution but a revolution. There has been some mental process such as the psychology of religion would call a conversion.[81]

The passage elaborates views Chambers had expressed more positively, some twenty-five years earlier, in the introduction to his *Timon of Athens* for the Red Letter Shakespeare ('. . . in the brain of Timon's creator some strange crisis is at hand. That the crisis took place is indisputable'). Now, in his most striking *rapprochement* with the mythos, he accepts a traditionary report which even the uncritical Lee had scorned: the assertion by the late seventeenth-century Gloucestershire clergyman Richard Davies that Shakespeare had 'died a Papist'—whatever that term meant to the Archdeacon.*

Chambers also furnishes the dramatist's last days with an imaginative setting—'the open fields and cool water-meadows and woodland of Stratford, and the great garden of New Place, where the mulberries he had planted were yet young'. We are not very far removed from the romantic world of that Edwardian sonnet in which Shakespeare prunes his rose and eats his pippin. Thus do past and present scholarly modes overlap in this most eminent of modern Shakespearians.

# 15

## *John Dover Wilson*

FOR eleven years Chambers had a junior colleague in the Technical branch who pursued a Shakespearian avocation with no less enthusiasm than his chief. Eventually he resigned from the Board to take an academic post, and years later he recalled bidding Chambers farewell:

Seeing my head poking round his door, 'What do *you* want?' he wearily asked. 'I've come to say good-bye,' I said. 'Oh, where are you off to?'—'London University'—

---

* To the laconic summation in his *Shakespeare* Chambers added a valuable gloss in a letter of 8 May 1932 addressed to Dover Wilson: '"Conversion" I used in the sense of the spiritual revolutions described in W. James's *Varieties of Religious Experience* and not in that of an Anglican or Catholic controversialist. But it might have been accompanied by some reversion to traditional observance which the informants of Davies called "papistry"' (uncalendared Dover Wilson papers in the National Library of Scotland).

'London University, umph; English, I suppose?'—'No,' I replied, hoping he would be pleased, 'Education.' At this he almost leapt from his chair, all lethargy vanished. 'Education,' he snorted, 'a *disgusting* subject!' He was then Second Secretary to the Board of Education.[82]

The junior official, a Provincial inspector, was John Dover Wilson.

For his achievement as editor of the Cambridge New Shakespeare—surely one of the most stimulating editions of the dramatist produced in this century—as well as for his textual and aesthetic criticism Wilson earned recognition as the doyen of Shakespearians long before his death in 1969 at the age of eighty-seven. We do not immediately think of him as a biographer, yet his *Essential Shakespeare* (1932) has perhaps reached as wide an audience as any modern reconstruction of the poet's career; few others can have achieved the tribute of translation into Serbo-Croatian. Readers put off by Chambers's frigid pages might well respond to a scholar who paints his subject with so much verve and colour (the palette does not exclude purple). Strictly speaking, *The Essential Shakespeare* is not a Life but a fantasia on the life or, to use Wilson's subtitle, *A Biographical Adventure*. '"Here, in a nutshell, is the kind of man I believe Shakespeare to have been", is what it [the title] is intended to convey,' he informs us in his Preface. 'I might perhaps have called it "A credible Shakespeare".'[83]

The germ of the book is to be found in Wilson's 1929 British Academy lecture, 'The Elizabethan Shakespeare', in which this St George of criticism waged battle with that hybrid monster, the Victorian conception of Shakespeare. This he saw as an amalgam of incompatibles. On the one hand the Olympian Bard, 'a great tragic poet, facing the vastidity of the universe, wrestling with the problems of evil and disaster—a man of brooding temper, of lofty thought, of grave demeanour, and, after passing through the fire, of joyful serenity of temper';[84] on the other, the mercantile genius, Lee's prosaic Shakespeare, writing masterpieces in order to keep the family cupboard well stored. It is interesting that as late as 1929 Wilson thought this conception sufficiently well entrenched to merit repudiation.

*The Essential Shakespeare* leaves the impression that little of Chambers's influence rubbed off on the younger man. Despite the assault on the Victorian heritage, a number of Wilson's opinions about issues great and small would not have discomfited his ancestors, as clearly they did discomfit Chambers. A well-to-do wool merchant, John Shakespeare was 'almost certainly' a Catholic recusant, and so ardent in his faith that (Wilson thinks) he preferred not to see his son educated at a local grammar school run by a Protestant clergyman; the only novelty here is the extremity to which a common view is pushed. In harmony with kindly Victorian sentiment, Wilson explains away the haste of Shakespeare's marriage by an assumed pre-contract. In London the dramatist began by reshaping the plays of established writers like Greene, who denounced the upstart crow for plagiarism (so much for Smart and Alexander). Despite these battered orthodoxies, however, Wilson has an intensely personal view of Shakespeare to promulgate. In so doing, he ranges himself against 'the scientific school of Shakespearean biography', which despite pretensions to detachment, is rich in 'suppressed hypotheses'. His own hypotheses Wilson makes no effort to suppress.

The frontispiece affords a clue to Wilson's programme. The cultivated youth portrayed by Grafton symbolizes a counterblast to the bourgeois conception of the poet fostered by the Folio engraving and the Stratford bust.[85] Wilson's first daring conjecture has to do with his subject's education: service as a singing-boy in the household of 'some great Catholic nobleman' prepared Shakespeare for his later acceptance as the darling of the curled darlings of the nation. For this Shakespeare is nothing if not aristocratic. As an actor-playwright for Lord Strange's Men, 'his facetious grace in writing' recommended him to persons of quality ('divers of worship'): first, Lord Strange, then Essex, and finally Southampton. All these great men cherished their friendship with Shakespeare, for he offered them sage counsel through his art, 'offered it respectfully, unobtrusively, but candidly and with admirable discretion'. Soon the poet entered Southampton's entourage, and during the plague closing of the theatres in the early 1590s, resided with him at Titchfield, the Earl's seat, perhaps in the capacity of tutor. There he became acquainted with the famous translator of Montaigne, for we know that Florio was a tutor in the same household. Maybe Shakespeare and Florio together toured Italy, where the former picked up his intimate knowledge of Venetian topography. While at Titchfield, Shakespeare mainly had to furnish dramatic entertainments. To this phase belongs *Love's Labour's Lost*, a comedy 'obviously' designed for private performance and 'undoubtedly' written for Southampton and his set. When the nobleman came of age he gave his protégé the munificient gift vouched for by tradition, although Wilson regards the rumoured figure of £1,000 as an exaggeration. 'Of the later relations between the dramatist and his patron', he writes, 'it is difficult to speak with any kind of assurance.'[86]—as though one could of the earlier.

Eventually Southampton fades from the poet's life (although *The Tempest* may glance at his interest in Virginia), replaced by a more dazzling presence. Essex was now receiving Shakespeare's dramatic counsel. In *The Merchant of Venice* he appealed subtly to the Earl to show mercy to Dr Lopez, a Portuguese Jew converted to Christianity and Elizabeth's personal physician, at whose treason trial Essex malignantly presided. The appeal was apparently too subtle, for the wretched Lopez was hanged, drawn, and quartered. Shakespeare did not give up. Through the character of Henry V he tried to make his friend into that kind of man, without notable success. In *Julius Caesar*, however, the playwright had the prescience to describe an Essex-like conspiracy. *Troilus and Cressida* represents a courageous attempt to goad the nobleman into a reconciliation with Elizabeth. Instead he made his abortive *putsch* on 8 February 1601; before the month ended he had joined Dr Lopez. The revival of *Richard II* on the eve of the rebellion might have proved calamitous for Shakespeare, but, remarks Wilson in one of his more inspired moments, 'if the Lord Chamberlain was a sensible man, a few words of explanation from the author of the play as to its exact tenour and purport should have placed matters in a proper light'.[87] Shakespeare went on to mourn his fallen idol, reincarnated as Prince of Denmark. 'And Hamlet's mystery?' Wilson asks. 'Hamlet's mystery is the mystery of Essex.'[88] Next question.

From Shakespeare's portrayal of 'the inner Essex', it follows that he loved the man. 'The rebellion and execution, followed by the rewriting of *Hamlet* as an everlasting memorial to his friend, were—can we doubt it?—the most profound experiences he had ever passed through.'[89] More profound, presumably, than marriage, or his liaison with the Dark Lady, or the loss of his only son. The accession of James failed to assuage Shakespeare's sense of loss, despite the favours lavished upon him and his troupe. A shadow of corruption fell across the land. With *King Lear* he came perilously close to madness, but gradually he recovered and renewed his attachment to Stratford, 'with its memories, its quiet pastures and wide skies, with all the wild life of bird and beast and flower, with the pleasant friendships and domesticities of the littletown, with his house and garden, with his own family, and especially perhaps with his younger daughter'.[90] (Yet the elder was clearly his favourite.) In the beatific vision of the enchanted island he celebrated his enfranchisement from the tragic mood and, with Prospero, bade the stage farewell.

It is rather as though a Frank Harris with professional know-how had chosen to cast Essex instead of Mary Fitton in the role of tragic catalyst, master-mistress of the poet's passion. (Incidentally, it is Wilson's development of this idea, not the idea itself, which has novelty: the Essex story is mentioned by Capell in his 1768 preface, taken up by Ten Brink in 1893, and elaborated upon by Brandes in 1896; a popular school primer of the 1890s speaks of a Shakespeare whose life suddenly grew dark after Essex perished on the scaffold.) Like Harris, Wilson succeeds in rescuing his subject from the pedants—his Shakespeare has glamour—but the price he pays is a book with the quality of romantic fiction. Despite all the comparisons made by Wilson between his subject and Shelley, Wordsworth, Dostoevsky, above all Keats, no credible portrait of the creating artist emerges. Despite the campaign against his Victorian predecessors, Wilson belongs squarely in the tradition of Furnivall, Dowden, and Brandes; his Shakespeare, like theirs, moves from the depths to a sentimental apotheosis.

Published to coincide with Shakespeare's birthday and the inauguration of
the Memorial Theatre (now the Royal Shakespeare Theatre) by the Prince of
Wales, *The Essential Shakespeare* met with instantaneous success. Praise came
to the author from eminent scholars—Harold Child, Lascelles Abercrombie,
A. W. Reed—but one suspects that he valued most of all the letter he received
from his former chief. He had read the book with 'a good deal of agreement',
Chambers wrote to Wilson on 8 May 1932. 'I think we have much in
common, so far as the interpretation of the significance of the plays is
concerned . . .'; and Chambers goes on to indulge in one of his favourite
fancies: 'Is it unreasonable to think that mental perturbation may come in
waves, and that a first wave in *Lear* may have been followed by a period of
comparative balance, and that again by a second wave in *Timon*?'[91] The
newspaper reviews were generally enthusiastic, although Allardyce Nicoll in
the *Yorkshire Post* shrewdly saw *The Essential Shakespeare* as representing a
new romanticism not so very different, after all, from the romanticism of the
Victorian era.

Wilson had to run a sterner gauntlet in the learned journals. A notice in
*Modern Language Review* (XXVII, 1932) courteously withheld approval, the
reviewer, C. J. Sisson, focusing especially on what he regarded as Wilson's
misreading of history. 'The book', Sisson concludes, 'haunts the imagination
of the reader, however much his reason may demur to it. It is perilous
reading.' Sportingly he sent Wilson a proof of the review, with an offer
to mend anything unfair. With equal chivalry the victim declined to interfere, but after the lapse of a few days he addressed a remarkable *apologia* to
Sisson:

Now the review has I hope been passed for press & is out of your hands, I may tell you,
sir, that it leaves me entirely unrepentant & proud to think that the book is having the
influence you speak of. I *want* it to, for the Shakespeare I was brought up on & all
young England has been brought up on is a thing unbelievable & repellent. I know
what I am saying: it is my business as professor of education to keep in touch with
English as she is taught in the schools. Moreover I have two children, now grown up,
who have boggled at the horror as I did myself. You say he is a 'man of straw'! Where
in heaven's name were you brought up? In the Islands of the Blessed?

And so I determined to give young people a different & more human Shakespeare.
It's one I entirely believe in myself—I don't say in every detail—but in essence. Of
course the scaffolding is hypothetical, as anything of Lee is, & as every 'Life' must be.
But I have at least not concealed that as Lee on every page attempts to do.

Whether Wilson actually posted the letter we do not know; a copy in his
hand is to be found among his uncatalogued papers in the National Library of
Scotland.*

* Sisson's personal papers, I have learned from his widow, were destroyed by a land-mine early in
the Second World War.

# 16

## Shakespeare's Mythical Sorrows

In any event *The Essential Shakespeare* and other biographical adventures continued to haunt Sisson's imagination. Before the British Academy at Burlington House on 25 April 1934 he delivered the annual Shakespeare Lecture. Entitled 'The Mythical Sorrows of Shakespeare', this performance would prove as momentous in its own way as Chambers's lecture, ten years previous, on the disintegration of Shakespeare's text.

Sisson writes with urbanity and good-natured wit (why not? he holds all the cards), and he carries his considerable learning lightly. A brief historical survey properly emphasizes the brothers Schlegel, Dowden, and Brandes; but Sisson also grasps the seminal importance of Hallam's brief evocation of Shakespeare's troubled heart, and is aware that behind Dowden's Four Periods obscurely lurks Coleridge's five Epochs. In Sisson's hands mere summary becomes a crushing weapon; he need only recapitulate Ten Brink to dispose of him: 'It appears that in 1607 Shakespeare's brother Edmund died, an event which helped to infuriate him. Fortunately, in 1608 his mother died, an event which restored him to a kindlier mood. So various are the effects of deaths in the family upon a great poet.'[92]

Early on, Sisson sets forth the 'four dogmas' on which subjective biography rests:

First, that the actual evolution of Shakespeare's personal life must be read into his poetic and dramatic work. Second, that dramatists write tragedies when their mood is tragic, and comedies when they are feeling pleased with life. Thirdly, that Shakespeare was so far a child of his own age that he faithfully reflected its spirit in his literary work, and fourthly, that the spirit of the age was heroic and optimistic under Elizabeth, degenerating towards the end of her reign into the cynicism, disillusionment, and pessimism which marked the reign of James the First.[93]

The task of refuting these fallacies, for so Sisson regards them, is complicated by the tendency of the offending biographers to shift the grounds of dialectic: when literary interpretation fails, one can always lose oneself in the clouds of *Zeitgeist* or embark on flights of intuitive fancy where certainty mocks reason.

Sisson nevertheless moves ahead on all four fronts. Unwilling to join Wilson in a chorus of the seventeenth-century blues, he enlarges upon points, first put forward in his review, with respect to achievements under James in colonization, diplomacy, and letters. Turning to Shakespeare, Sisson acknowledges the impossibility of disproving that a mood of disgust permeates a given play. On the other hand, proof that it does is equally elusive. In *Timon of Athens*, Brandes, Chambers, and Wilson discern tell-tale symptoms of nervous collapse, but for Dowden 'The impression which the play leaves is that of Shakespeare's sanity.' When psychiatrists disagree the layman must reserve judgement as to the diagnosis of the patient. Peculiarities of chronology, moreover, complicate the task of correlating the works with the life. Shakespeare went on writing comedies bubbling with the buoyancy of youth until almost forty—in other words, until middle age. Then, some seven years later, he appears before us in *The Tempest* in the twilight of life. 'There is something wrong', Sisson mildly observes, 'with either this youth, or this old age, or with both.'

The lecture concludes with a statement of Sisson's own credo. 'Shakespeare', he writes, 'was not stung into tragedy by any Dark Lady. He was not depressed into tragedy by the fall of Essex, who threatened revolution and chaos in England, to Shakespeare's horror and alarm; the cruelty of anarchy was a thought that haunted the poet like a nightmare. He did not degenerate into tragedy in a semi-delirium of cynicism and melancholy, ending in a religious crisis. Shakespeare *rose* to tragedy in the very height and peak of his powers, nowhere else so splendidly displayed, and maintained throughout his robust and transcendent faith in God and his creature Man.'[94] This position does not lack a subjective bias of its own. In fact we do not know how the dramatist reacted to the Essex rebellion, we can only guess; and that Shakespeare maintained at all times a transcendent faith in God and Man is merely one cloistered academic's rather too complacent inference from the same *œuvre* which has led other critics into making contrary inferences.

Yet no formal life of Shakespeare laying claim to serious regard can limit itself to the facts and to logical deductions from the facts alone; the writing of literary biography after all requires the play of literary imagination. Granted that Sisson directs most of his formidable ammunition against the abuses of interpretation which result not in legitimate biography but in oblique self-portraiture; still one may object that even in legitimate biography an element of self-portraiture is an inevitable concomitant of the writer's empathy with his subject. Sisson's limitation is that he fails to grasp the theoretical implications of his onslaught. He came to realize, however, that he had gone too far, for in 1950 he described his British Academy performance as expressing the 'extreme of resistance' to subjectivism, and he went on to acknowledge that, 'The pendulum, swinging from pole to pole, from Lee to Brandes, is gradually settling to rest in the centre, and few would now either accept the full lyrical interpretation of the plays, or deny to them their reflection of the growth and the increasing maturity of a poet's mind through which experience was transmuted into art as in a crucible.'*⁹⁵ But it is in the nature of extremes that they provoke extremes of reaction, and what Sisson had to say very much needed saying in 1934. At Burlington House he had a rendezvous with scholarly destiny.

No rebuttal followed. Instead, three years later, Sisson's long-time friend R. W. Chambers restated the case before the same forum with more words and less wit in 'The Jacobean Shakespeare and *Measure for Measure*'. But during the long hour (his lecture runs to twice the length of Sisson's) Chambers made some telling points. Surely the cruellest blow that can befall any man is the loss of his only son, yet Shakespeare turned out his most joyous work after the death of Hamnet in 1596; to this period belong Falstaff, Rosalind and Orlando, Beatrice and Benedick, and Sir Toby Belch. Then too, because of the disintegrators, success in denying Shakespeare all or part of the *Henry VI* trilogy and *Titus Andronicus*, the grim, even horrific, elements in the dramatist's early work (a phase condescendingly referred to as 'in the workshop') had been neglected and ignored. But the demonstration piece of his lecture is *Measure for Measure*, which Sir Edmund Chambers had taken, in the 1911 *Britannica*, as a testimonial to the author's 'profound disillusionment of spirit'. The other Chambers prefers to regard Isabella as a great exemplar of mercy rather than as a rancid virgin, and her play as an exalting drama of forgiveness having links with the late Romances. As for Shakespeare's supposed depression and nervous breakdown—R. W. Chambers can only report hearing a gentle voice from Stratford murmur, 'Good friend, for Jesus' sake forbear.'⁹⁶

It would be an exaggeration to claim that Sisson and R. W. Chambers put an end to biographical adventures inspired by Shakespeare. Some life-writers (in the popular camp) failed to sniff the wind of change altogether. Others, aware of theoretical objections, nevertheless insisted upon their prerogatives of interpretation; a few hit upon new—and dubious—techniques of extracting the life from the corpus. Yet at Burlington House in the early 1930s a critical turning-point was reached. The heyday of subjectivism was over.

# 17
## *Aspects of Shakespeare*

ALONG with general studies, there are the monographs on selected aspects of the poet's career, on his family and on the Warwickshire context. These

* His last statement on this theme, in a late unpublished lecture among the Sisson papers at the Shakespeare Institute in Birmingham, is even more concessive: '. . . it is a poor, cool, and distant critical embrace of literature that forgets, or ignores, the life-blood of a man that runs through the veins of his created works'.

treatises, sometimes more ambitious and illuminating than conventional Lives, are best reviewed not according to their chronological sequence of publication but with reference to the successive stages of their subject's history. If we can believe Arthur Gray of Jesus College, Cambridge (*A Chapter in the Early Life of Shakespeare*, 1926), Shakespeare did not grow up in Stratford after all—poor Fripp, how he wasted his energy!—but, a child of eight, was packed off by his father to Polesworth in northern Warwickshire, not far from Coventry. There, in the rambling manor house of Polesworth Hall, he served as a page in the ménage of Sir Henry Goodere, acquired his Latin, and was introduced to polite society. He may have stayed on for a time as tutor or schoolmaster before proceeding to London, where perhaps Henry Goodere the younger, having matriculated from St John's College, Cambridge, two years before Southampton, introduced him to the Earl. But who can believe the Master of Jesus College? Gray offers not a jot of supporting documentation; we await in vain a record, an allusion, a circumstance. This imagined chapter in Shakespeare's early life was evidently inspired by the example of his exact Warwickshire contemporary Michael Drayton, who indeed had employment as a page at Polesworth Hall, and in his writings more than once referred to his happy days there. It is astonishing that Gray managed to spin a whole book out of his crack-brained thesis; more astonishing still that a distinguished university press (Cambridge) should have published it.

Farmer in the eighteenth century had denigrated Shakespeare's classical attainments; J. Churton Collins exaggerated them in 1904 in an impressionistic and inexact essay, 'Shakespeare as a Classical Scholar', included in his *Studies in Shakespeare*. Despite important contributions by historians to the understanding of sixteenth-century education, the important subject of Shakespeare's schooling had still not received scientific investigation by the second quarter of the present century. Modern inquiry gets under way modestly with George A. Plimpton's *The Education of Shakespeare* (1933). In this general survey of grammar-school training, Plimpton devotes chapters to the principal teachers—Elyot, Ascham, Vives, Mulcaster, etc.—then goes on to the textbooks. Sections follow on writing, arithmetic, and geometry, Latin, Greek, and rhetoric, as well as on orthography, dictionaries, and letter-writing. All the chapters (except the one on Latin) are short, and amply stocked with quotations from primary sources. Bearing the date 'aet: 46–1610', a patently spurious portrait of Shakespeare, from Plimpton's own collection, serves as frontispiece. A discouraging preliminary, but for the rest *The Education of Shakespeare* is a decent, unpretentious introduction to a complicated field. Being by an amateur who makes no claim to Shakespearian expertise, it fails, however, to rate a mention in the imposing professional tomes of T. W. Baldwin.

One of the most tireless and productive of twentieth-century Shakespearians, Baldwin has explored all the highways, alternative routes, and indirect turnings of pedagogical practice in the poet's day. Unlike Dover Wilson, he assumes that his hero received a normal education, first at a Stratford petty school and then at the local grammar school—although Baldwin is of course conscious that no direct evidence of Shakespeare's attendance has ever been adduced (Rowe merely follows report) and that we have no first-hand information about the Stratford curriculum. He must work from the analogy of other schools and from inferences from Shakespeare's writings.

In *William Shakspere's Petty School* (1943) Baldwin describes the nature and scope of elementary education. The aim of this education is nicely summed up by the Norwich statute of 1566: to prepare the boy 'to say his catechism and to read perfectly both English and Latin, and to write competently'. Some schools also furnished instruction in the casting of accounts. The textbooks consisted of religious and moral material. That Shakespeare obtained this form of education seems reasonably clear from reflections of petty-school drilling, as cited by Baldwin, in the works. A final chapter, 'William Shakspere, Anglican', proposes that the dramatist was neither Catholic nor Calvinistic Puritan, but trained in and conforming with the Church of England.

*William Shakspere's Petty School* constitutes, in one reviewer's phrase, the porch to the temple, the latter edifice being *William Shakspere's Small Latine & Lesse Greeke* (1944). 'Whatever critics do', Baldwin pleads in a prefatory note, 'I hope they will not use the term "erudite" either of Shakspere's work or of mine upon it; neither is anything of the kind. Both merely use humdrum everyday routines of the most commonplace grammar school of Shakspere's day.' His petition fell on deaf ears; 'erudite' was precisely the word used to describe a work of overpowering fullness of coverage. The two volumes, amounting to 1,525 large pages, reflect monumental learning—and (it must be said) an equally monumental indifference to the frailty of the reader who undertakes to persevere through them.

Baldwin begins with Jonson's aphorism and follows the evolution of interpretation until, in the eighteenth century, 'small Latin' became canonized as 'little Latin', which is after all not quite the same thing. The development of the grammar-school curriculum in the sixteenth century is then traced. Baldwin starts with Erasmus's *De Ratione Studii*, examines the education of royalty and the systems at Winchester, Eton, Westminster, and elsewhere, and finishes this section with the theorists: Kempe, Brinsley, Hoole, and Clarke. The next segment takes up Shakespeare's presumed training. In the lower forms he would have received tuition in grammar from Lyly's *Short Introduction* and *Brevissima Institutio*, learned to construe from Cato, Aesop, and Terence (possibly with other supplements), and memorized

Withals, Stanbridge, or another to build a vocabulary. Using the Bible and perhaps also a collection of *sententiae*, he acquired the ability to write Latin grammatically; by means of colloquia he mastered Latin speaking. Pupils in the upper school were taught rhetoric, with enough logic to make it intelligible. Baldwin discusses exhaustingly the various rhetorical texts: Cicero's *Ad Herennium* and *Topica*, Susenbrotus, Erasmus's *Copia*, and Quintilian, the supreme authority. We see the boys writing epistles, then formal themes, and finally—the crown of grammar-school rhetoric—orations or declamations. The exercise of versifying followed, with Ovid, Virgil, and Horace as models; sometimes Juvenal and Persius too. The place of moral history and moral philosophy in the curriculum comes next in Baldwin's survey. In a chapter on Shakespeare's 'less Greek', Baldwin attempts to define the poet's possible acquirements; maybe he read in Greek part of the New Testament (*Acts* being the one book emphasized by all the authorities) in Beza's text, with the Genevan translation alongside.

'What of it?' Baldwin calls his last chapter. He emphasizes that in the society of Shakespeare's time only the grammar schools supplied formal literary instruction, university education being mainly professional. (True enough, but Baldwin exaggerates: Cambridge, as one reviewer pointed out, responded to intellectual flux, and the best tutors gave their charges a more humane reading diet.) 'If William Shakspere had the grammar school training of his day—or its equivalent—he had as good a formal literary training as had any of his contemporaries. At least, no miracles are required to account for such knowledge and techniques from the classics as he exhibits. Stratford grammar school will furnish all that is required.'[97] When Shakespeare began writing plays, he had a grasp of a rhetorical system, not merely of fragments from scattered sources. From rhetoric he derived literary construction and literary criticism; rhetoric gave him the theory and practice of imitation. His knowledge of the classics he derived from grammar-school sources; a knowledge coloured by the commentators he found in his editions: the *Plautus* and *Horace* of Lambinus, probably the *Terence* and *Virgil* of Willichius. None of the prose writers—Sallust, Livy, Cicero, etc.—seems to have left much of a mark on Shakespeare. Ovid appealed to him as an original and unpedantic poet; he preferred Virgil and Horace in their unscholarly moments. Finally Baldwin regards as plausible the tradition that Shakespeare taught school for a while before becoming attached to the theatre.

In his preface the scholar writes:

I have in no sense whatever written or attempted to write a book or books on Shakspere's education. I have attempted merely to present in as orderly fashion as my own nature and that of the materials would permit all the facts so far as I know them which appear to me to have a bearing upon the question of Shakspere's formal education, and to present my own conclusions upon those facts. I have in the present work attempted to discover exactly the what, how, and why of grammar school curriculum of Shakspere's day, and to evaluate as exactly the reflections of that curriculum in the work of Shakspere, whether he came into contact with it directly in grammar school or only indirectly from his environment as it were. It may be possible some centuries hence to write a nice little book on Shakspere's education, but such a book would be mere worthless dabble now.[98]

The austerity of the last sentence commands awe, perhaps not untinged with the suspicion that the professor doth underprotest too much.

Certainly the problem Baldwin has set for himself is fiercely difficult, and just as surely his work has flaws and limitations. The book fairly cries out for the editorial blue pencil: with cheerful prolixity Baldwin repeats material from chapter to chapter and sometimes even within the same chapter. Surmise figures more prominently in his argument than he may realize. For example, a passage in *Hamlet* ('Lay her i' th' earth, | And from her fair and unpolluted flesh | May violets spring!') seems to recollect that very difficult poet Persius ('nunc non e tumulo fortunataque favilla | nascentur violae?'). Perhaps, Baldwin suggests, Shakespeare made use of the commentary generally printed with the text of Mantuan—a line of whom the dramatist quotes in *Love's Labour's Lost*—and there found the passage from Persius which he later recalled. But all this impresses an eminent classicist, J. A. K. Thomson (*Shakespeare and the Classics*, 1952), as far too hypothetical; Laertes's fancy is so commonplace as to have the character of folklore. Throughout Baldwin deals not with what *was* but with what *may have been*. The conclusions arrived at through such a method must lack finality, and it is possible for Thomson to take a different view of Shakespeare's classical attainments.

Yet much of what Baldwin says in this *magnum opus* is reasonable and more than sufficiently supported by documentation. In a memorable essay on Shakespeare's reading F. P. Wilson distills Baldwin's contribution to irreducible essence in a single lucid sentence. 'Few who have read through T. W. Baldwin's treatise on *William Shakspere's Small Latine & Lesse Greeke*', Wilson writes, 'will have the strength to deny that Shakespeare acquired the grammar-school training of his day in grammar, logic, and rhetoric; that he could and did read in the originals some Terence and Plautus, some Ovid and Virgil; that possessing a reading knowledge of Latin all those short-cuts to learning in florilegia and compendia were at his service if he cared to avail himself of them; and that he read Latin not in the spirit of a scholar but of a poet.'[99] One may wish that Baldwin had cultivated a similar pithiness. Wilson's gifts are not every scholar's, however, and what counts is that, after almost two centuries, Farmer was at last properly answered.

The Lost Years have not surprisingly inspired a lion's share of theorizing, to which enthusiastic amateurs have generously contributed. In *Sergeant*

*Shakespeare* (1949) Duff Cooper develops an idea which first came to him amid the gas shells and stench of the Flanders trenches during the First World War. Shakespeare, he seeks 'to prove', enlisted in the army for service in the Low Countries in the campaigns of 1585–8 and was promoted to the rank of sergeant ('a man of his intelligence was not likely to remain for long a private soldier'). His enlistment came about when the Warwickshire lad, in hot water over the poaching incident, turned to the Earl of Leicester, then busy raising a force. Help was forthcoming only on condition of Shakespeare's joining up. As documentary evidence for this romance Cooper cites (as Thoms had cited in the previous century) Sir Philip Sidney's reference, in March 1586, to 'Will, my lord of Leicester's jesting player. . . . the knave [who] delivered the letters to my lady of Leicester'. Apparently the Earl used his foot-soldier as a postman, an employment requiring 'tact and discretion', if not necessarily military prowess. The biographer of Haig goes on to demonstrate Shakespeare's martial experience from episodes and allusions in the plays. The quality of Cooper's sensibility is illustrated by his explanation of that 'very curious interpolation', the quarrel between Brutus and Cassius in *Julius Caesar*, as 'a typical soldier's tiff'. His little book, Cooper modestly hoped, would amuse his wife, Lady Diana. Presumably it did.

Rather more eccentric views are expressed by William Bliss in *The Real Shakespeare* (1947), dedicated 'To all Shandeans and Pantagruelists, dead, living and yet to be born'. An authority on canoeing, Bliss proposes that Shakespeare at the age of thirteen ran away from home to circumnavigate the globe with Drake on the *Golden Hind*, as proved by the reference to 'remainder biscuit' in *As You Like It* (ship's biscuit being carried only on very long voyages). In 1585 the nautical Shakespeare sailed again, this time possibly as a mate or bos'n bound for the Levant (*vide* the Tiger in *Macbeth*). Shipwrecked upon the barren coast of Illyria, he eventually made his way to Venice, where he was brought to the Earl of Southampton, an English traveller willing and able to help out a distressed fellow-countryman. 'At any rate they met and straightway fell in love with each other.'[100]

To Shakespeare the soldier and Shakespeare the sailor we must now add Shakespeare the scrivener. This suggestion we owe to the ingenuity of E. B. Everitt, and may with diligence extricate it from *The Young Shakespeare* (1954), the prose of which imposes a heavy penance upon the well-wishing adventurer despite the author's apparently unironical claims to lucidity and brevity. Everitt identifies the dramatist as one of the 'sort of shifting companions' that Nashe complains (in his Epistle to Greene's *Menaphon*) have left the trade of noverint to busy themselves with the endeavours of art. So skilled a scrivener did Shakespeare become that he mastered no fewer than eight different styles of Elizabethan hand, meanwhile finding time to turn out *Edmund Ironside*, *The Troublesome Reign of King John*, and other plays not usually credited to him. Surprisingly, Everitt seems not to have won many converts. Nor for that matter have Cooper and Bliss, although in Eric Sams, the musicologist and retired civil servant, Everitt has found a mettlesome and voluble partisan three decades after the fact. Sams has renewed claims for Shakespeare's responsibility for *Edmund Ironside*, a manuscript play usually dated in the mid 1590s, of unknown theatrical auspices and a key part of Everitt's reconstruction of the young Shakespeare's playwriting career. Sam's views made the front page of London's *Sunday Times* (29 December 1985). Everitt, the *Sunday Times* reporter noted, 'was derided by the Shakespearian establishment, forced to print his edition of the play in Copenhagan and died deeply disappointed in 1981'. In 1986 Sams published his own elaborate four-hundred-page edition of *Shakespeare's Lost Play* (rev. 1985) about pre-Conquest Britain and the struggles between the English and the Vikings in the time of Canute. The Shakespearian establishment has remained unimpressed, although *Edmund Ironside* itself, after a lapse of almost four centuries, has now at last achieved stage representation on both sides of the Atlantic.

Even responsible professionals have their misguided moments. In a chapter of his *Organization and Personnel of the Shakespearean Company* (1929) confidently entitled 'Facing the Facts with Shakespeare', T. W. Baldwin manages to get his principal facts wrong. He offers the following theory: when Leicester's Men passed through or near Stratford in 1587, probably in May, Shakespeare signed on as an apprentice; so the evidence would 'strongly indicate'. A little over a year later the poet appeared in London with Leicester's company. After a lapse of seven years, the minimum period for formal apprenticeship, Shakespeare took up his freedom and became a full-fledged member of the troupe. Baldwin revives Malone's suggestion, first put forward by him in 1780 but later abandoned, that the novice's first theatrical employment was that of prompter's attendant—apprentice, Baldwin would say, maybe to Thomas Vincent, the early prompter. This reconstruction of a professional career has a beguiling simplicity, but unfortunately will not hold water. Baldwin presents no evidence to support his oracular pronouncement about Shakespeare's connection with Leicester's Men in 1588, and no evidence exists. More seriously, he has misundertsood the nature of Elizabethan guild organization and control. The Statute of Artificers of 1563 (5 Eliz. c. 4), cited by Baldwin, regulated only sixty-one mentioned artisan occupations, of which acting was not one. In any case, the Statute had little application to London. There was no guild of players, as Baldwin realizes, and his statement that 'the actors could secure boy labor only through the apprentice law, and they could admit as "masters" in their legal corporation only those who had conformed to that law', is untrue. Nor does the practice described by Baldwin have any sanction in custom.[101]

*In Shakespeare's Warwickshire and the Unknown Years* (1937), by Oliver Baker, is engagingly informative about sixteenth-century houses and domestic surroundings—stained and painted cloths provide matter for a lengthy discourse—and corrects misconceptions (J. Q. Adams's, for example) about the Hathaway cottage, Richard Shakespeare's house, etc., for the author has a rare knowledge of Elizabethan domestic architecture. But the book has chiefly attracted interest for a novel explanation—although less novel than Baker thinks—of the hiatus in the poet's career. In the Revd G. J. Piccope's *Lancashire and Cheshire Wills*, edited for the Chetham Society in 1860, Baker has found a will executed on 3 August 1581 by Alexander Hoghton of Lea, Lancs., in which that worthy shows concern for the well-being of two persons, presumably players, then residing with him. Hoghton leaves all his musical instruments and play clothes to his brother Thomas, but

if he will not keep and maintain players, then it is my will that Sir Thomas Heskethe, knight, shall have the same instruments and play clothes. And I most heartily require the said Sir Thomas to be friendly unto Foke Gyllome and William Shakeshafte now dwelling with me, and either to take them into his service or else to help them to some good master, as my trust is he will.[102]

Now, in the Snitterfield manor records the surname of the dramatist's grandfather has as one of its variants Shakstaff, which is not unlike Shakeshafte.* Although William is not known to have used either form, conceivably he altered his name slightly upon leaving home, possibly without parental consent, to attach himself to the Hoghton household as a player. Baker notes the frequency of the name Foke, or Fulke, in Warwickshire, and the plenitude of Gilloms at Bidford and Henley-in-Arden.

An intriguing possibility. Also intriguing is the date, 1923, when Baker says he came upon the will. For that year witnessed the publication of *The Elizabethan Stage*, and buried in a footnote (i. 280) is the crucial passage from Houghton's will, followed by the query, 'Was then William Shakshafte a player in 1581?' Why Chambers failed to discuss the question in his *William Shakespeare*, he afterwards confessed that he did not himself know, but he returned to the theme in a late essay, 'William Shakeshafte', included in his *Shakespearean Gleanings* (1944). Chambers gives an accurate text of the key provisions of the will—Baker had throughout displayed an amateur's casual-ness about spellings and transcriptions—and furnishes additional informa-tion about the Hoghton and Hesketh families. Chambers goes on to observe that the Stanleys had houses in Lancashire; Ferdinando, Lord Strange sojourned there often, and numerous records of visits by players, perhaps Lord Strange's Men, have come down. If, upon Alexander Hoghton's death, William Shakeshafte passed into the service of either Thomas Hoghton or Sir Thomas Hesketh, he might have readily thence entered Lord Strange's service, and, by an easy transition, made his entrée into the London

* Shakeschafte, thought to be an actual variant of Richard Shakespeare's name, turns out to be a misreading of Shakstaff in a (misdated) 1533 record.

theatrical world. It is a sobering reflection, however, that dozens of Shake-shaftes dwelt in Lancashire and Cheshire in the sixteenth century, although not all had the Christian name William.*

Some have sought to dispel, at least partially, the mystery surrounding the poet's first London years by investigating the topical bearings of an early comedy. As far back as 1903 Arthur Acheson had speculated (in *Shakespeare and the Rival Poet*) that the School of Night mentioned in *Love's Labour's Lost* existed as a clique, that Chapman's poem *The Shadow of Night* belonged to the 'School', and that Shakespeare satirized the coterie in his drama. Acheson's hint was handsomely followed up in the introduction to the New Shakespeare edition of *Love's Labour's Lost* (1923) and in M. C. Bradbrook's *The School of Night* (1936), but of greatest interest from the biographical standpoint is Frances Yates's *A Study of 'Love's Labour's Lost'* (1936). She sees Shakespeare as setting out in life as a pedagogue; not, *pace* Aubrey, in a country grammar school but, rather, in some secret Catholic foundation like the one run by Swithin Wells at Monkton Farleigh. Alternatively, he may have served as a tutor in the house of a Catholic nobleman, Southampton—Dover Wilson's suggestion—being a good bet. In any event Shakespeare early came within the Southampton ambience and met Essex and his two brilliant sisters, as well as Sidney's relations and friends. Thereby hangs a topical tale. In *Love's Labour's Lost* the dramatist sided with the Essex–Southampton faction against their political rival Ralegh, to whose set Chapman was poet-in-chief. Shakespeare also directs oblique satirical thrusts at Florio.

So far Yates offers no startling insights. (Nothing so original as, for example, Janet Spens's suggestion a few years earlier that in the original production Southampton acted the King, and Essex played Berowne.) But Yates goes on to propose that Shakespeare in this comedy defends Lady Penelope Rich—Essex's sister and Sidney's Stella—from indirect attacks, anti-Petrarchan in nature, made by the great Copernican and friend of Florio,

* In an article in *The Review of English Studies* for February 1970, 'Was William Shakespeare William Shakeshafte?', Douglas Hamer answered the question in the negative. The term *players* in the will, he concludes, more likely signifies musicians rather than actors, and the Shakeshafte referred to must have been a good deal older than seventeen, Shakespeare's age in 1581. In the 1970 article Hamer seemed to have made a crucial—indeed devastating—point respecting the bequests in the Hoghton will of 1581 (a sort of early non-subscription tontine arrangement) by which certain servants—eleven out of a total of thirty—were rewarded according to their seniority. The recipients drew annuities which increased with the death of each annuitant until the last drew the entire income for life. But in *Shakespeare: The 'Lost Years'* (Manchester, 1985), E. A. J. Honigmann noted that William Shakeshafte was one of four legatees bequeathed £2; only one of the thirty annuitants received a larger amount. Thomas Barton, Hoghton's steward, was awarded no specified sum, and—according to Hamer's interpretation—should therefore have been one of the younger legatees; but in 1587 Barton described himself as 'about fifty'. A faithful family retainer, Thomas Costom, left £1 per year, should have been older than Barton, but was in fact younger, being twenty-nine in 1581. Hamer's argument thus collapses, and the case for a Lancastrian connection for Shakespeare has been put forward by Honigmann with renewed vigour. Still, if Shakespeare was at seventeen in Hoghton's service, he would have had to be back in Stratford to woo, impregnate, and marry Anne Hathaway before his nineteenth birthday, not—on the face of it—the most plausible of scenarios.

Giordano Bruno, in his *Eroici Furori*. In *Love's Labour's Lost* Berowne's name and some of his characteristics, especially his predilection for astronomical metaphors and his combination of anti-feminism with a high-minded philosophy of love, are deliberately intended to recall Bruno to the spectators. Shakespeare also comes gallantly to the defence of Essex's other sister, the unhappily mated Countess of Northumberland, lately defamed by her husband (a member of the School of Night) in an unpublished essay 'On the Entertainment of a Mistress Being Inconsistent with the Pursuit of Learning'. Yates brought the manuscript to light at the Public Record Office. Both essay and play took their immediate inspiration from the Christmas 1594–5 Grays Inn Revels, which the Essex group planned against the Ralegh group. Shakespeare (who attended the Revels along with the various partisan lords and their ladies) sided with the former, Northumberland with the latter.

Such condensed summary of the argument fails to do justice to Yates's abstruse and eccentric scholarship. The upshot of it all reinforces her initial persuasion that *Love's Labour's Lost* expresses 'the spirit of aristocratic faction'. While no doubt congenial to those whose pulses quicken at mention of the *haut monde*, this esoteric reading of the play must spring from a curious notion of the nature of drama. What sort of audience, one wonders, could have grasped more than a small fraction of these workings of faction? An audience, one suspects, consisting primarily of Yateses. The whole super-structure of theory, as elaborated from Acheson to Yates, rests upon an insecure foundation. There is no evidence to link Chapman directly with Ralegh, and the phrase 'School of Night', spoken by the King of Navarre, may be a textual corruption, perhaps of 'suit of night'. From such fragile threads are recondite hypotheses spun.

Difficulties of this sort have not impeded subsequent pursuits of elusive topical meanings. In *Shakespeare's Rival* (1960) Robert Gittings assumes that *Love's Labour's Lost* was written for private performance before members of the Essex circle. Gittings makes 1597–8 Shakespeare's Living Year, and his special contribution is to suggest connections between the dramatist and an obscure poetaster, Gervase Markham, who (according to the author) figures as Armado in an hypothesized revision of the comedy, and also as the Rival Poet of the Sonnets. As one would expect of Gittings, the theories are gracefully presented, but he writes more persuasively when his subject is Keats.

Biographers of Southampton have understandably dwelt on Shakespeare's aristocratic connections. A. L. Rowse's *Shakespeare's Southampton, Patron of Virginia* (1965) is positive and superficial, G. P. V. Akrigg's *Shakespeare and the Earl of Southampton* (1968) less so. The latter cheerfully admits that conjecture, not fact, must be the interpretative historian's ultimate resort, and is prepared to make the best of it: 'Speculation, properly conducted, can be an exhilarating and exciting pastime.' Like any good student of Southampton, Akrigg identifies the Earl with the Friend of the Sonnets (Mr W.H. is Wriothesley, Henry), and just as naturally he regards *Love's Labour's Lost* as a coterie play written for private production before the nobleman's circle. More unconventional is Akrigg's willingness to see a homosexual strain (although not luridly pederastic) in the poet's love for the Fair Youth, a passion that may have been Shakespeare's greatest emotional experience. When the Earl betrayed him by taking up with the Dark Lady, he left his protégé emotionally maimed for life. In *All's Well That Ends Well* the Helena who loves Lord Bertram despite her inferior social status is the bisexual Shakespeare infatuated with 'that lascivious young boy'. The relationship takes other disguised forms in the plays. Bassanio in *The Merchant of Venice* is Southampton; Antonio is Shakespeare. Patron and poet lurk behind Prince Hal and Falstaff, and when Henry V bitterly excoriates Scroop and the other traitors, the dramatist is remembering his own betrayal. Hence too the 'compulsive theme' of betrayal throughout the canon, in *Julius Caesar, Troilus and Cressida*, and the rest. It is subjective biography, once again, that Akrigg offers, with Southampton cast in the role previously played by Mary Fitton and the Earl of Essex. Other students interpret the same literary evidence differently. Rowse, for example, puts Shakespeare firmly in the heterosexual camp.

Like the makers of myths and traditions, the writers of monographs tend to pass over in silence the middle years of the dramatist's life; they fail to show us the author of *Othello* and *King Lear* in the context of creation. But about his family and Warwickshire links they speak freely.

In *The Shakespeares and "The Old Faith"* (1946) John Henry de Groot subjects the Spiritual Last Will and Testament to the most thorough scrutiny it has ever received, and concludes that this document, except for the first two and a half articles, is authentic. Therefore John Shakespeare remained Catholic all his life. Brought up in a household infused with the spirit of the Old Faith but subjected to conflicting influences from school and church, the poet developed an attitude of religious tolerance which later manifested itself in his writings. A pleasing reconstruction, but much—too much—depends upon the *therefore*. Other John Shakespeares lived in Warwickshire, and the Stratford wheelwright was capable of forging a signature on an authentic document. The disappearance of the actual paper makes all conjecture uncertain, and this biographer has opted for a secular agnosticism as regards the Spiritual Last Will and Testament of John Shakespeare.

But the John Shakespeare Testament—if such it is—had earlier found sponsors in Father Herbert Thurston and the Comtesse de Chambrun (1873–1954), whose colourful life story has a Jamesian resonance. A prominent

Cincinatti heiress, the Comtesse—née Clara Langworth—flowered when the Gilded Age (the phrase is Mark Twain's) was losing its sheen. Enraptured with Shakespeare from the age of six, she went on to marry a French official and adopted his nationality. During the First World War she holed up at the Bibliothèque Nationale and other Parisian repositories; eventually she took a doctorate at the Sorbonne. Equally fluent in English and French, as her numerous publications testify, Chambrun maintained that she had at last 'definitely and permanently' demonstrated that the third Earl of Southampton was Shakespeare's lifelong patron. In her chapter on the Spiritual Testament, which more than once engaged her, in *Shakespeare Rediscovered by Means of Public Records, Secret Reports & Private Correspondence Newly Set Forth as Evidence on his Life & Work*—a work which delivers less than its title promises—Chambrun passionately maintains that the authenticity of the Testament now had 'full confirmation', a conclusion that did not entirely disarm scepticism.

The information adduced by E. A. B. Barnard and H. A. Shield in contributions of almost identical title has a limited antiquarian interest. In *New Links with Shakespeare* (1930), Barnard sheds light on Henry Condell from documents found in a large chest which had for years lain unopened at Hanley Court in Worcester. Also, in the inventory of Ralph Hubaud (at Birmingham Probate Registry), he has found a citation of £20 'Owing to Mr. Shakespre'. It will be recalled that Mr Shakespre had in 1605 paid Hubaud £440 for half of a leasehold interest in a parcel of Stratford tithes. Barnard tries less successfully to show that Shakespeare passed some of his early married years in the Broadway area of the Cotswolds—it is perhaps unnecessary to add that Barnard was a native of Broadway.

Some of the 'Links with Shakespeare' explored by Shield in a series of sixteen articles published in *Notes and Queries* between 1946 and 1958 are tenuous indeed. Patient genealogical research rubs shoulders with free-wheeling guesswork. Identifying the dramatist with William Shakeshafte, Shield rather discouragingly nominates a new Dark Lady, Jane Spencer, with whom Shakespeare had an affair at Rufford Hall, where he was acting in Sir Thomas Hesketh's troupe. The sheets must have shaken merrily at Rufford Hall, for Mistress Jane (a jolly matron) bore Sir Thomas's brother five illegitimate children. Her other qualifications for the role of Dark Lady are less conspicuous.

Even more specialized is Christopher Whitfield's *The Kinship of Thomas Combe II, William Reynolds and William Shakespeare* (1961), revised from an article in *Notes and Queries* (October 1961). An appreciative reader of Barnard, whom he corrects, Whitfield furnishes some fresh genealogical crumbs about the young Thomas Combe to whom Shakespeare left his sword. The rest is dry crusts of speculation. Whitfield believes that the poet's circle of acquaintance, which included many respected gentry 'all linked by a veritable spider's web of blood and marriage relationship with each other', extended south to Chipping Campden in North Gloucestershire, as well as to Broadway in Worcestershire, and the country south of the Avon. Shakespeare attended many a marriage and christening in those parts; so avers Whitfield, who himself hails from Broad Campden.

John Hall has been the subject of several studies. Frank Marcham's *William Shakespeare and His Daughter Susannah* (1931) prints the wills of Hall and his father William; also the texts of minor Chancery proceedings involving the physician in 1623–4 and his widow in 1636–7. More appealing is the idiosyncratic introduction in which Marcham accuses Halliwell-Phillipps and, by implication, Sidney Lee of inability to read Elizabethan handwriting. He also complains that Henry Clay Folger once refused him photocopies of some manuscripts; but, no man to nurse a grudge, Marcham supplies a frontispiece photograph of a benign looking Folger. In *Shakespeare's Son-in-Law* (1939) the unfortunate Arthur Gray, a two-time loser, triumphantly flourishes as new the William Hall will that Marcham printed eight years previously. Lastly, *The Shakespeare Circle: A Life of Dr. John Hall*, by C. Martin Mitchell, reflects much amateur enthusiasm and contains some quotations from unpublished manuscript material.

Of far greater importance than any of these efforts is Mark Eccles's *Shakespeare in Warwickshire* (1961). 'I have found no new records of Shakespeare', Eccles admits in his preface, 'but I have examined the manuscript sources and have given my interpretation of the evidence.'[103] In accordance with this modest programme Eccles sets forth the known facts about Shakespeare's ancestors, friends, and relations—above all, Shakespeare himself in his provincial setting: marrying and begetting, buying property, making his will. Eccles's presentation is stark. In a book that runs to under two hundred pages he can spare only a footnote to discuss, with sober dispassion, the vexed question of John Shakespeare's Testament. The style is correspondingly bare—bleak might be the better term for it. These pages bristle with names and dates, not all of them having the interest of direct relevance to the study of Shakespeare's life. The following passage on the occupants of the New House (not to be confused with New Place) is characteristic:

The New House in High Street was the home of Alderman William Smith, mercer and linendraper, who had been master of the gild in 1540. By his second wife, Alice, sister of John Watson, later Bishop of Winchester, he had six sons, all named in the will of John Bretchgirdle, vicar. After he retired and went to live at Worcester in 1577, the house was occupied by his son John, ironmonger, bailiff in 1604. Alice Smith's will bequeathed the glass and wainscot in the New House, and the wainscot still covers the walls.[104]

One hardly glimpses Shakespeare the artist; Eccles never permits himself a rhetorical flourish or a personal aside or a releasing flash of irony or humour. Nor does he attempt to evoke the atmosphere of place or to remind us that these men and women once had flesh and blood. 'The picture of Shakespeare's life in Warwickshire is a mosaic with most of the pieces missing,' he says, and refuses to restore any of them with the aid of imagination. Instead Eccles merely gives the facts and cites the documents. With this study objective biography reaches its logical extreme; this book is more uniformly austere even than Chambers, whose *William Shakespeare* it may be regarded as supplementing. 'In general', Macaulay once said, 'nothing is less attractive than an epitome.' *Shakespeare in Warwickshire* is an epitome of factual knowledge, and it is unattractive. It is also an exceedingly useful work which scholars would consult, and biographers exploit. This writer has found it an invaluable asset.

Eccles carries self-effacement too far. Despite the absence of any pretence to novelty, some of the facts he presents are new. To explain the significance of the second-best bed, source of so much prejudiced surmise, responsible scholars have examined other Elizabethan wills and extracted similar bequests; but Eccles produces the most apposite parallel of all. In the Prerogative Court of Canterbury he has uncovered the testament of William Palmer of Leamington, Glos., which was executed in 1573. Palmer leaves 'unto Elizabeth my wife all her wearing apparel and my second-best featherbed for herself, furnished, and one other meaner feather-bed, furnished, for her maid', besides doubling the marriage settlement, 'in consideration that she is a gentlewoman and drawing towards years, and that I would have her to live as one that were and had been my wife'.[105] In this case who can doubt that the bequest is intended as an affectionate gesture towards a cherished spouse? Characteristically, Eccles plays down his find by burying it in a note. But he does make part of his narrative his most suggestive discovery. Malone had thought that Shakespeare possibly joined one of the companies—Leicester's, Warwick's, or the Queen's—when they played at Stratford; Pollard favoured the Queen's Men, who stopped there in 1586 and 1587. A coroner's inquest at Thame in Oxfordshire revealed to Eccles that one William Knell had died of a sword wound in the throat on 13 June 1587. This Knell was a player in the Queen's company, which therefore lacked an actor when in the summer of 1587 they regaled the folk of Stratford with plays from their London repertory. Was this the cry of players that Shakespeare joined? A beguiling fancy, but unfortunately no more than that, for the Queen's Men might have touched down at Stratford any time between December 1586 and December 1587. And no evidence exists of any Elizabethan troupe ever having recruited while on the road. London was where the action was in show business, as in so much else.

# 18
## Leslie Hotson

CLEARLY the age of discovery is not past, despite the millions of documents which (so they claimed) passed through the hands of the Wallaces and despite the endless hours, more often than not fruitless, expended by countless other record-office toilers. This truism finds ample demonstration in the career of Leslie Hotson. He made his name in 1925 when he published his account of the circumstances leading up to the violent death of Christopher Marlowe among a seedy tavern crew whose very names smell of roguery: Skeres, Poley, and Frizer. This must surely be reckoned the premier biographical recovery—at least on the literary side—of modern times. It was no fluke; 'the alert ingenuity that detected and followed the clues removes the discovery from the class of happy accidents', the great Kittredge observed at the time, and his judgement was not long after confirmed by Hotson's retrieval, from the Public Records, of nine lost letters from Shelley to Harriet. Inevitably such a hunter will pursue the biggest game of all. In his Shakespeare quest Hotson has come up with nothing so intimate as the Shelley letters and nothing so spectacular as the Marlowe drama. But Hotson's sleuthing has not gone unrewarded.

In his *Memoranda* of 1884, designed to benefit succeeding generations, that old Chancery Lane hand Halliwell-Phillipps advised amateurs in the Record Office to sift the rolls of the Court of Queen's Bench. Acting upon this hint, Hotson found among the entries in the Controlment Roll for Michaelmas Term, 1596, the writ of attachment in which William Wayte craved the peace against William Shakspere, Francis Langley, and two unknown women. This Langley had in 1589 bought the manor of Paris Garden on the Bankside in Surrey, and there, six years later, he put up the Swan Theatre. Because the document is directed to the sheriff of Surrey, Hotson believes that Shakespeare then dwelt in Southwark, migrating across the river from Bishopsgate after his company began a stand at Langley's new playhouse.* Aware that petitions for sureties of the peace were often acts of retaliation, Hotson searched further and found an earlier writ in which Langley swore the peace against William Wayte and William Gardiner.

Hotson now closed in on the latter two quarries in *Shakespeare versus Shallow* (1931). He discovered that Gardiner was a well-known and widely detested Justice of the Peace with jurisdiction over the Bankside and Paris

---

* But such inferences are tricky. Did the Chamberlain's Men ever actually act at the Swan? In early 1596 they are referred to as playing at The Theatre; Pembroke's Men took over at the Swan in February 1597. And would Shakespeare have felt obliged to live so close to his place of employment? Heminges, Condell, and Burbage, we know, resided at some distance from theirs. Perhaps the quarrel, which chiefly involved Langley, took place near the Swan, with Shakespeare a visiting bystander rather than a resident of the district. Nevertheless it is a fact that Malone refers to a document, since vanished, which shows the poet as living in Southwark in 1596.

Garden; Wayte, his stepson and tool, is described as a 'loose person of no reckoning or value'. An earlier incarnation of the monstrous Sir Giles Overreach, Gardiner married a prosperous widow and proceeded to defraud her son and brothers and sisters. By means of other sharp practices, abetted by money-lending, he rose to wealth and civic dignity. In the spring of 1596 Gardiner quarrelled violently with Langley, who denounced him (with some cause) as 'a false perjured knave'. The Justice in turn sought to put the Swan out of business. It would be rash to identify Shakespeare as one of the principals in this feud. Only a single laconic entry mentions him.

But Hotson is not content to let matters rest. He goes on to suggest that in the figure of Justice Shallow the dramatist caricatured the rapacious Gardiner, who was permitted to impale upon his coat of arms—a golden griffin—three white luces (he had married a Lucy). Hence the allusion to white louses in *The Merry Wives of Windsor*. So too Hotson finds Gardiner in the Shallow of *2 Henry IV*, and he identifies Wayte with Abraham Slender. But Gardiner only occasionally blazoned his arms with luces—would any audience have understood the allusion, to which his surname gives no clue? A more serious objection is that the foolish, senile, and essentially harmless Shallow bears little resemblance to the overweening Gardiner. Hotson has a reply: 'To beknave him was to compliment him. But to stage the cunning justice as an imbecile fit only for inextinguishable laughter would flick him on the raw.'[106] This smacks of special pleading, and elsewhere Hotson exaggerates. He calls his first chapter 'Shakespeare's Quarrel', and in the course of his book more than once describes it as such; indeed, as Shakespeare's violent quarrel. But it was not Shakespeare's quarrel: there is no evidence that he ever set eyes on Gardiner. Speculation has a tendency to harden into illegitimate certainty. At the end Hotson refers to 'the ocular demonstration now given us of his [Shakespeare's] dramatic use of some of the life he knew', and in a later work he alludes to his theory as though it were established fact. Because of Hotson's positiveness, as well as his prestigious reputation, his Gardiner–Shallow identification has been taken more seriously than it deserves to be by later biographers—most recently, Irving Ribner in *William Shakespeare: An Introduction to His Life, Times, and Theatre* (1969).

In *I, William Shakespeare* (1937) Hotson sticks his neck out less in exploring the life of Thomas Russell, Esq., one of the two friends appointed by Shakespeare as overseers of his will, and the recipient of a £5 bequest. Hotson comes up with some interesting details about Russell, who was born in the village of Strensham, some miles down the Avon from Stratford, and later resided in his inherited manor of Alderminster, four miles below Clopton Bridge on the road that took Shakespeare to London. Russell's stepfather feuded with the Earl of Pembroke, a circumstance which (Hotson feels) makes more improbable than ever the identification of Mr W.H. with William Herbert. On the other hand, Thomas had a half-brother, Sir Maurice

Berkeley, who was associated with Southampton. Russell married Katherine Bampfield, whose sister became the wife. of William, elder brother of the Henry Willoughby who pined with love for Avisa and was the 'faithful friend' of W.S. in *Willobie His Avisa*. Later, a widower, he remarried with the mother of Leonard Digges. Probably through his stepfather Digges came to know Shakespeare, and thus to eulogize him in verse in the First Folio and the 1640 *Poems*. Of Shakespeare, Hotson writes, 'To learn what manner of men he singled out to share his affection is to penetrate farther into a knowledge of the man himself.'[107] One need not share this view—Shakespeare remains elusive—to be grateful for an attractive biography which rescues from oblivion some minor Elizabethans with Shakespearian connections.

*Shakespeare's Sonnets Dated and Other Essays* (1949) resuscitates a few others. Hotson identifies William Johnson, the citizen and vintner of London who took part with the poet in the purchase of the Blackfriars Gate-house, as the host of the Mermaid Tavern in Bread Street. Another trustee in this transaction, John Jackson, gent., turns up as one of the 'noble wits' at the monthly meetings at the Mermaid. This Jackson married the sister-in-law of Elias James, who brewed beer at the foot of Puddle Dock Hill and died, a rich bachelor, in 1610. He is the godly brewer memorialized in a manuscript epitaph which has been ascribed to Shakespeare. Thus does the expert researcher, attuned to the associations of names, fit together the pieces of the puzzle. One fact alone gives us pause, and that is the ordinariness of the name John Jackson. Hotson also finds suggestive the fact that one Thomas Savage, involved with Shakespeare and four of his colleagues in a tenancy agreement respecting the gardens and grounds of the Globe, left a bequest in his will (brought to light by Hotson) to the widow of Thomas Hesketh of Rufford—the same Hesketh to whom William Shakeshafte was recommended in 1581. Closer to home—literally as well as figuratively—is Hotson's discovery that in 1599, two years before his death, John Shakespeare sued John Walford in the Court of Common Pleas for payment for twenty-one tods of wool, or 588 lbs. (a tod is usually reckoned as coming to twenty-eight pounds), bought of him more than thirty years previously. Thus the old tradition of John Shakespeare the wool-driver, first recorded by Rowe, is not pure fantasy after all but derives from one of John's subsidiary business activities.

Of Hotson's more recent work only his *Mr W.H.* calls for brief notice in these pages. His candidate for the role of onlie begetter (equated with the Fair Youth) is William Hatcliffe, of Hatcliffe, Lincs., who was elected Prince of Purpoole for the Gray's Inn Revels of 1587-8. The argument surely falls or possibly stands upon Hotson's chronology, first promulgated in *Shakespeare's Sonnets Dated*. In *Mr W.H.* he asserts that his conviction (that Shakespeare wrote most of the poems at the age of twenty-five) has been 'widely accepted both in the literary press and by leading writers, scholars, and critics'. But Hotson exaggerates: his case for the dating has been enthusiastically rejected by most scholars. Nor does he demonstrate any connection between Shakespeare and Hatcliffe or between the poet and the Gray's Inn festivities. Instead he wanders down the dark path of cryptogram-hunting, wherein lies Baconianism and madness. A chief limitation of Hotson's later work, including this study, is his proneness to attach the same evidential significance to a literary inference (say, the interpretation of an ambiguous topical allusion) as to an irrefutable documentary discovery. Alas, as a Frenchman practising his English once said to me, 'They is not the same *même*.'

Hotson's literary personality contrasts revealingly with that of his fellow record-hunter Eccles. The two men are near contemporaries, and both began their sleuthing careers by following the spoor of Marlowe. But whereas Eccles persistently plays down his own contributions, Hotson maximizes them by dramatic inflation. The importance of his discoveries tends, indeed, to vary inversely with the lengths of the books in which they are presented (*The Death of Christopher Marlowe* required 76 pages; *Shakespeare versus Shallow*, 375, with more than half devoted to unjustified *pièces justificatives*). This reader must also confess impatience with the excessive heartiness of Hotson's style, all breathless with the thrill of discovery. But popularizers must sometimes pay this price, and Hotson's genuine exuberance has no doubt sustained him through many wearying hours in the isolation of the muniment room. For his additions to the Shakespeare life record—minor and sometimes tangential though they be—one's dominant response must be that of gratitude.

# 19
## *Journeys to the Unconscious*

SOME voyagers have preferred to take interior journeys of discovery. We tend to regard Caroline F. E. Spurgeon as an heroic, although not unfeminine, figure from an already somewhat remote past: a principal architect (with G. Wilson Knight and Wolfgang Clemen) of that profound shift in emphasis away from historicism and the study of the play as an action or an assemblage of characters or the embodiment of a philosophy, and towards a conception of the play as dramatic poem conveying meaning primarily through word-pictures. To think of Spurgeon is to think of her classifications of figures, and of the brightly coloured charts at the back of *Shakepeare's Imagery and What It Tells Us* (1935). We especially remember the second part of her study, 'The Function of the Imagery as Background and Undertone in Shakespeare's Art', with its sensitive discussion of 'the part played by recurrent images in raising and sustaining emotion, in providing atmosphere or in emphasising a theme'. That as far back as 1794 Walter Whiter in some ways anticipated her approach in his *Specimen of a Commentary* does not diminish appreciation for her extension of the frontiers of criticism. We may, however, forget that she also thought of her method as a means of widening the frontiers of biography; it would, she thought, enable us 'to get nearer to Shakespeare himself, to his mind, his tastes, his experiences, and his deeper thought than does any other single way I know of studying him'.[108] Part I of *Shakespeare's Imagery* concludes with a chapter entitled 'Shakespeare the Man'.

What manner of being emerges from the application to biography of this revolutionary method which unlocks the 'storehouse of the unconscious memory'? Shakespeare (we learn) had a healthy mind as well as body, and was cleanly in his habits, with a fastidious disdain for dirt and foul odours. A quiet chap, annoyed by noise, but practical rather than a dreamer: he made creative use of his silences, for he was busily absorbing impressions, registering them like a sensitive photographic plate. The countryside and fresh air appealed to him more than urban pleasures. Shakespeare enjoyed reading; horses he loved, as he did most animals, except spaniels and house dogs (probably because of the filthy practice of feeding them at table in those days). He played bowls with zest, and he was an expert archer. 'He was, indeed, good at all kinds of athletic sport and exercise, walking, running, dancing, jumping, leaping and swimming.'[109] A homy and domestical sort too, neat and handy when it came to household chores, especially if they involved carpentry. He preferred, however, to steer clear of such indoor nuisances as smoky chimneys, stopped ovens, and guttering candles. The attributes of the inner man Spurgeon sums up in five words: courage, sensitiveness, balance, humour, and wholesomeness. (The acute sensitivity, she grants, coexists oddly with the courage.) Of all the virtues, he most prized unselfish love; fear, rather than money, he regarded as the root of evil. If Shakespeare can be described in a word, it is *Christ-like* (perhaps that helps to explain his carpentering hobby). Thus do new critical methods furnish a modern variation on age-old bardolatry.

However pleasing to worshippers, this picture derives from an approach with the appearance but not the substance of objectivity. Spurgeon can be arbitrary in her classification of images. She refuses to allow, moreover, that a simple reference (distinguished from a figure by virtue of being conscious and deliberate) may have imaginative reverberations; conversely, she fails to take into account the overt artistic ends that imagery may serve—the whole rhetorical tradition lies outside her ken. One cannot, moreover, so easily distinguish between conscious and unconscious as Spurgeon seems to think. Emblematic stage imagery she ignores altogether. Nor does she attach due weight to the derivative, the proverbial, and the commonplace as imagistic components.*

Caroline Spurgeon's name appears more than once in Kenneth Muir and Sean O'Loughlin's *Voyage to Illyria* (1937), the title of which stands in conscious parallelism to *The Road to Xanadu*. Her approach is one of the five (along with 'false notes, recurrences, and fervours', Shakespeare's treatment of sources, etc.) that go into the making of what the authors describe as 'A New Study of Shakespeare', dedicated to a vindication of the personal heresy. Accepting Wordsworth's dictum, they look to the Sonnets as the key to the poet's heart. The experience of the cycle—the enslaved lover's detestation of female impurity, his Platonic adoration of the Fair Youth, the betrayal by Southampton—finds analogues in the plays. Valentine–Shakespeare forgiving Proteus–Southampton in the denouement of *The Two Gentlemen of Verona* dramatizes Sonnet 40 ('I do forgive thy robb'ry, gentle thief'). So too the Sonnets help to account for the theme of betrayal in *Julius Caesar* and Shakespeare's melancholy in *Hamlet*. With *King Lear* he reaches rock bottom; in *Timon of Athens* the flames at the centre of the Inferno sear him. Then, with *Antony and Cleopatra* comes the fusion of love and sex, and consequent regeneration, of which the late Romances are the testament. Despite novelties of method, this reconstruction of Shakespeare's internal life has, in its general outlines, a comfortably familiar ring: a synthesis of Campbell, Brandes, and Chambers, old wine in a new bottle. But *The Voyage to Illyria* has an engaging youthful *élan*.

A more pioneering work of imagistic criticism, Edward A. Armstrong's *Shakespeare's Imagination* (1946) illuminates the imaginative faculties of memory, emotion, and reason through an exploration of clusters of figures. (Not surprisingly, the author of *Bird Display* and *Birds of the Grey Wind* chooses images of birds and insects as *points d'appui* for his investigation.) Armstrong detects a peculiar obsessiveness, not necessarily morbid, in the poet's personality. He feels too that Shakespeare found the nucleus of Hamlet within himself, an *aperçu* calculated to startle few readers. Further carefully planned study might yield interesting results about the dramatist's buried life, Armstrong believes, but recognizing the high degree of autonomy of Shakespeare's imagination, he prudently leaves fulfilment of this task to others.

Study of the psychology of association and inspiration draws the investigator on to the subliminal springs of art; to those provinces beyond the threshold of consciousness where the psychoanalytic priesthood perform their mysterious rites. Armstrong himself has little taste for Freud: the latter's description of the imagination impresses him as defective, and even the term 'unconscious' rouses him to philosophical objection. Such scruples have not

---

* Most of these strictures, as well as others, are pronounced in the excellent critique of Spurgeon by R. A. Foakes, 'Suggestions for a New Approach to Shakespeare's Imagery', in *Shakespeare Survey* 5 (1952).

deterred the great host of the faithful who have, as it were, placed Shakespeare on the couch, and, in their professional journals, published the results of their free associations. In so doing they have produced an arcane literature of intimidating magnitude and sometimes stupefying jargon. Here only a brief sampling is possible of what is after all an esoteric branch of Shakespearian biography.

Sometimes the practitioner finds his way to Shakespeare's unconscious via the life record: rather a fragile vessel to bear so weighty a burden. As far back as 1907 Otto Rank observed at the Vienna Society, apropos of Shakespeare's move from Stratford to London, that many poets commence their poetic careers with a flight which for them signifies emancipation from parents and siblings. (This may be so, but it is hard to see how Shakespeare could have become a professional playwright without removing to the capital.) More usually the psychoanalyst links biographical data—the death of John Shakespeare, for example—with the *œuvre*; homosexuality in the Sonnets, and attitudes towards the father in *Hamlet* and towards the mother in *Coriolanus*, are favourite themes. The death of the father and of the father substitute (Essex) is conspicuous in the most celebrated essay in psychoanalytic criticism, Ernest Jones's *Hamlet and Oedipus* (1949), but Jones lays greater emphasis upon the Herbert–Fitton betrayal of the poet, a misfortune that the bisexual Shakespeare—so the analyst surmises on the basis of much clinical experience—helped to bring upon himself. Fortunately Shakespeare assuaged his suffering and preserved his sanity by writing *Hamlet*.

The contributions of Ella Freeman Sharpe exemplify a certain kind of Freudian biographical criticism. Praised by Freud's biographer Ernest Jones for her delicacy in following up slight clues, she suggests that in *Hamlet* Shakespeare dramatized his own 'regression' after the loss of his father and possibly of Mary Fitton as well. All the characters are the dramatist's 'introjections'; Shakespeare—himself the murdered majesty of Denmark, the slain Claudius, the water-logged Ophelia, etc.—ejects these phantasms symbolically and remains sane 'through a sublimation that satisfies the demands of the super-ego and the impulses of the id'. It is reassuring to learn that Shakespeare kept his sanity, although the reader of Sharpe's later essay, 'From *King Lear to The Tempest*', runs the risk of losing his. She finds manifested in *King Lear* a child's massive fantasies re-experienced with all their massive power in adult life. From the play we learn of two occasions when Lear's mother (that is, Shakespeare's mother) was pregnant. On the first, Shakespeare had not yet stabilized his sphincter control. At the time of the second pregnancy he was accustomed to walking about on his own and, when the baby arrived, he ran off. This episode took place, appropriately enough, in harvest time: 'He was found exhausted, dirty, and decked with flowers of late summer.' *King Lear* also reflects repressed childhood observations of female genitalia. 'Gloucester with a bandage over his bleeding eyes looks "like Goneril with a white beard"—telling us of repressed knowledge of menstruation, bandage, and pubic hair.' Sharpe's paper, one suspects, reveals more of her unconscious than of Shakespeare's.

Not all psychoanalytic criticism is so imaginative. In *The Personality of Shakespeare* (1953), Harold Grier McCurdy praises Sharpe on *King Lear* as 'sensitive and really admirable in its genre', but it is clear that she has failed to convince him. The balance displayed in McCurdy's book is quite beyond her. He discusses theory and method; he avoids the vanity of dogmatizing; he has troubled to acquaint himself with the principal critical documents, including Sisson on Shakespeare's mythical sorrows. Regarding the plays as wish-fulfilling fantasies of which the characters constitute the swarming components, McCurdy tallies the number of lines spoken by the various personages, weights them quantitatively, presents tables and graphs, and plots curves. He finds that one-sixth of Shakespeare's 'top-ranking' characters are 'gentle loyal women', and the remaining five-sixths 'proud undependable aggressive men'. He goes on to consider the secondary figures who help, oppose, or mate with those in the first rank. An analysis of themes follows, with tables of 'Sexual Betrayal, Real or Imagined' and the like. McCurdy also takes into consideration the facts of the life. For example, Edmund Lambert, John Shakespeare's brother-in-law, is seen as the chief villain in the events leading to the financial and social decline of the Shakespeares; he finds dramatic incarnation as Claudius.

From the quantitative data a portrait of the dramatist emerges. A bisexual personality, predominantly masculine, aggressive, prone to wide fluctuations of mood, this Shakespeare tried to suppress his feminine traits and to justify his existence by vigorous, even ruthless action. The homosexual tendency helps to explain the poet's paranoid suspiciousness, his jealous imagining that other men have been coming between him and the woman or women he loves. Over the years sensuality and self-indulgence grew more and more repugnant to Shakespeare, but he did not easily accept spiritualized emotions—'he could not admit the loving-kindness of Christian charity without feeling threatened with overwhelming weakness'. With McCurdy we travel some distance from Spurgeon's Christ-like Shakespeare.

What Spurgeon and McCurdy have in common is the illusory character of their objectivity. The psychologist's charts (like the image-classifier's) may dazzle the unwary with the imposing paraphernalia of statistics, but inevitably McCurdy's interpretation of character and theme is personal and reductive. Nor does his appendix comparing his line counts with those of W. J. Rolfe strengthen confidence. The divergences are rationalized as stemming from a different method of estimating part lines and prose passages, but one wonders whether such an explanation will account for a difference of about two hundred in the lines assigned to Hamlet. McCurdy's willingness to identify characters in the plays with actual persons whose lives are virtual blanks is aptly described by Norman N. Holland, applying Whitehead's formula, as 'the fallacy of misplaced concreteness'. The heterodoxy of the depth psychologists, who quarrel with each other over doctrinal issues, complicates the task of evaluation for a mere literary student. Sufficiently bemused by orthodox Freudians, this layman quails before the schismatics. But perhaps one too easily makes uneducated sport of a method operating according to its own laws, which are not those of critical or biographical evidence as they are ordinarily understood. For a more sympathetic response than that mustered in these pages, the reader will do well to consult the comprehensive chapter on 'Psychoanalysis and the Man' in Holland's *Psychoanalysis and Shakespeare* (1966).*

# 20

## *Pop Biography*

MEANWHILE the great tide of popular biography has surged on undiminished, and unruffled by Freudian perturbations. Ivor Brown, who with George Fearon wrote *Amazing Monument: A Short History of the Shakespeare Industry* (1939), has himself helped sustain that industry with *Shakespeare* (1949), *Shakespeare in his Time* (1960), *How Shakespeare Spent the Day* (1963), *The Women in Shakespeare's Life* (1968), and *Shakespeare and the Actors* (1970)—not to mention a one-act play, *William's Other Anne* (1937), dramatizing some of his pet notions. Other titles of the past quarter-century include Hazelton Spencer, *The Art and Life of William Shakespeare* (1947); Charles Norman, *So Worthy a Friend: William Shakespeare* (1947); Leonard Dobbs, *Shakespeare Revealed* (1948); Marchette Chute, *Shakespeare of London* (1949); Frank Ernest Hill, *To Meet Will Shakespeare* (1949); Hesketh Pearson, *A Life of Shakespeare* (1949); M. M. Reese, *Shakespeare: His World and His Work* (1953, rev. 1980); F. E. Halliday, *Shakespeare: A Pictorial Biography* (1956), *Shakespeare in His Age* (1960), and *The Life of Shakespeare* (1961); Hugh R. Williamson, *The Day Shakespeare Died* (1962); Peter Quennell,

---

* Solemn disquisitions, psychoanalytic or otherwise, may stir in the irreverent a craving for innocent diversion, which is happily forthcoming in Caryl Brahms and S. J. Simon's *No Bed for Bacon* (1941). Equipped with a 'WARNING TO SCHOLARS: *This book is fundamentally unsound*', it describes some curious goings-on. Henslowe indulges his mania for burning down his competitor's playhouses. The Immortal Bard, deeply perplexed about the spelling of his own name, makes baffled attempts to get started on *Love's Labour's Won*. At the Mermaid old Ben reads aloud the verse that will one day accompany the Droeshout engraving ('This figure that thou here seest put, | It was for gentle Shakespeare cut. . . .'). 'It is not good?' asks Jonson uneasily. 'It is good', says Shakespeare: 'But untimely'.

*Shakespeare: The Poet and His Background* (1963); and Roland Mushat Frye, *Shakespeare's Life and Times: A Pictorial Record* (1968). To this list, select rather than exhaustive, may be added two handy repositories of information, including the biographical: Halliday's *A Shakespeare Companion* (1952, rev. 1964, 1969) and *The Reader's Encyclopedia of Shakespeare* (1966), edited by Oscar James Campbell and Edward G. Quinn. The latter incorporates the charge brought against Susanna Shakespeare in 1606 for failing to receive the sacrament the previous Easter. This record is to be found in an act book of the Stratford Ecclesiastical Court, formerly among the family papers of the Sackvilles of Knole, and now deposited at the Kent Archive Office in Maidstone; the Assistant Archivist, Hugh A. Hanley, first brought it to light in an article in the *Times Literary Supplement* for 21 May 1964.

As a group these are books that the professional journals for the most part feel no compulsion to review. Most of the authors mine the same secondary sources—Chambers, Fripp, and (later) Eccles—in order to acquaint a heterogeneous audience of non-scholars with the facts and problems of Shakespeare's life. All document lightly when they document at all; none prints the records *in extenso*. Biography of this order requires no sojourns in the record office and no immersion in the Warwickshire ambience; Chute composed her entire Life without straying from the New York Public Library. Under the circumstances a certain amount of duplication is inevitable. One wonders, for example, why Frye should have troubled to compile his 'Pictorial Record' when Halliday's equally attractive (and cheaper) 'Pictorial Biography' was already available. But what astonishes us about these biographies is not how much they overlap but their variety of approach to their subject. In them are exhibited the techniques of journalism, fiction, and austerely factual narrative, of subjective and impersonal Life-writing; one encounters an historian's braggadocio, the encyclopaedist's urge to encompass Shakespeare's whole world, an eccentric's unhinged vision. How unlike (to illustrate the point) are the biographies published in the same year by Marchette Chute and Ivor Brown!

'I, for one,' Brown declares,

absolutely refuse to believe that he kept himself out of his writing, that his Sonnets are a formal exercise, and that his plays are examples of abstract and remote dramaturgy, from whose themes and persons and language he carefully withdrew all personal feeling and opinion. Only the dullest of study-bound professors could pretend that authorship so vibrant as Shakespeare's is possible on those lines. My own opinion (and certainly not mine alone) is that he portrayed and betrayed himself continually.[110]

Behind this manifesto lurks the ghost of Frank Harris. As a youth Brown read the puff for *The Man Shakespeare* by 'Jacob Tonson' (i.e., Arnold Bennett) in *The New Age*, and was thus introduced to the poet's 'tragic life story'. In his autobiography, *The Way of My World*, Brown long afterwards recalled the impact of Harris on his formative mind: 'this was a typically lush piece of writing, but it was a strong stimulant to one who had been bored almost into Bard-hatred by the academic mandarins. In persuading me that Shakespeare was a human being and therefore to be read with affection instead of studied with distaste, Tonson's service won, and has retained, my deep gratitude.'[111]

Despite his professional writer's disdain for dry academics, Brown brings to his task some of the academic virtues: a grounding in the best authorities (Gray, Smart, Alexander) and a sturdy common sense with respect to legendary accretions and the tantalizing blank spaces in Shakespeare's career. Moreover, as a drama critic with practical awareness of the machinery of stage production, he can venture upon his subject through the theatre door, a means of access usually denied university-tied scholars. But literary amateurism carries its penalties. Brown expresses some odd views. Unwilling to reduce Anne Whateley of Temple Grafton to the 'sad status of a shadow born of a misprint', he suggests that dashing Will simultaneously courted two Annes, that he took out a licence to marry the younger and prettier Whateley lass, but that Miss Hathaway, having the stronger claim by virtue of her belly's testimony, sent her champions posting, money in hand, to Worcester for the bond. Thus was poor Anne Whateley left at the church door. Brown regards the anti-Stratfordians seriously enough to entertain the notion that Shakespeare took over hints and ideas from his noble acquaintances, and sometimes even touched up their manuscripts, for they were busy writers who preferred to remain aloof from the degrading theatre; if *Love's Labour's Lost* was not inspired by Bacon, then maybe a member of the Southampton clique was the original begetter.

To these vagaries of speculation must be added excesses of style; too often animation deteriorates into purple journalese. Brown thus describes Shakespeare's transition from his Fourth Age of High-Fantastical to the Fifth Age of Bitter Comedy:

The ugliness of things leaped up at Shakespeare: like assassins surrounding a man from all quarters the deadly sins of envy, lust, jealousy and tyrannical ambition sprang out from behind the arras of his happy high-fantastical imaginings; they stabbed at his ecstasy in living. . . . the sun sank sharply in his sky. The pettiness of man and the frailty of woman began to obsess his meditation and obscure his laughing outlook on the glittering, turbulent panorama of the town. Nor was there the old comfort in the country. The earth which had been so fair a frame for meadows painted with delight became a pestilent congregation of vapours.[112]

Such a style is symptomatic of a simplistic critical vision. This Shakespeare turns out joyous plays when he is happy, tragedies when life oppresses him. By the time he wrote *Antony and Cleopatra* the Dark Lady was dead, and the dramatist could forgive her by making her an Egyptian lass unparalleled— 'The ecstasy and the agony were over,' Brown observes, the expected cliché at his fingertips. In the hands of such a critic a concordance becomes an instrument of medical diagnosis. From Shakespeare's references to ulcers, abscesses, boils, and plague sores, Brown surmises that 'in the early years of the century Shakespeare himself suffered from a severe attack of staphylococcic infection and was plagued with recurrent boils and even worse distresses of the blood and skin. . . .'[113]

Happily, Chute does not burden Shakespeare with boils. She refrains altogether from making biographical inferences from the works. The Sonnets, that irresistible quarry for subjective biographers, she relegates to an appendix (they receive barely a mention in the text). 'Many attempts have been made to interpret the sonnets as autobiographical', she concludes, 'and no doubt the desire to discover something about Shakespeare's private life is legitimate enough. But each reader finds a different story in the sonnets and reaches a different conclusion, and perhaps it is just as well. No single theory can safely be formed about them, and in the meantime William Shakespeare remains securely in possession of his privacy.'[114] The traditions and legends furnish matter for another appendix; here, rather than in the Life proper, one finds Fuller, Ward, Aubrey, Davies, and Rowe, the wit-combats with Jonson, the schoolteaching stint in the country, the fornication with Mistress Davenant, Southampton's largesse, the deer-poaching, and the merry meeting with Drayton and Jonson at which Shakespeare had one cup too many. Chute trusts to contemporary documents exclusively, the latest belonging to 1635, and this she accepts only because it includes the testimony of Cuthbert Burbage, Shakespeare's last surviving fellow. Nor does she attempt to round out her narrative with interpretative analysis of the canon—'The book is not a literary biography,' she insists. A wise prohibition, for her occasional critical remarks have a dispiriting *naïveté*. She lacks the equipment to evoke the genius and intellectual power of her subject; the climate of thought in which he breathed does not interest her. That with so many exclusions Chute should have been able to produce any biography at all is something of a triumph. That it should be as good as it is borders on the miraculous.

This legerdemain she accomplishes by dwelling upon Stratford and London, grammar school and Court, theatres and companies. A danger of the method is that background may become foreground, but somehow she overcomes this. Chute has a way of placing Shakespeare at the centre of Elizabethan circumstance; the poet (as one reviewer remarked) comes 'to dominate, and, enigma as he is, almost to take life at second-hand from the life around him'. Such an achievement—the creation of a biographical illusion—requires art. Yet, despite her exaltation of the fact, Chute does not entirely escape the temptation of conjecture: she postulates a break that 'evidently' took place between Anne and William Shakespeare within three years of their marriage, although no evidence of estrangement has come down.

Chute's total rejection of the Shakespeare–Mythos must have transient appeal to the historian confronted with incompatible and unverifiable traditions. But Malone long ago recognized that such data may harbour golden kernels of truth. 'What then are we to think of tradition as an element in the making of a biography?' Sir Edmund Chambers has asked more recently, and he goes on to supply a reasoned answer:

It is obviously far less reliable than record, which may be misinterpreted, but at least gives a germ of fact, which has to be worked in at its appropriate place. Tradition is attractive. It may deal with more picturesque and intimate matter than record. On the other hand, it may be due to invention. . . . In any case there is room for errors of transmission, either through inexact memories, or through the natural human instinct to leave a story better than you found it. Nevertheless, tradition cannot be altogether disregarded. A country neighbourhood is self-contained and tenacious of outstanding local personalities. . . . Our attitude towards tradition must therefore be one, neither of credulity nor of complete scepticism, but of critical balance. There are criteria to be borne in mind. Does the tradition arise early or late? Does it come from more than one independent source? Does it help to explain record, or contradict it?[115]

By failing to ask these questions, Chute betrays her inadequacy as a serious biographer. But perhaps this is (as Pope said in another connection) to try the citizen of one country by the laws of another. *Shakespeare of London* makes no pretences to being a critical Life, as the absence of any documentation mutely testifies. Chute is a responsible popularizer who has in one respect taken the easy way out. Within her self-imposed limits, she has produced an admirably sane book. That is no contemptible achievement.

Leonard Dobbs sorts oddly with Brown, Chute, and the rest, but then it is the eccentric's prerogative not quite to fit in with any company. His life was a succession of misfortunes. Dobbs sat for a research scholarship in science at Cambridge but was ill throughout the examination week and failed. When he finally took thirds in mathematics and science he concluded, not unreasonably, that he was not cut out for a life of research in pure science. Next Dobbs tried schoolteaching but was unable to enforce discipline; a sever motorcycle accident, which, his memorialist remarks, 'may not have been altogether unwelcome to him', terminated his pedagogical career. He took up sculpture but lacked the strength to work in stone. In Majorca, Dobbs made and sold marmalade but used too much sugar. A turning-point was his meeting with Hesketh Pearson just after (Pearson recalls) Leonard had 'cracked his head open by doing physical jerks in a room, the ceiling of which made it unsuitable for high jumping'. Pearson's *Life of Shakespeare* stimulated Dobbs, between moods of hilarity and dismal brooding, to expound upon the poet at a pub in Whatlington. The biographer encouraged him to put his views down on paper. 'New ideas poured from him in such abundance', Pearson recalls, 'that I began to get alarmed, begging him to steady himself and commence at once on the first volume of what promised to be a library if he went on like that.' After Dobbs's death in 1945, at the age of forty-two, *Shakespeare Revealed* was published through the good offices of Hugh Kingsmill. The latter sensibly withheld endorsement of the ideas developed in what was the last of Dobbs's calamities.

In these strange pages he works out his theory that Shakespeare used his fellow dramatists as models for the leading characters in his plays, which therefore become autobiographical allegories. Shakespeare represented Jonson as Falstaff, the rejection of the latter by Henry V corresponding to the separation of the two playwrights when Jonson left to join a rival company. In *Hamlet* Shakespeare expiated his guilty conscience over dethroning Marlowe with *Titus Andronicus*. 'This allegorical interpretation', Dobbs goes on, 'can, I think, be completed, if we regard the Queen, whom Claudius marries after killing her husband, as symbolizing the art which Marlowe was the first to possess, and which Shakespeare won from him as the seal of the kingship of the drama which he had usurped. Note the subtlety of this idea. . . .'[116] Alas, *Shakespeare Revealed* contains many such subtleties.

None of the remaining members of this heterogeneous crew aspires to match Dobbs for idiosyncratic originality, but their books contribute to the variety of biographical strategy already noted. These writers offer workaday Lives (Quennell), contentious Lives (Rowse), lightly fictionalized Lives (Hill); odd theories and old prejudices. According to Norman (*So Worthy a Friend*) Shakespeare addressed his Sonnets to both Southampton and Pembroke, in Cleopatra the Dark Lady and Mistress Davenant (in life separate) merge, Anne harboured bitter thoughts about the marital bed, and 'A scholarly thesis that the *Contention* and *True Tragedie* are merely surreptitious theater copies of Part Two and Part Three [of *Henry VI*] falls to pieces on examination', which is not forthcoming. In *The Day Shakespeare Died* Williamson presents 'a Catholic actor–playwright who was of no particular account to his contemporaries'. The same author reminds us, as we have not been reminded for half a century, that 'had Sidney Lee not abandoned his real name, Solomon Lazarus, he might have been less uncritically accepted in the first place as an authority on Anglicanism'.[117] The mean-spiritedness of this pointless remark (as though anybody did not know Lee was a Jew, whatever that has to do with his knowledge of Elizabethan Anglicanism) is alien to the temper of M. M. Reese, who accepts the serious obligation of the popularizer in an age when the advance of specialization has made information that should be available to all the province of the few. 'Scholarship', he writes, 'that will not mediate its conclusions in terms acceptable to ordinary men ends by strangling both itself and the object of its attentions.'[118] In *Shakespeare: His World and His Work* Reese mediates the findings of the specialists. To be sure his own scholarship is at second hand, he sometimes makes mistakes and sometimes displays greater confidence than the facts warrant, as when he assumes that Shakespeare sold his theatrical shares for a substantial capital sum. But one can hardly imagine a better ordered or more lucidly presented introduction to the dramatist's age, life, and work than this one, which fairly glows with reasonableness. Of the remaining popular biographies, Rowse's, having stirred the most controversy, most deserves comment here.

Fellow of All Souls, authority on the England of Elizabeth, versifier, and writer of best-selling autobiographical memoirs (*A Cornish Childhood, A Cornishman at Oxford*), Dr Rowse has produced a solid middle-brow biography in a hearty middle-brow style, all roast beef and Yorkshire pudding, to which his fondness for such archaisms as 'pshaw' lends a not unattractive period quaintness. In keeping with his style the author manifests a pre-Freudian innocence which puts him at a disadvantage in dealing with such matters as the sexuality of the Sonnets. No aroma of the record office scents this effort; as any good popularizer must, Rowse takes his facts from Chambers, Fripp, and Eccles. But although he offers not a single new document, Rowse does not disclaim originality. In the marriage (her second) of the Countess of Southampton to Sir Thomas Heneage on 2 May 1594, he discovers the occasion of *A Midsummer Night's Dream*. This suggestion had been put forward, but not pressed, by Stopes almost half a century previous in her life of Southampton.

Rowse maintains that the Sonnets were written between 1592 and 1595, when, for part of the time, plague closed the theatres; that Southampton is the Lovely Boy; that Marlowe furnished the pattern for the Rival Poet; and that Mr W.H.—Sir William Hervey (or Harvey, as Rowse prefers), the Countess of Southampton's third husband—handed over the Sonnets to the printers in 1609 and was in this sense alone the 'onlie begetter'. In presenting his case Rowse brings his formidable historical expertise to bear on the contemporary events alluded to in the poems. He argues forcefully, with little regard for the sensibilities of those who differ from him (Hotson's ideas are dismissed as, of all suggestions, 'the craziest'). But the language of verse, being imaginative and ambiguous, admits of more than one possible interpretation. Moreover, Marlowe, first suggested as the Rival as early as 1859, had no known connection with the Southampton set. It is in the nature of a fact that it puts an end to responsible speculation, but controversy about the Sonnets has gone on merrily since Rowse's biography, and few if any have had to change their allegiance. For the rest in this substantial book (484 pages), Rowse displays his customary magic with political and social history, Warwickshire and the Court; he has his finger on the pulse of the age, although on the theatres and companies he is negligible.

One can scarcely imagine better timed publication: October 1963, on the eve of the quatercentenary year. The appearance of the book was heralded by a series of four leader-page articles by Rowse in *The Times* for September 17–20, carried under the general title, 'Historian Answers Questions about Shakespeare'. An added headline in larger type proclaimed, over the first instalment, 'Only the Dark Lady Still a Mystery'. Early on in this piece Rowse declares. 'I am prepared to stake my reputation as an Elizabethan scholar on the claim that all the problems of the Sonnets save one—the identity of Shakespeare's mistress, the Dark Lady—are susceptible of solution, and that I have solved them.' These words forecast the tone of the book, in which Rowse emerges as the Sir Positive At-all of Shakespearian biography. '. . . I am overwhelmed by what historical investigation, by proper historical method, has brought to light. It has enabled me to solve, for the first time, and definitively, the problem of the Sonnets, which has teased so many generations and led so many people into a morass of conjecture.' Everything is 'now cleared up'. 'Now, for the first time, certainty as to dating has been achieved and the consequences are immeasurable.' '. . . the picture builds up gradually, inescapably, to certainty and conviction'. 'The game [of dating the Sonnets] has now come to an end, for good and all.' And so on; Rowse likes such words as 'fixes', 'confirms', and 'certainty'. Counterpointing the claims are sneers at the literary scholars, a miserable lot of dodos struggling along without Rowse's mastery of historical method. How this method differs from that of historically oriented literary students he nowhere explains, but I gather from my own conversations with him that the historian occupies himself with certainties, firmly stated, and not with speculations, in which the literary folk pointlessly indulge—a distinction made by Rowse in another *Times* article (17 January 1964). Apparently he speaks for himself rather than for his guild as a whole; other historians with whom I have discussed this question do not make positiveness an article of their creed. However this may be, Rowse's arresting pronouncements quickly gave his book notoriety.

A row inevitably followed; first in the correspondence columns of *The Times*, later in reviews and rejoinders to reviews. With the exquisite good taste for which it is celebrated, *Private Eye* on 18 October 1963 edified its readers with an anonymous broadside entitled, 'Swan of All Souls'. The piece quoted authentic samples of the Swan's verse ('The phallic crocuses are up and out | Standing on tiptoe as if to shout . . .'), as well as an extract from that 'new monument to egotistical niggling', *I Am Shakespeare*: 'Once the solution is stated categorically, it seems so obvious that one can only have contempt for those pseudo-experts, lamentably lacking my own historical insight, impeccable taste and implacable self-admiration, who have for so long perversely ignored all the clues which stare them in the face. What other English poet is there who combines in his verse the same admiration for the upper classes, the same interest in sex and the same feeling for nature?' Later the same year there appeared, in poignant contrast to both Rowse and *Private Eye*, Dover Wilson's urbane polemic, *An Introduction to the Sonnets of Shakespeare for the Use of Historians and Others*. A destructive notice of Rowse in the *TLS* ('Shadow and Substance in Shakespeare', 26 December 1963) drew an angry defence from the victim, finally giving the anonymous reviewer (John Crow) another chance to denounce 'Dr. Rowse's fat bad book'.

Since then thunderbolts hurled from All Souls have periodically resparked controversy; on 26 April 1969 a leader-page article, this one on Mr W.H., belatedly marked Shakespeare's birthday in *The Times* and elicited the usual spate of letters, all of them (except one from Rowse himself) unfavourable. Throughout this ordeal of public exposure he did not once relapse into diffidence. 'Really!' he exclaimed in *The Times* on 17 January 1964. 'What is the point of affecting a lack of confidence one does not feel? I should not have *dared* to put out such a book if I were not sure of my ground. No historian would dream of it, let alone a leading Elizbethan historian with much more to lose than a lot of reviewers, for the most part knowing little of the subject.'

Upon bringing out his Life, Rowse, like his great metaphysical poet predecessor (but without the word-play), found that when he had done he had not done: Rowse had claimed to have solved all the problems of the Sonnets save one; everything 'except for the identity of Shakespeare's mistress, which we are never likely to know'. Subsequently, at the Bodleian Library, while working his way through the case-books of Simon Forman, a contemporary of Shakespeare—Forman was born in 1552 and died in 1611—who enjoyed considerable success as physician, astrologer, and lecher, Rowse found his Dark Lady. She was Emilia Lanier, née Bassano, the daughter of Baptista Bassano and Margaret Johnson, who, although unmarried, lived together as man and wife. The Bassanos were a family of Court musicians who had come to England from Venice to serve Henry VIII. Their descendants stayed on at Court in the same capacity; Baptista's will describes him as 'the Queen's musician'. By the time she was seventeen in 1587, Emilia was an orphan with a dowry of £100, not a negligible sum in those days, but she was scarcely an heiress. She mended her fortune, however, by becoming the mistress of Henry Carey, 1st Lord Hunsdon, then well advanced in years. As Lord Chamberlain he supported the players in their sporadic skirmishes with the municipal authorities, and he was himself the patron of an acting company; for a while, just before his death, in 1596, he sponsored Shakespeare's company, the Chamberlain's Men. In 1593 Emilia consulted Dr Forman, who cast her horoscope, and said of her in his dairy that 'She is now very needy, in debt, and it seems for lucre's sake will be a good fellow, for necessity doth compel.' So she was promiscuous. On one occasion Forman, sent for by Emilia's maid, stayed all night with Emilia. In January 1600 she sent for Forman, who wondered 'whether she intendeth any more villainy'. By then he was finished with her.

His case established to his own satisfaction, Rowse recast his biography, mainly (one suspects) to give Emilia a showcase, and in 1973 published it as *Shakespeare the Man*. In the Preface to his emended second printing he claims that the resurfacing of Emilia 'has triumphantly vindicated the answers I have put forward all along, and the method by which they were found . . . The discovery of the Dark Lady completely corroborates, and puts the coping-stone on, my previous findings'—i.e., the chronology of the Sonnets and the identity of Fair Youth, Rival Poet, and Mr W.H. On 29 January 1973 *The Times* carried a feature article, headed 'Revealed at Last, Shakespeare's Dark Lady', by A. L. Rowse. Once published, *The Times* article was summarized in newspapers and magazines the world over, and for weeks afterward the correspondence columns reverberated with responses: heated, facetious, or merely informative. In the same year Rowse brought out a revised edition of the Sonnets, titled *Shakespeare's Sonnets: The Problems Solved*, complete with paraphrases for those who prefer to read their poetry as prose, and with sufficient reference to Emilia in the annotations.

But is she the Dark Lady? She was promiscuous, and her dates do accord with Rowse's dating of the Sonnets. Coming as she did from a musical family, she may well have been accomplished at the virginals. All this is in accord with what the poems suggest. Rowse observes that the husband's Christian name, William, makes an admirable basis for puns, lending another dimension to the word-play of the Will sonnets ('Whoever hath her wish, thou hast thy Will, | And Will to boot, and Will in over-plus.'). Rowse, however, is wrong about the name of the lady's husband: she married Alfonso—not such a good name for puns—not William, Lanier. And was Emilia dark? Stanley Wells was the first to look more closely at the passage in Forman's case-book, which Rowse conveniently reproduced in facsimile in his *Shakespeare the Man*. The key word is not 'brown' at all, but 'brave': she was very brave in youth, here meaning 'splendid', 'fine', 'showy'. We are left with a promiscuous lady, of whom there were others in Elizabethan London. *The Times*, which had announced in a front-page headline 'A. L. Rowse discovers Shakespeare's Dark Lady', quickly beat a prudent retreat, and the correspondence which followed used the non-committal heading, 'Another Dark Lady'. The next year, in 1974, Rowse had a chance to retrace his steps in *Simon Forman: Sex and Society in Shakespeare's Age*. Again he tells Emilia's story. She is now brave, not brown, and her husband's name is correctly given, in passing, as Alfonso. Nowhere does Rowse allude to past errors, and about his thesis he remains impenitent, all the more convinced 'that here in this Italianate woman we have the Dark Lady'. Others are less persuaded.

# 21
## *Journey's End*

THE year that *Shakespeare's Lives* made its first appearance, 1970, saw the publication of one of the most popular of popular Lives of recent decades, Anthony Burgess's *Shakespeare*, sumptuously embellished in both colour and black-and-white illustrations. In its Penguin Books incarnation (1972) this is an undeniably attractive oversized paperback calculated to seduce both Shakespeare and Burgess enthusiasts, two not entirely discrete constituencies, however more numerous the former may be. This *Shakespeare* testifies to the imaginitive resources—not universally congenial to the academically disciplined student—brought to bear on the task by a distinguished novelist who has elsewhere used the same gifts for explorations of James Joyce and Ernest Hemingway and his world.

'For my part', Burgess owns, 'I seize on the song in *Love's Labour's Lost* to visualize the sister Joan as a greasy girl who spends much of her time washing pots and pans in cold water.' (As for *Love's Labour's Lost* itself, this comedy is 'almost painfully aristocratic' and 'highly expressive' of the ethos of the Southampton circle, with 'many references' to the School of Night; in such matters Frances Yates, although not mentioned in this context, casts her long shadow.) Burgess sees Shakespeare's younger brother Gilbert as 'dully pious and possibly epileptic, the source of the falling sickness that comes in both *Julius Caesar* and *Othello*. I think of him as a stolid carver and snipper of gussets, a natural successor to his father in the glover's trade.' Needless to say, the documentary record vouchsafes no inkling that Joan was a greasy girl or Gilbert epileptic (we know only that he resided in both Stratford and London and that a record describes him not as a glover but a haberdasher of St Bride's parish).

So it goes. 'It is reasonable to believe', Burgess unreasonably declares, 'that Will wished to marry a girl named Anne Whateley.' (Shakespeare is customarily called Will in this Life, a cosiness which may not be to all readers' taste.) But there is no evidence for Burgess's reconstruction of events apart from the Worcester clerk's entry of the grant of a wedding licence in the Bishop's Register on 27 November 1582 (one day earlier than the marriage-licence bond itself), giving the bride's name as Anne Whateley of Temple Grafton. Actually we cannot even say with certainty that Anne Whateley ever actually existed. Behind Burgess's invention lurks, I suspect, Ivor Brown, two of whose books, including *The Women in Shakespeare's Life*, are duly cited in an end-note on the books which most helped the author in writing his own book.

One's reservations about the Burgess text apply to the illustrations as well. The views of the Globe playhouse consist of a detail from the Visscher panorama and a colour plate deriving from Visscher, whereas the much more reliable Wenceslaus Hollar panorama makes no appearance in these pages. The 1790 engraving—not, as here claimed, drawing—of New Place, supposedly presenting Shakespeare's house 'as it may well have looked', does no such thing, having only the doubtful authority of Jordan to vouch for it; Vertue's more responsible sketches of the frontage and plan of New Place go unreproduced and unreported. Near the end of the book two small illustrations depict a portion of the third page of Shakespeare's will and the record of his burial—regrettably, as the result of a printing error, these appear upside down and transposed, although an erratum slip compensates for some of the mischief. The caption refers to the poet's 'short, pullulating life'. Pullulating no doubt it was, and short by modern standards; but not all that short by the expectations of the age. Fletcher, who succeeded Shakespeare as the principal dramatist for the King's Men, died of the plague at forty-five while Beaumont, with whom in earlier days Fletcher had collaborated, gave up the ghost in his early thirties. Other examples are not far to seek.

Victoria Glendinning, Dame Rebecca West's biographer, has recently remarked apropos of A. N. Wilson—the novelist, journalist, and biographer of Tolstoy—'Fiction, like biography, is all lies' (*The Times*, 25 August 1988). Some lies are better than others. This reader prefers Burgess's unabashedly novelistic treatment of Shakespeare in the earlier *Nothing Like the Sun* (1964), in which the interior drama of the poet's love life is seen through a glass darkly. He is here not Will but WS. Seduced early one May morning while in a drunken stupor by an Anne Hathaway already bereft of her virgin-knot, WS loves (you guessed it) Anne Whateley of Temple Grafton, seventeen summers old, with a piping boy's voice and a girl's swelling young breasts. Untouched by men she would remain until placed between her lavender-scented sheets—but not by WS, who must marry his other Anne, as her swollen belly requires. Armed with dildo this Anne lures her boy-husband into strange sexual rites and later cuckolds him with his brother Richard on the second-best bed. Meanwhile, WS has been a Gloucestershire tutor (sacked for making a pederastic pass at a pupil), then a lawyer's clerk, and finally player and playwright in London. There the Earl of Southampton (Mr W.H.) tousles him and gives him £1,000 to buy theatre shares. There too he encounters the golden sorceress who had beckoned him once in a vision, then from an open brothel doorway in Bristol. Fatimah, called Lucy Negro, she comes from the East rather than West Indies; irresistible is the allure of her black eyes, thick lips, delicately splayed nose, and cool-warm brown shoulders. WS beds her and is rewarded with the pox. The progress of the spirochete is chronicled in horrifying detail. Hence, in the defiled city the poet's epiphany; WS will no longer be sweet Master Shakespeare. *Troilus and Cressida* now, with its delirium of coinages; Goneril and Timon; the final anger and pity.

An absurd gallimaufry of invention and (to put it midly) dubious biographical theorizing, but Burgess displays a redeeming Joycean gift for language. Diaries and stream-of-consciousness, the workaday life (with New Place the goal) and the dream life, blank verse, literary allusion, puns (Jakes peer), and snatches of song interweave to form a richly textured narrative. Normal syntax dissolves; the effect is Elizabethan yet the spirit modern. Informing it all is a vision of the libidinal springs of art, literature as 'an epiphenomenon of the action of the flesh'. One may not subscribe to this philosophy of creation. One may also discern in the sexual degradation of the protagonist a working out of the author's obsession rather than the fictionalization of fact; in contemplating the fate of WS we think of Crabbe or Hussey or Enderby, or of Spindrift in *The Doctor Is Sick*. For Burgess comes before us as novelist, not scholar, and he is entitled to the biographical irresponsibilities of art.

Meanwhile investigators continue to make responsible additions to knowledge. As an active townsman, John Shakespeare was—like others of his class—involved in litigation on a number of occasions. Four more cases involving him have recently come to light in the Exchequer court; these are chronicled by D. L. Thomas and N. E. Evans in 'John Shakespeare in the Exchequer', *Shakespeare Quarterly*, 35 [1984], 315–18. They reveal the Stratford glover as engaged in subsidiary wool dealings and money-lending transactions deemed illegal by the Tudor parliaments. In buying wool John had violated legislation restricting such activities to manufacturers or merchants of the Staple. (On John Shakespeare's subsidiary wool dealings, see Part III, Chapter 12, and Part VII, Chapter 18). In the sixteenth century—as in later periods—trade depended upon the availability of credit, but anachronistic statutes held usury 'a vice most odious and detestable', and subject to prosecution. Infractions unsurprisingly were widespread. Lacking means to enforce its regulations, the government relied upon private individuals to bring damaging information into the royal courts, the informers being rewarded with half the penalties assessed. Thus, in Hilary term 1570, one of the barons of the Exchequer exhibited an information by Anthony Harrison of Evesham in Worcestershire claiming that John Shakespeare of 'Stratford upon Haven' had lent the substantial sum of £100 to John Mussum of 'Woltun' in Warwickshire for £20 in interest. In the same term James Langrake of Whittlebury, Northamptonshire, reported that John 'Shappere alias Shakespere' of Stratford had lent John Musshem of Walton D'Eiville, a village not far from Stratford, the sum of £80, also with £20 due in interest. This Langrake, a professional informer, had been accused of raping one of his maidservants; he had been imprisoned for accepting considerations from the unfortunates he informed against, and was subsequently fined and prohibited from bringing further informations for a twelvemonth. In the first case there is no further record of proceedings after the recital of the accusation; in the second, John was fined 40s. The same pestiferous Langrake in 1572 informed the court that John 'Shaxspere' of 'Stretford super Haven' and John Lockesley of the same place had illegally bought 200 tods (i.e. 5,600 pounds) of wool, at a price of 14s. per tod, from Walter Newsam and others. Later that year John was accused—once again by Langrake—of buying 100 tods of wool at the same price from Edward and Richard 'Graunte' and other unidentified individuals. In both instances the accused apparently wound up compounding with the informer. Of John Shakespeare's parters in his transactions little is known, but Edward Grant was a Catholic landowner in Snitterfield who had a manor-house at nearby Northbrook. Recusants dangerously resorted to his house. His wife's nephew, John Somerville, plotted against the Queen in 1583, and his grandson John Grant would in the next reign be implicated in the Gunpowder Plot. John Shakespeare's usurious activities cast a curious sidelight on his playwright-son's *Merchant of Venice*, written a quarter of a century later.

Like father, like son, when it came to business dealings? One may ask—as E. A. J. Honigmann has asked, in his paper '"There is a world elsewhere": William Shakespeare, Businessman' at the World Shakespeare Congress in West Berlin (1986)—whether Richard Quiney was prepared to pay interest on the loan he solicited in his undelivered letter of 25 October 1598 to Shakespeare, and whether his 'loving good friend and countryman' would have expected to receive it. Like other prospective lenders, Shakespeare may well have. No explicit reference is made by Quiney to the payment of a fee, so we cannot certainly say whether any such payment would have been involved; but the possibility—to Professor Honigmann, the likelihood—is certainly there. Richard's father Adrian, a well-to-do mercer and Henley street neighbour of the Shakespeares, wrote to his son around 1 November, 'If you bargain with William Shakespeare and receive money there'—that is, at his hands—'bring your money home that you may buy such wares as you may sell presently to profit.' Friendly neighbours the Quineys and Shakespeare may have been, but neighbourly fellow-feeling need not—then as now—have sentimentally affected the outcome of business practicalities. The 'indifferent conditions' for which Abraham Sturley hoped may, in the event, have proved to be hard conditions.

In my account of William-Henry Ireland (see Part III, Chapters 8 to 10) in *William Shakespeare: Records and Images* (London and New York, 1981), are included facsimiles of items in the Folger Shakespeare Library, the British Library, and the Bodleian Library, as well as in the collection of Stuart B.

Schimmel of New York City and the collection of the Viscountess Eccles at Four Oaks Farm in New Jersey. Included is a photograph of Shakespeare's purported lock of hair, now also at Four Oaks Farm.

The chapter on 'Shakespeare Forgeries: Ireland and Collier', in the same book, recounts more briefly J. Payne Collier's almost century-long career (1789–1883), this time with a number of items reproduced in photographic facsimile. Some of these only passingly engage Dewey Ganzel in his richly readable *Fortune and Men's Eyes: The Career of John Payne Collier* (Oxford, 1982). In his preface, Ganzel allows that when he began his biography he was convinced of the validity of the traditional verdict on Collier's malfeasance, but as his research progressed, traditionalism slipped away, and what finally emerged was (as Ganzel himself puts it) a whodunit. Ganzel himself becomes counsel for the defence, unravelling an 'involved and complex story'. He offers new arguments favouring the genuineness of some at least of the readings purportedly introduced by the Old Corrector in his copy of the Second Folio, the chief exhibits being variants peculiar to the First Quarto of *Titus Andronicus* (1594), which did not resurface until after Collier's death. On close inspection, however, the arguments have their illusory aspect: see Arthur Freeman's devastating review of *Fortune and Men's Eyes* in the TLS (22 April 1983), 391–3. Among those Ganzel consulted when pursuing his research were Collier's great-granddaughter, the later Mrs Violet Cornelia Koop and her daughters. Koop furnished details of family history and gave Ganzel access to letters and manuscripts in her possession. Ganzel did not, however, seek to avail himself of Freeman's Collier collection, which contains unique items. Nor does he really confront, for example, the challenge offered by Joan Alleyn's letter of 21 October 1603, to her actor husband Edward, in the Henslowe Papers at Dulwich College. The postscript to the letter, despite the 'most decayed state' of the paper, as described by Collier, could never have contained the phrase 'Mr Shakespeare of the glob', which appears in Collier's transcription. A real-life Perry Mason, unlike his fictional counterpart, does not necessarily win all his cases. In *William Shakespeare: A Study of Facts and Problems*, E. K. Chambers furnishes an Appendix on 'Shakespearean Fabrications', of which the Collier inventory occupies eight pages of vol. ii. Not all of those items in the long list are indeed spurious: of the Simon Forman notes on Shakesperian performances Chambers says, 'These have been questioned, but are genuine.'

At the same time, Chambers's list is not complete. As I observe in 'Another Collier Forgery', *Notes and Queries*, n.s. 18 [1971], 155–6, there is a rambling epistle addressed by one Rychard Cockes to 'the right worshipfull Lady Lucy', which talks about venison and thus suggests that—contrary to Malone—Sir Thomas indeed had a deer-park in which Shakespeare might, after all, have poached, and as a result fled Stratford to avoid prosecution. Collier read his paper before the Society of Antiquaries on 2 December 1852, and it was published in the Society's journal, *Archaeologia*, the next year (35, 18–22). Unsurprisingly, the Cockes letter is nowhere to be found.

Further evidence of Collier's delinquency continues to surface, for example in Giles E. Dawson's painstaking essay, 'John Payne Collier's Great Forgery', *Studies in Bibliography*, 24 [1971], 1–26. There, Dawson demonstrates, from palaeographical evidence, that of Collier's notes on the eighty-three interpolated ballads in a manuscript commonplace-book at the Folger Shakespeare Library are products of the same hand. The normal inference is that the hand belongs to Collier, who once owned the item.

In 'Master W. H., R.I.P.', *PMLA*, 102 [1987], 42–54, Donald W. Foster maintains that, 'With but one unremarkable exception, nowhere do I find the word *begetter, father, parent*, or *sire* used to denote anyone but the person who wrote the work.' (The one exception to the poet-as-begetter metaphor appears in Samuel Daniel's prefatory sonnet to his *Delia* (1592) cycle, in which his verse is said to have been begotten by the Countess of Pembroke's hand.) 'By 1600, each role associated with the genesis of a printed text—author, editor, translator, patron, and printer—had already been assigned a conventional counterpart in the business of procreation.' As for Mr W.H., Foster suspects that he originated in a misprint, that Thorpe had written 'W. SH.' or W. Sh.', and that Eld's compositor omitted a letter—a mistake overlooked in a hasty proof-reading.

At around the same time, in 'Mrs Shakespeare', Barbara Everett of Somerville College, Oxford, had come up with a nomination of her own, Anne Hathaway's brother William, in *London Review of Books*, 8 [19 December 1986], 7–10. Shakespeare must have had lockable cupboards in his strong-built house, and in Anne he had a wife with keys to unlock them. Everett writes: 'Conceivable as a powerful, even attractively masculine woman, eight years older than the writer; one capable of obsessing her young husband for many miserable jealous years, then of maddening and amusing and at last ("second-best bed") boring him—it is believable that this perhaps ambitious, clever and wilful woman impatiently sent her brother off to London with the bundle of fair-copied, brilliant, confused poems which her obstinate husband wouldn't publish and which she in any case remembered, rightly or wrongly, as being mostly addressed to herself and therefore arguably her own.' Anne gave the manuscript poems to her brother, who took them to London and offered them to Thorpe. Hence the willing publisher's grateful dedication. A 'crossfire of letters' followed Everett's article—the phrase is Herbert Mitgang's in 'New Answers to Shakespeare Riddle', his summary account in the *New York Times* (3 March 1987), in which Foster and Everett are quoted as taking polite exception to one another. Also quoted is Dr Rowse, who has himself not been behindhand in contributing to the gaiety of nations with his own speculations. Rowse was less polite. 'Rubbish! Absolute rot!' he fulminated. 'Is there no end to human foolery?'

In *The Life and Times of William Shakespeare* (London, 1988), an attempt, for the general reader, to put Shakespeare's poetry 'in the context of his life and times', Everett's Oxford colleague, Peter Levi, the Oxford Professor of Poetry, scorns 'the recent crazy suggestion' that W.H. betokens William Hathaway whose 'very existence is pure conjecture' (it is not), and diffidently offers his own W.H.: William Hole, in 1618 appointed 'Head Sculptor of the iron for money in the Tower and Elsewhere for Life', the life terminating in 1624. Hole was an early English engraver of music who (Levi feels) 'almost certainly' knew Shakespeare. However that may be, this author fearlessly predicts that, whether or not there will be an end to human foolery, we have not heard the last of Mr W.H.

There are, unsurprisingly, other William Halls. John Boe ('Mr W.H.: A New Candidate,' *Shakespeare Quarterly*, 37 [1986], 97–8) offers the father of the physician who married Shakespeare's elder daughter, and himself apparently also a physician. Perhaps this Hall was one of the 'private friends', as Meres puts it, to whom Shakespeare's 'sugared sonnets' found their way, the manuscript being sold to Thorpe after Hall's death in 1607.

In *Shakespeare and the Public Records* (1985), with text and selection of documents by David Thomas, and a section by Jane Cox on the will and signatures, special interest attaches to Cox's discussion of the six Shakespeare signatures customarily regarded as authentic. As has often happened in the past, a sceptical inquirer has made necessary a re-examination of comfortable assumptions. She even queries the authenticity of the will signatures, although their shakiness points to a gravely ill signator holding the pen. 'Until the Statute of Frauds of 1667 there was no necessity for a will to bear the testator's signature at all. Manuals of the period indicate the form preferred by the doctors of civil law, namely that a will should be signed on every page and witnessed, but virtually any form was acceptable so long as it seemed to be a true representation of the dying man's wishes.' Even the names of the three witnesses to Shakespeare's will—Shaw, Robinson, and Sadler—look suspiciously similar to Cox. She finds numerous instances of 'forgeries' of witnesses' signatures among fifty-five wills proved in the Prerogative Court of Canterbury in the same month as Shakespeare's. Cox has deigned to milk a sacred cow.

'In 1594, when my story ends, 'Shakespeare turned thirty and more than half his life lay behind him.' Thus begins Russell Fraser's comparatively brief, yet complexly orchestrated narrative in *Young Shakespeare* (New York, 1988). Fraser has flashbacks to earlier times: to the Mystery cycles and to Moralities like *The Castle of Perseverance* with its personified Vices and Virtues; flash forwards: to such late Shakespeare as *The Tempest* and *Henry VIII*—even to *The Two Noble Kinsmen*, a collaboration with John Fletcher which may belong to Shakespeare's last year as a playwright for the King's Men. Stratford and London (and points between), the playhouses and audiences, Marlowe and the other University Wits, chronology, anti-Stratfordianism—Fraser sums up a lot of ground in expressive, at times eloquent prose. He takes a dim view of Shakespeare's marriage, which Fraser sees as 'pestered', the failure not owing to difference in years. Other students see the matter differently. 'No biography exists', Fraser maintains in his preface, 'that is simultaneously a comprehensive and scrupulous account of the life, and a consideration, worth having, of the art. This is the book I have sought to write.' Comprehensive *Young Shakespeare* is not, and it is not always scrupulous (there inevitably are lapses), but it is very much a book worth having.

Sallying forth once more unto the breach, Rowse—now well into his eighties—in 1989 gave us *Discovering Shakespeare: A Chapter in Literary History*. Unsurprisingly we have already heard much of what he tells us. The Sonnets constitute an autobiographical sequence which should be taken in *as a whole* (emphasis Rowse's) 'like a novel or a play'; the Dark Lady is Emilia Lanier; Simon Forman, medic and astrologer, makes his expected appearance; Thomas Thorpe 'without doubt' obtained the manuscript of the Sonnets from Sir William Hervey. But *Discovering Shakespeare* tells 'a shocking story' that gives this short book of fewer than two hundred pages its not inconsiderable piquance. Did not John Crow's long anonymous review in the *TLS*, now a quarter of a century old, malign Rowse's biography of Shakespeare? No matter that Crow is now dead and, we are assured, forgotten. But what can we expect from 'the shiftless society of today' or from the Shakespeare Trade Union, all those mean, unimaginative third-raters and their ilk? At least with Rowse readers always know where they stand; no obscurantist he. Never mind the occasional slip (*pace* Rowse, that most familiar of icons, the engraved Droeshout portrait of Shakespeare, appears not on the frontispiece to the 1623 Folio but the title-page): Rowse's prose has not lost its pungency.

# 22

# *Epilogue*

EARLY in the quatercentennial year someone remarked that the most sincere tribute to the National Poet might be 365 days of silence, but, as this suggestion appeared in a paper on Shakespeare read before a learned assembly, it failed to contribute to that laudable end. Although the quatercentenary yielded no major biography, not silence but an orgy of celebration greeted the anniversary: books, articles, lectures; ephemera, mostly pieties ground out for the occasion.

If the Shakespearian theme has lost none of its potency, the twentieth century, now in its last decade, yet lacks an authoritative Life conceived in the modern spirit. Looking back, one may perhaps wonder whether such a Life is possible. It is after all a melancholy truth that the three greatest of all contributions to the biographical tradition—Malone's posthumous biography, Halliwell-Phillipps's *Outlines*, Chambers's study of facts and problems—abandon continuous narrative. Perhaps we should despair of ever bridging the vertiginous expanse between the sublimity of the subject and the mundane inconsequence of the documentary record. What would we not give for a single personal letter, one page of diary! Hardy expressed what many have felt when he wrote:

> Bright baffling Soul, least capturable of themes,
> Thou, who display'dst a life of commonplace,
> Leaving no intimate word or personal trace
> Of high design outside the artistry
>     Of thy penned dreams,
> Still shalt remain at heart unread eternally.

A certain kind of literary biography, rich in detail about (in Yeats's phrase) the momentary self, is clearly impossible.

Yet the subject still beckons. Knight early and Lee late in the last century demonstrated the feasibility of a sustained, even shapely, biographical narrative; and Rowse, for all his limitations, has recently demonstrated it anew. Each generation must reinterpret the documentary record by its own lights and endeavour to sort out the relations of the man and the masks in the plays and Sonnets. Whatever we conclude in this regard, we may discern, in the *oeuvre* as a whole, the mysterious workings of a poet and dramatist's imagination; we can follow the development of mind and art, which, in the final resort, matter more to us than Shakespeare's private sorrows and ecstasies. We have too the advantages of a greater knowledge than our predecessors of the social, political, and cultural backgrounds, of the theatres and companies, and of the facts of the life themselves. What the greatest of Shakespearian scholars thought conceivable must surely now be more so than in Malone's time. Despite the barrenness of the present, every age craves its own synthesis; inevitably the attempt will be made. Meanwhile, Shakespeare abides.

# Notes

## Part I

1. Letter of William Hall to Edward Thwaites. Bodl. MS Rawlinson D. 377, fo. 90; printed by E. K. Chambers, *William Shakespeare: A Study of Facts and Problems* (Oxford, 1930), ii. 260.
2. Loc. cit.
3. Charles Douglas (*vere* Douglas Goldring), *Artist Quarter* (London, 1941), 341–2.
4. Jorge Luis Borges, 'Everything and Nothing', *Selected Poems 1923–1967*, ed. Norman Thomas DiGiovanni (New York, 1972), 259–60.
5. Richard Verstegan, *A Restitution of Decayed Intelligence* (Antwerp, 1605), 294.
6. For a compendious history of this structure, see Levi Fox, 'The Heritage of Shakespeare's Birthplace', *Shakespeare Survey 1* (Cambridge, 1948), 79–88. For my account of John Shakespeare and the Warwickshire context I have found especially helpful Mark Eccles, *Shakespeare in Warwickshire* (Madison, Wis., 1961).
7. Edgar I. Fripp, *Shakespeare, Man and Artist* (London, 1938), i. 43.
8. Chambers, *William Shakespeare*, ii. 41–2.
9. For a fairly recent and sensible treatment of the grant of arms, see C. W. Scott-Giles, *Shakespeare's Heraldry* (London, 1950), 27–41.
10. John Manningham, *Diary*, in British Museum. Harl. MS 5353; fo. 29ᵛ; printed by Chambers, *William Shakespeare*, ii. 212.
11. Patent Roll, I Jac. I, p. 2, m. 4; Gerald Eades Bentley, *Shakespeare: A Biographical Handbook* (New Haven, Conn., 1961), 91–2.
12. The complete text of the will is reproduced by Chambers, *William Shakespeare*, ii. 170–4.
13. *Greenes, Groats-worth of Witte* (London, 1592), sig. A3ᵛ.
14. Ibid. sigs. F1ᵛ–F2; reprinted by Chambers, *William Shakespeare*, ii. 188.
15. John Semple Smart, *Shakespeare: Truth and Tradition* (London, 1928), 196.
16. Thomas Nashe, *Pierce Penilesse His Supplication to the Diuell*, in *Works*, ed. Ronald B. McKerrow; rev. F. P. Wilson (Oxford, 1958), i. 154.
17. H[enry] C[hettle], *Kind-Harts Dreame* (London, n.d.), sigs. A3ᵛ–4; reprinted by Chambers, *William Shakespeare*, ii. 189.
18. *Kind-Harts Dreame*, sig. A4.
19. Francis Meres, *Palladis Tamia* (London, 1598), sig. Oo1ᵛ–Oo2; reprinted by Chambers, *William Shakespeare*, ii. 194, Bentley (*Shakespeare*, 199–203) takes the measure of Meres.
20. *Palladis Tamia*, sig. Oo2; Chambers, 194.
21. *Palladis Tamia*, sig. Oo4; Chambers, 195.
22. Two texts of the poem (which is not certainly Beaumont's) survive: Holgate MS, fo. 110, in the Pierpont Morgan Library, and British Museum Add. MS 30982, fo. 75ᵛ; Chambers (*William Shakespeare*, ii. 224), whom I quote, prints the Holgate text.
23. *Poems: Written by Wil. Shake-speare. Gent.* (London, 1640), sig. L1.
24. William Basse, 'On Mr. Wm. Shakespeare', British Museum MS Lansdowne 777, fo. 67ᵛ; printed by Chambers, *William Shakespeare*, ii. 226. Some of the variations between manuscripts are described by L. Toulmin Smith in *The Shakspere*

*Allusion-Book* ([London, 1909], i. 288), and the manuscripts themselves are listed by Ingleby (i. 289).

25. Ben Jonson, 'To the memory of my beloued, The Author Mr. William Shakespeare: And what he hath left vs', *Mr. William Shakespeares Comedies, Histories, and Tragedies*, sig. A4.

26. 'The Epistle Dedicatorie', sig. A2ᵛ.

27. *Timber*, in Ben Jonson, *Works*, ed. C. H. Herford and P. and E. Simpson (Oxford, 1925–52), viii. 583–4.

28. Jonson, 'To the memory of . . . Mr William Shakespeare', *Mr. William Shakespeares Comedies, Histories, and Tragedies*, sig, A4ᵛ.

29. Caroline F. E. Spurgeon, *Shakespeare's Imagery and What It Tells Us* (Cambridge, 1935), 202.

30. James Russell Lowell, 'Shakespeare Once More', in *Among My Books* (Boston, 1870), 200–1.

31. W. Carew Hazlitt, *Shakespeare: Himself and his Work: A Biographical Study* (London, 1914), 45.

32. William Shakespeare, *Venus and Adonis* (London, 1593), sig. A⁴.

33. William Shakespeare, *The Rape of Lucrece* (London, 1594), sig. A².

34. William Shakespeare, *Shake-speares Sonnets* (London, 1609), sig. A⁴.

35. Henry Willobie, *Willobie His Avisa* (London, 1594), c. xliv; reprinted by Chambers, *William Shakespeare*, ii. 191.

36. Thomas Heywood, *An Apology for Actors* (London, 1612), sig. G4ʳ⁻ᵛ; Chambers, *William Shakespeare*, ii. 218.

37. W. J. Fraser Hutcheson, *Shakespeare's Other Anne* (Glasgow, 1950), 68.

38. John F. Forbis, *The Shakespearean Enigma and the Elizabethan Mania* (New York, 1924), 200–1.

39. See Gerald Eades Bentley, *Shakespeare and Jonson: Their Reputations in the Seventeenth Century Compared* (Chicago, 1945); esp. ii. 130–40.

40. John Dryden, *The Life of Lucian*, in *The Works of Lucian* (London, 1711), i. 3–4.

# Part II

1. Thomas Fuller, *The History of the Worthies of England* (London, 1662), 'The Principality of Wales', 16.

2. The epitaph has only recently been noted. See the short article by Robert C. Evans, '"Whome None But Death Could Shake": An Unreported Epitaph on Shakespeare', *Shakespeare Quarterly*, 39 [1988], 60.

3. *A Banquet of Jeasts, or Change of Cheare* (London, 1630), no. 259; cited in *The Shakspere Allusion-Book*, i. 347.

4. Anon., Bodl. MS Ashmolean 38; printed by E. K. Chambers, *William Shakespeare: A Study of Facts and Problems* (Oxford, 1930), ii. 246.

5. Thomas Plume, Plume MS 25 in library at Maldon, Essex; printed by Chambers, *William Shakespeare*, ii. 247.

6. *Satiromastix*, I. ii. 366–7, in Thomas Dekker, *Dramatic Works*, ed. Fredson Bowers (Cambridge, 1953–61), i.

7. Bodl. MS Rawlinson Poet. 160, fo. 41; printed by Chambers, *William Shakespeare*, i. 551.

8. Leslie Hotson, 'Shakespeare Mourns a Godly Brewer', *Shakespeare's Sonnets Dated* (London, 1949), 111–24.

9. It is so described by Mr Evans the auctioneer, cited by J. O. Halliwell[-Phillipps], *Some Account of the Antiquities, Coins, Manuscripts, Rare Books, Ancient Documents, and Other*

*Reliques, Illustrative of the Life and Works of Shakespeare* (London, 1852), 33; a facsimile is provided on p. 35, and also in Halliwell[-Phillipps]'s edition of *The Works of Shakespeare*, i. 162. The epitaphs in this manuscript read:

> Shakspeare: An Epitaph on Sir Edward Standly.
> Ingraven on his tomb in Tong Church.

> Not monumental stones preserves our fame,
> Nor sky-aspiring pyramids our name;
> The memory of him for whom this stands
> Shall outlive marble and defacers' hands.
> When all to time's consumption shall be given,
> Standly for whom this stands shall stand in Heaven.

> *Idem, ibidem*: On Sir Thomas Standly.
> Ask who lies here but do not weep;
> He is not dead; he doth but sleep.
> This stony register is for his bones;
> His fame is more perpetual than these stones,
> And his own goodness with himself being gone,
> Shall live when earthly monument is none.

10. Quoted by John Quincey Adams, 'Shakespeare as a Writer of Epitaphs', *The Manly Anniversary Studies in Language and Literature* (Chicago, 1923), 87. I have profited from this essay.
11. Printed together with Patrick Hannay, *A Happy Husband* (London, 1619), sig. L2ᵛ. See also note 15.
12. *A Relation of a Short Survey . . .*, 77–8.
13. Anon., Bodl. MS Ashmolean 38; printed by Chambers, *William Shakespeare*, ii. 246.
14. From fo. 72 of a MS in the possession of B. Dobell, printed in the *Athenaeum*, 19 January 1901; see also Chambers, *William Shakespeare*, ii. 251.
15. John Aubrey, *Brief Lives*, Bodl. MS Aubrey 6, fo. 109. Printed by Chambers, *William Shakespeare*, ii. 253; see also *Aubrey's Brief Lives*, ed. Oliver Lawson Dick (London, 1949), 275.
16. Nicholas Rowe, 'Some Account of the Life &c. of Mr. William Shakspear', in Shakespeare, *Works*, ed. Rowe (London, 1709), vol. i, p. xxxvi; *Eighteenth Century Essays on Shakespeare*, ed. D. Nichol Smith (2nd edn.; Oxford, 1963), 20.
17. John Jordan, *Original Memoirs and Historical Accounts of the Families of Shakespeare and Hart*, ed. J. O. Halliwell[-Phillipps] (London, 1865), 32; reprinted by Chambers, *William Shakespeare*, ii. 294.
18. Francis Peck, *Explanatory and Critical Notes on Divers Passages of Shakespeare's Plays*, appended to *New Memoirs of the Life and Poetical Works of Mr. John Milton* (London, 1740), 222–3; reprinted by Chambers, *William Shakespeare*, ii. 273.
19. Jordan, *Original Memoirs*, 32; Chambers, *William Shakespeare*, 294.
20. Richard James, *Epistle*, Bodl. MS James 34; BM Grenville MS 35. Printed by Chambers, *William Shakespeare*, ii. 241–2.
21. Ibid.; Chambers, *William Shakespeare*, 242.
22. *The Life of Sir John Oldcastle*, prol., 6–8, 13–14, in Malone Society Reprint, ed. Percy Simpson (London, 1908).
23. For the former view see, for example, J. Q. Adams, *A Life of William Shakespeare* (London, 1923), 228; for the latter (a minority position), see Leslie Hotson, *Shakespeare versus Shallow* (Boston, 1931), 15, and 'The Earl of Essex and "Falstaff"', Shakespeare's Sonnets Dated, 150. Hotson suggests that the Brooke faction may

have commissioned *Sir John Oldcastle* in order to rehabilitate their slandered ancestor ('The Earl of Essex and "Falstaff"', 157).

24. Shakespeare, *Works*, ed. Rowe, vol. i, p. ix; Nichol Smith (ed.), *Eighteenth Century Essays*, 5.

25. The problem is ably discussed by William Green in his chapter 'The Brook–Broome Variant', *Shakespeare's 'Merry Wives of Windsor'* (Princeton, N.J., 1962), 107–20.

26. John Dennis, *The Comical Gallant* (London, 1702), sig. A2; reprinted by Chambers, *William Shakespeare*, ii. 263.

27. *The Person of Quality's Answer to Mr. Collier's Letter* (1704), reprinted in John Dennis, *Original Letters, Familiar, Moral and Critical* (London, 1721), ii. 232.

28. Shakespeare, *Works*, ed. Rowe, vol. i. pp. viii–ix; Nichol Smith (ed.) *Eighteenth Century Essays*, 5.

29. Charles Gildon, *Remarks on the Plays of Shakespear*, in *The Works of Mr William Shakespear* (London, 1710), vii. 291. This volume supplements Rowe's edition.

30. William Shakespeare, *A Collection of Poems* (London, n.d.), sig. A2ᵛ.

31. This identification was made by Oldys in a note in his copy of Fuller's *Worthies*. The copy has disappeared, but Steevens quotes Oldys's marginalia in his *Shakespeare* (London, 1778), i. 205.

32. Oldys's note is mentioned by Malone, *Supplementary Observations* (London, 1780), i. 44. For the rumour that Shakespeare 'spent at the rate of 1,000*l.* a year, as I have heard', see *Diary of the Rev. John Ward*, ed. Charles Severn (London, 1839), 183. The story of Shakespeare's bequest to his sister appears in Aubrey's memoranda; see Chambers, *William Shakespeare*, ii. 253.

33. Ibid. 279.

34. *The Lives and the Characters of the English Dramatick Poets. First Begun by Mr. Langbain, improv'd and continued down to this Time, by a Careful Hand* [i.e., Charles Gildon] (London, n.d.[?1698]), 126. The particular absurdity of this tradition is marked by G. E. Bentley (*Shakespeare: A Biographical Handbook*, 9–10).

35. Gildon, *Remarks . . .*, in Shakespeare, *Works*, ed. Rowe, vii. 404.

36. Charles Gildon, *Reflections on Rymer's Short View of Tragedy*, in *Miscellaneous Letters, and Essays, on Several Subjects. By Several Gentlemen and Ladies* (London, 1694), 89.

37. Lewis Theobald, *Double Falsehood; Or, the Distrest Lovers* (London, 1728), editor's preface.

38. Anon., *Essay against Too Much Reading* (London, 1728), pt. ii, p. 14.

39. Shakespeare, *Works*, ed. Rowe, vol. i, p. vi; Nichol Smith (ed.), *Eighteenth Century Essays*, 3.

40. The notes were first published by George Steevens in his edition of the *Plays of William Shakspeare* (London, 1778), i. 204.

41. Edward Capell, *Notes and Various Readings to Shakespeare* (London, 1774), I. i. 60.

42. Loc. cit.

43. R. B. Wheler recorded this story on 16 October 1807; a copy is in the Halliwell-Phillipps *Shakespeareana* in the Folger Library.

44. Numerous instances are collected in *The Shakspere Allusion-Book*, ii.

45. Nicholas L'Estrange, *Merry Passages and Jeasts*, BL MS Harleian 6395, fo. 2; printed in Jonson, *Works*, ed. Herford and Simpson, i. 186–7. By citing as his source 'Mr Dunn', L'Estrange gave rise to the curious idea that he had the story from Donne; but the Baronet, who never mentions Donne, elsewhere cites a Captain Duncomb as his authority.

46. Shakespeare, *Plays*, ed. Steevens, i. 204–5.

47. This information is offered anonymously in the *Town Jester*, around 1760, and reported in *The Life of William Shakespeare* (London, 1848), 187–8, by

Halliwell[-Phillipps], who says that he found substantially the same story in another collection, where it was ascribed to the poet Phillipps.

48. Anon., *The Life of That Reverend Divine, and Learned Historian, Dr. Thomas Fuller* (London, 1661), 61.

49. Fuller, *Worthies*, 2.

50. Ibid., 'Warwick–Shire', 126.

51. Shakespeare, *Works*, ed. Rowe, vol. i, pp. xii–xiii; Nichol Smith (ed.), *Eighteenth Century Essays*, 7.

52. Octavius Gilchrist, *An Examination of the Charges Maintained by Messrs. Malone, Chalmers, and others, of Ben Jonson's Enmity, &c. towards Shakspeare* (London, 1808), 1.

53. Shakespeare, *Works*, ed. Rowe, vol. i., p. xiii; Nichol Smith (ed.), *Eighteenth Century Essays*, 7.

54. Loc. cit.

55. John Dryden, *Essays*, ed. W. P. Ker (Oxford, 1900), ii. 18.

56. Shakespeare, *Works*, ed. Rowe, vol. i, p. xiv; Nichol Smith (ed.), *Eighteenth Century Essays*, 8.

57. The suggestion is made by Nichol Smith in *Eighteenth Century Essays*, 286.

58. Anon. (John Roberts), *An Answer to Mr. Pope's Preface to Shakespear* (London, 1729), 11.

59. Joseph Spence, *Observations, Anecdotes, and Characters of Books and Men, Collected from Conversation*, ed. James M. Osborn (Oxford, 1966), i. 23.

60. Ben Jonson, *Works*, ed. W. Gifford (London, 1816), viii. 181 n. 4.

61. The letter is printed by Anthony Powell, *John Aubrey and His Friends* (London, 1963), 183.

62. Quoted in Aubrey, *Brief Lives*, ed. Dick, p. lvii.

63. John Britton, *Memoir of John Aubrey* . . . (London, 1845), 15.

64. Chambers, *William Shakespeare*, ii. 254.

65. In his useful biography, *Sir William Davenant: Poet Venturer* (Philadelphia, 1935), Alfred Harbage writes: 'He [Aubrey] repented later and scored out this word [whore] very carefully, then the whole sentence' (p. 17). This may be a correct reconstruction, but I can find no authority for it, and we know that Wood bowdlerized material sent to him by Aubrey.

66. Langbaine, *Lives and Characters of the English Dramatick Poets*, ed. Gildon, 32.

67. *Remarks and Collections of Thomas Hearne*, ed. C. E. Doble (Oxford Historical Society; Oxford, 1885–1921), ii. 228. The same jest appears in the *Wit and Mirth* of the Water Poet ([London, 1629], no. 39, sig. B8; reprinted in *All the Workes of John Taylor* [London, 1630], 184), where, however, the child is unidentified and the godfather is 'goodman *Digland* the Gardiner'. Oldys, in his marginalia to Langbaine, not unreasonably suspected that Taylor was using a fictitious name.

68. Spence, *Observations, Anecdotes, and Characters* . . ., i. 184.

69. Shakespeare, *Plays*, ed. Steevens, i. 203–4.

70. Spence, *Observations, Anecdotes, and Characters* . . ., i. 185.

71. William Chetwood, *A General History of the Stage;* [(*More Particularly the Irish Theatre*) *from its Origin in Greece down to the Present Time*] (London, 1749), 21 n.

72. Walter Whiter, *A Specimen of a Commentary on Shakespeare* (London, 1794), 100 n.

73. Sir Walter Scott, *Waverley Novels* (Edinburgh, 1846), x. 226; cited by Harbage, *Sir William Davenant*, 4–5.

74. Sir William Davenant, *Madagascar; with Other Poems* (London, 1638), 37. The three-stanza poem is reprinted in Chambers, *William Shakespeare*, ii. 236–7.

75. John Dryden, *Works*, ed. Sir Walter Scott (Edinburgh, 1882), iii. 106.

76. John Downes, *Roscius Anglicanus, or, an Historical Review of the Stage* (London, 1708), 24.
77. Ibid. 21.
78. Shakespeare, *Works*, ed. Rowe, vol. i, p. x; Nichol Smith (ed.), *Eighteeenth Century Essays*, 5–6.
79. It is first attributed to Davenant, who 'us'd to tell the . . . whimsical Story', in a manuscript note, dated *c*.1748 by Halliwell-Phillipps, in Edinburgh University Library; see Chambers, *William Shakespeare*, ii. 284. Halliwell-Phillipps thought the handwriting resembled that of Oldys.
80. This work was published as by 'Mr. Cibber', but that Shiels was chiefly responsible we learn from Johnson and Boswell (see *Boswell's Life of Johnson*, ed. G. Birkbeck Hill; rev. L. F. Powell [Oxford, 1934], i. 187; iii. 29–31). Shiels provides the genealogy of the tradition ('William Shakespear', *Lives of the Poets*, i. 130).
81. See below, 436.
82. John Aubrey, *Brief Lives*, Bodl. MS Aubrey 6, fo. 109; Chambers, *William Shakespeare*, ii. 253.
83. Aubrey, *Brief Lives*, Bodl. MS Aubrey 6, fo. 109; Chambers, *William Shakespeare*, ii. 252.
84. Thomas Plume, Plume MS 25, fo. 161; Chambers, *William Shakespeare*, ii. 247.
85. Aubrey, *Brief Lives*, Bodl. MS Aubrey 6, fo. 109; Chambers, *William Shakespeare*, ii. 253.
86. The point is made by A. L. Rowse, *William Shakespeare: A Biography* (London, 1963), 48. Shakespeare movingly describes the slaughter of a calf in an elaborate simile in *2 Henry VI* (III. i. 210–16).
87. *The Charters of Endowment, Inventories, and Account Rolls, of the Priory of Finchale, in the County of Durham*, ed. James Raine (Surtees Soc. Pubs.; London, 1837), gloss., 441.
88. Shakespeare, *Works*, ed. Rowe, vol. i, pp. ii–iii; Nichol Smith (ed.), *Eighteenth Century Essays*, 1–2.
89. Chambers, *William Shakespeare*, ii. 259.
90. Shakespeare, *Works*, ed. Rowe, vol. i, p. v; Nichol Smith (ed.), *Eighteenth Century Essays*, 3. For this section Alice Fairfax-Lucy's attractive family chronicle, *Charlecote and the Lucys* (London, 1958), has furnished useful information and details.
91. Wood, *Athenae Oxonienses*, ed. Philip Bliss (London, 1813), vol. i, p. cxiii.
92. Corpus Christi College Library MS Fulman xv, 22, no. 7; Chambers, *William Shakespeare*, ii. 257.
93. William Shakespeare, *Plays and Poems*, ed. Edmond Malone (London, 1790), i. 1. 107.
94. Capell, *Notes*, ii. 75.
95. Loc. cit.
96. Cited by Steevens in his edition of Shakespeare's *Plays*, i. 223 (note to *Merry Wives*).
97. In his marginalia to Langbaine's *Account* (BM C. 28. g. 1), 455, Oldys says that the tradition derives from 'old Betterton'; but Betterton made inquiries in Stratford on behalf of Rowe when the latter was working on his 'Account'—see below, 204–5.
98. For instances of approbation see Adams, *Life of William Shakespeare*, 81.
99. Life of Shakespeare in *Biographia Brittannica* (London, 1763), vi. 1. 3627, signed 'P'.
100. 'Letter from the Place of Shakespeare's Nativity', *British Magazine Or Monthly Repository for Gentlemen and Ladies*, 3 [June 1762], 301.
101. MS in possession of Halliwell-Phillipps, and printed by him in *Outlines of the Life of Shakespeare* (7th edn.; London, 1887), ii. 326.
102. Loc. cit.

103. Robert Bell Wheler, *Collectanea de Stratford*, 202. in Shakespeare Birthplace Record Office.

104. Aubrey, *Brief Lives*, Bodl. MS 6, fo. 109; Chambers, *William Shakespeare*, ii. 254.

105. Chambers, *William Shakespeare*, ii. 259; Rowe, 'Account', p. vi.

106. William Shakespeare, *Plays*, ed. Samuel Johnson (London, 1765), vol. i, p. c.

107. David Lloyd, *Statesmen aad Favourites of England Since the Reformation* (London, 1665), 504.

108. Aubrey, *Brief Lives*, Bodl. MS 6, fo. 109; Chambers, *William Shakespeare*, ii. 253.

109. Shakespeare, *Plays and Poems*, ed. Malone, i. 2. 113.

110. Loc. cit.

111. Ward, *Diary*, 17.

112. Ibid. 184.

113. Ibid. 183.

114. W. L. Rushton, *Shakespeare An Archer* (London, 1897), 113.

115. Chambers, *William Shakespeare*, ii. 257.

116. Letter from Jordan to the editor of the *Gentleman's Magazine*, published by Halliwell-Phillipps in *Outlines*, ii. 399. On the Testament I have found the work of de Groot (see below, VII. 17) and McManaway most helpful.

117. The episode was the subject of a column by the Norfolk historian and journalist George Tucker in the *Virginian-Pilot and Ledger–Star* for 24 July 1988.

118. Edmond Malone, *An Inquiry into the Authenticity of Certain Miscellaneous Papers and Legal Instruments* (London, 1796), 198–9.

119. Anon. (John Roberts), *Answer to Mr. Pope's Preface*, 45, Chambers, *William Shakespeare*, ii. 272.

120. *Original Collections on Shakespeare and Stratford-on-Avon, by John Jordan, the Stratford Poet, Selected from the Original Manuscripts, Written about the Year 1780*, ed. J. O. Halliwell[-Phillipps] (London, 1864), 54; Chambers, *William Shakespeare*, ii. 292.

121. Shakespeare, *Plays and Poems*, ed. Malone, i. 1. 136 n. 7; Chambers, *William Shakespeare*, ii. 273.

122. Chambers, *William Shakespeare*, ii. 259.

123. *Lucan's Pharsalia*, trans. Nicholas Rowe (Dublin, 1719), pp. i–ii.

124. Edward Phillips, *Theatrum Poetarum* (London, 1675), 194.

125. Shakespeare, *Works*, ed. Rowe, vol. i, pp. i–ii; Nichol Smith (ed.), *Eighteenth Century Essays*, 1.

126. Dennis, *Original Letters*, i. 20.

127. *Gentleman's Magazine*, 100 [1830], 515.

128. Shakespeare, *Works*, ed. Rowe, vol. i, p. xxxiv; Nichol Smith (ed.), *Eighteenth Century Essays*, 19.

129. Shakespeare, *Works*, ed. Rowe, vol. i, p. x; Nichol Smith (ed.), *Eighteenth Century Essays*, 5.

130. See Abraham Fraunce, *The Arcadian Rhetoricke*, ed. Ethel Seaton (Lutrell Soc. Reprints, no. 9; Oxford, 1950), pp. xli–li.

131. Shakespeare, *Works*, ed. Rowe, vol. i, p. xxvi; Nichol Smith (ed.), *Eighteenth Century Essays*, 15.

132. Shakespeare, *Works*, ed. Rowe, vol. i, pp. viii, x; Nichol Smith (ed.), *Eighteenth Century Essays*, 4–5, 6.

133. William Ayre, *Memoirs of the Life and Writings of Alexander Pope, Esq.* (London, 1745), i. 209.

134. Dryden, *Essays*, ed. Ker, i. 175.

135. Shakespeare, *Works*, ed. Rowe, vol. i, p. xxxvii; Nichol Smith (ed.), *Eighteenth Century Essays*, 21.

136. Undated letter to Dennis (March ?1693/4) in *The Letters of John Dryden*, ed. Charles E. Ward (Durham, N.C., 1942), 71.

137. Samuel Johnson, *Lives of the English Poets*, ed. George Birkbeck Hill (Oxford, 1905), ii. 71.

138. Giles Jacob, *The Poetical Register: or, the Lives and Characters of the English Dramatick Poets. With an Account of Their Writings* (London, 1719), i. 228.

139. Loc. cit.

140. William Shakespeare, *Plays*, ed. G. Steevens (London, 1785), i. 194 n; cited by Smith, *Eighteenth Century Essays*, p. xxxviii.

141. Alexander Pope, *Imitations of Horace*, ed. John Butt (2nd edn.; London, 1953), 199 (The Twickenham Pope, iv).

142. William Shakespeare, *Works*, ed. Alexander Pope and William Warburton (London, 1747), vol. i, p. viii.

143. William Shakespeare, *Plays*, ed. Samuel Johnson (London, 1765), vol. i, pref., sig. C8.

144. The transcript and its provenance are fully described by Levi Fox in 'An Early Copy of Shakespeare's Will', *Shakespeare Survey* 4 (Cambridge, 1951), 69–77.

145. Letter of Joseph Greene, BL MS Lansdowne 721, fo. 2.

146. *Reliques of Ancient English Poetry. Consisting of Old Heroic Ballads, Songs, and other Pieces of Our Earlier Poets, together with Some Few of Later Date*, ed. Thomas Percy (4th edn.; London, 1794), i. 143.

147. William Guthrie, *A General History of Scotland, from the Earliest Accounts to the Present Time* (London, 1768), viii. 358.

148. Shakespeare, *Plays*, ed. Steevens [1778], i. 203.

149. John Taylor, *Records of My Life*, (London, 1832), i. 25.

150. Ibid. 28.

151. Anonymous (James Yeowell), *A Literary Antiquary. Memoir of William Oldys, Esq.* (London, 1862), p. xxxvii.

152. Alexander Chalmers, *The General Biographical Dictionary* (London, 1815), xxiii. 336.

153. Ibid. 339.

154. *Mr. William Shakespeare His Comedies, Histories, and Tragedies*, ed. Capell (London, 1767–8), vol. i, introd., 71–2.

155. Ibid. 72.

156. Ibid. 73–4.

# Part III

1. John Upton, *Critical Observations on Shakespeare* (London, 1746) 28.

2. 'Gleanings', *Universal Magazine*, 93 [1793], 184. My attention was drawn to this passage, as to several others, by Robert Whitbeck Babcock, *The Genesis of Shakespeare Idolatry 1766–1799* (Chapel Hill, NC, 1931), an unexciting but admirably detailed and informative monograph.

3. William Shakespeare, *Works*, ed. Alexander Pope (London, 1725), vol. i, pref., p. vi; *Eighteenth Century Essays on Shakespeare*, ed. D. Nichol Smith (2nd edn.; Oxford, 1963), 47.

4. William Shakespeare, *Plays*, ed. Samuel Johnson (London, 1765), vol. i, pref., sig. C1ᵛ; Nichol Smith (ed.), *Eighteenth Century Essays*, 125.

5. Maurice Morgann, *Essays on the Dramatic Character of Sir John Falstaff* (London, 1777) 105 n; Nichol Smith (ed.), *Eighteenth Century Essays*, 249 n.

6. Joseph Ritson, *Remarks, Critical and Illustrative* (London, 1783), pref., p. vi.

7. Martin Sherlock, *A Fragment on Shakspeare* (London, 1786), 13–14.

8. Shakespeare, *Works*, ed. Pope, vol. i, pref., pp. vi–vii; Nichol Smith (ed.), *Eighteeenth Century Essays*, 47.

9. Peter Whalley, *Enquiry into the Learning of Shakespeare* (London, 1748), pref., pp. iv–v.

10. Zachary Grey, *Critical, Historical, and Explanatory Notes on Shakespeare* (London, 1754), pref., pp. vi–vii.

11. Ibid. ii. 146.

12. Quoted by E. S. Shuckburgh, *Laurence Chaderton, D.D. . . . Richard Farmer, D.D.* (Cambridge, 1884), 48.

13. *George Steevens Correspondence*, Folger Shakespeare Library MS C. b. 10.

14. Richard Farmer, *An Essay on the Learning of Shakespeare*, (3rd edn.; London, 1789), 93–4; Nichol Smith (ed.), *Eighteenth Century Essays*, 201.

15. Nichol Smith (ed.), *Eighteenth Century Essays*, 152.

16. Farmer, *Essay*, pref., n.p.; Nichol Smith (ed.), *Eighteenth Century Essays*, 151.

17. James Northcote, *The Life of Sir Joshua Reynolds* (2nd edn.; London, 1818), i. 152.

18. *Boswell's Life of Johnson*, ed. G. Birkbeck Hill; rev. L. F. Powell (Oxford, 1934), iv. 18.

19. Farmer, *Essay*, pref., n.p.; Nichol Smith (ed.), *Eighteenth Century Essays*, 152.

20. Farmer, *Essay*, pref., n.p.; Nichol Smith (ed.), *Eighteenth Century Essays*, 153.

21. *Universal Magazine*, 55 [1769], 159.

22. Charles Dibdin, *The Professional Life of Mr. Dibdin Written by Himself* (London, 1803), i. 77.

23. See Mollie Sands, *The Eighteenth Century Pleasure Gardens of Marylebone* (Society for Theatre Research; London, 1987), esp. 78–9.

24. Edmond Malone, *Supplement to the Edition of Shakespeare's Plays Published in 1778 by Samuel Johnson and George Steevens* (London, 1780), ii. 369 n.

25. *Lichtenberg's Visits to England*, trans. and ed. Margaret L. Mare and W. H. Quarrell (Oxford, 1938), 105.

26. T. B., *Gentleman's Magazine*, 39 [1769], 345.

27. Anon., *Shakespeare's Jests, or the Jubilee Jester* . . . (London, n.d.), 1, 1–2, 17. This was apparently the first jestbook to be published in America (New York, 1774).

28. Ibid. 13.

29. Ibid. 61.

30. 'On Literary Fame, and the Historical Characters of *Shakspeare*', *Essays by a Society of Gentlemen, at Exeter* (London, [1796]), 249.

31. John Taylor, *Records of My Life* (London, 1832), ii. 155.

32. James Boswell, 'A Biographical Memoir, &c.', in Shakespeare, *Plays and Poems*, ed. Boswell and Malone (London, 1821), vol. i, p. iv.

33. James Prior, *Life of Edmund Malone* (London, 1860), 9.

34. Ibid. 19.

35. Joseph Farington, *The Farington Diary*, ed. James Greig (London, 1922–8), ii. 186; entry for 29 Jan. 1804. In 1795, Boswell informed Farington, Malone offered himself to Maria Bover, a celebrated wit and beauty, but was rejected. He has never been a favourite of the ladies,' observed Boswell, expert in such matters, 'He is too soft in his manners' (*Farington Diary*, i. 88).

36. Prior, *Malone*, 34.

37. *Farington Diary*, i. 44.

38. Prior, *Malone*, 56.

39. *Manuscripts and Correspondence of James, First Earl of Charlemont* (Historical Manuscripts Commission, 12th Rep., app., pt. x; 13th Rep., app., pt. viii; London, 1891–4), i. 338–9.

40. See *Boswell's Life of Johnson*, iii. 359 n. 2; iv. 36 n. 3, 51 n. 2.

41. William Shakespeare, *Plays*, ed. S. Johnson and G. Steevens (2nd edn.; London, 1778), I. 280 n. o.

42. See G. B. Harrison, 'A Note on *Coriolanus*', *Joseph Quincy Adams Memorial Studies*, ed. James G. McManaway, Giles E. Dawson, and Edwin E. Willoughby (Washington, DC, 1948), 239–40.

43. William A. Ringler, Jun., 'Spenser, Shakespeare, Honor, and Worship', *Renaissance News*, xiv (1961), 159–61.

44. Gildon, *Remarks*, in Shakespeare, *Works*, ed. Rowe, vii. 447.

45. Malone, *Supplement*, i. 653. Italics Malone's.

46. *The Modern Universal British Traveller; or, A New, Complete, and Accurate Tour through England, Wales, Scotland, and the Neighbouring Islands. Comprising All That Is Worthy of Observation in Great Britain* (London, 1779), 180. The articles on England were contributed by Charles Burlington.

47. Malone, *Supplement*, i. 654. In a note to this note Steevens takes advantage of the occasion to declare his disbelief in the Combe epitaph.

48. Ibid. 657.

49. Gildon, *Remarks . . .*, in Shakespeare, *Works*, ed. Rowe, vii. 446.

50. William Shakespeare, *Plays*, ed. Johnson and Steevens (London, 1793), vol. i, advertisement, pp. vii–viii.

51. *Manuscripts and Correspondence of James, First Earl of Charlemont*, i. 387.

52. For more detailed discussion, see S. Schoenbaum, *Internal Evidence*, 22–4.

53. *Letters of James Boswell*, ed. Chauncey Brewster Tinker (Oxford, 1924), ii. 381.

54. Boswell, *Life of Johnson*, i. 7.

55. William Shakespeare, *Plays and Poems*, ed. E. Malone (London, 1790), i. 2. 180.

56. Shakespeare Birthplace Record Office MS 15, fos. 5ᵛ–6; *Original Letters from Edmund Malone, the Editor of Shakespeare, to John Jordan, the Poet*, ed. J. O. Halliwell[-Phillipps] (London, 1864), 12.

57. Frank Simpson, 'New Place: The Only Representation of Shakespeare's House, from an Unpublished Manuscript', *Shakespeare Survey* 5 (Cambridge, 1952), 55–7.

58. Shakespeare, *Plays and Poems*, ed. Malone, i. 1. 118 n. 6.

59. Ibid. i. 2. 1.

60. Letter from Malone to Warton, dated 17 Aug. 1789; British Library MS Add. 42561, fo. 208ᵛ.

61. Letter from Burke to Malone, MS in Beinecke Rare Book Library of Yale University.

62. Shakespeare, *Plays and Poems*, ed. Malone, i. 1. 388.

63. Ibid. 414.

64. Letter from Malone to Charlemont, dated 15 November 1793, in *Manuscripts and Correspondence of James, First Earl of Charlemont*, ii. 221.

65. Chambers, *William Shakespeare*, ii. 184.

66. Victor Hugo, *William Shakespeare*, trans. Melville B. Anderson (Chicago, 1887), 35.

67. Shakespeare Birthplace Record Office MS 122, fo. 46ᵛ. Letter dated 29 April 1790.

68. *Original Collections on Shakespeare and Stratford-on-Avon, by John Jordan, the Stratford Poet, Selected from the Original Manuscripts, Written about the Year 1780*, ed. J. O. Halliwell[-Phillipps] (London, 1864), 44.

69. William Henry Ireland, *Confessions*, (London, 1805), 21.

70. Samuel Ireland, *Picturesque Views on the Upper, or Warwickshire Avon* (London, 1795), 189.

71. William Henry Ireland, *A Full and Explanatory Account of the Shakespearian Forgery*, 65 (MS in the possession of Mrs Donald F. Hyde at Four Oaks Farm, NJ). In a prefatory statement Ireland says that his 'hasty Notes' were written at various periods; a late entry (219) gives his age as 26, which would make the year of composition 1803 by

his own reckoning, although the date 1800 appears in the margin. What is clear, despite these confusions, is that the *Full and Explanatory Account* antedates Ireland's printed *Confessions*. From the latter he discreetly omitted any reference to his father cursing during the visit to New Place.

72. W. H. Ireland, *Confessions*, 30–2.

73. A transcription of the Wheler paper appears in vol. xx of the Halliwell[-Phillipps] scrap-books in the Folger Shakespeare Library (MS W. b. 223). A MS note in the British Library copy of William-Henry Ireland's *Authentic Account of Shakespearian Manuscripts* (shelf mark: 642. d. 29) describes Williams as 'a country wit who amused himself with telling the story in order to ridicule Mr. Ireland'. Grebanier's suggestion that Jordan had a share in the deception is unsupported.

74. MS note (1834) in *Ireland's Shaksperian Fabrications* in Harvard College Library (TS 680. 23. 5F).

75. W. H. Ireland, *Confessions*, 7.

76. Herbert Croft, *Love and Madness* (London, 1780), 138; cited, not entirely accurately, by Grebanier (*Great Shakespeare Forgery*, 66–7).

77. W. H. Ireland, *Confessions*, 11.

78. Ibid. 45.

79. Ibid. 64.

80. Ibid. 50–1.

81. Ibid. 53.

82. James Boaden, *Memoirs of the Life of John Philip Kemble, Esq.* (London, 1825), ii. 163; cited, along with the following, by Grebanier (*Great Shakespeare Forgery*, 24–5).

83. Ibid. 164.

84. For texts of the Ireland fabrications I have used (unless otherwise stated) the texts published by his father in *Miscellaneous Papers and Legal Instruments under the Hand and Seal of William Shakespeare* (London, 1796), but I have followed the capitalization of the facsimiles.

85. W. H. Ireland, *Confessions*, 86.

86. Ibid. 57.

87. Ibid. 68. Italics Ireland's.

88. W. H. Ireland, *Full and Explanatory Account*, 104.

89. W. H. Ireland, *Confessions*, 197.

90. Lancelot Andrewes, *A Sermon Preached before the Kings Maiestie . . .* (London, 1606), 14. Copy in the collection of Mr Stuart B. Schimmel.

91. Letter from 'Mr. H.' to S. Ireland, BL MS Add. 30346, fo. 47ᵛ.

92. Letter from 'Mr. H.' to S. Ireland, ibid., fos. 54–5.

93. Letter from S. Ireland to 'Mr. H.', ibid., fo. 66.

94. W. H. Ireland, *Confessions*, 77. Italics Ireland's.

95. Ibid. 96.

96. Letter from Francis Webb to the Revd Dr Jackson, BL MS Add. 30346, fo. 99.

97. Letter from Joseph Ritson to George Paton, National Library of Scotland, Adv. MS 29. 5. 8, fo. 99.

98. W. H. Ireland, *Confessions*, 228.

99. S. Ireland, *Miscellaneous Papers and Legal Instruments*, 2 (Huntington Library copy).

100. W. H. Ireland, *Confessions*, 231.

101. S. Ireland, *Miscellaneous Papers and Legal Instruments*, 3 (Huntington Library copy).

102. W. H. Ireland, *Full and Explanatory Account*, 143.

103. They were published for the first time by J. O. Halliwell[-Phillipps] in *A Descriptive Catalogue of a Collection of Shakespeariana . . . Including a Remarkable Series of the Ireland*

*Forgeries* (London, 1866). This volume, limited to an edition of 50 copies, is apparently unknown to Mair and Grebanier.

104. Letter from Ritson to Paton, National Library of Scotland, Adv. MS 29. 5. 8, fo. 100.

105. Transcript by Samuel Ireland in British Library MS Add. 30346, fo. 69.

106. Letter from Malone to Byng, BL MS Add. 30346, fo. 70.

107. Letter from Byng to Malone, dated 8 March [1795]; Bodl. MS Malone 38, fo. 163; see too the letter from Byng (also dated 8 March) which follows, fo. 164.

108. Edmond Malone, *An Inquiry into the Authenticity of Certain Miscellaneous Papers and Legal Instruments* (London, 1796), 166.

109. Loc. cit.

110. James Boaden, *A Letter to George Steevens, Esq.*, (London, 1796), 21 n.

111. Ibid. 42.

112. Philalethes, *Shakspeare's Manuscripts, in the Possession of Mr. Ireland, Examined, Respecting the Internal and External Evidence of Their Authenticity* (London, 1796), 12, 24.

113. G. M. Woodward, *Familiar Verses from the Ghost of Willy Shakspeare to Sammy Ireland* (London, 1796), 6.

114. Malone, *Inquiry*, 360.

115. Ibid. 354.

116. Ibid. 198–9.

117. Ibid. 215–16.

118. Ibid. 167 n.

119. S. Ireland, *Mr. Ireland's Vindication of His Conduct* (London, 1796), 48.

120. Folger Shakespeare Library MS S. b. 119. The letter, which is pasted into a scrapbook, is there described by William as a transcript but is in fact in Samuel Ireland's hand.

121. Shakespeare, *Plays and Poems*, ed. Boswell and Malone, vol. i, p. lxx.

122. Letter of Joseph Ritson, MS R–V autogrs. Misc-English, MA 2515, in the Pierpont Morgan Library; printed in Ritson, *Letters*, ed. J. Frank (London, 1833), ii. 141.

123. *Manuscripts and Correspondence of James, First Earl of Charlemont*, ii. 268.

124. *Collections on the Ireland Forgeries*, Shakespeare Birthplace Record Office MS 114.

125. Letter of W. H. Ireland to John Byng, Shakespeare Birthplace Record Office MS 114, fo. 47a.

126. Anon., 'The Shakespeare Ireland Forgeries', *London Review*, 1 [1860], 395.

127. Reprinted in *Chalmeriana: or A Collection of Papers Literary and Political, Entitled, Letters, Verses, &c. Occasioned by Reading a Late Heavy Supplemental Apology for the Believers in the Shakespeare Papers by George Chalmers, F.R.S.S.A.* (London, 1800), 1.

128. George Chalmers, *An Apology for the Believers in the Shakspeare-Papers* . . . (London, 1797), 51–2.

129. George Chalmers, *A Supplemental Apology for the Believers in the Shakspeare-Papers* . . . (London, 1799), p. 55.

130. Ibid. 59 n. y.

131. Ibid. 52.

132. *Monthly Review*, 31 [1800], 189; cited in Variorum *Sonnets*, ed. Rollins, ii. 249.

133. Letter dated 3 July 1790, in *Original Letters from Edmund Malone . . . to John Jordan* . . ., 26.

134. See Chambers, *William Shakespeare*, ii. 96, 98.

135. *The Correspondence of Edmond Malone, the Editor of Shakespeare, with the Rev. James Davenport, D.D., Vicar of Stratford-on-Avon*, ed. J. O. Halliwell[-Phillipps] (London, 1864), 56–7.

136. Ibid. 81–2.

137. Ibid. 82.
138. Malone, *Inquiry*, pp. 3–4.
139. *The Correspondence of Thomas Percy & Edmond Malone*, ed. Arthur Tillotson (Louisiana State University Press, 1944), 93 (The Percy Letters, i).
140. *Autograph Letters of James Boswell 1759–1794*, fo. 77 (V8A. MA. 981), in the Pierpont Morgan Library.
141. Letter to Malone from Hunt and Hobbes, MS in the Beinecke Rare Book Library of Yale University.
142. *Correspondence of Edmund Malone . . . with the Rev. James Davenport*, 83.
143. Ibid. 84.
144. *Correspondence of Thomas Percy & Edmond Malone*, 242.
145. *Original Letters from Edmund Malone . . . to John Jordan*, 49–50.
146. John Dryden, *Prose Works*, ed. Malone (London, 1800), vol. i, pt. 1, p. vi.
147. Shakespeare, *Plays and Poems*, ed. Boswell and Malone, vol. i, p. xx.
148. Ibid. 279.
149. Ibid. 119.
150. *Correspondence of Thomas Percy & Edmond Malone*, 60; see also Malone's letter to Charlemont, 15 November 1793, in *Manuscripts and Correspondence of James, First Earl of Charlemont*, ii. 221.
151. Shakespeare, *Plays and Poems*, ed. Boswell and Malone, ii. 609 n.

# Part IV

1. I am here indebted to Richard Altick who, in *Lives and Letters: A History of Literary Biography in England and America* (New York, 1965), 77–8, gives an amusing and informative description of the range of biographical expression in this period.
2. Thomas Carlyle, 'Goethe', *Critical and Miscellaneous Essays*, in *Works* (London, 1898–1901), xxvi. 245, 246.
3. 'Shakspeare a Tory and a Gentleman' (1828) in Hartley Coleridge, *Essays and Marginalia* (London 1851), i. 115. My attention was drawn to Coleridge's essay, as well as to Very and Carlyle's 'Goethe', by M. H. Abrams, *The Mirror and the Lamp: Romantic Theory and the Critical Tradition* (New York, 1953), 248; see also his excellent brief discussion, 'The Paradox of Shakespeare', 244–9.
4. Jones Very, 'Shakespeare', *Essays and Poems* (Boston, 1839), 61.
5. Ibid. 46; 'Hamlet', *Essays and Poems*, 86 ff.
6. August Wilhelm Schlegel, *Sämmtliche Werke*, ed. E. Böcking (Leipzig, 1846), vii. 38 n.
7. A. W. Schlegel, *A Course of Lectures on Dramatic Art and Literature*, trans. John Black; rev. A. J. W. Morrison (London, 1846), 352. It should be borne in mind that the printed text represents an expansion of the lectures delivered by Schlegel in 1808.
8. Ibid. 369.
9. *Coleridge's Shakespearean Criticism*, ed. Thomas Middleton Raysor (London, 1930), ii. 30 n. 3.
10. Ibid. ii. 266.
11. Ibid. ii. 119.
12. Ibid. ii. 356.
13. Ibid. ii. 333.
14. Ibid. ii. 91.
15. Benjamin Bailey, quoted by Walter Jackson Bate, *John Keats* (Cambridge, Mass., 1963), 218.
16. *The Letters of John Keats 1814–1821*, ed. Hyder Edward Rollins (Cambridge, Mass., 1958), ii. 67.

17. Ibid. ii. 115–16.
18. Ibid. i. 193.
19. Ibid. i. 194.
20. William Hazlitt, *Lectures on the English Poets* (London, 1818), 92–3.
21. *The Letters of Charles Armitage Brown*, ed. Jack Stillinger (Cambridge, Mass., 1966), 323–4.
22. Charles Armitage Brown, *Shakespeare's Autobiographical Poems, Being His Sonnets Clearly Developed: with His Character Drawn Chiefly from His Works* (London, 1838), 104.
23. Ibid. 67.
24. Ibid. 98–9.
25. Thomas Carlyle, 'The Hero as Poet', *On Heroes, Hero-Worship, & the Heroic in History* (London, 1841), 178.
26. Thomas Carlyle, 'Jean Paul Friedrich Richter', *Critical and Miscellaneous Essays*, in *Works*, xxvii. 101.
27. Carlyle, 'The Hero as Poet', 175.
28. Ibid. 179.
29. Ibid. 180.
30. Robert Bell Wheler, *History and Antiquities of Stratford-upon-Avon* (Stratford-upon-Avon, n.d.), 131.
31. Shakespeare Birthplace Record Office MS 32, 53–4.
32. Keats, *Letters*, i. 239–40.
33. John Mason Good, *The Study of Medicine* (4th edn.; London, 1840), iii. 322–3.
34. Nathan Drake, *Shakspeare and His Times* (London, 1817), vol. i, pref., pp. iii–iv.
35. Ibid. ii. 613.
36. Ibid. ii. 73.
37. Ibid. ii. 62.
38. William Shakespeare, *Plays and Poems*, ed. E. Malone and J. Boswell (London, 1821), xx. 249. Boswell's comment does not escape the Variorum editor (*Sonners*, ed. Rollins, ii. 186).
39. Drake, *Shakspeare and His Times*, ii. 65. The italics are Drake's.
40. Ibid. ii. 69.
41. *Gentleman's Magazine*, 88 [1818]. 2. 241–2.
42. John Dover Wilson, *An Introduction to the Sonnets of Shakespeare for the Use of Historians and Others* (Cambridge, 1964), 63.
43. James Boaden, *On the Sonnets of Shakespeare: Identifying the Person to Whom They Are Addressed; and Elucidating Several Points in the Poet's History* (London, 1837), 59.
44. For Steeven's arguments, see Shakespeare, *Plays and Poems*, ed. Malone and Boswell, ii. 663–75.
45. Brown, *Shakespeare's Autobiographical Poems*, 42.
46. 'Illustrations of the Sonnets', *The Pictorial Edition of the Works of Shakspere*, ed. C. Knight (London, n.d.), Poems, 126–7.
47. James Boaden, *An Inquiry into the Authenticity of Various Pictures and Prints . . . Offered to the Public as Portraits of Shakspeare* (London, 1824), 27.
48. James Granger, *A Biographical History of England, from Egbert the Great to the Revolution* (5th edn.; London, 1824), i. 310 n.
49. On Taylor see Mary Edmond, 'The Chandos Portrait: A Suggested Painter', *Burlington Magazine*, 124 (1982), 146–9.
50. William Shakespeare, *Plays*, ed. S. Johnson and G. Steevens (London, 1793), vol. i. advertisement, p. iv.
51. Boaden, *Inquiry*, 41.

52. Ibid. 49. For my discussion of this portrait I have once again found M. H. Spielmann's work invaluable.
53. Quoted by William Jaggard, *Shakespeare Bibliography* (Stratford-upon-Avon, 1911), 172. Despite much effort, I have been unable to trace to its source this passage, which Jaggard surely errs in attributing to Johnson. Jennens is similarly characterized, but with less economy, in John Nichols's *Literary Anecdotes of the Eighteenth Century* (London, 1812), iii. 120–4.
54. Boaden, *Inquiry*, 47.
55. Ibid. 50–1.
56. *The Critical Review*, 30 [1770], 439.
57. Ibid. 31 [1771], 84.
58. Printed in 'Shakespeare', *European Magazine*, 26 [1794], 278. Steevens is responsible for this anonymous article.
59. Loc. cit.
60. Ibid. 279.
61. Ibid. 279 n. 7.
62. M. H. Spielmann, 'The Portraits of Shakespeare', in Shakespeare, *Works* (Stratford-upon-Avon, 1904–7), x. 397–8.
63. Boswell's *Life of Johnson*, ed. G. Birkbeck Hill; rev. L. F. Powell (Oxford, 1934), iv. 178.
64. Shakespeare, *Plays and Poems*, ed. Boswell and Malone, vol. i., p. xxvii.
65. Ibid., vol. i, p. xxviii.
66. Letter from Charles Lamb to Barron Field, dated 22 September 1822, in *The Works of Charles and Mary Lamb*, ed. E. V. Lucas (London, 1905), vii. 573–4. In his note Lucas gives a history of the portrait.
67. Ben Jonson, *Works*, ed. W. Gifford (London, 1816), vol. i., p. ccli.
68. See S. Schoenbaum, 'Shakespeare and Jonson: Fact and Myth', *The Elizabethan Theatre*, ed. David Galloway (1970), 1–19, and T. J. B. Spencer's rejoinder, 'Ben Jonson on his beloved, The Author Mr. William Shakespeare', *The Elizabethan Theatre IV*, ed. George Hibbard (1974), 22–40.
69. Ibid. pp. lxv–lxvi.
70. I. A. Shapiro, 'The "Mermaid Club"', *MLR*, 45 [1950], 6–17, is an authoritative study to which I am much indebted. Shapiro correctly cites Gifford as the source of the legend, but somewhat misleadingly refers to the 1875 reprint of his edition of Jonson.
71. John Ward, *Diary*, ed. C. Severn (London, 1839), 60–1.
72. Ibid. 183. I have modernized spelling.
73. David Hume, *The History of England* (London, 1763), vi. 131.
74. Loc. cit.
75. Loc. cit.
76. Shakespeare, *Plays and Poems*, ed. Malone, i. 1. 176 n. 6.
77. Henry Hallam, *Introduction to the Literature of Europe in the Fifteenth, Sixteenth, and Seventeenth Centuries* (London, 1837–9), ii. 176.
78. Ibid. iii. 85. The significance of this passage for future biography is not lost upon C. J. Sisson; see *The Mythical Sorrows of Shakespeare* (Proceedings of the British Academy, xx; London, 1934), 7–8.
79. Augustine Skottowe, *The Life of Shakspeare; Enquiries into the Originality of His Dramatic Plots and Characters; and Essays on the Ancient Theatres and Theatrical Usages* (London, 1824), i. 1–2.
80. Ibid. i. 65.
81. Ibid. i. 9.

82. *The Monthly Review*, 105 [1824], art. 11, p. 412.

83. Charles Symmons, 'The Life of William Shakespeare', in Shakespeare, *Dramatic Works*, ed. Samuel Weller Singer (Chiswick, 1826), i. 41.

84. Brown, *Shakespeare's Autobiographical Poems*, p. 215 n.

85. 'William Shakespear' (1564–1616)', *The Cabinet Cyclopaedia*, ed. Dionysius Lardner (London, 1830–49), 'Lives of the Most Eminent Literary and Scientific Men of Great Britain', ii. 83.

86. Ibid. 99.

87. Ibid. 100.

88. Ibid. 102.

89. Ibid. 130.

90. 'Notices of the Life of Lord Byron', in George Gordon, Lord Byron, *Works*, ed. Thomas Moore (London, 1832), iii. 136 n.

91. Richard Ryan, *Dramatic Table Talk; or Scenes, Situations, & Adventures, Serious & Comic, in Theatrical History & Biography* (London, 1825), ii. 156–7. My analysis of this passage is anticipated by Gerald Eade Bentley, *Shakespeare: A Biographical Handbook* (New Haven, Conn., 1961), 9.

92. Anna Jameson, *Memoirs of the Loves of the Poets. Biographical Sketches of Women Celebrated in Ancient and Modern Poetry* (2nd edn.; London, 1831), i. 240.

93. Ibid. 246. Mrs Jameson saw no need to correct the later impressions of her book.

94. Ibid. 246–7.

95. Anon., 'The Confessions of William Shakspeare', *New Monthly Magazine*, 2nd ser., xliii (1835), 309. My attention was drawn to this obscure title by the Hyder Rollins Variorum edition of the *Sonnets*, which I have found an invaluable guide to the labyrinthine literature on the subject.

96. Ibid. 2.

97. William Shakespeare, *Dramatic Works*, introd. John Britton (Chiswick, 1814), vol. i, p. xvii.

98. William Shakespeare, *Dramatic Works*, ed. Thomas Campbell (London, 1838), p. xvi.

99. Ibid. p. lxiii.

100. Letter to Campbell to Hunter in the collection of Louis Marder of Evanston.

101. For a compendious account of the origin and development of encyclopaedias, see Robert Collison, *Encyclopaedias: Their History throughout the Ages* (London, 1964), and esp. ch. 7, 'The Nineteenth Century', 174–98.

102. See above, II. 93.

103. See Edward Hughes, 'The Authorship of the "Discourse of the Commonwealth"', *Bulletin of the John Rylands Library*, 21 [1937], 167–75.

104. 'Shakespeare [William]', *Biographia Britannica* (London, 1763), vi. 3,630–1.

105. John Berkenhout, *Biographia Literaria; or a Biographical History of Literature* (London, 1777), 401.

106. 'Shakespeare or Shakspeare (William)', *Encyclopaedia Britannica* (3rd edn.; Edinburgh, 1797), xvii. 331.

107. Ibid. xvii. 332.

108. John Aiken *et al.*, *General Biography* (London, 1799–1815), ix. 122.

109. 'Shakspere, William', *The Penny Cyclopaedia of The Society for the Diffusion of Useful Knowledge* (London, 1841), xxi. 339.

110. Thomas De Quincey, *Works* (4th edn.; Edinburgh, 1863), vol. xv, pref., pp. vi–vii.

111. Ian Jack, *English Literature 1815–1832* (Oxford, 1963), 299. The lapse is only momentary, for Jack mentions the *Britannica* article in his bibliography (p. 525).

112. Thomas De Quincey, 'Shakspeare', *Encyclopaedia Britannica* (7th edn.; Edinburgh, 1842), xx. 175.

113. Ibid. xx. 177.

114. Anon., 'Life of Mr. Edward Alleyn, Comedian', *The Theatrical Review; or, Annals of the Drama*, 1 [1763], 64.

115. Richard Fenton, *A Tour in Quest of Genealogy* (London, 1811), 29–30.

116. Ibid. 190.

117. Ibid. 192.

118. C. M. Ingleby, 'The Literary Career of a Shakespeare Forger', *Shakespeare: The Man and the Book* (London, 1877–81), pt. 2, p. 146 n.

119. Chambers, *William Shakespeare*, ii. 384.

120. The newspaper accounts quoted in this paragraph were transcribed by G. Hilder Libbis and are to be found in vol. ii of his *Ireland Shakespeare Fabrications*, a miscellaneous collection of unpublished notes and other materials now in the Huntington Library. To Libbis must go the credit for fitting together these pieces of the jigsaw puzzle. That this was not accomplished sooner is odd, for a couple of the newspaper items appear in the volume of Ireland cuttings in the British Library (MS Add. 30349).

121. J. O. Halliwell[-Phillipps], *Letters of Authors*, xxiii. 28, in Edinburgh University Library.

122. John Payne Collier, *An Old Man's Diary, Forty Years Ago; for the Last Six Months of 1832* (London, 1871), vol. ii, pref., p. ii.

123. Quoted by Henry B. Wheatley, *Notes on the Life of John Payne Collier; with a Complete List of His Works, and an Account of Such Shakespeare Documents as Are Believed to Be Spurious* (London, 1884), 6.

124. J. P. Collier, *Autobiography*, Folger Shakespeare Library MS M. a. 230, p. 150.

125. J. P. Collier, *The History of English Dramatic Poetry to the Time of Shakespeare: Annals of the Stage to the Restoration* (London, 1831), i. 331.

126. Ibid. i. 403 n.

127. Ibid. i. 298.

128. J. P. Collier, *Diary*, xvii. 69 (Folger Shakespeare Library MS M. a. 33).

129. J. P. Collier, *Mr. J. Payne Collier's Reply to Mr. N. E. S. A. Hamilton's 'Inquiry' into the Imputed Shakespeare Forgeries* (London, 1860), 35. In the margin alongside this passage of his copy (now in the Rare Book Room of the Newberry Library) Howard Staunton, Shakespeare editor and celebrated chess player, wrote: '*much more*'.

130. J. P. Collier, *New Facts Regarding the Life of Shakespeare* (London, 1835), 55.

131. Ibid. 22–3.

132. Ibid. 33.

133. Ibid. 41.

134. Ibid. 48–9.

135. J. P. Collier, *Farther Particulars Regarding Shakespeare and His Works* (London, 1839), 56.

136. *Memoirs of Edward Alleyn, Founder of Dulwich College: Including Some New Particulars Respecting Shakespeare, Ben Jonson, Massinger, Marston, Dekker, &c.*, ed. J. P. Collier (London, 1841), 68 (Shakespeare Soc. Pubs., 2).

137. Ibid. 91.

138. *The Life of William Shakespeare*, in Shakespeare, *Works*, ed. J. P. Collier (London, 1842–53), vol. i., p. cxxxii.

139. Loc. cit.

140. Ibid. p. cxlvii.

141. The letter, from W. W. Williams, is preserved in a copy of vol. i of Collier's 1842–53 edition of Shakespeare in the BL (shelf mark C. 134. fo. 1).

142. Quoted by Charles Knight, *Passages of a Working Life During Half a Century: with a Prelude of Early Reminiscences* (London, 1864), ii. 296.
143. C. Knight, *William Shakspere: A Biography* (London, 1843). Knight does not name Croker as his correspondent, but merely refers to 'a gentleman of great critical sagacity' (p. 500).
144. Joseph Hunter, *New Illustrations of the Life, Studies, and Writings of Shakespeare. Supplementary to All the Editions* (London, 1845), i. 68.
145. Ibid. 73.
146. Joseph Hunter, British Library MS Add. 24497, fo. 47.
147. J. O. Halliwell[-Phillipps], *Observations on the Shaksperian Forgeries at Bridgewater House; Illustrative of a Facsimile of the Spurious Letter of H. S.* (London, 1853), 7.
148. Ibid. 8.
149. Alexander Dyce, *Remarks on Mr. J. P. Collier's and Mr. C. Knight's Editions of Shakespeare* (London, 1844), pref., p. vi.
150. J. P. Collier, *Notes and Emendations to the Texts of Shakespeare's Plays, from Early Manuscript Corrections in a Copy of the Folio, 1632, in the Possession of J. Payne Collier* (London, 1852), intro., p. vi.
151. Reproduced by N. E. S. A. Hamilton, *An Inquiry into the Genuineness of the Manuscript Corrections in Mr. J. Payne Collier's Annotated Shakspere, Folio, 1632* (London, 1860), 16–17.
152. Collier, *Autobiography*, 120–1. It is fair to note that there are places in the *Autobiography* where Collier confuses Lord Campbell with Thomas Campbell, the poet and editor whom he also knew.
153. Letter of Frederic Madden to C. M. Ingleby in Folger Shakespeare Library MS C. a. 14, fo. 1.
154. Frederic Madden, *Journal* for 1859, p. 185 (Bodl. MS C. 169).
155. Ibid. 227.
156. Letter of J. P. Collier to S. Leigh Sotheby in Folger Shakespeare Library (MS C. b. 8, fo. 2).
157. Hamilton, *An Inquiry into the Genuineness of the Manuscript Corrections . . .*, 82.
158. C. M. Ingleby, Folger Shakespeare Library MS M. b. 25.
159. C. M. Ingleby, *Supplement to "A Complete View of the Shakespeare Controversy"*, MS (unpaginated) in the possession of Louis Marder of Evanston.
160. Collier, *Diary*, xv. 7–8 (Folger Shakespeare Library MS M. a. 31).
161. Ibid., xii. 36–7 (Folger Shakespeare Library MS M. a. 29).
162. Letter preserved in the Huntington Library copy of *An Old Man's Diary*, pt. 1 (shelf no. 18141).
163. Collier, *Autobiography*, 146–7. This passage has, I believe, been previously noticed only by Giles Dawson, in an unpublished address.
164. Ibid. 148.
165. Letter, dated 18 April 1878, of J. P. Collier to an unknown correspondent; in the possession of Dr Arthur Freeman of Cambridge, Mass.
166. Collier, *Diary*, xxvi. 40–1 (Folger Shakespeare Library MS M. a. 40).
167. Hunter, *New Illustrations*, i. 24.
168. Ibid. 27.
169. Ibid. 51.
170. Ibid. 115.

# Part V

1. Charles Knight, *Passages of a Working Life During Half a Century; with a Prelude of Early Reminiscences* (London, 1864[63]–5), ii. 19.

2. 'Brougham, Henry Peter', *DNB* vi. 456.

3. Thomas De Quincey, 'Shakespeare', *Encyclopaedia Britannica* (7th edn.; Edinburgh, 1842), xx. 179.

4. *The Pictorial Edition of the Works of Shakspeare*, ed. C. Knight (London, 1839–42), Comedies, ii. 192.

5. Knight, *Passages of a Working Life . . .*, ii. 294–5.

6. Ibid. ii. 303.

7. Knight, *William Shakspere*, 286–7.

9. Ibid. , Advertisement, unpaged.

9. See above, IV. 8.

10. Knight, *William Shakspere*, 21. The capitals are his.

11. Ibid. 526.

12. Ibid. 521.

13. Ibid. 112.

14. J. O. Halliwell[-Phillipps], *The Working Life of William Shakespeare. Including Many Particulars Respecting the Poet and His Family Never before Published* (London, 1848), pref., vii.

15. Charles Knight, *William Shakspere; Biography* (rev. edn.; London, 1850), 174.

16. Halliwell's undated letter is in the University College, London, Library; for Collier's reply (May 1840) see Halliwell[-Phillipps], *Letters of Authors*, iii. fo. 35, in Edinburgh University Library.

17. J. O. Halliwell[-Phillipps], *Statement in Answer to Reports Which Have Been Spread Abroad Against Mr. James Orchard Halliwell* (London, 1845), 4.

18. Letter of J. O. Halliwell[-Phillipps], printed by A. N. L. Munby, *The Family Affairs of Sir Thomas Phillipps* (Cambridge, 1952), 41 (*Phillipps Studies*, no. 2).

19. Letter of Henry Ellis to J. O. Halliwell[-Phillipps], *Statement*, 3.

20. The facts of the case have been exhaustively and impartially investigated by D. A. Winstanley, who concludes that, 'after giving him [Halliwell] the benefit of every doubt it is impossible not to believe that he stole the manuscripts from the college library' ('Halliwell Phillipps [sic] and Trinity College Library', *Library*, 5th ser., 2 [1948], 277).

21. Quoted by Winstanley, 264.

22. E. V. Lucas, *Reading, Writing, and Remembering: A Literary Record* (New York, 1932), 48.

23. In fairness to Munby it should be noted that the Life has been pieced together by Nicolas Barker from the five volumes of Munby's *Phillipps Studies*. These I have found invaluable.

24. Anon. letter to T. Phillipps, printed by Munby, *Family Affairs . . .*, 47.

25. Ibid. 52.

26. Ibid. 78.

27. Loc. cit.

28. J. O. Halliwell[-Phillipps], *A Calendar of the Shakespearean Rarities, Drawings and Engravings, Preserved at Hollingbury Copse, Near Brighton* (London, 1887), pref., 5.

29. 'In Shakespeare's Country', *Gentleman's Magazine*, NS, 14 [1875], 451.

30. J. O. Halliwell[-Phillipps], *A List of Works Illustrative of the Life and Writings of Shakespeare . . .* (London, 1867), 29.

31. Halliwell[-Phillipps], *Life of William Shakespeare*, 175.

32. Ibid. p. x.

33. Ibid. 297–8.

34. See above, IV. 7.

35. Halliwell[-Phillipps], *Life of William Shakespeare*, 151.

36. William Shakespeare, *Works*, ed. J. O. Halliwell[-Phillipps] (London, 1853–65), vol. i, pref., p. iv.
37. *A Life of William Shakespeare*, in Shakespeare, *Works*, ed. Halliwell[-Phillipps), i. 151.
38. The passage is quoted above, I. 2.
39. Letters from F. W. Fairholt, dated 2 November 1865, in Halliwell[-Phillipps], *Letters of Authors*, vol. 101, fo. 32, in Edinburgh University Library.
40. Henrietta Halliwell-Phillipps, *Diary*, Halliwell-Phillipps Collection 330, p. 160 (Edinburgh University Library).
41. F. Madden, *Journals* for 1870–2, Bodl. MS Eng. hist. C. 182, 421; quoted by Munby, *Family Affairs of Sir Thomas Phillipps*, 105.
42. Henrietta Halliwell-Phillipps, *Diary*, Halliwell-Phillipps Collection, 330, p. 203 (Edinburgh University Library).
43. Halliwell[-Phillipps], *List of Works*, 55.
44. Letter of J. O. Halliwell-Phillipps, Birmingham Shakespeare Library MS S849F, fo. 71.
45. Letter of J. O. Halliwell-Phillipps, National Library of Scotland MS 2632.
46. Letter of J. O. Halliwell-Phillipps, Birmingham Shakespeare Library MS S849F, fo. 73.
47. J. O. Halliwell-Phillipps, *Calendar of the Shakespeare Rarities*, title-page.
48. J. O. Halliwell-Phillipps, *Outlines of the Life of Shakespeare* (7th edn.; London, 1887), vol. i, pref., p. xvii. The preface is dated March 1884.
49. Halliwell-Phillipps, *Outlines*, i. 234.
50. Holograph letter of Halliwell-Phillipps to Justin Winsor, in *Controversy between J. O. Halliwell-Phillipps and F. J. Furnivall 1881* in Harvard College Library (11485. 8F*).
51. Letter of J. O. Halliwell-Phillipps in the possession of Dr Arthur Freeman of London.
52. *Some Account of the Life of Shakespeare*, in William Shakespeare, *Works*, ed. A. Dyce (London, 1857), vol. i, p. xv n. 34.
53. Ibid., pref., p. xvi.
54. William Watkiss Lloyd, *The Life of William Shakespeare*, in William Shakespeare, *Dramatic Works*, ed. S. W. Singer (London, 1856), vol. i., p. xcvii.
55. Richard Grant White, *Memoirs of the Life of William Shakespeare, with an Essay toward the Expression of His Genius, and an Account of the Rise and Progress of the English Drama* (Boston, 1865), 147.
56. Karl Elze, *William Shakespeare. A Literary Biography*, trans. L. Dora Schmitz (London, 1901), 81–2.
57. François Pierre Guillaume Guizot, *Shakespeare and His Times* (London, 1852), 23–4.
58. Victor Hugo, *William Shakespeare*, trans. M. B. Anderson (Chicago, 1887), 10–11.
59. Ibid. 11.
60. *Memoir of Shakespeare*, in William Shakespeare, *Poems*, ed. A. Dyce (Aldine Poets; London, 1832), p. lxxvi.
61. William Shakespeare, *Works*, ed. H. N. Hudson (Boston, 1856), xi. 118. My attention was drawn to this passage, as to others cited in this section, by Rollins's Variorum edition of the Sonnets.
62. Ralph Waldo Emerson, *Representative Men*, in *Complete Works* (Boston, 1903–15), iv. 209.
63. David Masson, 'Shakespeare and Goethe', *Essays Biographical and Critical: Chiefly on English Poets* (Cambridge, 1856) 12.
64. D. Masson, *Autobiography of Shakespeare*, Folger Shakespeare Library MS S. a. 169, i. 57–8.
65. Thomas Kenny, *The Life and Genius of Shakespeare* (London, 1864), 53–4.

66. Halliwell-Phillipps, *Outlines . . .*, i. 174.
67. Ebenezer Forsyth, *Shakspere: Some Notes on His Character and Writings* (Edinburgh, 1867), 23.
68. Léon de Wailly, 'Sonnets de Shakspeare', *Revue des deux mondes*, 3rd ser., iv (1834), 688.
69. Friedrich Nietzsche, *Morgenröthe*, aphorism 76, trans. J. M. Kennedy, in *Complete Works*, ed. Oscar Levy (New York, 1911), ix. 77.
70. Philarète Chasles, 'Shakspeare's Sonnets', *Athenaeum* (16 Feb. 1867), 223.
71. Charles Edmonds, 'A Shakspearean Discovery', *Athenaeum* (22 Nov. 1873), 661.
72. Philarète Chasles, 'Hints for the Elucidation of Shakespeare's Sonnets', *Athenaeum* (25 Jan. 1862), 116.
73. Walter Bourchier Devereux, *Lives and Letters of the Devereux, Earls of Essex . . .* (London, 1853), i. 145.
74. Oscar Wilde, *The Portrait of Mr W. H.*, in *Blackwood's Edinburgh Magazine*, 146 [1889], 13.
75. Ibid. 19.
76. Ibid. 21.
77. Hesketh Pearson, *The Life of Oscar Wilde* (London, 1954), 142–3.
78. Richard Ellmann, *Oscar Wilde* (1988), 297–8 *et passim*. Ellmann mistakenly says that both Shakespeare and Wilde had two children.
79. Charles Ricketts, *Oscar Wilde: Recollections* (London, 1932), 48.
80. Oscar Wilde, *The Portrait of Mr W.H.* (London, 1921), 49.
81. Ibid. 53.
82. Ellmann, *Oscar Wilde*, 296.
83. James Joyce, *Ulysses* (New York, 1934), 196.
84. Julien Green, *Personal Record, 1928–1939*, trans. Jocelyn Godefroi (New York, 1939), 66.
85. Samuel Butler, *Shakespeare's Sonnets Reconsidered, and in Part Rearranged . . .* (London, 1899), 122.
86. Ibid. 86–7.
87. Henry Festing Jones, *Samuel Butler Author of* Erewhon *(1835–1902), A Memoir* (London, 1919), ii. 313.
88. Quoted by Philip Henderson, *Samuel Butler: The Incarnate Bachelor* (London, 1953), 215.
89. Ibid.
90. Gerald Massey, *Shakspeare's Sonnets Never before Interpreted* (London, 1866), 436.
91. Cired by Rollins, Variorum *Sonnets*, ii. 261.
92. Bernard Shaw, *Misalliance . . .* (London, 1914), 104, 107.
93. William Shakespeare, *Sonnets*, ed. Thomas Tyler (London, 1890), 87.
94. *The Sonnets of William Shakspere, Rearranged and Divided into Four Parts*, ed. Robert Cartwright (London, 1859), 8–9; see also Cartwright, *The Footsteps of Shakspere* (London, 1862), 154–5.
95. William J. Thoms, 'Was Shakespeare Ever a Soldier?' *Three Notelets on Shakespeare* (London, 1865), 136. As originally published in *Notes and Queries*, 2nd ser., 7 [1859], 330–3 and 351–5, Thom's article lacks this sentence.
96. John Campbell, *Shakespeare's Legal Acquirements Considered* (London, 1859), 30.
97. Ibid. 23.
98. 'The Shakespeare Death-Mask', *Antiquary*, 2 [1860], 65.
99. Frederick J. Pohl, 'The Death-Mask', *Shakespeare Quarterly*, 12 [1961], 115–25.
100. Charles Dickens, *Letters*, ed. G. Howarth and M. Dickens (London, 1893), 173.
101. J. Parker Norris, 'Shaksperian Gossip', *American Bibliopolist*, 8 [1876], 40.

102. Washington Irving, *The Sketch Book of Geoffrey Crayon, Gent.* (New York, 1819–20), no. vii. 65.
103. Anon., 'Shakespeariana', *Monthly Magazine*, 45 [1818], 2.
104. This witness was James Hare, who many years later described his experiences in the *Birmingham Weekly Post*. The account was privately reprinted in 1955 in a pamphlet entitled *The Dust of Shakespeare*.
105. Letter of Dowden to Ingleby, dated 20 Aug. 1883, Folger Shakespeare Library MS W. a. 74 (6).
106. Letter of Newman to Ingleby, dated 16 Oct. 1883, Folger Shakespeare Library MS W. a. 74 (24).
107. 'Shakespeare—The Individual', *The Collected Works of Walter Bagehot*, ed. Norman St John-Stevas (Cambridge, Mass., 1965), i. 213–14. I have profited from the prefatory biographical memoir to this edition.
108. Kenny, *Life and Genius of Shakespeare*, 341.
109. Ibid. 68.
110. Ibid. 89–90.
111. John A. Heraud, *Shakspere: His Inner Life as Intimated in His Works* (London, 1865), 8.
112. Ibid. 26.
113. S. W. Fullom, *History of William Shakespeare, Player and Poet: with New Facts and Traditions* (London, 1862), 202.
114. Ibid. 81.
115. Ibid. 109.
116. Ibid. 149.
117. Ibid. 232, 234.
118. Ibid. 186.
119. In this section I reproduce, with slight alteration, some passages from my *Internal Evidence and Elizabethan Dramatic Authorship*, 38 ff., in which I give an account of the Society's assault on the Shakespeare canon.
120. 'Furnivall, Frederick James', *DNB*, 2nd supp., ii. 65.
121. *The New Shakspere Society's Transactions* (London, 1874), pp. vi–vii. In fairness to Furnival it should be noted that this passage is from a shorthand report of his opening speech to the Society; but he supervised publication.
122. F. G. Fleay, *Shakespeare Manual* (London, 1876), 108.
123. A. H. Bullen, pref. to H. Dugdale Sykes, *Sidelights on Shakespeare* (London, 1919), p. vii.
124. F. G. Fleay, *A Chronicle History of the Life and Work of William Shakespeare Player, Poet, and Playmaker* (London, 1886), 16.
125. Ibid. 23.
126. Charlotte Carmichael Stopes, *The Life of Henry, Third Earl of Southampton, Shakespeare's Patron* (Cambridge, 1922), pref., p. vi.
127. Dowden, *Shakspere*, 38.
128. Ibid. pref., p. v.
129. Ibid. 5.
130. Ibid. 163.
131. Ibid. 33.
132. *Letters of Edward Dowden and His Correspondents*, ed. Elizabeth D. and Hilda M. Dowden (London, 1914), 69–70.
133. William Shakespeare, *Sonnets*, ed. Dowden (London, 1881), introd., 6.
134. Ibid. 11.
135. This observation I owe to Professor Terence Spencer.

136. Sidney Lee, *The Impersonal Aspect of Shakespeare's Art* (London, 1909), 7.
137. Ibid. 14.
138. Edward Dowden, *Essays Modern and Elizabethan* (London, 1910), 263.
139. Georg Brandes, *William Shakespeare: A Critical Study*, trans. William Archer and Mary Morison (London, 1898), i. 179.
140. Ibid. ii. 327.
141. Ibid. ii. 233.
142. Ibid. ii. 27.
143. Ibid. i. 280–1.
144. Ibid. ii. 134.
145. The passages cited from *Ulysses* (New York, 1934) appear, respectively, on 202, 192–3, 203, 188; Brandes's comment on *Venus and Adonis* occurs in *Shakespeare*, i. 68. For a fuller discussion, of especial interest to the student of Joyce, see William M. Schutte, *Joyce and Shakespeare* (New Haven, Conn., 1957).
146. Leslie Stephen, *Some Early Impressions* (London, 1924), 160.
147. Sidney Lee, *Stratford-on-Avon from the Earliest Times to the Death of William Shakespeare* (London, 1885), 2. Lee revised this work for new editions in 1900 and 1907.
148. Sidney Lee, 'Shakespeare, William', *DNB* li. 363. Lee's turnabouts are chronicled in detail by Rollins in vol. ii of the Variorum *Sonnets*.
149. Sidney Lee, 'The Perspective of Biography', *Elizabethan and Other Essays*, ed. Frederick S. Boas (Oxford, 1929), 63.
150. Lee, 'Shakespeare, William', 379.
151. Ibid. 352.
152. Ibid. 384.
153. Logan Pearsall Smith, *On Reading Shakespeare* (London, 1933), 149–50.
154. Sidney Lee, *A Life of William Shakespeare* (London, 1898), pref., p. vi.
155. Leslie Stephen, 'Shakespeare as a Man', in *Studies of a Biographer* (2nd ser.; London, 1902), iv. 3.
156. Lee, *A Life of William Shakespeare*, p. 265.
157. Ibid. 176–7.
158. Ibid. 109.
159. Ibid. 154.
160. Ibid. 92.
161. Sidney Lee, *A Life of William Shakespeare* (4th rev. edn.; London, 1925), 117.
162. For Lee's discussion see the revised *Life*, 474–81; for an authoritative account of the episode see Chambers, *William Shakespeare*, ii. 141–52.
163. See Lee, 307–15.

# Part VI

1. Letter of Delia Bacon to Mrs J. H. Woodruff, printed by Vivian C. Hopkins, *Prodigal Puritan: A Life of Delia Bacon* (Cambridge, Mass., 1959), 176–7; Theodore Bacon, *Delia Bacon: A Biographical Sketch* (Boston, 1888), 62.
2. *New Letters of Thomas Carlyle*, ed. Alexander Carlyle (London, 1904), ii. 150.
3. Letter of Thomas Carlyle to Edward Everett, dated 22 December 1854; printed by Hopkins, *Prodigal Puritan*, 187. The letter does not appear in Rusk's edition of *Letters of Ralph Waldo Emerson*, cited by Miss Hopkins in her notes.
4. Francis Bacon, *The Advancement of Learning*, bk. i.
5. Letter of Delia Bacon to Charles Butler, printed by LeRoy Elwood Kimball, 'Miss Bacon Advances Learning', *Colophon*, NS, 2 [1937], 342–3; also Hopkins, *Prodigal Puritan*, 184.

6. Theodore Bacon, *Delia Bacon*, 76.

7. Delia Bacon, 'William Shakespeare and His Plays; and Inquiry Concerning Them', *Putnam's Monthly Magazine*, 7 [1856], 19.

8. Ibid. 9.

9. Ibid.

10. Ibid. 14.

11. Ibid. 1.

12. Letter of Delia Bacon to Nathaniel Hawthorne, printed by Hopkins, *Prodigal Puritan*, 200.

13. Hopkins, 205.

14. Ibid. 237.

15. Ibid. 243.

16. Delia Bacon, *The Philosophy of the Plays of Shakspere Unfolded* (London, 1857), p. xxii.

17. Ibid. pp. xvii–xviii.

18. Ibid. 579.

19. Letter of Delia Bacon to David Rice, printed by Hopkins, *Prodigal Puritan*, 248.

20. James Townley, *High Life below Stairs* (London, 1759), 38.

21. [Herbert Lawrence], *The Life and Adventures of Common Sense: An Historical Allegory* (London, 1769), i. 146.

22. Ibid. 146–9.

23. Anon., *The Story of the Learned Pig* (London, 1786), 37–8.

24. Ibid. 38.

25. James Corton Cowell, *Some Reflections on the Life of William Shakespeare: A Paper Read before the Ipswich Philosophic Society*, MS 294 in the University of London Library, 67–8.

26. Joseph C. Hart, *The Romance of Yachting: Voyage the First* (New York, 1848), pref., 5. No chronicle of a second voyage appeared.

27. Ibid. 228.

28. Ibid. 211–12.

29. 'Who Wrote Shakespeare?', *Chambers's Edinburgh Journal* (7 Aug. 1852), 88.

30. William Henry Smith, *Bacon and Shakespeare. An Inquiry Touching Players, Playhouses, and Play-writers in the Days of Elizabeth* (London, 1857), 2.

31. George Henry Townsend, *William Shakespeare Not an Imposter* (London, 1857), 122.

32. W. H. Smith, *Bacon and Shakespeare: William Shakespeare: His Position as Regards the Plays, etc.* (London, 1884), 6.

33. Ignatius Donnelly, *The Great Cryptogram: Francis Bacon's Cipher in the So-Called Shakespeare Plays* (Chicago, 1888), introd., pp. v–vi.

34. H. N. Gibson, *The Shakespeare Claimants* (London, 1962), 71.

35. Donnelly, *The Great Cryptogram*, 663.

36. These deficiencies are pointed out and illustrated by W. F. and E. Friedman in *The Shakespearean Ciphers Examined* (Cambridge, 1957), 38–9.

37. John T. Doyle, Folger Shakespeare Library TS S. a. 123, 14.

38. Ignatius Donnelly, *Diary*, quoted by Martin Ridge, *Ignatius Donnelly: The Portrait of a Politician* (Chicago, 1962), 243.

39. Edwin Bormann, *Francis Bacon's Cryptic Rhymes and the Truth They Reveal* (London, 1906), 212, 213. Wadsworth savours this passage as Bormann's 'most brilliant stroke', which indeed it is.

40. Henry James, *Letters*, ed. Percy Lubbock (London, 1920), i. 432.

41. Copyright © 1970 by The Mark Twain Company.

42. 'In the Tribunal of Literary Criticism. Bacon *vs.* Shakespeare', *Arena*, 6 [1892], 706.

43. *Arena*, 7 [1893], 734.

44. Orville W. Owen, *Sir Francis Bacon's Cipher Story* (Detroit, 1893), 25. For my account of Owen I have profited greatly from the authoritative chapter on him by W. F. and E. Friedman in *The Shakespearean Ciphers Examined*, 63–76.

45. Frederick W. Mann, *The Owen Cipher: A Paper Read before the Witenagemote* (New York, 1894), 16. Quotation and example are furnished by the Friedmans, 69.

46. Owen, *Sir Francis Bacon's Cipher Story*, 3.

47. Ibid. 97–8.

48. Ibid. ii. introd., n.p.

49. Loc. cit.

50. Kate Prescott, *Reminiscences of a Baconian* (The Haven Press, 1949), 73.

51. Burrell F. Ruth, 'Dr. Orville W. Owen: Recent Recollections', *American Baconiana*, 1 [1924], 17.

52. For my account of the Gallup ciphers I am again much indebted to the Friedmans (pp. 188–278), who had the incalculable advantage of personal acquaintance with this lady. All discussions of Mrs Gallup must now derive from *The Shakespearean Ciphers Examined*. I must add that Mrs Friedman generously agreed to be interviewed by me on this subject.

53. Elizabeth Wells Gallup, *The Bi-literal Cipher* . . . (Detroit, 1899), 23.

54. Elizabeth Wells Gallup, 'The Bi-literal Cipher of Sir Francis Bacon. A New Light on a Few Old Books', *Pall Mall Magazine*, 26 [1902], 398; reprinted in Gallup, *Concerning the Bi-literal Cypher of Francis Bacon . . . Pros and Cons of the Controversy* . . . (Detroit, n.d.), 60.

55. 'Mrs. Gallup's Biliteral Cipher', *Baconiana*, NS, 9 [1901], 5.

56. Elizabeth Wells Gallup, *The Bi-literal Cypher of Sir Francis Bacon . . . Part III. The Lost Manuscripts. Where They Were Hidden* (London, 1910), 2–3.

57. Ibid. 6.

58. Ibid. 13.

59. Ibid. 16.

60. B. G. Theobald, 'Mrs. Gallup's Competence', *Baconiana*, 3rd ser., 22 [1936], 126.

61. W. F. and E. Friedman, *The Shakespearean Ciphers Examined*, 216–44.

62. Natalie Rice Clark, *Hamlet on the Dial Stage* (Paris, 1931), pref., 7.

63. Ibid. 8.

64. Edwin Durning-Lawrence, *Bacon Is Shake-speare* (London, 1910), 102.

65. William Stone Booth, *Some Acrostic Signatures of Francis Bacon* (Boston, 1909), 47.

66. For an expert critique, see W. F. and E. Friedman, *The Shakespearean Ciphers Examined*, 114–30.

67. W. S. Booth, *Sublte Shining Secrecies* . . . (Boston, 1925), introd., 3–4.

68. Walter Conrad Arensberg, *The Baconian Keys* (rev. edn.; Pittsurgh, 1928), pref., 10.

69. Mrs Henry Pott, *Francis Bacon and His Secret Society: An Attempt to Collect and Unite the Lost Links of a Long and Strong Chain* (2nd edn.; London, 1911), pref., 5.

70. W. C. Arensberg, *The Secret Grave of Francis Bacon at Lichfield* (San Francisco, 1923), 36.

71. Ibid. 39.

72. W. C. Arensberg, *The Burial of Francis Bacon and His Mother in the Lichfield Chapter House: An Open Communication to the Dean and Chapter of Lichfield Concerning the Rosicrucians* (Pittsburgh, 1924), 19.

73. Appleton Morgan, *The Shakespearean Myth: William Shakespeare and Circumstantial Evidence* (Cincinnatti, 1881), 304–5.

74. Ibid. 284.

75. John H. Stotsenburg, *An Impartial Study of the Shakespeare Title* (Louisville, Ky., 1904), 510.

76. George Greenwood, *The Shakespeare Problem Restated* (London, 1908), pref., p. x. My attention was drawn to this note by Frank W. Wadsworth (*The Poacher from Stratford* (Berkeley, 1958), 97–8).

77. Gilbert Slater, *Seven Shakespeares: A Discussion of the Evidence for Various Theories with Regard to Shakespeare's Identity* (London, 1931), 220–1.

78. Alden Brooks, *Will Shakespeare Factotum and Agent* (New York, 1937), 373.

79. Alden Brooks, *This Side of Shakespeare* (New York, 1964), 129–30.

80. J. Thomas Looney, *'Shakespeare' Identified in Edward De Vere the Seventeenth Earl of Oxford* (London, 1920), introd., 16.

81. Ibid. 57. I owe this insight to R. C. Churchill, *Shakespeare and his Betters* (London, 1958), 197.

82. Letter of J. Thomas Looney to Charles Wisner Barrell, dated 6 June 1937; printed in *The Shakespeare Fellowship Quarterly*, 5 [1944], 21.

83. Looney, *'Shakespeare' Identified*, 118–19.

84. Ibid. 131.

85. *Aubrey's Brief Lives*, ed. O. L. Dick (London, 1949), 305. This episode, which has escaped the noses of the Oxfordians, is cited by Wadsworth (*The Poacher from Stratford*, 111).

86. Looney, *'Shakespeare' Identified*, 385.

87. Ibid. 211.

88. Ibid. 413.

89. Ibid. 530.

90. Ibid. 509.

91. Eva Turner Clark, *Hidden Allusions in Shakespeare's Plays: A Study of the Oxford Theory Based on the Records of Early Court Revels and Personalities of the Times* (New York, 1931), 603.

92. Dorothy and Charlton Ogburn, *This Star of England: 'William Shakespeare' Man of the Renaissance* (New York, 1952), 1004. This particular aberration is cited by Giles E. Dawson in a withering review; see *Shakespeare Quarterly*, 4 [1953], 165–70.

93. Gerald H. Rendall, *Shakespeare Sonnets and Edward De Vere* (London, 1930), 210.

94. Percy Allen, *Talks with Elizabethans Revealing the Mystery of 'William Shakespeare'* (London, n.d.), 40.

95. Ibid. 32.

96. Ibid. 72–3.

97. Ernest Jones, *Sigmund Freud: Life and Work* (London, 1953–7), i. 24.

98. Ibid. iii. 460.

99. 'Address Delivered in the Goethe House at Frankfort', *The Standard Edition of the Complete Psychological Works of Sigmund Freud*, trans. James Strachey, Anna Freud, Alix Strachey, and Alan Tyson (London, 1953– ), xxi. 211.

100. Jones, *Sigmund Freud*. iii. 488.

101. Quoted by A. Bronson Feldman, 'The Confessions of William Shakespeare', *American Imago*, 10 [1953], 165.

102. Freud, *Complete Psychological Works*, xxiii. 192 n. 1.

103. Norman N. Holland. *Psychoanalysis and Shakespeare* (New York, 1966), 59.

104. Harry Trosman, 'Freud and the Controversy over Shakespearean Authorship', *Journal of the American Psychoanalytic Association*, 13 [1965], 492.

105. Calvin Hoffman, *The Man Who Was 'Shakespeare'* (London, 1955), 121. The title of the English edition, from which I quote, differs slightly from that of the American edition.

106. R. C. Churchill, *Shakespeare and His Betters* (London, 1958), 117.

107. *The Autobiography of Malcolm X* (New York, 1965), 187. The *Autobiography* was ghost-written by Alex Haley, who would go on to achieve celebrity as the author of *Roots*.

# Part VII

1. Richard Ellmann, 'The Two Faces of Edward', *Edwardians and Late Victorians: English Institute Essays, 1959*, ed. Ellmann (New York, 1960), 188–210. For the Elinor Glyn reference (which led me to *Romantic Adventure*) and other suggestions, I am indebted to Samuel Hyne's able book, *The Edwardian Turn of Mind* (Princeton, N.J., 1968). Barbara Tuchman's *The Proud Tower* (London, 1966), although highly impressionistic, also directed me to sources I might otherwise have missed.

2. Charles Isaac Elton, *William Shakespeare: His Family and Friends* (London, 1904), 129. Halliwell-Phillipps (*Outlines*, ii. 185) furnishes the Henslowe example.

3. Ibid. 39.

4. Ibid. 349–50.

5. David Masson, *Shakespeare Personally* (London, 1914), 147.

6. Ibid. 190.

7. Charlotte Carmichael Stopes, *Shakespeare's Warwickshire Contemporaries* (2nd edn.; London, 1907), 33.

8. C. C. Stopes, *Shakespeare's Family, Being a Record of the Ancestors and Descendants of William Shakespeare, with Some Account of the Ardens* (London, 1901), 63.

9. C. C. Stopes, *Shakespeare's Environment* (2nd edn.; London, 1918), 155.

10. Envelope 3, box I, Wallace Papers, Huntington Library. This collection has not yet been catalogued.

11. Letter to Professor Förster; item iv, no. 7, box I, Wallace Papers.

12. C. W. Wallace, item A–v–2, box I, Wallace Papers.

13. C. W. Wallace, *Advance Sheets from Shakespeare, the Globe, and Blackfriars* (Stratford-upon-Avon, 1909), 3.

14. C. W. Wallace, envelope 27, box II, Wallace Papers.

15. C. W. Wallace, 'New Shakespeare Discoveries: Shakespeare as a Man among Men', *Harper's Magazine*, 120 [1910], 490.

16. C. W. Wallace, 'Feelings', envelope 27, box II, Wallace Papers.

17. C. C. Stopes, *Burbage and Shakespeare's Stage* (London, 1913), pref., p. ix.

18. Stopes, *Shakespeare's Environment*, 205.

19. Albert Feuillerat, 'The History of the First Blackfriars Theatre', *Athenaeum* (9 Nov. 1912), 563.

20. Joseph William Gray, *Shakespeare's Marriage, His Departure from Stratford, and Other Incidents in His Life* (London, 1905), 38.

21. Ibid. 51

22. Reported by H. M. Spielmann, 'The "Grafton" and "Sanders" Portraits of Shakespeare', *Connoisseur*, 23 [1909], 100.

23. John Semple Smart, *Shakespeare: Truth and Tradition* (London, 1928), 12. I am here paraphrasing a note supplied by Smart's former colleague W. Macneile Dixon.

24. John Dover Wilson, *The Essential Shakespeare: A Biographical Adventure* (Cambridge, 1932), 7–8.

25. 'Shakespeare's Marriage', *Spectator*, 95 [29 July 1905], 153. The review is unsigned.

26. Ibid. 154.

27. Ibid. 153.

28. Michael Holroyd, *Lytton Strachey: A Critical Biography* (London, 1967), i. 88.

29. Lytton Strachey, 'Shakespeare's Final Period', *Books and Characters French & English* (London, 1922), 63.
30. Ibid. 60.
31. Ibid. 48.
32. *The Letters of Oscar Wilde*, ed. Rupert Hart-Davis (London, 1962), 608. This letter, dated 13 June 1897, is frequently quoted by Harris's biographers.
33. This anecdote, which is told by more than one of Harris's biographers, appears in A. I. Tobin and Elmer Gertz, *Frank Harris: A Study in Black and White* (Chicago, 1931), 118.
34. Frank Harris, *The Man Shakespeare and His Tragic Life-story* (London, 1909), introd., pp. ix–x.
35. *Frank Harris to Arnold Bennett: Fifty-eight Letters, 1908–1910*, ed. Mitchell Kennerley (Merion Station, Pa., 1936), 14.
36. Ibid. 23.
37. *New York Times*, quoted by Tobin and Gertz, *Frank Harris*, 168.
38. Arnold Bennett, quoted by E. Merrill Root, *Frank Harris* (New York, 1947), 176.
39. Tobin and Gertz, *Frank Harris*, 163–4.
40. Root, *Frank Harris*, 176.
41. Harris, *The Man Shakespeare . . .*, 231.
42. Ibid. 55.
43. Ibid. 141.
44. Ibid. 403.
45. Ibid. 372–3.
46. Ibid. introd., p. xi.
47. Georg Brandes, *William Shakespeare: A Critical Study*, trans. W. Archer and M. Morison (London, 1898), ii. 412.
48. Arnold Bennett, *Letters*, ed. James Hepburn (London, 1968), ii. 240.
49. Frank Harris, *The Women of Shakespeare* (London, 1911), 121.
50. Frank Harris, *Shakespeare and His Love* (London, 1910), introd., p. xi.
51. *Daily News* (London), 23 Nov. 1910.
52. Bernard Shaw, *Misalliance . . .* (London, 1914), 140.
53. Ibid. pref., 108.
54. Ibid. pref., 126.
55. W. Carew Hazlitt, *Shakespear: Himself and His Work: A Biographical Study* (London, 1914), 45.
56. Sarah Taylor Shatford, *Shakespeare's Revelations by Shakespeare's Spirit* (New York, 1919), 45–6.
57. S. T. Shatford, *My Proof of Immortality with My Demiséd Act* (New York, 1924), 76.
58. S. T. Shatford, *This Book for Him I Name for Jesus' Sake* (New York, [1920]), 7.
59. Hubert Ord, *Chaucer and the Rival Poet in Shakespeare's Sonnets* (London, 1921).
60. Charles Creighton, *Shakespeare's Story of His Life* (London, 1904), 6.
61. Arthur Acheson, *Shakespeare's Sonnet Story* (London, 1922), 133.
62. Clara Longworth de Chambrun, *Two Loves I Have: The Romance of William Shakespeare* (Philadelphia, 1934), 306.
63. Elizabeth Goudge, *The Golden Skylark and Other Stories* (London, 1941), 157–8.
64. G. B. Harrison, *Shakespeare under Elizabeth* (New York, 1963), 64.
65. Simone Arnaud, 'Les Sonnets de Shakespeare', *Nouvelle Revue*, 71 [1891], 537.
66. Edgar I. Fripp, *Shakespeare, Man and Artist* (London, 1938), i, pref., p. ix.
67. Ibid. i. 186.
68. Ibid. i. 29.
69. J. Q. Adams, *A Life of William Shakespeare* (London, 1923), pref., p. ix.

70. Ibid. 354.
71. Ibid. 95.
72. Smart, *Shakespeare: Truth and Tradition*, 50.
73. Ibid. 87.
74. Robert Morant, quoted by F. P. Wilson in 'Sir Edmund Kerchever Chambers 1866–1954', *Proceedings of the British Academy*, 42 [1956], 273.
75. E. K. Chambers, *Carmina Argentea* (n.p., 1918), 25.
76. E. K. Chambers, *William Shakespeare: A Study of Facts and Problems* (Oxford, 1930), i, pref., p. ix.
77. Ibid. i. 18; ii. 52.
78. Ibid. i. 26.
79. Herbert Thurston, 'The "Mr. W.H." of Shakespeare's Sonnets', *Month*, 156 [1930], 427.
80. Chambers, *William Shakespeare*, i. 22.
81. Ibid. i. 86.
82. J. D. Wilson, quoted in F. P. Wilson, 'Chambers', 274.
83. J. D. Wilson, *Essential Shakespeare*, pref., p. viii.
84. J. D. Wilson, *The Elizabethan Shakespeare* (London, 1929), 19.
85. See above, VII. 5.
86. J. D. Wilson, *Essential Shakespeare*, 67.
87. Ibid. 103.
88. Ibid. 106.
89. Ibid. 107.
90. Ibid. 136–7.
91. Uncatalogued letter of E. K. Chambers among the John Dover Wilson papers in the National Library of Scotland.
92. C. J. Sisson, *Mythical Sorrows of Shakespeare* (London, 1934), 9.
93. Ibid. 5.
94. Ibid. 27–8.
95. C. J. Sisson, 'Studies in the Life and Environment of Shakespeare since 1900', *Shakespeare Survey 3* (Cambridge, 1950), 9. As a review of scholarship, this essay is superficial.
96. R. W. Chambers, *The Jacobean Shakespeare and* Measure for Measure (London, 1937), 56.
97. T. W. Baldwin, *William Shakspere's Small Latine & Lesse Greeke* (Urbana, 1944), ii. 663.
98. Ibid. vol. i, pp. ix–x.
99. F. P. Wilson, 'Shakespeare's Reading', *Shakespeare Survey 3*, 15.
100. William Bliss, *The Real Shakespeare: A Counterblast to Commentators* (London, 1947), 111.
101. For a fuller and authoritative discussion of this issue, see E. K. Chambers, *William Shakespeare*, ii. 82–4.
102. Oliver Baker, *In Shakespeare's Warwickshire and the Lost Years* (London, 1937), 298. I have corrected Baker's sometimes faulty transcription, and I have modernized spelling and punctuation.
103. Mark Eccles, *Shakespeare in Warwickshire* (Madison, Wis., 1961), pref., p. vi.
104. Ibid. 46.
105. Ibid. 164–5. I have modernized spelling and punctuation.
106. Leslie Hotson, *Shakespeare versus Shallow* (London, 1931), 110.
107. Leslie Hotson, *I, William Shakespeare Do Appoint Thomas Russell, Esq. . . . .* (London, 1937), 12.

108. Caroline F. E. Spurgeon, *Shakespeare's Imagery and What It Tells Us* (Cambridge, 1935), pref., p. x.
109. Ibid. 204–5. For Ms Spurgeon's description of Shakespeare's physique, see above, p. 546.
110. Ivor Brown, *Shakespeare* (London, 1949), 155.
111. Ivor Brown, *The Way of my World* (London, 1954), 138–9.
112. Brown, *Shakespeare*, 164, 165.
113. Ibid. 218.
114. Marchette Chute, *Shakespeare of London* (London, 1951), 300–1.
115. E. K. Chambers, *Sources for a Biography of Shakespeare* (London, 1946), 66, 67.
116. Leonard Dobbs, *Shakespeare Revealed* (London, n.d.), 119.
117. Hugh Ross Williamson, *The Day Shakespeare Died* (London, 1962), 12.
118. M. M. Reese, *Shakespeare: His World and His Work* (London, 1953), pp. viii–ix.

# Index